RECURSION THEORY

PROCEEDINGS OF SYMPOSIA
IN PURE MATHEMATICS
Volume 42

RECURSION THEORY

AMERICAN MATHEMATICAL SOCIETY
PROVIDENCE, RHODE ISLAND

PROCEEDINGS OF SYMPOSIA IN PURE MATHEMATICS
OF THE AMERICAN MATHEMATICAL SOCIETY
VOLUME 42

PROCEEDINGS OF THE AMS-ASL SUMMER INSTITUTE
ON RECURSION THEORY
HELD AT CORNELL UNIVERSITY
ITHACA, NEW YORK
JUNE 28–JULY 16, 1982

EDITED BY
ANIL NERODE AND RICHARD A. SHORE

Prepared by the American Mathematical Society
with partial support from National Science Foundation grant MCS 8120074

1980 *Mathematics Subject Classification*. Primary 03D25, 03D30, 03D45, 03D55, 03D60, 03D70, 03D80, 03E05, 03E10, 03E35, 03E45, 03E47, 03F30, 03G05, 13E05, 14M05, 28A12, 90D05, 90D13.

Library of Congress Cataloging in Publication Data

Main entry under title:

Recursion theory.

(Proceedings of symposia in pure mathematics; v. 42)

Papers presented at the 1982 AMS Summer Research Institute held at Cornell University from June 28 to July 16, 1982, and co-sponsored by the ASL.

Bibliography: p.

1. Recursion theory--Addresses, essays, lectures. I. Nerode, Anil, 1932– . II. Shore, Richard A., 1946– . III. American Mathematical Society. IV. AMS Summer Research Institute (1982: Cornell University) V. Series.

QA9.6.R4 1984 511.3 84-18525

ISBN 0-8218-1447-8

The paper used in this book is acid-free and falls within the guidelines established to insure permanence and durability.

Table of Contents

Preface

The 1982 AMS Summer Research Institute was cosponsored by the ASL and was devoted to Recursion Theory. It met at Cornell University from June 28 to July 16. Our intention was to consider recursion theory in the broadest sense. This is reflected in the lists of participants and lectures (Appendices I and II) as well as in the contents of this Proceedings volume. The hour talks were roughly grouped around seven short courses—two in Classical Recursion Theory and one each in Generalized Recursion Theory, Fine Structure of L, Descriptive Set Theory, Effective Mathematics and Complexity Theory (Computer Sciences). These series correspond to the sections of this volume except that we have put the two set theoretic subjects into one section and have combined the papers on the foundational topics with those on computer science. Both of these are natural alignments since the talks in Descriptive Set Theory dealt mainly with the structure of $L(\mathbf{R})$ and the papers in complexity theory are strongly related to classical undecidability and incompleteness results.

This volume contains contributions representing most of the short courses and hour lectures given at the Institute. The papers representing the talks by Martin, Steel and Woodin in Descriptive Set Theory have however appeared in *Cabal Seminar*, 1979–81 Springer Lecture Notes in Mathematics no. 1019. In addition to the courses and hour lectures there were many shorter talks given by the participants and organized in sections. Extended abstracts for thirty-two of these talks appeared as a special bound issue of the Recursive Function Theory Newsletter edited by Iraj Kalantari under the auspices of Western Illinois and Cornell Universities.

As Cochairmen of the organizing committee we would like to thank our fellow committee members, S. Feferman, Y. Moschovakis, H. Putnam, G. Sacks, J. Shoenfield and R. Soare for their advice and encouragement. The excellent logistical support which helped make the Institute a success was provided by Peter Fejer who was the local coordinator and especially by Dottie Smith of the AMS who helped run just about everything. Both deserve our thanks.

Anil Nerode

Richard A. Shore

I. CLASSICAL RECURSION THEORY

Proceedings of Symposia in Pure Mathematics
Volume **42**, 1985

REA Operators, R. E. Degrees and Minimal Covers

CARL G. JOCKUSCH, JR.[1] AND RICHARD A. SHORE[1]

We will sketch a proof that every degree $\geqslant \mathbf{0}^{(\omega)}$ is a minimal cover. In a nutshell, the idea of the proof is to show that the minimal degree construction of Sacks (viewed as an operator which takes each set A to its Sacks minimal cover $M(A)$) has certain features in common with the ω-jump operator. These common features are sufficient to make it possible to prove that every degree $\geqslant \mathbf{0}^{(\omega)}$ has a representative in the range of M along the same lines as the known corresponding result for the ω-jump. The Sacks minimal degree construction is related to the ω-jump (rather than some other iteration of the jump) because the Sacks construction (relative to A) yields a set at level ω of the difference hierarchy for $\Delta_2^{0,A}$-sets. A refinement of the main result shows that the set of degrees of arithmetical sets is definable in the structure $\langle \mathscr{D}, \leqslant \rangle$, where $\mathscr{D}$ is the set of degrees and $\leqslant$ is its usual partial ordering. This, in turn, yields a number of corollaries about definability, homogeneity and automorphisms of $\langle \mathscr{D}, \leqslant \rangle$ when combined with work of Nerode and Shore.

Before tackling generalizations of the ω-jump, we consider generalizations of the jump itself. Let $J_e(A) = A \oplus W_e^A$. An operator on ${}^{\omega}2$ of the form J_e for some e is called an REA operator or, sometimes, a pseudo-jump operator. (Here "REA" stands for r.e. in and above.) The jump operator is an REA operator, up to recursive isomorphism. Existence proofs in the theory of r.e. degrees give many more examples of REA operators. Also analogues of the Friedberg completeness criterion and the existence of a nonrecursive r.e. set with $A' \equiv_T K$ hold with the jump operator replaced by an arbitrary REA operator. This gives a fairly general method for using known constructions in the theory of r.e. degrees to obtain new results or new proofs of old results. Full proofs of the results in this paper for

1980 *Mathematics Subject Classification*. Primary 03D30; Secondary 03G10.
Key words and phrases. Degrees of unsolvability, minimal covers.
[1] Research partially supported by grants from the National Science Foundation.

iterations of REA operators, as well as extensions, generalizations and other applications are given in **[6]** while those for the noniterated case are given in **[5]**.

We use standard recursion-theoretic notation. Usually the letters A, B, C, etc. represent subsets of $\omega = \{0, 1, 2, \dots\}$, while f, g, h represent total functions from ω into ω. Other lowercase Roman letters usually represent elements of ω. Let $A \oplus B = \{2n\colon n \in A\} \cup \{2n+1\colon n \in B\}$, and let ${}^{\omega}2$ be the power set of ω. We identify sets $A \subseteq \omega$ with their characteristic functions so that $n \in A$ if and only if $A(n) = 1$. We use the letters σ, τ for strings, i.e. functions from finite initial segments of ω into $\{0, 1\}$, or alternatively finite sequences of 0's and 1's. The length of σ, denoted $|\sigma|$, is the length of σ as a finite sequence. We say that τ extends σ (denoted $\tau \supseteq \sigma$) if $|\tau| \geqslant |\sigma|$ and $\sigma(n) = \tau(n)$, all $n < |\sigma|$. Let W_e^{σ} be the set of x such that the eth Turing machine with oracle converges in at most $|\sigma|$ steps with argument x and oracle σ. Thus we define W_e^A (the eth set r.e. in A) to be $\bigcup\{W_e^{\sigma}\colon \sigma \subseteq A\}$. (Here $\sigma \subseteq A$ means that the characteristic function of A extends σ.) For strings σ, τ we write $\sigma^\frown\tau$ for the usual concatenation of σ, τ viewed as finite sequences. If $i \in \{0, 1\}$, we write i for the corresponding string of length 1. Also we write 0^n1 for the string of length $n + 1$ which corresponds to a sequence of n 0's followed by a 1. Let $\langle \cdot\,, \cdot \rangle$ be a recursive pairing function from ω^2 onto ω, and let $A^{[e]} = \{x\colon \langle e, x\rangle \in A\}$. $A \leqslant_T B$ means that A is recursive in B, and $A \equiv_T B$ means that A and B are recursive in each other. A' is the jump of A and $A^{(n)}$ is the nth jump of A. $A^{(\omega)}$ is such that $(A^{(\omega)})^{[n]} = A^{(n)}$ for all n.

The following prototype result extends the Friedberg completeness criterion to REA operators. The proof is virtually the same as for the Friedberg completeness criterion. Recall that $J_e(A) = A \oplus W_e^A$.

THEOREM 1. *For every $e \in \omega$ and every set C with $0' \leqslant_T C$, there is a set A such that $J_e(A) \equiv_T A \oplus 0' \equiv_T C$.*

We consider two kinds of refinements of Theorem 1. The first kind of refinement **[5]** involves requiring A in Theorem 1 to be r.e. when $C \equiv_T 0'$. It extends Friedberg's result that there is a low nonzero r.e. degree and is proved by a finite injury priority argument which is quite similar to that used to prove the latter, although additional positive conditions must be interleaved to ensure that $0' \leqslant_T J_e(A)$.

THEOREM 2. $(\forall e)(\exists A)$[*A is r.e. and nonrecursive and $J_e(A) \equiv_T 0'$*].

A construction of an r.e. set W_e naturally corresponds, when relativized, to the REA operator J_e. Thus Theorem 2 converts a construction of an r.e. degree with a given property $\mathscr{P}$ to a construction of an r.e. degree **a** such that **0**$'$ has $\mathscr{P}$ relative to **a**. For instance, Theorem 2 converts the construction of a low nonzero r.e. degree (itself a special case of Theorem 2) to a construction of an incomplete high r.e. degree. Iteration of these arguments yields a *finite injury* proof of G. Sacks' result **[12]** that there are r.e. degrees at all levels of the high$_n$, low$_n$ hierarchy. (An r.e. degree **a** is called low$_n$ if $\mathbf{a}^{(n)} = \mathbf{0}^{(n)}$ and high$_n$ if $\mathbf{a}^{(n)} = \mathbf{0}^{(n+1)}$.) An application of the recursion theorem to the uniform relative version of Theorem 2 immediately yields a number e so that, for all $A \subseteq \omega$, $A <_T J_e(A)$ and $J_e(J_e(A)) \equiv_T A'$. Thus J_e is a sort of "square root" of the jump operation, although we do not know whether such an operator can be degree invariant. It is easily seen that, for this e, the degree of $J_e(\varnothing)$ is an r.e. degree which is not high_n or low_n for any $n < \omega$. (Earlier proofs of the existence of such a degree were given by Martin, Lachlan and Sacks; see [**12**].) The next result is an extension of Theorem 2 which applies to constructions of comparable pairs of r.e. degrees in a way that Theorem 2 applies to constructions of a single r.e. degree. The proof is again a finite injury priority argument similar to that for Theorem 2.

THEOREM 3. *Let e and i be numbers with $J_i(A) \leqslant_T J_e(A)$ for all A. Then there exists a nonrecursive r.e. set A such that $J_i(A)$ has r.e. degree and $J_e(A)$ has degree* $\mathbf{0}'$.

When applied to the relativized version of A. H. Lachlan's construction [**8**] of r.e. degrees **c**, **d** such that $\mathbf{c} < \mathbf{d}$ and **d** cannot be split by incomparable r.e. degrees $> \mathbf{c}$, Theorem 3 produces an r.e. degree **b** such that $\mathbf{0}'$ cannot be so split over **b**. (Such a **b** was first shown to exist by L. Harrington [**2**].)

The second direction in which Theorem 1 is generalized [**6**] involves finding and applying analogues for REA operators of the iterated Friedberg jump theorem [**9**]. (The latter asserts that for any recursive ordinal α, every degree $\geqslant \mathbf{0}^{(\alpha)}$ is the αth jump of some degree.) Here we consider analogues of the case $\alpha \leqslant \omega$ only, although the general case for (notations for) arbitrary recursive ordinals is handled in [**6**]. For $n < \omega$, an operator J on ${}^{\omega}2$ is called n-REA if J is a composition $J_{e_n} \circ \cdots \circ J_{e_1}$ of n REA operators. An operator J is called ω-REA if there is a recursive function f such that $(J(A))^{[e]} = (J_{f(e)} \circ J_{f(e-1)} \circ \cdots \circ J_{f(0)})(A)$ for all $A \subseteq \omega$ and all $e \in \omega$. Of course, for $\alpha \leqslant \omega$, the α-fold Turing jump operator is an α-REA operator (up to recursive isomorphism), and it is the most fundamental example of such an operator. Later we shall see that the minimal degree construction of Sacks is also on ω-REA operator (up to Turing degree) but we first obtain the promised extension of the iterated Friedberg jump theorem to α-REA operators.

THEOREM 4. *If $\alpha \leqslant \omega$, J is an α-REA operator, and $0^{(\alpha)} \leqslant_T C$, then there is a set A such that $J(A) \equiv_T 0^{(\alpha)} \oplus A \equiv_T C$.*

The proof of Theorem 4 is a modification of the forcing argument used by MacIntyre to prove iterated forms of Friedberg's completeness criterion [**9**] and is well illustrated by the case $\alpha = 2$. Say that $J(A) = J_b(J_a(A))$. The idea is to construct a set A which is "generic" for all statements of the form $k \in J_b(J_a(A))$ (rather than for *all* two-quantifier sentences of arithmetic as in the proof of the iterated Friedberg completeness criterion). The method of coding C is just as in the proof of the Friedberg completeness criterion. The first step in the proof is the construction of a perfect tree $T_0 \leqslant_T 0'$ each branch of which is "generic" for all statements of the form $k \in J_a(A)$. To specify T_0 inductively, suppose σ is a node at level k on T_0. If there is a string $\tau \supseteq \sigma^\frown 0$ with $k \in J_a(\tau)$, then the "left" immediate successor of σ at level $k+1$ of T_0 is the first such τ. (Here $J_a(\tau) = \{k: \tau(k) = 1\} \oplus W_a^\tau$.) If there is no such τ, the left immediate successor of σ on T_0 is simply $\sigma^\frown 0$. The right immediate successor of σ on T_0 is defined similarly, with $\sigma^\frown 1$ in place of $\sigma^\frown 0$. With each node τ on T_0 associate a string $I_0(\tau)$ which gives the information about $J_a(A)$ "forced" by τ, so that $J_a(A) \supseteq I_0(\tau)$ whenever $A \supseteq \tau$. Specifically, if τ occurs at level $k+1$ of T_0, then $I_0(\tau)$ has length $k+1$ and, for $j \leqslant k$, $I_0(\tau)(j) = 1$ if and only if $j \in J_a(\tau)$.

Now within T_0 one defines a perfect tree $T_1 \leqslant_T 0''$ such that all branches of T_1 are "generic" for $J_b \circ J_a$. The construction of T_1 is the same as that of T_0 except that τ is required to lie on T_0, the strings $\sigma^\frown 0$, $\sigma^\frown 1$ are replaced by the immediate successors of σ on T_0, and the condition "$k \in J_a(\tau)$" is replaced by "$k \in J_b(I_0(\tau))$". Finally, let A be that branch of T_1 which goes right at level k of T_1 if and only if $k \in C$, all $k \in \omega$.

We sketch the verification that $J(A) \equiv_T C$, which is essentially as in the Friedberg completeness criterion. By the construction, for all k, $k \in J(A)$ if and only if $k \in J_b(I_0(\tau))$, where τ is the unique string of level $k+1$ on T_1 extended by A. This string τ may be computed effectively from C because C describes the path of A through T_1 and $T_1 \leqslant_T 0'' \leqslant_T C$. Also $I_0 \leqslant_T 0' \leqslant_T C$. It follows that $J(A) \leqslant_T C$. To verify that $C \leqslant_T J(A)$, first compute from an oracle for $J_a(A)$ which strings extended by A lie on T_0. (Here the point is that the string τ mentioned in the definition of T_0 exists if and only if $k \in J_a(A)$, and when τ exists it may be found by an effective search.) Having done this one may then carry out the corresponding computation for T_1 with an oracle for $J_b(J_a(A))$ and along the way determine which way A branches at each node of T_1. Therefore $C \leqslant_T J(A)$.

The next result gives an interesting example of an ω-REA operator. A degree **b** is called a *minimal cover* of a degree **a** if **b** $>$ **a** and no degree **c** satisfies **b** $>$ **c** $>$ **a**.

THEOREM 5. *There is an ω-REA operator J such that, for all A, the degree of $J(A)$ is a minimal cover of that of A.*

The operator J asserted to exist by Theorem 5 is, up to Turing degree, the Sacks minimal degree construction [**11**]. Let $M(A)$ be the set obtained from the Sacks minimal degree construction relative to A. Thus $M(A) \leqslant_T A'$ and the degree of $M(A)$ is a minimal cover of that of A. We claim that there is an ω-REA operator J such that $J(A) \equiv_T M(A)$ for all A. To prove this we need to know not only that $M(A) \leqslant_T A'$ but that $M(A)$ occurs by level ω in the difference hierarchy of $\Delta_2^{0,A}$-sets in a uniform way. This means that there is a number m and a recursive function g such that, for all A and n,

(i) $\{m\}^A(\cdot, \cdot)$ is total,

(ii) $M(A)(n) = \lim_s \{m\}^A(n, s)$, and

(iii) $|\{s: \{m\}^A(n, s) \neq \{m\}^A(n, s+1)\}| \leqslant g(n)$.

The existence of such m and g may be seen quite easily by analysing the Sacks minimal degree construction relative to A **[11]** and its natural A-recursive approximation. (In fact, one may take $g(n) = 2^n$.)

Call a set B ω-r.e. if there exist recursive functions f, g such that for all n, $B(n) = \lim_s f(n, s)$ and $|\{s: f(n, s) \neq f(n, s + 1)\}| \leqslant g(n)$. (It is easy to see that B is ω-r.e. if and only if $B \leqslant_{tt} K$ if and only if $B \leqslant_w K$. See **[1]** for further information.) If we show uniformly that for every ω-r.e. set B there is an ω-REA operator J with $B \equiv_T J(\varnothing)$, we can then conclude by relativizing the argument that the existence of m and g for M as above implies that there is an ω-REA operator J with $M(A) \equiv_T J(A)$ for all A. This will conclude the proof of Theorem 5. Assume now that B is ω-r.e. with f, g as above. Let

$$k(n, s) = |\{t < s: f(n, t) \neq f(n, t + 1)\}|.$$

We first define a set E of the same degree as B such that $E^{[0]}$ is r.e. and, for all i, $E^{[i+1]}$ is r.e. in $E^{[i]}$, uniformly in i. Let E be the unique set such that, for all i,

$$E^{[i]} = \{\langle n, s\rangle : k(n, s) = g(n) - i \text{ and } f(n, s) \neq B(n)\}.$$

It is clear that $E \leqslant_T B$. Since $k(n,0) = 0$, the equivalence

$$\langle n,0\rangle \in E^{[g(n)]} \Leftrightarrow f(n,0) \neq B(n)$$

holds. We assume without loss of generality that f is 0-1 valued, so the above equivalence shows that $B \leqslant_T E$. Also $E^{[0]} = \varnothing$ and, for all n, s, and i,

$$\langle n, s\rangle \in E^{[i+1]} \Leftrightarrow k(n, s) = g(n) - i - 1 \,\&\, (\exists t > s)[f(n, t) \neq f(n,s) \,\&\, k(n, t) = g(n) - i \,\&\, \langle n, t\rangle \notin E^{[i]}].$$

To prove this, observe that if $f(n, s) \neq B(n)$ then there exists $t > s$ with $f(n, t) \neq f(n, s)$ and the least such t satisfies $k(n, t) = k(n, s) + 1$. (Also use the assumption that f is 0-1 valued.) The above equivalence implies that $E^{[i+1]}$ is r.e. in $E^{[i]}$, uniformly in i. Finally, let G be the unique set such that $G^{[0]} = \varnothing \oplus E^{[0]} = \varnothing$ and, for all i, $G^{[i+1]} = G^{[i]} \oplus E^{[i+1]}$. Then $G \equiv_T E \equiv_T B$, and G is easily seen to be ω-REA from the facts that $E^{[0]}$ is r.e. and $E^{[i+1]}$ is uniformly r.e. in $E^{[i]}$. This concludes our sketch of the proof of Theorem 5.

REMARK. The above shows that every ω-r.e. set B has the same degree as $J(\varnothing)$ for some ω-REA operator J. For $k < \omega$, a set B is called weakly k-r.e. if there is a recursive function f such that, for all n, $B(n) = \lim_s f(n, s)$, and

$$|\{s: f(n, s) \neq f(n, s + 1)\}| \leqslant k.$$

A very similar argument shows that every weakly k-r.e. set has the same degree as $J(\varnothing)$ for some k-REA operator J. (See **[6]** for applications of this and treatment of levels beyond ω.)

In §3 of Harrington and Kechris [3] it was shown that every degree $\geqslant$ the degree of Kleene's $\mathcal{O}$ is a minimal cover (of some lower degree). The next corollary refines this result from the degree of Kleene's $\mathcal{O}$ to $\mathbf{0}^{(\omega)}$. It is best possible with respect to the hyperarithmetic hierarchy since Jockusch and Soare [7] proved that $\mathbf{0}^{(n)}$ fails to be a minimal cover for each $n < \omega$.

COROLLARY 6. *Every degree* $\geqslant \mathbf{0}^{(\omega)}$ *is a minimal cover.*

Corollary 6 follows at once from Theorems 4 and 5.

It follows from Corollary 6 and the cited result of Jockusch and Soare that the set of bases of cones of minimal covers separates the set $\mathscr{A}$ of degrees of arithmetical sets from the degrees $\geqslant \mathbf{0}^{(\omega)}$, and this separating set is obviously definable in $(\mathscr{D}, \leqslant)$. A refinement of Corollary 6 along the lines of Harrington and Shore [**4**] will now be used to show that $\mathscr{A}$ is actually definable in $(\mathscr{D}, \leqslant)$. As in [**4**], let $\mathscr{C}$ be the downward closure in the degrees of $\mathscr{C}_0$, where

$$\mathscr{C}_0 = \{\mathbf{d}\colon (\forall \mathbf{a})[\mathbf{a} \cup \mathbf{d} \text{ is not a minimal cover of } \mathbf{a}]\}$$
$$(= \{\mathbf{d}\colon (\forall \mathbf{b})(\forall \mathbf{a})[\mathbf{b} \geqslant \mathbf{d} \;\&\; \mathbf{b} \text{ a minimal cover of } \mathbf{a} \rightarrow \mathbf{a} \geqslant \mathbf{d}]\}).$$

Then $\mathscr{C}$ is obviously definable in $(\mathscr{D}, \leqslant)$, and it is shown in [**4**] that $\mathscr{A} \subseteq \mathscr{C} \subseteq \mathscr{H}$, where $\mathscr{H}$ denotes the set of degrees of hyperarithmetical sets. No minimal cover is in $\mathscr{C}_0$, so Corollary 6 implies that no degree $\geqslant \mathbf{0}^{(\omega)}$ is in $\mathscr{C}$.

THEOREM 7. $\mathscr{A} = \mathscr{C}$.

The inclusion $\mathscr{C} \subseteq \mathscr{A}$ of Theorem 7 is a consequence of the next result which is a "cone avoidance" refinement of Corollary 6.

THEOREM 8. *If* $\mathbf{c} \geqslant \mathbf{0}^{(\omega)}$ *and* $\mathbf{d}$ *is not arithmetical, then there exists a degree* $\mathbf{a} \ngeqslant \mathbf{d}$ *such that* $\mathbf{c}$ *is a minimal cover of* $\mathbf{a}$.

PROOF. The conclusion of Theorem 8 is immediate from Corollary 6 unless $\mathbf{c} \geqslant \mathbf{d}$. Hence we assume $\mathbf{c} \geqslant \mathbf{d}$.

Let $M(A)$ be the set obtained from the Sacks minimal degree construction relative to A. Thus the degree of $M(A)$ is a minimal cover of that of A and, as remarked in the proof of Theorem 5, $M(A)$ is uniformly ω-r.e. in A. Let C and D be sets of degree $\mathbf{c}$ and $\mathbf{d}$, respectively. Suppose we construct a set A such that $M(A) \equiv_T A \oplus D \equiv_T C$. Then the degree $\mathbf{a}$ of A will satisfy the conclusion of Theorem 8. (If $\mathbf{a} \geqslant \mathbf{d}$, the $M(A) \leqslant_T A \oplus D \leqslant_T A$, contradicting $A <_T M(A)$.) To motivate the construction of A, we first prove a simpler join theorem in which M is replaced by the jump operator.

PROPOSITION 9. *If* D *is not recursive and* $0' \oplus D \leqslant_T C$, *then there exists a set* A *such that* $A' \equiv_T A \oplus D \equiv_T C$.

PROOF. This result is a special case of the Posner-Robinson Join Theorem [**10**, Theorem 3] but is proved here by a modification of the trick used in [**10**, Theorem 1]. We may assume without loss of generality that $\overline{D}$ is not r.e. by replacing D by $D \oplus \overline{D}$. The characteristic function of A is obtained as $\bigcup_k \sigma_k$, where $\sigma_0 \subseteq \sigma_1 \subseteq \cdots$ are strings. Let $\sigma_0 = \varnothing$, the empty string.

Stage $2k + 1$. Let $\sigma_{2k+1} = \sigma(k)\frown C(k)$.

Stage $2k + 2$. Let $S_k = \{m\colon (\exists \tau \supseteq \sigma_{2k+1}\frown 0^n 1)[k \in W_k^\tau]\}$.

Then S_k is r.e. so $S_k \neq \overline{D}$. Let n_k be the least number n such that $n \in S_k$ if and only if $n \in D$. If $n_k \in S_k$, let σ_{2k+2} be the least $\tau \supseteq \sigma_{2k+1}\frown 0^{n_k} 1$ with $k \in W_k^\tau$. Otherwise let σ_{2k+2} be $\hat{\sigma}_{2k+1} 0_1^{n_k}$.

Let $f(k) = \sigma_k$. Then f is recursive in each of the three sets A', $A \oplus D$, C. For instance, to show that f is recursive in $A \oplus D$, note first that the odd stages may be carried out recursively in A. For the even stages, suppose σ_{2k+1} is given. Then n_k may be found from A as the unique n such that $A \supseteq \sigma_{2k+1}^\frown 0^n 1$. If $n_k \in D$, σ_{2k+2} is the least $\tau \supseteq \sigma_{2k+1}^\frown 0^{n_k} 1$ with $k \in W_k^\tau$ and this τ may be found by an effective search. If $n_k \notin D$, then $\sigma_{2k+2} = \sigma_{2k+1}^\frown 0^{n_k} 1$.

On the other hand, each of the three sets A', $A \oplus D$, C is recursive in f. For instance, $k \in A'$ if and only if (by definition) $k \in W_k^A$ if and only if $k \in W_k^{\sigma_{2k+2}}$. Also $C(k) = \sigma_{2k+1}(|\sigma_{2k+1}| - 1)$, so $C \leqslant_T f$. Finally, $A \oplus D \leqslant_T f$ since $A' \leqslant_T f$ and $D \leqslant_T C \leqslant_T f$. Therefore $A' \equiv_T A \oplus D \equiv_T C$.

REMARK. We conjecture that the following weak analogue of Proposition 9 holds for ω-jump: If D is not arithmetical, there exists a set A with $A^{(\omega)} \leqslant_T A \oplus D$. This conjecture implies Theorem 7 as follows. Suppose $\mathbf{d}$ is not arithmetical. Let $\mathbf{e}$ be a degree with $\mathbf{e}^{(\omega)} \leqslant \mathbf{e} \cup \mathbf{d}$. By Corollary 6, relativized to $\mathbf{e}$, $\mathbf{e} \cup \mathbf{d}$ is a mimimal cover of some degree $\mathbf{a} \geqslant \mathbf{e}$. Then $\mathbf{a} \ngeqslant \mathbf{d}$ since otherwise $\mathbf{a} \geqslant \mathbf{e} \cup \mathbf{d}$. It follows that $\mathbf{d} \notin \mathscr{C}_0$. Hence $\mathscr{C}_0 \subseteq \mathscr{A}$, from which it follows that $\mathscr{C} \subseteq \mathscr{A}$ because $\mathscr{A}$ because $\mathscr{A}$ is closed downwards. However, the conjecture itself remains open. The following special case of the conjecture is easy to verify: If f is a function not majorized by any arithmetic function, then there exists a set A with $A^{(\omega)} \leqslant_T A \oplus f$. The set A will be Cohen generic for arithmetic so that $A^{(\omega)} \leqslant_T A \oplus 0^{(\omega)}$. The characteristic function of A is obtained as the union of an ascending sequence of strings $\{\sigma_k\}$ such that $\sigma_{2k+1} = \sigma_{2k}^\frown 0^{(\omega)}(k)$ and σ_{2k+2} forces the kth sentence of arithmetic or its negation. In addition, the length of σ_{2k+2} is required to be $f(n)$, where n is the unique number such that $\sigma_{2k+1} \supseteq \sigma_{2k+1}^\frown 0^n 1$. The existence of σ_{2k+2} follows from the hypothesis on f and standard basic properties of forcing. The sequence $\{\sigma_k\}$ is recursive in $A \oplus f$, so

$$A^{(\omega)} \leqslant_T A \oplus 0^{(\omega)} \leqslant_T \{\sigma_k\} \leqslant_T A \oplus f.$$

(This proof is virtually identical to that of the Posner-Robinson theorem [**10**].)

We return now to the task of constructing a set A with $M(A) \equiv_T A \oplus D \equiv_T C$, where $M(A)$ is the Sacks minimal cover of A. The only properties of M needed for this construction are that A is recursive in $M(A)$ and that $M(A)$ is ω-r.e. in A, both uniformly in A.

The construction is along the lines of the proof of Proposition 9 but the forcing used there is replaced by the more complicated forcing from the proof of Theorem 4. We obtain A as the unique common branch of the trees $T_0, T_1, \ldots$, where each T_i is an arithmetical perfect tree and T_{i+1} is a subtree of T_i. We now view a tree T as a mapping from strings to strings. A string is on T if and only if it is in the range of T, and the left, right immediate successors of $T(\sigma)$ on T are $T(\sigma^\frown 0)$, $T(\sigma^\frown 1)$, respectively. The full subtree of T above the node $T(\sigma)$ is the tree $\tilde{T}$ given by $\tilde{T}(\tau) = T(\sigma^\frown \tau)$.

Let g be the recursive bounding function which witnesses that $M(A)$ is uniformly ω-r.e. in A, as in the proof of Theorem 5. The relativization of the proof of Theorem 5 easily yields, given $A \subseteq \omega$ and $k \in \omega$, a sequence of sets $C_k^{0,A}$, $C_k^{1,A},\ldots,C_k^{g(k),A}$ such that uniformly in i, k, and A: $C_k^{0,A}$ is r.e. in A, $C_k^{i+1,A}$ is r.e. in $A \oplus C_k^{i,A}$ for $i < g(k)$, $C_k^{i,A}$ is recursive in $M(A)$, and $k \in M(A)$ if and only if $0 \in C_k^{g(k),A}$. (Let $C_k^{i,A} = \{s\colon\ k^A(k,s) = g(n) - i$ and $f^A(k,s) \neq M(A)(k)\}$, where k^A, f^A are the relativized versions of k, f, respectively, from the proof of Theorem 5 with B now replaced by $M(A)$. To obtain that $k \in M(A)$ if and only if $0 \in C_k^{g(k),A}$ we are assuming, without loss of generality, that $f^A(k,0) = 0$ for all k.)

To begin the construction of the sequence $\{T_i\}$ of trees, let T_0 be the identity tree, so $T_0(\sigma) = \sigma$ for all σ.

Stage $2k + 1$. Let T_{2k+1} be the full subtree of T_{2k} above $T_{2k}(C(k))$.

Stage $2k + 2$. Let the sets $C_k^{0,A}$, $C_k^{1,A},\ldots,C_k^{g(k),A}$ be as previously specified. Let $\hat{C}_k^{0,A} = A \oplus C_k^{0,A}$ and for $0 \leqslant i < g(k)$, let $\hat{C}_k^{i+1,A} = \hat{C}_k^{i,A} \oplus C_k^{i+1,A}$. Let $e_0,\ldots,e_{g(k)}$ be numbers such that, for all A, $\hat{C}_k^{0,A} = J_{e_0}(A)$ and, for $0 \leqslant i < g(k)$, $\hat{C}_k^{i+1,A} = J_{e_{i+1}}(\hat{C}_k^{i,A})$. Such numbers may be found effectively from k.

For our fixed k and each n, consider the following process: Let U_0^n be the full subtree of T_{2k+1} above the string $T_{2k+1}(0^n1)$. Within U_0^n form the subtree $U_1^n \leqslant_T (U_0^n)'$ such that all branches of U_1^n are "generic" for J_{e_0}, as in Theorem 4. Then within U_1^n form the subtree $U_2^n \leqslant_T (U_1^n)'$ such that all branches of U_2^n are "generic" for $J_{e_1}J_{e_0}$, again as in Theorem 4. Continue in this way until $U_{g(k)}^n$ is formed. Finally let $T_{2k+2}^{\langle n\rangle}$ (the "nth candidate for T_{2k+2}") be the full subtree of $U_{g(k)}^n$ above $U_{g(k)}^n(0^\frown 0)$. Let $\hat{J}(A) = J_{e_{g(k)}} \cdots J_{e_0}(A)$. By construction, if two branches of $U_{g(k)}^n$ extend the same node at level $i + 1$ of $U_{g(k)}^n$ then both or neither satisfy $i \in \hat{J}(A)$. Note that $k \in M(A)$ if and only if $0 \in C_k^{g(k),A}$ if and only if $1 \in \hat{C}_k^{g(k),A} = \hat{J}(A)$. Also all branches of $T_{2k+2}^{\langle n\rangle}$ extend the same node at level 2 of $U_{g(k)}^n$. Hence all or no branches A of $T_{2k+2}^{\langle n\rangle}$ satisfy $k \in M(A)$. Let

$$S_k = \left\{n\colon \text{all branches } A \text{ of } T_{2k+2}^{\langle n\rangle} \text{ satisfy } k \in M(A)\right\}.$$

Then S_k is arithmetic, indeed $S_k \leqslant_T T_{2k+1}^{(g(k)+1)}$. Therefore $S_k \neq \overline{D}$. Let n_k be the least number n such that $n \in S_k$ if and only if $n \in D$. Finally set $T_{2k+2} = T_{2k+2}^{\langle n_k\rangle}$. This completes the construction of the trees T_k and hence of A.

Let $f_k(i)$ be the pair $(\sigma, T_k(\sigma))$ which is uniquely determined by the conditions $|\sigma| = i$ and $A \supseteq T_k(\sigma)$. Let $f(k, i) = f_k(i)$. Then f is recursive in each of the sets $M(A)$, $A \oplus D$, and C. This is proved by establishing that f_k is recursive in each of these sets by a uniform induction on k. The main point here is that, as in Theorem 4, once it is known whether $k \in M(A)$ it is possible to compute f_{2k+2} from f_{2k+1}. Also as in Proposition 9, if f_{2k+1}, A and D are given, one may determine n_k and whether $k \in M(A)$. In the other direction, it is not hard to see that each of the three sets $M(A)$, $A \oplus D$, C is recursive in f. We omit further details.

A number of results about definability, homogeneity and automorphisms for $(\mathscr{D}, \leqslant)$ follow from Corollary 6 by work of Nerode and Shore. In particular, the results in §§3 and 4 of Shore [**13**] (also stated in Harrington and Shore [**4**]) now hold with "hyperarithmetical" replaced everywhere by "arithmetical". For instance, by Theorem 4.5 of [**13**], every automorphism of $(\mathscr{D}, \leqslant)$ is the identity on a cone of degrees with an arithmetic base. (See also Shore's paper [**14**] in this volume.)

Let $\mathscr{B}$ be the set of degrees $\mathbf{d}$ such that every $\mathbf{c} \geqslant \mathbf{d}$ is a minimal cover. $\mathscr{B}$ is obviously closed upwards, and Corollary 6 asserts that $\mathbf{0}^{(\omega)} \in \mathscr{B}$. By [7], no arithmetical degree is in $\mathscr{B}$. However, little more is known about $\mathscr{B}$. It is conceivable, but seems very unlikely, that $\mathscr{B} = \mathscr{D}(\geqslant \mathbf{0}^{(\omega)})$ (the set of degrees $\geqslant \mathbf{0}^{(\omega)}$). Assuming that $\mathscr{B} \neq \mathscr{D}(\geqslant \mathbf{0}^{(\omega)})$, it is reasonable to ask whether $\mathscr{B}$ contains a degree $< \mathbf{0}^{(\omega)}$, a degree $\ngeqslant \mathbf{0}'$, a minimal degree, etc.

REFERENCES

1. R. Epstein, R. Haas and R. Kramer, *Hierarchies of sets and degrees below* **0**′, Logic Year 1979–80 (M. Lerman, J. H. Schmerl and R. I. Soare, eds.), Lecture Notes in Math., vol. 859, Springer-Verlag, Berlin, Heidelberg and New York, 1981, pp. 32–48.

2. L. Harrington, *Understanding Lachlan's monster paper*, handwritten notes, 1980.

3. L. Harrington and A. Kechris, *A basis result for* Σ^0_3 *sets of reals with an application to minimal covers*, Proc. Amer. Math. Soc. **53** (1975), 445–448.

4. L. Harrington and R. Shore, *Definable degrees and automorphisms of* $\mathscr{D}$, Bull. Amer. Math. Soc. (N. S.) **4** (1981), 97–100.

5. C. Jockusch and R. Shore, *Pseudo jump operators*. I: *The r.e. case*, Trans. Amer. Math. Soc. **275** (1983), 599–609.

6. ______, *Pseudo jump operators*. II: *Transfinite iterations, hierarchies and minimal covers*, J. Symbolic Logic (to appear).

7. ______, *Minimal covers and arithmetical sets*, Proc. Amer. Math. Soc. **25** (1970), 856–859.

8. A. H. Lachlan, *A recursively enumerable degree which will not split over all lesser ones*, Ann. Math. Logic **9** (1975), 307–365.

9. J. MacIntyre, *Transfinite extensions of Friedberg's completeness criterion*, J. Symbolic Logic **42** (1977), 1–10.

10. D. Posner and R. Robinson, *Degrees joining to* **0**′, J. Symbolic Logic **46** (1981), 714–722.

11. G. E. Sacks, *Degrees of unsolvability*, Ann. of Math. Studies, No. 55, Princeton Univ. Press, Princeton, N. J., 1966.

12. ______, *On a theorem of Lachlan and Martin*, Proc. Amer. Math. Soc. **18** (1967), 140–141.

13. R. Shore, *The degrees of unsolvability*: *Global results*, Logic Year 1979–80 (M. Lerman, J. H. Schmerl and R. I. Soare, eds.), Lecture Notes in Math., vol. 859, Springer-Verlag, Berlin, Heidelberg and New York, 1981, pp. 283–301.

14. ______, *The structure of the degrees of unsolvabilty*, these PROCEEDINGS, pp. 33–51.

DEPARTMENT OF MATHEMATICS, UNIVERSITY OF ILLINOIS, URBANA, ILLINOIS 61801

DEPARTMENT OF MATHEMATICS, CORNELL UNIVERSITY, ITHACA, NEW YORK 14853

Proceedings of Symposia in Pure Mathematics
Volume **42**, 1985

The Embedding Problem for the Recursively Enumerable Degrees

MANUEL LERMAN[1]

1. History. Let $\mathscr{L} = \langle L, \leqslant, \vee, \wedge \rangle$ be a finite lattice with least element 0 and greatest element 1. $\leqslant$, $\vee$ and $\wedge$ are the ordering, join and meet operations of $\mathscr{L}$, respectively. If $\mathscr{L}$ is a sublattice of the set $\mathscr{R}$ of r.e. degrees, then we write $\mathscr{L} = \langle \mathbf{L}, \leqslant, \cup, \cap \rangle$ and note that $\mathbf{0}$ and $\mathbf{0}'$ are, respectively, the least and greatest elements of $\mathscr{R}$.

We will use several languages to talk about lattices. Each will consist of the pure predicate calculus together with some extra symbols used to interpret elements, relations and functions of $\mathscr{L}$; these symbols are confused with the corresponding symbols for $\mathscr{L}$. The nonlogical symbols for the respective languages are

$$\mathbb{L}_0\colon \{\leqslant, \vee, \wedge, 0\}, \quad \mathbb{L}_1\colon \{\leqslant\}, \quad \mathbb{L}_2\colon \{\leqslant, \vee, 0, 1\}, \quad \mathbb{L}_3\colon \{\leqslant, \vee, \wedge, 1\}.$$

The Embedding Problem (EP) for $\mathscr{R}$ asks for a characterization of the finite lattices which are embeddable into $\mathscr{R}$; the embedding is required to preserve ordering, meets, joins and 0. The formulation of this problem is closely tied to Shoenfield's Conjecture (SC) **[Sh1]** which was stated in 1963 at a symposium in Berkeley. Shoenfield was interested in the Decision Problem (DP) for the elementary theory of the r.e. degrees, and his conjecture yields, as a corollary, the decidability of this elementary theory.

SHOENFIELD'S CONJECTURE. *If* $\bar{\mathbf{a}}$ *is a vector of r.e. degrees which satisfies the diagram* $D(\bar{x})$ *in the language* $\mathbb{L}_2$, *and* $D^*(\bar{x}, y)$ *is any consistent upper semilattice (usl) diagram extending* D, *then there is an r.e. degree* $\mathbf{b}$ *such that* $\mathscr{R} \models D^*(\bar{\mathbf{a}}, \mathbf{b})$.

1980 *Mathematics Subject Classification.* Primary 03D25, 03G10.
[1] Research supported by the National Science Foundation through grant number MCS78-01849.

At the time, the Sacks Density Theorem [**Sa**] had been proved, and implied all instances of SC in which the diagram D^* was linearly ordered.

SC was first refuted through an embedding result due, independently, to Lachlan [**La1**] and Yates [**Y**]. Lachlan and Yates showed that there is a *minimal pair* of r.e. degrees, i.e., a pair $\mathbf{a}, \mathbf{b}$ of incomparable r.e. degrees such that $\mathbf{a} \cap \mathbf{b} = \mathbf{0}$. Thus not every realization of D in the figure can be extended to a realization of D^* (dotted lines imply that meets are not specified).

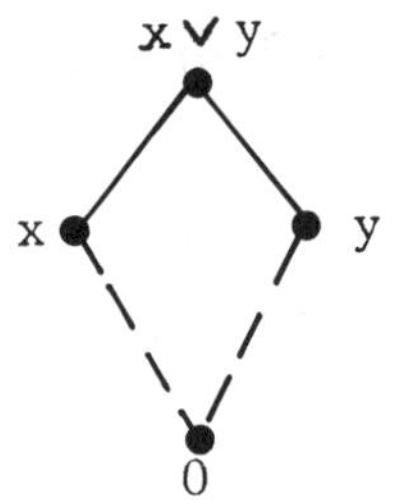

D

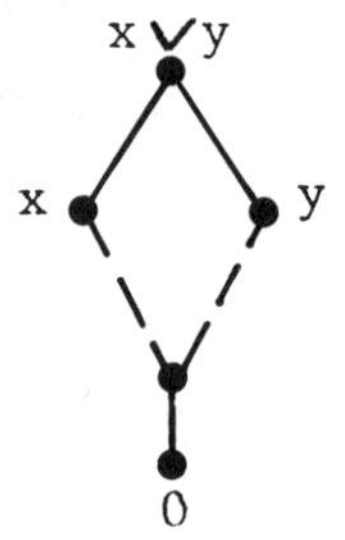

D*

Thus SC was of no help in solving the DP. At that time, all that was known about the DP was that the existential theory of $\mathscr{R}$, $\exists_1 \cap \mathrm{Th}(\mathscr{R})$, in $\mathbb{L}_1$ is decidable, a result which easily follows from the techniques of Friedberg [**F**] and Muchnik [**M**].

The next natural step in attacking the DP is to try to determine whether $\forall_2 \cap \mathrm{Th}(\mathscr{R})$ is decidable in the language $\mathbb{L}_1$. Since a solution to the EP would characterize a natural fragment of $\forall_2 \cap \mathrm{Th}(\mathscr{R})$, it is not surprising that this problem was attacked first. Shore has made good progress towards a decision procedure for $\forall_2 \cap \mathrm{Th}(\mathscr{R})$, but a major obstacle to extending Shore's results to larger classes of sentences is the need for a solution to the EP. Some results related to the EP for the language $\mathbb{L}_3$ have been obtained by Lachlan [**La1**] and more recently by Ambos-Spies [**A**], but these results lag behind those for the language $\mathbb{L}_0$.

The next result concerning the EP was obtained independently by Thomason [**T**] and Lerman, who showed that every finite distributive lattice is embeddable into $\mathscr{R}$. In the case of distributive lattices, there is no real conflict between meet preservation requirements and join preservation requirements. The techniques which were introduced deal with meet preservation for nontrivial meets of the lattice.

Join preservation is more complicated for nondistributive lattices. Lachlan [**La2**] introduced a technique for preserving joins, and used this technique to embed the two five-element nondistributive lattices into $\mathscr{R}$. Lachlan's technique was quickly noticed to be extendable to any lattice in which all meets are trivial. However, the conflicts between the join preservation techniques of Lachlan and the meet preservation techniques of Lerman and Thomason needed to be resolved.

The EP was given further stimulus by Shoenfield [**Sh2**] in a 1974 talk. There, Shoenfield conjectured that all finite lattices are lattice embeddable into $\mathscr{R}$ through an embedding which preserves 0. There seemed to be, at that time, almost unanimous agreement that Shoenfield was correct. I disagreed, and conjectured that the lattices S_8 (pictured in the diagram) was not embeddable into $\mathscr{R}$. The reasons for my conjecture were the following:

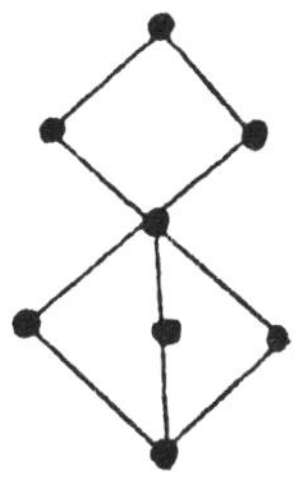

S_8

(1) I had earlier tried to embed every finite lattice into $\mathscr{R}$, and S_8 was the simplest lattice to describe for which the embedding techniques of that time failed, and no reasonable modification seemed possible.

(2) No attempt had yet been made at proving lattices nonembeddable. Thus the nonembeddability of S_8 seemed to me to be at least as likely as its embeddability.

(3) If S_8 were found to be embeddable, everyone would forget my conjecture, thinking it to be frivolous. Otherwise, my prophecy would stand out. In fact, my conjecture was accurate, as Lachlan and Soare [**LaS**] later showed.

Another question about $\mathscr{R}$ which had been raised by Jockusch is the Categoricity Problem (CP): Is Th($\mathscr{R}$) $\aleph_0$-categorical? The general feeling was that Th($\mathscr{R}$) was not $\aleph_0$-categorical. To prove this, one needed to exhibit as $n > 0$ such that Th($\mathscr{R}$) realizes infinitely many n-types. Lachlan noted that if all finite lattices were embeddable into $\mathscr{R}$, then since there are infinitely many finite lattices with three generators, the non-$\aleph_0$-categoricity of Th($\mathscr{R}$) would follow. Lerman, Shore and Soare [**LSS**] followed up on this idea, embedding infinitely many nonisomorphic *partial lattices* (only certain join and meet relationships are specified on the universe of the structure) with three generators into $\mathscr{R}$. Their proof extended the Lachlan technique for embedding N_5 (the five-element nonmodular lattice pictured) to cases where nontrivial meets exist. Subsequently, Harrington and Shelah [**HS**] proved that Th($\mathscr{R}$) is not decidable (and not $\aleph_0$-categorical) using monstrous injury priority argument techniques.

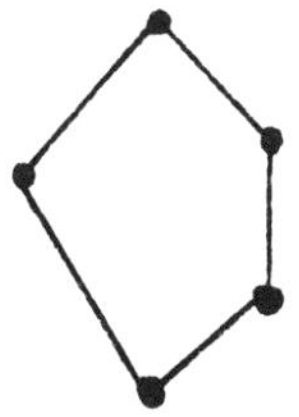

N_5

Ongoing work by Ambos, Lerman and Soare has resulted in both embedding and nonembedding conditions which subsume all known results. This work analyzes lattices in terms of decompositions of parts of the lattice by prime usl filters. Both embedding and nonembedding conditions are obtained, but these conditions are not complementary. There are two types of conflicts between meet preservation requirements and joint preservation requirements. One type of conflict has been resolved; a resolution of the second type of conflict will hopefully be found in the not too distant future, and it is also hoped that the two ways of resolving conflicts can coexist.

2. Embedding techniques. We wish, at least partially, to describe the evolution of techniques used to embed finite lattices into $\mathscr{R}$. We will not discuss techniques for proving lattices nonembeddable, except for a few comments which we make in this paragraph. These latter techniques are based on those used by Lachlan [**La1**] to prove the nondiamond theorem: If **a** and **b** are incomparable r.e. degrees and $\mathbf{a} \cup \mathbf{b} = \mathbf{0}'$, then $\mathbf{a} \cap \mathbf{b} \neq \mathbf{0}$. The nonembedding techniques and the embedding techniques are almost complementary in that a strategy for defeating one construction can almost invariably be turned into an extension of the previous strategy for the other construction. This interplay has been especially useful for determining whether particular lattices are embeddable, and has frequently been responsible for pointing the way to proofs of stronger results.

We first consider the basic technique for constructing a minimal pair $\mathbf{a}_0$ and $\mathbf{a}_1$ of r.e. degrees. We build r.e. sets A_0 and A_1 of degree $\mathbf{a}_0$ and $\mathbf{a}_1$, respectively, which satisfy the following requirements ($\{\Phi_e\colon e \in N\}$ is a recursive enumeration of all partial recursive functionals):

$$P_{e,i}\colon \Phi_e(A_i) \neq A_{1-i},$$
$$Q_{e,j}\colon \Phi_e(A_0) = \Phi_j(A_1) \text{ total } \quad \Phi_e(A_0) \text{ recursive.}$$

$P_{e,i}$ is attacked by appointing a follower x to witness its satisfaction. Thus we try to guarantee that either $\Phi_e(A_0; x)\uparrow$ (diverges) or $\Phi_e(A_0; x) \neq A_1(x)$. This is accomplished by waiting until $\Phi_e(A_0; x)$ converges. If and when this happens, we try to preserve the computation $\Phi_e(A_0; x)$ and place $x \in A_1$ if necessary, to achieve the inequality. Conflicts between requirements are resolved through the use of priorities as in Friedberg [**F**] and Muchnik [**M**].

We satisfy $Q_{e,j}$ by computing $\Phi_e(A_0)$ at carefully chosen stages. The choice of stages will depend on the action of higher priority requirements, which is determined by a finite amount of information. Once such requirements reach their final states, we look for a stage such that $\Phi_e(A_0; y) = \Phi_j(A_1; y)$ for all sufficiently small y, say all $y \leqslant x$. If, at that stage, $\Phi_e(A_0; x) = z$, we guarantee that all later stages, as we approximate to A_0 via $\{A_0^s\}$ and to A_1 via $\{A_1^s\}$, it will be the case that either $\Phi_e(A_0^s; x) = z$ or $\Phi_j(A_1^s; x) = z$. Thus if, in fact, $\Phi_e(A_0) = \Phi_j(A_1)$, then $\Phi_e(A_0; x) = z$. Each $Q_{e,j}$ must act to restrain followers, allowing injuries to only one of the computations $\Phi_e(A_0; x) = z$ or $\Phi_j(A_1; x) = z$ at a time, with no injuries allowed to the other computation until the first has recovered (and hence both, again, are equal). Each $Q_{e,j}$ will exercise only finite permanent restraint, permitting the eventual satisfaction of all requirements.

Let $\mathscr{L} = \langle L, \leqslant, \vee, \wedge \rangle$ be an arbitrary finite lattice. We wish to define an embedding taking each $a_i \in L$ to an r.e. set A_i. The following requirements must be satisfied:

$$S_{i,j}: a_i \leqslant a_j \Rightarrow A_i \leqslant_T A_j,$$
$$P_{e,i,j}: a_i \nleqslant a_j \Rightarrow \Phi_e(A_j) \neq A_i,$$
$$Q_{e,m,i,j}: a_i \wedge a_j = a_k \Rightarrow \big(\Phi_e(A_i) = \Phi_m(A_j) \text{ total} \Rightarrow \Phi_e(A_i) \leqslant_T A_k\big),$$
$$R_{i,j}: a_i \vee a_j = a_k \Rightarrow A_k \leqslant_T A_i \oplus A_j.$$

We satisfy $S_{i,j}$ by imposing the constraint that whenever $a_i \leqslant a_j$ and $x \in A_i^{s+1} - A_i^s$, then $x \in A_j^{s+1} - A_j^s$. $P_{e,i,j}$ is attacked as before, except that we now *target* the follower x for some A_k with $a_k \leqslant a_i$. If $\mathscr{L}$ is a distributive lattice, then $\{a_m: a_m \leqslant a_i \ \&\ a_m \nleqslant a_j\}$ has a least element a_k, and we target x for A_k instead of A_i. Thus if $a_u \vee a_v = a_w$ and $a_w \geqslant a_k$, then either $a_u \geqslant a_k$ or $a_v \geqslant a_k$ (else $a_w = a_u \vee a_v \leqslant a_j \ngeqslant a_k$). $R_{u,v}$ will then automatically be satisfied, as if $x \in A_w^{s+1} - A_w^s$, then $x \in A_u^{s+1} - A_u^s$ or $x \in A_v^{s+1} - A_v^s$.

We attempt to satisfy $Q_{e,m,i,j}$ as we previously satisfied $Q_{e,j}$. Thus we choose an appropriate stage at which

$$\Phi_{e,s}(A_i^s; x) = \Phi_{m,s}(A_j^t; x) = z \tag{1}$$

and try to guarantee that either $\Phi_{e,t}(A_i^t; x) = z$ or $\Phi_{m,t}(A_j^t; x) = z$ for all $t \geqslant s$. This may no longer be possible, as some follower x may enter A_k injuring both computations simultaneously. (Recall that $a_k = a_i \wedge a_j$.) We arrange the construction so that any follower which injures both computations of (1) is appointed as a follower before (1) is specified. This allows us to use an A_k oracle to determine whether the computation in (1) will be preserved. Thus $\Phi_e(A_i) \leqslant_T A_k$.

Join preservation requirements introduce new difficulties when nondistributive lattices are considered. Consider the nonmodular lattice N_5 pictured in the diagram. Suppose that we try to satisfy the requirement $\Phi_e(A_2) \neq A_3$ with the follower x targeted for A_3. Since $a_2 \vee a_4 = a_1 > a_3$, we must determine whether or not $x \in A_3$ through the use of A_2 and A_4 oracles. Since the computation $\Phi_e(A_2; x)$ must be preserved once it is found, we initially rely on an A_4 oracle. Thus we assign a trace x_1 for x, target x_1 for A_4, and require that if $x \in A_3^s$, then $x_1 \in A_4^r$ for some $r \leqslant s$. Since $a_3 \vee a_4 = a_0$, we may have to separate x and x_1 in order to avoid conflicts with meet preservation requirements. But once we discover a computation $\Phi_{e,s}(A_2^s; x)$, we can appoint a trace x_2 for x, target x_2 for A_2, and guarantee that x_2 is sufficiently large so as not to interfere with the computation just discovered. x and x_2 are free to enter their targeted sets simultaneously. Thus to decide if $x \in A_3$, we ask the A_4 oracle to tell us whether or not $x_1 \in A_4$. If $x_1 \notin A_4$ then $x \notin A_3$. If $x_1 \in A_4$, then we find the stage at

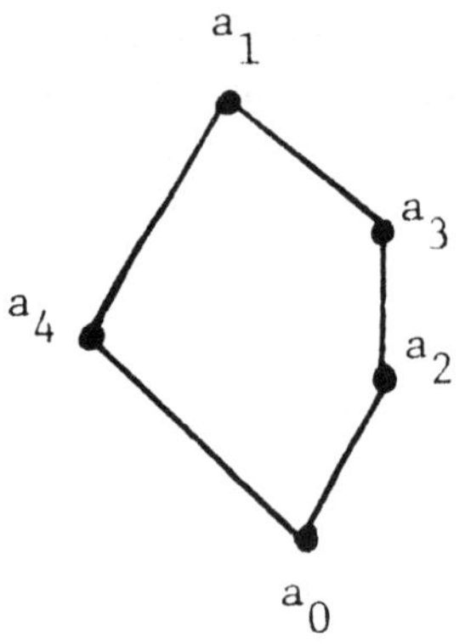

N_5

which x_1 entered A_4. If x does not enter A_4 at that same stage, then x_2 has already been appointed, and $x \in A_3 \Leftrightarrow x_2 \in A_2$.

The construction which embeds the five-element modular nondistributive lattice M_5 (see the diagram) into $\mathscr{R}$ requires a potentially infinite list of traces. For suppose that we try to satisfy the requirement $\Phi_e(A_3) \neq A_2$ with the follower x. Since $a_3 \vee a_4 \geqslant a_2$, $x = x_0$ will need a trace x_1 targeted for A_4. But $a_2 \vee a_3 \geqslant a_4$, so x_1 will need a trace targeted for A_2, etc. (A_3 targets cannot be used until we find a computation $\Phi_{e,s}(A_3^s; x)$.) If and when a computation $\Phi_{e,s}(A_3^s; x)$ is found, we may be prohibited by the fact that $a_2 \wedge a_4 = a_0$ from allowing x and all its traces to enter their target sets simultaneously, else the corresponding meet preservation requirement could be injured. Thus we arrange for each trace of x to be permitted by $N_{e,m,2,4}$ to enter its target set, one trace at a time, with traces appointed later entering their target sets first. Once these traces are separated, each x_k requires a new trace. But we are now free to assign traces targeted for A_3, at least for this one meet preservation requirement. We proceed through all meet preservation requirements of sufficiently high priority, lower priority requirements first, following a similar procedure of trace separation and appointment of new traces for each requirement. If each of these meet preservation requirements has reached its final state, x, all its traces, all their traces, etc. will enter their respective target sets, lower priority traces first, without injuring any meet preservation requirements.

The use of traces seems to be necessary for constructions embedding nondistributive lattices into $\mathscr{R}$. If such a lattice also has nontrivial meets, then we face the problem of having traces, rather than followers, injure meet preservation requirements. Since traces may be appointed relatively late in the construction, conflicts with meet preservation requirements arise. For the procedure to preserve meets relies on the ability to identify all followers and traces which may injure a computation $\Phi_e(A_i; x) = z$ at the time when the computation is specified. Such is the case with S_8, and the severity of these conflicts allowed Lachlan and Soare **[LaS]** to prove that S_8 is not embeddable into $\mathscr{R}$.

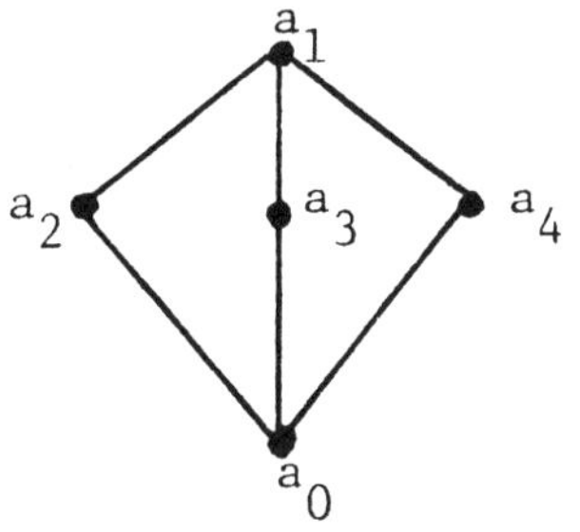

M_5

The solution to the EP will ultimately be based on an algebraic framework. A careful analysis of the sets for which traces for x must be appointed when x is targeted for A_k leads us to consider prime usl filters for the lattice. A *usl filter* F for $\mathscr{L}$ is an upwards closed subset of $\mathscr{L}$. F is *prime* if whenever $a \vee b \in F$, then either $a \in F$ or $b \in F$. When x is targeted for A_k, x must have traces targeted for A_m (or some A_r such that $a_r \leqslant a_m$) for every a_m in some prime usl filter F containing a_k (x is a trace of x). Given $a \in L$, $L[a]$ is the sublattice of L with universe $\{x \in L: x \leqslant a\}$. The nonembedding condition states that $\mathscr{L}$ is not embeddable into $\mathscr{R}$ if there are $a, b \in L$ with $b \leqslant a$ such that for every prime filter F for $L[a]$, if $F \cap L[b] = \varnothing$ then F is nonprincipal and contains an element c which is the meet of elements $p, q \in L$ with $p \nleqslant a$. This condition exactly reflects the situation where x will require the continuous appointment of traces targeted below A_j ($c = a_j$), some of which will be appointed too late for consideration in meet preservation. If F is principal, then its least point requires no traces, so this situation will not occur.

The other problem which arises in the embedding construction is the effect of the collapse of x and its traces into entities which enter their target sets separately. This requires a new assignment of traces, and may also necessitate the late appointment of injurious traces. This is the conflict which still remains to be resolved. We have a condition, EC, which allows us to resolve this conflict, but it is technical in nature, so we choose not to state it here. It is hoped that the final solution to the EP will be in terms of a simple and natural algebraic condition.

References

[A] K. Ambos-Spies, *On the structure of the recursively enumerable degrees*, Inaugural-Dissertation, Ludwigs-Maxmilians-Universitat Munchen, 1980.

[F] R. M. Friedberg, *Two recursively enumerable sets of incomparable degrees of unsolvability*, Proc. Nat. Acad. Sci. U.S.A. **43** (1957), 236–238.

[HS] L. Harrington and S. Shelah, *The undecidability of the recursively enumerable degrees*, Bull. Amer. Math. Soc. (N.S.) **6** (1982), 79–80.

[La1] A. H. Lachlan, *Lower bounds for pairs of recursively enumerable degrees*, Proc. London Math. Soc. **16** (1966), 537–569.

[La2] ______, *Embedding nondistributive lattices in the recursively enumerable degrees*, Conference in Mathematical Logic (London, 1970), Lecture Notes in Math., vol. 255, Springer-Verlag, Berlin, Heidelberg and New York, 1972, pp. 149–77.

[**LaS**] A. H. Lachlan and R. I. Soare, *Not every finite lattice in embeddable in the recursively enumerable degrees*, Adv. in Math. **37** (1980), 74--82.

[**LSS**] M. Lerman, R. A. Shore and R. I. Soare, *The elementary theory of the r. e. degrees is not $\aleph_0$-categorical*, Adv. in Math. (to appear).

[**M**] A. A. Muchnik, *On the unsolvability of the problem of reducibility in the theory of algorithms*, Dokl. Akad. Nauk SSSR **108** (1956), 194–197. (Russian)

[**Sa**] G. E. Sacks, *The recursively enumerable degrees are dense*, Ann. of Math. (2) **80** (1964), 300–312.

[**Sh1**] J. R. Shoenfield, *Applications of model theory to degrees of unsolvability*, Theory of Models (Proc. Internat. Sympos., Berkeley, 1963), North-Holland, Amsterdam, 1965, pp. 359–363.

[**Sh2**] ______, *The decision problem for recursively enumerable degrees*, Bull. Amer. Math. Soc. **81** (1975), 973–977.

[**T**] S. K. Thomason, *Sublattices of the recursively enumerable degrees*, Z. Math. Logik Grundlag. Math. **17** (1971), 273–280.

[**Y**] C. E. M. Yates, *A minimal pair of r. e. degrees*, J. Symbolic Logic **31** (1966), 159–168.

DEPARTMENT OF MATHEMATICS, UNIVERSITY OF CONNECTICUT, STORRS, CONNECTICUT 06268

Proceedings of Symposia in Pure Mathematics
Volume **42**, 1985

Major Subsets and Automorphisms of Recursively Enumerable Sets

WOLFGANG MAASS[1]

1. Introduction. For recursively enumerable (r.e.) sets M and A with $M \subseteq A$ one says that *M is a major subset of A* ($M \subset_m A$) if $A - M$ is infinite and if $A \cup W = N \Rightarrow M \cup W =^* N$ for every r.e. set W ($=^*$ means equality up to finitely many numbers, N is the set of all natural numbers).

According to Lachlan [**4**] every nonrecursive r.e. set A has a major subset M. An abundance of results has been proved about major subsets (see Soare [**17**] for a survey) since they are of critical importance for questions concerning decidability and automorphisms of the lattice of r. e. sets $\mathcal{E}^*$.

In Maass and Stob [**13**] it was shown that for any $M \subset_m A$ and $\tilde{M} \subset_m \tilde{A}$ there is an effective isomorphism between the intervals $\mathcal{E}^*(A - M)$ and $\mathcal{E}^*(\tilde{A} - \tilde{M})$. In this paper we point out some consequences of this result and answer two related questions.

The Leitmotiv of a large part of this paper is the structural resemblance between semilow$_{1.5}$ sets and sets in an interval bounded by a major subset. One calls an r.e. set D *semilow*$_{1.5}$ if there is a total recursive function f s.t., for all $e \in N$,

$$W_e \cap \overline{D} \text{ infinite} \Leftrightarrow W_{f(e)} \text{ infinite.}$$

If we consider sets $M \subset_m A$ and D with $M \subseteq D \subset_\infty A$, then $D \subset_m A$ and therefore the effective isomorphism from [**13**] supplies a recursive function f s.t., for all $e \in N$,

$$W_e \cap \overline{D} \cap (A - M) \text{ infinite} \Leftrightarrow W_{f(e)} \cap (A - M) \text{ infinite.}$$

1980 *Mathematics Subject Classification*. Primary 03D25; Secondary 03D30.

[1]During preparation of this paper the author was supported by the Heisenberg Programm der Deutschen Forschungsgemeinschaft, West Germany.

The outer splitting property is another common feature of semilow$_{1.5}$ sets and sets in an interval bounded by a major subset [**10**, **13**].

In §1 we use the splitting property from [**12**] in order to give a negative answer to a question of M. Lerman: Assuming $M \subset_m A$ and $\tilde{M} \subset_m A$, is there an isomorphism Φ: $\mathscr{E}^*(A) \to \mathscr{E}^*(A)$ with $\Phi(M^*) = \tilde{M}^*$?

In §2 we consider automorphisms of $\mathscr{M}^* := \mathscr{E}^*(A - M)$ for $M \subset_m A$. We generalize the notions promptly simple and semilow to $\mathscr{M}^*$ and show that promptly simple semilow sets are automorphic in $\mathscr{M}^*$.

In §3 we use another slight extension of the isomorphism construction from [**13**] in order to derive a few basic facts about the homomorphisms of Boolean algebras that are generated in $\mathscr{E}^*$ by major subsets.

In §4 we provide the antithesis to the previously exploited lowness properties of major subsets by showing that no major subset (and thus no r-maximal set) is semi-low$_2$.

For an arbitrary set $S \subseteq N$ one defines

$$\mathscr{E}(S) := \{ W \cap S | W \text{ r.e.} \}.$$

$\mathscr{E}(S)$ is a lattice under set-theoretic union and intersection.

We write

$$\mathscr{E}_C(S) := \{ U | U \in \mathscr{E}(S) \text{ and } S - U \in \mathscr{E}(S) \}$$

for the sublattice of complemented elements in $\mathscr{E}(S)$. Obviously $\mathscr{E}_C(S)$ is a Boolean algebra.

We write T^* for the equivalence class of T w.r.t. the equivalence relation $=^*$. We write $\mathscr{E}^*(S)$, $\mathscr{E}_C^*(S)$ for the corresponding quotient lattices.

CONVENTION. (1) Capital letters denote r.e. sets (unless we say "an arbitrary set").

(2) We say that an r.e. set A is semilow, semilow$_{1.5}$, etc. instead of saying that the complement of A has these properties as in the original definitions in Soare [**16**] and Bennison and Soare [**1**].

I would like to thank Alistair Lachlan, Bob Soare, Michael Stob and Martin Ziegler very much for stimulating discussions on the subject of this paper.

2. The splitting property for major subsets. M. Lerman has raised the following question. Assume that M and $\tilde{M}$ are two major subsets of an r.e. set A. Does there exist an automorphism Φ of $\mathscr{E}^*(A)$ with $\Phi(M^*) = \tilde{M}^*$?

It is tempting to believe that such an automorphism exists. Soare [**15**] has shown that any two maximal sets are automorphic in $\mathscr{E}^*$ and the construction of a major subset of a given set A can be arranged to look very similar to the construction of a maximal set inside the universe A. In both cases, $A - M$ consists of the final resting places of infinitely many markers Γ_e, $e \in N$, that seek to maximalize their e-state w.r.t. certain arrays of r.e. sets (see Soare [**17**, Theorem 8.2]). Nevertheless, the answer to the question above is no.

THEOREM 2.1. *Assume A is nonrecursive. Then there are major subsets M, $\tilde{M}$ of A such that $\Phi(M^*) \neq \tilde{M}^*$ for every automorphism Φ of $\mathscr{E}^*(A)$.*

In order to prove this theorem we consider the following property.

DEFINITION 2.2. Assume $B \subseteq D \subseteq A$. We say that *$D$ has the splitting property in $A - B$* if one can split every r.e. set $W \subseteq A$ into r.e. sets W', W'' s.t. $W' \subseteq D$ and

$$(A - W) \cup B \text{ not r.e.} \Rightarrow (A - W') \cup B, (A - W'') \cup B \text{ not r.e.}$$

If we take $A := N$ and $B := \varnothing$ this definition coincides with the weak splitting property from Maass, Shore and Stob [**12**].

Notice that one can define the splitting property in $A - B$ by an elementary definition over $\mathscr{E}^*(A - B)$. We show in Lemma 2.3 that every nonrecursive set A has a major subset M that has the splitting property in A, and in Lemma 2.5 that every nonrecursive set A has, as well, a major subset $\tilde{M}$ that has not the splitting property in A. This will finish the proof of Theorem 2.1. Observe that this argument exploits the structural resemblance to low sets, where the splitting property can be used to show that not all low sets are automorphic in $\mathscr{E}^*$.

LEMMA 2.3. *Assume A is not recursive. Then there is a major subset $M \subset_m A$ s.t. M is promptly simple in A and therefore has the splitting property in A.*

PROOF. A standard construction of a major subset $M \subset_m A$ as e.g. in Soare [**17**, Theorem 8.2] can obviously be combined with the satisfaction of the standard finitary positive requirements that make M promptly simple in A. According to Theorem 2.2 in [**12**] prompt simplicity implies the splitting property and this still holds if we substitute the universe N by A.

LEMMA 2.4. *Assume $B \subseteq D \subseteq A$, $\mathscr{E}^*(A - D)$ is not a Boolean algebra and the Turing degree of D is half of a minimal pair. Then D does not have the splitting property in $A - B$.*

PROOF. This is shown in Theorem 3.1 in Maass, Shore and Stob [**12**] for the case $A = N$ and $B = \varnothing$. The same argument works as well in this more general situation.

LEMMA 2.5. *Assume A is not recursive. Then there is a major subset $M \subset_m A$ s.t. M has not the splitting property in A.*

PROOF. Let **h** be a high Turing degree that is half of a minimal pair (see Lachlan [**2**]). Since **h** is high there exists a major subset $M \subset_m A$ of degree **h** (Lerman [**5**]). We then apply Lemma 2.4 with $B := \varnothing$ and $D := M$, and see that M does not have the splitting property in A.

One could as well construct directly a set M with the desired properties.

REMARK 2.6. Assume $M \subset_m A$. One can use the splitting property in $A - M$ in order to show, for various properties P, that not all sets D in $\mathscr{E}(A - M)$ with property P are automorphic in $\mathscr{E}^*(A - M) =: \mathscr{M}^*$. For many P it is easy to construct a set D with property P that has, in addition, the splitting property in

$A - M$. If there exist, in addition, sets D with property P in every high degree, one can use Lemma 2.4 with $B := M$ in order to get some D with property P that does not have the splitting property in $A - M$.

3. Automorphisms of $\mathcal{M}^*$. It is not so easy to find sets that are automorphic in $\mathcal{M}^*$, since there are no sets with few supersets in $\mathcal{M}^*$ (like e.g. maximal sets in $\mathcal{E}^*$). We show here that there are analogies of promptly simple and semilow sets in $\mathcal{M}^*$ and that sets with both properties are automorphic in $\mathcal{M}^*$.

Since it is easy to make sets in $\mathcal{M}^*$ promptly simple and semilow, and since all these sets realize the same 1-type in $\mathcal{M}^*$, we suggest as the next step to look for a decision procedure for $\forall\exists$-sentences in $\mathcal{M}^*$, which have a promptly simple semilow set in $\mathcal{M}^*$ as parameter (a decision procedure for $\forall\exists$-sentences without parameters in $\mathcal{M}^*$ is given in Stob [**18**]). This will supply valuable experience towards a decision procedure for the $\exists\forall\exists$-theory of $\mathcal{M}^*$.

DEFINITION 3.1. Consider sets B, D, A with $B \subseteq D \subseteq A$.

(a) We say that D *is promptly simple in* $A - B$ if $A - D$ is infinite and if there is a recursive function f and an enumeration of all r.e. subsets $(U_i)_{i \in N}$ of A s.t.

$$\forall j \in N\big(U_j \cap (A - B) \text{ infinite} \Rightarrow \exists x, s\big(x \in U_{j,s} - U_{j,s-1} \wedge x \notin B \wedge x \in D_{f(s)}\big)\big).$$

(b) We say that D *is semilow in* $A - B$ if there is an enumeration of D and all r.e. subsets $(U_i)_{i \in N}$ of A s.t.

$$\forall i((U_i \setminus D) \cap (A - B) \text{ infinite} \Rightarrow (U_i - D) \cap (A - B) \text{ infinite}).$$

REMARK 3.2. Because of Theorem 5.1 below not every characterization of semilow can be generalized to an interval $A - M$ with $M \subset_m A$. Of course the preceding definition is equivalent to the standard definition of semilow if $A - B = N$. Note that sets in $\mathcal{E}_C(A - B)$ are always semilow in $A - B$.

THEOREM 3.3. *Assume $M \subset_m A$. Then there is a set D with $M \subseteq D \subseteq A$ s.t. D is promptly simple in $A - M$ and D is semilow in $A - M$.*

PROOF. We use similar arguments as in Stob [**18**].

We construct a recursive enumeration of a set D s.t. $M \subseteq D \subseteq A$ and s.t. with respect to the standard enumeration of $(W_e)_{e \in N}$ the following requirements are satisfied for all $k, i, n, j \in N$:

$$N_k: |A - D| \geqslant k,$$

$$N_{i,n}: (W_i \setminus D) \cap (A - M) \text{ infinite} \Rightarrow |(W_i - D) \cap (A - M)| \geqslant n$$

and

$$P_j: W_j \cap (A - M) \text{ infinite} \Rightarrow \exists x, s\big(x \in W_{j,s} \cap (A_s - W_{j,s-1}) \cap A_{s-1} \wedge x \notin M \wedge x \in D_s\big).$$

We define in the usual fashion (see [13]) nondecreasing recursive functions ('movable markers') $\Gamma_{N_k}(s)$, $\Gamma_{N_{i,n}}(s)$, $\Gamma_{P_j}(s)$ s.t.

$$\lim_{s\to\infty} \Gamma_{N_k}(s) = \infty \quad \text{iff} \quad |A - D| < k,$$

$$\lim_{s\to\infty} \Gamma_{N_{i,n}}(s) = \infty \quad \text{iff} \quad \text{the conclusion of } N_{i,n} \text{ does not hold}$$

and

$$\lim_{s\to\infty} \Gamma_{P_j}(s) = \infty \quad \text{iff} \quad \text{the conclusion of } P_j \text{ does not hold.}$$

THE CONSTRUCTION. We fix enumerations of A and M. For $x \in A$ we define $s_x := \mu s(x \in A_s)$.

Stage s. We consider all $x \in A_s$. We place x in D if $x \in M_s$ or if there exists some $j \in N$ s.t.

$$\begin{aligned} &x \in W_{j,s} \cap (A_s - W_{j,s-1}) \cap A_{s-1}, \\ &\Gamma_{P_j}(s_x) > x, \\ &\Gamma_{N_k}(s_x) \leqslant x \quad \text{for all } k \leqslant j, s \\ &\Gamma_{N_{i,n}}(s_x) \leqslant x \quad \text{for all } \langle i, n\rangle \leqslant j \text{ with } x \in W_{i,s-1} \end{aligned}$$

(in this case we say that P_j forces x into D).

LEMMA 3.4. *Every requirement P_j forces only finitely many elements of $A - M$ into D.*

PROOF. If P_j forces some element of $(A - M)$ into D then $\lim_{s\to\infty} \Gamma_{P_j}(s) < \infty$.

LEMMA 3.5. *Every requirement N_k is satisfied.*

PROOF. Otherwise, $\Gamma_{N_k}(s_x) > x$ for almost all $x \in A - M$.

LEMMA 3.6. Every requirement $N_{i,n}$ is satisfied.

PROOF. Assume $N_{i,n}$ is not satisfied. Thus $(W_i \setminus D) \cap (A - M)$ is infinite and $\lim_{s\to\infty} \Gamma_{N_{i,n}}(s) = \infty$. Since $M \subset_m A$ we have $\Gamma_{N_{i,n}}(s_x) > x$ for almost all $x \in A - M$. Therefore, almost all of the infinitely many elements of $(W_i \setminus D) \cap (A - M)$ can only be forced into D by P_j with $j < \langle i, n\rangle$. Thus $N_{i,n}$ is satisfied since these P_j together force only finitely many elements of $A - M$ into D according to Lemma 3.4, a contradiction.

LEMMA 3.7. *Every requirement P_j is satisfied.*

PROOF. This follows easily from Lemmas 3.5 and 3.6.

This finishes the proof of Theorem 3.3.

THEOREM 3.8. *Assume $M \subset_m A$ and $\tilde{M} \subset_m \tilde{A}$. Further assume that $M \subseteq D \subseteq A$, $\tilde{M} \subseteq \tilde{D} \subseteq \tilde{A}$ and D, $\tilde{D}$ are promptly simple and semilow in $A - M$, resp. $\tilde{A} - \tilde{M}$. Then there is an isomorphism*

$$\Phi\colon \mathscr{E}^*(A - M) \to \mathscr{E}^*(\tilde{A} - \tilde{M}) \quad \textit{with } \Phi(D^*) = \tilde{D}^*.$$

PROOF. First we notice that $D \subset_m A$ and $\tilde{D} \subset_m \tilde{A}$. Therefore there exists an effective isomorphism [13]

$$\psi: \mathcal{E}^*(A - D) \to \mathcal{E}^*(\tilde{A} - \tilde{D}).$$

More exactly, there are simultaneously recursively enumerable arrays $(U_i)_{i\in N}$, $(\hat{V}_i)_{i\in N}$, $(\hat{U}_i)_{i\in N}$ and $(V_i)_{i\in N}$ s.t., for every $i \in N$, $U_i = {}^*W_i = {}^*V_i$ and s.t., for every state ν (see [10] for notation), infinitely many elements of $A - D$ have state ν w.r.t. $(U_i)_{i\in N}$, $(\hat{V}_i)_{i\in N}$ iff infinitely many elements of $\tilde{A} - \tilde{D}$ have state ν w.r.t. $(\hat{U}_i)_{i\in N}$, $(V_i)_{i\in N}$.

It follows from Lemma 5.5 in [13] that, in addition, the following weak covering property (∗) holds (notation as in [13]).

(∗) For every state ν: if infinitely many elements of $A - D$ have state ν w.r.t. $(U_i)_{i\in N}$, $(\hat{V}_i)_{i\in N}$ at some point of the enumeration then there is a state $\nu_1 \leqslant \nu$ s.t. infinitely many elements of $\tilde{A} - \tilde{D}$ have state ν_1 w.r.t. $(\hat{U}_i)_{i\in N}$, $(V_i)_{i\in N}$ at some point of the enumeration + symmetrical counterpart.

We proceed then analogously as for low sets in Maass [11]. We relativize the prompt and low shrinking property of [11] to the interval $A - M$. Since D is promptly simple and semilow in $A - M$, D has the prompt and low shrinking property in $A - M$ (this is proved exactly as in the unrelativized case [11]). Analogously $\tilde{D}$ has the prompt and low shrinking property in $\tilde{A} - \tilde{M}$. This implies via a relativized version of the Shrinking Lemma in [11] that we can shrink the sets $\hat{V}_i$ to sets $V_i' \subseteq \hat{V}_i$ and the sets $\hat{U}_i$ to sets $U_i' \subseteq \hat{U}_i$ s.t.

$$V_i' \cap (A - D) =^* \hat{V}_i \cap (A - D), \qquad U_i' \cap (\tilde{A} - \tilde{D}) =^* \hat{U}_i \cap (\tilde{A} - \tilde{D}),$$

and s.t., in addition, the covering property (∗∗) of the following Extension Theorem for $\mathcal{M}$ is satisfied. The conclusion of the Extension Theorem for $\mathcal{M}$ implies that we can continue the isomorphism $\psi: \mathcal{E}^*(A - D) \to \mathcal{E}^*(\tilde{A} - \tilde{D})$ to an automorphism $\Phi: \mathcal{E}^*(A - M) \to \mathcal{E}^*(\tilde{A} - \tilde{M})$ with $\Phi(D^*) = \tilde{D}^*$.

THEOREM 3.9 (EXTENSION THEOREM FOR $\mathcal{M}$). *Assume $M \subset_m D$ and $\tilde{M} \subset_m \tilde{D}$. Further assume that there is a simultaneous enumeration of M, D, $\tilde{M}$, $\tilde{D}$ and arrays*

$$(U_i)_{i\in N}, (V_i')_{i\in N}, (U_i')_{i\in N}, (V_i)_{i\in N} \text{ s.t. for every } i \in N$$
$$D \searrow V_i' = \varnothing \quad \text{and} \quad \tilde{D} \searrow U_i' = \varnothing,$$

and s.t. the following covering property (∗∗) holds.

(∗∗) *For every state ν: if infinitely many elements of $D - M$ enter D in state ν w.r.t. $(U_i)_{i\in N}$, $(V_i')_{i\in N}$ then there is a state $\nu_1 \leqslant \nu$ s.t. infinitely many elements of $\tilde{D} - \tilde{M}$ enter $\tilde{D}$ in state ν_1 w.r.t. $(U_i')_{i\in N}$, $(V_i)_{i\in N}$ + symmetrical counterpart.*

Then one can extend the sets V_i', U_i' to sets $\tilde{V}_i$, $\tilde{U}_i$ s.t., for every $i \in N$,

$$\tilde{V}_i \cap \overline{D} = V_i' \cap \overline{D}, \qquad \tilde{U}_i \cap \overline{\tilde{D}} = U_i' \cap \overline{\tilde{D}},$$

and s.t. for every state ν infinitely many elements of $D - M$ have state ν w.r.t. $(U_i)_{i\in N}$, $(\tilde{V}_i)_{i\in N}$ iff infinitely many elements of $\tilde{D} - \tilde{M}$ have state ν w.r.t. $(\tilde{U}_i)_{i\in N}$, $(V_i)_{i\in N}$.

PROOF. The construction is a trivial extension of the construction in [**13**]. The only difference in the proof occurs at the beginning of the verification of Claim 3 in the proof of [**13**, Lemma 5.5]. At this point one uses covering property $(**)$ instead of the argument: "Of course $\sigma_2 \neq \varnothing \ldots$".

4. On the structure of subalgebras and homomorphisms of Boolean algebras generated by major subsets. If S, $\tilde{S}$ are two arbitrary infinite sets with $S \subset \tilde{S}$, then the Boolean algebra B' of intersection with S of r.e. sets that look recursive on $\tilde{S}$ (i.e. $\{(R \cap S)^* | R^* \in \mathscr{E}^*(\tilde{S})$ and $(\tilde{S} - R)^* \in \mathscr{E}^*(\tilde{S})\}$) is a subalgebra of the Boolean algebra B of r.e. sets that look recursive on S (i.e. $\{U^* | U^* \in \mathscr{E}^*(S)$ and $(S - U)^* \in \mathscr{E}^*(S)\} =: \mathscr{E}_C^*(S)$). In general when $\tilde{S}$ grows, fewer sets look recursive on $\tilde{S}$ and thus the subalgebra $B' \hookrightarrow B$ shrinks.

We consider here only the case where $\tilde{S} := N$ (thus the sets that look recursive on $\tilde{S}$ are the real recursive sets) and $S := A - M$ for r.e. sets A and M.

Discussions with M. Stob led to the observation that the Owings Splitting Theorem restricts the subalgebras $B' \hookrightarrow B$ that arise in this way: if $B' \hookrightarrow B$ is represented as above then one can split every element of $B - B'$ into two elements of $B - B'$. It is not difficult to see that there are embeddings of Boolean algebras into the countable atomless Boolean algebra which do not have this special property.

The characterization of all subalgebras of Boolean algebras that arise in $\mathscr{E}^*$ in this way is an interesting although quite difficult project.

We restrict our attention here to those subalgebras that arise in the case where $M \subset_m A$. It is well known that the subalgebra $B' \hookrightarrow B$ that is represented by $\tilde{S} := N$ and $S := A - M$ (with M, A r.e.) as above satisfies

$$\forall b \in B(b \neq 0 \Rightarrow \exists a \in B(a < b \wedge a \notin B'))$$

iff $M \subset_m A$ (this follows as well from Corollary 4.5). In particular, B is always the atomless countable Boolean algebra if $M \subset_m A$.

In the case $M \subset_m A$ we have the special property that every recursive set R has (up to $=^*$) a unique continuation from $\bar{A}$ onto $\bar{M}$. Thus the function

$$H_{A,M}: \mathscr{E}_C^*(\bar{A}) \to \mathscr{E}_C^*(A - M)$$

defined by $(R \cap \bar{A})^* \mapsto (R \cap (A - M))^*$ for recursive sets R is independent of the choice of R and therefore well defined. Obviously $H_{A,M}$ is a homomorphism of Boolean algebras where Range($H_{A,M}$) is the previously considered subalgebra B'. By choosing suitable A we thus get some control over the subalgebra B'. Observe that $H_{A,M}$ is 1-1 if M is a small major subset of A and Range($H_{A,M}$) is just the 0-1 Boolean algebra if M is an r-maximal major subset of A.

These maps $H_{A,M}$ are of interest with regard to the characterization of orbits of r.e. sets under automorphisms of $\mathscr{E}^*$. The characterization of the orbit, in which an r.e. sets A lies, requires a full understanding of the function

$$\mathscr{E}^*(\bar{A}) \ni (W \cap \bar{A})^* \mapsto \{X^* \mid X \subseteq A, X \text{ r.e. and } X \cup (W - A) \text{ r.e.}\} \subseteq \mathscr{E}^*(A).$$

The preceding homomorphisms $H_{A,M}$ are embryos of these functions, where we consider only recursive sets W and restrict our attention to the smaller lattice $\mathscr{E}^*(A - M)$ (instead of $\mathscr{E}^*(A)$) where the continuation of the recursive set W is still unique.

THEOREM 4.1. *Assume $M \subset_m A$, $\tilde{M} \subset_m \tilde{A}$ and, for every recursive set R,*

$$R \cap (A - M) =^* \varnothing \Leftrightarrow R \cap (\tilde{A} - \tilde{M}) =^* \varnothing.$$

Then there is an isomorphism

$$\Phi: \mathscr{E}^*(A - M) \to \mathscr{E}^*(\tilde{A} - \tilde{M})$$

s.t., for every recursive set R,

$$\Phi((R \cap (A - M))^*) = (R \cap (\tilde{A} - \tilde{M}))^*.$$

PROOF. This is another inessential extension of the isomorphism construction in Maass and Stob [**13**]. We fix a simultaneous enumeration of an array $(R_i)_{i \in N}$ that contains exactly the recursive sets: at Stage s we enumerate x in R_i if

$$\phi_{i,s}(x) \simeq 1 \quad \text{and} \quad \forall y < x(\phi_{i,s}(y)\downarrow \text{ and } \phi_{i,s}(y) \in \{0,1\}).$$

LEMMA 4.2. *Assume $e \in N$. Almost all $x \in A - M$ have the property*

$$\{i \leqslant e \mid x \in R_i\} = \{i \leqslant e \mid x \in R_{i,t_x}, \text{ where } t_x := \mu t(x \in A_t)\}.$$

The same holds for $\tilde{A} - \tilde{M}$.

PROOF. If $R_i \cap (A - M)$ is infinite, then $x \in R_{i,t_x}$ for almost all $x \in R_i \cap (A - M)$.

We need an extension of the construction in Maass and Stob [**13**] similar to the one that was needed for the proof of the Extension Theorem for $\mathscr{M}$. We have to take into account that elements are already in certain sets R_i before they appear in the construction. But unlike the situation in the Extension Theorem, we do not even have to extend these R_i to get an exact matching of states on both sides. Lemma 4.2 enables us to perform the construction of [**13**] simultaneously but separately inside every e-state of the R_i.

States ν are now 4-tuples $\langle e, \sigma, \tau, \rho \rangle$ (instead of tuples $\langle e, \sigma, \tau \rangle$ as in [**13**]) with $\sigma, \tau, \rho \subseteq e + 1$.

For the construction in $A - M$ we say that $x \in A$ has state $\langle e, \sigma, \tau, \rho \rangle$ at Stage s if

$$\sigma = \{i \leqslant e \mid x \in U_{i,s}\}, \qquad \tau = \{i \leqslant e \mid x \in \hat{V}_{i,s}\}$$

and

$$\rho = \{i \leqslant e \mid x \in R_{i,t_x}\},$$

where t_x is the stage where the simultaneous enumeration function g enumerated x into A.

For the construction in $\tilde{A} - \tilde{M}$ we say that $\tilde{x} \in \tilde{A}$ has state $\langle e, \sigma, \tau, \rho \rangle$ at Stage s if

$$\sigma = \{i \leqslant e | x \in \hat{U}_{i,s}\}, \qquad \tau = \{i \leqslant e | x \in V_{i,s}\}$$

and

$$\rho = \{i \leqslant e | x \in R_{i,t_{\tilde{x}}}\},$$

where $t_{\tilde{x}}$ is the stage where $\tilde{x}$ entered $\tilde{A}$.

For $\nu = \langle e, \sigma, \tau, \rho \rangle$ and $\nu' = \langle e', \sigma', \tau', \rho' \rangle$ we say that $\nu \preccurlyeq \nu'$ (ν' extends ν) if $e \leqslant e'$, $\sigma = \sigma' \cap (e + 1)$, $\tau = \tau' \cap (e + 1)$ and $\rho = \rho' \cap (e + 1)$.

We say that $\nu \geqslant \nu'$ (ν covers ν') if $e = e'$, $\sigma \supseteq \sigma'$, $\tau \subseteq \tau'$ and $\rho = \rho'$.

The goal of the construction is to get sets U_i, $\hat{V}_i$, $\hat{U}_i$, V_i s.t., for every $i \in N$,

$$U_i =^* W_i \cap (A - M) \quad \text{and} \quad V_i =^* W_i \cap (\tilde{A} - \tilde{M})$$

and s.t. for every state $\nu = \langle e, \sigma, \tau, \rho \rangle$ infinitely many $x \in A - M$ have final state ν iff infinitely many $\tilde{x} \in \tilde{A} - \tilde{M}$ have final state ν.

Of course, a stream $\mathscr{S}(X)$ consists now of 4-tuples $\nu = \langle e, \sigma, \tau, \rho \rangle$ and boxes B_ν, a well as the function $q(s, \nu)$, are now defined for these extended states ν.

The only change in the proof of [**13**] occurs in the verification of Claim 3 in the proof of [**13**, Lemma 5.5]. One argues now as follows.

By contradiction fix $\nu_2 = \langle e, \sigma_2, \tau_2, \rho_2 \rangle$ such that the claim fails for ν_2, σ_2 is minimal and τ_2 is minimal for σ_2. Assume first that $\sigma_2 = \varnothing$. Since ν_2 occurs infinitely often in $\mathscr{T}$, there are infinitely many $\hat{y} \in \tilde{A} - \tilde{M}$ s.t.

$$\rho_2 = \{i \leqslant e | \hat{y} \in R_{i,t_{\hat{y}}}\}.$$

By the preceding lemma this implies, for the recursive set $R := \bigcap\{R_i | i \in \rho_2\}$, that $R \cap (\tilde{A} - \tilde{M})$ is infinite. By the assumption of the theorem this implies that $R \cap (A - M)$ is as well infinite. Thus, by Lemma 4.2,

$$S := \{x \in A - M | \{i \leqslant e | x \in R_{i,t_x}\} = \rho_2\}$$

is infinite. All $x \in S$ are in some state $\nu' = \langle e, \sigma', \tau', \rho_2 \rangle$ with $\tau' = \varnothing$ when they run for the first time over track $\mathscr{G}$. We have $\nu' \geqslant \nu$ for such a state ν' and therefore Claim 3 holds for ν_2, a contradiction. One argues then for $\sigma_2 \neq \varnothing$ as in [**13**]. The analogous versions of Lemmas 5.6 and 5.7 from [**13**] show that the construction meets our previously mentioned goal.

This finishes the proof of Theorem 4.1.

COROLLARY 4.3. *Assume M and $\tilde{M}$ are major subsets of A s.t. the maps*

$$H_{A,M}: \mathscr{E}_C^*(\bar{A}) \to \mathscr{E}_C^*(A - M) \quad \text{and} \quad H_{A,\tilde{M}}: \mathscr{E}_C^*(\bar{A}) \to \mathscr{E}_C^*(A - \tilde{M})$$

have the same kernel. Then $H_{A,M}$ and $H_{A,\tilde{M}}$ are identical up to isomorphism, i.e. there exists an isomorphism Φ: $\mathscr{E}^*(A - M) \to \mathscr{E}^*(A - \tilde{M})$ s.t., for all recursive R,

$$H_{A,\tilde{M}}((R \cap \bar{A})^*) = \Phi(H_{A,M}((R \cap \bar{A})^*)).$$

PROOF. This follows immediately from Theorem 4.1 and the definition of $H_{A,M}$ before Theorem 4.1.

REMARK 4.4. The previous corollary shows that for a fixed A the homomorphisms $H_{A,M}$ for $M \subset_m A$ are already determined by their kernel. This becomes different if we let A vary. The following is an example of sets A, $\tilde{A}$, where both $\mathcal{E}^*(\overline{A})$ and $\mathcal{E}^*(\overline{\tilde{A}})$ are the countable atomless Boolean algebra, and of sets $M \subset_m A$ and $\tilde{M} \subset_m \tilde{A}$ s.t. both $H_{A,M}$ and $H_{\tilde{A},\tilde{M}}$ are 1-1 embeddings of the countable atomless Boolean algebra into the countable atomless Boolean algebra $\mathcal{E}^*_C(A - M)$, resp. $\mathcal{E}^*_C(\tilde{A} - \tilde{M})$, but where the Boolean algebra $\mathcal{E}^*_C(A - M)$ together with the distinguished subalgebra Range($H_{A,M}$) has a different structure than $\mathcal{E}^*_C(\tilde{A} - \tilde{M})$ together with the subalgebra Range($H_{\tilde{A},\tilde{M}}$). We take both A and $\tilde{A}$ to be atomless hyperhypersimple. We choose A semilow$_2$ and $\tilde{A}$ s.t. it does not have an r-maximal major subset (see Lerman, Shore and Soare [**6**]). We let M ($\tilde{M}$) be any small major subset of A ($\tilde{A}$). This implies immediately that both $H_{A,M}$ and $H_{\tilde{A},\tilde{M}}$ are 1-1 (thus they have the same kernel). Further, one can split every nonzero element of $\mathcal{E}^*_C(\tilde{A} - \tilde{M})$ by an element in Range($H_{\tilde{A},\tilde{M}}$) (i.e. by a recursive set) whereas there is a nonzero element in $\mathcal{E}^*_C(A - M)$ which cannot be split by elements in Range($H_{A,M}$) (actually below every nonzero element in $\mathcal{E}^*_C(A - M)$ there is one with this property).

The following corollary shows that for $M \subset_m A$ the homomorphism $H_{A,M}$ has a characteristic homogeneity property.

COROLLARY 4.5. *Assume $M \subset_m A$. Then the homomorphism*

$$H_{A,M}: \mathcal{E}^*_C(\overline{A}) \to \mathcal{E}^*_C(A - M)$$

has the following homogeneity property: one can split every nonzero element $T^* \in \mathcal{E}^*_C(A - M)$ into two pieces $T^*_1, T^*_2 \in \mathcal{E}^*_C(A - M) -$ Range$(H_{A,M})$ s.t., for every $i \in \{1,2\}$, the homomorphism

$$H^{T_i}_{A,M}: \mathcal{E}^*_C(\overline{A}) \to \mathcal{E}^*_C(T_i)$$

with $H^{T_i}_{A,M}(b) = H_{A,M}(b) \cap T^*_i$ for $b \in \mathcal{E}^*_C(\overline{A})$ has the same structure as the homomorphism

$$H^{T}_{A,M}: \mathcal{E}^*_C(\overline{A}) \to \mathcal{E}^*_C(T)$$

with $H^{T}_{A,M}(b) = H_{A,M}(b) \cap T^*$ for $b \in \mathcal{E}^*_C(\overline{A})$, i.e. there is an isomorphism

$$\Phi_i: \mathcal{E}^*_C(T_i) \to \mathcal{E}^*_C(T)$$

s.t., for all $b \in \mathcal{E}^*_C(\overline{A})$,

$$H^{T}_{A,M}(b) = \Phi_i\big(H^{T_i}_{A,M}(b)\big).$$

PROOF. Every nonzero element $T^* \in \mathcal{E}^*_C(A - M)$ has the form $(W - M)^*$ for some r.e. W with $M \subseteq W \subseteq A$ and $M \cup \overline{W}$ not r.e. According to the Owings Splitting Theorem [**17**] we can split W into r.e. sets W_1, W_2 such that, for all r.e. U and $i \in \{1,2\}$,

$$(U - W_i) \cup M \text{ r.e.} \Rightarrow (U - W) \cup M \text{ r.e.}$$

We define $T_i := W_i - M$.

We show that for every recursive R we have, for $i \in \{1,2\}$,

$$(*) \qquad R \cap (W - M) \text{ infinite} \Leftrightarrow R \cap (W_i - M) \text{ infinite}.$$

Assume R is recursive and $R \cap (W_i - M)$ is finite. Then $(R - W_i) \cup M$ is r.e. and thus $(R - W) \cup M$ is as well r.e. The union of the r.e. sets $(R - W) \cup M$ and $\overline{R} \cup W$ is equal to N. We apply reduction to these sets and get a recursive R_0 with $R_0 \cap \overline{W} = R \cap \overline{W}$ and $R_0 \cap (W - M) = \varnothing$. Since $M \subset_m W$, this implies that $R \cap (W - M)$ is finite.

We get $T_i^* \notin \text{Range}(H_{A,M})$ directly from $(*)$. Further with $(*)$ we get the desired isomorphism Φ_i from Theorem 4.1.

5. Major subsets are not semilow$_2$. An r.e. set B is called semilow$_2$ if $\{e | W_e \cap \overline{B} \text{ infinite}\} \leqslant_T 0''$. Many classes of r.e. sets (e.g. atomless hyperhypersimple sets) that consist only of sets of high degree contain some particularly well-behaved representatives that are semilow$_2$. These often have interesting special properties (e.g. all semilow$_2$ atomless hyperhypersimple sets are automorphic **[9]**). We show below that the class of major subsets (and thus the class of r-maximal sets that are not maximal) does not contain semilow$_2$ sets. On the contrary, all major subsets are as far away from being semilow$_2$ as possible. It follows from **[13]** that the Turing degree of the set

$$\{e | W_e \cap (A - M) \text{ infinite}\}$$

does not depend on the choice of A, M with $M \subset_m A$. We show here that this degree is equal to $0'''$. It is obvious that $\{e | W_e \cap (A - M) \text{ infinite}\}$ is recursive in $\{e | W_e \cap \overline{M} \text{ infinite}\}$.

THEOREM 5.1. *Assume $M \subset_m A$. Then*

$$\deg\{e | W_e \cap (A - M) \text{ infinite}\} = 0'''.$$

PROOF. Let $S_3 \in 0'''$ be a Σ_3^0 set. It is easy to construct an r.e. set W_{e_0} and a recursive function p s.t.

$$\forall e \in N\big(e \in S_3 \Leftrightarrow W_{p(e)} - W_{e_0} \text{ finite}\big).$$

If $M \subset_m A$ then M has the outer splitting property in A **[13]**. Thus there is a recursive function g s.t. the sets $W_{g(i)}$ are pairwise disjoint,

$$A = \bigcup_{i \in N} W_{g(i)}$$

and

$$\forall i \in N\big(W_{g(i)} \cap (A - M) \text{ is finite and nonempty}\big).$$

We use this to embed $\mathscr{E}^*$ effectively into $\mathscr{E}^*(A - M)$. There is a recursive function k s.t., for all $e, e' \in N$,

$$W_e - W_{e'} \text{ finite} \Leftrightarrow \big(W_{k(e)} - W_{k(e')}\big) \cap (A - M) \text{ finite}$$

(set $W_{k(e)} := \bigcup_{i \in W_e} W_{g(i)}$).

We define $\tilde{M} := W_{k(e_0)} \cup M$. Then, for all $e \in N$,

$$e \in S_3 \Leftrightarrow W_{p(e)} - W_{e_0} \text{ finite} \Leftrightarrow \left(W_{k(p(e))} - W_{k(e_0)}\right) \cap (A - M) \text{ finite}$$

$$\Leftrightarrow W_{k(p(e))} \cap (A - \tilde{M}) \text{ finite}.$$

Thus S_3 is recursive in $\{e | W_e \cap (A - \tilde{M}) \text{ finite}\}$ and since $\tilde{M} \subset_m A$ this set has the same degree as

$$\left\{e | W_e \cap (A - M) \text{ finite}\right\}.$$

REFERENCES

1. V. L. Bennison and R. I. Soare, *Some lowness properties and computational complexity sequences*, Theoret. Comput. Sci. **6** (1978), 233–254.

2. A. H. Lachlan, *Lower bounds for pairs of recursively enumerable degrees*, Proc. London Math. Soc. **3** (1966), 537–569.

3. ______, *On the lattice of recursively enumerable sets*, Trans. Amer. Math. Soc. **130** (1968), 1–37.

4. ______, *The elementary theory of recursively enumerable sets*, Duke Math. J. **35** (1968), 123–146.

5. M. Lerman, *Some theorems on r-maximal sets and major subsets of recursively enumerable sets*, J. Symbolic Logic **36** (1971), 193–215.

6. M. Lerman, R. A. Shore and R. I. Soare, *r-maximal major subsets*, Israel J. Math. **31** (1978), 1–18.

7. M. Lerman and R. I. Soare, *d-simple sets, small sets and degree classes*, Pacific J. Math. **87** (1980), 135–155.

8. W. Maass, *Recursively enumerable generic sets*, J. Symbolic Logic **47** (1982), 809–823.

9. ______, *On the orbits of hyperhypersimple sets*, J. Symbolic Logic **49** (1984), 51–62.

10. ______, *Characterization of recursively enumerable sets with supersets effectively isomorphic to all recursively enumerable sets*, Trans. Amer. Math. Soc. **279** (1983), 311–336.

11. ______, *Variations on promptly simple sets*, J. Symbolic Logic (to appear).

12. W. Maass, R. A. Shore and M. Stob, *Splitting properties and jump classes*, Israel J. Math. **39** (1981), 210–224.

13. W. Maass and M. Stob, *The intervals of the lattice of recursively enumerable sets determined by major subsets*, Ann. Pure Appl. Logic **24** (1983), 189–212.

14. M. Rubin, *The theory of Boolean algebras with a distinguished subalgebra is undecidable*, Ann. Sci. Univ. Clermont-Ferrand II Math. **13** (1976), 129–133.

15. R. I. Soare, *Automorphisms of the lattice of recursively enumerable sets. Part* I: *Maximal sets*, Ann. of Math. (2) **100** (1974), 80–120.

16. ______, *Computational complexity, speedable and levelable sets*, J. Symbolic Logic **42** (1977), 545–563.

17. ______, *Recursively enumerable sets and degrees*, Bull. Amer. Math. Soc. **84** (1978), 1149–1181.

18. M. Stob, Doctoral Dissertation, University of Chicago, Chicago, 1980.

DEPARTMENT OF MATHEMATICS, UNIVERSITY OF CHICAGO, CHICAGO ILLINOIS 60637

Current address: Department of Mathematics, Statistics and Computer Science, University of Illinois at Chicago, Illinois 60680

Proceedings of Symposia in Pure Mathematics
Volume **42**, 1985

The Structure of the Degrees of Unsolvability

RICHARD A. SHORE

Abstract. Our goal in these lectures is to survey the results to date on the structure of the degrees of unsolvability ordered by Turing reducibility. We also hope to give some indications of the methods of proof used for the various types of theorems that will be discussed. We will start with local and first order facts, both old and new, and then move on to global and second order results including ones on definability, automorphisms and homogeniety problems. Although we will mainly be concerned with the structure of the degrees $\mathscr{D}$ as a whole, most of what we say will apply to the standard substructures of $\mathscr{D}$ as well: $\mathscr{D}(\leqslant \mathbf{0}')$, the degrees below $\mathbf{0}'$; $\mathscr{A}$, the degrees of the arithmetic sets; $\mathscr{H}$, the degrees of the hyperarithmetic sets; etc. Much of it will also be applicable to other reducibility orderings and degrees from 1-1 to arithmetic. In addition to the papers that we will cite, an excellent source for nearly all of what we will cover (and more) is Lerman [**1983**].

1. Local and first order results. The notion of relative computability first appears in Turing [**1939**] while the study of the r.e. degrees was intiated in Post [**1944**] and of the structure of the degrees as a whole in Kleene and Post [**1954**]. We will begin with some of the basic facts established in the 1954 paper.

FACT 1.1. $\mathscr{D}$ is an upper semilattice with least element $\mathbf{0}$, the degree of the recursive sets. Note that $\mathbf{a} \vee \mathbf{b} = \deg(A \oplus B)$. (We use $A, B, C, \ldots$ to stand for subsets of ω, and $\mathbf{a}, \mathbf{b}, \mathbf{c}, \ldots$ for the corresponding degrees and set $A \oplus B = \{2x | x \in A\} \cup \{2x + 1 | x \in B\}$.)

FACT 1.2. There are no maximal elements in $\mathscr{D}$.

PROOF. Either by a direct *diagonalization* against the countably many sets recursive in a given one A or by the effective version of this argument:

Jump operator: $\mathbf{A}' = \deg(A')$, where $A' = \{e | \phi_e^A(e)\downarrow\}$ and $\mathbf{a} < \mathbf{a}'$.

REMARK 1.3. Iterating the jump gives a linearly ordered strictly increasing sequence of degrees which can be continued into the transfinite through the recursive ordinals by taking recursive joins at limit levels, e.g. $\varnothing^{(\omega)} = \bigoplus_{n \in \omega} \varnothing^{(n)} = \{\langle n, x\rangle | x \in \varnothing^{(n)}\}$. At ordinals beyond ω_1^{ck} the situation becomes more complex and is intimately connected with the fine structure of L and master codes for

1980 *Mathematics Subject Classification.* Primary 03D30, 03–02.

countable levels of the constructible hierarchy. There has been a long line of work in this area starting with Mostowski, Kleene and Post with major contributions made later by Putnam and his coworkers. As we are not primarily interested in the jump operator we will just refer the reader to some of the more recent work along these lines: Jockusch and Simpson [**1976**] and Hodes [**1978**, **1980**, **1983**].

FACT 1.4. Every degree has at most countably many predecessors as there are only countably many reduction procedures.

FACT 1.5. $\mathscr{D}$ is not linearly ordered; in fact,

THEOREM 1.6 (KLEENE AND POST [**1954**]). *Every countable partial ordering can be embedded in* $\mathscr{D}$.

PROOF. As for the r.e. degrees we divide the goal into countably many *requirements* but now no priorities are needed. We meet each requirement in turn by a *finite extension* of the finite approximations to the sets being constructed which have been determined so far. □

It is frequently useful and usually suggestive to think of degree constructions as forcing arguments in settings smaller than set theory. Thus the method of Kleene and Post corresponds to Cohen forcing with finite conditions with the requirements corresponding to the list of sentences to be decided (by forcing them or their negations).

There are two important phenomena which we can illustrate with this construction:

Relativization. By replacing the list of partial recursive functions by the list of ones partial recursive in A in the above construction and then joining each of the sets constructed with A we see that we can embed any countable p.o. in $\mathscr{D}(\geqslant \mathbf{a})$, the degrees above $\mathbf{a}$ for any degree $\mathbf{a}$.

Localization. A careful examination of this construction gives bounds on where the degrees constructed lie. Basically given the p.o. the only questions are whether or not there exists a finite extension of a given finite set forcing some convergence fact—a Σ_1 question and so recursive in one jump. Combining this calculation with the above relativization we have

THEOREM 1.6a. *Every countable p.o. can be embedded in* $\mathscr{D}[\mathbf{a}, \mathbf{a}']$ *for every* $\mathbf{a}$.

Note that a priori we seem to need to work above the presentation of the given p.o. This problem is avoided here by observing that there is an $\aleph_1$-universal recursively presented p.o. (the atomless Boolean algebra). Equivalently, it suffices to embed a countable set S of independent degrees, i.e., no element of S is recursive in the join of any finite subset of other elements of S. It will be important later, however, that no such trick is available for embeddings preserving more structure, e.g., as upper semilattices or as initial segments.

This embedding theorem alone suffices to determine the one-quantifier theory of $\mathscr{D}$.

COROLLARY 1.7. *For every* **a**, *any existential sentence is true in* $\mathscr{D}(\geqslant \mathbf{a})$ *iff it is consistent with the axioms for a partial order. Thus the* 1-*quantifier theory of* $\mathscr{D}(\geqslant \mathbf{a})$ *is decidable and independent of* **a**.

The next step up from the embedding problem is that of extension of embeddings. Again the notion of independence supplies an easy formulation of such a result.

THEOREM 1.8 (KLEENE AND POST [**1954**]). *For each* $n, m \in \omega$ *we have that* $\forall A_1 \cdots \forall A_m \exists B_1 \cdots \exists B_n$ *s.t.*

$$\forall k \leqslant n \left[B_k \nleqslant \bigoplus_{j \leqslant m} A_j \oplus \bigoplus_{k \neq i \leqslant n} B_i \right] \quad \textit{and} \quad \forall i \leqslant m \left[A_i \nleqslant \bigoplus_{i \neq j \leqslant m} A_j \oplus \bigoplus_{k \leqslant n} B_k \right].$$

PROOF. *Finite forcing.*

Relativization. Just add A to the set of the A_i.

Localization. One can get all the $B_k \leqslant (\oplus_{j \leqslant m} A_j)'$.

We call this an extension of embeddings result since it easily implies

COROLLARY 1.9. *If* $P_1 \subseteq P_2$ *are finite lattices (or equivalently upper semilattices) with* 0 *s.t. no element of* $P_2 - P_1$ *is below any element of* P_1 *and* $f\colon P_1 \to \mathscr{D}$ *is an embedding, then we can find a* g *to complete the diagram*:

$$\begin{array}{ccc} P_1 & \xrightarrow{f} & \mathscr{D} \\ \downarrow & \nearrow_{g} & \\ P_2 & & \end{array}$$

The obvious question now is whether the restriction on P_2 in the corollary is necessary. The simplest case here asks if there is a minimal degree. This question was left open by Kleene and Post but answered affirmatively in Spector [**1956**].

THEOREM 1.10. $\exists$ *a minimal degree, i.e. an* **a**, *such that the only degree* $\mathbf{b} < \mathbf{a}$ *is* $\mathbf{0}$.

PROOF. *Forcing with recursive perfect trees.*

Relativization. 1.10a. Every degree **a** has a *minimal cover* **b**, i.e. $(\mathbf{a}, \mathbf{b}) = \varnothing$.

Localization. Here the sentences that must be decided are 2-quantifier ones and so Spector's minimal degree is recursive in $\mathbf{0}^{(2)}$.

There then followed a long series of extensions, improvements and refinements characterizing initial segments of $\mathscr{D}$ by many workers. We mention three important ones.

THEOREM 1.11 (LACHLAN [**1968**]). *Every countable distributive lattice* $\mathscr{L}$ *with* 0 *and* 1, *is isomorphic to an initial segment of* $\mathscr{D}$.

THEOREM 1.12 (LERMAN [**1971**]). *Every finite lattice* $\mathscr{L}$ *is isomorphic to an initial segment of* $\mathscr{D}$.

THEOREM 1.13 (LACHLAN AND LEBEUF [**1976**]). *Every countable upper semilattice with* 0 *is isomorphic to an initial segment of* $\mathscr{D}$.

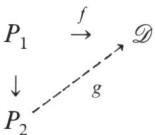

PROOFS. Quite considerable elaborations of the minimal degree construction. One still forces with recursive perfect trees but they must now be constructed with various homogeneity and uniformity properties that essentially guarantee that the forcing conditions deciding the required two quantifier sentences are still dense. These restrictions correspond to algebraic refinements of the lattice-theoretic representation theorems that had to be proved to actually produce even a manageable list of requirements for the construction.

The relativizations of these theorems is as usual essentially immediate. Localization computations again require two jumps but as there is no recursively presented universal lattice (or even u.s.l.) we also need a presentation of the given lattice $\mathscr{L}$.

THEOREM 1.11a. *If $\mathscr{L}$ is a countable distributive lattice presentable recursively in* $\mathbf{c} \geqslant \mathbf{a}''$ *then $\mathscr{L}$ is isomorphic to initial segments of* $\mathscr{D}[\mathbf{a}, \mathbf{c}]$ *and* $\mathscr{D}(\leqslant \mathbf{c})$.

THEOREM 1.12a. *For every* $\mathbf{a}$, *every finite lattice is isomorphic to an initial segment of* $\mathscr{D}[\mathbf{a}, \mathbf{a}'']$.

Now Theorem 1.12a shows that the requirements of Corollary 1.9 are, in fact, necessary. This suffices to decide the 2-quantifier theory of $\mathscr{D}$.

COROLLARY 1.14 (SHORE [**1978**] AND LERMAN [**1983**]). *For every* $\mathbf{a}$ *the* 2-*quantifier theory of* $\mathscr{D}(\geqslant \mathbf{a})$ *is decidable and independent of* $\mathbf{a}$.

PROOF. One reduces any $\forall\exists$ sentence to a finite set of extension of embeddings problems which are then answered one way or the other by Theorem 1.12a and Corollary 1.9.

On the other hand, since the theory of distributive lattices is undecidable, Theorem 1.12a proves

THEOREM 1.15. (LACHLAN [**1968**]). *For each* $\mathbf{a}$ *the theory of* $\mathscr{D}(\geqslant \mathbf{a})$ *is undecidable.*

In fact, decidability stops at the 2-quantifier level and the facts we have so far suffice to prove

THEOREM 1.16 (SCHMERL; SEE LERMAN [**1983**]). *For each* $\mathbf{a}$ *the* 3-*quantifier theory of* $\mathscr{D}(\geqslant \mathbf{a})$ *is undecidable.*

Such results and the method of proof by relativization suggested

ROGER'S HOMOGENEITY PROBLEM. Is Th($\mathscr{D}(\geqslant \mathbf{a})$) independent of $\mathbf{a}$? Or even: Are $\mathscr{D}(\geqslant \mathbf{a})$ and $\mathscr{D}(\geqslant \mathbf{b})$ isomorphic for every $\mathbf{a}$ and $\mathbf{b}$?

The solutions will come later.

Two questions about improvements of these embedding results come immediately to mind. First, can the localization results be improved from two jumps to one?

THEOREM 1.17 (SACKS [**1963**]). *Every degree* $\mathbf{a}$ *has a minimal cover below* $\mathbf{a}'$.

PROOF. A combination of a recursive approximation construction and a priority argument. (A recursive approximation construction is one in which one recursively defines a function $f(x, s)$ such that $\forall x[\lim_{s\to\infty} f(x, s)$ exists]. The limit function f is then recursive in $\mathbf{0}'$ by the limit lemma.)

Again there followed a sequence of improvements by Yates, Cooper, Epstein, etc. culminating in Lerman [**1983**] who proved, for example,

THEOREM 1.18. *If $\mathscr{L}$ is a countable upper semilattice presentable recursively in* $\mathbf{0}'$ *then $\mathscr{L}$ is isomorphic to an initial segment of* $\mathscr{D}(\leqslant \mathbf{0}')$.

The proof is too complicated for us to even name the methods except by indirection: forcing with partial trees, infinite injury arguments and more.

For a survey of other types of results on the structure of $\mathscr{D}(\leqslant \mathbf{0}')$ see Posner [**1980**].

The second direction of generalization is into the uncountable. Of course by Fact 1.4 we can only consider locally countable partial orderings or lattices (i.e. ones in which every element has only countably many predecessors) of size at most $c = 2^{\aleph_0}$.

THEOREM 1.19 (SACKS [**1963**, **1966**]). (1) *If $\mathscr{P}$ is a locally countable p.o. of size at most c in which every element has at most $\aleph_1$ many successors, then $\mathscr{P}$ is embeddable in $\mathscr{D}$.*

(2) *If $\mathscr{P}$ is a locally finite p.o. of size at most c, then $\mathscr{P}$ is embeddable in $\mathscr{D}$.*

Of course Theorem 1.19(1) entirely settles the problem of embedding partial orderings in $\mathscr{D}$ if CH holds. For initial segments the problem has now been solved assuming CH. Abraham and Shore [**1985**] have shown that every locally countable upper semilattice (with 0) of size $\aleph_1$ is embeddable as an initial segment of $\mathscr{D}$. (Rubin had announced such a result for distributive lattices in Rubin [**1979a**, **1979b**].) Without CH the situation is unsettled. For initial segment, lattice and u.s.l. embeddings, Groszek and Slaman [**1983**] have an independence result.

THEOREM 1.20 (GROSZEK AND SLAMAN). Con(ZF) $\to$ Con(ZFC + *there is a locally finite u.s.l. of size $\aleph_2 < c$ which cannot be embedded in $\mathscr{D}$*).

Although the situation for p.o.'s is still completely open they have a result which shows that the known methods cannot work.

THEOREM 1.21 (GROSZEK AND SLAMAN). Con(ZF) $\to$ Con(ZFC + *there is a maximal independent set in $\mathscr{D}$ of size $< c$*).

2. Some global and second order facts. Of course, by considering embeddings of infinite and especially uncountable structures in $\mathscr{D}$ we have already moved from first order and local properties to second order and global ones. We now want to consider some other related second order properties that will have both first order and global consequences.

Perhaps the algebraically most natural second order objects in a lattice or u.s.l. are the ideals (i.e. the sets closed downward and under join). Of course there are always the principal ideals $I_{\mathbf{x}} = \{\mathbf{y} | \mathbf{y} \leqslant \mathbf{x}\}$ which are always countable. It is a somewhat surprising fact (and a quite important one) that every countable ideal is the intersection of two principal ones.

THEOREM 2.1 (SPECTOR [**1956**]). *For every countable ideal I in $\mathcal{D}$ there are two degrees* $\mathbf{x}$ *and* $\mathbf{y}$ *called an exact pair for I such that $I = I_{\mathbf{x}} \cap I_{\mathbf{y}}$.*

PROOF. *Forcing with infinite-coinfinite conditions.* One adds on infinitely many new elements to code in the next degree in I and makes finite extensions à la Kleene and Post to keep all unwanted sets out of the intersection. □

As usual, relativization is routine and the theorem holds in $\mathcal{D}(\geqslant \mathbf{a})$ for every $\mathbf{a}$. The calculation for localization is that one must have a listing of sets $\{A_i\}_{i<\omega}$ representing the elements of I and then ask finite extension questions about finite subsets of this list.

THEOREM 2.1a. *If $\{A_i | i < \omega\}$ is a list of sets with degrees in an ideal I such that each element of I contains some set in the list, then there is an exact pair* $\mathbf{x}$ *and* $\mathbf{y}$ *for I with $\mathbf{x}, \mathbf{y} \leqslant \oplus_{n<\omega}(\oplus_{i<n} A_i)'$.*

Greater care and a new coding scheme can be combined to improve this calculation but it is only needed to work inside substructures not closed order jump such as $\mathcal{D}(\leqslant \mathbf{0}')$ as in Shore [**1981**].

An immediate first order application is

COROLLARY 2.2 (KLEENE AND POST [**1981**]). *$\mathcal{D}$ is not a lattice.*

PROOF. No exact pair for any strictly ascending sequence of degrees such as the $\mathbf{0}^{(n)}$ can have a greatest lower bound. □

As a hint of things to come about Th($\mathcal{D}$), note that this theorem gives us a way of translating a second order quantifier (over countable ideals) into a first order one over pairs of degrees. Before considering the resulting characterization of Th($\mathcal{D}$) we want to touch on an entirely different set of second order notions—those of measure and category.

The typical result in this area says that in some sense almost all degrees have some local or first order property. We will briefly mention the common measures used and a few of the known results along this line.

2.3. *Cardinality.* The measure here is given by countability versus either uncountability or cardinality $c = 2^{\aleph_0}$.

Sample: *Minimal degrees.* (a) One can show that there are uncountably many minimal degrees by proving that one can diagonalize against any given countable list simply by adding on the appropriate requirements to the standard construction.

(b) One can prove that there are c many minimal degrees by observing that the class of sets of minimal degrees is analytic and as it is uncountable a general descriptive set theory theorem shows that I has cardinality c (Sacks [**1966**]).

(c) One can directly construct a perfect set of minimal degrees. One simply forces with trees of trees which at the end (upon intersection) give a perfect tree, every path through which is of minimal degree (Lacombe, unpublished, and Sacks [**1961**]).

2.4. *Baire category*. Of course this approach is closely connected with the finite extension method and Cohen forcing. Thus, for example, an arithmetic set of degrees is comeager iff it contains a real which is Cohen generic for arithmetic. We recommend Jockusch [**1980**] for a survey of this area and only present a couple of examples from that paper.

(a) $\{\mathbf{a} \mid \mathscr{D}(\leqslant \mathbf{a})$ is a lattice$\}$ is meager. Indeed, by applying a general theorem of Martin one can conclude that $\{\mathbf{a} \mid \exists \mathbf{b} \leqslant \mathbf{a}\ (\mathbf{0} < \mathbf{b}\ \&\ \mathscr{D}(\leqslant \mathbf{b})$ is a lattice$)\}$ is meager as is, in particular, the class of degrees with a minimal predecessor.

(b) The set of degrees $\mathbf{a}$ with the cupping property, i.e., $\forall \mathbf{b} \geqslant \mathbf{a}\ \exists \mathbf{c}(\mathbf{a} \vee \mathbf{c} = \mathbf{b})$, is meager.

(c) The set of degrees which are r.e. in some smaller degree is comeager.

As one should expect the proofs in this area are usually arguments about finite extension methods or, equivalently, Cohen forcing for various subsystems of arithmetic.

2.5. *Lebesgue measure*. (a) (Sacks [**1963**]) The set of minimal degrees has measure 0.

(b) (Paris [**1977**]) The set of degrees with a minimal predecessor has measure 0.

(c) (Kurtz [**1981**]) The set of degrtees r.e. in some lower degree has measure 1.

Typically the proofs for measure are considerably more difficult than for category. Basically, one often uses complicated approximations in measure coupled with the 0-1 law and at times Fubini's theorem. As for category, result (c) can be combined with the embedding theorems below an r.e. degree (relativized) in Shore [**1981**] to show, for example, that $\mathscr{D}(\leqslant \mathbf{a})$ need not be isomorphic to $\mathscr{D}(\leqslant \mathbf{b})$ even if both are generic or random.

A more unusual measure-theoretic result is

(d) (Stillwell [**1972**]) The almost all theory of degrees is decidable.

2.6. *Martin measure*. Here we say that a set is large if it contains a cone, i.e., $\mathscr{D}(\geqslant \mathbf{a})$ for some $\mathbf{a}$. This measure is closely connected to infinite games and the axiom of determinacy. Indeed, it was brought to light by Martin's lemma that a Turing degree game is determined iff the winning set or its complement contains a cone. Thus, by Martin [**1975**] every Borel set of degrees or its complement contains a cone. Typically, one starts with a first order property of a degree $\mathbf{a}$. If it depends only on $\mathscr{D}(\leqslant \mathbf{a})$, for example, $\mathbf{a}$ is a minimal cover, then it is Borel and so it or its complement contains a cone. In this example the relativization of Spector's construction (Theorem 1.10a) shows that this set is contained in $\mathscr{D}$ and hence

(a) (Jockusch [**1973**]) There is a cone of minimal covers.

Clearly, one can give analogous arguments for many other properties.

In general, such results on measure and category relativize to $\mathscr{D}(\geqslant \mathbf{a})$ straightforwardly. On the other hand, it is unclear what localization might mean here except perhaps in the case of Martin measure where the obvious problem is to calculate the base of an appropriate cone or equivalently the strategy for the associated game.

One very simple example is the Friedberg Completeness Theorem which shows that every degree above $\mathbf{0}'$ is the jump of some degree below it and has the cupping property. In fact, iterating the completeness theorem of Friedberg [**1957a**] provides calculations of the bases of cones for several properties.

THEOREM 2.7 (SELMAN [**1972**]). *For each $n \geqslant 1$ and any degree $\mathbf{c} \geqslant \mathbf{d}^{(n)}$ there are degrees $\mathbf{a}, \mathbf{b} \geqslant \mathbf{d}$ such that*

$$\mathbf{a} \vee \mathbf{b} = \mathbf{c} = \mathbf{a}^{(n)} = \mathbf{a} \vee \mathbf{d}^{(n)} = \mathbf{b}^{(n)} = \mathbf{b} \vee \mathbf{d}^{(n)}.$$

The direct proof in Selman [**1972**] is by a forcing argument with finite conditions for n-quantifier arithmetic. Simpler indirect arguments are possible (Jockusch [**1974**]).

A more interesting question is presented by the set of minimal covers. We begin with a lower bound.

THEOREM 2.8 (JOCKUSCH AND SOARE [**1970**]). *No $\mathbf{0}^{(n)}$ is a minimal cover.*

PROOF. By induction using the relativization of the old result proved by a standard priority argument that no r.e. degree is minimal (Friedberg [**1957**]). □

These results also imply an interesting corollary.

COROLLARY 2.9. *$\mathscr{D}$ is not elementarily equivalent to $\mathscr{A}$, the degrees of the arithmetic sets.*

In the other direction one can try to calculate a strategy for some appropriate game. The obvious one is Σ^0_5 but Jockusch first proved the theorem using only Σ^0_4-determinacy. Even this, however, does not give a strategy provably in analysis. Later Harrington and Kechris [**1975**] showed that one could get by with a Σ^0_3 game by using Sacks' construction of a minimal degree below $\mathbf{0}'$ in place of Spector's below $0''$. They also proved that Σ^0_1-determinacy is sufficient to determine such games and that one can compute a strategy for a type of associated game, and so a base for the cone of minimal covers, recursively in Kleen's $\mathscr{O}$ (the complete Π^1_1 set). This left a gap between $0^{(\omega)}$ and $0^{(\omega_1^{\mathrm{ck}})} = \mathscr{O}$, which has recently been filled in.

THEOREM 2.10 (JOCKUSCH AND SHORE [**1984**]). *$\mathbf{0}^{(\omega)}$ is the base of a cone of minimal covers.*

The proof is by a type of completeness theorem for operators generalizing the jump and its iterates. Thus it is a type of nested forcing argument in which one nests perfect trees each of which has every path generic for a particular sequence of sentences derived from the analysis of Sacks' minimal degree as not just Turing reducible to $\mathbf{0}'$, but as actually lying at the ωth level of the difference hierarchy. (See also Jockusch and Shore [these PROCEEDINGS].)

COROLLARY 2.11. *$\mathscr{A}$ is not elementarily equivalent to $\mathscr{H}$ (the degrees of the hyperarithmetic sets) nor to any ideal containing* $0^{(\omega)}$.

We will soon see more applications of a strengthening of this theorem but first we shall characterize the theory of $\mathscr{D}$ and its jump ideals (at least up to recursive isomorphism).

3. Characterizing Th($\mathscr{D}$). We have seen that Th($\mathscr{D}$) is undecidable (Theorem 1.15). Our goal now is the result of Simpson [**1977**] that it is in fact recursively isomorphic to $\mathrm{Th}^2(N)$, the theory of true second order arithmetic. It is of course clear that $\mathrm{Th}(\mathscr{D}) \leqslant_{1\text{-}1} \mathrm{Th}^2(N)$. The proof of the other direction that we will sketch is from Nerode and Shore [**1979**].

THEOREM 3.1. *The theory of quantification over all symmetric binary relations (i.e. graphs) on a countable domain is* 1-1 *equivalent to* $\mathrm{Th}^2(N)$.

PROOF. This is essentially the old fact that every relation can be coded by a graph (Rabin and Scott [**n.d.**] or Lavrov [**1963**]).

THEOREM 3.2 (NERODE AND SHORE [**1979**]). *The theory of countable distributive lattices with quantification over ideals is recursively isomorphic to* $\mathrm{Th}^2(N)$.

PROOF. The idea is that in each member of a finitely axiomatizable subclass of distributive lattices we can code every s.i.b. relation on the set of minimal elements of the lattice by an ideal. The trick is to require, for each pair of minimal elements x and y, that there exists a unique element of the lattice $c(x, y)$ which is strictly above $x \vee y$ but not above any other minimal elements. As these code elements are join irreducible they also form an independent set of elements. Thus an ideal generated by a subset of these codes can be generated by only one such set. Quantification over subsets of the codes is therefore easily replaced by quantification over ideals of the lattice. Of course the subsets of the codes correspond to the subsets of pairs of distinct minimal elements and so code all s.i.b. relations on them.

THEOREM 3.3 (SIMPSON). $\mathrm{Th}(\mathscr{D}) \equiv_{1\text{-}1} \mathrm{Th}^2(N)$.

PROOF. All countable distributive lattices are represented as initial segments of $\mathscr{D}$ by Theorem 1.11 while we can quantify over ideals in such lattices by first order quantification over pairs of degrees by Theorem 2.1. □

In the spirit of localization a careful examination of what is really needed to prove the above result will give the corresponding theorems for small substructures of $\mathscr{D}$. To begin with, one does not need all relations to code arithmetic. We can surely get by with, say, $+$, $\times$ and $\leqslant$. These can be coded by three ideals in our lattice which we may as well assume to be principal. We can then further refine our subclass of lattices by adding on some finite axiomization for these relations. Thus we can code in a definable way some finite axiomization of arithmetic into distributive lattices and hence in $\mathcal{D}$. We can easily get recursively presented lattices which code models of arithmetic in this way. Thus by a local version of the embedding results (Theorem 1.11a) they will be available in any ideal containing, say, $\mathbf{0}''$. The next step is to pick out the codes of standard models of (true) arithmetic (i.e. the ones isomorphic to N). We can guarantee that a model is standard by requiring that every proper initial segment (in some set of initial segments of the model which includes the one consisting of all the standard integers) has a top element. Initial segments (like all subsets of the degrees representing the integers of some model in some initial segment of $\mathcal{D}$) are coded by ideals, and so exact pairs. Thus, for the translation of the sentence asserting that all proper initial segments have top elements to guarantee that a model is standard when interpreted inside some substructure of $\mathcal{D}$, it suffices for the substructure to contain an exact pair for the ideal generated by the interpretations of the standard integers in any initial segment coding a model of arithmetic. As one can easily list the degrees representing the standard integers arithmetically in the top element of the associated initial segment, the local version of Spector's Theorem 2.1a guarantees that an exact pair for the required ideal always exists in any ideal of $\mathcal{D}$ closed under jump. □

We have thus shown that Th(N) is 1-1 reducible to Th($\mathcal{C}$) for any jump ideal $\mathcal{C} \subseteq \mathcal{D}$. To characterize Th($\mathcal{C}$) one must now calculate exactly which subsets of any model coded in $\mathcal{C}$ have exact pairs in $\mathcal{C}$. Again elementary calculations and Theorem 12.1a show that they are exactly C^*, the sets with degrees in $\mathcal{C}$.

THEOREM 3.4 (NERODE AND SHORE [**1980**]). *If $\mathcal{C} \subseteq \mathcal{D}$ is a jump ideal, then* $\mathrm{Th}(\mathcal{C}) \equiv_{1\text{-}1} \mathrm{Th}(N, \mathcal{C}^*, +, \times, \leqslant, \in)$ *(the theory of true arithmetic with quantification over sets with degrees in $\mathcal{C}$).*

Typical examples of jump ideals are, of course, $\mathcal{A}$ and $\mathcal{H}$ or, more generally, any of the $\mathcal{D}_n$ = the degrees of the Δ^1_n sets, $n \geqslant 0$. As the reduction is uniform, i.e. it proceeds by a transformation of sentences which is independent of $\mathcal{C}$, we can also use this characterization to prove that various degree structures are not elementarily equivalent when we can distinguish between the corresponding models of second order arithmetic.

COROLLARY 3.5 (NERODE AND SHORE [**1980**]). (1) *For every n, $\mathcal{D}_n \not\equiv \mathcal{D}$.*
(2) $\mathcal{D}_0 \neq \mathcal{D}_1$.
(3) *If $V = L$ or PD holds, then $\mathcal{D}_n \not\equiv \mathcal{D}_m$ if $n \neq m$.*

PROOF. For each pair one can point to some set which is definable in one associated model of second order arithmetic but not in the other. (The set-theoretic hypotheses are used to derive selection theorems which then define the strength of quantification over the appropriate level of the analytic hierarchy.)

If one carries the localization of these arguments to an extreme form one can prove, for example,

THEOREM 3.6 (SHORE [**1981**]). $\mathrm{Th}(\mathscr{D}(\leqslant \mathbf{0}')) \equiv_{1\text{-}1} \mathrm{Th}(N) \equiv_{1\text{-}1} \mathrm{Th}\,\mathscr{D}(\leqslant \mathbf{a})$ *for any r.e.* **a**.

On the other hand it is not at all clear what the relativization of these results should say. The ingredients of the proof that $\mathrm{Th}^2(N) \leqslant_{1\text{-}1} \mathrm{Th}(\mathscr{D})$ all relativize and so $\mathrm{Th}^2(N) \leqslant_{1\text{-}1} \mathrm{Th}(\mathscr{D}(\geqslant \mathbf{a}))$ for every **a**, but what of the other direction? The usual flavor of relativization would tend towards the guess that $\mathrm{Th}(\mathscr{D}(\geqslant \mathbf{a}))$ is recursively isomorphic not to $\mathrm{Th}^2(N)$ but to the theory of second order arithmetic with some additional predicate or constant for **a** or some similar sort of claim. If so, the validity of such a relativized form would contradict the general principle embodied in the homogeneity problems since for sufficiently different degrees **a** the theories of arithmetic with an additional predicate associated with them should be different. Thus in either case we have here the first hint of a difficulty with the principle of relativization.

Our proof of Theorem 3.3 can also be applied to other reducibility orderings satisfying the appropriate embedding and exact pair theorems.

THEOREM 3.7 (NERODE AND SHORE [**1979**]). *The theories of* 1-1, *m*-1, *wtt, tt and arithmetic degrees are all recursively isomorphic to* $\mathrm{Th}^2(N)$

Elaborations of these ideas as in Shore [**1982**] can be used to distinguish between some of these theories not previously known to be distinct.

THEOREM 3.8.(a) (SHORE [**1982a**]) *The truth-table degrees are not elementarily equivalent to* $\mathscr{D}$.

(b) (SHORE [**1982b**]) *The arithmetic degrees are not elementarily equivalent to* $\mathscr{D}$.

(We should note that Mohrherr [**1982**] has found a direct proof of (a).)

4. Definability, automorphisms and homogeneity. Most of the basic questions in these areas were first raised by Rogers [**1967** and **1967a**] and have received considerable attention since then. Progress towards answering such questions was first made in the setting of the degrees with the jump operator $\mathscr{D}'$. We will first mention a few of the workers in this area: Feiner [**1970**], and then Yates [**1972**], refuted the homogeneity conjecture (in its isomorphism version for $\mathscr{D}'$). Results restricting the possible automorphisms of $\mathscr{D}'$ were obtained in Yates [**1972**], Jockusch and Solovay [**1977**], Richter [**1979**] and Epstein [**1979**]. Definability results for $\mathscr{D}'$ can also be found in Jockusch and Simpson [**1976**] and Simpson [**1977**].

More recently, considerable progress has been made on such questions for $\mathscr{D}$ beginning with the papers of Nerode and Shore [**1979** and **1980**] and including Shore [**1979**, **1981**, **1981a** and **1982**] as well as Harrington and Shore [**1981**] and Jockusch and Shore [**1984** and these PROCEEDINGS]. As is often the case, the best approach is to start at the end with the strongest result. We therefore start with Jockusch and Shore [**1984** and these PROCEEDINGS] and questions of definability in $\mathscr{D}$. The approach is through the minimal cover problem.

Recall that Jockusch and Soare [**1970**] shows that no $\mathbf{0}^{(n)}$ is a minimal cover. In fact, they show that $\mathbf{0}^{(n)} \vee \mathbf{a}$ is not minimal over $\mathbf{a}$ for any $\mathbf{a}$. This observation leads one to consider two classes of degrees, $\mathscr{C}_0$ (the strongly nonminimal covers) $= \{\mathbf{c} | \forall \mathbf{a} (\mathbf{a} \vee \mathbf{c}$ is not a minimal cover of $\mathbf{a})\}$ and its downward closure $\mathscr{C} = \{\mathbf{c} | \exists \mathbf{b} \geqslant \mathbf{c}\ \forall \mathbf{a} (\mathbf{a} \vee \mathbf{c}$ is not a minimal cover of $\mathbf{a})\}$. As shown in Harrington and Shore [**1981**] the ideas of the proof in Jockusch and Soare [**1970**] show that $\mathscr{C}_0$ is closed under jump and join and so $\mathscr{C}$ is a jump ideal:

Suppose $\mathbf{c} \in C_0$ but $\mathbf{c}' \vee \mathbf{a}$ is a minimal cover of $\mathbf{a}$. If $\mathbf{c} \not\leqslant \mathbf{a}$ then $\mathbf{a} < \mathbf{c} \vee \mathbf{a} \leqslant \mathbf{c}' \vee \mathbf{a}$ and so $\mathbf{c} \vee \mathbf{a}$ is a minimal cover of $\mathbf{a}$, for a contradiction. On the other hand, if $\mathbf{c} \leqslant \mathbf{a}$ then $\mathbf{c}' \vee \mathbf{a}$ is r.e. in $\mathbf{a}$ and so, by the relativization of the theorem that no r.e. degree is minimal, we again have a contradiction. The proof for join is similar.

The crucial point now is that one can also prove that if $\mathbf{c}$ is not arithmetic then there is an $\mathbf{a}$ such that $\mathbf{c} \vee \mathbf{a}$ is a minimal cover of $\mathbf{a}$ (Jockusch and Shore [**1984** or these PROCEEDINGS]). (Harrington and Shore [**1981**] had earlier proved that $\mathscr{C} \subseteq \mathscr{H}$ by a game-theoretic argument. The construction of $\mathbf{a}$ for a nonarithmetic $\mathbf{c}$ combines the methods used to show that $\mathbf{0}^{(\omega)}$ is a base of a cone of minimal covers with a new proof of the join theorem of Posner and Robinson [**1981**].)

THEOREM 4.1. *$\mathscr{C} = \mathscr{A}$ and so the degrees of the arithmetic sets are definable in $\mathscr{D}$.*

The ingredients of this proof all relativize, and so if we let

$$\mathscr{C}^{\mathbf{a}} = \left\{ \mathbf{x} \geqslant \mathbf{a} | \mathscr{D}(\geqslant \mathbf{a}) \vDash \exists \mathbf{z} \geqslant \mathbf{x}\, \forall \mathbf{y}\, (\mathbf{y} \vee \mathbf{z} \text{ is not a minimal cover of } \mathbf{y}) \right\},$$

then we have

THEOREM 4.1a. *$\mathscr{C}^{\mathbf{a}}$ is the set of all degrees $\geqslant \mathbf{a}$ of sets arithmetical in $\mathbf{a}$.*

This application of relativization coupled with one other gives an immediate counterexample to the homogeneity conjecture which was originally suggested by this very principle. In fact very few cones can be isomorphic:

THEOREM 4.2. *If $\mathscr{D}(\geqslant \mathbf{a}) \cong \mathscr{D}(\geqslant \mathbf{b})$, then $\mathbf{a}$ and $\mathbf{b}$ are of the same arithmetic degree.*

PROOF. For any A, one can easily construct a distributive lattice $\mathscr{L}_A$ arithmetic in A which arithmetically codes $\mathbf{a}$ (i.e., such that A is arithmetic in any presentation of $\mathscr{L}_A$). By Theorem 1.11a, $\mathscr{L}_A$ is isomorphic to an initial segment of $\mathscr{D}(\geqslant \mathbf{a})$ with top element arithmetic in A and hence in $\mathscr{C}^{a}$. Now if $\mathscr{D}(\geqslant \mathbf{a}) \cong \mathscr{D}(\geqslant \mathbf{b})$, then $\mathscr{L}_A$ is isomorphic to an initial segment of $\mathscr{D}(\geqslant \mathbf{b})$ with top element in $\mathscr{C}^{\mathbf{b}}$ (i.e., arithmetic in $\mathbf{b}$). Thus A is arithmetic in $\mathbf{b}$. □

Indeed, by applying the analysis used to localize the characterization of Th($\mathscr{D}$) to Th($\mathscr{C}$) for jump ideals $\mathscr{C}$, we can refute the elementary equivalence version of the homogeneity conjecture as well.

THEOREM 4.3. *If $\mathscr{D}(\geqslant \mathbf{a}) \equiv \mathscr{D}$, then $\mathbf{a}$ is arithmetic.*

PROOF. If $\mathscr{D}(\geqslant a) \equiv \mathscr{D}$, then $\mathscr{C}^{\mathbf{a}} \equiv \mathscr{C}$ but the characterization of theories gives a uniform reduction of the theory of second order arithmetic with quantification over sets with degrees in $\mathscr{C}$ (over sets recursive in those with degrees in $\mathscr{C}^{\mathbf{a}}$) to Th($\mathscr{C}$) (Th($\mathscr{C}^{\mathbf{a}}$)). These two models of arithmetic are elementarily equivalent iff **a** is arithmetic. (Otherwise the sentence saying there is a nonarithmetic set distinguishes them.) □

The sentence exhibiting the elementary difference between $\mathscr{D}$ and $\mathscr{D}(\geqslant \mathbf{a})$ guaranteed by this theorem is quite complex. Note, however, that proof of the theorem itself fails to relativize. The proof really uses the fact that **0** is definable in second order arithmetic and so the best one can do in the way of relativization is

THEOREM 4.3a. *If* **c** *is definable in second order arithmetic and* $\mathscr{D}(\geqslant \mathbf{a}) \equiv \mathscr{D}(\geqslant \mathbf{c})$, *then* **a** *and* **c** *are of the same arithmetic degree.*

Moreover, there is little hope of improving on this situation since

PROPOSITION 4.4. $PD \Rightarrow \exists \mathbf{c} \forall \mathbf{a} \geqslant \mathbf{c}[\mathscr{D}(\geqslant \mathbf{a}) \equiv \mathscr{D}(\geqslant \mathbf{c})]$.

PROOF. Intersect the cones on which each sentence is true or false.

On the other hand,

PROPOSITION 4.5. $V = L \Rightarrow \forall \mathbf{a}, \mathbf{b}[\mathscr{D}(\geqslant \mathbf{a}) \equiv \mathscr{D}(\geqslant \mathbf{b}) \to$ **a** *and* **b** *are of the same arithmetic degree*].

PROOF. The least pair of sets forming an exact pair for the degrees arithmetic in **a**(**b**) is definable uniformly in each structure and so the same for both.

From our definition of the arithmetic degrees in $\mathscr{D}$ as $\mathscr{C}$ we can also give a definition of $\mathscr{H}$, the hyperarithmetic ones, by using various special properties of $\mathscr{H}$ (Jockusch and Shore [**1984**]), but much more is true. The translation of second order arithmetic into $\mathscr{D}$ used to localize the results in Th($\mathscr{D}$) provides a wholesale method of transforming definitions in second order arithmetic into ones in $\mathscr{D}$ once we are given the starting point of a definable jump ideal such as $\mathscr{A}$. The only restriction is that the relation must be "arithmetically invariant".

THEOREM 4.6. *Any relation R on degrees which is invariant under joining with arithmetic degrees, i.e.,*

$$\forall \mathbf{x}_1, \ldots, \mathbf{x}_n \forall \mathbf{a} \in \mathscr{A}[R(\mathbf{x}_1, \ldots, \mathbf{x}_n) \Leftrightarrow R(\mathbf{x}_1 \vee \mathbf{a}, \ldots, \mathbf{x}_n \vee \mathbf{a})],$$

is definable in $\mathscr{D}$ *iff it is definable in second order arithmetic.*

COROLLARY 4.7. *The following are definable in* $\mathscr{D}$:

(1) $\mathbf{a} \equiv_T \mathbf{b}^{(\omega)}$.

(2) **a** *is* Δ^1_n *in* **b** *for each* $n \geqslant 0$.

(3) $\mathbf{a} \equiv_T \mathcal{O}^{\mathbf{b}}$.

(4) **a** *is constructible from* **b**.

(5) $\mathbf{a} \equiv_T \mathbf{b}^{\#}$.

(6) *Any relation on degrees above $\mathscr{A}$ which is definable in second order arithmetic, and so, in particular, all the different notions of upper bounds for jump ideals (as, for example, Jockusch and Simpson* [**1976**] *or Hodes* [**1978, 1980**]).

PROOF OF THEOREM 4.6. Consider the special case of defining a single degree $\mathbf{s}$ (necessarily above $\mathscr{A}$) which is definable in second order arithmetic.

(1) By Theorem 2.7, $\mathbf{s}$ is, for each $n \in \omega$ and $\mathbf{a} \in \mathscr{A}$, the least upper bound of $\{\mathbf{x} \geqslant \mathbf{a} | \mathbf{x}^{(n)} \leqslant \mathbf{s}\}$.

(2) The calculations for the local versions of the characterization of Th($\mathscr{D}$) (Theorem 3.4) show that the sets coded by an exact pair below $\mathbf{x}$ in a standard model of arithmetic below $\mathbf{x}$ are all recursive in $\mathbf{x}^{(n)}$ for some fixed n.

(3) The localization of Spector's theorem guarantees that we can code $X \in \mathbf{x}$ by an exact pair below $\mathbf{x} \vee \mathbf{0}''$ in a model of arithmetic below $\mathbf{x} \vee \mathbf{0}''$.

Mixing these results together we can show that $\mathbf{s}$ is the least upper bound of various sets of degrees below which there are only exact pairs for sets in associated models of arithmetic which have nth jumps below $\mathbf{s}$. The point is that as $\mathbf{s}$ is definable we can say that $\mathbf{x}^{(n)} \leqslant \mathbf{s}$ in second order arithmetic and so require this of sets coded in models of arithmetic in $\mathscr{D}$. (See Jockusch and Shore [**1984**] for a complete proof of this theorem.) □

Turning now to automorphisms of $\mathscr{D}$ we first note that Theorem 4.2 automatically implies that if $\phi: \mathscr{D} \to \mathscr{D}$ is an automorphism, then $\phi(\mathbf{x})$ is of the same arithmetic degree as $\mathbf{x}$ for every $\mathbf{x}$. We can actually do much better by a direct argument which also gives a much more direct solution to the homogeneity problem. Let $\phi: \mathscr{D}(\geqslant \mathbf{a}) \to \mathscr{D}(\geqslant \mathbf{b})$ be an isomorphism.

LEMMA 4.8. *If* $\mathbf{x} \geqslant \phi^{-1}(\mathbf{b}^{(2)})$, *then* $\phi(x) \leqslant \mathbf{x}^{(5)}$.

PROOF. As $\phi(\mathbf{x}) \geqslant \mathbf{b}^{(2)}$, Theorem 1.11a says that the lattice $L_{\phi(\mathbf{x})}$ is isomorphic to an initial segment of $[\mathbf{b}, \phi(\mathbf{x})]$ and hence to one of $[\mathbf{a}, \mathbf{x}]$. As the decoding of $\phi(\mathbf{x})$ from $L_{\phi(\mathbf{x})}$ and so from $[\mathbf{a}, \mathbf{x}]$ is arithmetic, $\phi(\mathbf{x}) \leqslant \mathbf{x}^{(5)}$. (A straightforward quantifier count shows that five jumps suffice.) □

To solve the homogeneity problem we first prove

LEMMA 4.9. $\phi^{-1}(\mathbf{b}^{(2)}) \leqslant \phi(\mathbf{a}^{(2)})^{(7)}$.

PROOF. Again by Theorem 1.11a, $L_{\phi^{-1}(\mathbf{b}^{(2)})}$ is isomorphic to an initial segment of $[\mathbf{a}, \phi^{-1}(\mathbf{b}^{(2)}) \vee \mathbf{a}^{(2)}]$ and so by the isomorphism to one of $[\mathbf{b}, \mathbf{b}^{(2)} \vee \phi(\mathbf{a}^{(2)})]$. By decoding, $\phi^{-1}(\mathbf{b}^{(2)}) \leqslant (\mathbf{b}^{(2)} \vee \phi(\mathbf{a}^{(2)}))^5$ but as $\mathbf{b} \leqslant \phi(\mathbf{a}^{(2)})$, $\phi^{-1}(\mathbf{b}^{(2)}) \leqslant (\phi(\mathbf{a}^{(2)})^{(2)})^{(5)} = \phi(\mathbf{a}^{(2)})^{(7)}$.

THEOREM 4.10. *The homogeneity conjecture fails.*

PROOF. Suppose, for example, that $\mathbf{b}$ is any base of a cone of minimal covers in $\mathscr{D}(\geqslant \mathbf{a}^{(2)})$ (as given by relativizing Jockusch [**1973**] or Harrington and Kechris [**1975**]). Thus $\phi(\mathbf{a}^{(2)})^{(7)} \geqslant \mathbf{b}$ is also the base of a cone of minimal covers in

$\mathscr{D}(\geqslant \mathbf{a}^{(2)})$. By the isomorphism, $\phi(\phi(\mathbf{a}^{(2)})^{(7)})$ is one in $\mathscr{D}(\geqslant \phi(\mathbf{a}^{(2)}))$, but, by Lemmas 4.8 and 4.9,

$$\phi\left(\phi(\mathbf{a}^{(2)})^{(7)}\right) \leqslant \left(\phi(\mathbf{a}^{(2)})^{(7)}\right)^{(5)} = \phi(\mathbf{a}^{(2)})^{(12)}.$$

As this degree is not a minimal cover in $\mathscr{D}(\geqslant \phi(\mathbf{a}^{(2)}))$ by relativizing Theorem 2.8 (Jockusch and Soare [**1970**]) we have the required contradiction. In particular, we may choose $\mathbf{a} = \mathbf{0}$ and $\mathbf{b} = \mathscr{O}$ to see that $\mathscr{D} \not\cong \mathscr{D}(\geqslant \mathscr{O})$. □

Alternatively we can improve upon Lemma 4.8 in another direction to get our result on automorphisms.

THEOREM 4.11. *ϕ is the identity on a cone with a base which is the join of an element arithmetic in* **a** *and one arithmetic in* **b**.

PROOF. For notational convenience let $\mathbf{c} = \phi^{-1}(\mathbf{b}^{(2)})$. If $\mathbf{x} \geqslant \mathbf{c}^{(5)}$, then by the Freidberg Completeness Theorem 2.7 there is a $\mathbf{y} \geqslant \mathbf{c}$ with $\mathbf{y}^{(5)} = \mathbf{y} \vee \mathbf{c}^{(5)} = \mathbf{x}$. Thus $\phi(\mathbf{x}) = \phi(\mathbf{y}) \vee \phi(\mathbf{c}^{(5)})$, and so if $\mathbf{x} \geqslant \phi(\mathbf{c}^{(5)}) \vee \mathbf{c}^{(5)}$ we can apply Lemma 4.8 to conclude that $\phi(\mathbf{x}) \leqslant \mathbf{y}^{(5)} \vee \phi(\mathbf{c}^{(5)}) = \mathbf{x}$. If we now run the same argument for $\phi^{-1}\colon \mathscr{D}(> \mathbf{b}) \to \mathscr{D}(\geqslant \mathbf{a})$, we see that if $\mathbf{x} \geqslant \phi(\mathbf{d}^{(5)}) \vee \mathbf{d}^{(5)}$ then $\phi^{-1}(\mathbf{x}) \leqslant \mathbf{x}$, where $\mathbf{d} = \phi(\mathbf{a}^{(2)})$. Combing the two results shows that if

$$\mathbf{x} \geqslant \phi(\mathbf{c}^{(5)}) \vee \mathbf{d}^{(5)} \vee \mathbf{c}^{(5)} \vee \phi(\mathbf{d}^{(5)}),$$

then $\phi(\mathbf{x}) = \mathbf{x}$. Now ϕ maps the degrees arithmetic in **a** onto those arithmetic in **b** by Theorem 4.1a (and similarly for ϕ^{-1}). As $\mathbf{b}^{(2)}$ is arithmetic in **b**, $\mathbf{c} = \phi^{-1}(\mathbf{b}^{(2)})$ is arithmetic in **a** and so then is $\mathbf{c}^{(5)}$. One more step shows that $\phi(\mathbf{c}^{(5)})$ is arithmetic in **b**. Similarly, $\mathbf{d}^{(5)}$ is arithmetic in **b** and $\phi(\mathbf{d}^{(5)})$ is arithmetic in **a**. Thus we see that **x** is the join of a degree in $\mathscr{C}^{\mathbf{a}}$ with one in $\mathscr{C}^{\mathbf{b}}$ the required bound.

COROLLARY 4.12. *Every automorphism of $\mathscr{D}$ is the identity on every degree above all the arithmetic ones.*

Another line of investigation that has been pursued in the study of automorphisms of $\mathscr{D}$ has been that of determining automorphism bases; that is, sets S of degrees such that if $\phi \restriction S = \mathrm{id} \restriction S$ for any automorphism ϕ, then $\phi = \mathrm{id}$ on $\mathscr{D}$. Thus, for example, the minimal degrees form an automorphism basis since they generate $\mathscr{D}$ (with $\vee$ and $\wedge$) by Jockusch and Posner [**1982**] which gives many other results along these lines. One of their results combined with Corollary 4.12 shows there is a cone (with base recursive in $\mathbf{0}^{(\omega)}$) which is an automorphism basis for $\mathscr{D}$ (though, of course, it cannot generate $\mathscr{D}$).

5. Open questions. We want to close with a few open questions and some speculations. First a couple of obvious ones of long standing.

Q.5.1. Are there any nontrivial automorphisms of $\mathscr{D}$ (Rogers)?

And, in the same vein:

Q.5.2. Are there distinct degrees **a** and **b** with $\mathscr{D}(\geqslant \mathbf{a}) \cong \mathscr{D}(\geqslant \mathbf{b})$?

Q.5.3. Is the jump operator definable in $\mathscr{D}$ (Kleene and Post)?

It is worth noting that in the context of degrees of constructibility C, not only can one duplicate the main result for Turing degrees but these questions can also be answered.

THEOREM 5.4 (FARRINGTON [**1984**]). *If* $\aleph_1^{L(r)} < \aleph_1$ *for every real* r, *then*:

(1) $\mathrm{Th}(C) \equiv_{1\text{-}1} \mathrm{Th}^2(N)$.

(2) *The sharp operator is definable in* C

Using Farrington's lemmas, Grosek [**1984**] *has observed that one can code enough in* C *to also prove*:

(3) *There are no nontrivial automorphisms of* C.

Indeed Farrington [**1983**] shows that (3) holds assuming only that $\aleph_2^L < \aleph_1$.

The essential relevant difference between C and $\mathscr{D}$ is that one can often code and recover sets in the C-degrees below $\mathbf{x}$ constructibly in $\mathbf{x}$ but, in general, only recursively $\mathbf{x}^{(3)}$ in the T-degrees below $\mathbf{x}$.

Q.5.5. Can one give degree theoretically natural definitions (e.g. ones not using codings of arithmetic) for natural relations on degrees the way Theorem 4.1a gives for "arithmetic in"? Jockusch and Shore [**1984**] give one such for "hyperarithmetic in" and a couple of other examples. A reasonable test case might be the degree $\mathbf{0}^{(\omega)}$.

$\mathbf{0}^{(\omega)}$ is, of course, definable in $\mathscr{D}$ but it is naturally definable in $\mathscr{D}'$. Indeed, if one adds on the jump operator many more notions have natural definitions—see Jockusch and Simpson [**1976**] and Hodes [**1983**].

Along the lines of merely taking one more step towards solutions of these problems we suggest refining the characterization of $\mathscr{C}$ to answer:

Q.5.6. Is $\mathscr{C}_0 = \{\mathbf{c} \mid \forall \mathbf{a}{:}(\mathbf{a} \vee \mathbf{c}$ is not a minimal cover of $\mathbf{a})\}$ actually equal to the union over n of the sets of n-REA degrees? (The proof of Theorem 4.1 shows that for each n, $\mathscr{C}_0$ contains all of the n-REA degrees and is closed under the relation "n-REA in".)

If the answer to Q.5.4 is yes, then one can use the results of Shore [**1981**] on $\mathscr{D}(\leqslant \mathbf{0}')$ to show for example that $\mathbf{0}^{(3)}$ is definable in $\mathscr{D}$, the triple jump is invariant and most likely it is even definable in $\mathscr{D}$. Other results would then follow as in Theorem 4.6 or Nerode and Shore [**1980**, Theorem 2.8].

For our last question we want to return to the homogeneity conjecture and the characterization of $\mathrm{Th}(\mathscr{D})$ that here gave us the first inkling of its failure. We know that for each $\mathbf{a}$, $\mathrm{Th}^2(N) \leqslant_{1\text{-}1} \mathrm{Th}(\mathscr{D}(\geqslant \mathbf{a}))$ uniformly (i.e. independently of $\mathbf{a}$) and, of course, for every $\mathbf{a}$, $\mathrm{Th}(\mathscr{D}(\geqslant \mathbf{a})) \leqslant_{1\text{-}1} \mathrm{Th}^2(N, \mathbf{a})$—the theory of second order arithmetic with a predicate for $\mathbf{a}$—via a uniform reduction.

Theorem 4.1a and the results of §3 show that $\mathrm{Th}^2(N, \mathscr{A}_a) \leqslant_{1\text{-}1} \mathrm{Th}(\mathscr{D}(\geqslant \mathbf{a}))$ uniformly, where $\mathscr{A}_\mathbf{a}$ is a predicate for the sets arithmetic in $\mathbf{a}$. Now if, for example, there is well ordering of the reals definable in second order arithmetic, then $\mathbf{a}$ is definable from $\mathscr{A}_\mathbf{a}$ and so we have

THEOREM 5.7. *If there is a well-ordering of the reals definable in second order arithmetic, then* $\mathrm{Th}(\mathscr{D}(\geqslant \mathbf{a})) \equiv_{1\text{-}1} \mathrm{Th}^2(N, \mathbf{a})$ *for every* $\mathbf{a}$.

Now this theorem not only has an extra set-theoretic hypothesis but the ordering is used in such a way that one introduces parameters to pick out **a** from $\mathscr{A}_{\mathbf{a}}$. Thus one loses uniformity as well. If one could get the result of Theorem 5.7 uniformly (i.e. with reduction procedures independent of **a**) then, still assuming a definable well-ordering of the reals, one could not only prove that there are no nontrivial automorphisms of $\mathscr{D}$ but even that $\mathscr{D}(\geqslant \mathbf{A}) \equiv \mathscr{D}(\geqslant \mathbf{b})$ if and only if $\mathbf{a} = \mathbf{b}$.

Thus perhaps the ultimate application and simultaneously the strongest possible refutation of the homogeneity principle would be a positive answer to

Q.5.8. Is, for every **a**, $\mathrm{Th}(\mathscr{D}(\geqslant \mathbf{a})) \equiv_{1\text{-}1} \mathrm{Th}^2(N, \mathbf{a})$ and can the reductions be given uniformly, that is independently of **a**?

References

U. Abraham and R. A. Shore, [**1985**], *Initial segments of the Turing degrees of size* $\aleph_1$ (to appear).

R. Epstein [**1979**], *Degrees of unsolvability: Structure and theory*, Lecture Notes in Math., vol. 759, Springer-Verlag, Berlin.

P. Farrington [**1983**], *Hinges and automorphisms of the degrees of non-constructibility*, J. London Math. Soc. (2) **28**, 193–202.

______ [**1984**], *The first order theory of the c-degrees* (to appear).

L. Feiner [**1970**], *The strong homogeneity conjecture*, J. Symbolic Logic **35**, 375–377.

R. M. Friedberg [**1957**], *The fine structure of degrees of unsolvability of recursively enumerable sets*, Seminars of Cornell Institute for Symbolic Logic, pp. 404–406.

______ [**1957a**], *A criterion for completeness of degrees of unsolvability*, J. Symbolic Logic **22**, 159–160.

M. Groszek [**1984**], *Applications of iterated perfect set forcing* (to appear).

M. J. Groszek and T. A. Slaman, [**1983**], *Independence results on the global structure of the Turing degrees*, Trans. Amer. Math. Soc. **277**, 579–588.

L. Harrington and A. Kechris [**1975**], *A basis result for* Σ_3^0 *sets of reals with an application to minimal covers*, Proc. Amer. Math. Soc. **53**, 445–448.

L. Harrington and R. A. Shore [**1981**], *Definable degrees and automorphisms of* $\mathscr{D}$, Bull. Amer. Math. Soc. (N.S.) **4**, 97–99.

H. Hodes [**1978**], *Uniform upper bounds on ideals of Turing degrees*, J. Symbolic Logic **43**, 601–612.

______, [**1980**], *Jumping through the transfinite: The master code hierarchy of Turing degrees*, J. Symbolic Logic **45**, 204–220.

______ [**1983**], *More on uniform upper bounds on ideals of Turing degrees*, J. Symbolic Logic **48**, 441–457.

C. G. Jockusch, Jr. [**1973**], *An application of* Σ_4^0 *determinacy to the degrees of unsolvability*, J. Symbolic Logic **38**, 293–294.

______, [**1974**], Review of Selman [**1972**] in Math. Reviews **45**, #3155.

______, [**1980**], *Degrees of generic sets*, Recursion Theory: Its Generalizations and Applications (F. R. Drake and S. S. Warner, ed.), London Math. Soc. Lecture Notes, No. 45, Cambridge Univ. Press, London and New York.

C. G. Jockusch, Jr. and D. Posner [**1982**], *Automorphism bases for the degrees of unsolvability*, Israel J. Math. **40**, 150–164.

C. G. Jockusch, Jr. and R. A. Shore, *REA operators, r.e. degrees and minimal covers*, these PROCEEDINGS.

______ [**1984**], *Pseudo-jump operators. II: Transfinite iterations, hierarchies and minimal covers*, J. Symbolic Logic **49**, 183–214.

C. G. Jockusch, Jr. and S. G. Simpson [**1976**], *A degree theoretic definition of the ramified analytical hierarchy*, Ann. Math. Logic **10**, 1–32.

C. G. Jockusch, Jr. and R. I. Soare [**1970**], *Minimal covers and arithmetical sets*, Proc. Amer. Math. Soc. **25**, 856–859.

C. G. Jockusch, Jr. and R. M. Solovay [**1977**], *Fixed points of jump preserving automorphisms of degrees*, Israel J. Math. **26**, 91–94.

S. C. Kleene and E. L. Post [**1954**], *The upper semi-lattice of degrees of unsolvability*, Ann. of Math. (2) **59**, 379–407.

S. A. Kurtz [**1981**], *Randomness and genericity in the degrees of unsolvability*, Ph. D. Thesis, University of Illinois.

A. H. Lachlan [**1968**], *Distributive initial segments of the degrees of unsolvability*, Z. Math. Logik Grundlag. Math. **14**, 457–472.

A. H. Lachlan and R. Lebeuf [**1976**], *Countable initial segments of the degrees of unsolvability*, J. Symbolic Logic **41**, 289–300.

I. A. Lavrov [**1963**], *Effective unseparability of the sets of identically true formulae and finitely refutable formulae for certain elementary theories*, Algebra i. Logika **2**, 5–18.

M. Lerman [**1971**], *Initial segments of the degrees of unsolvability*, Ann. of Math. (2) **93**, 365–389.

______ [**1983**], *The degrees of unsolvability*, Springer-Verlag, Berlin.

D. A. Martin [**1975**], *Borel determinacy*, Ann. of Math. (2) **102**, 363–371.

J. L. Mohrherr [**1982**], Ph. D. Thesis, University of Illinois.

A. Nerode and R. A. Shore [**1979**], *Second order logic and first order theories of reducibility orderings*, The Kleene Symposium (J. Barwise, J. Keisler and K. Kunen, eds.), North-Holland, Amsterdam.

______ [**1980**], *Reducibility orderings: Theories, definability and automorphisms*, Ann. Math. Logic **18**, 61–89.

J. B. Paris [**1977**], *Measure and minimal degrees*, Ann. Math. Logic **11**, 203–216.

D. B. Posner [**1980**], *A survey of the non-RE degrees* $\leqslant \mathbf{0}'$, Recursion Theory: Its Generalisations and Applications (F. R. Drake and S. S. Warner, eds.), London Math. Soc. Lecture Notes, No. 45, Cambridge Univ. Press, London and New York, pp. 52–109.

D. B. Posner and R. W. Robinson [**1981**], *Degrees joining to* $\mathbf{0}'$, J. Symbolic Logic **46**, 714–722.

E. L. Post [**1944**], *Recursively enumerable sets of positive integers and their decision problems*, Bull. Amer. Math. Soc. **50**, 284–316.

N. Rabin and D. Scott [**n.d.**], *The undecidability of some simple theories*, mimeographed notes.

L. J. C. Richter [**1979**], *On automorphisms of the degrees that preserve jumps*, Israel J. Math. **32**, 27–31.

H. Rogers, Jr. [**1967**], *Theory of recursive functions and effective computability*, McGraw-Hill, New York.

______, [**1967a**], *Some problems of definability in recursive function theory*, Set Models and Recursion Theory, Proc. Summer School in Math. Logic and Tenth Logic Colloq. (Lercester, August–September 1965; J. N. Crossley, ed.), North-Holland, Amsterdam.

J. M. Rubin [**1979a**], *The existence of an* ω_1 *initial segment of Turing degrees*, Notices Amer. Math. Soc. **26**, A-425.

______ [**1979b**], *Distributive uncountable initial segments of the degrees of unsolvability*, Notices Amer. Math. Soc. **26**, A-619.

G. E. Sacks [**1961**], *On suborderings of degrees of recursive unsolvability*, Ph. D. Thesis, Cornell University, New York.

______ [**1963**], *On the degrees less than* $\mathbf{0}'$, Ann. of Math. (2) **77**, 211–231.

______ [**1966**], *Degrees of unsolvability*, 2nd ed., Ann. of Math. Studies, no. 55, Princeton Univ. Press, Princeton, N. J.

A. L. Selman [**1972**], *Applications of forcing to the degree theory of the arithmetic hierarchy*, Proc. London Math. Soc. (3) **25**, 586–602.

R. A. Shore [**1978**], *On the* $\forall\exists$ *sentences of* α*-recursion theory*, Generalized Recursion Theory. II (J. Fenstad et al., eds.), North-Holland, Amsterdam, pp. 331–353.

______ [**1979**], *The homogeneity conjecture*, Proc. Nat. Acad. Sci. U.S.A. **76**, 4218–4219.

______ [**1981**], *The degrees of unsolvability: Global results*, Logic Year 1979-80 (M. Lerman, J. Schmerl and R. I. Soare, eds.), Lecture Notes in Math., Springer-Verlag, Berlin and New York.

______ [**1981a**], *The theory of the degrees below* $\mathbf{0}'$, J. London Math. Soc. **24**, 1–14.

______ [**1982**], *On homogeneity and definability in the first order theory of the Turing degrees*, J. Symbolic Logic **47**, 8–16.

______ [**1982a**], *The Turing and truth-table degrees are not elementarily equivalent*, Logic Colloquium '80 (D. van Dolen, ed.), North-Holland, Amsterdam.

______ [**1984**], *The arithmetic degrees are not elementarily equivalent to the Turing degrees*, Arch. Math. Logik **24**.

S. G. Simpson [**1977**], *First order theory of the degrees of recursive unsolvability*, Ann. of Math. (2) **105**, 121–139.

______ [**1977a**], *Degrees of unsolvability: A survey of results*, Handbook of Mathematical Logic (J. Barwise, ed.), North-Holland, Amsterdam.

C. Spector [**1956**], *On degrees of recursive unsolvability*, Ann. of Math. (2) **64**, 581–592.

J. Stillwell [**1972**], *Decidability of the almost all theory of degrees*, J. Symbolic Logic **37**, 501–506.

A. M. Turing [**1939**], *Systems of logic based on ordinals*, Proc. London Math. Soc. (2) **45**, 161–228.

C. E. M. Yates [**1972**], *Initial segments and implications for the structure of degrees*, Conference in Mathematical Logic (London, 1970; W. Hodges, ed.), Lecture Notes in Math., no. 255, Springer-Verlag, Berlin, pp. 305–335.

DEPARTMENT OF MATHEMATICS, CORNELL UNIVERSITY, ITHACA, NEW YORK, 14853

Proceedings of Symposia in Pure Mathematics
Volume **42**, 1985

Tree Arguments in Recursion Theory and the $0'''$-Priority Method

ROBERT I. SOARE[1]

Table of Contents

Introduction. This Summer Research Institute in Recursion Theory marks the 25th anniversary of a similar Institute for Symbolic Logic held during the summer of 1957, also at Cornell. Since recursion theory began as a branch of mathematics approximately 25 years earlier with the work of Gödel, Church, Kleene, Post and Turing, it seems fitting to review the subject at least once every quarter of a century. The period just before the 1957 Cornell Summer Institute was particularly important for recursion theory because it was then that Friedburg [**1957a**]

1980 *Mathematics Subject Classification.* Primary 03D25, 03G10; Secondary 03–03.

[1]Portions of the material were presented by the author in a short course on recursively enumerable sets and degrees in July, 1982, during the AMS Summer Research Institute on Recursion Theory at Cornell University. The remainder was worked out and presented in a series of lectures for a seminar at the University of Leeds during the fall of 1982 while the author was partially supported by the British Science and Engineering Research Council. The author is grateful to some of the seminar participants: K. Ambos-Spies. S. B. Cooper, F. R. Drake, C. G. Jockusch and S. S. Wainer for helpful comments and suggestions. The author is also greateful to L. Harrington, and particularly to T. Slaman for many lengthy discussions on the $0'''$-priority method which are reflected in the paper, and to M. Lerman, R. A. Shore and Steffen Lempp, who made helpful suggestions after reading the manuscript.

0082-0717/85 $1.00 + $.25 per page

and independently, Muchnik [**1956**] solved Post's problem [**1944**] by introducing the finite injury priority method to construct a nonrecursive recursively enumerable (r.e.) set which is incomplete. The priority method, together with its many extensions and generalizations, has played a fundamental role in recursion theory ever since, not only in recursion theory on the integers, but in generalized recursion theory, including recursion theory on certain ordinals, E-recursion theory, and in recursive model theory, the effective content of mathematics, computational complexity, and other areas. During the 1957 Cornell meeting, Friedberg [**1957b**] further developed the priority method and applied it to questions about r.e. degrees, and Friedberg [**1958**] also solved a problem of Myhill arising from Post's work by constructing a maximal set. This maximal set construction has proved very important in most of the later work on the algebraic structure of the r.e. sets, particularly their structure as a distributive lattice.

The priority method should be viewed as an effective version of the well-known Baire Category Theorem (see Kelley [**1955**, p. 200]), which asserts that in a suitable topological space $\mathscr{S}$ (say a compact Hausdorff space or a complete metric space) a countable sequence of dense open sets $\{D_n\colon n \in \omega\}$ has nonempty intersection. For example, let $\mathscr{S}$ be the Cantor space 2^ω with the usual topology. Let $2^{<\omega}$ denote the set of finite sequences of 0's and 1's. To exhibit $f \in 2^\omega$, $f \in \bigcap_n D_n$, construct $f = \bigcup_n \sigma_n$, for $\sigma_n \in 2^{<\omega}$ as follows. Let $\sigma_{-1} = \lambda \in 2^{<\omega}$, the empty sequence. Given σ_{n-1}, find a string σ_n extending σ_{n-1} (written $\sigma_{n-1} \subseteq \sigma_n$) such that D_n contains the basic open set $\{g\colon g \in 2^\omega$ and $\sigma_n \subset g\}$, which is possible since D_n is dense and open.

Post [**1944**] gave the first intuitive treatment of r.e. sets and the role they played in Gödel's incompleteness theorem [**1931**]. For $A \subseteq \omega$ let $\deg(A)$ denote the Turing degree of A, namely $\{B\colon A \equiv_T B\}$. Let $\mathbf{0} = \deg(\varnothing)$ and $\mathbf{0}' = \deg(K)$, where K (also written $\varnothing'$) is the complete r.e. set $\{e\colon e \in W_e\}$, and where $\{W_e\colon e \in \omega\}$ is a standard ordering of all r.e. sets. Post raised the question subsequently known as *Post's problem* of whether there exists an r.e. set A such that A is *incomplete*, namely $\mathbf{0} < \deg(A) < \mathbf{0}'$. Post hoped to construct an incomplete r.e. set A by finding a property of the complement $\bar{A}$ compatible with A being nonrecursive and which implies that $\bar{A}$ is sufficiently "thin" with respect to containment of certain r.e. sets so that $K \not\leqslant_T A$. While this approach succeeded in producing r.e. sets incomplete with respect to certain other reducibilities (see Odifreddi [**1981**]), it did not solve Post's problem for *Turing* reducibility.

The next major development was the result by Kleene and Post [**1954**] that there exist sets A and $B \subseteq \omega$ of degree $\leqslant \mathbf{0}'$ which are Turing incomparable ($A \not\leqslant_T B$ and $B \not\leqslant_T A$) and hence incomplete. Of course, A and B are not r.e. but they are recursive in $\varnothing'$ which is fairly close. More important, Kleene and Post proved the theorem by decomposing a complicated condition, $A \not\leqslant_T B$, into an infinite sequence of simpler conditions called *requirements* of the form R_{2e}: $A \neq \{e\}^B$, $e \in \omega$, where $\{e\}^X$ denotes the eth Turing reduction (partial recursion functional) with oracle X, and similarly R_{2e+1}: $B \neq \{e\}^A$ for $B \not\leqslant_T A$. Because of the continuity of $\{e\}^X$, viewed as a partial functional on the Cantor space, the requirement R_e can be satisfied using the "finite extension" method of the Baire Category Theorem, but relativized now to the oracle $\varnothing'$. This decomposition of a complicated condition into simpler requirements which may be satisfied by the Baire Category method has also led to dramatic advances elsewhere, for example, in Paul Cohen's [**1963**] invention of "forcing" to prove the independence of the continuum hypothesis.

Within three years of the Kleene-Post result, Friedberg and, independently, Muchnik, had each solved Post's problem by replacing the $\varnothing'$-oracle construction of Kleene and Post by a recursive construction proceeding through infinitely many stages to insure that the previous sets A and B are now in addition r.e. At each stage s in the recursive construction, one has only approximations A_s, B_s, $\{e\}_s^{A_s}$ and $\{e\}_s^{B_s}$ to the sets being constructed and the partial recursive (p.r.) functionals. A requirement R_n has *higher priority* (also called *stronger* priority) than R_m if $n < m$. At a stage s in the recursive construction, we may take some action in the style of the Baire Category method to meet some requirement R_m. At a later stage $t > s$, new information may become available concerning some higher priority requirement R_n, $n < m$, causing us to act to satisfy R_n, thereby possibly "injuring" R_m by undoing the effect of the previous action. In this case we must begin anew to satisfy R_m. As long as R_m is injured at most finitely often, its action eventually remains undisturbed and R_m remains satisfied. This is the essence of the finite injury priority method.

The finite injury priority method was further developed, but it soon proved inadequate to handle more difficult questions. It was superseded by the infinite injury priority method in which a given requirement R_n may be injured infinitely often. The first example of this method was a weak form of the thickness lemma by Shoenfield [**1961**] in connection with representability of recursive functions in incomplete theories. Independently Sacks [**1963a**, **1963c**, **1964a**, **1964b**] developed a different and more powerful version of the infinite injury priority method, and he used it to prove a wealth of results about r.e. sets and degrees. Yates [**1966a**, **1966b**] and also Lachlan [**1966b**], further extended and applied the method, and each independently solved a problem of Sacks by inventing the important minimal pair method which is another kind of infinite injury method and will be described later.

By this time, the arguments had become technically very complicated and difficult to follow. Various attempts were made by Sacks [**1963a**, Chapter 4], Lachlan [**1967**] and others, to give a common framework for priority arguments which would make them more intelligible. These attempts included a game theoretic approach (Lachlan [**1970**]) as well as a topological approach (Lachlan [**1973a**, **1973b**]). The latter ideas were further developed in Soare [**1976**] to give a unified and comprehensible treatment of the infinite injury method yielding as corollaries most of the infinite injury results by Sacks, Yates and others, and in Soare [**1980a**] to give Lachlan's "nested strategies" version of the minimal pair method, and other important methods for constructing r.e. degrees

The major new development in r.e. degrees during the last few years not covered in Soare [**1976**, **1978a**, or **1980b**] is the use of trees in priority constructions and the $0'''$-priority method which requires the use of these trees. A priority argument is classified as $0'$, $0''$ or $0'''$ according to whether one requires a $\varnothing'$-, $\varnothing''$- or $\varnothing'''$-oracle at the end of the construction to determine exactly how each requirement was satisfied. For example, finite injury priority arguments are of level $0'$, while the infinite injury and minimal pair arguments are level $0''$. The use of trees (even in $0''$-priority arguments) and the $0'''$-priority method, originally introduced by Lachlan, have become increasingly important in recent years. (For example, Harrington and Shelah [**1982a**] have announced the undecidability of the elementary theory of the r.e. degrees using the latter method.) The present paper is an attempt to lay a comprehensible framework for these new methods, analogously to Soare [**1976** and **1980a**] for the $0''$-priority methods. The exposition of the $0'''$-priority method will be further developed in Slaman and Soare [**ta**] from which several of the ideas in the present paper are derived. The nature of this approach is to return to the spirit of the Baire Category Theorem but without the topological machinery of Lachlan [**1973a**]. One first isolates the strategy for meeting a single requirement R_n. One then considers the necessary modifications so that if R_n treats the lower priority requirements in a certain fashion, he will be treated by the higher priority requirements in the same fashion and will therefore be given at least the minimal environment necessary for his strategy to succeed.

In §1 we give a very brief history of r.e. sets and degrees concentrating primarily on recent results not already covered in our survey (Soare [**1978b**]). In §2 we review the Friedberg-Muchnik Theorem and then present it as a tree argument in order to become familiar with the use of trees in a very elementary setting. In §3 we elaborate upon this example to discuss the general use of trees in priority arguments. In §4 we review the minimal pair construction to remind the reader of the "nested strategies" method which is necessary for tree arguments and for the $0'''$-priority method. In §5 we combine the minimal pair method with trees to produce an example where trees are needed in a $0''$-priority construction. In §6 we present the nonbounding theorem of Lachlan [**1979**] which we believe to be the easiest example of a $0'''$-priority argument. We close in §§7 and 8 with some general remarks about how the method may be adapted and applied to other problems.

We use the standard notation from Rogers [**1967**] and Soare [**1978b**]. We deal with sets and functions over the nonnegative integers $\omega = \{0, 1, 2, \ldots\}$. We identify a set A with its characteristic function χ_A and write $A(x)$ instead of $\chi_A(x)$. Let $f \restriction x$ denote the restriction of f to arguments $y < x$, and $A \restriction x$ denote $\chi_A \restriction x$. Let $A \subseteq {}^*B$ denote that $A \subseteq B$ except for finitely many elements and $A = {}^*B$ denote that $A \subseteq {}^*B$ and $B \subseteq {}^*A$. Let $\langle x, y\rangle$ denote the integer which is the image of the ordered pair (x, y) under the standard recursive pairing function $\langle \cdot , \cdot \rangle$ from $\omega \times \omega$ one-one onto ω. For $A \in \omega$, let $A^{[y]} = \{\langle x, z\rangle : \langle x, z\rangle \in A \;\&\; z = y\}$, namely the yth row of A which is viewed (by identification under the pairing function) as a subset of $\omega \times \omega$. Let $\{e\}_s^A(x){\downarrow} = y$ denote that the eth partial recursive functional with oracle A and input x converges in $\leqslant s$ steps, and $\{e\}_s^A(x)\uparrow$ denote divergence. In the former case we let $u(A; e, x, s)$ denote the largest element used in the computation and we assume that the steps are arranged so that if $\{e\}_s^A(x) = y$, then x, y and $u(A; e, x, s)$ are all $\leqslant s$. In any recursive construction in which we are constructing an r.e. set A, we let A_s denote the elements enumerated in A by the end of stage s. If Λ is any set, then Λ^ω denotes the set of functions $f: \Lambda \to \omega$; $\Lambda^{<\omega}$ denotes the set of finite strings of elements of Λ; lower case Greek letters $\alpha, \beta, \gamma, \eta, \xi, \zeta$ range over $\Lambda^{<\omega}$; $\alpha \subseteq \beta$ ($\alpha \subset \beta$) denotes that β extends (properly extends α), and $\alpha \subset f$ that f extends α; $\alpha {}^\wedge \beta$ represents the concatenation of α followed by β, $\lambda \in \Lambda$ is the empty string and $\langle a \rangle$, for $a \in \Lambda$, the string of length 1 consisting of a; $|\sigma|$ denotes the length of string σ. Further notation about trees and strings is given in Definition 2.2. In addition to the usual logical symbols the quantifiers $(\exists^\infty x)R(x)$ and $(\text{a.e.}x)R(x)$ denote "there exist infinitely many x such that $R(x)$," and "for almost every x" respectively. When we write $a \leqslant b, c$ we mean $a \leqslant b$ and $a \leqslant c$, and similarly $a \leqslant b, c \leqslant d$ denotes that $a \leqslant b \leqslant d$ and $a \leqslant c \leqslant d$. (*Warning*: all degrees from now on will be r.e. unless explicitly mentioned otherwise. In particular, all quantifiers of theorems in §1 range over **R**.)

1. A brief survey of recursively enumerable sets and degrees. This is intended to be only a very brief account of some of the main results about r.e. sets and degrees, particularly those not found in our previous survey Soare [**1978b**]. Many more results not mentioned are contained in papers listed in the extensive bibliography but even this is not exhaustive.

1.1. *The upper semilattice of the r.e. degrees.* A (Turing) degree is r.e. if it contains an r.e. set. Like the degrees as a whole, the r.e. degrees **R** form an upper semilattice being closed under supremum (sup) $\mathbf{a} \cup \mathbf{b}$ but not necessarily under infimum (inf) $\mathbf{a} \cap \mathbf{b}$ ([Lachlan, **1966b**]). Let $\mathbf{R}^+ = \mathbf{R} - \{\mathbf{0}\}$.

After Friedberg and Muchnik constructed incomparable r.e. degrees, the finite injury method was developed to perhaps its strongest form by Sacks [**1963b**] to prove

1.1. THEOREM (SACKS SPLITTING THEOREM).

$$(\forall \mathbf{a} > \mathbf{0})(\exists \mathbf{b}_0, \mathbf{b}_1)[\mathbf{a} = \mathbf{b}_0 \cup \mathbf{b}_1 \;\&\; \mathbf{b}_0 | \mathbf{b}_1],$$

where $\mathbf{b}_0 | \mathbf{b}_1$ *denotes that* $\mathbf{b}_0$ *and* $\mathbf{b}_1$ *are incomparable.*

To prove this theorem, Sacks invented a new technique for meeting a negative requirement of the form N_e: $\{e\}^{B_i} \neq A$ in which he attempted to preserve *agreement* between both sides of the equation, rather than disagreement as one might have attempted. Sacks combined this idea with his version of the infinite injury priority method to prove numerous other theorems about **R** including the jump theorem [**1963c**] and particularly the elegant density theorem [**1964a**].

1.2. THEOREM (SACKS DENSITY THEOREM).

$$(\forall \mathbf{a}, \mathbf{c})[\mathbf{c} < \mathbf{a} \Rightarrow (\exists \mathbf{b})[\mathbf{c} < \mathbf{b} < \mathbf{a}]].$$

Yates [**1966b** and **1969**] also developed a similar version of the infinite injury method and used it to obtain a number of results about the classification in the arithmetical hierarchy of index sets of r.e. sets, and to give a different proof of the density theorem.

For each $n \geqslant 0$ define the subclasses of $\mathbf{R}$,

$$\mathbf{H}_n = \{\mathbf{d}\colon \mathbf{d} \in \mathbf{R} \text{ and } \mathbf{d}^{(n)} = \mathbf{0}^{(n+1)}\} \quad \text{and} \quad \mathbf{L}_n = \{\mathbf{d}\colon \mathbf{d} \in \mathbf{R} \text{ and } \mathbf{d}^{(n)} = \mathbf{0}^{(n)}\},$$

where $\mathbf{d}^{(0)} = \mathbf{d}$, and $\overline{\mathbf{L}}_n = \mathbf{R} - \mathbf{L}_n$. (Note that $\mathbf{R}^+ = \overline{\mathbf{L}}_0$.) The degrees in $\mathbf{H}_1(\mathbf{L}_1)$ are called *high* (*low*) since they have the highest (lowest) possible jump. An r.e. set A is *high* (*low*) if $\deg(A) \in \mathbf{H}_1(\mathbf{L}_1)$.

It was natural to ask to what extent these two theorems of Sacks could be combined. Using the recursion theorem Robinson [**1971a**, Corollary 9] extended the Sacks splitting theorem (which constitutes the case $\mathbf{c} = \mathbf{0}$ below).

1.3. THEOREM (ROBINSON SPLITTING THEOREM).

$$(\forall \mathbf{a})(\forall \mathbf{c} \in \mathbf{L}_1)[\mathbf{c} < \mathbf{a} \Rightarrow (\exists \mathbf{b}_0, \mathbf{b}_1)[\mathbf{a} = \mathbf{b}_0 \cup \mathbf{b}_1 \textit{ and } \mathbf{c} < \mathbf{b}_i < \mathbf{a} \textit{ for } i < 2]].$$

However, Lachlan [**1975a**] *proved that this could not be done for every* $\mathbf{c} \in \mathbf{R}$.

1.4. THEOREM (LACHLAN NONSPLITTING THEOREM).

$$(\exists \mathbf{a}, \mathbf{c})[\mathbf{c} < \mathbf{a} \,\&(\forall \mathbf{b}_0, \mathbf{b}_1)[[\mathbf{c} \leqslant \mathbf{b}_0 \,\&\, \mathbf{c} \leqslant \mathbf{b}_1 \,\&\, \mathbf{a} \leqslant \mathbf{b}_0 \cup \mathbf{b}_1] \Rightarrow [\mathbf{a} \leqslant \mathbf{b}_0 \vee \mathbf{a} \leqslant \mathbf{b}_1]]].$$

To prove this theorem, Lachlan introduced what we now call the $0'''$-priority method (see §6). (The method appeared so difficult that the result and proof were sometimes referred to as the "monster theorem" and "monster method".) An exposition of this theorem will be presented in Slaman and Soare [**ta**] using the same framework as in §6 for the $0'''$-priority method.

The uniformity of structure of $\mathbf{R}$ exhibited by the splitting and density theorems of Sacks led Shoenfield [**1965**] to conjecture that $\mathbf{R}$ is a dense structure as an upper semilattice analogously as the rationals are dense as a linear ordering. (The conjecture asserts that if $\vec{\mathbf{a}} \in R$ satisfies a diagram $D(\vec{x})$ and $D_1(\vec{x})$ is any consistent diagram in the language $L(\leqslant, \cup, \mathbf{0}, \mathbf{0}')$ extending D, then there exists $\mathbf{b} \in \mathbf{R}$ such that $D_1(\vec{\mathbf{a}}, \mathbf{b})$.) Shoenfield listed two consequences of his conjecture:

(1.1) If $\mathbf{a}, \mathbf{b}$ are incomparable then they have no infimum in $\mathbf{R}$;

(1.2) given degrees $\mathbf{0} < \mathbf{b} < \mathbf{a}$, there exists $\mathbf{c} < \mathbf{a}$ such that $\mathbf{a} = \mathbf{b} \cup \mathbf{c}$.

Unfortunately, both consequences are false but each led to the development of important new ideas and methods of proof.

We say that incomparable degrees $\mathbf{a}, \mathbf{b} \in \mathbf{R}$ form a *minimal pair* if the infimum $\mathbf{a} \cap \mathbf{b}$ exists and is $\mathbf{0}$. (Note that throughout this paper we use "minimal pair" to refer only to r.e. degrees, although it is often used more generally for non-r.e. degrees.) Yates [**1966a**] and, independently, Lachlan [**1966b**] refuted Shoenfield's conjecture by constructing a minimal pair (as we prove in §4). This led to the question of which degrees $\mathbf{c} \in \mathbf{R}^+$ *bound* a minimal pair $\mathbf{a}, \mathbf{b}$ in the sense that $\mathbf{c} > \mathbf{a}$ and $\mathbf{c} \geqslant \mathbf{b}$. Cooper [**1974b**] proved that every $\mathbf{c} \in \mathbf{H}_1$ does so, but Lachlan

[1979] proved that not every $\mathbf{c} \in \mathbf{R}^+$ does so. In §6 we present our version of the proof of this Lachlan nonbounding theorem as perhaps the easiest example of a proof using the $0'''$-priority method.

Consequence (1.2) of the Shoenfield conjecture was also refuted and gave rise to a number of results about cupping and capping in $\mathbf{R}$. A degree $\mathbf{a} \in \mathbf{R}$ caps (cups) if there is a degree $\mathbf{b} \in \mathbf{R}$, $\mathbf{0} < \mathbf{b} < \mathbf{0}'$, such that $\mathbf{a} \cap \mathbf{b} = \mathbf{0}$ ($\mathbf{a} \cup \mathbf{b} = \mathbf{0}'$). A degree $\mathbf{a} \in \mathbf{R}$ has the *anticupping* (a.c.) *property* if there is a nonzero r.e. degree $\mathbf{b} < \mathbf{a}$ such that for no r.e. $\mathbf{c} < \mathbf{a}$ does $\mathbf{a} = \mathbf{b} \cup \mathbf{c}$. Consequence (1.2) asserts that no $\mathbf{a} \in \mathbf{R}$ has the a.c. property. Lachlan **[1966c]** disproved (1.2), and Ladner and Sasso **[1975]** showed that every $\mathbf{a} \in \mathbf{R}^+$ has a predecessor $\mathbf{b} \in \mathbf{L}_2 \cap \mathbf{R}^+$ with the a.c. property. Yates and later Cooper **[1974a]** proved that $\mathbf{0}'$ has the a.c. property, and Harrington **[1976]** proved that every $\mathbf{a} \in \mathbf{H}_1$ has it. D. Miller **[1981a]** gives a good exposition of this proof and obtains certain extensions **[1981b]**. The proof is interesting in that it goes beyond the usual infinite injury method because we may have $\liminf_s r(e, s) = \infty$ where $r(e, s)$ is the restraint function for a requirement R_e. However, the proof does not involve either trees or the $0'''$-priority method, and will not be discussed here. Harrington **[1978]** did, however, use the $0'''$-priority method to prove that not every $\mathbf{a} \in \mathbf{R}^+$ has the a.c. property, and indeed, that there exits $\mathbf{a} \in \mathbf{R}^+$ such that every $\mathbf{b} < \mathbf{a}$ can be nontrivially cupped to every $\mathbf{d} \geqslant \mathbf{a}$.

1.5. THEOREM (PLUS CUPPING THEOREM—HARRINGTON).

$$(\exists \mathbf{a} > \mathbf{0})(\forall \mathbf{d} \geqslant \mathbf{a})(\forall \mathbf{b})_{\mathbf{0}<\mathbf{b}<\mathbf{a}}(\exists \mathbf{c} < \mathbf{d})[\mathbf{b} \cup \mathbf{c} = \mathbf{d}].$$

Fejer and Soare **[1981]** showed that the special case where $\mathbf{d} = \mathbf{0}'$ can be proved much more simply using only an $0''$-priority argument, and this serves as a good introduction to the gap-cogap method which appears in $0'''$-priority constructions. They showed also that this special case has as immediate corollaries the following two very pleasing results, each originally proved by Harrington (unpublished) which assert that every $\mathbf{a} \in \mathbf{R}$ either cups or caps and that some degrees do both.

1.6. COROLLARY (CUP OR CAP THEOREM—HARRINGTON).

$$(\forall \mathbf{a})(\exists \mathbf{b})_{\mathbf{0}<\mathbf{b}<\mathbf{0}'}[\mathbf{a} \cup \mathbf{b} = \mathbf{0}' \textit{ or } \mathbf{a} \cap \mathbf{b} = \mathbf{0}].$$

1.7. COROLLARY (CUP AND CAP THEOREM—HARRINGTON).

$$(\exists \mathbf{a})(\exists \mathbf{b}, \mathbf{c})_{\mathbf{0}<\mathbf{b},\mathbf{c}<\mathbf{0}'}[\mathbf{a} \cup \mathbf{b} = \mathbf{0}' \textit{ and } \mathbf{a} \cap \mathbf{c} = \mathbf{0}].$$

In spite of those failures of Shoenfield's conjecture, recent results have shown surprising algebraic uniformity of structure of $\mathbf{R}$, in the spirit of the Sacks density and splitting theorems. Let $\mathbf{M} \subseteq \mathbf{R}$ consist of $\mathbf{0}$ and those degrees which form half of a minimal pair. Let $\mathbf{NC} = \mathbf{R} - \mathbf{M}$. The degrees $\mathbf{a} \in \mathbf{M}$ are called *cappable* and those in $\mathbf{NC}$ *noncappable*. Ambos-Spies, Jockusch, Shore and Soare **[1984]** showed that $\mathbf{M}$ is an ideal in $\mathbf{R}$ and that $\mathbf{NC}$ is a strong filter (namely closed upwards), and such that if $\mathbf{a}, \mathbf{b} \in \mathbf{NC}$ there exists $\mathbf{c} \in \mathbf{NC}$, $\mathbf{c} \leqslant \mathbf{a}, \mathbf{b}$. This gives the first algebraic decomposition of **R** into the disjoint union of an ideal and a filter. Furthermore, **NC** coincides with five other apparently unrelated classes of r.e. degrees: **ENC**, the effectively noncappable degrees; **LC**, the degrees cuppable to $\mathbf{0}'$ by some $\mathbf{b} \in \mathbf{L}_1$; **PS**, the degrees of the promptly simple sets of subsection 1.3; **G**, the degrees of r.e. sets in the orbit of the r.e. generic set under automorphisms of $\mathscr{E}$ (see subsection 1.3); and $\mathbf{SP\overline{H}}$, the degrees of non-*hh*-simple sets having a certain splitting property studied by Maass, Shore and Stob [**1981**]. Recently Schwarz [**1982** and **1984**] has studied the quotient stucture **R** modulo the ideal **M**, and has shown, for example, that the Sacks splitting theorem holds there.

A degree $\mathbf{a} \in \mathbf{R}$ is *branching* if there exist incomparable degrees $\mathbf{b}, \mathbf{c} \in \mathbf{R}$ such that $\mathbf{a} = \mathbf{b} \cap \mathbf{c}$. For example, the existence of a minimal pair shows that **0** is branching. Fejer [**1980** and **1983**] and Slaman [**ta**] each extended the Sacks density theorem by proving the density of the nonbranching degrees and the density of the branching degrees respectively. (The latter improves an earlier result of Fejer [**1982**].)

With regard to the structure of degrees r.e. relative to a given degree, Shore [**1982**] refuted a conjecture of Sacks [**1966**] by showing that the degrees r.e. in and above a degree **d** are not in general isomorphic to **R**. The proof uses the embedding method of Lerman, Shore and Soare [**1984**]. New, simple proofs of various results on r.e. degrees are given by Jockusch and Shore [**1983**] by introducing the notion of a pseudo-jump operator and proving certain completeness-type theorems. For example, Jockusch and Shore [**1983**] prove that for any e and $i \in \omega$ there is a nonrecursive r.e. set A such that $W_e^A \oplus A$ has r.e. degree and $W_e^A \oplus W_i^A \oplus A$ has degeee $\mathbf{0}'$. This enables a very simple proof using Theorem 1.4 of Harrington's result [**1980**] that the upper degree **a** of the Lachlan nonsplitting theorem 1.4 may be $\mathbf{0}'$. On the other hand, Soare and Stob [**1982**] have shown for any degree $\mathbf{c} \in \mathbf{R}^+$ there is a degree $\mathbf{a} \geqslant \mathbf{c}$, such that **a** is r.e. in **c** but $\mathbf{a} \notin \mathbf{R}$. The Jockusch-Shore and Soare-Stob results move in opposite directions to one another, and each limits the amount to which the other can be extended or made uniform.

1.2. *The elementary theory of* **R** *and embedding lattices into* **R**. Let Th(**R**) be the elementary theory of the structure $(\mathbf{R}, \leqslant, \cup, \mathbf{0}, \mathbf{0}')$. Lerman, Shore and Soare [**1984**] proved that Th(**R**) is not $\aleph_0$-categorical by proving that there are infinitely many 3-types because there are infinitely many nonisomorphic finite "partial lattices" generated by three elements, all of which can be embedded into **R**. Harrington and Shelah [**1982a**] have announced that Th(**R**) is undecidable by showing that the theory of partial orderings is reducible to it. Their proof [**1982b**] uses the $0'''$-priority method, and also shows that Th(**R**) is not $\aleph_0$-categorical.

Closely related to deciding fragments of Th(**R**) is the important embedding question of which finite lattices can be embedded in **R** as lattices (i.e. preserving sups and infs). (For example, this question must be answered before one can decide even the $\exists$-sentences of $L(\leqslant, \cup, \cap, 0)$ true in **R**.) By extending the minimal pair method, Thomason [**1971**], Lachlan [**1972**] and, independently,

Lerman (unpublished) showed that all countable distributive lattices could be embedded in **R** as lattices, and Lachlan [**1972**], by a more difficult method, showed that the two nondistributive five element lattices can also be so embedded. However, Lachlan and Soare [**1980**] refuted the Embedding Conjecture by proving that not all finite lattices can be so embedded. Further work has been done by Ambos-Spies, Lerman and Soare.

Also of interest is the question of which lattices can be lattice embedded in **R** by mappings which preserve least and *greatest* elements. Lachlan's nondiamond theorem [**1966b**] established that the diamond lattice (i.e. the four element Boolean algebra) could not be so embedded. From this it was conjectured that if $\mathbf{a}, \mathbf{b} \in \mathbf{R}$ satisfy $\mathbf{a} \cup \mathbf{b} = \mathbf{0}'$, then **a** and **b** have no infimum in **R**. This was refuted by Shoenfield and Soare [**1978**] and, independently, by Lachlan [**1980**]. Indeed, Lachlan proved the following more general and pleasing result.

1.8. THEOREM (LACHLAN SPLITTING THEOREM).

$$(\forall \mathbf{a} > \mathbf{0})(\exists \mathbf{b}_0, \mathbf{b}_1, \mathbf{c})[\mathbf{a} = \mathbf{b}_0 \cup \mathbf{b}_1 \,\&\, \mathbf{b}_0 \mid \mathbf{b}_1 \,\&\, \mathbf{c} = \mathbf{b}_0 \cap \mathbf{b}_1].$$

The proof uses a new technique for preserving infimums which may have other applications. Further work related to the nondiamond theorem has been done by Ambos-Spies [**1980** and **1984a**], who proves that in Theorem 1.8 we may replace the last clause by "$\mathbf{b}_0 \cap \mathbf{b}_1$ does not exist".

1.3. *The lattice of r.e. sets.* Let $\mathscr{E}$ denote the lattice of r.e. sets under inclusion, $\mathscr{F}$ the ideal of finite sets and $\mathscr{E}^*$ the quotient lattice of $\mathscr{E}$ modulo $\mathscr{F}$. In his attempt to construct an incomplete r.e. set, Post defined a coinfinite r.e. set A to be *simple* if $\bar{A}$ contains no infinite r.e. set. Dekker [**1954**] proved that simplicity will not guarantee incompleteness because the degrees of the simple sets are precisely $\mathbf{R}^+$. Myhill proposed an r.e. set A with a still thinner complement $\bar{A}$ and called A *maximal* if A^* is a coatom of $\mathscr{E}^*$, i.e. if $A \in \mathscr{E}$ and $\bar{A}$ is infinite but cannot be split into two infinite pieces by any r.e. set W_e. A maximal set was first constructed by Friedberg [**1958**]. Sacks [**1964b**] constructed an incomplete maximal set, Yates [**1965**] a complete one and Martin [**1966b**] showed that the degrees of maximal sets are precisely $\mathbf{H}_1$.

For any set $S \subseteq \omega$ (not necessarily r.e.) let $\mathscr{E}(S)$ denote the lattice $\{W_e \cap S: e \in \omega\}$ under inclusion. Lachlan [**1968a**, **1968c**, **1968d** and **1970**] extensively studied $\mathscr{E}$ and classified those Boolean algebras which can occur as $\mathscr{E}(\bar{A})$ for $A \in \mathscr{E}$. Soare [**1974b**] answered a question of Martin and Lachlan by developing some machinery for generating automorphisms of $\mathscr{E}$, and proved that for any two maximal sets A and B there is an automorphism of $\mathscr{E}$ carrying A to B (i.e. A and B are *automorphic*). He also showed that the group of automorphisms Aut $\mathscr{E}^*$ is k-ply transitive on its coatoms for any $k \in \omega$. Soare [**1982a**] later showed that if A is r.e. coinfinite and low (or even if $\bar{A}$ is semilow, namely $\{e: W_e \cap \bar{A} \neq \varnothing\} \leqslant_T \phi'$) then $\mathscr{E}^*(\bar{A}) \simeq^{\text{eff}} \mathscr{E}^*$, where $\simeq^{\text{eff}}$ denotes that the isomorphism is effective on indices of r.e. sets. Soare [**1977**] showed that an r.e. set A has $\bar{A}$ semilow iff A has a certain computational complexity property called nonspeedability.

Maass [**1982**] introduced a different computational complexity property called prompt simplicity, and showed that if two r.e. sets are both promptly simple and nonspeedable, then they are automorphic. He constructed an r.e. generic set, which has roughly all the properties which a finite injury priority argument could guarantee, and he showed that each such set has both the complexity properties above. For a survey of the results about the computational complexity of r.e. sets, see Soare [**1982b**]. Promptly simple sets had surprising consequences both for the lattice $\mathscr{E}$ (see Maass, Shore and Stob [**1981**]) and for the r.e. degrees **R** (see Ambos-Spies, Jockusch, Shore and Soare [**1984**]). Schwarz [**1982, 1984**] classified the index sets of the promptly simple sets and degrees.

Maass [**1983**] also simplified the automorphism machinery of Soare and obtained an exact characterization of those coinfinite r.e. sets A such that $\mathscr{E}^*(\overline{A}) \simeq^{\text{eff}} \mathscr{E}^*$. Maass and Stob [**ta**] further extended the automorphism machinery to prove that if $A_1, B_1, A_2, B_2 \in \mathscr{E}$ and for $i = 1, 2$, A_i is a major subset of B_i (in the sense of Lachlan [**1968d**]), then $\mathscr{E}^*(A_1 - B_1) \simeq \mathscr{E}^*(A_2 - B_2)$, and furthermore that the $\vec{\forall}\vec{\exists}$-theory of $\mathscr{E}^*(A_1 - B_1)$ is decidable. This is a very difficult and pleasing result. Nowhere simple sets and their properties in $\mathscr{E}^*$ were studied by Shore [**1978**], and an algebraic characterization of $\mathscr{E}^*(\overline{K})$ was given. Other results about $\mathscr{E}$ were obtained by D. Miller [**1981b**] and Stob [**1979, 1982b, ta**]. Lachlan [**1968d**] proved the equidecidability of the elementary theories of $\mathscr{E}$ and $\mathscr{E}^*$ and gave a decision procedure for the $\vec{\forall}\vec{\exists}$-sentences true in $\mathscr{E}^*$. Lerman and Soare [**1980b**] extended this to the sentences in an expanded language with additional predicates for identifying maximal and *hh*-simple sets. There have been recent attempts to prove the undeciability of the theory of $\mathscr{E}^*$. Herrmann [**ta**] and Harrington have each announced its undecidability.

There are many other interesting and important recent results on r.e. sets and degrees too numerous to describe in detail. We refer the reader particularly to Arslanov [**1981**], Arslanov, Nadirov and Salovev [**1977**], Bickford [**1983**], Bickford and Mills [**ta**], Fejer [**1980**], Herrman [**1978, 1981, ta**], Jockusch [**1981**], Lerman [**1977, 1978**], Lermann and Remmel [**ta**], Lerman, Shore and Soare [**1978**], Lerman and Soare [**1980a**], Maass [**ta**], Maass and Homer [**ta**], D. Miller [**1981b**], D. Miller and Remmel [**ta**], Mohrherr [**1982**], Odifreddi [**1981**], Schwarz [**1982**], Shore [**1977**], Soare [**1980b, ta**], Stob [**1979**] and Welch [**1981**].

2. The 0′-priority method using trees. We first give a standard proof (from Soare [**ta**, Chapter VII]) of the Friedberg-Muchnik Theorem, and then we recast the proof using trees. In this case there is no particular advantage in using trees, but it will serve to introduce the reader to tree constructions in a very simple setting.

2.1. THEOREM (FRIEDBERG - MUCHNIK). *There exist r.e. sets A and B such that $A \not\leq_T B$ and $B \not\leq_T A$ (and hence $\varnothing <_T A, B <_T \varnothing'$).*

STANDARD PROOF OF THEOREM 2.1. It suffices to recursively enumerate A and B to meet, for all e, the requirements

$$R_{2e}: A \neq \{e\}^B, \qquad R_{2e+1}: B \neq \{e\}^A.$$

The strategy (called the *basic module*) for meeting a single such requirement R_{2e} is to attach to R_{2e} a *candidate* $x \in \omega^{[e]}$, which has not yet been enumerated in A, and to wait for a stage $s+1$ such that

$$(2.1) \qquad \{e\}_s^{B_s}(x)\downarrow = 0.$$

(If no such stage exists, we do nothing and R_{2e} is automatically satisfied by x because $A(x) = 0$ and either $\{e\}^B(x)\uparrow$ or $\{e\}^B(x)\downarrow \neq 0$.) If $s+1$ exists, we say R_{2e} *requires attention* at stage $s+1$. Now R_{2e} *receives attention* and we (1) enumerate x in A; and (2) define the *restraint function* (i.e. the "wall") $r(2e, s+1) = s+1$ (which by our convention is greater than the greatest element $u(B_s; e, x, s)$ used in the computation), and we attempt (with priority R_{2e}) to restrain any numbers $y \leqslant r = r(2e, s+1)$ from later entering B. If we achieve the latter objective, then $B \upharpoonright r = B_s \upharpoonright r$. Hence, by the use principle,

$$\{e\}^B(x) = \{e\}^{B\upharpoonright r}(x) = \{e\}^{B_s\upharpoonright r}(x) = 0.$$

However, $A(x) = 1$ so requirement R_{2e} is satisfied. (The strategy for R_{2e+1} is the same but with the roles of A and B reversed.)

To accommodate all requirements simultaneously, we must occasionally, but only finitely often, change the candidate $x(e)$ for requirement R_e. Let $x(e, s)$ denote the current candidate for R_e at the end of stage s and $x(e) = \lim_s x(e, s)$. To keep candidates for different requirements distinct, we choose all candidates for R_e from $\omega^{[e]}$. If R_e receives attention at stage $s+1$, then for all $i > e$ we cancel the witness $x(i, s)$ and redefine $x(i, s+1)$ to be some $y > r = r(e, s+1)$. Thus, only requirements R_i of *stronger priority* than R_e (namely $i < e$) can later *injure* R_e by contributing some $x < r$ to A or B. After every R_i, $i < e$, has ceased to receive attention, R_e will receive attention at most once, at which time it will become satisfied and will remain satisfied forever.

Construction of A and B. Let A_s (B_s) denote the set of elements enumerated into A (B) by the end of stage s. To *initialize* R_e *at state* s means to cancel the previous candidate $x(e, s-1)$ (if $s > 0$); to define $x(e, s)$ to be the least $y \in \omega^{[e]}$, $y > s$, $y \notin A_s \cup B_s$, and $y >$ any previous candidate for R_e; and to define $r(e, s) = -1$ (signifying that R_e may later require attention).

Stage $s = 0$. Let $A_0 = B_0 = \varnothing$, and initialize R_e for every e.

Stage $s+1$. Requirement R_{2e} *requires attention* if

$$(2.2) \qquad \{e\}_s^{B_s}(x(2e, s))\downarrow = 0 \quad \text{and} \quad r(2e, s) = -1,$$

and R_{2e+1} *requires attention* if (2.2) holds with A_s and $2e+1$ in place of B_s and $2e$. Choose the least $i \leqslant s$ such that R_i requires attention. We say that R_i *receives attention* or *acts*. Set $r(i, s+1) = s+1$; enumerate $x(i, s)$ in A if i is even and in B if i is odd; for all $k > i$ initialize R_k; and for all $j < i$, leaves the values of the current candidate and restraint unchanged. If i fails to exist go to stage $s+2$.

LEMMA. *For every i, requirement R_i acts at most finitely often and is met.*

PROOF. Fix i and assume the lemma by induction for all $j < i$. Choose s minimal so that no R_j, $j < i$, receives attention at any stage $t > s$. Hence, $r(i, s) = -1$, and for all $t \geqslant s$, $x(i, t) = x(i, s) = x(i)$. Suppose $i = 2e$. (The case of i odd is similar.) If R_{2e} never receives attention after stage s, then $A(x(2e)) = 0$ and $\{e\}^B(x(2e)) \neq 0$. If R_{2e} receives attention at some stage $t + 1 > s$, then $\{e\}_t^{B_t}(x(2e)) = 0$, $x(2e) \in A_{t+1} - A_t$ and $B_t \upharpoonright t = B \upharpoonright t$ so $\{e\}^B(x(2e)) = 0 \neq A(x(2e))$. In the latter case, R_{2e} never receives attention at any stage $v \geqslant t + 1$. (Note that each requirement R_i can be injured at most $2^i - 1$ times.) □

We now wish to recast the same construction using trees. Let Λ be a countable (usually finite) setwith a linear ordering $<_\Lambda$, and define T to be the tree $\Lambda^{<\omega}$, i.e. the finite sequences of elements of Λ. Let lower case Greek letters $\alpha, \beta, \gamma, \ldots,$ range over $\Lambda^{<\omega}$ and f and g range over Λ^ω. Let $|\alpha|$ denote the length of α. Let $\alpha \subseteq \beta$ ($\alpha \subset \beta$) denote that string β extends (properly extends) α.

2.2. DEFINITION. Let $\alpha, \beta \in T$.

(i) α is to the *left* of β ($\alpha <_L \beta$) if

$$(\exists a, b \in \Lambda)(\exists \gamma \in T)\left[\gamma\hat{\ }\langle a\rangle \subseteq \alpha \,\&\, \gamma\hat{\ }\langle b\rangle \subseteq \beta \,\&\, a <_\Lambda b\right].$$

(ii) $\alpha \leqslant \beta$ if $\alpha <_L \beta$ or $\alpha \subseteq \beta$.

(iii) $\alpha < \beta$ if $\alpha \leqslant \beta$ and $\alpha \neq \beta$.

Note that $\alpha \leqslant \beta$ is a kind of modified Kleene-Brouwer ordering. If $\alpha \subset \beta$ then α is a *predecessor* of β and β is a *successor* of α. (Thus, we view the tree T as growing downward with λ as the top node.)

For this section we fix $\Lambda = 2 = \{0, 1\}$ with the usual order, and $T = 2^{<\omega}$. For each $\alpha \in T$ we have an α-strategy which is a version of the basic module above, and which is designed to attempt to satisfy requirement R_i if $|\alpha| = i$. Namely, α has a candidate $x(\alpha)$ and a restraint $r(\alpha)$ whose values at the end of stage s are denoted by $x(\alpha, s)$ and $r(\alpha, s)$. Now α *requires* attention where $|\alpha| = 2e$ if (2.2) holds with $|\alpha|$ in place of $2e$ (and in addition with A_s in place of B_s if $|\alpha| = 2e + 1$). At some stage $s + 1$ a certain α which requires attention and whose guess "seems correct" will *receive attention* (*act*), at which time we enumerate $x(\alpha, s)$ in A if $|\alpha| = 2e$ (in B if $|\alpha| = 2e + 1$), set $r(\alpha, s + 1) = s$ to preserve the computation and initialize (therefore injure) all nodes $\beta > \alpha$, since these are the nodes of T of weaker priority than α.

The difference between this and the previous construction is that previously some R_e may act as many as 2^e times, since R_e begins anew after each injury. Here, each of the 2^e strings α of length e acts at most once, since α is equipped with a "guess" about the outcomes of attempts to satisfy R_i, for all $i < e$, and α will only act when this guess "seems correct". Requirement R_e will ultimately be met by the unique α of length e with the correct guess, namely $\alpha = f \upharpoonright e$ where $f \in 2^\omega$ is the *true path* defined as follows.

2.3. DEFINITION. The *true path* $f \in 2^\omega$ through T (for the construction below) is defined by induction on n. Given $\alpha = f \upharpoonright n$, define

$$f(n) = \begin{cases} 0 & \text{if } (\exists s)[R_\alpha \text{ acts at stage } s], \\ 1 & \text{otherwise.} \end{cases}$$

Although f is not recursive, clearly $f \leqslant_T \varnothing'$ and indeed, $f(n) = \lim_s \delta_s(n)$ where we define the recursive sequence $\{\delta_s: s \in \omega\}$ of strings in T, such that $|\delta_s| = s$, as follows by induction on $n < s$. Given $\alpha = \delta_s \upharpoonright n$, if $n < s$ define

$$\delta_s(n) = \begin{cases} 0 & \text{if } (\exists t \leqslant s)[R_\alpha \text{ acts at stage } t], \\ 1 & \text{otherwise.} \end{cases}$$

Notice that for each n, $\delta_{s+1} \upharpoonright n \leqslant \delta_s \upharpoonright n$ (i.e. as s increases δ_s never moves rightward) so $\lim_s \delta_s(n)$ exists and equals $f(n)$. (In the more general $0''$ or $0'''$ tree constructions we can only guarantee that $f(n) = \liminf_s \delta_s(n)$ in the sense of §3, namely f is the leftmost path visited infinitely often by the recursive sequence $\{\delta_s: s \in \omega\}$.)

Thus each $\alpha \in T$ may be identified with a guess about which $\beta \subset \alpha$ will eventually act, namely α guesses that $\beta = \alpha \upharpoonright k$ will act iff $\alpha(k) = 0$. If $\alpha \subseteq \delta_s$ then at stage $s + 1$ α's guess *seems correct* and α is eligible to act. (Note that at such a stage, α is guessing that the only strings $\beta \subset \alpha$ which will ever act have already done so.)

TREE PROOF OF THEOREM 2.1.

Construction of A and B. For each $\alpha \in T$ we have parameters $x(\alpha)$ and $r(\alpha)$, whose values at the end of stage s are denoted by $x(\alpha, s)$ and $r(\alpha, s)$. When a parameter is assigned a value, it retains that value until a new value is assigned. Let $\omega^{[\alpha]} = \alpha^{[h]}$ when n is the code number for α in some effective coding of T. To *initialize* α at stage s means to define $r(\alpha, s) = -1$, and $x(\alpha, s)$ to be the least $y \in \omega^{[\alpha]}$, $y \notin A_s \cup B_s$, and $y >$ all previous values of parameter $x(\alpha)$.

Stage $s = 0$. Initialize all $\alpha \in T$ and set $A_0 = B_0 = \varnothing$.

Stage $s + 1$. Let α be the $\subset$-minimal $\gamma \subseteq \delta_s$ which requires attention. We say α *acts*. Set $r(\alpha, s + 1) = s$, enumerate $x(\alpha, s)$ in A if $|\alpha|$ is even and in B if $|\alpha|$ is odd, and initialize all γ, $\alpha < \gamma$.

LEMMA. *For every i requirement R_i is satisfied by $\alpha = f \upharpoonright i$.*

PROOF. Fix i, let $\alpha = f \upharpoonright i$, and assume the lemma for all $j < i$. Choose s minimal such that $\alpha \subseteq \delta_t$ for all $t \geqslant s$. Now $r(\alpha, s) = -1$ and α is never initialized after stage s. Let $x = \lim_t x(\alpha, t) = x(\alpha, s)$. Suppose $i = 2e$. If α never acts then $\{e\}^B(x) \neq 0$. If α acts at some stage $t > s$, then $\{e\}_t^{B \upharpoonright t}(x)\downarrow = 0$, and $A_t(x) = 1$, but $x(\beta, t) > t$ for all $\beta > \alpha$ so β cannot later injure α by contributing some $y \leqslant t$ to $A \cup B$. Furthermore, if $\beta < \alpha$ then β will never act after stage t because $\alpha \subset f$. In either case $\{e\}^B(x) \neq A(x)$. □

Notice that there is no injury in the construction along the true path. Namely, if $\alpha \subset \beta \subset f$ then α never injures β, because β has a correct guess about whether α will ever act and β only acts after its guess seems correct. (This is as close to a 0-injury solution to Post's problem as is known.)

3. The tree method in priority arguments and the classification of $0'$-, $0''$ and $0'''$-priority arguments. With the simple tree proof of §2 in mind, we now make some general remarks on the use of trees in the priority method before we study some more difficult examples. Some of these remarks were suggested by Harrington [**1982**] and in private conversations, although the use of trees was introduced by Lachlan [**1975a**] and used in Lachlan [**1979**]. We adopt some of the notation, terminology and conventions of the latter which we find particularly convenient, although our proof in §6 of the nonbounding theorem will be different.

To prove a theorem in recursion theory we first write down a list of requirements, R_i, $i \in \omega$, sufficient to establish it. We then formulate a strategy called the *basic module* for satisfying a single such requirement in isolation. The basic module may require infinitely many actions and may have several possible final outcomes (possibly infinitely many).

Let Λ be a coding of the set of outcomes, and $<_\Lambda$ an appropriate ordering of Λ, generally chosen so that if $a <_\Lambda b$ and a "appears correct" infinitely often, then b is not the correct final outcome. (For example, in §2, $\Lambda = \{0, 1\}$ where outcome 1 denotes the outcome that $\{e\}^B(x) \neq 0$, and outcome 0 denotes that $\{e\}^B(x) = 0$ and we enumerate x in A.)

If there are several different types of requirements (e.g. positive and negative) we may need a different type of basic module for each, and the outcome set Λ will vary according to the type of the requirement. (For simplicity, we consider only those cases where there is only one type of basic module and a single outcome set Λ. The other types can easily be combined with this method in the obvious way.)

Next we define the priority tree $T = \Lambda^{<\omega}$, and use the notation and terminology about T defined in §2, especially Definition 2.2. For each $\alpha \in T$ we define a α-strategy for meeting requirement R_i, where $i = |\alpha|$. The α-strategy is merely the natural adaptation of the basic module but takes into account that for each $k < |\alpha|$, $\alpha(k) = a \in \Lambda$ means that α "guesses" that the outcome of the β-strategy for $\beta = \alpha \restriction k$ is a. (This gives α a considerable advantage over the standard linear version of the same argument where R_i has no information about the outcomes of strategies for higher priority requirements R_j, $j < i$.) We defer for the moment the specification of the whole construction and exactly when the α-strategy is allowed to act.

At the end of the construction, we define the *true path* $f \in \Lambda^\omega$ by induction on n as in §2. Namely, if $\alpha = f \restriction n$, let $f(n) \in \Lambda$ be the final outcome of the α-strategy. In presenting the full recursive construction, $\mathscr{C} = \bigcup\{\mathscr{C}_\alpha \colon \alpha \in T\}$, where $\mathscr{C}_\alpha$ is that portion of the construction performed by the α-strategy; our main objective is to insure that what Harrington [**1982**] calls the *construction along the true path*, namely $\hat{\mathscr{C}} = \bigcup\{\mathscr{C}_\alpha \colon \alpha \subset f\}$, succeeds because at the end we shall verify

$$(3.1) \qquad (\forall i)[\alpha = f \restriction i \Rightarrow \alpha \text{ satisfies requirement } R_i].$$

Of course, $\hat{\mathscr{C}}$ is not really a "construction" in any effective sense, since f is not recursive, but $\hat{\mathscr{C}}$ is merely a part of the whole recursive construction $\mathscr{C}$. Nevertheless, $\hat{\mathscr{C}}$ is the only essential part of $\mathscr{C}$ because by (3.1) the only α's which will matter at the end are those $\alpha \subset f$. However, since we cannot recursively identify those $\alpha \subset f$ during the construction, we must specify possible action $\mathscr{C}_\alpha$ for each $\alpha \in T$, such that if $\alpha \subset f$ then α is given at least the minimal environment required by the original basic module to succeed.

How well can we identify f during the construction? In our example of §2, the outcomes $\Lambda = \{0, 1\}$ were each finitary in nature. Namely, we could have defined a recursive function $H(\alpha, s)$ such that $H(\alpha, s) = 1$ if α has not yet acted by the end of stage s, and $= 0$ otherwise; and we could have proved that $\lim_s H(\alpha, s)$ exists for all α. In an infinite injury (namely $0''$) construction, the outcomes will usually include infinitary as well as finitary outcomes. For example, in the thickness lemma (see Soare [**1976**, p. 520]), we have positive requirements of the form P_e: $B^{[e]} =^* A^{[e]}$ where we are constructing an r.e. set A; and B is a given r.e. set such that $B^{[e]}$ is either $\omega^{[e]}$ or is finite. If $|\alpha| = i$ and $R_i = P_e$, then the final outcomes for α are $\Lambda = \{0, 1\}$ where 1 denotes that R_i contributes at most finitely many elements to A, while 0 denotes that R_i contributes infinitely many. In other cases, such as the minimal pair (§4) or the thickness lemma, R_i may be a negative requirement N_e with an associated recursive restraint function $r(e, s)$ such that $\liminf_s r(e, s)$ exists. The set of outcomes for N_e will be $\Lambda = \omega$ where outcome k denotes that $\liminf_s r(e, s) = k$.

We shall attempt to define a recursive approximation $\{\delta_s \colon s \in \omega\}$ to f, where $\delta_s \in 2^{<\omega}$ and $|\delta_s| = s$. Fix s and define $\delta_s(n)$ by induction on n for $n < s$. Suppose $\alpha = \delta_s \restriction n$. Using the Thickness Lemma example above, if R_n is N_e, define $\delta_s(n) = r(e, s)$. If R_n is P_e then define $\delta_s(n) = 0$ if $A_s^{[e]} - A_t^{[e]} \neq \varnothing$, where t is the greatest stage $< s$ such that $\alpha \subseteq \delta_t$, and $\delta_s(n) = 1$ otherwise. The crucial point is that f is the leftmost path visited infinitely often by $\{\delta_s \colon s \in T\}$, namely $f \restriction n = \liminf_s \delta_s \restriction n$ in the sense of (3.2) and (3.3) below, i.e. if $\alpha = f \restriction n$ then

$$(3.2) \qquad (\exists^{<\infty} s)[\delta_s <_L \alpha]$$

and

$$(3.3) \qquad (\exists^{\infty} s)[\alpha \subseteq \delta_s].$$

Furthermore, we shall arrange that

$$(3.4) \qquad (\forall \beta)[\delta_s <_L \beta \Rightarrow \beta \text{ is initialized at the end of stage } s],$$

where β is *initialized at stage* s if as in §2 all its parameters are reset and any pending β-action is cancelled. At stage $s + 1$, those $\alpha \subseteq \delta_s$ are said to *appear correct*, and s is called an *α-stage* for such α. Furthermore, α is only eligible to act at an α-stage, namely

$$(3.5) \qquad (\forall \alpha)(\forall s)[\alpha \text{ acts at stage } s \Rightarrow \alpha \subseteq \delta_s].$$

(The picture is that at the end of stage s, those β to the right of δ_s, i.e. $\delta_s <_L \beta$, are initialized and any pending β-action or restraint is cancelled. During stage $s + 1$, those $\alpha \subseteq \delta_s$ are eligible to act because their "guess" is in agreement with δ_s. Those $\beta <_L \delta_s$ are regarded as "asleep" during stage $s + 1$ while waiting for some later stage t such that $\beta \subseteq \delta_t$. While β is asleep, any pending β-action or restraint is preserved but no new β-action is taken. If $\alpha = f \upharpoonright n$ then, like the princess who wakes only once every hundred years, α will awaken at each of the infinitely many α-stages, possibly take some action, and will go to sleep until the next α-stage, during which interval α's action is preserved. During the recursive construction $\mathscr{C}$ none of all the β, $|\beta| = i$, working on R_i know which β will be chosen for royalty (namely $\alpha = f \upharpoonright i$) and will succeed in satisfying R_i. Thus each β simply "does his duty to logic" by doing his best to satisfy R_i. Those $\beta <_L \alpha$ will act finitely often, not often enough to satisify R_i, while those β, $\alpha <_L \beta$, will be initialized infinitely often, thus also failing to satisfy R_i, while α alone succeeds.)

Fix $\alpha = f \upharpoonright n$ and consider the construction from the point of view of α. As in Harrington [**1982**] decompose $T - \{\alpha\}$ into the disjoint sets of nodes to the left, right, above and below α:

(3.6)
$$L = \{\beta: \beta <_L \alpha\}, \quad R = \{\beta: \alpha <_L \beta\}, \quad A = \{\beta: \beta \subset \alpha\}, \quad B = \{\beta: \alpha \subset \beta\},$$

and let $\mathscr{C}_X = \{\mathscr{C}_\beta: \beta \in X\}$ for $X \subseteq T$. By (3.2) and (3.5) there are only finitely many stages when any $\beta <_L \alpha$ acts, say none after s_0, so $\mathscr{C}_L$ has only a finite effect upon α. Those $\beta \in R$ will be initialized just before any α-stage, and so $\mathscr{C}_R$ will never give any interference when α wants to act. Finally, any action performed by α at an α-stage $> s_0$ will remain in force until the next α-stage. In most $0''$-constructions (such as the minimal pair of §4), we can show that the basic module can be adapted to succeed if played on an infinite recursive set of stages (the α-stages $> s_0$) in place of *all* stage $s \in \omega$, and so the α-strategy will succeed. (This modification is the main contribution of the version of the minimal pair proof which we present in §4, and is an essential feature of any tree argument at level $0''$ or $0'''$.) For $0'''$-constructions the α-strategy requires not only this but also the active "cooperation" by $\mathscr{C}_A$ as we describe later in subsection 6.4. The guiding principle is what Harrington [**1982**] calls the *golden rule*, namely that α must behave towards $\beta \in R$ (respectively $\beta \in B$) as α would have $\mathscr{C}_L$ (respectively $\mathscr{C}_A$) behave towards α, providing that β's guess appears correct often enough. Thus, α achieves the cooperation of $\mathscr{C}_A$ by actively cooperating with certain $\beta \supset \alpha$ when β seems correct (see subsection 6.4).

The tree method suggests a classification of priority arguments into either $0'$-, $0''$- or $0'''$-arguments as follows. We define a construction $\mathscr{C}$ to be of level $0^{(n)}$ if it requires a $\varnothing^{(n)}$-oracle to compute (at the end of the construction) exactly how each requirement R_i was satisfied. In all tree arguments (even $0'''$) the approximation $\{\delta_s(n): s \in \omega\}$ will be recursive as a function of s and n. For the finite injury constructions, as in §2, we have $f(n) = \lim_s \delta_s(n)$ so $f \leqslant_T \varnothing'$, while in the infinite injury constructions we merely have $f \upharpoonright n = \liminf_s \delta_s \upharpoonright n$ as in (3.2) and (3.3) so $f \leqslant_T \varnothing''$. But if $\alpha = f \upharpoonright n$, then $f(n)$ is the true outcome of the α-strategy, and α satisfies R_n by (3.1). Hence the finite injury arguments have level $0'$ and the infinite injury arguments level $0''$.

But in all known tree arguments, the true path $f \leqslant_T \varnothing''$ because $f \restriction n = \liminf_s \delta_s \restriction n$ as above. Thus there is a mystery of how any construction $\mathscr{C}$ can be classified as having level $0'''$. The answer is that (3.1) no longer holds. Namely, in a typical $0'''$-construction (for example in §6) the requirements R_i are more complicated than before being usually of the logical form

$$(3.7)\qquad R_i \equiv P_{i,1} \vee P_{i,2} \vee \cdots \vee P_{i,m} \vee (\exists X_i)\left[N_i(X_i) \& (\forall j) R_{i,j}(X_i)\right],$$

where X_i is some object being built during the construction such as an r.e. set (as in (3.8)) or a Turing reduction as in Lachlan [**1975a**] or Slaman [**ta**].

For example, in the Lachlan nonbounding theorem of §6, to construct a nonrecursive r.e. set C whose degree does not bound a minimal pair we must meet for all i the requirement

$$(3.8)\qquad R_i\colon \Phi_i^C \neq A_i \vee \Psi_i^C \neq B_i \vee A_i \text{ is recursive} \vee B_i \text{ is recursive}$$
$$\vee (\exists D_i)\left[D_i \leqslant_T A_i \& D_i \leqslant_T B_i \& (\forall j)\left[D_i \neq \overline{W}_j\right]\right].$$

where $\{(A_i, B_i, \Phi_i, \Psi_i)\colon i \in \omega\}$ is a standard enumeration of all 4-tuples (A, B, Φ, Ψ) such that A, B are r.e. sets and Φ, Ψ are p.r. functionals, and where D_i is an r.e. set. (Note that (3.8) is a special case of (3.7) where $m = 4$, $P_{i,k}$, $k \leqslant 4$, is the kth clause of (3.8), N_i asserts "$D_i \leqslant_T A_i$ and $D_i \leqslant_T B_i$" and the subrequirement $R_{i,j}$ asserts "$D_i \neq \overline{W}_j$".)

Since the overall requirement R_i is too complicated to attack all at once, we begin to construct D_i and describe a basic module to satisfy $R_{i,j}$ for a fixed i and j. For simplicity, assume that Λ, the set of outcomes of this basic module, is the same for all i and j. Define the priority tree $T = \Lambda^{<\omega}$, and the recursive functions i, j: $T \to \omega$ by $\langle i(\alpha), j(\alpha)\rangle = m$ where $|\alpha| = \langle m, p\rangle$. Assign α to work on subrequirement $R_{i(\alpha),j(\alpha)}$. (Thus for all i and j there are infinitely many n such that all α of length n are assigned to $R_{i,j}$.)

The set of outcomes Λ will be decomposed into disjoint sets Λ_1 and Λ_2. (For example in §6, $\Lambda_1 = \{s, w\}$ and $\Lambda_2 = \{g1, w1, g2, w2\}$.) For every $\alpha \in T$ and $\alpha \in \Lambda$ we shall associate a subrequirement $M(\alpha, a)$ which is $R_{i(\alpha),j(\alpha)}$ if $a \in \Lambda_1$ and is $P_{i',k}$ for some $i' \leqslant i(\alpha)$, $k \leqslant 4$, if $a \in \Lambda_2$. Now (3.1) for $0''$-arguments is replaced by the (3.9) and (3.10) for $0'''$-arguments,

$$(3.9)\qquad \alpha = f \restriction n \& a = f(n) \in \Lambda_2 \Rightarrow \text{the } \alpha\text{-strategy satisfies } M(\alpha, a) = P_{i',k}$$

and

$$(3.10)\qquad \alpha = f \restriction n \& a = f(n) \in \Lambda_1 \Rightarrow$$
$$\text{the } \alpha\text{-strategy satisfies } M(\alpha, a) = R_{i(\alpha),j(\alpha)} \text{ or } i \in S(\alpha),$$

where

$$(3.11)\qquad S(\alpha) =_{\text{dfn}} \{i'\colon (\exists\beta)(\exists a)(\exists k)\left[\beta^\wedge\langle a\rangle \subseteq \alpha \& M(\beta, a) = P_{i',k}\right].$$

Namely, if $\alpha = f \restriction na = f(n)$ and $a \in \Lambda_2$ then α satisfies $P_{i',k'}$, and hence $R_{i'}$, for some $i' \leqslant i$, and it is unnecessary to consider subrequirement $R_{i',j'}$, for any j'. Now $S(\alpha)$ denotes the set of i' such that $R_{i'}$ has been satisfied in this way by some $\beta \subset \alpha$. If no $\beta \subset \alpha$ has already satisfied R_i and $a \in \Lambda_1$, then (3.10) asserts that α satisfies the subrequirement $R_{i(\alpha),j(\alpha)}$. We shall also insure that if α satisfies R_i then no β, $\alpha \subset \beta \subset f$, also satisfies R_i, namely

$$(3.12)\quad (\forall \alpha, \beta)(\forall i', a, b, k, l)\big[\big[\alpha \subset \beta \subset f \,\&\, f(|\alpha|) = a \,\&\, f(|\beta|) = b \,\&\, M(\alpha, a) = P_{i',k}\big] \Rightarrow M(\beta, b) \neq P_{i',l}\big].$$

Lines (3.9) to (3.12) represent the heart of the $0'''$ overall strategy.

Now requirement R_i is eventually satisfied as follows. From (3.9) and (3.12) we can define for each i

$$(3.13)\quad \tau(f, i) = (\mu\beta \subset f)(\forall n \geqslant |\beta|)[\alpha = f \restriction n \,\&\, i(\alpha) \leqslant i \Rightarrow f(n) \in \Lambda_1],$$

Namely,

$$(3.14)\quad \tau(f, i) = (\mu\beta \subset f)(\forall \alpha)(\forall ek \leqslant i) \cdot [\beta \subset \alpha \subset f \Rightarrow [e \in S(\alpha) \Leftrightarrow e \in S(\beta)]].$$

Now if there is some $\alpha \subseteq \tau(f, i)$ with $a = f(|\alpha|)$ and $M(\alpha, a) = P_{i,k}$ then R_i is satisfied by $P_{i,k}$. Otherwise, for all $\alpha \subset f$, $i \notin S(\alpha)$. For each j, choose n sufficiently large that $\tau(f, i) \subset \alpha = f \restriction n$, $i(\alpha) = i$ and $j(\alpha) = j$. Now by (3.10), subrequirement $R_{i,j}$ is satisfied by α. Note that since $f \leqslant_T \varnothing''$ it requires a $\varnothing'''$-oracle to decide whether R_i is satisfied by the first alternative or the second, namely whether

$$(3.15)\quad (\exists n)[\alpha = f \restriction n \,\&\, (\exists k \leqslant 4)[M(\alpha, f(n)) = P_{i,k}]].$$

(This explains the mystery of how a construction can be classified as $0'''$ even though $f \leqslant_T \varnothing''$.)

It has been suggested by Harrington that a $0'''$-construction can be viewed as an $0'$-construction over a $0''$-construction. We can explain this by referring to the example above. Fix R_i. The last clause of (3.8) requires that if no subrequirement $P_{i,k}$ is satisfied then we must construct an r.e. set D_i meeting subrequirement N_i and subrequirements $R_{i,j}$ for all $j \in \omega$. For each $\alpha \in T$ we have a possible candidate $D_{\alpha,i}$ for D_i as follows. We begin with a certain candidate $D_{\lambda,i}$ for D_i. In passing from α to $\beta = \alpha\hat{\ }\langle a\rangle$, we replace the old set $D_{\alpha,i}$ by a new version $D_{\beta,i}$ just if $i' \in S(\beta) - S(\alpha)$ for some $i' \leqslant i$, namely if (at least if $\beta \subset f$) β satisfies $R_{i'}$ for some $i' \leqslant i$. This is regarded as an "injury" to the final specification of D_i, but is justified by the more important progress made via $P_{i',k}$ on $R_{i'}$. Since this injury occurs finitely often along the true path f, there is a final value $\tau = \tau(f, i)$ such that $D_{\tau,i} = D_{\alpha,i}$ for all α, $\tau \subseteq \alpha \subset f$, and $D_{\tau,i}$ is the r.e. set D_i which we present to satisfy the final clause of R_i. We say that D_i *originates* at node $\tau(f, i)$. (As we saw in §2 in $0'$- or $0''$-constructions there is *no* injury along the true path because if $\alpha \subset f$, then α has a correct guess about each $\beta \subset \alpha$ and can never be injured by such a β.)

Recursive constructions are often thought of as performed by a "clerk". Here we give the clerk an $\varnothing''$-oracle so that he can compute f, and we think of the clerk as performing the $0''$-construction along the true path $\hat{\mathscr{C}}$. As the clerk descends along f and replaces $D_{\alpha,i}$ by a new version $D_{\beta,i}$ where $\beta = \alpha\hat{\ }\langle a\rangle \subset f$, because

$i' \in S(\beta) - S(\alpha)$ for $i' < i$, we call this *injury along the true path* because it appears to the $\varnothing''$-clerk exactly as the usual finite injury appears to the recursive clerk. In other $0'''$-constructions, such as the Lachlan nonsplitting theorem **[1975]** or the density of branching degrees by Slaman **[ta]**, the object X_i of (3.7) which we must construct may be a Turing reduction Ψ_i. For every $\alpha \in T$ we have a candidate $\Psi_{\alpha,i}$ which may be replaced by a new version $\Psi_{\beta,i}$ as we descend along f. The final candidate Ψ_i will be $\Psi_{\tau(f,i),i}$.

From a practical point of view the main complications in the $0'''$-method arise from the fact that subrequirements $R_{i,j}$ for some R_i lie at infinitely many levels of the tree. If $\alpha \subset \beta$ then in the usual $0''$-constructions, $i(\alpha) < i(\beta)$ and hence α can carry out his behaviour without regard to β. It is β's task to guess about the outcome of the α-strategy and to act accordingly. In the $0'''$-constructions we may have $i(\alpha) > i(\beta)$ so that α must consciously cooperate with β.

Namely in order to succeed, α needs an environment similar to that of the basic module and hence α needs the cooperation of $\mathscr{C}_A$ of (3.6). By the golden rule (following (3.6)) α can be assured of this cooperation provided he cooperates with those $\beta \supset \alpha$ when β seems correct. Thus, the fundamental steps in the α-strategy will be the same as those of the original basic module, but in addition α will often have to: (1) allow various $\beta \supset \alpha$ to initiate some actions like those which α itself needs to initiate; and (2) allow those same β to complete certain of these actions if the β-actions affect some R_i for $i < i(\alpha)$, before α is allowed to complete its own pending action. The resulting final α-strategy, called the *modified α-module*, takes into consideration the α-delays caused by (2). The precise specification of these modifications for a given construction is what causes all the extra difficulties in the $0'''$-method as opposed to the $0''$-method (see subsection 6.4).

4. A standard $0''$-argument, the minimal pair method. Recall that nonzero r.e. degrees $\mathbf{a}, \mathbf{b}$ form a *minimal pair* if $\mathbf{a} \cap \mathbf{b} = \mathbf{0}$. Yates **[1966a]** and, simultaneously, Lachlan **[1966b]** each proved the existence of minimal pairs thereby refuting Shoenfield's conjecture.

The proof uses negative requirements N_i with associated recursive restraint functions $r(i, s)$ such that $\liminf_s r(i, s) < \infty$ even though perhaps $\limsup_s r(i, s) = \infty$. The main difficulty is to insure that the restraint functions drop back simultaneously, so that $\liminf_s \tilde{r}(i, s) < \infty$ where $\tilde{r}(i, s) = \max\{r(j, s): j \leqslant i\}$. Lachlan **[1973a]** suggested a new way to achieve this by having the strategy σ_1 for N_1 broken into infinitely many different strategies σ_1^j, one for each guess j about $\liminf_s r(0, s)$. Requirement N_1 is finally satisfied by σ_1^k where k is the true value of $\liminf_s r(0, s)$. Furthermore, σ_1^k is only allowed to act when his guess "seems correct", namely at the stages in $S^k = \{s: r(0, s) = k\}$. This idea is the forerunner of the tree method of §3. Moreover, the idea that strategy σ_1^k can succeed, even if it plays only at the stages in S^k, is crucial to the tree method because $\alpha \in T$ can play only at the α-stages described in §3.

The following argument can be viewed as a tree argument where $R_i = N_i$, and the set of outcomes is $\Lambda = \omega$, where outcome k of R_i denotes that $\liminf_s r(i, s) = k$. However, for simplicity we choose to suppress the tree here because the outcomes for R_j, $j < i$, can be incorporated into a single number $r(i-1, s)$ which is the only information N_i needs. In §4 and §5 we give tree versions of a $0''$-arguments.

4.1. THEOREM. (LACHLAN - YATES) *There exist nonzero r.e. degrees* **a** *and* **b** *which form a minimal pair, namely* $\mathbf{a} \cap \mathbf{b} = \mathbf{0}$.

PROOF OF THEOREM 4.1. It suffices to construct r.e. sets A and B satisfying, for all e, i, j, the requirements

$$P_{2e}: A \neq \{e\}, \qquad P_{2e+1}: B \neq \{e\}$$

and

$$N'_{\langle i,j\rangle}: \{i\}^A = \{j\}^B = f \text{ total} \Rightarrow f \text{ is recursive.}$$

The following remark allows us to simplify the form of the negative requirements.

4.2. REMARK (POSNER). To satisfy all $N'_{\langle i,j\rangle}$, $i, j \in \omega$, it suffices to satisfy, for all e, the requirement

$$N_e: \{e\}^A = \{e\}^B = g \text{ total} \Rightarrow g \text{ is recursive.}$$

PROOF. We may assume without loss of generality that we can arrange that $A \neq B$, say $n_0 \in A - B$. For each i and j there is an index e such that

$$\{e\}^X(x) = \begin{cases} \{i\}^X(x) & \text{if } n_0 \in X, \\ \{j\}^X(x) & \text{if } n_0 \notin X. \end{cases}$$

The remark follows immediately. □

From now on we shall replace all occurrences of negative requirements similar to $N'_{\langle i,j\rangle}$ by equivalent requirements N_e.

Given $\{A_t: t \leqslant s\}$ and $\{B_t: t \leqslant s\}$ we define as usual the functions

$$\text{(length function)} \quad l(e,s) = \max\left\{x: (\forall y < x)\left[\{e\}_s^{A_s}(y)\downarrow = \{e\}_s^{B_s}(y)\right]\right\},$$

$$\text{(maximum length function)} \quad m(e,s) = \max\{l(e,t): t \leqslant s\}.$$

A stage s is called 0-*expansionary* if $l(0,s) > m(0, s-1)$. Define the *restraint function*

$$r(0,s) = \begin{cases} 0 & \text{if } s \text{ is 0-expansionary,} \\ \text{the greatest 0-expansionary stage } t < s, & \text{otherwise.} \end{cases}$$

(Notice that as in §2 we can define the restraint function in terms of a stage s, rather than an element z used in a computation at stage s since $z \leqslant u(A_s; e, z, s) \leqslant s$ by our convention.)

The *basic module*, namely the strategy σ_0 for meeting a *single* negative requirement N_0, is to allow x to enter $A \cup B$ at stage $s+1$ only if $x \geqslant r(0,s)$, and even then at most *one* of the sets A, B receives an element x at any stage. Thus, if x destroys one of the computations $\{0\}_s^{A_s}(p) = q$ or $\{0\}_s^{B_s}(p) = q$ for some $p < l(0,s)$, say $\{0\}_s^{A_s}(p)$, then the other computation $\{0\}_s^{B_s}(p) = q$ will be preserved until the A-computation is restored, and both sides output q again. In this way, if

$\{0\}^A = \{0\}^B = g$ is a total function, then g is recursive. (To compute $g(p)$ we find the least s such that $p < l(0, s)$ and we set $g(p) = \{0\}_s^{A_s}(p)$.) Furthermore, $\liminf_s r(0, s) < \infty$, since $\liminf_s r(0, s) = 0$ unless there is a largest 0-expansionary stage t in which case $r(0, s) = t$ for all $s \geqslant t$.

This fundamental strategy of having one side or the other hold the computation at all times is applied to the other negative requirements N_e, $e > 0$, but with some crucial modifications to force the negative restraints to drop back simultaneously, thus creating "windows" through the restraints. For example, to drop back simultaneously with N_0, N_1 must guess the value of $k = \liminf_s r(0, s)$. Thus, N_1 must simultaneously play infinitely many strategies σ_1^k, $k \in \omega$, one for each possible value of k. Each strategy σ_1^k is played like σ_0 but with $S^k = \{s: r(0, s) = k\}$ in place of ω as the set of stages during which it is active, and through which its length functions l and m are defined. This allows σ_1^k to open its window more often since its length functions ignore the stages in $\omega - S^k$. Strategy σ_1^k still succeeds providing any restraint it imposes is maintained during intermediate stages $s \notin S^k$ while σ_1^k is dormant. Thus, at stage s if $k = r(0, s)$, we play σ_1^k; maintain the restraints previously imposed by the dormant σ_1^i, $i < k$; and discard any restraints imposed by σ_1^j, $j > k$. Thus, if $k = \liminf_s r(0, s)$, then (1) strategy σ_1^k succeeds in meeting N_1; (2) the strategies σ_1^i, $i < k$, impose finitely many restraints over the whole construction; and (3) the strategies σ_1^j, $j > k$, drop all restraint at each stage $s \in S^k$. Thus the entire restraint $r(1, s)$ imposed by N_0 and N_1 *together*, has $\liminf_s r(1, s) < \infty$.

Construction of A and B.

Stage $s = 0$. Do nothing.

Stage $s + 1$. Given A_s and B_s define the restraint function $r(e, s)$ for N_e by induction on e as follows. Define $r(0, s)$ as above. A stage s is $(e + 1)$-*expansionary* if

$$(\forall t < s)[r(e, t) = r(e, s) \Rightarrow l(e + 1, t) < l(e + 1, s)].$$

Define

$$r'(e + 1, s) = \begin{cases} \text{maximum } (e + 1)\text{-expansionary stage } t < s \\ \text{such that } r(e, t) = r(e, s), \\ \text{if } s \text{ is not } (e + 1)\text{-expansionary and } t \text{ exists}, \\ 0 \quad \text{otherwise.} \end{cases}$$

Define $r(e + 1, s)$ to be the maximum of

(i) $r(e, s)$,

(ii) those t such that $r(e, t) < r(e, s)$ and $t < s$,

(iii) $r'(e + 1, s)$.

(See the tree version of the definition of $r(e, s)$ following the proof of Lemma 3.)

Requirement P_{2e} *requires attention* if

(4.1) $$\neg(\exists y)[y \in A_s \& \{e\}_s(y) = 0]$$

and

(4.2) $$(\exists x)\big[\{e\}_s(x) = 0 \,\&\, r(2e, s) < x \,\&\, x \in \omega^{[e]}\big],$$

and likewise P_{2e+1} with B in place of A and $r(2l+1, s)$ in place of $r(2e, s)$. Choose the highest priority requirement P_e which requires attention and the least x satisfying (4.2) for that e. Enumerate x in A if e is even (in B if e is odd).

LEMMA 1. $(\forall e)[\liminf_s r(e, s) < \infty]$.

PROOF. We first prove the case $e = 0$. Suppose $\limsup_s r(0, s) = \infty$. Then there are infinitely many 0-expansionary stages, so by definition there are infinitely many s such that $r(0, s) = 0$, and hence $\liminf_s r(0, s) = 0$. For the inductive step, fix e and assume $k = \liminf_s r(e, s)$. Then there are only finitely many stages s such that $r(e, s) < k$. Let t be the largest. Let $S = \{s\colon r(e, s) = k\}$. Either there are infinitely many $(e+1)$-expansionary stages in S, in which case $\liminf_s r(e+1, s) = \max\{t, k\}$, or else there is a largest $(e+1)$-expansionary stage $v \in S$, in which case $\liminf_s r(e+1, s) = \max\{t, k, v\}$.

LEMMA 2. *Every positive requirement is satisfied and acts at most once.*

PROOF. Consider requirement P_{2e} (since P_{2e+1} is similar). If $A = \{e\}$, then $A \cap \omega^{[e]} = \varnothing$ so there exists

$$x \in \bar{A} \cap \omega^{[e]}, \quad \{e\}(x)\downarrow = 0, \quad \text{and} \quad x > \liminf_s r(2e, s).$$

Hence, some such x is eventually enumerated into A satisfying P_{2e}.

LEMMA 3. $(\forall e)$ *requirement* N_e *is met.*

PROOF. Fix e and let $k = \liminf_s r(e-1, s)$ and $S = \{s\colon r(e-1, s) = k\}$. (If $e = 0$ let $S = \omega$ and $k = 0$.) Choose s' such that no P_i, $i < e$, acts after stage s' and $r(e-1, s) \geqslant k$ for all $s \geqslant s'$. Now assume that $\{e\}^A = \{e\}^B = g$ is a total function. To recursively compute $g(p)$, $p \in \omega$, find an e-expansionary stage $s'' \in S$, $s'' > s'$, such that $l(e, s'') > p$. There are infinitely many e-expansionary stages because $\{e\}^A = \{e\}^B$. Let $q = \{e\}_{s''}^{A_{s''}}(p) = \{e\}_{s''}^{B_{s''}}(p)$. We shall prove by induction on t that for all $t \geqslant s''$ either

(1) $\{e\}_t^{A_t}(p) = q$ or

(2) $\{e\}_t^{B_t}(p) = q$,

and hence, that $g(p) = q$. Let $s_1 < s_2 < \cdots$ be the e-expansionary stages in S which are greater or equal to s''. Both (1) and (2) hold for $t = s_1$ since $s_1 = s''$. Fix n and assume by induction that both (1) and (2) hold for $t = s_n$ (and that either (1) or (2) holds at every t, $s_1 \leqslant t \leqslant s_n$). Now at stage $s_n + 1$ at most one element enters $A \oplus B$ so at most one of the computations (1), (2) for $t = s_n$ is destroyed, and the other, say (1), holds for $t = s_n + 1$. Now by choice of s'',

$$(\forall t)\big[[s_n < t < s_{n+1} \,\&\, t \notin S] \Rightarrow r(e-1, t) > k\big],$$

and hence

$$(\forall t)\big[[s_n < t < s_{n+1}] \Rightarrow r(e, t) \geqslant s_n\big],$$

by clauses (ii) and (iii) in the definition of $r(e, t)$, namely by (ii) for $t \notin S$, and by (iii) for $t \in S$. (If $e = 0$, only the latter clause applies.) Now

$$s_n \geqslant u(A_{s_n}; e, p, s_n) = u(A_{s_n+1}; e, p, s_n + 1) = u(A_{s_{n+1}}; e, p, s_{n+1})$$

and hence (1) holds for $t = s_{n+1}$. But s_{n+1} is e-expansionary, so the A and B computations must converge and agree. Thus, both (1) and (2) hold for $t = s_{n+1}$. □

It is instructive to define the restraint function $r(e, s)$ using the tree $T = \Lambda^{<\omega}$ where $\Lambda = \omega$ with the usual order. We shall define a recursive sequence of strings $\{\delta_t: t \in \omega\}$ which approximate the true path f as in §3. For $\alpha \in T$ a stage s is an *α-stage* if $\alpha \subseteq \delta_s$ or $s = 0$. The α-strategy is the same as the basic module but with the set S^α of α-stages in place of ω. Namely, define $l(\alpha, s) = l(e, s)$, where $e = |\alpha|$, and

$$m(\alpha, s) = \max\{l(\alpha, t): t \leqslant s \,\&\, t \in S^\alpha\}.$$

An α-stage s is *α-expansionary* if $s = 0$ or $l(\alpha, s) > m(\alpha, s - 1)$. The restraint function for the α-strategy alone is

$$(4.3) \qquad r(\alpha, s) = \begin{cases} \text{the greatest } \alpha\text{-stage } t < s, & \text{if } s \text{ is not an } \alpha\text{-stage,} \\ 0 \quad \text{if } s \text{ is } \alpha\text{-expansionary,} \\ \text{the greatest } \alpha\text{-expansionary stage } t < s & \text{otherwise,} \end{cases}$$

For $s \geqslant 0$ define $\delta_s \in T$, $|\delta_s| = s$, as follows. Given $\alpha = \delta_s \upharpoonright e$, $e < s$, define

$$\delta_s(e) = r(e, s) =_{\text{dfn}} \max\{r(\beta, s): \beta \leqslant \alpha\}.$$

(Note that for $\beta <_L \alpha$, $r(\beta, s)$ is defined by the first clause of (4.3) so $\delta_s \upharpoonright e$ suffices to define $r(\beta, s)$ for all $\beta \leqslant \alpha$.) Now the true path $f \in \Lambda^\omega$ is defined by $f(e) = r(e) =_{\text{dfn}} \liminf_s r(e, s)$, and has the property that $f \upharpoonright e = \liminf_s \delta_s \upharpoonright e$ in sense of (3.2) and (3.3).

This definition produces essentially the same restraint function $r(e, s)$ as the first definition, but it more clearly illustrates how the α-strategies are combined. The reason that the use of trees can be eliminated in the proof as first given, is that this argument resembles a Markov process in that the only necessary bit of information for the α-strategy to know is $\alpha(e - 1) = r(e - 1, s)$, where $e = |\alpha|$, because we have insured that $r(e - 1, s) \geqslant r(j, s)$ for all $j \leqslant e - 1$. This applies to many other constructions also, such as the plus cupping theorem (see Fejer and Soare [**1981**]). However, certain other constructions are best done on a tree if they have both a minimal pair type restraint function $r(e, s)$ and also infinitary positive requirements, as we now see in §5 where we examine the most typical such construction.

5. The tree method with $0''$-priority arguments, a minimal pair of high degrees. Lachlan [**1966b**] constructed a pair of maximal sets whose degrees form a minimal pair. Martin [**1966b**] proved that the degrees of maximal sets are precisely the high r.e. degrees. Hence, Lachlan's result is equivalent to the existence of a minimal pair of high r.e. degrees.

We are interested in this result because in the proof we shall construct an r.e. set A where we must guess not only about the value of $\liminf_s r(e, s)$ as in §4, but also about which rows $A^{[e]}$ are infinite. For this section, we shall use the priority tree,

$$(5.1) \qquad T = \{\alpha: \alpha \in \omega^{<\omega} \ \& \ (\forall e)[\alpha(2e + 1) \in \{0,1\}]\}.$$

The true path f through T will be such that

$$(5.2) \qquad f(2e) = r(e) =_{\text{dfn}} \liminf_s r(e,s)$$

and

$$(5.3) \qquad f(2e + 1) = \begin{cases} 0 & \text{if } A^{[e]} \text{ is infinite,} \\ 1 & \text{if } A^{[e]} \text{ is finite.} \end{cases}$$

The method used here can be adapted to give tree proofs for virtually all the standard infinite injury constructions, for example in Soare [**1976**]. More importantly, the tree method also applies to all those $0''$-arguments requiring both infinitary positive requirements (as in the usual infinite injury) and also a minimal pair type of restraint function $r(e, s)$ as in §4, rather than the type of restraint function for avoiding an upper cone as in a typical infinitary injury argument such as the Sacks density theorem, or jump theorem. In the latter two cases the restraint functions can be made to drop back simultaneously using the "true stages" method (see Soare [**1976**]). The minimal pair type restraint function requires the nested strategies method of §4 to make the restraint functions drop back simultaneously, and hence requires the tree method when combined with infinitary positive requirements. Thus, this section should be viewed as a more powerful method than Soare [**1976**] for handling infinite injury arguments.

5.1. THEOREM (LACHLAN). *There exists a minimal pair of high r.e. degrees.*

PROOF. We shall construct an r.e. set C such that the degrees of $A = \bigcup_y C^{[2y]}$ and $B = \bigcup_y C^{[2y+1]}$ form a minimal pair by satisfying for all e the negative requirement as in §4,

$$N_e: \{e\}^A = \{e\}^B = g \text{ total} \Rightarrow g \text{ is recursive.}$$

Now fix (as in Soare [**1976**, p. 521] an r.e. set D such that for all e,

$$(5.4) \qquad e \in \varnothing'' \Rightarrow D^{[2e]} \text{ and } D^{[2e+1]} \text{ are finite}$$

and

$$(5.5) \qquad e \notin \varnothing'' \Rightarrow D^{[2e]} =^* \omega^{[2e]} \text{ and } D^{[2e+1]} =^* \omega^{[2e+1]}.$$

To insure that A and B have high degree, we shall construct C a *thick subset* of D, namely $C \subseteq D$ and meeting for each e the positive requirement,

$$P_e: C^{[e]} =^* D^{[e]}.$$

Note that P_e insures that $\varnothing'' \leqslant_T A'$, and $\varnothing'' \leqslant_T B'$ by the limit lemma (see Shoenfield [**1971**, p. 29]), because (5.4) and (5.5) guarantee that

$$\varnothing''(e) = 1 - \lim_x A(\langle x, 2e\rangle) = 1 - \lim_x B(\langle x, 2e+1\rangle).$$

Let T be as in (5.1). For each s we shall define $\delta_s \in T$, $|\delta_s| = 2s$, such that for all n, $f \restriction n = \liminf_s \delta_s \restriction n$ as in §3. For $\alpha \in T$, a stage s is an *α-stage* if $\alpha \subseteq \delta_s$ or $s = 0$. Let S^α be the set of α-stages.

5.2. DEFINITION. An A-computation $\{e\}_s^{A_s}(x) = y$ is *α-correct* if

$$(5.6) \qquad (\forall i < e)[\alpha(2i+1) = 0$$
$$\Rightarrow (\forall z)[\alpha(2i) < z \leqslant u(A_s; e, x, s) \,\&\, z \in \omega^{[i]} \Rightarrow z \in C_s^{[i]}]],$$

and similarly for a B-computation (with B_s in place of A_s).

(The intuition is that by (5.2) α guesses that $\alpha(2i) = r(i) =_{\text{dfn}} \liminf_s r(i, s)$. Also by (5.3), P_e, and $D^{[e]}$, if $\alpha(2i+1) = 0$, then α guesses that $C^{[i]} =^* \omega^{[i]}$ and indeed, α believes that all numbers $z > r(i)$, $z \in \omega^{[i]}$, will eventually enter $C^{[i]}$. Thus, α does not "believe in" a computation until these elements have entered C.)

We now adapt the basic module to $\alpha \in T$, $|\alpha| = 2e$, by defining the recursive functions

$$l(\alpha, s) = \max\{x \colon (\forall y < x)[\{e\}_s^{A_s}(y)\downarrow = \{e\}_s^{B_s}(y)$$
$$\text{and these computations are } \alpha\text{-correct}]\},$$

$$m(\alpha, s) = \max\{l(\alpha, t) \colon t \leqslant s \,\&\, t \in S^\alpha\}.$$

A stage s is *α-expansionary* if $s = 0$ or $s > 0$, s is an α-stage, and $l(\alpha, s) > m(\alpha, t)$, where t is the greatest α-stage $< s$. (If $|\alpha|$ is odd, say $2e+1$, these three definitions have no meaning, since α is assigned to P_e.)

Construction of A and B.

Stage $s = 0$. Let $C_0 = \varnothing$, and $r(\alpha, -1) = 0$ for all $x \in T$, and $r(\alpha, s) = 0$ for all α such that $|\alpha|$ is odd and for all s.

Stage $s+1$. First define $\delta_s \in T$, $|\delta_s| = 2s$, by performing the steps $e < s$ below in increasing order of e. Next, choose the least integer $x = \langle y, e\rangle < s$ such that $x \in D_s^{[e]} - C_s^{[e]}$ and $x > r(e, s) = \delta_s(2e)$. Enumerate x in C. If no such x exists, do nothing.

Step e, $0 \leqslant e < s$. Given $\alpha = \delta_s \restriction 2e$, define $\delta_s(2e)$ and $\delta_s(2e+1)$ as follows. For $\beta \leqslant \alpha$, $|\beta|$ even, define

$$(5.7) \qquad r(\beta, s) = \begin{cases} 0 & \text{if } \beta \subseteq \alpha \text{ and } s \text{ is } \beta\text{-expansionary}, \\ \text{the greatest } \beta\text{-expansionary stage } t < s, & \text{otherwise.} \end{cases}$$

Define

$$\delta_s(2e) = r(e, s) = \max\{r(\beta, s) \colon \beta \leqslant \alpha\}.$$

Now let $\gamma = \delta_s \restriction 2e+1$. Define

$$\delta_s(2e+1) = \begin{cases} 0 & \text{if } |C_s^{[e]}| > |C_t^{[e]}| \text{ where } t \text{ is the maximum } \gamma\text{-stage} < s, \\ 1 & \text{otherwise.} \end{cases}$$

(The intuition is that γ is assigned to P_e and γ "wakes up" only at γ-stages. If the cardinality of $C^{[e]}$ has increased since γ was last awake, then our current guess is that $C^{[e]}$ will be infinite, so we set $\delta_s(2e+1) = 0$ according to (5.3), and $= 1$ otherwise.)

If $e < s - 1$, go to step $e + 1$. Otherwise, define $r(\beta, s) = r(\beta, s - 1)$ for all β such that $r(\beta, s)$ has not yet been defined and go to stage $s + 2$.

LEMMA 1. *There exists $f \in \Lambda^\omega$, called the* true path, *such that*
(i) $(\forall n)[f \upharpoonright n = \liminf_s \delta_s \upharpoonright n$ *as defined in* (3.2) *and* (3.3)],
(ii) $(\forall e)[f(2e) = r(e) =_{\text{dfn}} \liminf_s r(e, s)]$,
(iii) $(\forall e)[[f(2e+1) = 0 \Leftrightarrow |C^{[e]}| = \infty] \& [f(2e+1) = 1 \Leftrightarrow |C^{[e]}| < \infty]]$.

PROOF. We prove (i), (ii) and (iii) simultaneously by induction. Fix $e \geqslant 0$, let $\alpha = f \upharpoonright 2e$, and assume (i) for $n = 2e$, and (ii) and (iii) for all $e' < e$. Choose t minimal such that there is no $s > t$, $\delta_s <_L \alpha$. Let $k = \max\{r(j): j < e\}$. Now, as in Lemma 1 of §4, if there are infinitely many α-expansionary stages, then $r(\alpha, s) = 0$ at each and $f(2e) = r(e) = k$. Otherwise, there is a largest such stage v, $r(\alpha, s) = v$ for almost all s, and $r(e) = \max\{k, v\}$. In either case, we have (i) for $n = 2e + 1$ and (ii) for e. Now let $\gamma = f \upharpoonright 2e + 1$. If $C^{[e]}$ is finite then $\delta_s(2e+1) = 1$ for almost all s. If $|C^{[e]}| = \infty$ then there are infinitely many s such that $\delta_s \supset \gamma^\wedge\langle 0\rangle$ because there are infinitely many γ-stages. In either case, we have (i) for $n = 2e + 2$ and (iii) for e.

LEMMA 2. $(\forall e)[C^{[e]} =^* D^{[e]}]$. (More precisely, for all e, $C^{[e]} \subseteq D^{[e]}$ and $D^{[e]} \subseteq \{0, 1, \ldots, r(e)\} \cup C^{[e]}$.)

PROOF. Choose any $x \in D^{[e]}$, $x > r(e)$. Choose s_0 such that for all $y < x$, $y \in C$ iff $y \in C_{s_0}$. Now at the next stage, $s + 1 > s_0$ such that $r(e, s) = r(e)$, x is enumerated in C_{s+1} if $x \notin C_s$ already.

LEMMA 3. $(\forall e)[\{e\}^A = \{e\}^B = g$ *total* $\Rightarrow g$ *is recursive*].

PROOF. Assume that $\{e\}^A = \{e\}^B = g$ is a total function. Let $\alpha = f \upharpoonright 2e$. We shall show that the α-strategy satisfies N_e. By Lemma 1(iii), α has the correct guess about $|C^{[i]}|$, the cardinality of $C^{[i]}$, for all $i < e$, namely $\alpha(2i+1) = 0$ iff $|C^{[i]}| = \infty$. By Lemma 1(i) choose an α-stage t such that there is no $s \geqslant t$ with $\delta_s <_L \alpha$. Note that by Lemma 1(ii) we have

$$(5.8) \qquad (\forall i < e)(\forall s \geqslant t)[r(i) \leqslant r(i, s)],$$

and hence

$$(5.9) \qquad (\forall i < e)\left[C_t^{[i]} \upharpoonright r(i) + 1 = C^{[i]} \upharpoonright r(i) + 1\right].$$

By Lemma 1(iii), we also have

$$(5.10) \qquad (\forall i < e)\left[C^{[i]} \text{finite} \Rightarrow C_t^{[i]} = C^{[i]}\right],$$

because if $C^{[i]}$ is finite, then $\alpha(2i+1) = 1$, but if some x enters $C^{[i]}$ at a stage $v > t$ then at some stage $s \geqslant v > t$ we must have $\delta_s \supseteq (\alpha \upharpoonright 2i)^\wedge\langle 0\rangle$, so $\delta_s <_L \alpha$ contrary to the choice of t.

To recursively compute $g(p)$ find the least α-expansionary stage $v > t$ such that $l(\alpha, v) > p$. Let $g(p) = \{e\}_v^{A_v}(p)$. Note that v exists because if $\{e\}^A(p) = q$ with use u, then the computation $\{e\}_s^{A_s}(p) = q$ will be α-correct infinitely often, namely for all $s > t$ such that $z \in C_s$ for all $z \in \omega^{[i]}$ if $i < e$, $\alpha(2i+1) = 0$, and $r(i) < z \leqslant u$. As in Theorem 4.1, it now follows by induction on $s \geqslant v$ that for all $s \geqslant v$ either

(5.11) $\{e\}_s^{A_s}(p) = q$ is an α-correct computation,

or

(5.12) $\{e\}_s^{B_s}(p) = q$ is an α-correct computation.

The extra point here, beyond those in Lemma 3 of §4, is that by (5.9), (5.10) and the definition of α-correct computation, we cannot have $z \in C_{s+1}^{[i]} - C_s^{[i]}$ if $i < e$, $s > t$, and z is used in an α-correct computation $\{e\}_s^{A_s}(p) = q$ or $\{e\}_s^{B_s}(p) = q$. As usual, the restraint function $r(e, s)$ prevents any $z \in \omega^{[i]}$ for $i \geqslant e$ from entering C and destroying a computation (5.11)or (5.12), unless s is an α-expansionary stage, in which case *both* (5.11) and (5.12) hold for s. □

5.3. REMARK. In the proof of Theorem 5.1 and in essentially all $0''$-tree constructions, we could have modified the tree so that there is no injury in the "construction" along the true path $\hat{\mathscr{C}} = \bigcup\{\mathscr{C}_\alpha : \alpha \subset f\}$. (Namely, if $\beta \subset \alpha \subset f$ then the α-strategy is never injured by β in the following sense.)

Let $\alpha = f \restriction 2e$. Define a slightly modified restraint function $\tilde{r}(\beta, s)$ as in (5.7) but in the third clause of (5.7) replace t by u, the greatest element used in an A or B computation at step t on any argument $x < l(\alpha, t)$. (Note that $\tilde{r}(\alpha, s) < r(\alpha, s)$ and is sufficient restraint to show that the α-strategy succeeds.) We say that α is *injured at stage* $s+1$ if $z \in C_{s+1} - C_s$ for some $z \leqslant \tilde{r}(\alpha, s)$. Furthermore, if $z \in \omega^{[i]}$ and $\beta = \delta_s \restriction 2i+1$, we say β *injures* α.

Now α can be injured finitely often by those $\beta <_L \alpha$. In the proof of Theorem 5.1, α can be also injured finitely often by those $\beta \subset \alpha$ if $|\beta| = 2i+1$ and $\alpha(2i+1) = 1$ because α knows that $D^{[i]}$ is finite but does not know its exact members. This can be remedied by replacing the tree T of (5.1) by $T = \omega^{<\omega}$ and (5.3) by

$$f(2i+1) = \begin{cases} 0 & \text{if } |C^{[i]}| = \infty, \\ k+1 & \text{if } C^{[i]} = D_k, \end{cases}$$

where $\{D_k\}_{k \in \omega}$ is the usual canonical indexing of finite sets. Now if $\alpha = f \restriction 2e$ then α has complete information about $C^{[i]}$ for all $i < e$, and defines α-correct computations accordingly, so that α is never injured by any $\beta \subset \alpha$.

5.4. REMARK. In the above construction we could have let the priority tree be simply $T = 2^{<\omega}$. Now $f(2e+1)$ is defined by (5.3) as before. If $\alpha = f \restriction 2e$, then $f(2e) = 0$ if there are infinitely many α-expansionary stages and $f(2e) = 1$ otherwise.

The construction and proof are carried out as above, except that in the definition of α-correct computation in (5.6) we replace "$\alpha(2i) < z$" by "$r(i, s) < z$", define $r(e, s)$ as above, but define $\delta_s(2e)$ as follows. If $\alpha = \delta_s \restriction 2e$, let $\delta_s(2e) = 0$ if s is α-expansionary, and $= 1$ otherwise. Lemma 1(ii) is modified by letting $r(e)$ be as previously, but $f(2e)$ as above. In the proofs of the lemmas no information is lost, because if $\delta_s(2e) = 1$ and $r(e, s) = k$, this is equivalent to the former case where $\delta_s(2e) = k$. In particular, Lemma 3 and (5.8)–(5.10) all hold as before.

This approach has some advantages because from now on, particularly in the $0'''$-constructions, it is convenient to use only finitely branching priority trees T. (However, α does not have complete information about all $\beta \subset \alpha$ as in Remark 5.3, and so α may be injured by such β.)

6. An $0'''$-priority argument, the Lachlan nonbounding theorem.

6.1. *Preliminaries.* It is an open question to decide exactly which degrees $\mathbf{a} \in \mathbf{R}^+$ bound minimal pairs. Cooper [**1974a**] showed that any high degree $\mathbf{a} \in \mathbf{H}_1$ will suffice, but in the following theorem, Lachlan shows that not every $\mathbf{a} \in \mathbf{R}^+$ will suffice.

6.1. THEOREM (LACHLAN NONBOUNDING THEOREM). *There exists an r.e. degree* $\mathbf{c} > \mathbf{0}$ *with no r.e. minimal pair below it.*

For this section we now adopt some notation of Lachlan [**1979**] which is convenient for this and other $0'''$-arguments. Upper case Greek letters Φ^A, $\Psi^A, \ldots$, denote A-partial recursive (p.r.) functions previously denoted by $\{e\}^\theta$, $e \in \omega$. When we define an r.e. set or p.r. functional Φ, during a recursive construction, then we use $A[s]$, $\Phi[s]$ to denote the result by the end of stage s of the construction (formerly denoted by A_s, $\{e\}_s$, etc.). If m is a parameter then $m[s]$ denotes the value at the end of stage s. We also allow whole expressions to be qualified by "$[s]$". For example, $\Phi^A(x)[s]$ denotes $\Phi[s]^{A[s]}(x)$. The advantage is that during the construction we regard A, Φ, m and so on as in a state of formation. During a given stage (which may involve many substages) we can allow the notation A to denote the set of elements so far enumerated in A, m to denote the current value of the parameter m and so on. When necessary to avoid confusion we append s and write $A[s]$, $m[s]$ or $\Phi[s]$ to denote this result by the *end* of stage s.

PROOF OF THEOREM 6.1. We must construct an r.e. set C satisfying, for all i, the requirements

$$P_i\colon C \neq \overline{W_i},$$

$$R_i\colon \Phi_i^C = A_i \;\&\; \Psi_i^C = B_i \;\&\; A_i \text{ is not recursive} \;\&\; B_i \text{ is not recursive}$$
$$\Rightarrow (\exists \text{r.e. } D)[D \text{ nonrecursive} \;\&\; D \leqslant_T A_i \;\&\; D \leqslant_T B_i],$$

where $\{(A_i, B_i, \Phi_i, \Psi_i)\colon i \in \omega\}$ is a standard indexing of all 4-tuples (A, B, Φ, Ψ) where A, B are r.e. sets and Φ, Ψ are p.r. functionals. Suppose that A_i, B_i, Φ_i and

Ψ_i are supplied with recursive enumerations in a uniformly effective fashion. Without loss of generality, we may assume for all x and s,

$$x \in A_i[s] - A_i[s-1] \Rightarrow \Phi_i^C(x)[s] = 1 \tag{6.1}$$

and

$$x \in B_i[s] - B_i[s-1] \Rightarrow \Psi_i^C(x)[s] = 1. \tag{6.2}$$

This is achieved by withholding numbers from A_i and B_i until Φ_i^C and Ψ_i^C take the value 1. (This does no harm since we are only concerned with $(A_i, B_i, \Phi_i, \Psi_i)$ if $A_i = \Phi_i^C$ and $B_i = \Psi_i^C$.) Define

$$l(\Phi_i^C, A_i)[s] = \max\{x\colon (\forall y < x)[A_i(y)[s] = \Phi_i^C(y)[s]]\}, \tag{6.3}$$

and similarly $l(\Psi_i^C, B_i)[s]$, with B_i and Ψ_i in place of A_i and Φ_i. The advantage of (6.1) and (6.2) is that at some stage if $x < l(\Phi_i^C, A_i)$ we can preserve $A_i \upharpoonright x$ by preserving $C \upharpoonright u$ where u is the greatest element used in the computation, and likewise for $B_i \upharpoonright x$.

6.2. *The basic module for meeting a subrequirement*. To understand our strategy, it is convenient to rewrite requirement R_i in the form of (3.7) as follows:

$$R_i\colon \Phi_i^C \neq A_i \vee \Psi_i^C \neq B_i \vee A_i \text{ recursive} \vee B_i \text{ recursive}$$
$$\vee (\exists \text{r.e. } D_i)\left[D_i \leqslant_T A_i \,\&\, D_i \leqslant_T B_i \,\&(\forall j)\left[D_i \neq \overline{W_j}\right]\right]. \tag{6.4}$$

Let $P_{i,k}$, $k \leqslant 4$, denote the kth disjunct of (6.4), and define the subrequirement

$$R_{i,j}\colon D_i \neq \overline{W}_j.$$

Our strategy is to attempt to satisfy R_i by constructing an r.e. set D_i to satisfy the last disjunct of R_i. The basic module below attempts to meet only a single subrequirement $R_{i,j}$ (consistent with making $D_i \leqslant_T A_i$, B_i uniformly). It either achieves that objective, or else satisfies $P_{i,k}$ for some $k \leqslant 4$, in which case R_i is satisfied at once and no further subrequirements $R_{i,j'}$, $j' \neq j$, need be examined. (The following basic module is not the same as in Lachlan [**1979**], but more similar to that hinted at by Harrington [**1982**], although the latter gives no details. This basic module gives rise to a simpler tree of outcomes than in Lachlan, although the initial module is slightly more complicated, having two types of gaps instead of one.)

Fix i and j and drop these subscripts from A_i, B_i, Φ_i and Ψ_i. We have parameters x, r_1, r_2, and r, which should be viewed as functions of the stage s. A parameter once given a value retains that value until a new value is assigned. Now x is the current candidate to satisfy $R_{i,j}$, r_1 and r_2 are restraint functions which tend to prevent elements from entering C, in order to preserve $A \upharpoonright x$ and $B \upharpoonright x$ respectively, and $r = \max\{r_1, r_2\}$. To *reset* the candidate x means to cancel the old candidate $x[s]$ and to let $x[s+1]$ be the least $y \in \omega^{[\langle i,j\rangle]}$, $y > s$ and greater than any previous candidates. To *initialize* (the basic module) means to reset x and to set $r_1 = r_2 = 0$. We initialize at stage $s = 0$ and at any later stage when some $z \leqslant r$ enters C (due to the action for some requirement P_k in which case the basic module begins all over on the new candidate x.

The basic module consists of the following steps. (See Figure 6.1).

Step 1. Wait for s such that $x \in W_j[s]$. At stage $s+1$ *open an A-gap* by setting $r_1[s+1] = 0$ and go to Step 2.

Step 2. Wait for the least $t \geqslant s+1$ such that $l(\Phi^C, A)[t] > x$. At stage $t+1$ *close the A-gap* and perform Step 2(a) or 2(b) according to which case applies.

Step 2(a) (*successful close*). Suppose $A[s] \upharpoonright x \neq A[t] \upharpoonright x$. *Open a B-gap* by defining $r_2[t+1] = 0$. (Note that r_1 remains 0 so $r[t+1] = 0$.) Go to Step 3.

Step 2(b) (*unsuccessful close*). Suppose $A[s] \upharpoonright x = A[t] \upharpoonright x$. Define $r_1[t+1] = t$ (to preserve $A[t] \upharpoonright x$), reset x and go to Step 1.

Step 3. Wait for the least $v \geqslant t+1$ such that $l(\Psi^C, B)[v] > x$. At stage $v+1$ *close the B-gap* and perform Step 3(a) or 3(b) according to which case applies.

Step 3(a) (*successful close*). Suppose $B[v] \upharpoonright x \neq B[t] \upharpoonright x$. Enumerate x in D.

Step 3(b) (*unsuccessful close*). Suppose $B[v] \upharpoonright x = B[t] \upharpoonright x$. Define $r_2[v+1] = v$ (to preserve $B[v] \upharpoonright x$), reset x and go to Step 1.

(Note that at the opening of an A-gap (B-gap), the dropping of the restraint r_1 (r_2) will tend to allow other positive requirements to enumerate elements into C, thereby causing Φ^C (Ψ^C) to change and thus $l(\Phi^C, A)$ ($l(\Psi^C, B)$) to drop down.)

First note that if the four hypotheses of R_i hold, then $D \leqslant_T A$ and $D \leqslant_T B$ uniformly because we require both A and B permission (*namely* changes in $A \upharpoonright x$ and $B \upharpoonright x$ before x enters D. Now $R_{i,j}$ is satisfied if we either wait forever for s such that $x \in W_j[s]$, or if some B-gap is closed successfully, in which case x enters D and $D \cap W_j \neq \varnothing$. If we remain in an A-gap (B-gap) forever waiting to close, then $\Phi^C \neq A$ ($\Psi^C \neq B$), and hence R_i is satisfied by the first (second) clause of (6.4).

Suppose that there are infinitely many B-gaps, say begun at stages $\{t_n: n \in \omega\}$ and closed (necessarily unsuccessfully) at stages $\{v_n: n \in \omega\}$, where $t_0 < v_0 < t_1 < v_1 < \cdots$. The intervals of stages $\{s: t_n \leqslant s < v_n\}$ are called *B-gaps*, since $r_2 = 0$ and we are free to enumerate any number into C, while the intervals $\{s: v_n \leqslant s < t_{n+1}\}$ are called *cogaps*.

Let us suppose that the basic module is *injured* only finitely often, namely that there exists s_0 such that

$$(6.5) \qquad (\forall s > s_0)(\forall z)[z \in C[s+1] - C[s] \Rightarrow z > r[s]].$$

Then we can show that B is recursive as follows. To compute $B(p)$ choose $t > s_0$ such that $x = x[t] > p$, and a B-gap is begun at stage $t+1$. Since the B-gap is closed unsuccessfully at $v+1 > t+1$, $B[t] \upharpoonright x = B[v] \upharpoonright x$. If the next B-gap is begun at stage $t'+1 > v+1$, then the C-restraint r_2 insures that $B[v] \upharpoonright x = B[t'] \upharpoonright x$. Now the argument continues with x replaced by $x' = x[t'+1] > x$, and establishes that $B[t] \upharpoonright x = B \upharpoonright x$. Hence $p \in B$ iff $p \in B[t]$. Note that $\lim_t x[t] = \infty$ since each new candidate is chosen to be greaterthan any previous candidates.

If there are only finitely many B-gaps but infinitely many A-gaps, then A is recursive by the same argument with r_1, A and Φ in place of r_2, B and Ψ, respectively, and with $s_1 > s_0$ large enough to satisfy (6.5) and greater than all the stages when the B-gaps were closed (so that after s_1 every A-gap is closed unsuccessfully).

Finally, note that $\liminf_s r[s] < \infty$. If there are only finitely many A and B gaps then $\liminf_s r[s]$ exists. If there are infinitely many B-gaps then $\liminf_s r[s] = 0$, since $r[s] = 0$ if a B-gap is begun at stage s. If there are finitely many B-gaps and infinitely many A-gaps, then $k = \lim_s r_2[s]$ exists, and so $r[s] = k$ for all $s > s_1$ such that an A-gap is begun at stage s, because $r_1[s] = 0$.

It is useful to view the basic module as a kind of finite automaton M, whose states are $\{s, w2, w1, w\}$ arranged in order from left to right as listed, and whose transitions between states are those given by Steps 1, 2(a), 2(b), 3(a) and 3(b) of the basic module. Namely, M begins in the starting state w, attempting to move leftward to the final state s. While M is in state w ($w1$, $w2$) M waits for the exit condition associated with the corresponding Step 1 (2, 3) to be satisfied. Then M moves left or right to a new state, according to which transition step applies (see Figure 6.1).

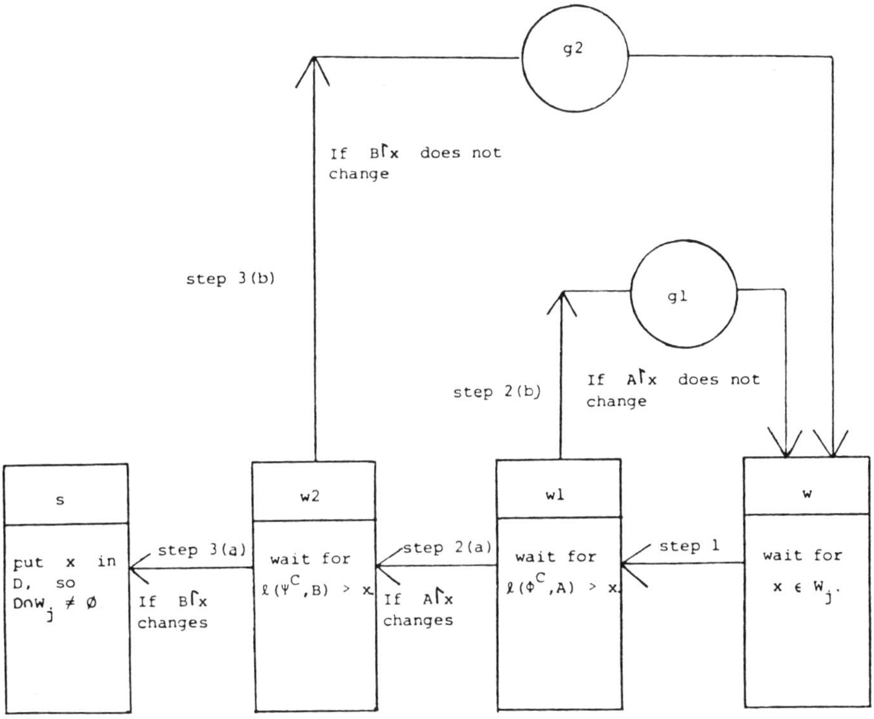

FIGURE 6.1. Diagram of the basic module as a finite automaton.

Note that the desired leftward progress towards the final state s may be interrupted by an unsuccessful close of an A-gap (B-gap), and therefore a return to the starting state w. This may cause M to perform infinite cycles returning rightward infinitely often from state $w1$ or $w2$ or both.

Let $F[s]$ denote the state of M at the end of stage s. The possible *outcomes* of the basic module are

$$\Lambda = \{s, g2, w2, g1, w1, w\}, \tag{6.6}$$

with ordering $<_\Lambda$ as listed. These are divided into the finitary ($0'$) outcomes $\Lambda^{\text{Fin}} = \{s, w2, w1, w\}$ and the infinitary ($0''$) outcomes $\Lambda^{\text{Inf}} = \{g1, g2\}$.

We define the outcome as follows. If $\lim_s F[s] = a$, then $a \in \Lambda^{\text{Fin}}$ and the outcome is a. If $\lim_s F[s]$ does not exist, then the outcome is $g2$ if there are infinitely many B-gaps (infinitely many right exits from $w2$), and the outcome is $g1$ otherwise (in which case there are infinitely may right exits from $w1$).

The states of M are shown in Figure 6.1 as rectangles. The outcomes $g1$ and $g2$ are not themselves states of M, but represent infinite cycles and are pictured as circles in Figure 6.1 where the rightward motion of the cycle begins. (Thus the outcome of the basic module is the $<_\Lambda$ least (i.e. leftmost) $a \in \Lambda$ such that M either is in or passes through rectangle a or circle a at infinitely many stages.)

The following table summarizes the progress made on R_i or $R_{i,j}$ by the basic module for each outcome $a \in \Lambda$.

Outcome	s	$g2$	$w2$	$g1$	$w1$	w
Effect on $R_{i,j}$	win	none	none	none	none	win
Effect on the first four clauses of R_i, $P_{i,k}$, $k \leqslant 4$	none	B_i recursive ($P_{i,4}$)	$\Psi_i^C \neq B_i$ ($P_{i,2}$)	A_i recursive ($P_{i,3}$)	$\Phi_i^C \neq A_i$ ($P_{i,1}$)	none

FIGURE 6.2. Analysis of outcome.

(Fixing i and repeating the basic module for all $j \in \omega$, it is now clear that there is a nonrecursive r.e. set C which satisfies a single requirement R_i of (6.4) and such that D_i can be found uniformly recursive in each of A_i and B_i. It then follows, using the recursion theorem, that there is no uniform effective procedure for passing from the index of a nonrecursive r.e. set C to indices of r.e. sets $A, B \leqslant_T C$ whose degrees form a minimal pair together with indices of reductions Φ, and Ψ from C to A and B. This was first announced by Lachlan and Ladner.)

6.3. *The priority tree*. Let Λ be as in (6.6) with the ordering $<_\Lambda$ as listed. (This ordering is chosen so that if $a <_\Lambda b$ and a "appears correct" infinitely often, then a witnesses that b is not the correct outcome. Roughly, a "appears correct" at s if $a \in \Lambda^{\text{Fin}}$ and $F[s] = a$, or $g1$ ($g2$) and M exits right at stage s from $w1$ ($w2$).)

6.2. DEFINITION. (i) Let the *priority tree* be $T = \Lambda^{<\omega}$.

(ii) Define recursive functions i, j: $T \to \omega$ as follows. If $\alpha \in T$, $|\alpha| = \langle m, n\rangle$ and $m = \langle i, j\rangle$ then $i(\alpha) = i$ and $j(\alpha) = j$.

We shall denote the sets $A_{i(\alpha)}$, $B_{i(\alpha)}$ and $W_{j(\alpha)}$ by A_α, B_α, W_α and the functions $\Phi_{i(\alpha)}$ and $\Psi_{i(\alpha)}$ by Φ_α, Ψ_α. For each $\alpha \in T$ we have an associated r.e. set D_α which is the set D with which we are trying to satisfy the last clause of $R_{i(\alpha)}$. (Note that unlike A_α the set D_α is indexed by $\alpha \in T$, not merely by $i(\alpha)$). See the discussions about D_α in §§3, 6.3, 6.4, and Lemma 6 of §6.6. For every $\alpha \in T$ we have a strategy whose first aim is to satisfy requirement P_α: $C \neq \overline{W}_\alpha$, and whose second aim is to satisfy the subrequirement R_α: $D_\alpha \neq \overline{W}_\alpha$.

To achieve the latter, we have a variant of the basic module called the *α-module* which attempts to satisfy subrequirement $R_{i(\alpha),j(\alpha)}$. Let $F(\alpha)[s]$ denote the state of the α-module at the end of stage s. (Note that Definition 6.2(ii) insures that for all i and j, there are infinitely many k, such that for all α of length k, $i(\alpha) = i$ and $j(\alpha) = j$. We use the tree notation as in §§2 and 3.) The set of outcomes for each α-module will be Λ.

6.3. DEFINITION. The *true path* $f \in \Lambda^\omega$ is defined by induction on n as follows. Let $\alpha = f \upharpoonright n$. Define

$$f(n) = \begin{cases} a & \text{if } \lim_s F(\alpha)[s] = a \ (\text{necessarily } a \in \Lambda^{\text{Fin}}), \\ g2 & \text{if } \alpha \text{ opens infinitely many } B_\alpha\text{-gaps}, \\ g1 & \text{if } \alpha \text{ opens finitely many } B_\alpha\text{-gaps and infinitely many } A_\alpha\text{-gaps}. \end{cases}$$

(Note that $f(n)$ is the $<_\Lambda$ least $a \in \Lambda$ which "appears correct" infinitely often.) As in the earlier theorems, we shall have a recursive sequence $\{\delta[s]: s \in \omega\}$ of strings, such that $\alpha \subseteq \delta[s]$ for any α which acts at stage s, and we must prove (in the Leftmost Path Lemma (see subsection 6.6)) that $f \upharpoonright n = \liminf_s \delta[s] \upharpoonright n$.

If $\alpha \subset f$, then the analysis of outcomes in Figure 6.2 still applies, except that if the outcome for α is $w1$ or $w2$, then we can only conclude that $\Phi_k^C \neq A_k$ or $\Psi_k^C \neq B_k$ for *some* $k \leqslant i(\alpha)$, not necessarily $k = i(\alpha)$. (This is because of the necessary modifications to the α-module to be described in subsection 6.4.) In this case there will be a fixed $k \leqslant i(\alpha)$ associated with α, and we say that α *blames* k (for its failure to satisfy its subrequirement). We insure that if $\alpha \subset \beta \subset f$, and α blames $k \leqslant i$ then β does not blame the same k.

Furthermore, if $\alpha^\wedge\langle gk\rangle \subseteq \beta$, $k \in \{1,2\}$ and $i = i(\alpha) = i(\beta)$, then β will never be allowed to open a gap (since if $\alpha^\wedge\langle gk\rangle \subset f$, then A_i or B_i is recursive and there is no need for β ever to act).

6.4. DEFINITION. For each $\alpha \in T$ and $i \in \omega$ define

$$\tau(\alpha, i) = (\mu\beta \subseteq \alpha)(\forall\gamma)_{\beta\subseteq\gamma\subset\alpha}\left[i(\gamma) \leqslant i \,\&\, \gamma^\wedge\langle a\rangle \subseteq \alpha \Rightarrow a \in \{s, w\}\right].$$

Hence for all i there are only finitely many $\alpha \subset f$, $i(\alpha) \leqslant i$ and $\alpha^\wedge\langle a\rangle \subset f$ for $a \in \{g2, g1, w2, w1\}$. Thus, for all i, $\tau(f, i) = \lim\{\tau(\alpha, i): \alpha \subset f\}$ exists and has the same meaning as in (3.13) and (3.14) with $\Lambda_1 = \{s, w\}$ here.

As explained in §3, it is useful to view D_i as obtained as follows. For each $\alpha \in T$ and $i \in \omega$ we have a version $D_{\alpha,i}$ for the final r.e. set D_i. Let $\beta = \alpha^\wedge\langle a\rangle$. If $a \in \Lambda_1$ or $i(\alpha) > i$, we let $D_{\alpha,i} = D_{\beta,i}$. Otherwise, we replace $D_{\alpha,i}$ by a new version $D_{\beta,i}$ because α witnessed that some $R_{i'}$, $i' \leqslant i$, was satisfied by one of the first four clauses of (6.4) for R_i. The replacement process occurs for finitely many $\alpha \subset f$, and $D_{\tau(f,i),i}$ is the final version of D_i presented to satisfy R_i. For each $\alpha \in T$ we think of $\tau(\alpha, i)$ as the node $\beta \subseteq \alpha$ at which the set $D_{\alpha,i}$ originates.

This picture of approximations $D_{\alpha,i}$, $\alpha \subset f$, is the most useful for other $0'''$-arguments. However, for this theorem the skeptical reader may think of D_i as follows. For each α we may regard D_α in a purely formal way as a set of cardinality $\leqslant 1$, because it contains at most one element x, and x must be enumerated by the α-module. The final set D_i for R_i is then

$$(6.7) \qquad D_i = \bigcup\{D_\alpha : \alpha \in E_i\},$$

where as in (3.15) and (6.24)

$$(6.8) \qquad E_i = \{\alpha : \tau(f, i) \subseteq \alpha \,\&\, \tau(f, i) = \tau(\alpha, i) \,\&\, i = i(\alpha)\}.$$

6.4. *The modifications in the α-module for the sake of $\beta \supset \alpha$.* (This subsection is not necessary to the formal proof, but contains the motivation for it. We use the notation from subsections 6.5.) The most important part of an $0'''$-argument concerns the modifications which must be made in the basic module to obtain the *α-module* for $\alpha \in T$ so that if $\beta = \alpha\hat{\ }\langle a\rangle$ and $\beta \subset f$ then the β-strategy can succeed. For $a \in \Lambda^{\mathrm{Fin}}$, this is relatively easy. We now consider $a \in \Lambda - \Lambda^{\mathrm{Fin}}$, namely $a \in \{g1, g2\}$. Fix $\alpha \in T$ and $\beta = \alpha\hat{\ }\langle g2\rangle$ (since the case of $g1$ is similar).

First we should let β open gaps while α is in a B_α-gap (since at these stages $r(\alpha)[s] = 0$). (In this way we can hope to achieve $\liminf_s \tilde{r}(\gamma)[s] < \infty$, for $\gamma \subset f$ where $\tilde{r}(\gamma)[s] = \max\{r(\xi)[s] \colon \xi \leqslant \gamma\}$.) To accomplish this, it is convenient to have substages $t \leqslant s$ for each stage s, so that more than one γ can act at each stage. For example, if $|\alpha| < s$ and α opens a B_α-gap at substage t of stage s, then it is reasonable to let β open a gap at substage $t + 1$, because if we force β to wait until stage $s + 1$, then α may be ready to close the B_α-gap and so β may never be able to act.

At the end of each substage t of a given stage $s + 1$, we define in (6.12) a set of *accessible nodes* AC_t consisting of those γ whose guess currently "seems correct", and thus for which it is reasonable to let γ act at substage $t + 1$. If $\alpha \in \mathrm{AC}_t$ and $|\alpha| < s$, we should certainly add $\alpha\hat{\ }\langle F_t(\alpha)\rangle$ to AC_t, where $F_t(\alpha)$ is the value of $F(\alpha)$ at the end of substage t. However, $\alpha\hat{\ }\langle a\rangle$ for $a \in \{g1, g2\}$ will never be added in this way. As a first approximation we could add $\alpha\hat{\ }\langle g2\rangle$ to AC_t merely whenever α *opens* a B_α-gap, but it is better for later purposes to allow $\alpha\hat{\ }\langle g2\rangle \in \mathrm{AC}_t$ whenever $\alpha \in \mathrm{AC}_t$ and α is *in* a B_α-gap (namely whenever $\alpha\hat{\ }\langle w2\rangle \in \mathrm{AC}_t$). This extra accessibility is acceptable from β's viewpoint because the condition it desires, namely $r(\alpha)[s] = 0$, holds not just when α *opens* the B-gap, but also at all stages *during* that gap.

However, this extra accessibility of $\beta = \alpha\hat{\ }\langle g2\rangle$ may allow β to act more often than it deserves. Suppose, for example, that $\alpha = f \restriction n$ and $\lim_s F(\alpha)[s] = w2$ so that $f(n) = w2$. Then β will be accessible at almost every stage when α is accessible, but β cannot be allowed to act infinitely often because $\beta <_L f$. The *control function* $\mathrm{CO}(\alpha)[s]$ in (6.11) is designed to insure that such nodes β act at most finitely often. Intuitively, a node γ gets a new *token* (also called a "chip") whenever $\mathrm{CO}(\gamma)$ increases. Now the γ-module requires a fresh token for each exit from the starting state w (by condition (6.16c)) although γ may perform any further actions on the same token, until $F(\gamma) = w$ again. Thus the control function insures that:

(6.9) If α opens finitely many A_α-gaps (B_α-gaps) then for $k = 1$ ($k = 2$) $(\exists^{<\infty} s)(\exists \gamma \supseteq \alpha^\wedge\langle gk\rangle)[\gamma$ acts at $s]$.

Hence, $\delta[s] <_L \alpha^\wedge\langle w2\rangle$ for only finitely many s. However, at the same time

$$(6.10) \qquad \alpha \subset f \Rightarrow \lim_s \mathrm{CO}(\alpha)[s] = \infty.$$

However, all of this so far could have arisen in some $0''$-construction. The heart of the difficulty for an $0'''$-construction lies in the following delay, which represents the active cooperation of α with various $\gamma \supset \alpha$, and which distinguishes the $0'''$ from the $0''$-method. When $F_t(\alpha) = a$ and $F_{t+1}(\alpha) = b \neq a$, then we initialize all γ, $\alpha^\wedge\langle a\rangle \leqslant \gamma$. Hence, at any moment in time there is at most one $a \in \Lambda^{\mathrm{Fin}}$ such that one (or more) $\gamma \supseteq \alpha^\wedge\langle a\rangle$ has an open A_γ-gap or B_γ-gap (namely $F(\gamma) = w1$ or $F(\gamma) = w2$), which we refer to as *pending γ-action*. Also, if $F(\alpha)$ moves from $w1$ to $w2$, we initialize all γ, $\alpha^\wedge\langle w2\rangle <_L \gamma$, including $\gamma \supseteq \alpha^\wedge\langle g1\rangle$ since the latter guess "appears false". However, when $F(\alpha)$ moves *right* from $w2$ it may leave pending γ-action for $\gamma \supseteq \alpha^\wedge\langle g2\rangle$. Later γ may open an A_γ-gap, close it unsuccessfully moving *right* again, and leave pending ξ-action for $\xi \supseteq \alpha^\wedge\langle g1\rangle$ as well. This γ-action and ξ-action may remain pending forever, even if γ and ξ are ready to close their gaps, because perhaps γ and ξ are never again accessible.

This problem becomes acute in Lemma 7 of subsection 6.6 where we must show that $D_i \leqslant_T A_i$ and $D_i \leqslant_T B_i$. To do this we must show that almost every A_α-gap or B_α-gap once opened by some $\alpha \in E_i$ of (3.15) or (6.24) is eventually closed or cancelled, at which point we can effectively determine whether x enters D_i. Now suppose that $\beta = \alpha^\wedge\langle g2\rangle$, $i(\alpha) > i(\beta) = i$, and $\alpha^\wedge\langle w\rangle \subset f$ so $\beta <_L f$. Perhaps while α was previously in a B_α-gap, α allowed β to also open a gap. Later α closed its B_α-gap, $F(\alpha)$ returned to w forever, leaving β inaccessible and therefore unable to close its gap even if $\Phi_i^C = A_i$ and $\Psi_i^C = B_i$. If this is repeated over infinitely many nodes $\alpha^n \subset f$, $n \in \omega$, then we get infinitely many nodes $\beta^n <_L f$, $i(\beta^n) = i$, such that β remains in a gap forever, and we fail.

To remedy this, we have the delay function $d(\alpha, k)$ (see (6.15)) which forces α to cooperate with certain $\beta \supseteq \alpha^\wedge\langle gk\rangle$. Suppose α is accessible (i.e. $\alpha \in \mathrm{AC}_t$), α is in a B_α-gap (i.e. $F_t(\alpha) = w2$), $\beta \supseteq \alpha^\wedge\langle g2\rangle$, $i(\beta) < i(\alpha)$ and β is in a gap (i.e. $F(\beta) = w2$ or $w1$). Then we define $d(\alpha, 2) = 1$, which (because of (6.18c)) prevents α from closing its gap until β closes first. (If $i(\beta) > i(\alpha)$ then there is no prohibition on α and α may close its gap leving pending β-action.) Now β may be delaying closing its gap because $d(\beta, k') = 1$ on account of some $\gamma \supseteq \beta^\wedge\langle gk'\rangle$, with $i(\gamma) < i(\beta)$, and so on. However, since $i(\gamma)$ decreases as we move down through such γ, we must eventually come to some $\xi \supseteq \alpha^\wedge\langle g2\rangle$ such that ξ is in a gap, say an A_ξ-gap, and $d(\xi, i) = 0$ (namely, ξ is not delaying for any $\gamma \supset \xi$ but ξ is waiting for $l(\Phi_\xi, A_\xi) > x(\xi)$). (See the proof of Lemma 4.) While ξ is waiting, we can view α as *blaming* $i = i(\xi)$ for its failure to close its gap. So long as this condition persists, we do not allow any ζ', $\xi <_L \zeta'$, such that $i(\zeta') = i$ to open new A_i-gaps (see (6.16e)) because ξ is temporarily witnessing that $\Phi_i^C \neq A_i$. Hence no $\alpha' \supset \alpha$, such that $\alpha'^\wedge\langle wk'\rangle \subset f$ can blame the same i.

(As an alternative to this delay function, we can insert extra outcomes in the α-module, where $F(\alpha)$ waits while delaying for some $\beta \supseteq \alpha^\wedge\langle gk\rangle$. This complicates the α-module and the tree, but gives simple statements of certain lemmas. See §7.)

Some such delay or cooperation by α for the sake of $\beta \supset \alpha$ always arises in $0'''$-arguments, because although α has higher *local priority*, namely $|\alpha| < |\beta|$, β may be working on a requirement R_i of higher *global priority*, if $i(\beta) < i(\alpha)$. (In most $0''$-arguments, $|\alpha| < |\beta|$ iff $i(\alpha) < i(\beta)$, so the two priorities coincide.)

6.5. *The construction*. During the construction we regard A_α, B_α, Φ_α, Ψ_α, C, W_j and so on as in a state of formation. At a given point during the construction if we write A_α we mean the finite set of elements so far enumerated in A_α, and similarly for Φ_α. When necessary to avoid confusion, we append $[s]$ and write $A_\alpha[s]$ or $\Phi_\alpha[s]$ to denote the result by the end of stage s. However, in the verification after the construction, we also use thenotations A_α, Φ_α and so on to denote the completed sets and functionals.

To state the construction, it is useful to have the following parameters (as in the basic module) which have roughly the following meanings:

$x(\alpha) =$ the current candidate for the α-module,

$r(\alpha, 1) =$ the restraint imposed on C to preserve $A_\alpha \restriction x(\alpha)$,

$r(\alpha, 2) =$ the restraint imposed on C to preserve $B_\alpha \restriction x(\alpha)$,

$r(\alpha) = \max\{r(\alpha, 1), r(\alpha, 2)\}$,

$F(\alpha) =$ the current state $a \in \Lambda^{\text{Fin}} = \{s, w2, w1, w\}$ of the α-module,

$m(\alpha, 1) =$ the number of times α has exited rightward from state $w1$ (i.e. has unsuccessfully closed an A_α-gap),

$m(\alpha, 2) =$ the number of times α has exited rightward from state $w2$ (i.e. has unsuccessfully closed a B_α-gap).

The values of the parameters may change during the construction. A parameter once given a value retains that value until redefined. Let $\omega^{[\alpha]}$ denote $\omega^{[n]}$ where n is the code number for α in some effective coding of T. To *reset the candidate* $x(\alpha)$ at stage s means to cancel the old candidate $x(\alpha)[s-1]$ (if $s > 0$), and to define $x(\alpha)[s]$ to be the least $y \in \omega^{[\alpha]}$, $y < s$ and y greater than all previous candidates. To *initialize node* α means to reset $x(\alpha)$, to set $F(\alpha) = w$ unless $F(\alpha) = s$, to set $r(\alpha, 1) = r(\alpha, 2) = r(\alpha) = 0$, and to cancel any A_α- or B_α-gaps.

Stage $s = 0$. Initialize all nodes $\alpha \in T$. Define $\delta[0] = \lambda$.

At the end of stage s we are given $m(\alpha, k)[s]$ for $\alpha \in T$ and $k \in \{0, 1\}$. Define the *control function* $\text{CO}(\alpha)[s]$ by induction on $|\alpha|$ as follows:

(6.11) $$\text{CO}(\lambda)[s] = s,$$

$$\text{CO}(\alpha^\wedge\langle a\rangle)[s] = \begin{cases} \text{CO}(\alpha)[s] \dot{-} 1 & \text{if } a \in \Lambda - \{g1, g2\}, \\ \min\{\text{CO}(\alpha)[s] \dot{-} 1, m(\alpha, k)\} & \text{if } a = gk, k \in \{1, 2\}. \end{cases}$$

(Note that for each $\alpha \in T$, the function $\text{CO}(\alpha)[s]$ is nondecreasing in s.)

Stage $s + 1$. The construction proceeds by substages $t \leqslant s$. We refer to substage t of stage $s + 1$ as *stage* $(s + 1, t)$. The value of parameter p at the end of substage t will be denoted by p_t. At the end of substage t we shall have $\delta_t \in T$ such that $\delta_t \subseteq \delta_{t+1} \subseteq \delta[s + 1]$, and only those $\alpha \subseteq \delta[s + 1]$ may act during stage $s + 1$.

Substage $t = 0$. Define $\delta_0 = \lambda$.

Substage $t + 1 \leqslant s + 1$. We are given δ_t, and for all $\alpha \in T$, $F_t(\alpha) \in \{s, w2, w1, w\}$, the current state of the α-module.

Define AC_t (sometimes written $\mathrm{AC}_t[s + 1]$) the *accessible nodes* at stage $(s + 1, t + 1)$, by the following closure properties:

$$(6.12)\qquad \begin{array}{l} \lambda \in \mathrm{AC}_t, \qquad \alpha \in \mathrm{AC}_t \,\&\, |\alpha| \leqslant s \Rightarrow \alpha\hat{\ }\langle F_t(\alpha)\rangle \in \mathrm{AC}_t, \\ \alpha\hat{\ }\langle wk\rangle \in \mathrm{AC}_T \Rightarrow \alpha\hat{\ }\langle gk\rangle \in \mathrm{AC}_t \quad \text{for } k \in \{1, 2\}. \end{array}$$

We say that α *requires attention at stage* $(s + 1, t + 1)$ if $\alpha \in \mathrm{AC}_t$ and one of the following conditions holds (as defined precisely later):

(6.13)
ready to satisfy P_α,
ready to begin an A_α-gap,
ready to close an A_α-gap,
ready to close a B_α-gap.

Let α be the least $\leqslant$ -least $\gamma \supseteq \delta_t$ such that γ requires attention. If the first clause of (6.13) holds for α then perform the action to satisfy P_α, and let $\delta_{t+1} = \alpha$. Otherwise, exactly one of the last three clauses of (6.13) holds. Perform the indicated action. If α opens a gap, let $\delta_{t+1} = \alpha\hat{\ }\langle gk\rangle$ where $k = 1$ if it is an A_α-gap and $k = 2$ if it is a B_α-gap. If α does not open a gap, let $\delta_{t+1} = \alpha\hat{\ }\langle F_{t+1}(\alpha)\rangle$. Go to stage $s + 2$ if either: α acted to satisfy P_α; $t = s$; or α fails to exist. Otherwise, go to substage $t + 2$.

Define $\delta[s + 1] \in T$ of length $s + 1$ as follows. Choose t maximal such that δ_t was defined. Let $\delta(n)[s + 1] = \delta_t(n)$ for all $n < |\delta_t|$. For n, $|\delta_t| \leqslant n \leqslant s$, let $\alpha = \delta[s + 1] \upharpoonright n$ and define $\delta(n)[s + 1] = F_t(\alpha)$. Initialize all γ, $\delta[s] <_L \gamma$.

6.5. Definition. Let $S^\alpha = \{s \colon \alpha \subseteq \delta[s]\}$. The stages in S^α are called *α-stages*.

Note that α only acts at an α-stage. Furthermore, if α acts, $F(\alpha)[s] = a$ and $F(\alpha)[s + 1] = b$, then all $\gamma \geqslant \alpha\hat{\ }\langle a\rangle$ will be initialized. In addition, by Steps 2(b) and 3(b), if $b <_\Lambda a$ then all γ, $\alpha\hat{\ }\langle b\rangle <_L \gamma$, will also be initialized. We say that α is *in a gap forever* if $\lim_s F(\alpha)[s] = wk$, $k \in \{1, 2\}$, and α is *accessible infinitely often* if $\alpha \in \mathrm{AC}_t[s]$ for infinitely many stages (s, t).

(6.14)

Ready to satisfy P_α:

(a) $C \cap W_\alpha = \varnothing$,

(b) $(\exists y)\big[y \in W_\alpha \cap \omega^{[\alpha]} \,\&\, \gamma > \tilde{r}(\alpha)$, where $\tilde{r}(\alpha) =_{\mathrm{dfn}} \max\{r(\beta) \colon \beta \leqslant \alpha\}$.

Action. Enumerate the least such y in C. Initialize all $\beta \geqslant \alpha$.

The next three cases (6.16), (6.17) and (6.18) resemble Steps 1, 2 and 3 of the basic module but with the modifications explained in subsection 6.4. (We label the corresponding actions as Steps 1, 2(a), 2(b), 3(a) or 3(b) to correspond to Figure 6.1.) The most important modification involves the following *delay function* defined at the end of stage $(s+1, t)$ for all $\alpha \in T$ and $k \in \{1,2\}$,

$$(6.15)\quad d_t(\alpha, k) = \begin{cases} 1 & \text{if } \alpha \notin AC_t \text{ or if } F_t(\alpha) = wk \,\&(\exists\beta \supseteq \alpha^\wedge\langle gk\rangle) \\ & [\beta \in \mathrm{AC}_t \,\&\, i(\beta) < i(\alpha) \,\&[F_t(\beta) = w1 \text{ or } = w2]], \\ 0 & \text{otherwise} \end{cases}$$

(6.16) Ready to open an A_α-gap:

(a) $F_t(\alpha) = w$,

(b) $x(\alpha) \in W_\alpha$,

(c) $\mathrm{CO}(\alpha) >$ the number of A_α-gaps previously begun,

(d) $\neg(\exists\beta)(\exists k)[i(\beta) = i(\alpha) \,\&\, \beta^\wedge\langle gk\rangle \subseteq \alpha]$,

(e) $\neg(\exists\beta)[\beta < \alpha \,\&\, i(\beta) = i(\alpha) \,\&\, \beta \in \mathrm{AC}_t$
$\&(\exists k)[F_t(\beta) = wk \,\&\, d_t(\beta, k) = 0]]$.

Action. (This action is called Step 1.) Open an A_α-gap, define $F(\alpha) = w1$ and $r(\alpha, 1) = 0$. Initialize all γ such that $\alpha^\wedge\langle w\rangle \leqslant \gamma$.

(6.17) Ready to close an A_α-gap:

(a) $F_t(\alpha) = w1$,

(b) $l(\Phi_\alpha^C, A_\alpha) > x(\alpha)$,

(c) $d_t(\alpha, 1) = 0$.

Action. Let $v + 1 < s + 1$ be the stage when the current A_α-gap was opened and let $x = x(\alpha)[s]$. Close the A_α-gap.

Step 2(a) (*left exit from* w1). If $A_\alpha[s] \restriction x \neq A_\alpha[v] \restriction x$ then open a B_α-gap, define $F(\alpha) = w2$ and $r(\alpha, 2) = 0$. (Hence, $r(\alpha) = 0$ also.) Initialize all γ, $\alpha^\wedge\langle w2\rangle <_L \gamma$.

Step 2(b) (*right exit from* w1). If $A_\alpha[s] \restriction x = A_\alpha[v] \restriction x$, define $F(\alpha) = w$, $m(\alpha, 1) = m(\alpha, 1) + 1$ and $r(\alpha, 1) = s$ (to preserve $A_\alpha \restriction x$), reset $x(\alpha)$, and initialize all γ, $\alpha^\wedge\langle w1\rangle \leqslant \gamma$.

(6.18) Ready to close a B_α-gap:

(a) $F_t(\alpha) = w2$,

(b) $l(\Psi_\alpha^C, B_\alpha) > x(\alpha)$,

(c) $d_t(\alpha, 2) = 0$.

Action. Let $v + 1 < s + 1$ be the stage when the current B_α-gap was opened, and let $x = x(\alpha)[s]$. Close the B_α-gap.

Step 3(a) (*left exit from* w2). If $B_\alpha[s] \restriction x \neq B_\alpha[v] \restriction x$, then enumerate x in D_α, and define $F(\alpha) = s$. Initialize all γ, $\alpha^\wedge\langle s\rangle <_L \gamma$.

Step 3(b) (*right exit from* w2). If $B_\alpha[s] \restriction x = B_\alpha[v] \restriction x$, then define $F(\alpha) = w$, $m(\alpha, 2) = m(\alpha, 2) + 1$ and $r(\alpha, 2) = s$ (to preserve $B_\alpha \restriction x$). Reset $x(\alpha)$, and initialize all γ, $\alpha^\wedge\langle w2\rangle \leqslant \gamma$.

6.6. *The verification.* Let f be the true path of Definition 6.3. We first prove that f is the leftmost path visited by $\{\delta[s]: s \in \omega\}$ infinitely often, and that each $\alpha \subset f$ has the necessary conditions to succeed.

LEMMA 1 (LEFTMOST PATH LEMMA). Let $\alpha = f \upharpoonright n$. For all n,

(i) $\alpha = \liminf_s \delta[s] \upharpoonright n$,

(ii) $\lim_s \mathrm{CO}(\alpha)[s] = \infty$,

(iii) $\lim\{\tilde{r}(\alpha)[s]: s \in S^{\alpha+}\}$ *exists where* $\alpha^+ = \alpha^\wedge\langle f(n)\rangle$, S^α *is defined in Definition* 6.5 *and*

$$\tilde{r}(\alpha)[s] =_{\mathrm{dfn}} \max\{(\beta)[s]: \beta \leqslant \alpha\}.$$

PROOF. Clearly (i) and (ii) hold for $n = 0$. Fix $n \geqslant 0$, assume (i) and (ii) for n and (iii) for all $m < n$. Let $\beta = f \upharpoonright n$, $a = f(n)$ and $\alpha = \beta^\wedge\langle a\rangle = \beta^+$. We prove (i) and (ii) for α and (iii) for β.

To prove (i) for α assume for a contradiction that $\delta[s] <_L \alpha$ for infinitely many s. Then for some $b <_\Lambda a$, $\delta[s] \supseteq \gamma = \beta^\wedge\langle b\rangle$ for infinitely many s, and hence γ is accessible infinitely often. By the definition (6.12) of AC_t, and the fact that $a = \liminf_s F(\beta)[s]$, this can only happen if $b = gk$ and $a = wk$ for some $k \in \{1,2\}$. But if $a = wk$, then $\lim_s F(\beta)[s] = wk$, so β exits from state wk only finitely often, and $\lim_s m(\beta, k)[s] < \infty$. Hence, by the definition (6.11) of the control function, $\lim_s \mathrm{CO}(\xi)[s] < \infty$ for all $\xi \supseteq \gamma$ and for almost all $\xi \supseteq \gamma$ $\mathrm{CO}(\xi)[s] = 0$ for all s. Thus, only finitely many $\xi \supseteq \gamma$ ever act and each acts at most finitely often because by (6.16)(c) ξ can open a new A_ξ-gap at most $\lim_s \mathrm{CO}(\xi)[s]$ times, and ξ can act only finitely often before it returns to state ω and must open a new A_ξ-gap. Thus, $\delta[s] <_L \alpha$ finitely often.

To prove that $\alpha \subseteq \delta[s]$ infinitely often, note that $\beta \subseteq \delta[s]$ infinitely often, and so it follows for α if $a \in \Lambda^{\mathrm{Fin}}$. Suppose $a = gk$, say $k = 2$ (since the case $k = 1$ is similar). Then β opens infinitely many B_β-gaps. However, if β opens a B_β-gap at stage (s, t) for $s > |\beta|$, then $\delta_{t+1} = \beta^\wedge\langle g2\rangle$ and $\delta_{t+1} \subseteq \delta[s]$. This proves (i) for α.

To prove (ii) for α we assume (ii) for β. Now, if $a \in \Lambda^{\mathrm{Fin}}$, then $\lim_s F(\beta)[s] = a$, so (ii) holds for α. If $a = g2$, then β opens infinitely many B_β-gaps, necessarily closing each unsuccessfully and exiting right from state $w2$, so $\lim_s m(\beta, 2)[s] = \infty$ and (ii) holds for α. If $a = g1$, then β opens infinitely many A_β-gaps, closes almost all unsuccessfully (with a right exit from $w1$), so $\lim_s m(\beta, 1)[s] = \infty$ and (ii) holds for α.

To prove (iii) for β, let β^- be the predecessor (if any) of β, and

$$\tilde{r}(\beta^-) = \lim\{\tilde{r}(\beta^-, s): s \in S^\beta\}.$$

Choose v such that for all $s \geqslant v$, $\beta \leqslant \delta[s]$ and if $s \in S^\beta$ then $\tilde{r}(\beta^-, s) = \tilde{r}(\beta^-)$. Now as in the basic module $r(\beta) =_{\mathrm{dfn}} \liminf_s r(\beta)[s]$ exists and $r(\beta)[s] = r(\beta)$ at almost every s such that $\alpha \subseteq \delta[s]$ (since if $a = gk$ or wk then $r(\beta, k)[s] = 0$ at such an s). Hence, there exists $t \geqslant v$ and an integer $\tilde{r}(\beta)$ such that $\tilde{r}(\beta)[s] = \tilde{r}(\beta)$ for all $s \in S^\alpha$, $s \geqslant t$.

LEMMA 2. *C is nonrecursive.*

PROOF. Note that $\bar{C}$ is infinite because if we choose α such that $W_\alpha = \emptyset$ then $C \cap \omega^{[\alpha]} = \emptyset$. Assume for a contradiction that $\bar{C}$ is r.e., say $\bar{C} = W_j$. Choose α, y and s such that $j(\alpha) = j$, $\alpha \subset f$, $y \in \omega^{[\alpha]}$, $y \in W_j[s-1]$, $\alpha^+ \subseteq \delta[s]$, $\tilde{r}(\alpha)[s] = \tilde{r}(\alpha) < y$ (using Lemma 1), and for all $\beta \subset \alpha$, if β ever acts to satisfy P_β (i.e., (6.14)), then β has already done so before stage s. Hence, $C[s] \cap W_j[s-1] \neq \emptyset$, because if $C[s-1] \cap W_j[s-1] = \emptyset$, then α will be allowed to receive attention to satisfy P_α at stage s and will contribute y (or some other $z \in W_j[s-1]$) to C at stage s.

6.6 DEFINITION. For $\alpha \in T$ and $k \in \{1,2\}$ let $d(\alpha, k) = \liminf_{s,t} d_t(\alpha, k)[s]$.

(Namely, $d(\alpha, k) = 0$ if there are infinitely many stages (s, t) such that $d_t(\alpha, k)[s] = 0$ and $d(\alpha, k) = 1$ otherwise. Thus, $d(\alpha, k) = 0$ only if there are infinitely many stages (s, t) when $\alpha \in AC_t[s]$ and α is not delaying for some $\beta \supseteq \alpha^\wedge\langle gk\rangle$, $i(\beta) < i(\alpha)$, $\beta \in AC_t[s]$ but β is permanently in a gap.)

LEMMA 3 (DELAY FUNCTION LEMMA). *Suppose $\beta <_L f$ or $\beta^\wedge\langle wk\rangle \subset f$, and β is accessible infinitely often. Choose n maximal such that $\beta \supseteq \gamma = f \upharpoonright n$. Let $\gamma^+ = \gamma^\wedge\langle f(n)\rangle$. Then there exists s_0 such that for all $s \geq s_0$, and $t \leq s$,*

(i) $\beta \in AC_t[s] \Leftrightarrow \gamma \in AC_t[s] \Leftrightarrow \gamma^+ \in AC_t[s]$, and

(ii) $\gamma \in AC_t[s] \Rightarrow d_t(\beta, k)[s] = d(\beta, k)$ for $k \in \{1,2\}$.

PROOF. Clearly, if $\beta \in AC_t[s]$ then $\gamma \in AC_t[s]$. Now by definition of γ, either there exists $k' \in \{1,2\}$ such that $\gamma^+ = \gamma^\wedge\langle wk'\rangle$ and $\beta \subseteq \gamma^\wedge\langle gk'\rangle$; or $\beta = \gamma$ and $\gamma^+ = \gamma^\wedge\langle wk'\rangle$ for $k' = k$. By Lemma 1, choose s_0 such that for all $s \geq s_0$, $\gamma^+ \leq \delta[s]$ and $F_t(\gamma)[s] = wk'$ for all $t \leq s$, and no P_α, $\alpha \leq \gamma^+$, acts at any stage $s \geq s_0$. Hence, for $s \geq s_0$, $\gamma \in AC_t[s]$ iff $\gamma^+ \in AC_T[s]$. Also no $\zeta <_L \gamma^+$ acts at any $s \geq s_0$ so we must have

(6.19) $$(\forall \zeta <_L \gamma^+)(\forall s \geq s_0)(\forall t \leq s)\left[F_t(\zeta)[s] = F(\zeta)[s]\right].$$

Let $\xi = \gamma^\wedge\langle gk'\rangle$. By (6.12) and (6.19) if any of γ, γ^+, and ξ is in $AC_t[s]$, for $s \geq s_0$, $t \leq s$, then all are in $AC_t[s]$. Now for any $\zeta \supseteq \xi$, if ζ is accessible infinitely often for $s \geq s_0$, $t \leq s$, then $\zeta \in AC_t[s]$ iff $\xi \in AC_t[s]$, since the former depends only upon whether $\xi \in AC_t[s]$ and upon the value of $F_t(\zeta)[s]$, $\zeta \supseteq \xi$, which is constant after s_0 by (6.19). Hence,

(6.20)
$$(\forall \zeta \supseteq \xi)(\forall s \geq s_0)(\forall t \leq s)[\zeta \text{ is accessible infinitely often} \Rightarrow [\zeta \in AC_t[s] \Leftrightarrow \gamma \in AC_t[s]]].$$

This establishes (i) since either $\beta = \gamma$ or $\beta \supseteq \xi$, and since β is accessible infinitely often. Now (ii) follows since the value of $d_t(\beta, k)[s]$ depends only upon the value of $F_t(\zeta)[s]$ and whether $AC_t[s]$ for certain $\zeta \supseteq \xi$, all of which is constant by (6.19) and (6.20) for all $\zeta \supseteq \xi$, on those stages (s, t), $s \geq s_0$ and $t \leq s$, such that $\xi \in AC_t[s]$. Indeed we have

(6.21)
$$(\forall \zeta)[[\zeta = \xi \text{ or } \zeta = \gamma] \Rightarrow (\forall s \geq s_0)(\forall t \leq s)(\forall k'')\left[\gamma \in AC_t[s] \Rightarrow d_t(\zeta, k'')[s] = d(\zeta, k'')\right]].$$

LEMMA 4 (FINITE INJURY ALONG THE TRUE PATH LEMMA).
$(\forall i)(\exists m)(\forall \beta \supseteq f \upharpoonright m)[[\beta$ *is in a gap forever and* β *is accessible infinitely often*$]$
$\Rightarrow i(\beta) > i]$.

PROOF. Proceed by induction on i. Assume the statement for all $i' < i$ with some $m' \geqslant 0$. Fix some $\beta \subseteq f \upharpoonright m'$ such that $\lim_s F(\beta)[s] = wk$, $i(\beta) = i$, and β is accessible infinitely often, if such β exists. Let $n \geqslant m'$ be maximal such that $\beta \supseteq \gamma = f \upharpoonright n$. Since any β with $f <_L \beta$ would cancel its β-gap eventually we must have $\beta <_L \gamma$ or $\beta = \gamma$ and $\beta \,\hat{}\, \langle wk \rangle \subset f$ so Lemma 3 applies. Let γ^+, ξ, and s_0 be as in the proof of Lemma 3. Now (6.19), (6.30), (6.21) and the inductive hypothesis on m' imply

$$(6.22) \qquad (\forall s \geqslant s_0)(\forall t \leqslant s)[\gamma \in AC_t[s] \Rightarrow d_t(\beta, k)[s] = 0],$$

since otherwise there exists $\zeta \supseteq \beta \,\hat{}\, \langle gk \rangle$ with $i(\zeta) \leqslant i(\beta)$, ζ permanently in a gap, and ζ accessible infinitely often contrary to the choice of m'. Now no $\zeta \supseteq \gamma^+$, with $i(\zeta) = i(\beta)$ can open an A_ζ-gap at stage $(s, t+1)$, $s \geqslant s_0$, since if so then $\zeta \in AC_t[s]$, so $\gamma \in AC_t[s]$, $\beta \in AC_t[s]$ and $d_t(\beta, k)[s] = 0$ by Lemma 3 contradicting (6.16)(e). Hence, $m = s_0$ works for β since any ζ in a gap at s_0 has $|\zeta| < s_0$.

LEMMA 5 (TRUTH OF OUTCOME LEMMA). *Let* $\alpha = f \upharpoonright n$.

(i) *If* $f(n) = s$, *then* $D_\alpha \cap W_\alpha \neq \varnothing$.

(ii) *If* $f(n) = gk$, *then* A_α *is recursive if* $k = 1$, *and* B_α *is recursive if* $k = 2$.

(iii) *If* $f(n) = wk$, *and* $d(\alpha, k) = 0$, *then* $\Phi_\alpha^C \neq A_\alpha$ *if* $k = 1$, *and* $\Psi_\alpha^C \neq b_\alpha$ *if* $k = 2$.

(iv) *If* $\beta \leqslant \alpha$, $\lim_s F(\beta)[s] = wk$, $d(\beta, k) = 0$, *then* $\Phi_\beta^C \neq B_\beta$ *if* $k = 2$.

PROOF.

(i) If $f(n) = s$ then at some stage the α-strategy enumerated into D_α some $x(\alpha)$ from W_α.

(ii) Suppose $f(n) = g2$. Let $\alpha^+ = \alpha \,\hat{}\, \langle g2 \rangle$. Choose s_0 such that $\alpha^+ \leqslant \delta[s]$ for all $s \geqslant s_0$, and for all $\gamma \subseteq \alpha^+$, P_γ is satisfied before stage s_0 if at all. Now (6.5) holds for s_0 with $r(\alpha)[s]$ in place of $r[s]$ so the same proof as in the basic module shows that B_α is recursive.

(iii) This is a special case of (iv).

(iv) Since $\lim_s F(\beta)[s] = wk$, β remains in a gap forever, and $x = \lim_s x(\beta)[s]$ exists. Now β opens finitely many gaps, so $\lim_s CO(\beta)[s] < \infty$, and hence finitely many $\xi \supseteq \beta \,\hat{}\, \langle gk \rangle$ ever act, and each acts at most finitely often. But $d(\beta, k) = 0$ so there is no $\xi \supseteq \beta \,\hat{}\, \langle gk \rangle$, such that ξ is in a gap forever, $i(\xi) < i(\beta)$ and ξ is accessible infinitely often. Hence, there exists s_0 such that for all $s \geqslant s_0$ if $\beta \in AC_t[s]$ then $d_t(\beta, k)[s] = 0$, and $F_t(\beta)[s] = wk$, namely (6.17)(a) and (c) hold.

Fix k, say $k = 1$. Suppose for a contradiction that $\Phi_\beta^C = A_\beta$, so that (6.17)(b) holds for almost all s. Choose $s_1 \geqslant s_0$ such that for all $s \geqslant s_1$, $f \upharpoonright (|\beta| + 1) \leqslant \delta[s]$, and no P_α, $\alpha \leqslant f \upharpoonright (|\beta| + 1)$, acts at s; no $\xi \leqslant \beta$ acts to satisfy P_ξ at stage s; and all the conditions of (6.17) are satisfied if $\beta \in AC_t[s]$. Let $s \geqslant s_1$ be any α-stage $\geqslant s_1$.

Case 1. $\beta \subseteq \alpha$.

Choose t maximal such that $\delta_t[s] \subseteq \beta$. Then either $\delta_t[s] \subset \beta$ or $\beta \in AC_t[s]$ and hence $\delta_{t+1}[s] = \beta$. Thus β acts to chose the A_β-gap either at stage (s, t) or $(s, t+1)$, contrary to hypothesis on s_0.

Case 2. $\beta <_L \alpha$.

Choose $n < |\alpha|$ maximal such that $\alpha \upharpoonright n \subseteq \beta$. Let $\gamma = \alpha \upharpoonright n$. Choose t maximal such that $\delta_t[s] \subseteq \gamma$. Now $\gamma \in AC_t[s]$, so by Lemma 3 $\beta \in AC_t[s]$ also. Thus, β acts to close the a_β-gap at stage $(s, t+1)$.

LEMMA. $(\forall i)(\exists m)(\forall n \geqslant m)[[\alpha = f \upharpoonright n \,\&\, i = i(\alpha)] \Rightarrow f(n) \in \{s, w\}]$.

PROOF. Fix i. Choose m' satisfying Lemma 4. Now if $\alpha = f \upharpoonright p$ for $p \geqslant m'$ and $i(\alpha) = i$ then $f(p) \notin \{w1, w2\}$ since α is accessible whenever $\alpha \subseteq \delta[s]$, which happens infinitely often by Lemma 1(i). Now suppose $f(p) = gk$, $k \in \{1, 2\}$, for some such α and p. Then by (6.16)(d) no $\gamma \supseteq \alpha\hat{}\langle gk \rangle = \alpha^+$ such that $i(\gamma) = i$ can ever open a gap. Hence the lemma follows by choosing $m \geqslant \max\{m', p+1\}$.

LEMMA 7. *Fix i such that $\Phi_i^C = A_i$, $\Psi_i^C = B_i$, A^i is nonrecursive, and B_i is nonrecursive. Then there exists a nonrecursive r.e. set D such that $D \leqslant_T A_i$ and $D \leqslant_T B_i$.*

PROOF. By Lemma 5, choose $\eta = f \upharpoonright p$ with p minimal such that

$$(\forall n \geqslant p)\left[\alpha = f \upharpoonright n \,\&\, i(\alpha) \leqslant i \Rightarrow f(n) \in \{s, w\}\right].$$

Define as in (6.7)

$$E = \{\alpha : \eta \subseteq \alpha \,\&\, i(\alpha) = i \,\&\, \tau(\alpha, i) = \eta\}, \text{ and} \tag{6.24}$$

$$D = \bigcup \{D_\alpha :: \alpha \in E\}. \tag{6.25}$$

Clearly D is r.e. since E is recursive. (Note that E contains nodes $\alpha <_L f$ as well as $\alpha \subset f$ and $f <_L \alpha$.)

To see that D is nonrecursive, suppose for a contradiction that $D = \overline{W_j}$. Fix $n \geqslant p$, such that $i(\alpha) = i$ and $j(\alpha) = j$ for $\alpha = f \upharpoonright n$. By choice of p, $f(n) \in \{s, w\}$. If $f(n) = s$, then $D_\alpha \cap W_j \neq \varnothing$, by Lemma 5. Suppose $f(n) = w$. Choose s_0 such that for all $s \geqslant s_0$, $F(\alpha)[s] = w$, $\alpha\hat{}\langle w \rangle \leqslant \delta[s]$, and

$$(\forall \beta \subseteq \alpha)\left[\beta \text{ does not satisfy } P_\beta \text{ during stage } s\right],$$

$$(\forall \beta \subset \alpha)\left[\text{if } \beta \text{ acts during stage } s \text{ then } (\exists k \in \{1, 2\})\left[\beta\hat{}\langle gk \rangle \subset \alpha\right]\right].$$

Let $x = x(\alpha)[s_0]$. Then $x = x(\alpha)[s]$ for all $s \geqslant s_0$ since $x(\alpha)$ is never reset after s_0. Now by the hypotheses of Lemma 7 and by Lemma 5, there can be no $\beta < \alpha$ as mentioned in (6.16)(d) or (6.16)(e) which would infinitely often prevent α from opening an A_α-gap. Since $x \in W_j$ and $\lim_s CO(\alpha)[s] = \infty$ by Lemma 1, α eventually opens such a gap after stage s_0, contrary to the hypothesis. Hence, D is nonrecursive.

Let m be as in Lemma 4, and $q = \max\{m, p\}$. Let $\xi = f \upharpoonright q$. To prove that $D \leqslant_T A_i$ and $D \leqslant_T B_i$, it suffices to prove that for almost all $\alpha \in E$, any α-gap once opened is later closed or cancelled. We shall prove this for all $\beta \in E$, $\beta \supseteq \xi$.

(This suffices since finitely many $\beta <_L \xi$ ever act and all $\alpha, f <_L \alpha$, are initialized infinitely often.)

Suppose $\beta \in E$, $\beta \supseteq \xi$, and β opens a gap at stage (s_0, t_0) which is never closed or cancelled, say $\lim_s F(\beta)[s] = wk$. Then $\beta <_L f$. Now by Lemma 3(ii), by Lemma 4, and $q \geq m$, $d(\beta, k) = 0$. Furthermore, $\Phi_i^C = A_i$ and $\Psi_i^C = B_i$, so β is ready to close the gap at almost every stage. To obtain the contradiction it suffices to show that β is accessible infinitely often, indeed for all (s, t), $s > s_0$, $\beta \in AC_s[t]$ iff $\xi \in AC_s[t]$.

If this is false, choose $s > s_0$ and then $t \leq s$ minimal such that $\xi \in AC_s[t]$ and $\beta \notin AC_s[t]$. Now at stage (s, t), some α must have acted where $\xi \subseteq \alpha \subset \beta$, $\alpha^\wedge\langle a\rangle \subseteq \beta$, $b = F_t(\alpha)$, If $b <_\Lambda a$ then the β-gap would have been cancelled, so $a <_\Lambda b$, namely α closes some gap unsuccessfully and moves right. Since β was not cancelled, $a = gk'$, $k' \in \{1, 2\}$. But then $\beta \in E$ and $\alpha^\wedge\langle gk'\rangle \subseteq \beta$ imply $i(\alpha) > i$. However, in this case $d_{t-1}(\alpha, k')[s] = 1$ because at stage (s, t), β is in a gap and $i(\beta) < i(\alpha)$. Thus, α could not have closed its gap at stage (s, t). (This is the whole point of the delay function.) □

7. Expanding the tree of outcomes to gain more information. A disadvantage of the method in §6, is that if $\alpha = f \restriction n$ remains forever in a gap, say $f(n) = wk$, then $f(n)$ does not tell us *which* $i \leq i(\alpha)$ is to blame for α's failure to close the gap. This can be remedied by adding to Λ^{Fin} extra "waiting states" v_i^k, $i < i(\alpha)$, $k \in \{1, 2\}$, such that $F(\alpha)$ lies in the state v_i^k if α is currently in a gap and blames i. This complicates the motion of the α-module and the priority tree, but simplifies the Truth of Outcome Lemma and certain clauses in the construction.

More formally, let

$$\Lambda = \left\{s, g2, w2, \ldots, v_1^2, v_0^2, g1, w1, \ldots, v_1^1, v_0^1, w\right\}$$

with the order $<_\Lambda$ as listed. Define recursive functions i, j: $\Lambda^{<\omega} \to \omega$ as in subsection 6.2. Define the priority tree T by $\lambda \in T$; and $\alpha \in T$ implies $\alpha^\wedge\langle a\rangle \in T$ for

$$a \in \left\{s, g2, w2, v_{i(\alpha)-1}^2, \ldots, v_0^2, g1, w1, v_{i(\alpha)-1}^1, \ldots, v_0^1, w\right\}.$$

In addition to the function $d_t(\alpha, k)$ of (6.15), we have $d_t(\alpha, k, i)$ for each $i < i(\alpha)$ which $= 1$ if $d_t(\alpha, k) = 1$ via some β satisfying (6.15) with $i(\beta) = i$, and which $= 0$ otherwise. Now while α is in a gap, say an A_α-gap so $k = 1$, $F(\alpha) = v_i^1$ where i is minimal such that $d(\alpha, 1, i) = 1$ and $F(\alpha) = w1$ if there is no such i. While α remains in the gap, $F(\alpha)$ can move along the states $\{w1\} \cup \{v_i^1\colon i < i(\alpha)\}$, but departure from $w1$ does not count as a right exit from $w1$ and does not increase the value of $m(\alpha, 1)$. Hence, if α remains in the gap forever, $\lim_s m(\alpha, 1)[s] < \infty$, $\lim_s \mathrm{CO}(\alpha^\wedge\langle g1\rangle)[s] < \infty$ and so after some stage s_0 no $\beta \supseteq \alpha^\wedge\langle g1\rangle$ ever acts. In this case $\lim_s F(\alpha)[s]$ exists because $d(\alpha, 1, i)$ can only change value when some such β acts. In the third clause of the definition (6.12) AC_t, we replace $\alpha^\wedge\langle wk\rangle$ by $\alpha^\wedge\langle a\rangle$ for some $a \in \{wk\} \cup \{v_i^k\colon i < i(\alpha)\}$ where a is the current value of $F(\alpha)$. We also make the obvious modifications in the rest of the construction.

The advantage is that in the Truth of Outcome Lemma (§ 6, Lemma 3) we can replace (iv) by (v) by

$$(7.1) \qquad f(n) = wk \Rightarrow \Phi_i^C \neq A_i \vee \Psi_k^C \neq B_i$$

and

$$(7.2) \qquad f(n) = v_{i'}^k \Rightarrow \Phi_{i'}^C \neq A_{i'} \vee \Psi_{i'}^C \neq B_{i'}.$$

Thus, by analogy with (3.11) we can define $S(\alpha)$ as the set of i such that for some $\beta \subset \alpha$ R_i has already been satisfied by β (if $\beta \subset f$), namely

$$(7.3) \qquad S(\alpha) = \{i: (\exists \beta \subset \alpha)(\exists k)[\beta\,\hat{}\,\langle v_i^k\rangle \supseteq \alpha$$
$$\vee [i = i(\beta) \& [\beta\,\hat{}\,\langle wk\rangle \subseteq \alpha \vee \beta\,\hat{}\,\langle gk\rangle \subseteq \alpha]]]\}.$$

In the construction of subsection 6.5 we can replace (6.16d) and (6.16e) by simply

$$(7.4) \qquad i(\alpha) \notin S(\alpha).$$

Once the reader is familiar with tree constructions, he will usually prefer the extra information offered by the extra nodes of this method over that of §6. It should be viewed as merely a refinement of the latter where the state wk is "spread out" into the states $\{wk\} \cup \{v_i^k: i < i(\alpha)\}$.

8. A framework for $0'''$-priority arguments. The machinery of §6 provides a useful general framework for other $0'''$-constructions such as the Lachlan non-splitting theorem [**1975a**] as will be explained in Slaman and Soare [**ta**]. One begins with a basic module for handling a certain subrequirement $R_{i,j}$. (In general, R_{ij} may also represent the resolution of the conflict between one requirement S_j versus one opposing requirement R_i.) As in subsection 6.2, the basic module is represented as a kind of finite automaton which moves through states $\Lambda_j^{\text{Fin}} = \{s, wj, w(j-1), \ldots, w0, w\}$ beginning in the starting state and satisfying $R_{i,j}$ iff it reaches the final state s. While in state wi, it waits for a certain even t to occur, usually $l(\Phi_i, \Psi_i) > x$ where Φ_i and Ψ_i are p.r. functionals (or perhaps r.e. sets) which are being played by the "opponent". When the event occurs, it is classified as "successful" or "unsuccessful" (analogously to the successful or unsuccessful closing of the gaps in subsection 6.2, according to the situation at the time, and the module moves left to $w(i+1)$ (or s) or right to w (or perhaps to some wi', $i' < i$) according to whether the event is successful or not.

The transition diagram of Figure 6.1 is now expanded to include a rectangle for each $a \in \Lambda_j^{\text{Fin}}$. Above each rectangle wi is also a circle labelled gi and corresponding to the possibility of infinitely many right exits from state wi. The infinitary outcomes are $\Lambda_j^{\text{Inf}} = \{gi: i \leqslant j\}$. The set of all outcomes is $\Lambda_j = \Lambda_j^{\text{Fin}} \cup \Lambda_j^{\text{Inf}}$. To obtain the priority tree t as in subsection 6.3, let

$$\Lambda = \{s, \ldots, g2, g1, w1, g0, w0, w\}$$

with the order $<_\Lambda$ as listed. For $\alpha \in \Lambda^{<\omega}$ define $i(\alpha)$ and $j(\alpha)$ as in subsection 6.3. Define T by $\lambda \in T$; and

$$\alpha \in T \,\&\, j = j(\alpha) \Rightarrow \alpha^\wedge\langle a\rangle \in T \quad \text{for all } a \in \Lambda_j.$$

Now fix $\alpha \in T$, and let $i = i(\alpha)$ and $j = j(\alpha)$. The first approximation to the α-module is simply the basic module for $R_{i,j}$ modified in the obvious way for those $i' \leqslant i(\alpha)$, $i' \in S(\alpha)$ (defined analogously to (3.11) as the set of i' such that for some $\beta \subset \alpha$ if $\beta \subset f$ then β satisfies the overall requirement $R_{i'}$). We can also define the control function $\mathrm{CO}(\gamma)[s]$ as in (6.11) where $m(\gamma, k)$ is now the number of times γ has exited right from state wk. Thus if α exits right from wk finitely often, then at only finitely many stages s can some $\gamma \subseteq \alpha^\wedge\langle gk\rangle$ act, since we shall always require as in (6.16c) that γ cannot exit from the start state w unless $\mathrm{CO}(\gamma)$ exceeds the number of previous such exits.

As in subsection 6.4, the critical part is now to decide when during α's moves, α should allow those $\beta \supseteq \alpha^\wedge\langle a\rangle$ to play. This part is peculiar to each construction and cannot be described generally, but only by the golden rule that if $\beta \subset f$ then α must supply β with the same "environment" which α would need to succeed. Usually the most difficult case is $\beta = \alpha^\wedge\langle gk\rangle$ for some k. In this case, if β's guess is correct then $F(\alpha)$ will pass through the cycle of states $\{w, w_0, w1, \ldots, wk\}$ infinitely often, and α may let β act at any convenient time during the cycle.

For example, in the Lachlan nonsplitting theorem (see Slaman and Soare [**ta**]) we must construct r.e. sets $A >_T C$ such that $\deg(A)$ does not split over $\deg(C)$. The notion of accessibility in subsection 6.5 must be refined to allow (A, γ) stages (and (C, γ) stages) when γ is free to enumerate some element into $A(C)$. The (A, γ) stages are disjoint from the (C, γ) stages. For certain k it is convenient to give $\beta = \alpha^\wedge\langle gk\rangle$ an (A, β) stage when $F(\alpha)$ enters wk and a (C, β) stage when $F(\alpha)$ exits right from wk. For other values of k or other types of action, it might be convenient to allow β to act when $F(\alpha) = w$ and the state last occupied by α was wk. In general α should allow $\beta = \alpha^\wedge\langle gk\rangle$ to act after $\mathrm{CO}(\beta)$ is large, and particularly (1) when the action is convenient from β's point of view (so β's action will not injure α); and (2) when the state of $F(\alpha)$ is favourable from β's point of view (for example when α drops restraint, or in general when β's guess about all $\gamma \subseteq \alpha$ has just appeared correct).

Finally, after allowing $\beta \supseteq \alpha^\wedge\langle gk\rangle$ to act, α must delay (as in subsection 6.4) until β completes any pending action on any overall requirement $R_{e'}$ of higher priority than the requirement R_e upon which α is now waiting. (For example, if $i = i(\alpha)$, and α is now waiting at node we, $e < i$, for $l(\Phi_e, \Psi_e) > x(\alpha)$, we regard α as currently "waiting upon" R_e.) While α is in this delay phase, say $F(\alpha) = wk$, it may be convenient to allow β to be accessible whenever α is accessible, so that β can complete its action. Thus, it is this delay which leads in general to a *tree* of accessible nodes rather than merely a linearly ordered set as in the $0''$-tree construction.

The construction must be arranged so that we can prove the appropriate Leftmost Path Lemma (like Lemma 1 of §6) which asserts that each $\alpha \subset f$ is given the necessary environment to succeed, and the Truth of Outcome Lemma (like Lemma 3 of §6) which asserts that if $\alpha = f \upharpoonright n$ and $f(n) = a$ then the subrequirement corresponding to α and a is indeed satisfied. The other lemmas will be peculiar to the given construction.

Bibliography

K. Ambos-Spies

[1980] *On the structure of the recursively enumerable degrees*, Dissertation, University of Munich.

[1984a] *An extension of the non-diamond theorem in classical and α-recursion theory*, J. Symbolic Logic **49**, 586–607.

[1984b] *On pairs of recursively enumerable degrees*, Trans. Amer. Math. Soc. **383**, 507–531.

[1985] *Anti-mitotic recursively enumerable sets*, Proc. 1st Workshop on Foundations of Theoretical Computer Science GTI, Paderborn, 1982.

[ta] *Automorphism bases for the recursively enumerable degrees.*

[ta] *Generators of the recursively enumerable degrees.*

K. Ambos-Spies, C. G. Jockusch, Jr., R. A. Shore and R. I. Soare

[1984] *An algebraic decomposition of the recursively enumerable degrees and the coincidence of several degree classes with the promptly simple degrees*, Trans. Amer. Math. Soc. **281**, 109–128.

K. Ambos-Spies and P. A. Fejer

[ta] *Degree theoretic splitting properties of recursively enumerable sets.*

M. M. Arslanov

[1968] *Two theorems on recursively enumerable sets*, Algebra and Logic **7**, No. 3, 4–8.

[1981] *On some general theorems about fixed points*, Math. Univ. News, **228**, No. 5, 9–16.

M. M. Arsanov, R. F. Nadirov and V. D. Solovev

[1977] *Completeness criteria for recursively enumerable sets and some general theorems on fixed points*, Math. Univ. News, **179**, No. 4, 3–7.

V. L. Bennison

[1976] *On the computational complexity of recursively enumerable sets*, Ph. D. Dissertation, University of Chicago.

[1980] *Recursively enumerable complexity sequences and measure independence*, J. Symbolic Logic **45**, 417–438.

V. L. Bennison and R. I. Soare

[1978] *Some lowness properties and computational complexity sequences*, Theoret. Comput. Sci. **6**, 233–254.

M. Bickford

[1983] Ph. D. Dissertation, University of Wisconsin.

M. Bickford and C. F. Mills

[ta] *Lowness properties of r.e. sets.*

P. J. Cohen

[1963] *The independence of the continuum hypothesis*. I, Proc. Nat. Acad. Sci. U.S.A. **50**, 1143–1148.

[1966] *Set theory and the continuum hypothesis*, Benjamin, New York.

P. Cohen and C. G. Jockusch, Jr.

[1975] *A lattice property of Post's simple set*, Illinois J. Math. **19**, 450–453.

S. B. Cooper

[1972a] *Sets recursively enumerable in high degrees*, Notices Amer. Math. Soc. **19**, A–20.

[1972b] *Minimal upper bounds for sequences of recursively enumerable degrees*, J. London Math. Soc. **5**, 445–450.

[1972c] *Degrees of unsolvability complementary between recursively enumerable degrees*. I, Ann. Math. Logic **4**, 31–73.

[1972d] *Jump equivalence of the Δ_2^0 hyperhyperimmune sets*, J. Symbolic Logic **37**, 598–600.

[1974a] *On a theorem of C. E. M. Yates*, handwritten notes.

[1974b] *An annotated bibliography for the structure of the degrees below* $\mathbf{0}'$ *with special reference to that of the recursively enumerable degrees*, Recursive Funct. Theory News. **5**, 1–15.

M. Davis

[**1958**] *Computability and unsolvability*, McGraw-Hill, New York.

[**1973**] *Hilbert's tenth problem is unsolvable*, Amer. Math. Monthly **80**, 233–269.

M. David, (editor)

[**1965**] *The undecidable. Basic papers on undecidable propositions, unsolvable problems, and computable functions*, Raven Press, Hewlitt, New York.

A. N. Degtev

[**1971**] *Hypersimple sets with retraceable complements.* Algebra i Logika **10**, 235–246.

J. Dekker

[**1953**] *Two notes on recursively enumerable sets*, Proc. Amer. Math. Soc. **4**, 495–501.

[**1954**] *A theorem on hypersimple sets*, Proc. Amer. Math. Soc. **5**, 791–796.

[**1955**] *Productive sets*, Trans. Amer. Math. Soc. **78**, 129–149.

J. C. E. Dekker and J. Myhill

[**1958a**] *Some theorems on classes of recursively enumerable sets*, Trans. Amer. Math. Soc. **89**, 25–29.

[**1958b**] *Retraceable sets*, Canad. J. Math. **10**, 357–373.

Y. L. Ershov

[**1964**] *Decidability of the elementary theory of relatively complemented distributive lattices and of the theory of filters*, Algebra and Logic **3**, 17–38. (Russian)

[**1968a**] *A hierarchy of sets. Part* I, Algebra and Logic **7**, 24–43.

[**1968b**] *A hierarchy of sets. Part* II, Algebra and Logic **7**, 212–232.

[**1970**] *A hierarchy of sets. Part* III, Algebra and Logic **9**, 20–31.

P. A. Fejer

[**1980**] *The structure of definable subclasses of the recursively enumerable degrees*, Ph. D. Dissertation, University of Chicago.

[**1982**] *Branching degrees above low degrees*, Trans. Amer. Math. Soc. **173**, 157–180.

[**1983**] *The density of the nonbranching degrees*, Ann. Math. Logic **24**, 113–130.

P. A. Fejer and R. I. Soare

[**1981**] *The plus-cupping theorem for the recursively enumerable degrees*, Proc. Logic Year 1979–80 (M. Lerman, J. H. Schmerl and R. I. Soare, eds.), Lecture Notes in Math., vol. 859, Springer-Verlag, Berlin and New York, pp. 49–62.

R. M. Friedberg

[**1957a**] *Two recursively enumerable sets of incomparable degrees of unsolvability*, Proc. Nat. Acad. Sci. U.S.A. **43**, 236–238.

[**1957b**] *The fine structure of degrees of unsolvability of recursively enumerable sets*, Summaries of Cornell University Summer Institute of Symbolic Logic, pp. 404–406.

[**1958**] *Three theorems on recursive enumeration*: I. *Decomposition*, II. *Maximal set*, III. *Enumeration without duplication*, J. Symbolic Logic **23**, 309–316.

[**1975**] *A criterion for completeness of degrees of unsolvability*, J. Symbolic Logic **22**, 159–160.

R. M. Friedberg and H. Rogers, Jr.

[**1959**] *Reducibility and completeness for sets of integers*, Z. Math. Logik Grundlag. Math. **5**, 117–125.

K. Gödel

[**1931**] *Über formal unentscheidbare sätze der Principia Mathematica und verwandter systeme.* I, Monatsch. Math. Phys. **38**, 173–178.

[**1934**] *On undecidable propositions of formal mathematical systems*, Lectures at the Institute for Advanced Study, Princeton, New Jersey (Notes by S. C. Kleene and Barkley Rosser); reprinted in M. Davis [**1965**].

L. Harrington

[**1976**] *On Cooper's proof of a theorem of Yates. Parts* I *and* II, handwritten notes.

[**1978**] *Plus-cupping in the recursively enumerable degrees*, handwritten notes.

[**1980**] *Understanding Lachlan's monster paper*, handwritten notes.

[**1982**] *A gentle approach to priority arguments*, Lecture, A.M.S. Summer Research Institute in Recursion Theory, Cornell University.

L. Harrington and S. Shelah

[**1982a**] *The undecidability of the recursively enumerable degrees* (research announcement), Bull. Amer. Math. Soc. (N.S.) **6**, 79–80.

[**1982b**] *The undecidability of the recursively enumerable degrees*, handwritten notes.

L. Hay

[**1973a**] *The halting problem relativized to complements*, Proc. Amer. Math. Soc. **41**, 583–587.

[**1973b**] *The class of recurisvely enumerable subsets of a recursively enumerable set*, Pacific J. Math. **46**, 167–183.

[**1975**] *Spectra and the halting problem*, Z. Math. Logik Grundlag. Math. **21**, 167–176.

E. Herrmann

[**1978**] *Der Verband der rekursiv aufzählbaren Mengen (Entscheidungsproblem)*, Seminarbericht Nr. 10, Berlin.

[**1981**] *Die Verbandsiegenschaften der rekursiv aufzahlbaren Mengen*, Seminarbericht Nr. 36, Berlin.

[**ta**] *Orbits of hyperhypersimple sets and the lattice of* Σ_3^0 *sets*, J. Symbolic Logic.

[**ta**] *Definable structues in the lattice of recursively enumerable sets.*

[**ta**] *Definable Boolean pairs in the lattice of recursively enumerable sets.*

M. Ingrassia

[**1981**] *P-genericity for recursively enumerable sets*, Doctoral Dissertation, University of Illinois at Urbana-Champaign.

[**ta**] *P-generic r.e. degrees are dense.*

[**ta**] *Restricted notions of P-genericity for r.e. sets.*

C. G. Jockusch, Jr.

[**1972**] *Degrees in which the recursive sets are uniformly recursive*, Canad. J. Math. **24**, 1092–1099.

[**1974**] Π_1^0 *classes and Boolean combinations of recursively enumerable sets*, J. Symbolic Logic **39**, 95–96.

[**1981**] *Three easy construcions of recursively enumerable sets*, Proc. Logic Year 1979–80 (M. Lerman, J. H. Schmerl and R. I. Soare, eds.) Lecture Notes in Math., vol. 859, Springer-Verlag, Berlin, pp. 83–91.

C. G. Jockusch, Jr. and R. A. Shore

[**1983**] *Pseudo jump operators* I: *The R.E. case*, Trans. Amer. Math. Soc. **275**, 599–609.

[**ta**] *REA operators, r.e. degrees, and minimal covers*, these PROCEEDINGS.

C. G. Jockusch, Jr. and R. I. Soare

[**1971**] *A minimal pair of* Π_1^0 *classes*, J. Symbolic Logic **36**, 66–78.

[**1972a**] *Degrees of members of* Π_1^0 *classes*, Pacific J. Math. **40**, 605–616.

[**1972b**] Π_1^0 *classes and degrees of theories*, Trans. Amer. Math. Soc. **173**, 33–56.

[**1973**] *Post's problem and his hypersimple set*, J. Symbolic Logic **38**, 316–326.

S. Kallibekov

[**1971**] *Index sets of degrees of unsolvability*, Algebra i Logika **10**, 316–326. (Russian)

J. L. Kelley

[**1955**] *General topology*, van Nostrand, Princeton, N. J.

S. C. Kleene and E. L. Post

[**1954**] *The upper semi-lattice of degrees of recursive unsolvability*, Ann. of Math. (2) **59**, 379–407.

A. H. Lachlan

[**1965a**] *Some notions of reducibility and productiveness*, Z. Math. Logik Grundlag. Math. **11**, 17–44.

[**1965b**] *On a problem of G. E. Sacks*, Proc. Amer. Math. Soc. **16**, 972–979.

[**1966a**] *A note on universal sets*, J. Symbolic Logic **31**, 573–574.

[**1966b**] *Lower bounds for pairs of recursively enumerable degrees*, Proc. London Math. Soc. **16**, 537–569.

[**1966c**] *The impossibility of finding relative complements for recursively enumerable degrees*, J. Symbolic Logic **31**, 434–454.

[**1967**] *The priority method.* I, Z. Math. Logik Grundlag. Math. **13**, 1–10.

[**1968a**] *Degrees of recursively enumerable sets which have no maximal superset*, J. Symbolic Logic **33**, 431–443.

[**1968b**] *Distributive initial segments of the degrees of unsolvability*, Z. Math. Logik Grundlag. Math. **14**, 457–472.

[**1968c**] *On the lattice of recursively enumerable sets*, Trans. Amer. Math. Soc. **130**, 1–37.

[**1968d**] *The elementary theory of recursively enumerable sets*, Duke Math. J. **35**, 123–146.

[**1968e**] *Complete recursively enumerable sets*, Proc. Amer. Math. Soc. **19**, 99–102.

[1970] *On some games which are relevant to the theory of recursively enumerable sets*, Ann. of Math (2) **91**, 291–310.

[1972] *Embedding nondistributive lattices in the recursively enumerable degrees*, (Conference in Math. Logic, London, 1970), Lecture Notes in Math., vol. 255, Springer-Verlag, New York, pp. 149–177.

[1973a] *The priority method for the construction of recursively enumerable sets*, (Proc. Cambridge Summer School in Logic, 1971), Lecture Notes in Math., vol. 337, Springer-Verlag, Berlina and New York.

[1973b] *Recursively enumerable degrees*, Lectures, Warsaw, April, 1973, handwritten notes.

[1975a] *A recursively enumerable degree which will not split over all lesser ones*, Ann. Math. Logic **9**, 307–365.

[1975b] *Uniform enumeration operations*, J. Symbolic Logic **40**, 401–409.

[1975c] *wtt-complete sets are not necessarily tt-complete*, Proc. Amer. Math. Soc. **48**, 429–434.

[1979] *Bounding minimal pairs*, J. Symbolic Logic **44**, 626–642.

[1980] *Decomposition of recursively enumerable degrees*, Proc. Amer. Math. Soc. **79**, 629–634.

A. H. Lachlan and R. I. Soare

[1980] *Not every finite lattice is embeddable in the recursively enumerable degrees*, Adv. in Math. **37**, 74–82.

R. E. Ladner

[1973a] *Mitotic recursively enumerable sets*, J. Symbolic Logic **38**, 199–211.

[1973b] *A completely mitotic nonrecursive recursively enumerable degree*, Trans. Amer. Math. Soc. **184**, 479–507.

R. E. Ladner and L. P. Sasso

[1975] *The weak truth table degrees of recursively enumerable sets*, Ann. Math. Logic **4**, 429–448.

M. Lerman

[1970a] *Recursive functions modulo co-r-maximal sets*, Trans. Amer. Math. Soc. **148**, 429–444.

[1970b] *Turing degrees and many-one degrees of maximal sets*, J. Symbolic Logic **35**, 29–40.

[1971] *Some theorems on r-maximal sets and major subsets of recursively enumerable sets*, J. Symbolic Logic **36**, 193–215.

[1973] *Admissible ordinals and priority arguments*, (Proc. Cambridge Summer School in Logic, 1971), Lecture Notes in Math., Vol. 337, Springer-Verlag, New York.

[1976] *Congruence relations, filters, ideals and definability in lattices of α-recursively enumerable sets*, J. Symbolic Logic **41**, 405–418.

[1978] *Lattices of α-recursively enumerable sets*, Generalized Recursion Theory. II, (J. E. Fenstad, R. O. Gandy and G. E. Sacks, eds.) North Holland, Amsterdam, 1978, 223–238.

[1977] *Automorphism bases for the semilattice of recursively enumerable degrees*, Notices Amer. Math. Soc. **24**, A-251. Abstract #77T-E10

[1978] *On elementary theories of some lattices of α-recursively enumerble sets*, Ann. Math. Logic **14**, 227–272.

[1980] *The degrees of unsolvability: some recent results*, Recursion Theory: Its Generalizations and Applications, (F. Drake and S. S. Wainer, eds.) London Math. Soc. Lecture Notes Ser. No. 45, Cambridge Univ. Press, Cambridge, England, pp. 140–157.

[1983a] *Degrees of unsolvability*, Perspectives in Mathematical Logic, Omega Series, Springer-Verlag, Berlin and New York.

[1983b] *The structures of recursion theory*, (Proc. First Southeast Asia Conf. in Math Logic), (C. T. Chong and Wicks, eds.), Studies in Logic and Foundations of Math. III, North-Holland and Amsterdam, pp. 77–95.

[ta] *The embedding problem for the r.e. degrees*, these PROCEEDINGS.

M. Lerman and J. Remmel

[ta] *The universal splitting property*. I (Proc. Logic Colloq. '80) (D. Van Dalen, Lascar and Smiley, eds.) Studies in Logic and the Foundations of Math., vol. 108, North-Holland, N. Y., pp. 181–208.

[ta] *The universal splitting property*. II, J. Symbolic Logic.

M. Lerman, R. A. Shore and R. I. Soare

[1978] *r-maximal major subsets*, Israel J. Math. **31**, 1–18.

[1984] *The elementary theory of the recursively enumerable degrees is not $\aleph_0$-categorical*, Adv. in Math.

M. Lerman and R. I. Soare

[1980a] *d-simple sets, small sets, and degree classes*, Pacific J. Math. **87**, 135–155.

[1980b] *A decidable fragment of the elementary theory of the lattice of recursively enumerable sets*, Trans. Amer. Math. Soc. **257**, 1–37.

W. Maass

[1982] *Recursively enumerable generic sets*, J. Symbolic Logic, **47**, 809–823.

[ta] *On the orbits of hyperhypersimple sets*, J. Symbolic logic.

[1983] *Characterization of recursively enumerable sets with supersets effectively isomorphic to all recursively enumerable sets*, Trans. Amer. Math. Soc. **279**, 311–336.

[ta] *Variations on promptly simple sets.*

[ta] *Major subsets and automorphisms of recursively enumerable sets*, this PROCEEDINGS.

W. Maass and S. Homer

[ta] *Oracle dependent properties of the lattice of NP sets.*

W. Maass, R. A. Shore and M. Stob

[1981] *Splitting properties and jump classes*, Israel J. Math. **39**, 210–224.

W. Maass and M. Stob

[ta] *Intervals of the lattice of recursively enumerable sets determined by major subsets.*

S. S. Marchenkov

[1976] *A class of partial sets*, Mat. Zametki **20**, 473–478.

I. Marques

[1973] *Complexity properties of recursively enumerable sets*, Ph. D. Dissertation, University of California, Berkeley.

[1975] *On degrees of unsolvability and complexity properties*, J. Symbolic Logic **40**, 529–540.

D. A. Martin

[1966a] *Completeness, the recursion theorem, and effectively simple sets*, Proc. Amer. Math. Soc. **17**, 838–842.

[1966b] *Classes of r.e. sets and degrees of unsolvability*, Z. Math. Logik Grundlag. Math. **12**, 295–310.

[1966c] *A theorem on hyperhypersimple sets*, J. Symbolic Logic **28**, 273–278.

[1966d] *On a question of G. E. Sacks*, J. Symbolic Logic, **31**, 66–69.

D. A. Martin and M. B. Pour-el

[1970] *Axiomatizable theories with few axiomatizable extensions*, J. Symbolic Logic **35**, 205–209.

Y. Matijasevic

[1970] *Enumerable sets are diophantine*, Dokl. Akad. Nauk SSSR **191**, 279–282. (Russian)

[1971] *Diophantine representation of enumerable predicates*, Izv. Akad. Nauk SSSR Ser. Math. **35**, 3–30. (Russian)

T. G. McLaughlin

[1965] *On a class of complete simple sets*, Canad. Math. Bull. **8**, 33–37.

[1966] *Retraceable sets and recursive permutations*, Proc. Amer. Math. Soc. **17**, 427–429.

Y. T. Medvedev

[1955] *On nonisomorphic recursively enumerable sets*, Dokl. Akad. Nauk SSSR **102**, 211–214. (Russian)

D. Miller

[1981a] *High recursively enumerable degrees and the anti-cupping property*, (Proc. Logic Year 1979–80), (M. Lerman, J. H. Schmerl and R. I. Soare, eds.), Lecture Notes in Math., vol. 859, Springer-Verlag, Berlin and New York, pp. 230–267.

[1981b] *The relationship between the structure and degrees of recursively enumerable sets*, Ph. D. Dissertation, University of Chicago.

D. Miller and J. B. Remmel

[ta] *Effectively nowhere simple sets*, J. Symbolic Logic.

W. Miller and D. A. Martin

[1968] *The degree of hyperimmune sets*, Z. Math. Logik Grundlag. Math. **14**, 159–166.

J. Mohrherr

[1982] *Index sets and truth-table degrees*, Ph. D. Dissertation, University of Illinois at Chicago Circle.

[ta] *A refinement of* low_n *and* $high_n$ *for the r.e. degrees.*

[ta] *Kleene index sets and functional m-degrees*, J. Symbolic Logic.

M. D. Morley and R. I. Soare

[1975] *Boolean algebras, splitting theorems and* Δ_2^0 *sets*, Fund. Math. **90**, 45–52.

P.H. Morris

[1974] *Complexity theoretic properties of recursively enumerable sets*, Ph. D. Dissertation, University of California, Irvine.

A. A. Muchnik

[1956] *On the unsolvability of the problem of reducibility in the theory of algorithms*, Dokl. Akad. Nauk SSSR **108**, 194–197. (Russian)

J. Myhill

[1955] *Creative sets*, Z. Math. Logik Grundlag. Math. **1**, 97–108.

[1956] *The lattice of recursively enumerable sets*, J. Symbolic Logic **21**, 220 (abstract).

P. Odifreddi

[1981] *Strong reducibilities*, Bull. Amer. Math. Soc. (N.S.) **4**, 37–86.

J. C. Owings, Jr.

[1967] *Recursion metarecursion, and inclusion*, J. Symbolic Logic **32**, 173–178.

[1970] *Review of Lachlan* [**1968c, 1968d**] *and Robinson* [**1967a, 1967b**], J. Symbolic Logic **35**, 153–155.

E. L. Post

[1944] *Recursively enumerable sets of integers and their decision problems*, Bull. Amer. Math. Soc. **50**, 285–316.

H. G. Rice

[1953] *Classes of recursively enumerable sets and their decision problems*, Trans. Amer. Math. Soc. **74**, 358–366.

[1956a] *Recursive and recursively enumerable orders*, Trans. Amer. Math. Soc. **83**, 277–300.

[1956b] *On completely recursively enumerable classes and their key arrays*, J. Symbolic Logic **21**, 304–308.

R. W. Robinson

[1966a] *The inclusion lattice and degrees of unsolvability of the recursively enumerable sets*, Ph. D. Dissertation, Cornell University.

[1966b] *Recursively enumerable sets are not contained in any maximal set*, Notices Amer. Math. Soc. **13**, 325. Abstract #632–4.

[1967a] *Simplicity of recursively enumerable sets*, J. Symbolic Logic **32**, 162–172.

[1967b] *Two theorems on hyperhypersimple sets*, Trans. Amer. Math. Soc. **128**, 531–538.

[1968] *A dichotomy of the recursively enumerable sets*, Z. Math. Logic Grundlag. Math. **14**, 339–356.

[1971a] *Interpolation and embedding in the recursively enumerable degrees*, Ann. of Math. (2) **93**, 285–314.

[1971b] *Jump restricted interpolation in the r.e. degrees*, Ann. of Math. (2) **93**, 586–596.

H. Rogers, Jr.

[1959] *Computing degrees of unsolvability*, Math. Ann. **138**, 125–140.

[1967] *Theory of recursive functions and effective computability*, McGraw-Hill, New York.

G. E. Sacks

[1961] *A minimal degree less than* 0′, Bull. Amer. Math. Soc. **67**, 416–419.

[1963a] *Degrees of unsolvability*, Ann. of Math. Studies, No. 55, Princeton Univ. Press, Princeton, N. J; see rev. ed., 1966.

[1963b] *On degrees less than* 0′, Ann. of Math. (2) **77**, 211–231.

[1963c] *Recursive enumerability and the jump operator*, Trans. Amer. Math. Soc. **108**, 223–239.

[1964a] *The recursively enumerable degrees are dense*, Ann. of Math. (2) **80**, 300–312.

[1964b] *A maximal set which is not complete*, Michigan Math. J. **11**, 193–205.

[1964c] *A simple set which is not effectively simple*, Proc. Amer. Math. Soc. **15**, 51–55.

[1966] *Degrees of unsolvability*, rev. ed., Ann. of Math. Studies, No. 55, Princeton Univ. Press, Princeton, N. J.

[1967] *On a theorem of Lachlan and Martin*, Proc. Amer. Math. Soc. **18**, 140–141.

[1977] *RE sets higher up*, Logic, Foundations of Mathematics and Computability Theory, (D. Reidel, ed.), Dordrecht, Holland, pp. 173–194.

L. P. Sasso

[1974] *Deficiency sets and bounded information reducibilities*, Trans. Amer. Math. Soc. **200**, 267–290.

S. Schwarz

[**1982**] *Index sets of recursively enumerable sets, quotient lattices, and recursive linear orderings*, Ph. D. Dissertation, Univeesity of Chicago.

[**1984**] *The quotient semilattice of the recursively enumerable degrees modulo the cappable degrees*, Trans. Amer. Math. Soc. **283**, 315–328.

[**ta**] *Index sets related to prompt simplicity*.

[**ta**] *Index sets related to the high-low hierarchy*.

J. R. Shoenfield

[**1957**] *Quasicreative sets*, Proc. Amer. Math. Soc. **8**, 964–967.

[**1958**] *The class of recursive functions*, Proc. Amer. Math. Soc. **9**, 690–692.

[**1959**] *On degrees of unsolvability*, Ann. of Math. (2) **69**, 644–653.

[**1960**] *Degrees of models*, J. Symbolic Logic **25**, 233–237.

[**1961**] *Undecidable and creative theories*, Fundamenta Mathematicae **49**, 171–179.

[**1965**] *Application of model theory to degrees of unsolvability*, Proc. Sympos. Theory of Models (J. W. Addison, L. Henkin and A. Tarski, eds.), North-Holland, Amsterdam, pp. 359–363.

[**1966**] *A theorem on minimal degrees*, J. Symbolic Logic **31**, 539–544.

[**1971**] *Degrees of unsolvability*, North-Holland, Amsterdam.

[**1975**] *The decision problem for recursively enumerable degrees*, Bull. Amer. Math. Soc. **81**, 973–977.

[**1976**] *Degrees of classes of r.e. sets*, J. Symbolic Logic **41**, 695–696.

J. R. Schoenfield and R. I. Soare

[**1978**] *The generalized diamond theorem*, Recursive Funct. Theory News. **19** Abstract #219

R. A. Shore

[**1977**] *Determining automorphisms of the recursively enumerable sets*, Proc. Amer. Math. Soc. **65**, 318–326.

[**1978**] *Nowhere simple sets and the lattice of recursively enumerable sets*, J. Symbolic Logic **43**, 322–330.

[**1980**] *$L^*(K)$ and other lattices of the recursively enumerable sets*, Proc. Amer. Math. Soc. **80**, 143–146.

[**1982**] *Finitely generated codings and the degrees r.e. in a degree* **d**, Proc. Amer. Math. Soc. **84**, 256–263.

[**1981**] *The theory of the degrees below* **0′**, J. London Math. Soc. **24**, 1–14.

S. G. Simpson

[**1977**] *Degrees of unsolvability: A survey of results*, Handbook of Mathematical Logic (J. Barwise, ed.), North-Holland, Amsterdam.

T. Slaman

[**ta**] *The density of the nonbranching degrees*.

T. Slaman and R. I. Soare

[**ta**] *The Lachlan nonsplitting theorem and the* $0'''$ *priority method*.

R. M. Smullyan

[**1964**] *Effectively simple sets*, Proc. Amer. Math. Soc. **15**, 893–895.

R. I. Soare

[**1972**] *The Friedberg-Muchnik theorem re-examined*, Canad. J. Math. **24**, 1070–1078.

[**1974a**] *Automorphisms of the lattice of recursively enumerable sets*, Bull. Amer. Math. Soc. **80**, 53–58.

[**1974b**] *Automorphisms of the lattice of recursively enumerable sets*. Part I: *Maximal sets*, Ann. of Math. (2) **100**, 80–120.

[**1976**] *The infinite injury priority method*, J. Symbolic Logic **41**, 513–530.

[**1977**] *Computational complexity, speedable and levelable sets*, J. Symbolic Logic **42**, 545–563.

[**1978a**] *Recursive enumerability*, Proc. Internat. Congress Math. (1978), pp. 275–280.

[**1978b**] *Recursively enumerable sets and degrees*, Bull. Amer. Mth. Soc. **84**, 1149–1181.

[**1980a**] *Fundamental methods for constructing recursively enumerable degrees*, Recursion Theory: Its Generalisations and Applications (Proc. Logic Colloq. '79, Leeds, August 1979) (F. R. Drake and S. S. Wainer, eds.) Lecture Notes, vol. 45, Cambridge University Press, pp. 1–51.

[**1980b**] *Constructions in the recursively enumerable degrees*, C. I. M. E. Conf. on Recursion Theory and Computational Complexity (Bressanone, Italy, June 1979).

[**1982a**] *Automorphisms of the lattice of recursively enumerable sets*. Part II: *Low sets*, Ann. Math. Logic **22**, 69–107.

[**1982b**] *Computational complexity of recursively enumerable sets*, Inform. and Control **52**, 8–18.

[**ta**] *Recursively enumerable sets and degrees*, Omega Series in Logic, Springer-Verlag, Berlin and New York.

R. I. Soare and M. Stob

[**1982**] *Relative recursive enumerability*, Proc. Herbrand Sympos. Logic Colloq. '81, (J. Stern, ed.) North-Holland, Amsterdam, pp. 299–324.

C. Spector

[**1956**] *On degrees of recursive unsolvability*, Ann. of Math. (2) **64**, 581–592.

M. Stob

[**1979**] *The structure and elementary theory of the recursively enumerable sets*, Ph. D. Thesis, University of Chicago.

[**1982a**] *Invariance of properties under automorphisms of the lattice of recursively enumerable sets*, Pacific J. Math. **100**, 445–471.

[**1982b**] *Index sets and degrees of unsolvability*, J. Symbolic Logic **47**, 241–248.

[**1983**] *wtt-degrees and T-degrees of recursively enumerable sets*, J. Symbolic Logic **48**, 921–930.

[**ta**] *Major subsets and the latice of recursively enumerable sets*, these PROCEEDINGS.

A. Tarski, A. Mostowski and R. M. Robinson

[**1953**] *Undecidable theories*, North-Holland, Amsterdam.

S. Tennenbaum

[**1961**] *Degrees of unsolvability and the rate of growth of functions*, Notices Amer. Math. Soc. **8**, 608.

[**1962**] *Degrees of unsolvability and the rate of growth of functions*, Proc. Sympos. Math. Theory of Automata, Microwave Res. Inst. Sympos. Ser., vol. 12, Polytechnic Press, Brooklyn, New York.

S. K. Thomason

[**1971**] *Sublattices of the recursively enumerable degrees*, Z. Math. Logik Grundlag. Math. **17**, 272–280.

R. E. Tulloss

[**1971**] *Some complexities of simplicity: concerning grades of simplicity of recursively enumerable sets*, Ph. D. Dissertation, Univ. of California, Berkeley.

A. M. Turing

[**1936**] *On computable numbers, with an application to the Entscheidungs problem*, Proc. London Math. Soc. **42**, 230–265; **43**, 544–546.

V. A. Uspenskiĭ

[**1957**] *Some notes on recursively enumerable sets*, Z. Math. Logik Grundlag. Math. **3**, 157–170; English transl., Amer. Math. Soc. Transl. **23**, 80–101.

L. Welch

[**1981**] *A hierarchy of families of recursively enumerable degrees and a theorem on bounding minimal pairs*, Ph. D. Dissertation, University of Illinois at Urbana-Champaign.

C. E. M. Yates

[**1962**] *Recursively enumerable sets and retracing functions*, Z. Math. Logik Grundlag. Math. **8**, 331–345.

[**1965**] *Three theorems on the degrees of recursively enumerable sets*, Duke Math. J. **32**, 461–468.

[**1966a**] *A minimal pair of recursively enumerabledegrees*, J. Symbolic Logic **31**, 159–168.

[**1966b**] *On the degrees of index sets*, Trans. Amer. Math. Soc. **121**, 309–328.

[**1967**] *Recursively enumerable degrees and the degrees less than* $0'$, Sets, Models, and Recursion Theory, North-Holland, Amsterdam, pp. 264–271.

[**1969**] *On the degrees of index sets*. II, Trans. Amer. Math. Soc. **135**, 249–266.

[**1970**] *Initial segments of the degrees of unsolvability*. Part I, Mathematical Logic and the Foundations of Set Theory (Jerusalem), North-Holland, Amsterdam, pp. 63–83.

[**1972**] *Initial segments and implications for the structure of degrees* (Conf. in Math. Logic, London, 1970), Lecture Notes in Math., vol. 255, Springer-Verlag, Berlin and New York, pp. 305–335.

[**1974**] *Prioric games and minimal degrees below* 0′, Fund. Math **82**, 217–237.

[**1976**] *Banach-Mazur games, comeager sets, and degrees of unsolvability*, Math. Proc. Cambridge Philos. Soc.

DEPARTMENT OF MATHEMATICS, UNIVERSITY OF CHICAGO, CHICAGO, ILLINOIS 60617

Proceedings of Symposia in Pure Mathematics
Volume **42**, 1985

Major Subsets and the Lattice of Recursively Enumerable Sets

MICHAEL STOB[1]

Abstract. Maass and Stob have recently shown that the intervals of the lattice $\mathscr{E}$ of recursively enumerable sets determined by major subsets are isomorphic. The importance of this theorem for the study of the structure of $\mathscr{E}$ is discussed. A construction for the requirement of the theorem is described in detail in order to emphasize the similarities to and differences from other automorphism results. The two crucial properties of major subsets are isolated.

Introduction. If X is any subset of the set N of natural numbers, let $\mathscr{E}(X)$ denote the lattice formed by the sets $\{W \cap X \mid W \text{ r.e.}\}$ under inclusion. Let $\mathscr{E}^*(X)$ denote $\mathscr{E}(X)$ modulo the ideal of finite subsets of X. Let $\mathscr{E}(\mathscr{E}^*)$ abbreviate $\mathscr{E}(N)$ $(\mathscr{E}^*(N))$. Further, if $A \in \mathscr{E}(X)$, let $A^* \in \mathscr{E}^*(X)$ denote the equivalence class of A and $A =^* B$ $(A \subseteq^* B)$ denote $A^* = B^*$ $(A^* \subseteq B^*)$.

DEFINITION (LACHLAN). If $A, B \in \mathscr{E}$, then B is a *major subset* of A $(B \subset_m A)$ if $B \subseteq A$, $A - B$ is infinite, and for all $W \in \mathscr{E}$, if $A \cup W = N$ then $B \cup W =^* N$.

Major subsets were introduced by Lachlan [**2**] in the course of his study of the lattice structure of $\mathscr{E}$. Since their introduction, they have proved to be of considerable importance in studying the lattice structure of $\mathscr{E}$, the elementary theory of $\mathscr{E}$, and the relationship of an r.e. set to its Turing degree.

Recently, Maass and Stob [**11**] established the following result.

THEOREM. *If* $B \subset_m A$ *and* $\hat{B} \subset_M \hat{A}$, *then* $\mathscr{E}^*(A - B) \cong \mathscr{E}^*(\hat{A} - \hat{B})$.

The proof of this Theorem is a modification and considerable extension of the automorphism machinery of Soare. A description and simplification of that machinery as well as a unified presentation of the various automorphism results

1980 *Mathematics Subject Classification*. Primary 03D25.
[1] The research was partially supported by NSF Grants MCS 80-02937 and MCS 82-00032.

may be found in the paper by Maass [**8**]. In this paper then we will not give the proof to the Theorem. Instead, we will first (in §1) give some background information on the importance of major subsets in the study of the structure of $\mathscr{E}^*$ as well as on the significance of this theorem and other recent research in that program. We will also mention some fruitful areas for further study. Then (in §2) we will discuss in some detail the properties of major subsets which play a crucial role in the proof of the theorem and show how these properties combine to allow us to meet a simplified version of one requirement of the Theorem. This paper is intended to be a companion to that of Maass; we hope it will provide some insight into automorphism constructions in general, besides providing a useful survey of major subsets.

1. Major subsets and the study of $\mathscr{E}$. A major program in the study of the structure of $\mathscr{E}^*$ has been the study of those lattices which arise as $\mathscr{E}^*(A - B)$ for r.e. sets $B \subseteq A$, particularly in the case $A = N$. It is easy to see that $\mathscr{E}^*(A - B)$ is isomorphic to the interval determined by B and A. Thus the classification of lattices which arise in this way is of importance for the decision problem for $\mathscr{E}^*$ and for studying the automorphism types of elements of $\mathscr{E}^*$. It is this program of characterizing isomorphism types of $\mathscr{E}^*(\overline{B})$ for which Lachlan in [**2**] first used major subsets and for which major subsets are of crucial importance.

Lachlan gave a complete characterization of the Boolean algebras $\mathscr{B}$ for which $\mathscr{B} \cong \mathscr{E}^*(\overline{B})$ for some r.e. B. (In addition, he showed that the sets B for which $\mathscr{E}^*(\overline{B})$ is a Boolean algebra are precisely the hyperhypersimple sets.) He then used major subsets to give examples of sets B for which $\mathscr{E}^*(\overline{B})$ is not a Boolean algebra. By our Theorem, there is a lattice $\mathscr{M}^*$ such that $\mathscr{M}^* \cong \mathscr{E}^*(A - B)$ whenever $B \subset_m A$. Lachlan essentially noticed the following properties of $\mathscr{M}^*$.

FACT 1. $\mathscr{M}^*$ is not a Boolean algebra.

FACT 2. $\mathscr{M}^*$ is densely ordered. (A proof of this is found in §2.)

FACT 3. Every nontrivial interval of $\mathscr{M}^*$ is again isomorphic to $\mathscr{M}^*$ (for if $B \subseteq C \subseteq D \subseteq A$ and $B \subset_m A$, then $C =^* D$ or $C \subset_m D$).

FACT 4. The sublattice of $\mathscr{M}^*$ of complemented elements is the countable atomless Boolean algebra.

$\mathscr{M}^*$ is therefore an example of a lattice which is not a Boolean algebra but for which $\mathscr{M}^* \cong \mathscr{E}^*(\overline{B})$ for some B. Let $\hat{B} \subset_m A$, let f be a 1-1 recursive function from N onto A, and let $B = f^{-1}(\hat{B})$. Then $\mathscr{E}^*(\overline{B}) \cong \mathscr{E}^*(A - \hat{B}) \cong \mathscr{M}^*$.

Lachlan then used major subsets to give examples of sets B for which $\mathscr{E}^*(\overline{B})$ is not the 2-element Bolean algebra and does not have *any* nontrivial complemented elements. Such sets B are called r-maximal since no recursive set splits $\overline{B}$ nontrivially.

Lachlan noticed that if B is r-maximal, $B \subset_m A$ for every coinfinite superset A. (Thus, if B is r-maximal, every *proper* initial segment of $\mathscr{E}^*(\overline{B})$ is isomorphic to $\mathscr{M}^*$.) Lachlan gave two different examples of r-maximal sets. Let M be a maximal set. (M is maximal if M is not split nontrivially by any r.e. set or, equivalently,

$\mathcal{E}^*(\overline{M})$ is the two-element Boolean algebra.) Then Lachlan showed that any major subset B of M is r-maximal. In fact, using our theorem, it is possible to describe the isomorphism type of $\mathcal{E}^*(\overline{B})$ for all r-maximal sets which arise in this way. $\mathcal{E}^*(\overline{B})$ is simply formed by adjoining to $\mathcal{M}^*$ one more element which is greater than all the elements of $\mathcal{M}^*$. (If $B \subset_m M$ and M is maximal, then if $B \subseteq C$, either $C \subseteq^* M$ or $C =^* N$.) We will call this lattice $\mathcal{M}^{*+}$. Although there is only one isomorphism type among lattices $\mathcal{E}^*(\overline{B})$ for major subsets B of maximal sets, it turns out that there are many different 1-types realized by such sets B and so such sets fall into many distinct orbits under automorphisms of $\mathcal{E}^*$. This stands in opposition to Soare's first use of the automorphism machinery [**12**] where he showed that the maximal sets, those for which $\mathcal{E}^*(\overline{A}) \equiv 2$, form an orbit in $\mathcal{E}^*$. Since Soare's theorem, much evidence has been adduced to support the following

CONJECTURE 1.1. *The sets A such that $\mathcal{E}^*(\overline{A}) \cong \mathcal{L}$ form an orbit under the group of automorphisms of $\mathcal{E}^*$ if and only if $\mathcal{L}$ is a finite Boolean algebra.*

This evidence may be found in Maass, Shore, and Stob [**10**] where it is shown that if $\mathcal{L}$ is not a Boolean algebra and if for every high r.e. degree **a**, there is a set A such that $\mathcal{E}^*(\overline{A}) \cong \mathcal{L}$ and $\deg(A) = \mathbf{a}$, then $\mathcal{E}^*(\overline{A}) \cong \mathcal{L}$ does not determine the orbit of A. The elementary property distinguishing such sets A with $\mathcal{E}^*(\overline{A}) \cong \mathcal{L}$ is the splitting property.

DEFINITION 1.2. An r.e. set A has the *weak splitting property* if for every r.e. B there are sets B_0 and B_1 such that $B_0 \cup B_1 = B$, $B_0 \cap B_1 = \varnothing$, $B_0 \subseteq A$, and if B is nonrecursive then B_0 and B_1 are nonrecursive.

Maass, Shore and Stob show that there are high degrees **a** which contain no sets with the weak splitting property other than hyperhypersimple sets. However, they also show that for every lattice $\mathcal{L}$ such that $\mathcal{L} \cong \mathcal{E}^*(\overline{A})$ for some r.e. set A, A can be chosen to have the weak splitting property. Thus for the lattice $\mathcal{M}^{*+}$, since major subsets of maximal sets exist in every high degree, there are sets A with $\mathcal{E}^*(\overline{A}) \cong \mathcal{M}^{*+}$ which have the splitting property and some which do not.

A second example of an r-maximal set due to Lachlan is that of an r-maximal set A with no maximal superset. For such a set A, $\mathcal{E}^*(\overline{A})$ is dense, has no complemented elements, and has every initial segment isomorphic to $\mathcal{M}^*$. Robinson had previously given an example of such a set, and, as Lerman and Soare [7] notices, a genuinely different example. Lachlan's and Robinson's examples are such that the corresponding lattices $\mathcal{E}^*(\overline{A})$ are not isomorphic for, in Lachlan's example, $\mathcal{E}^*(\overline{A})$ satisfies the sentence $(\exists x)[x \neq 0 \wedge x \neq 1 \Rightarrow (\forall y)[y \cap x = 0 \Rightarrow y = 0]]$ and Robinson's example does not. It is not known how many isomorphism types there are among the lattices $\mathcal{E}^*(\overline{A})$ for $\overline{A}$ r-maximal with no maximal superset.

The theorem of Lachlan that most clearly indicates the importance of major subsets for the program of studying intervals of $\mathcal{E}^*$ is one which he only added in proof to [**2**].

THEOREM 1.3. *If $B \subset A$ and $B \cup \bar{A}$ is not r.e., then there is an r.e. set C such that $B \subseteq C \subset_m A$.*

It is easy to see that this theorem implies the following fact about $\mathscr{E}^*$.

FACT 5. Every interval of $\mathscr{E}^*$ which is not a Boolean algebra has a subinterval isomorphic to $\mathscr{M}^*$.

Obviously, Fact 5 means that any classification of the isomorphism types of intervals of $\mathscr{E}^*$, other than those which are Boolean algebras, implies an understanding of the structure of $\mathscr{M}^*$.

In view of Fact 5 and its implications for the study of the lattice structure, and so the decision problem, of $\mathscr{E}^*$, two different programs have been proposed. On the one hand, we could restrict our attention to the structure and decision problem for $\mathscr{M}^*$. After all, if the theory of $\mathscr{M}^*$ were proved undecidable so would be the theory of $\mathscr{E}^*$. In favor of this program is the large amount of information we already have about $\mathscr{M}^*$ and the power of the techniques for working with major subsets that have been developed. An exhaustive catalog of properties of $\mathscr{M}^*$ may be found in Herrmann [**1**]. (Herrmann and Lerman, independently, are responsible for conjecturing that our theorem is true.) A test question in this program is whether infinitely many different 1-types are realized in $\mathscr{M}^*$.

A program which might be viewed as diametrically opposed to the first was that suggested by Lerman. In [**4**], Lerman noticed that the relation $A \sim B$ iff $A \cap B \subset_m A \cup B$ is an equivalence relation on $\mathscr{E}^*$. Thus $\mathscr{E}^*/\sim$ is again a distributive lattice with 0 and 1. What is important is that in passing from $\mathscr{E}^*$ to $\mathscr{E}^*/\sim$, intervals determined by major subsets are collapsed. Thus there is a chance that the structure of $\mathscr{E}^*/\sim$ is simpler than that of $\mathscr{E}^*$; in this program Fact 5 might be viewed as an obstruction to understanding $\mathscr{E}^*$. Again, a decision procedure, undecidability theorem, characterization of filters, or description of 1-types for $\mathscr{E}^*/\sim$ would have direct implications for the study of $\mathscr{E}^*$.

While our Theorem and the preceding discussion emphasized the similarity of structure among major subsets, elementary differences between major subsets have also been exploited to gain further information about $\mathscr{E}^*$. One elementary type of major subset that has been used in this way is the r-maximal major subset of Lerman, Shore and Soare [**7**]. B is an r-maximal major subset of A if $B \subset_m A$ and no recursive set splits $A - B$ nontrivially. Lerman, Shore and Soare showed that there are r.e. sets A such that $\mathscr{E}^*(\bar{A})$ is the countable atomless Boolean algebra and some such have r-maximal major subsets while others do not. This, of course, rules out the countable atomless Boolean algebra as a counterexample to Conjecture 1.1. A second elementary type of major subset is the small major subset of Lachlan which played a crucial role in his decision procedure for the $\forall\exists$-theory of $\mathscr{E}^*$.

DEFINITION 1.4. If $B \subseteq A$, then B is small in A ($B \subset_s A$) if, for every pair of r.e. sets U, V,

$$U \supset V \cap (A - B) \Rightarrow U \cup (V - A) \quad \text{is r.e.} \tag{1.1}$$

An r.e. set B is small if $B \subset_s A$ for some coinfinite r.e. set A.

The name is appropriate; for instance, if $C \subset B \subset_s A$ or $C \subset_s B \subseteq A$ then $C \subset_s A$, and if $C \subset_s A$ and A is nonrecursive there is a set B such that $C \subseteq B \subseteq A$ but C is not small in B or B in A. Lachlan [**3**] showed that every nonrecursive r.e. set has a small major subset.

An illuminating way of viewing one of the differences between the r-maximal and small major subsets is due to Maass. Maass was interested in attempting to represent Boolean algebras with distinguished subalgebras in $\mathscr{E}^*$. For any X, the complemented elements of $\mathscr{E}^*(X)$ form a Boolean algebra, $\mathscr{R}^*(X)$. If $X = A - B$ and $B \subset_m A$, Fact 4 says that $\mathscr{R}^*(X)$ is the countable atomless Boolean algebra. Maass [**9**] noticed that there was a natural homomorphism $H_{A,B}$: $\mathscr{R}^*(\bar{A}) \to \mathscr{R}^*(A - B)$ whenever $B \subset_m A$, defined by

$$(1.2) \qquad (R \cap \bar{A})^* \to (R \cap (A - B))^*,$$

where R ranges over recursive sets. $B \subset_m A$ guarantees that the map defined by (1.2) is well defined, i.e., independent of the choice of R. Thus the range of $H_{A,B}$ is a subalgebra of $\mathscr{R}^*(A - B)$ and is a quotient algebra of $\mathscr{R}^*(\bar{A})$.

It turns out that if $B \subset_{sm} A$, then $H_{A,B}$ is 1-1, but if $B \subset_{rm} A$, then $H_{A,B}$ is a trivial map; the range of $H_{A,B}$ is the two-element Boolean algebra. Conceivably, the maps $H_{A,B}$ might provide enough complexity to conclude that the theory of $\mathscr{E}^*$ is undecidable. Maass has shown at least that two distinct embeddings of the countable atomless Boolean algebra into itself can be represented in this way. Maass and Stob [**11**] have given a complete characterization of the factor lattices of $\mathscr{E}^*(\bar{A})$ which can arise from such maps in terms of an effective presentation of $\mathscr{E}^*(\bar{A})$.

2. One requirement. The automorphism machinery of Soare is a general method for constructing isomorphisms from $\mathscr{E}^*(X)$ to $\mathscr{E}^*(\hat{X})$ for sets X and $\hat{X}$ satisfying certain minimal conditions. In all the variations of the construction, four recursive arrays of r.e. sets $\{U_i\}_{i\in N}$, $\{\hat{V}_i\}_{i\in N}$, $\{\hat{U}_i\}_{i\in N}$ and $\{V_i\}_{i\in N}$ are enumerated. The "forth" part of the intended isomorphism is the map which sends $U_i \cap X$ to $\hat{U}_i \cap \hat{X}$ for every i. (Thus for every r.e. set W there must be an i such that $W \cap X = U_i \cap X$.) The "back" part of the isomorphism is the map sending $V_i \cap \hat{X}$ to $\hat{V}_i \cap X$ for ever i. The requirement that these maps actually define an isomorphism is the following:

(2.1) For every Boolean combination B of finitely many of the sets U_i and $\hat{V}_i$, infinitely many elements of X are in B if and only if infinitely many elements of $\hat{X}$ are in the corresponding Boolean combination $\hat{B}$ of the sets $\hat{U}_i$ and V_i.

We will consider just one pair of r.e. sets U and $\hat{U}$ to see what problems arise in attempting to meet (2.1). In this case, (2.1) reduces to

$$(2.2) \qquad U \cap X \text{ infinite} \quad \text{iff} \quad \hat{U} \cap \hat{X} \text{ is infinite},$$

$$(2.3) \qquad \bar{U} \cap X \text{ infinite} \quad \text{iff} \quad \bar{\hat{U}} \cap \hat{X} \text{ is infinite}.$$

We will suppose that an enumeration of U is given; in many constructions $\{U_i\}_{i \in N}$ is simply chosen to be any fixed array containing all the r.e. sets. (In Soare's original use of the method, however, $\{U_i\}_{i \in N}$ was specially chosen so that the simultaneous enumeration of the sets $\{U_i\}_{i \in N}$ had certain properties.)

Surely one of the major obstacles in meeting (2.2) is the complexity of the hypothesis

$$U \cap X \text{ is infinite.} \tag{2.4}$$

If X is r.e., $\{e | W_e \cap X \text{ is infinite}\}$ is Π_2. In this case there is a relatively straightforward strategy which suffices to meet (2.2) uniformly in an index for U. One simply enumerates another element in $\hat{U} \cap \hat{X}$ each time a new element appears in $U \cap X$. Suppose X is not r.e., however, say $X = \bar{A}$ for some r.e. set A. In this case $\{e | W_e \cap X \text{ is infinite}\}$ is Π_3 and, as such, the hypothesis (2.4) is extremely difficult to use in a recursive construction. The difficulty with the straightforward strategy described above for the case X r.e. is, of course, that $U \cap X$ may "look" infinite by virtue of there being many stages s such that $U_s \cap X_s$ has large cardinality, where U_s and X_s are the recursive approximations to U and X at stage s. It is this obstacle, for instance, that Maass faced when he showed **[8]** that if A is semilow$_{1.5}$, then $\mathscr{E}^*(\bar{A}) \cong \mathscr{E}^*$. However, it is precisely the hypothesis on A that allowed Maass to meet requirement (2.2); $\{e | W_e \cap X \text{ is infinite}\}$ is Π_2 if $X = \bar{A}$ and A is semilow$_{1.5}$. Thus Maass could "count" the elements of $U \cap X$ with an r.e. set W, i.e. W satisfies $U \cap X$ infinite iff W infinite. Maass could then use the same strategy as before.

This strategy is not available, however, if $X = A - B$ and $B \subset_m A$. In fact, Maass has shown that if $B \subset_m A$, $\{e | W_e \cap (A - B) \text{ is infinite}\} \equiv_T \mathbf{0}'''$, so we cannot hope that the hypothesis (2.4) can be verified in a Π_2 way. However, a more delicate analysis of the straightforward strategy and hypothesis (2.4) leads to the fundamental observation enabling us to meet (2.2). Hypothesis (2.4) can be viewed as the conjunction of the infinitely many hypotheses

$$|U \cap X| \geqslant k. \tag{$(2.5)_k$}$$

If X, and hence $U \cap X$, is r.e., the enumeration of $U \cap X$ provides us at each stage s with a guess as to which of the hypotheses $(2.5)_k$ are correct. Namely we guess that $|U \cap X| \geqslant k$ just if $|U_s \cap X_s| \geqslant k$, and our strategy is to ensure that $|\hat{U} \cap \hat{X}| \geqslant k$ if we guess that $(2.5)_k$ is correct. The relevant properties of the guesses are that we guess correctly about $(2.5)_k$ at almost every stage s and, of course, for the purpose of doing a recursive construction, our guesses are uniform in k (and U). Notice that each hypothesis $(2.5)_k$ is a Σ_1-predicate of U. Now for the case $X = A - B$, $B \subset_m A$, $\hat{X} = \hat{A} - \hat{B}$, $\hat{B} \subset_m \hat{A}$, we have only that the hypotheses $(2.5)_k$ are Σ_2 in U. It is easy to see that it is not possible to guess at $(2.5)_k$ uniformly in k and U and still guess correctly at almost every stage. What we do instead to meet (2.2) is essentially to associate the guessing at $(2.5)_k$ with the elements x of $\hat{A}$ rather than the stages s of the construction. The crucial lemma is the following

LEMMA 2.1 (Π_2 GUESSING LEMMA FOR MAJOR SUBSETS). *Let $B \subset_m A$ and let P be a Π_2 predicate. Then, uniformly in an index for P, we can assign to every $x \in A$ a guess at the truth of P such that almost every $x \in A - B$ guesses correctly. Furthermore, if $\{A_s\}_{s \in N}$ is a fixed recursive enumeration of A, the guesses can be assigned to each x at the stage s such that $x \in A_s - A_{s-1}$.*

PROOF. Let $\{A_s\}_{s \in N}$ be a fixed enumeration of A. Uniformly in P there is a recursive function f such that $f(s) \leqslant f(s+1)$ for all s and such that $\lim_s f(s) < \infty$ iff P is false. Fix P and therefore such an f. For each x in A let

$$s(x) = (\mu s)[x \in A_s - A_{s-1}].$$

Assign to x the guess that P is true iff $x < f(s(x))$. Obviously if P is false, $\lim_s f(x) < \infty$. So, for almost every $x \in A$, $x > f(s(x))$ and each such x guesses that P is false. On the other hand, if P is true, $\lim_s f(s) = \infty$. Let

$$W = \{x | (\exists s)[x \in \bar{A}_s \text{ and } x < f(s)]\}.$$

Obviously $W \supseteq \bar{A}$ and so, since $B \subset_m A$, $W^* \supseteq A - B$. For any $x \in W \cap (A - B)$, $x < f(s(x))$ since f is increasing and there is a stage s such that $x < f(s)$ while x is still in $\bar{A}$. Thus each $x \in W \cap (A - B)$ is assigned a guess that P is true. $\square$

The lemma essentially solves the problem of $U \cap (A - B)$ becoming too large. The elements of $\hat{A}$ are assigned guesses at each of the sentences $(2.5)_k$,

$$\text{“}|U \cap (A - B)| \geqslant k\text{”.}$$

Only those elements $x \in \hat{A}$ which guess that $|U \cap (A - B)|$ is large can be used to make $\hat{U} \cap (\hat{A} - \hat{B})$ large. Formally, for each k we need only arrange that only finitely many of the elements of $\hat{A}$ which guess that

$$\text{“}|U \cap (A - B)| \geqslant k\text{”}$$

is false will be enumerated in $\hat{U}$. For instance, we could arrange that at any stage s, only the k least elements of $\hat{U}_s \cap (\hat{A}_s - \hat{B}_s)$ guess that $|U \cap (A - B)| < k$. This imposes a finitary restraint on finding the $(k+1)$st element of $\hat{U}$ if $U \cap (A - B)$ really is infinite because, in this case, almost every $x \in \hat{A} - \hat{B}$ guesses that $|U \cap (A - B)| \geqslant k$. An enumeration procedure for $\hat{U}$ based on this lemma, which suffices to meet (2.2), is the following. If $x \in \hat{A}_s - \hat{B}_s$, then enumerate x in $\hat{U}$ at stage s if and only if there is an integer j such that there are fewer than j elements of $\hat{U}_s \cap (\hat{A}_s - \hat{B}_s)$ which are $\leqslant x$ but x guesses $|U \cap (A - B)| \geqslant j$.

While it is straightforward to check that the above construction suffices to meet (2.2), further complications occur when attempting to meet (2.3). Of course, (2.2) and (2.3) conflict; an element enumerated in $\hat{U}$ for the purpose of meeting (2.2) cannot be used to make $\bar{\hat{U}} \cap (\hat{A} - \hat{B})$ large. What could happen is this. Suppose we are trying to insure that $|\bar{\hat{U}} \cap (\hat{A} - \hat{B})| \geqslant j$. Since elements of $\hat{A}$ can leave $\hat{A} - \hat{B}$ by being enumerated in $\hat{B}$, the element we designate for this purpose may be so enumerated. There is nothing to prevent this from happening infinitely often if our choice for the witness to $|\bar{\hat{U}} \cap (\hat{A} - \hat{B})| \geqslant j$ is injudicious. This would result in our failing to meet $|\overline{\hat{U}} \cap (\hat{A} - \hat{B})| \geqslant j$ and perhaps enumerating each element of $\hat{A} - \hat{B}$ into $\hat{U}$.

Again, Maass faced the same problem in showing that if A is semilow$_{1.5}$, then $\mathscr{E}^*(\overline{A}) \cong \mathscr{E}^*$. Since A is r.e., elements leave $\overline{A}$. Maass solved the problem by introducing a new property of semilow$_{1.5}$ sets, the outer splitting property. We then generalized this property to solve the problem in our case.

DEFINITION. X has the *finite splitting property* if there are recursive functions f_0 and f_1 such that, for every $e \in N$,

(i) $W_{f_0(e)} \cap W_{f_1(e)} = \varnothing$,

(ii) $W_e \supseteq W_{f_0(e)} \cup W_f f_1(e) \supseteq X \cap W_e$,

(iii) $W_{f_1(e)} \cap X$ is finite,

(iv) $W_e \cap X$ infinite $\rightarrow W_{f_1(e)} \cap X \neq \varnothing$.

We will call $W_{f_1(e)}$ the *critical part* of W_e. If $X = \overline{A}$, we say that A has the *outer splitting* property.

Maass' Theorem was that if A is semilow$_{1.5}$, A has the outer splitting property. For major subsets, we have the following

THEOREM 2.2. *If $B \subset_m A$, then $A - B$ has the finite splitting property.*

PROOF. Fix e. We show how to enumerate $W_{f_0(e)}$ and $W_{f_1(e)}$. By the recursion theorem we may assume that we know the index of $W_{f_1(e)}$. Let P be the Π_2 sentence "$W_{f_1(e)} \cap (A - B) = \varnothing$". By Lemma 2.1 we may assume that every $x \in A$ is equipped with a guess at the truth of P such that almost every $x \in A - B$ guesses correctly. If $x \in W_e \cap A$, then enumerate $x \in W_{f_1(e)}$ if x guesses P is true; otherwise enumerate $x \in W_{f_0(e)}$. By the enumeration procedure, (i) and (ii) obviously hold. For (iii), suppose $W_{f_1(e)} \cap (A - B) \neq \varnothing$. Then almost every $x \in A - B$ guesses that P is false and no such x can be enumerated in $W_{f_1(e)}$. Thus $W_{f_1(e)} \cap (A - B)$ is finite. For (iv), suppose that $W_e \cap (A - B)$ is infinite and, for a contradiction, that $W_{f_1(e)} \cap (A - B) = \varnothing$. Then almost every $x \in A - B$ guesses that P is true. Hence, almost every $x \in W_e \cap (A - B)$ is enumerated in $W_{f_1(e)}$, contradicting the assumption that $W_{f_1(e)} \cap (A - B)$ is empty. □

What the finite splitting property provides is the following. Suppose that we need one "witness" x of $\hat{A} - \hat{B}$ to satisfy a given requirement and suppose that there is an r.e. set W of possible candidates for that witness. For instance, the requirements might be $|\overline{\hat{U}} \cap (\hat{A} - \hat{B})| \geqslant j$ and the set of candidates

$$W = \{x \in \hat{A} \mid x \textit{ guesses } |U \cap (A - B)| \geqslant j\}.$$

Let C be the critical part of W. Then if we use only the candidates in C, we are only committing finitely many elements of $\hat{A} - \hat{B}$ to the requirement by (iii). By (iv), if $W \cap (\hat{A} - \hat{B})$ is infinite, at least one candidate $x \in C$ will be true (i.e., $x \in \hat{A} - \hat{B}$). Further, we can decide if $x \in C$ as soon as we have decided if $x \in W$ ((i), (ii)). By way of summarizing the preceding discussion, we can now show how to meet requirements (2.2) and (2.3).

THEOREM 2.3. *Suppose that* $B \subset_m A$, $\hat{B} \subset_m \hat{A}$ *and* U *is r.e. Then there is an r.e. set* $\hat{U}$, *an index for which can be produced effectively from one for* U, *such that*

$$U \cap (A - B) \text{ is infinite} \quad \text{iff} \quad \hat{U} \cap (\hat{A} - \hat{B}) \text{ is infinite},$$
$$\overline{U} \cap (A - B) \text{ is infinite} \quad \text{iff} \quad \overline{\hat{U}} \cap (\hat{A} - \hat{B}) \text{ is infinite}.$$

PROOF. The sentences $|U \cap (A - B)| \geqslant k$ and $|\overline{U} \cap (A - B)| \geqslant k$ are Σ_2. By Lemma 2.1 we can associate guesses with elements of $\hat{A}$ at each of these sentences such that, for any one of these, almost every $x \in \hat{A}$ guesses correctly. We define a sequence of sets as follows. Let

$$T_1 = \{x \in \hat{A} \mid x \text{ guesses } |U \cap (A - B)| \geqslant 1\}.$$

Given T_{2i-1}, define T_{2i} and T_{2i+1} by

$$T_{2i} = \{x \in \hat{A} | x \text{ guesses } |\overline{U} \cap (A - B)| \geqslant i,\ x \geqslant 2i, \text{ and}$$
$$x \text{ is not in the critical part of } T_j, \text{ for any } j < 2i\},$$
$$T_{2i+1} = \{x \in \hat{A} | x \text{ guesses } |U \cap (A - B)| \geqslant i + 1,\ x \geqslant 2i + 1, \text{ and}$$
$$x \text{ is not in the critical part of } T_j, \text{ for any } j < 2i + 1\}.$$

We may assume that $T_2 \supseteq T_4 \supseteq T_6 \supseteq \cdots$ and $T_1 \supseteq T_3 \supseteq T_5 \supseteq \cdots$. (Notice that for any $x \in \hat{A}$ we can determine which of the sets T_j that x is in, if any, at the stage at which x enters $\hat{A}$ and for each such T_j we can then also determine at that stage whether x is in the critical part of T_j.) We now describe how to enumerate $\hat{U}$.

If $x \in \hat{A}$, then enumerate x in $\hat{U}$ only if x is in the critical part of T_{2i+1} for some i, or x is not in the critical part of any set T_j but there is a k such that x guesses $|U \cap (A - B)| \geqslant k$ and $|\overline{U} \cap (A - B)| < k$.

Suppose $U \cap (A - B)$ is infinite. Then for every i, T_{2i+1} satisfies $T_{2i+1} \cap (\hat{A} - \hat{B}) =^* (\hat{A} - \hat{B})$ because almost every $x \in \hat{A} - \hat{B}$ guesses $|U \cap (A - B)| \geqslant i + 1$ and only finitely many elements of $\hat{A} - \hat{B}$ can be in the critical parts of the sets T_j, $j < 2i + 1$. Thus the critical part of T_{2i+1} contains an element of $\hat{A} - \hat{B}$ which is enumerated in $\hat{U}$. Since the critical parts of the sets T_{2i+1} are disjoint, $\hat{U} \cap (\hat{A} - \hat{B})$ is infinite. A similar argument applied to the sets T_{2i} shows that if $\overline{U} \cap (A - B)$ is infinite, so is $\overline{\hat{U}} \cap (\hat{A} - \hat{B})$.

Suppose that $U \cap (A - B)$ is finite. Then almost every set T_{2i+1} is finite on $\hat{A} - \hat{B}$, so only finitely many elements of $\hat{A} - \hat{B}$ can be enumerated in $\hat{U}$ by virtue of being in the critical parts of such sets. Now there is a k such that almost every $x \in \hat{A} - \hat{B}$ guesses $|U \cap (A - B)| < k$ but $|\overline{U} \cap (A - B)| \geqslant k$. Thus only finitely many $x \in \hat{A} - \hat{B}$ can be enumerated in $\hat{U}$ by virtue of the other clause in the condition on enumeration in $\hat{U}$. Thus $\hat{U} \cap (\hat{A} - \hat{B})$ is finite. A similar argument shows that if $\overline{U} \cap (A - B)$ is finite, then almost every $x \in \hat{A} - \hat{B}$ is enumerated in $\hat{U}$. $\square$

REFERENCES

1. E. Herrmann, *The lattice of recursively enumerable sets*, Seminarbericht, No. 10, Humboldt University, Berlin, 1978. (German)

2. A. H. Lachlan, *On the lattice of recursively enumerable sets*, Trans. Amer. Math. Soc. **130** (1968), 1–37.

3. ______, *The elementary theory of recursively enumerable sets*, Duke Math. J. **35**, (1968), 123–146.

4. M. Lerman, *Congruence relations, filters, ideals and definability in lattices of α-recursively enumerable sets*, J. Symbolic Logic **41** (1976), 405–418.

5. M. Lerman and R. I. Soare, *A decidable fragment of the elementary theory of the lattice of recursively enumerable sets*, Trans. Amer. Math. Soc. **257** (1980), 1–37.

6. ______, *d-simple sets, small sets, and degree classes*, Pacific J. Math. **87** (1980), 135–155.

7. M. Lerman, R. A. Shore and R. I. Soare, *r-maximal major subsets*, Israel J. Math. **31** (1978), 1–18.

8. W. Maass, *Characterization of recursively enumerable sets with supersets effectively isomorphic to all recursively enumerable sets*, Trans. Amer. Math. Soc., **279** (1983), 311–336.

9. ______, *Major subsets and automorphisms of recursively enumerable sets*, these PROCEEDINGS.

10. W. Maass, R. A. Shore and M. Stob, *Splitting properties and jump classes*, Israel J. Math. **39** (1981), 210–244.

11. W. Maass and M. Stob, *The intervals of the lattice of recursively enumerable sets determined by major subsets*, Ann. Pure Appl. Logic **24** (1983), 189–212.

12. R. I. Soare, *Automorphisms of the lattice of recursively enumerable sets. Part* I. *Maximal sets*, Ann. of Math. (2) **100** (1974), 80–120.

DEPARTMENT OF MATHEMATICS, CALVIN COLLEGE, GRAND RAPIDS, MICHIGAN 49506

II. GENERALIZED RECURSION THEORY

Proceedings of Symposia in Pure Mathematics
Volume **42**, 1985

Unimonotone Functions of Finite Types (Recursive Functionals and Quantifiers of Finite Types Revisited IV)

S. C. KLEENE

I have reduced the "RFQFTR IV" to a subtitle to emphasize that the present paper can be read as a separate paper, indeed serving as an introduction to the RFQFTR series, of which it is the fourth. I do tie the present treatment in with RFQFTR I-III, citing them as (Kleene) [**1978, 1980, 1982**] (see the references at the end).[1]

12. Summary of the project.

12.1. At the Summer Institute for Symbolic Logic at Cornell University twenty-five years before the present Summer Institute, I presented material from my paper RFQFT I, which I had sent to *Transactions of the American Mathematical Society* a few weeks earlier, and which was published there in [**1959**], as was RFQFT II in [**1963**]. In those two papers, the notion of partial recursiveness was extended to functions of variables of finite types, and the resulting theory was developed.

12.2. *The* [**1959**] *types and schemata.* Type 0 = the natural numbers $\{0, 1, 2, \ldots\}$, for which I use as variables $a, b, c, \ldots, \alpha^0, \beta^0, \gamma^0, \ldots$. For $j > 0$, type j = {the total one-place functions from type $j-1$ into type 0}, for which I use as variables $\alpha^j, \beta^j, \gamma^j, \ldots$, and for $j = 1$ $\alpha, \beta, \gamma, \ldots$.

A function $\phi(\mathfrak{A})$, where $\mathfrak{A}$ is $n_0, \ldots, n_r$ variables of types $0, \ldots, r$, is *partial* (*primitive*) *recursive* iff $\phi = \phi_p$ where each of $\phi_1, \ldots, \phi_p$ is defined by one of the following schemata S1–S9 (S1–S8) outright or from earlier functions in the list.

S1 $\quad \phi(a, \mathfrak{B}) \simeq a' \quad [= a + 1], \qquad \langle 1, n \rangle.$

S2 $\quad \phi(\mathfrak{A}) \simeq q, \qquad \langle 2, n, q \rangle.$

S3 $\quad \phi(a, \mathfrak{B}) \simeq a, \qquad \langle 3, n \rangle.$

1980 *Mathematics Subject Classification*. Primary 03-xx; Secondary 03D65.

[1]See Footnote 2 of [**1982**] for some corrections and clarifications to [**1978, 1980**]. Also in [**1980**, p. 4 l. 2], for "1978" read "1959".

S4 $\phi(\mathfrak{A}) \simeq \psi(\chi(\mathfrak{A}), \mathfrak{A})$, $\langle 4, n, g, h\rangle$.

S5 $\begin{cases} \phi(0, \mathfrak{B}) \simeq \psi(\mathfrak{B}), \\ \phi(a', \mathfrak{B}) \simeq \chi(a, \phi(a, \mathfrak{B}), \mathfrak{B}), \end{cases}$ $\langle 5, n, g, h\rangle$.

S6 $\phi(\mathfrak{A}) \simeq \psi(\mathfrak{A}_1)$ where $\mathfrak{A}$ is $\mathfrak{A}_1$ with the $k+1$-*st* type-j variable moved to the front, $\langle 6, n, j, k, g\rangle$.

S7 $\phi(\alpha^1, \alpha^0, \mathfrak{B}) \simeq \alpha^1(\alpha^0)$, $\langle 7, n\rangle$.

S8 $\phi(\alpha^j, \mathfrak{B}) \simeq \alpha^j(\lambda\alpha^{j-2}\chi(\alpha^j, \alpha^{j-2}, \mathfrak{B}))$ for $j \geqslant 2$, $\langle 8, n, j, h\rangle$.

S9 $\phi(a, \mathfrak{B}, \overline{\mathfrak{A}}) \simeq \{a\}(\mathfrak{B})$, $\langle 9, n, \bar{n}\rangle$.

The functions are indexed at the right; e.g. $\langle 4, n, g, n\rangle$ $(= 2^4 \cdot 3^n \cdot 5^g \cdot 7^h)$ indexes the function ϕ introduced by an application of S4 where $n = \langle n_0, \ldots, n_r\rangle$ $(= p_0^{n_0} \cdot \cdots \cdot p_r^{n_r})$ and g, h are the indices of ψ, χ as previously defined by applications of the schemata. In S9, $\{a\}(\mathfrak{B})$ is $\phi^{(a)}(\mathfrak{B})$ where $\phi^{(a)}$ (temporary notation) is the function of $\mathfrak{B}$ indexed by a (or, if a is not an index of a function of $\mathfrak{B}$, is the totally undefined function of $\mathfrak{B}$).

12.3. You will observe that in S8 I provided for the substitution of a λ-functor $\lambda^{j-2}\chi(\alpha^j, \alpha^{j-2}, \mathfrak{B})$ directly as the argument of a type-j variable α^j $(j \geqslant 2)$. Rather laboriously this substitutivity was generalized to hold in the following qualified form [**1959**, XXII].

Substitutivity of λ-functors (*restricted form*). If $\phi(\mathfrak{A}, \sigma^n, \mathfrak{B})$ and $\theta(\mathfrak{A}, \mathfrak{B}, \tau^{n-1})$ are partial recursive, there is a partial recursive function $\phi(\mathfrak{A}, \mathfrak{B})$ such that $\phi(\mathfrak{A}, \mathfrak{B}) = \phi(\mathfrak{A}, \lambda\tau^{n-1}\theta(\mathfrak{A}, \mathfrak{B}, \tau^{n-1}), \mathfrak{B})$ for values of $\mathfrak{A}$, $\mathfrak{B}$ for which $\lambda\tau^{n-1}\theta(\mathfrak{A}, \mathfrak{B}, \tau^{n-1})$ is completely defined and $\phi(\mathfrak{A}, \lambda\tau^{n-1}\theta(\mathfrak{A}, \mathfrak{B}, \tau^{n-1}), \mathfrak{B})$ is defined $(n \geqslant 1)$.

12.4. *The first recursion theorem in ordinary recursion theory* [**1952**, Theorem XXVI, p. 348]. When $\psi(\eta; \mathfrak{A})$ is a partial recursive functional (η being a variable for a function of the variables $\mathfrak{A}$), the least solution for ϕ of

$$\phi(\mathfrak{A}) \simeq \psi(\phi; \mathfrak{A})$$

is partial recursive. ▽

To entertain the first recursion theorem in the context of [**1959**], we need to have the function variable η, and for repeated applications of it (and maybe for other purposes) a list of function variables (or of assumed functions) $\theta_1, \ldots, \theta_l = \Theta$ where θ_t is a function of $m_{t_0}, \ldots, m_{t_r}$ variables of types $0, \ldots, r$. Write $m_t = \langle m_{t_0}, \ldots, m_{t_r}\rangle$ and $m = \langle m_1, \ldots, m_l\rangle$.

The nine [**1959**] schemata displayed above can be rewritten replacing the functions $\phi(\mathfrak{A})$, etc. by functionals $\phi(\Theta; \mathfrak{A})$, etc., as follows, e.g.:[2]

S1 $\phi(\Theta; a, \mathfrak{B}) \simeq a'$ $[= a + 1]$, $\langle 1, m, n\rangle$.

[2] It is convenient to call $\phi(\mathfrak{A})$ a "function" and $\phi(\Theta; \mathfrak{A})$ (when the Θ are variable) a "functional", even though when $\mathfrak{A}$ includes variables of types $j > 0$ $\phi(\mathfrak{A})$ is strictly speaking a functional (as is used in the titles "RFQFT"), and indeed the type-j objects α^j themselves are functionals for $j \geqslant 2$ (cf. [**1978**, 1.1]).

S4 $\phi(\Theta; \mathfrak{A}) \simeq \psi(\Theta; \chi(\Theta; \mathfrak{A}), \mathfrak{A})$, $\langle 4, m, n, g, h \rangle$.

S8

$$\phi(\Theta; \alpha^j, \mathfrak{B}) \simeq \alpha^j(\lambda\alpha^{j-2}\chi(\Theta; \alpha^j, \alpha^{j-2}, \mathfrak{B})) \quad \text{for } j \geqslant 2, \qquad \langle 8, m, n, j, h \rangle.$$

S9

$$\phi(\Theta; a, \mathfrak{B}, \bar{\mathfrak{A}}) \simeq \{a\}^{\Theta}(\mathfrak{B}), \qquad \langle 9, m, n, \bar{n} \rangle.$$

Furthermore, a schema S0.t to introduce each one θ_t of the l assumed functions Θ can be prefixed. In stating this, I followed the model of S8, which in introducing a function variable α^j for $j \geqslant 2$ provided for substitution of a λ-functor. For illustration (with $m_t = \langle 1, 1, 1 \rangle$, so θ_t is $\theta_t(c, \alpha^1, \alpha^2)$ say):
S0.t

$$\phi(\Theta; c, \mathfrak{B}) \simeq \theta_t(c, \lambda s\chi_2(\Theta; s, \mathfrak{B}), \lambda\sigma\chi_3(\Theta; \sigma, \mathfrak{B})), \quad \langle 0, m, n, t, h_2, h_3 \rangle.$$

With these ten schemata, the first recursion theorem was established to hold for a special class of the functionals $\psi(\eta; \mathfrak{A})$ [**1963**, LXIV], and likewise with η, Θ instead of just η [**1963**, LXIV*].

As I showed by examples in [**1963**, LVI and LXVI], neither the substitutivity of λ-functors nor the first recursion theorem hold without restriction in the [**1959**] theory.

12.5. In [**1978**], the first of these RFQFT Revisited papers (from the Second Symposium on Generalized Recursion Theory, at Oslo in 1977), I changed the ten schemata for the definition of partial recursive functions (S1–S9 relativized plus S0.t ($t = 1, \ldots, l$)) to provide without restriction both the substitutivity of λ-functors and the first recursion theorem. They are provided directly by new schemata S4.j for $j > 0$ and S11, respectively. I am restricting our attention now to functions of variables of the types 0, 1, 2, 3 (so $r = 3$).

Of course, having S4.j for $j > 0$, the special cases of substitutivity of λ-functors in S8 and S0.t become redundant. So I simplify S0.t to S0 as follows; and S7.j for $j = 2, 3$ take over the remaining role of S8. Furthermore S11, when supplemented in the context of the other schemata by two new initial functions $a \dot{-} 1$ and cs (Schemata S1.1 and S5.1, "cs" for "case"), enables the schema S5 of primitive recursion to be derived [**1978**, (XII)]; so I omit it. Likewise, S9 can be derived using S11, etc. [**1978**, (XIX)]. And as a minor detail, I use just 0 rather than any constant q in S2. Thus we arrive at:

The [**1978**] *schemata.*

S0 $\phi(\Theta; \mathfrak{B}, \mathfrak{C}) \simeq \theta_t(\mathfrak{B})$, $\langle 0, m, n, t \rangle$.

S1.0 $\phi(\Theta; a, \mathfrak{B}) \simeq a'$ $[= a + 1]$, $\langle 1, m, n, 0 \rangle$.

S1.1 $\phi(\Theta; a, \mathfrak{B}) \simeq a \dot{-} 1$ $\left[= \begin{cases} a - 1 & \text{if } a > 0 \\ 0 & \text{if } a = 0 \end{cases}\right]$, $\langle 1, m, n, 1 \rangle$.

S2.0 $\phi(\Theta; \mathfrak{A}) \simeq 0$, $\langle 2, m, n \rangle$.

S3 $\phi(\Theta; a, \mathfrak{B}) \simeq a$, $\langle 3, m, n \rangle$.

S4.0 $\phi(\Theta; \mathfrak{A}) \simeq \psi(\Theta; \chi(\Theta; \mathfrak{A}), \mathfrak{A})$, $\langle 4, m, n, 0, g, h\rangle$

S4.j $(j = 1, 2, 3)$ $\phi(\Theta; \mathfrak{A}) \simeq \psi(\Theta; \lambda\beta^{j-1}\chi(\Theta; \beta^{j-1}, \mathfrak{A}), \mathfrak{A})$,
$\langle 4, m, n, j, g, h\rangle$

S5.1 $\phi(\Theta; a, b, c, \mathfrak{B}) \simeq \mathrm{cs}(a, b, c) \left[= \begin{cases} b & \text{if } a = 0 \\ c & \text{if } a > 0 \end{cases}\right]$, $\langle 5, m, n, 1\rangle$.

S6.j $(j = 0, 1, 2, 3)$ $\phi(\Theta; \mathfrak{A}) \simeq \psi(\Theta; \mathfrak{A}_1)$ where $\mathfrak{A}$ is $\mathfrak{A}_1$
with the $k + 1$-st type-j variable moved to the front, $\langle 6, m, n, j, k, g\rangle$.

S7.j $(j = 1, 2, 3)$ $\phi(\Theta; \alpha^j, \alpha^{j-1}, \mathfrak{B}) \simeq \alpha^j(\alpha^{j-1})$, $\langle 7, m, n, j\rangle$.

S11 $\phi(\Theta; \mathfrak{A}) \simeq \psi(\lambda\mathfrak{A}\phi(\Theta; \mathfrak{A}), \Theta; \mathfrak{A})$
$[\simeq \psi(\phi, \Theta; \mathfrak{A})$ briefly$]$, $\langle 11, m, n, g\rangle$.

12.6. At the same time, I changed the rules for computation. Let us compare how computation went in [**1959**] and in [**1978**].

Computation. Computation is defined relative to a given partial recursive derivation $\phi_1, \ldots, \phi_p$ (of ϕ_p) from assumed functions $\theta_1, \ldots, \theta_l = \Theta$ (perhaps empty, i.e. $l = 0$). First, formation rules are stated to describe the 0-*expressions*, which syntactically are compatible with representing type-0 objects under a given assignment Ω of interpretations to the free variables and the Θ; and similarly *j-expressions* for $j > 0$ [**1978**, 2.2].[3]

To illustrate computation, suppose we have the following series of schema applications introducing successively $\rho, \psi, \ldots, \chi, \phi$:

$$\rho(a, b) \simeq a \quad \text{by S3.}$$

$$\psi(b, a) \simeq \rho(a, b) \quad \text{by S6.0.}$$

$$\vdots$$

$$\chi(a) \simeq \cdots$$

$$\phi(a) \simeq \psi(\chi(a), a) \quad \text{by S4.0.}$$

To compute $\phi(a)$ under the assignment Ω of a natural number r as value of the variable a, we begin by using the application of S4.0 which introduced ϕ. This gives us $\psi(\chi(a), a)$. Then in my [**1959**] theory,[4] we started a subcomputation of the part $\chi(a)$; and only if we were able to complete it, say obtaining the value n, did we continue the computation of $\phi(a)$ along the "principal branch" by using the application of S6.0 which introduced ψ. I flag a numeral completing a

[3] If ϕ_i with its arguments is $\phi_i(\Theta_i; \mathfrak{A}_i)$, we can write it simply "$\phi_i(\mathfrak{A}_i)$", since each Θ_i is determined by the partial recursive derivation $\phi_1, \ldots, \phi_p$ (with its analysis), or by the index of ϕ_p.

[4] I am presenting both versions of computation here "syntactically", in the manner of [**1978**], etc. In [**1959**], "syntactical considerations" were avoided by using instead $r + 2$-tuples $(z, a, \alpha^1, \ldots, \alpha^r)$ consisting of an index z and $1 + r$ contracted arguments for it [**1959**, 5.3].

computation (principal or subordinate). Thus we obtained the completed computation:

1959
$$\phi(a) \text{ --- } \psi(\chi(a), a) \text{ --- } \psi(N, a) \text{ --- } \rho(a, N) \text{ --- } a \text{ --- } R\dagger.$$
$$\chi(a) \cdots N\dagger$$

$N = 0'^{\cdots\prime}$ with n accents; $R = 0'^{\cdots\prime}$ with r accents.

In [**1978**], instead I authorized plowing right ahead from $\psi(\chi(a), a)$ by using the schema application introducing ψ, carrying along the $\chi(a)$ unevaluated, and only asking for it to be computed if and when we come to need a value for it. (Shall we say that I abandoned the caution of youth for the impetuosity of middle age?) In the present example, we don't need a value of $\chi(a)$:

1978
$$\phi(a) \text{ --- } \psi(\chi(a), a) \text{ --- } \rho(a, \chi(a)) \text{ --- } a \text{ --- } R\dagger.$$

As another illustration, suppose that $\phi(\alpha^2, a)$ is introduced by the [**1959**] schema S8 for $j = 2$, thus:

1959
$$\phi(\alpha^2, a) \simeq \alpha^2(\lambda\alpha^0\chi(\alpha^2, \alpha^0, a)) \quad \text{by S8.}$$

In [**1959**], I computed it as follows. Only if all the subcomputations of $\chi(\alpha^2, \alpha^0, a)$ with $0, 1, 2, \ldots$ as the values of α^0 could be completed, determining a type-1 object $\alpha^1 = \{\langle 0, n_0\rangle, \langle 1, n_1\rangle, \langle 2, n_2\rangle, \ldots\}$—only then could the step "____ $M\dagger$" for $m = \alpha^2(\alpha^1)$, where α^2 is the function interpreting the variable α^2, be taken:

1959
$$\phi(\alpha^2, a) \text{ --- } \alpha^2(\lambda\alpha^0\, \chi(\alpha^2, \alpha^0, a)) \text{ --- } M\dagger.$$
$$\vdots$$
$$\chi(\alpha^2, \alpha^0, a) \cdots \Omega_2 \cdots N_2^\dagger.$$
$$\chi(\alpha^2, \alpha^0, a) \cdots \Omega_1 \cdots N_1^\dagger$$
$$\chi(\alpha^2, \alpha^0, a) \cdots \Omega_0 \cdots N_0^\dagger$$

$m = \alpha^2(\alpha^1)$ for $\alpha^1 = \{\langle 0, n_0\rangle, \langle 1, n_1\rangle, \langle 2, n_2\rangle, \ldots\}$.

We do not have Schema S8 in the [**1978**] theory; but we have the more general S4.j for $j > 0$. In the same manner as with S4.0, if $\phi(a)$ has been introduced thus,

1978
$$\phi(a) \simeq \psi(\lambda\beta^0\chi(\beta^0, a), a) \quad \text{by S4.1,}$$

we plow right ahead using the schema which introduced ψ, instead of first asking for values of $\chi(\beta^0, a)$ for each of $0, 1, 2, \ldots$ as the value of β^0. Only if and when such values are needed do we seek to compute them. Why worry about what you may not need to know until the need becomes present?

12.7. These changes in [**1978**] altered the complexion of computation. Formerly, at each step in a completed computation we dealt with expressions interpretable using our types.[4] For example, if a 0-expression $\phi_i(A, B_1, B_2, C)$, where A, B_1, B_2, C are 0-, 1-, 1-, 2-expressions, occurred in a completed computation, each of A, B_1, B_2, C was interpretable by a type-0, -1, -1, -2 object, respectively. Now our carrying forward arguments of functions, without trying to compute them until the need arises, results in our computations being governed by formal rules without the expressions we are manipulating always being interpretable (in completed computations), under the assignment Ω currently in force to the free variables and the Θ, as objects of our types 0, 1, 2, 3.

So in my second paper [**1980**] (from the 1978 Kleene Symposium at Madison) of the RFQFTR series, I addressed the problem of modifying my type structure so as to clothe all the expressions that can arise in computation with interpretations—to provide a "semantics". Indeed, for establishing some of the properties of our functions and functionals listed in [**1978**, §3], such clothing was needed.

In retrospect, is it not a little anomalous that, in my 1959 theory of partial functions, their function arguments α^j $(j > 1)$ were restricted to be total? (Somewhat obliquely this was remarked in [**1963**, end 9.3])

12.8. To free ourselves from this restriction, the obvious first step is to add to type 0 indefinition or ignorance to obtain type $\dot{0}$. Then for $j > 0$ we replace type j by type $j^{\cdot}$, as follows.

The [**1980**] *Types* $\dot{0}$, $\dot{1}$, $\dot{2}$, $\dot{3}$. Type $\dot{0} = \{\mathrm{u}, 0, 1, 2, \ldots\}$ where u is "undefined". For $j > 0$, type $j^{\cdot} = \{$"suitable" partial one-place functions from type $(j-1)^{\cdot}$ into type $0\}$. Variables for objects of type $j^{\cdot}$ $(j = 0, 1, 2, 3)$ are like those for type j except dotted. This applies to the interpretation. For the definition of "j-expression", we have only one kind of variables for each of $j = 0, 1, 2, 3$, which may be written either dotted or undotted.

12.9. We do not want our type $j^{\cdot}$ (for $j > 0$) to comprise all partial one-place functions from type $(j-1)^{\cdot}$ into type 0, but only ones which behave suitably, i.e. in accordance with our prejudices on this matter.

"Suitably"—I think all will agree—should include the following. Suppose that in determining a function value $\dot{\alpha}^j(\dot{\alpha}^{j-1})$ (where $j \geqslant 2$) we have taken into account the values of $\dot{\alpha}^{j-1}$ only for certain arguments. Maybe $\dot{\alpha}^{j-1}$ is undefined for all other arguments, or maybe we just did not know or have not used such other values as it has. Then if we replace $\dot{\alpha}^{j-1}$ by a function extending it to more arguments (or wake up to $\dot{\alpha}^{j-1}$ having other values), our previous result (i.e. the value of $\dot{\alpha}^j(\dot{\alpha}^{j-1})$) should not have to be reconsidered, let alone revised. I formulate this as "monotonicity", as follows.

Consider our (partial) functions as sets of ordered pairs with second members $\in$ type 0. For $j > 1$, $\dot{\alpha}^j$ is *monotone*, iff, whenever $\overline{\dot{\alpha}^{j-1}} \supset \dot{\alpha}^{j-1}$ (set-theoretically) and $\dot{\alpha}^j(\dot{\alpha}^{j-1})$ is defined, then $\dot{\alpha}^j(\overline{\dot{\alpha}^{j-1}})$ is defined with the same value. And $\dot{\alpha}^1$ is *monotone*, iff, if $\dot{\alpha}^1(\mathrm{u})$ is defined, then $\dot{\alpha}^1(r)$ is defined with the same value for each $r \in$ type 0. (The foregoing holds for $j = 1$, if we put $\overline{\dot{\alpha}^0} \supset \dot{\alpha}^0 \equiv \overline{\dot{\alpha}^0} = \dot{\alpha}^0 \vee \dot{\alpha}^0 = \mathrm{u}$.)

12.10. However in [**1980**] I did not stop at requiring the functions encompassed in type j for $j > 1$ to be monotone. A coherent theory can be developed more specifically in more than one direction. My aim in the present paper is to show how the direction which I indicated in [**1980**], further developed in [**1982**] (from the Logic Symposion and A. S. L. Summer Meeting at Patras, Greece in 1980), and have brought to a certain conclusion by work now in a manuscript,[5] is quite natural and compelling from a certain point of view.

In the cited papers it may have seemed somewhat ad hoc; and indeed I arrived at it only after making a misstep in the preliminary manuscript of [**1978**], which was revealed to me early in 1978 by an example of David Kierstead, who subsequently (in a 1979 Wisconsin PhD thesis, and in [**1980** and **1983**]) developed a different symantics.[6]

13. A new approach to unimonotone functions $\dot{\alpha}^j$ ($j = 1, 2, 3$).

13.1. I focus now on the problem: How should our type-j functions $\dot{\alpha}^j$ gather information about their type-$(j-1)$ arguments $\dot{\alpha}^{j-1}$ in undertaking to answer a question we put to them "What is $\dot{\alpha}^j(\dot{\alpha}^{j-1})$?" (briefly, "$\dot{\alpha}^j(\dot{\alpha}^{j-1})$?" or even "$\dot{\alpha}^{j-1}$?"). It is for $j > 1$ that there really is a problem.

I shall model our $\dot{\alpha}^j$'s behavior on the way computation works. It will be computation of such a character that, whenever we succeed in completing a computation, no step will have been wasted. Of, if "$\dot{\alpha}^j(\dot{\alpha}^{j-1})$?" is answered, each question asked by $\dot{\alpha}^j$ of $\dot{\alpha}^{j-1}$ in progressing toward the answer elicits some necessary new information about $\dot{\alpha}^{j-1}$.[6]

As to how computation works, I assume initially only the most rudimentary features of it. A computation will, as illustrated in 12.6, be constructed as a (horizontal) tree branching toward the right, with the 0-expression being computed at its initial (leftmost) vertex, and with at most one step possible at each moment to add a vertex (bearing another 0-expression) horizontally or downward rightward from a vertex already incorporated into the tree. Along each path in the tree running horizontally rightward, our aim (which motivates our choice of the formal computation rules for adding vertices) is to reach a (flagged) numeral (ending that path) which expresses the common value of all the 0-expressions along it, using the assignment Ω of interpretations, to (at least) the free variables of its 0-expressions and to the functions Θ if any, which is in force along it. A vertex introduced downward rightward from a given vertex starts a subcomputation (under a like assignment $\bar{\Omega}$)[7] of the 0-expression placed there, the value of

[5] In particular, I have formulated explicitly the rule E4.3 for the computation of expressions of the form $\gamma^3(B)$ and the case of E7 for $\dot{\theta}_t(\cdots)$ with 2-expressions included as arguments, and (with the present paper) have established #A–#D of [**1980**, p. 26] extended to include $j = 3$. A principal step is to prove a general monotonicity theorem for computations, from which some other results quickly follow, e.g. that the values computed are independent of the choice of the oracles used (cf. [**1980**, 7.5]).

[6] See my [**1980**, 8.2 and Footnote 9 and **1982**, 11.4].

[7] As we shall not actually see until 13.3e, the 0-expression beginning a subcomputation may have an additional free variable, so the assignment has to be expanded to interpret that. Thus, at a vertex not on the principal branch running horizontally rightward from the initial vertex, the assignment may not be just the Ω in effect there.

which (hopefully to be obtained as a flagged numeral horizontally rightward from there) is required for (hopefully) making the next step horizontally rightward from the given vertex.

I think we can agree that the definition of type $\dot{0}$ as $\{\mathrm{u}, 0, 1, 2, \ldots\}$ mandates that in computing we handle a type-$\dot{0}$ variable γ^0 as follows. (Those of our formal computation rules not affected by the change from types 0, 1, 2, 3 to types $\dot{0}$, $\dot{1}$, $\dot{2}$, $\dot{3}$, which are straightforward, are given in [**1978**, 2.4].)

Rule for computing γ^0 under an assignment Ω including an interpretation $\dot{\alpha}^0$ for γ^0. In pursuing a computation, if at a given moment we have just arrived at a vertex bearing γ^0 (we say γ^0 has *surfaced* there, or that we have there a *surfacing of* γ^0 or a γ^0-*face* γ^0), then the computation cannot be continued or can be continued horizontally rightward from that vertex by "—— $R\dagger$" (where R is the numeral for r), according as $\dot{\alpha}^0 = \mathrm{u}$ (*Case* 1^0) or $\dot{\alpha}^0 = r \in$ type 0 (*Case* 2^0).

13.2a. *Computation of* E_{γ^0} *under* $\Omega_{\dot{\alpha}^0}$.[8] Take a 0-expression E_{γ^0}; and consider how, when we compute it under an assignment $\Omega_{\dot{\alpha}^0}$ consisting of $\dot{\alpha}^0$ as the interpretation of γ^0 and a fixed Ω as the assignment to the other free variables of E_{γ^0} (at least) and to the Θ, information about $\dot{\alpha}^0$ is used.

Case $\bar{1}^1$: In computing E_{γ^0} under $\Omega_{\dot{\alpha}^0}$, we neither obtain a value nor encounter a surfacing of γ^0. So E_{γ^0} is undefined whatever the interpretation $\dot{\alpha}^0$ of γ^0, and thus $\lambda\gamma^0 E_{\gamma^0}$ represents under Ω the totally undefined function $\lambda\dot{\alpha}^0\mathrm{u}$.

Case $\bar{2}^1$: We obtain a value say n for E_{γ^0} without γ^0 surfacing. So $\lambda\gamma^0 E_{\gamma^0}$ under Ω represents the constant function $\lambda\dot{\alpha}^0 n$.

Case $\bar{3}^1$: In computing E_{γ^0} under $\Omega_{\dot{\alpha}^0}$, we come to a γ^0-face γ^0. Now if $\dot{\alpha}^0$ is u, by Case 1^0 (in the preceding paragraph) we can't continue. If $\dot{\alpha}^0$ is an $r \in$ type 0, by Case 2^0 we continue from this γ^0-face by the step "—— $R\dagger$". Doing the same (with this r) from every subsequent surfacing of γ^0, we may fail to obtain a value for E_{γ^0}, or we may eventually obtain a value n for E_{γ^0}, under $\Omega_{\dot{\alpha}^0}$, the result (no value, or what value n) depending (in general) on the r.

13.2b. The computation of E_{γ^0} under $\Omega_{\dot{\alpha}^0}$ just described exemplifies a specific agency (fixed by E_{γ^0} and Ω) for operating on any type-$\dot{0}$ object $\dot{\alpha}^0$ to produce u or an $n \in$ type 0. Now I abstract from those agencies (for various E_{γ^0} and Ω) to formulate the kind of operation performed by each of them. In doing this, I have enjoyed using the imagery of "oracles" (introduced by Turing in [**1939**], but as we shall proceed greatly extended and elaborated).[9] I call such an agent as I have just exemplified an "$\dot{\alpha}^1$-oracle".

$\dot{\alpha}^1$-*oracles*. An oracle for a function $\dot{\alpha}^1$ (an $\dot{\alpha}^1$-oracle) operates as follows. To ask her "What is $\dot{\alpha}^1(\dot{\alpha}^0)$?", we put $\dot{\alpha}^0$ in a closed envelope if $\dot{\alpha}^0$ is an $r \in$ type 0, or leave the envelope empty if $\dot{\alpha}^0$ is u, and present the envelope to her.

---

[8] The case numbers $\bar{1}^1$, $\bar{2}^1$ and $\bar{3}^1$ match those of [**1980**, 6.2], while Cases 1^1, 2^1 and 3^1 were used in 5.3 there.

[9] For the history before 1978, cf. [**1978**, Footnote 2].

Case $\bar{1}^1$: The $\dot{\alpha}^1$-oracle does not open the envelope, and does not answer our question. So $\alpha^1(\dot{\alpha}^0) = \mathrm{u}$. Indeed, since the $\dot{\alpha}^1$-oracle did not allow herself to learn the contents (or emptiness) of the envelope, this is the case for all $\dot{\alpha}^0$ (not just the one we used); so $\dot{\alpha}^1 = \lambda\alpha^0\mathrm{u}$.

Case $\bar{2}^1$: Without opening the envelope, the $\dot{\alpha}^1$-oracle pronounces that $\dot{\alpha}^1(\dot{\alpha}^0) = n$. So $\dot{\alpha}^1 = \lambda\dot{\alpha}^0 n$.

Case $\bar{3}^1$: The $\dot{\alpha}^1$-oracle opens the envelope. If she finds it empty, she does not answer our question. So $\dot{\alpha}^1(\dot{\alpha}^0) = \mathrm{u}$. (This is when $\dot{\alpha}^0$ is u. $\dot{\alpha}^1(\mathrm{u})$ can be defined only by Case $\bar{2}^1$ holding.) If she finds inside a natural number r, then, depending (in general) on the r, she may fail to answer, so $\dot{\alpha}^1(\dot{\alpha}^0) = \mathrm{u}$, or she may answer that $\dot{\alpha}^1(\dot{\alpha}^0) = n$ for some $n \in$ type 0 depending (in general) on the r. $\triangledown$

This definition of an "$\dot{\alpha}^1$-oracle" may be used in either of two situations. We may start with a given partial one-place function $\dot{\alpha}^1$ from type $\dot{0}$ into type 0; then the answers the oracle gives to be an "$\dot{\alpha}^1$-oracle" must be correct for that function. Or we may start with an oracle that behaves as just described, and discover the values (when defined) of such a function $\dot{\alpha}^1$ (for which she is an "$\dot{\alpha}^1$-oracle") through her answers.

13.2c. *Definition of type* $\dot{1}$. Now I can define *type* $\dot{1}$ (compatibly with my preliminary stipulations in 12.8, 12.9) to consist of the partial one-place functions $\dot{\alpha}^1$ from type $\dot{0}$ into type 0 for which there are $\dot{\alpha}^1$-oracles, provided each such $\dot{\alpha}^1$ is monotone (cf. 12.9). But indeed it is, since $\dot{\alpha}^1(\mathrm{u})$ is defined only under Case $\bar{2}^1$, in which case our oracle also makes $\dot{\alpha}^1(r)$ defined with the same value for every $r \in$ type 0.

13.2d. $\lambda\gamma^0 E_{\gamma^0}$ *under* Ω *represents a type-*$\dot{1}$ *object* $\dot{\alpha}^1$ such that, for each $\dot{\alpha}^0 \in$ type $\dot{0}$, computing E_{γ^0} under $\Omega_{\dot{\alpha}^0}$ gives $\dot{\alpha}^1(\dot{\alpha}^0)$.

13.2e. I have not yet said how a type-$\dot{1}$ variable γ^1 is to be managed in computation. But now, I submit, our notion of an $\dot{\alpha}^1$-oracle makes it clear what to do.

Rule for computing $\gamma^1(B)$ under an assignment Ω including an interpretation $\dot{\alpha}^1$ for γ^1 with a choice of an $\dot{\alpha}^1$-oracle. At a *surfacing of* γ^1 (or γ^1-*face*) $\gamma^1(B)$, we shall proceed as follows, according to the case in 13.2b applicable to the $\dot{\alpha}^1$-oracle. In *Case* $\bar{1}^1$, the computation cannot be continued after reaching the γ^1-face. In *Case* $\bar{2}^1$, it is continued from the γ^1-face $\gamma^1(B)$ by "——$N\dagger$". In *Case* $\bar{3}^1$, the computation is continued by starting a subcomputation of B to the γ^1-face $\gamma^1(B)$ (under the assignment $\bar{\Omega}$ in effect at the γ^1-face):[7]

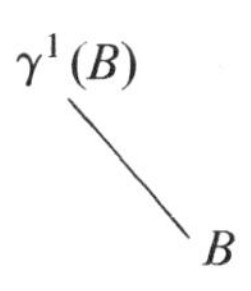

If this subcomputation can be completed to give a value $r \in$ type 0 for B and $\dot{\alpha}^1(r) = n \in$ type 0, the part of the computation begun with the γ^1-face $\gamma^1(B)$ can be completed thus:

$$\begin{array}{lll} \gamma^1(B) & \text{———} & N\dagger. \\ & \diagdown & \\ & & B \cdots R\dagger \end{array}$$

Otherwise, the computation cannot be continued horizontally rightward from the γ^1-face $\gamma^1(B)$.

13.3a. *Computation of* E_{γ^1} *under* $\Omega_{\dot{\alpha}^1}$. Consider how, in the computation of a 0-expression E_{γ^1} under $\Omega_{\dot{\alpha}^1}$ ($\dot{\alpha}^1$ interpreting γ^1) with a given $\dot{\alpha}^1$-oracle, information about $\dot{\alpha}^1$ is used.

Cases $\bar{1}^2$ and $\bar{2}^2$ (analogous to Cases $\bar{1}^1$ and $\bar{2}^1$ in 13.2a) are obvious: $\lambda\gamma^1 E_{\gamma^1}$ under Ω represents, respectively, $\lambda\dot{\alpha}^1 \text{u}$ or $\lambda\dot{\alpha}^1 m$.

Case $\bar{3}^2$: We come to a surfacing $\gamma^1(B)$ of γ^1. If Case $\bar{1}^1$ (in 13.2b) applies to the $\dot{\alpha}^1$-oracle (giving *Subcase* $\overline{3.1}^2$ here), by 13.2e the computation cannot be continued from the γ^1-face $\gamma^1(B)$. If Case $\bar{2}^1$ applies (*Subcase* $\overline{3.2}^2$), it is continued from the present γ^1-face by "—— $N\dagger$" and similarly with the same n from each subsequent surfacing of γ^1. Proceeding in this manner, we may fail to complete the computation of E_{γ^1}, or we may succeed with result say m. If Case $\bar{3}^1$ applies (*Subcase* $\overline{3.3}^2$), we may have a nonempty series of γ^1-faces $\gamma^1(B_0), \gamma^1(B_1), \ldots, \gamma^1(B_\zeta), \ldots$ (B_0 is the above B) with respective subcomputations of $B_0, B_1, \ldots, B_\zeta, \ldots$ under respective assignments[7] $\Omega^0_{\dot{\alpha}^1}, \Omega^1_{\dot{\alpha}^1}, \ldots, \Omega^\zeta_{\dot{\alpha}^1}, \ldots$ giving distinct values $r_0, r_1, \ldots, r_\zeta, \ldots$ (besides maybe some others giving the same r's as earlier ones) for which $\dot{\alpha}^1(r_0) = n_0$, $\dot{\alpha}^1(r_1) = n_1, \ldots$, $\dot{\alpha}^1(r_\zeta) = n_\zeta, \ldots$, so that from the γ^1-faces we can pass horizontally to $N_0^\dagger, N_1^\dagger, \ldots, N_\zeta^\dagger, \ldots$, after all of which (say for $\zeta < \xi$) the computation of E_{γ_1} is completed with a value say m being obtained. Or the process may end short of this result by the subcomputation of B_0 under $\Omega^0_{\dot{\alpha}^1}$ not leading to a value r_0, or by $\dot{\alpha}^1(r_\zeta) = \text{u}$ for some $\zeta \geqslant 0$, or after all $\zeta < \xi$ (with $\xi > 0$) by our then getting neither a γ^1-face for which the subcomputation gives a new r nor a value m for E_{γ^1}.

13.3b. Now we have been led to a notion of an "$\dot{\alpha}^2$-oracle" who operates in the same manner, but acting on her own rather than on the basis of what happens in computing a given 0-expression E_{γ^1} under $\Omega_{\dot{\alpha}^1}$ with a given Ω.

$\dot{\alpha}^2$-*oracles*. We ask an $\dot{\alpha}^2$-oracle "$\dot{\alpha}^2(\dot{\alpha}^1)$?" by handing her a closed envelope containing an $\dot{\alpha}^1$-oracle.

Cases $\bar{1}^2$ and $\bar{2}^2$ should be clear: $\dot{\alpha}^2 = \lambda\dot{\alpha}^1 \text{u}$ or $\dot{\alpha}^2 = \lambda\dot{\alpha}^1 m$, respectively. In *Case* $\bar{3}^2$, the $\dot{\alpha}^2$-oracle opens our envelope, and initially questions the $\dot{\alpha}^1$-oracle who steps out of it with an empty envelope ($\dot{\alpha}^0 = \text{u}$). For an $\dot{\alpha}^1$-oracle under Case $\bar{1}^1$ of 13.2b (*Subcase* $\overline{3.1}^2$), the $\dot{\alpha}^2$-oracle stands mute ($\dot{\alpha}^2(\dot{\alpha}^1) = \text{u}$). Under Case $\bar{2}^1$ (*Subcase* $\overline{3.2}^2$), she either stands mute ($\dot{\alpha}^2(\lambda\dot{\alpha}^0 n) = \text{u}$) or declares that $\dot{\alpha}^2(\dot{\alpha}^1) = m$ ($\dot{\alpha}^2(\lambda\dot{\alpha}^0 n) = m$). For an $\dot{\alpha}^1$-oracle under Case $\bar{3}^1$ (*Subcase* $\overline{3.3}^2$), the $\dot{\alpha}^2$-oracle may stand mute, or she may ask of the $\dot{\alpha}^1$-oracle a nonempty series of distinct questions

$$r_0, r_1, \ldots, r_\zeta, \ldots$$

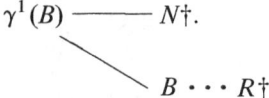

(i.e. "$\dot{\alpha}^1(r_0)$?", "$\dot{\alpha}^1(r_1)$?", etc.), being answered with

$$n_0, n_1, \ldots, n_\zeta, \ldots.$$

Each r_ζ is determined by the $\dot{\alpha}^2$-oracle on the basis of the information about the $\dot{\alpha}^1$-oracle to date, i.e. that the $\dot{\alpha}^1$-oracle opens envelopes and that $\dot{\alpha}^1 \supset \{\langle r_\eta, n_\eta\rangle | \eta < \zeta\}$. This process continues until it is ended by the $\dot{\alpha}^1$-oracle for some $\zeta \geqslant 0$ not answering the question "$\dot{\alpha}^1(r_\zeta)$?" by an n_ζ ($\dot{\alpha}^1(r_\zeta) = \mathrm{u}$), or, after questions have been asked and answered for all $\zeta < \xi$ with $\xi > 0$, by the $\dot{\alpha}^2$-oracle on the basis of the information to date not asking a next question but either standing mute or declaring that $\dot{\alpha}^2(\dot{\alpha}^1) = m$. If it is ended other than in the last fashion just listed, $\dot{\alpha}^2(\dot{\alpha}^1) = \mathrm{u}$. ▽

Just as with $\dot{\alpha}^1$-oracles, we may have an "$\dot{\alpha}^2$-oracle" who answers correctly for an already-known partial one-place function $\dot{\alpha}^2$ from type $\dot{1}$ into type 0, or we may discover the values (when defined) of such a function $\dot{\alpha}^2$ through the oracle's behavior. For the latter, we need to know that the oracle's answers depend only on $\dot{\alpha}^1$ as a type-$\dot{1}$ object and not on the choice of the $\dot{\alpha}^1$-oracle with which we question her. But an $\dot{\alpha}^1$ of type $\dot{1}$ has a unique oracle (in her behavior if not in her person), except in the case $\dot{\alpha}^1 = \lambda\dot{\alpha}^0\mathrm{u}$, in which she can have an oracle under Case $\overline{1}^1$ (who does not open the envelope) or one under Case $\overline{3}^1$ who opens the envelope but then invariably stands mute (there being no $r \in$ type 0 we can put in it that will please her). But for $\dot{\alpha}^1 = \lambda\dot{\alpha}^0\mathrm{u}$, an $\dot{\alpha}^2$-oracle, as just described, can answer our question "$\dot{\alpha}^2(\dot{\alpha}^1)$?" (and E_{γ_1} can be computed under $\Omega_{\dot{\alpha}^1}$) only under Case $\overline{2}^2$, and thus independently of the $\dot{\alpha}^1$-oracle.

Preparatory to using the $\dot{\alpha}^2$-oracles to define type $\dot{2}$, I ask whether the function $\dot{\alpha}^2$ whose values are revealed to us by an $\dot{\alpha}^2$-oracle is necessarily monotone. That is, will it be true that, for all $\dot{\alpha}^1$, $\overline{\dot{\alpha}^1} \in$ type $\dot{1}$, ($\dot{\alpha}^2(\dot{\alpha}^1)$ is defined, say $= m$) & $\dot{\alpha}^1 \subset \overline{\dot{\alpha}^1} \to (\dot{\alpha}^2(\overline{\dot{\alpha}^1})$ is defined and $= m$)? By examining the various cases, we quickly see that this will be true, except possibly when the $\dot{\alpha}^2$-oracle gives the value m under Subcase $\overline{3.3}^2$ on the basis of knowing that the $\dot{\alpha}^1$-oracle opens envelopes and $\dot{\alpha}^1$ has the value n for each r_ζ with $\zeta < \xi$. So she gives this value m for all such $\dot{\alpha}^1$'s, in particular for $\dot{\alpha}^1 = \{\langle r_\zeta, n\rangle | \zeta < \xi\}$, which has no other values. In this case, monotonicity requires that she gives the same value m under Subcase $\overline{3.2}^2$ for $\dot{\alpha}^1 = \lambda\dot{\alpha}^0 n$ (whose oracle declares that $\dot{\alpha}^1(\dot{\alpha}^0) = n$ without opening the envelope). I now mandate that an $\dot{\alpha}^2$-oracle must so conduct herself (i.e. be "monotonicity preserving") in order to be "licensed".

13.3c. *Definition of type* $\dot{2}$. Type $\dot{2}$ shall consist of the partial one-place functions $\dot{\alpha}^2$ from type $\dot{1}$ into type 0 whose values are revealed by licensed (i.e. monotonicity preserving) $\dot{\alpha}^2$-oracles.

13.3d. $\lambda\gamma^1 E_{\gamma^1}$ *under* Ω *represents a type-*$\dot{2}$ *object* $\dot{\alpha}^2$ such that, for each $\dot{\alpha}^1 \in$ type $\dot{1}$, computing E_{γ^1} under $\Omega_{\dot{\alpha}^1}$ gives $\dot{\alpha}^2(\dot{\alpha}^1)$. The only problem in inferring this from 13.3a—13.3c is whether, in mandating that the $\dot{\alpha}^2$-oracles preserve monotonicity, I imposed a requirement not necessarily observed in the computation of an E_{γ^1} under $\Omega_{\dot{\alpha}^1}$. One is tempted to deal with this by asking whether, when the computation of E_{γ^1} under $\Omega_{\dot\alpha^1}$ is completed under Subcase $\overline{3.3}^2$ with all the subcomputations of B resulting in the same value n (thus in particular for $\dot\alpha^1 = \{\langle r_\zeta, n\rangle | \zeta < \xi\}$), then simply by excising all those subcomputations of B we will not have in our hands a computation completed under Subcase $\overline{3.2}^2$ (for $\dot\alpha^1 = \lambda\dot\alpha^0 n$) with the same result. Actually, the situation may not be quite this simple, because e.g. there may be only in the excised subcomputations some surfacings of c for an application of the rule we are to state in 13.3e to some $\gamma^2(B)$ to the left of the excisions, the loss of which might call for changing the applicable subcase there. These reflections illustrate the kind of reasoning by which it can be established that the computation of E_{γ^1} under $\Omega_{\dot\alpha^1}$ (and computations generally) preserve monotonicity (monotonicity having been imposed on the oracles for the interpretations of the type-j variables for $j \geqslant 2$ and the $\dot\theta_t$'s).[5]

13.3e. For the next rule, the m obtained is exactly what the $\dot\alpha^2$-oracle gives when questioned with the $\dot\alpha^1$-oracle whom we abstracted in 13.2b from our analysis in 13.2a of the computation of E_{γ^0} under $\Omega_{\dot\alpha^0}$ (from which the Cases $\overline{2}^1$, $\overline{3}^1$ here come) when E_{γ^0}, γ^0 are the present $B(c)$, c.

Rule for computing $\gamma^2(B)$ under an assignment Ω including an interpretation $\dot\alpha^2$ for γ^2 with a choice of an $\dot\alpha^2$-oracle. Suppose we have come to a *surfacing of* γ^2 or γ^2-*face* $\gamma^2(B)$. For brevity, I proceed at once to *Subcase* $\overline{3.2}^2$ in which (*Case* $\overline{3}^2$) the $\dot\alpha^2$-oracle authorizes starting a subcomputation to $\gamma^2(B)$ of $B(c)$ (c a type-$\dot 0$ variable not occurring free in B) and (Case $\overline{2}^1$) without c surfacing this leads to the value n, whereupon, according as $\dot\alpha^2(\lambda\dot\alpha^0 n) = \text{u}$ or $\dot\alpha^2(\lambda\dot\alpha^0 n) = m$, the computation ends or the part begun with $\gamma^2(B)$ is completed thus:

(i)

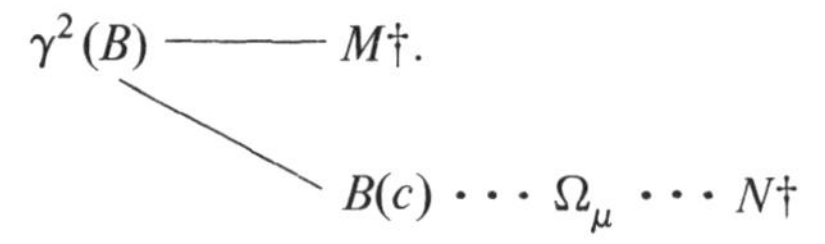

In *Subcase* $\overline{3.3}^2$ (*Case* $\overline{3}^2$ now with $\overline{3}^1$), if the computation of $\gamma^2(B)$ can be completed it will look like this:

(ii)

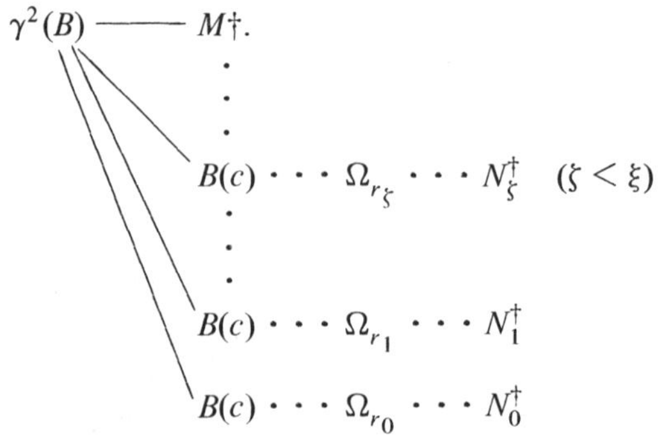

In the first (lowest) subcomputation of $B(c)$, begun without a value for c, c surfaces (Case $\bar{3}^1$), whereupon (if the computation can be completed) the $\dot{\alpha}^2$-oracle supplies r_0 as the value of c. On completing this subcomputation, with result n_0, then (if $\xi > 1$) the $\dot{\alpha}^2$-oracle authorizes a next higher subcomputation of $B(c)$ with c interpreted by r_1, etc., until a family of subcomputations has accumulated, representing a function $\dot{\rho} = \{\langle r_\zeta, n_\zeta \rangle | \zeta < \xi\}$ $(\xi > 0)$ such that ($\dot{\alpha}^1$ opens envelopes) & $\dot{\alpha}^1 \supset \dot{\rho}^1 \to \dot{\alpha}^2(\dot{\alpha}^1) = m$, whereupon the step "——$M\dagger$" is taken at the top.

This rule introduces the possibility of computation steps after an infinity of preceding steps. As an example, we might want our $\dot{\alpha}^2$-oracle to answer "$\dot{\alpha}^2(\dot{\alpha}^1)$?" only on the basis of knowing the values of $\dot{\alpha}^1$ on all the natural numbers, thus with $r_0, r_1, \ldots, r_\zeta, \ldots$ being $0, 1, 2, \ldots$ ($\xi = \omega$ in (ii)). This was the case in [**1978**] with $\alpha^2(\alpha^1)$ (undotted); and in our present semantics using the types $\dot{0}$, $\dot{1}$, $\dot{2}$, $\dot{3}$ we shall want the former theory with the types 0, 1, 2, 3 to be represented. Or we might e.g. want the $\dot{\alpha}^2$-oracle to explore the values of $\dot{\alpha}^1$ fully on the even numbers and then on some of the odd numbers (so $\xi > \omega$, with ω a value of ζ). The possibility of computation steps with infinitely many preceding steps thus arises from our unwillingness to restrict our $\dot{\alpha}^2$-oracles to asking only finitely many questions.

In a completed computation of E_{γ^1} under $\Omega_{\dot{\alpha}^1}$ (in 13.3a) with an E_{γ^1} not having in it a type-$\dot{2}$ variable γ^2 interpreted in $\Omega_{\dot{\alpha}^1}$ by an $\dot{\alpha}^2$ with an $\dot{\alpha}^2$-oracle, or something with a similar potency, there are only finitely many B_ζ's $(\xi < \omega)$.

For definiteness in the face of this possibility, we shall agree that, when we have a well-ordered family of (uncompleted) computations, each extending the preceding ones with no last, then the union of them shall be a computation iff it is a *well-founded* tree, i.e. a tree with only finite branches.

This convention is used, e.g., when in building toward (ii) we have a family of computations, comprising one of the form

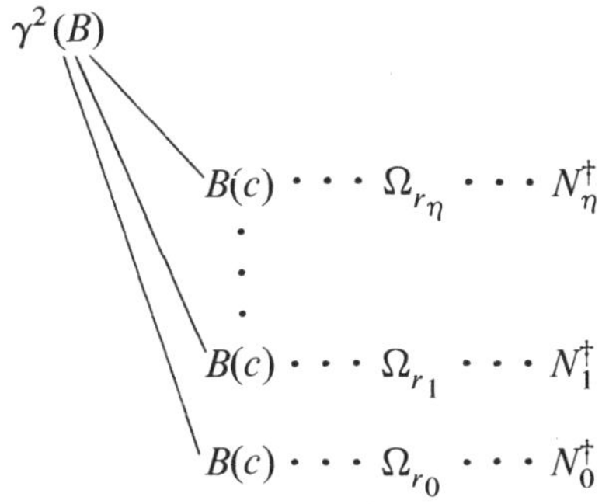

for each $\eta <$ a limit ordinal ζ. From the union of this family, thus accepted as a computation, we may be able to make the step to start a next higher subcomputation of $B(c)$ under Ω_{r_ζ} ($\zeta <$ the ξ of (ii)) or to add "——$M\dagger$" rightward from $\gamma^2(B)$ (so ζ is the ξ of (ii)).

In the case just illustrated, the union will always be well-founded; but cases occur when it is not [**1978** (XIV)].

13.4a. *Computation of* E_{γ^2} *under* $\Omega_{\dot\alpha^2}$. Again we are using a given $\dot\alpha^2$-oracle for the interpretation $\dot\alpha^2$ of γ^2. I pass at once to the principal case (*Subcase* $\overline{3.3^3}$), i.e. that (*Case* $\bar{3}^3$) γ^2 surfaces as $\gamma^2(B)$ and (Case $\bar{3}^2$ in 13.3b) the $\dot\alpha^2$-oracle opens envelopes. So a subcomputation of $B(c)$ to $\gamma^2(B)$ is started under the assignment in force there, and likewise at each subsequent surfacing of γ^2. What, if anything, can happen?

I shall analyze the computation with respect to *stages*, indexed by ordinals, each stage constituting its status immediately after new information has been called for from the $\dot\alpha^2$-oracle and has been provided. Of course, Stage 0 will not be reached if neither of the two events which first calls for information from the $\dot\alpha^2$-oracle occurs (namely, the subcomputation of some $B(c)$ being completed with result n without c having surfaced, or c surfacing), or if the $\dot\alpha^2$-oracle declines to supply the information (respectively, a value m for $\dot\alpha^2(\lambda\dot\alpha^0 n)$, or an interpretation r_0 for c). Similarly, subsequently to reaching a given stage, a next stage may not be reached, either because the computation does not progress to the point of calling for new information, or because the $\dot\alpha^2$-oracle does not supply it.

Supposing that we do reach at least a first stage (Stage 0), we will have the following picture. At any stage σ we have reached, in the computation of the moment there will be surfacings $\gamma^2(B)$ of γ^2 with various B's under respective assignments (maybe several with the same B and assignment). Each of these γ^2-faces with its subcomputations will be of the form (i) or (ii) displayed above, except that (ii) may be in various stages of incompletion, and except that the $M\dagger$ in (i) or in a completed (ii), or the top $B(c)$ in an uncompleted (ii), may have been authorized but not yet put there. (Our stages are counted from the moments when new information has just been called for and received, ready to be acted upon.)

Each of the γ^2-faces (with the 0-expression whose addition is authorized, but maybe not yet made) exhibits a budget of information about the $\dot\alpha^2$-oracle under questioning with a certain $\dot\alpha^1$-oracle ("What is $\dot\alpha^2(\dot\alpha^1)$?"). Several surfacings (perhaps with different B's or assignments) may exhibit the same budget. The distinct budgets can be shown on a chart by "branches" of the following forms (1), (2a), (2b) and (2c). The branches are indexed by ordinals in the order in which *new* functions $\dot\alpha^1$ enter, not just further information about old ones. The "o" says that $\dot\alpha^1$ (or more precisely, its oracle) opens envelopes.

(1) $\lambda\alpha^0 n : m\surd$

(2a) o $\varnothing : r_0?$

(2b) o $\{\langle r_\eta, n_\eta\rangle | \eta < \zeta\} : r_\zeta?$

(2c) o $\{\langle r_\zeta, n_\zeta\rangle | \zeta < \xi\} : m_\surd$

It would be a mistake to suppose that each of (2) must be completed with an "$m\surd$" (i.e. brought to the form (2c)) before further (1)'s and (2)'s are started. For, in computing E_{γ^2} under $\Omega_{\dot\alpha^2}$, if we come to a γ^2-face $\gamma^2(B)$ whose B in turn includes free occurrences of γ^2, there may be γ^2-faces within a subcomputation of

$B(c)$ to that γ^2-face. The subcomputations to those will have to be completed before the subcomputation of that $B(c)$ can be.

Without trying to analyze such interrelations between different γ^2-faces, we can simply say that chart branches may be extended and new ones started, skipping around from one to another.[10]

$$\circ \{\langle r_0, n_0\rangle, \langle r_1, n_1\rangle, \langle r_2, n_2\rangle\}: m\checkmark \qquad 0$$
$$\circ \{\langle r_0, n_0\rangle, \langle r_1, n_1\rangle, \langle r_2, \overline{n}_2\rangle, \langle \overline{r}_3, \overline{n}_3\rangle\}: \overline{r}_4? \qquad 1$$
$$\lambda\alpha^0 \overline{\overline{n}} : \overline{\overline{m}}\checkmark \qquad 2$$

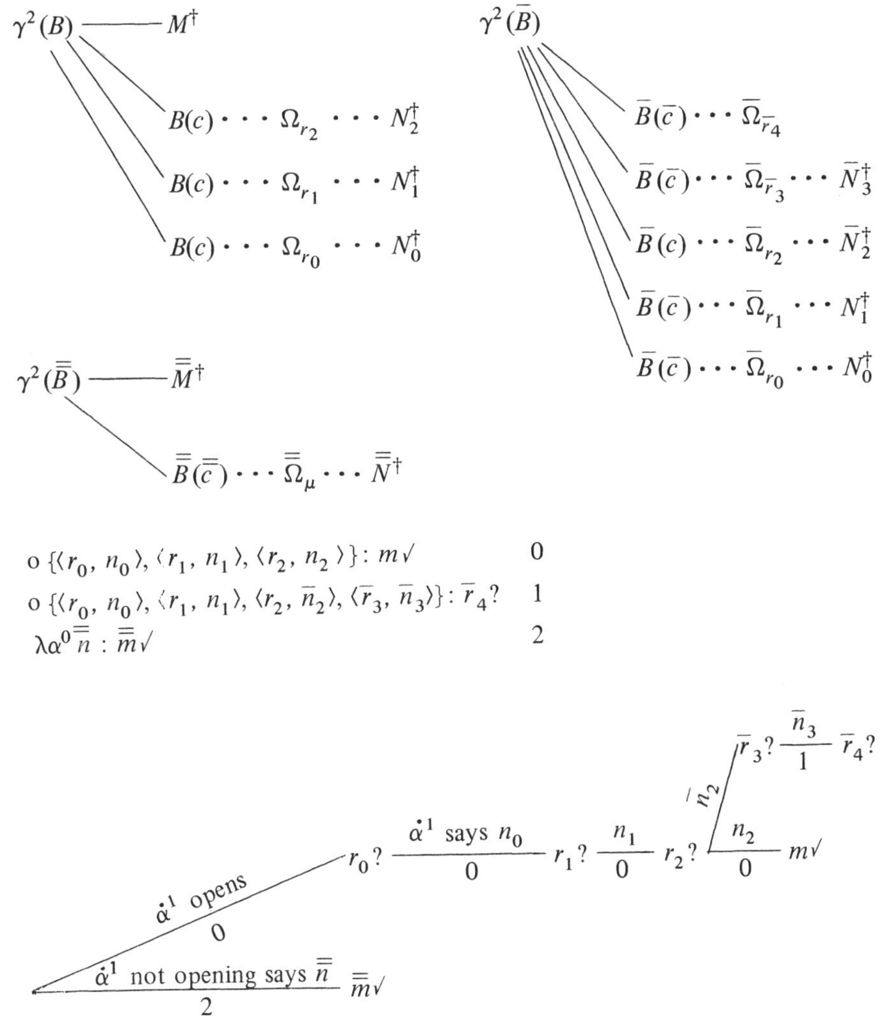

FIGURE 1. Samples of γ^2-faces and corresponding chart branches.

---

[10] The present approach to $\dot{\alpha}^3$-oracles (in 13.4b) makes the corresponding feature of them rather obvious, while in writing [**1978**] I only arrived at it through reflecting on Kierstead's example.[6] Formalizing his example, we do get an E_{γ^2} whose computation under $\Omega_{\dot{\alpha}^2}$ with a suitable $\dot{\alpha}^2$ (e.g. $\dot{\alpha}_1^2$ of [**1980**, 8.2]) contains subcomputations of $\overline{B}(\overline{c})$ within a subcomputation of $B(c)$.

I chose the term "branches" to fit what is a better format for presenting the information about $\dot{\alpha}^2$ questioned with $\dot{\alpha}^1$'s, in which we have a tree, with the vertices (after the first and at least before the last) along each branch representing arguments r of a function $\dot{\alpha}^1$, and the segments (after the first) representing values n of $\dot{\alpha}^1$ on the just preceding arguments. Two branches diverge after the first vertex for which the functions represented by them behave differently.

To illustrate, I show in Figure 1 three γ^2-faces with their corresponding branches, first in the format of (1)—(2c) and then in the tree representation. I have written B, $\overline{B}$ and $\overline{\overline{B}}$ in the γ^2-faces, but two or all three of them might be the same 0-expression computed under different assignments (the assignments in force where they arise).

The chart in Figure 1 is a Stage-6 chart, each of its seven line segments (in the tree representation) having been added to reach a stage. The branches are indexed by 0, 1, 2. I am taking it that the $\gamma^2(B)$ shown at the upper left was the first surfacing of γ^2 in the computation of E_{γ^2}, and the subcomputations for it were carried through the third (under Ω_{r_2} with result n_2) before the third subcomputation to the $\gamma^2(\overline{B})$ at the upper right was completed (under $\overline{\Omega}_{r_2}$ with result $\overline{n}_2 > n_2$).[11] The branch indexed by 0 belongs to the $\gamma^2(B)$ (and maybe to some other γ^2-faces $\gamma^2(\tilde{B})$ with $\tilde{B}(\tilde{c})$ receiving the same respective values n_0, n_1, n_2 under $\tilde{\Omega}_{r_0}$, $\tilde{\Omega}_{r_1}$, $\tilde{\Omega}_{r_2}$); while Branch 1, belonging to the $\gamma^2(\overline{B})$, diverges from it after the "r_2?". Branch 2 arose after the computation of $\overline{B}(\overline{c})$ under $\overline{\Omega}_{r_2}$.

Suppose we have run through a series of stages indexed by the ordinals $\tau <$ a limit ordinal σ. These stages correspond to moments in pursuing the computation of E_{γ^2} under $\Omega_{\dot{\alpha}^2}$ with no last. If the union of these momentary computations is a well-founded tree, then, as we said in 13.3e, we take the union to be a computation; otherwise, we shall not continue toward a Stage σ.

Suppose the union is well founded. In the union of the corresponding charts, there may be indexed branches with no last status (being extended after each stage $\tau < \sigma$). I call these *aspiring* branches. Each correspond to one or more γ^2-faces $\gamma^2(B)$ as in (ii) but uncompleted. For each, new information from the $\dot{\alpha}^2$-oracle will be needed sooner or later (but we know it now), if the computation can be completed. I say "sooner or later", because, if this happens (in a given act of unionization) simultaneously with several branches, the next computation steps for their γ^2-faces (under our rule in 13.3e for computing $\gamma^2(B)$) will have to be done in series (maybe not even consecutively). So, for each of them, we question the $\dot{\alpha}^2$-oracle with the $\dot{\alpha}^1$ represented by the newly unionized branch, hopefully obtaining an "r?" or an "$m\surd$", which we then write at the end of the aspiring branch (on a vertex at a limit position!). If we can thus finish each aspiring branch (or if there are no aspiring branches) in the unionized chart, we then have

[11]As dealt with in [**1982**, 11.2], the tree we draw on the chart at each stage is an initial subtree of the "$\dot{\alpha}^2$-oracle tree", which (as described in 10.1) gives full information about the $\dot{\alpha}^2$-oracle. At a vertex where branches diverge for $n = 0, 1, 2, \ldots$ as the values of $\dot{\alpha}^1(r)$, we draw them in order from the bottom up.

the Stage σ chart, and we construe the unionized computation as Stage σ of the computation.

That the unionization might set the stage for steps associated with several γ^2-faces is a consequence of our entertaining the possibility of skipping from working on one subcomputation to working on another under the computation rule of 13.4e for $\gamma^3(B)$ with an $\dot{\alpha}^3$-oracle (and under the like case for $\dot{\theta}_t(\ldots B\ldots)$).[5]

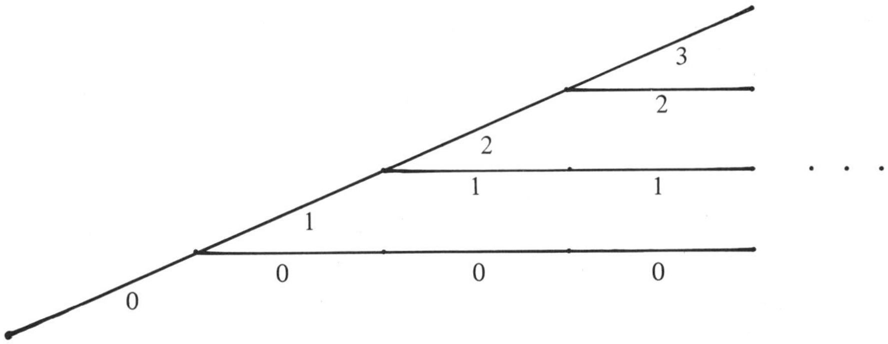

FIGURE 2. Example of an unindexed chart branch after unionization.

Figure 2 shows how an unindexed branch might arise from the unionization. The branches which are indexed by $0, 1, 2, \ldots$ each run horizontally rightward after a finite number of line segments slanting upwards. But the branch slanting upward ad infinitum has no index. (An indexed branch will show its index on all segments from a certain one on, or on a final vertex at an infinite position.) It could be that this unindexed branch does not describe the subcomputations for any γ^2-face $\gamma^2(B)$ (present or future), though each finite initial segment of it runs with those for γ^2-faces going with indexed branches which began to run horizontally further out. Or there may be, now or later, γ^2-faces $\gamma^2(B)$ for which it does represent the family of subcomputations as they accumulated up through the stages $\tau < \sigma$, or as they will accumulate later. (The subcomputations for these γ^2-faces simply ran slowly enough in the computation of E_{γ^2} that at each stage $\tau < \sigma$ the requisite information from the $\dot{\alpha}^2$-oracle had already been obtained in pursuing a subcomputation for another γ^2-face which at a later stage $\tau < \sigma$ led to a different n.) In the latter case, we will eventually need to query the $\dot{\alpha}^2$-oracle with the function $\dot{\alpha}^1$ represented by the unindexed infinite chart branch, namely when it first happens in the pursuit of the computation that the next step (if one is possible) is to be from the incomplete (ii) for a γ^2-face represented by the branch. If we get an "r?" or an "$m\surd$", we then put that at the end of the branch, and index the branch (showing the index on the vertex thus added at an infinite position) with the next ordinal not already used as an index.

The computation of E_{γ^2} under $\Omega_{\dot{\alpha}^2}$ may be completed with result s, of course only if the chart at the latest stage has only branches ending with an "$m\surd$". (A branch ending with an "r?" would correspond to an uncompleted application of the rule for computing $\gamma^2(B)$, which would render the whole computation uncompleted.)

13.4b. $\dot{\alpha}^3$-*oracles.* In the manner already illustrated at types $\dot{1}$ and $\dot{2}$, this analysis of the role of the $\dot{\alpha}^2$-oracle in the computation of E_{γ^2} under $\Omega_{\dot{\alpha}^2}$ leads us to a formulation of how $\dot{\alpha}^3$-oracles shall operate.

The key observation is that in such a computation, for a given fixed E_{γ^2} and fixed assignment Ω to the free variables of E_{γ^2} other than γ^2 and to the Θ, however the $\dot{\alpha}^2$ assigned to γ^2 and its oracle may vary, the following is the case. In the principal case (*Subcase* $\overline{3.3^3}$), at each moment of the computation what exists is the same for each $\dot{\alpha}^2$-oracle compatible with the chart at the last stage reached (i.e. such that that chart states true information about her), or, after a family of stages $\tau <$ *a* limit ordinal σ, compatible with the charts at all those stages τ. For indeed, as we follow the progress of the computation, the chart (updated from stage to stage) logs all the information thus far elicited about the $\dot{\alpha}^2$-oracle, and everything else in the situation is fixed by E_{γ^2} and Ω.

We get an $\dot{\alpha}^3$-oracle by letting her, in the principal case, simply from contemplating the chart of the moment (or in approaching a limit stage, the union of the preceding charts), ask the $\dot{\alpha}^2$-oracle for new information, instead of continuing the computation of an E_{γ^2} under $\Omega_{\dot{\alpha}^2}$ from the corresponding stage to determine her next question. Then in the tree format, each chart branch represents at each stage an $\dot{\alpha}^1$, which is a creature of the $\dot{\alpha}^3$-oracle (who determines what is written on the segments, in particular the n's) used by her in probing the $\dot{\alpha}^2$-oracle (who supplies the r's and m's written on the vertices).

Branches of the form (1) (as I said in [**1982**, 11.2], where the $\dot{\alpha}^3$-oracles are described in full detail) are of the *first kind*; of the form (2a), (2b) or (2c) of the *second kind*. When the $\dot{\alpha}^3$-oracle does her part toward reaching a stage (either initially toward reaching Stage 0, or after reaching Stage σ toward reaching Stage $\sigma + 1$), I say she is using one of "Options 1–5". This corresponds to the progress of the computation of E_{γ^2} under $\Omega_{\dot{\alpha}^2}$ toward turning up the need for new information about $\dot{\alpha}^2$. In particular, when she seeks to acquire a branch (1), I say she is using *Option* 1 toward the next stage in her questioning of the $\dot{\alpha}^2$-oracle; when she seeks to acquire a branch (2a) (only once), *Option* 2; when she seeks to extend an existing branch of the second kind by a next segment, *Option* 3; when she seeks to add a segment diverging from an existing branch of the second kind, *Option* 4; and when she picks up an infinite unindexed branch, *Option* 5.

The final result, namely that $\dot{\alpha}^3(\dot{\alpha}^2) = s$, if reached, is decided upon by the $\dot{\alpha}^3$-oracle contemplating a chart in a stage in which each branch has some "$m_\kappa \surd$" at its end and represents say $\dot{\rho}_\kappa$ ($\kappa < \mu$), where κ indexes the branch, and ($\dot{\alpha}^2$ opens envelopes) & ($\dot{\alpha}^2 \supset \{\langle \dot{\rho}_\kappa, m_\kappa \rangle | \kappa < \mu\}) \rightarrow \dot{\alpha}^3(\dot{\alpha}^2) = s$.

It is not as simple as for $\dot{\alpha}^2$-oracles to get that the result is independent of the choice of the oracle used in our questioning of the $\dot{\alpha}^3$-oracle. There is a greater range of possibilities for having various oracles for a given type-$\dot{2}$ object $\dot{\alpha}^2$ than was the case for a type-$\dot{1}$ object $\dot{\alpha}^1$. As the matter is dealt with in 11.2, I get this independence, and simultaneously the monotonicity of the function $\dot{\alpha}^3$ given by the $\dot{\alpha}^3$-oracle's performance, by legislating that an $\dot{\alpha}^3$-oracle to be licensed must observe the following version of monotonicity: *For given* $\dot\alpha^2$- *and* $\overline{\dot\alpha^2}$-*oracles with* $\dot\alpha^2 \subset \overline{\dot\alpha^2}$, *if the* $\dot\alpha^3$-*oracle declares that* $\dot\alpha^3(\dot\alpha^2) = s$, *she must also declare that* $\dot\alpha^3(\overline{\dot\alpha^2}) = s$.[12]

13.4c. *Definition of type* $\dot 3$. Type $\dot 3$ shall comprise the partial one-place functions $\dot\alpha^3$ from type $\dot 2$ into type 0 whose values are revealed by licensed $\dot\alpha^3$-oracles.

13.4d. $\lambda\gamma^2 E_{\gamma^2}$ *under* Ω *represents a type-*$\dot 3$ *object* $\dot\alpha^3$ such that, for each $\dot\alpha^2 \in$ type $\dot 2$, computing E_{γ^2} under $\Omega_{\dot\alpha^2}$ gives $\dot\alpha^3(\dot\alpha^2)$. That the $\dot\alpha^3$-oracle which we obtain in 13.4b by abstracting from 13.4a is properly licensed can be established as a corollary of our general monotonicity result for computations.[5]

13.4e. *Rule for computing* $\gamma^3(B)$ under an assignment Ω which includes an interpretation $\dot\alpha^3$ for γ^3 with a choice of an $\dot\alpha^3$-oracle. It is straightforward to formulate this rule in detail,[5] just as we went from "computation of E_{γ^1} under $\Omega_{\dot\alpha^1}$" in 13.3a to "$\dot\alpha^2$-oracles" in 13.3b to the "rule for computing $\gamma^2(B)$" in 13.3e. The s given by the rule, or there being no result, is exactly what the $\dot\alpha^3$-oracle gives when questioned with the $\dot\alpha^2$-oracle who is abstracted in 13.3b from the computation of E_{γ^1} under $\Omega_{\dot\alpha^1}$ in 13.3a when E_{γ^1}, γ^1 are the present $B(\gamma)$, γ. Indeed, in the principal case (*Subcase* $\overline{3.3}^3$), we will have the same succession of Stage σ charts.[13] There will be subcomputations of $B(\gamma)$ corresponding to each chart branch, with γ interpreted in the end (when the chart branch is completed with an "$m_{\surd}$") by the $\dot\alpha^1 = \dot\rho$ represented by that branch. Thus (if a value s is reached) we will have a family of subcomputations of $B(\gamma)$ under $\Omega_{\dot\rho_\kappa}$, which (unlike those of $B(c)$ under Ω_{r_ξ} for computing $\gamma^2(B)$) are not necessarily completed in the order in which they were started.

13.5. *Unimonotone functions.* I have called the functions $\dot\alpha^j$ comprised in type $j^{\cdot}$ for $j = 1, 2, 3$ *unimonotone*, meaning that they are *monotone* and, for each $\dot\alpha^{j-1}$ for which $\dot\alpha^j(\dot\alpha^{j-1})$ is defined, there is a *un*ique *i*ntrinsically determined minimum *basis* $\dot\beta^{j-1}$ with $\dot\beta^{j-1} \subset \dot\alpha^{j-1}$ and $\dot\alpha^j(\dot\beta^{j-1})$ defined and $= \dot\alpha^j(\dot\alpha^{j-1})$. For $j > 1$, this $\dot\beta^{j-1}$ is the type-$(j-1)^{\cdot}$ subfunction of $\dot\alpha^{j-1}$ which has been explored intrinsically (i.e. from within) in determining that $\dot\alpha^j(\dot\alpha^{j-1})$ is defined and its value. For $j = 1$, unimonotonicity is synonymous with monotonicity (i.e. the *un.-i.*-basis part is automatic).

Although this semantics is undoubtedly complicated, it seems to me in a way very natural; and the fact that it works is a fundamental property of our type-$\dot 0$, -$\dot 1$, -$\dot 2$, -$\dot 3$ objects that may conceivably have interesting applications.

---

[12]An alternative formulation is also given there, in which this requirement (called there (4)) is broken down into three others, of which (1) separately requires the result s, if given, to be independent of the choice of the $\dot\alpha^2$-oracle, and (2) and (3) mandate special cases of monotonicity.

[13]The following is the case, as we should expect. Applying the present rule to the computation of simply $\gamma^3(\gamma^2)$ (i.e. taking B to be just a variable γ^2, with Ω including interpretations $\dot\alpha^3$, $\dot\alpha^2$ of γ^3, γ^2 with given $\dot\alpha^3$-, $\dot\alpha^2$-oracles), we get the same result s or no result, and in the principal case the same series of charts, as in questioning the $\dot\alpha^3$-oracle with the $\dot\alpha^2$-oracle.

This follows (by the proposition just stated in the text) from the like proposition about the computation of $\gamma^2(\gamma^1)$, by which the $\dot\alpha^2$-oracle abstracted from it is the $\dot\alpha^2$-oracle we started with.

This in turn follows from the like for the computation of $\gamma^1(\gamma^0)$.

References

Kierstead, David Philip

[**1980**] *A semantics for Kleene's j-expressions*. The Kleene Symposium (J. Barwise, H. J. Keisler and K. Kunen, eds.), North-Holland, Amsterdam, New York and Oxford, pp. 353–336.

[**1983**] *Syntax and semantics in higher-type recursion theory*, Trans. Amer. Math. Soc. **276**, 67–105.

Kleene, Stephen Cole

[**1952**] *Introduction to metamathematics*, North-Holland, Amsterdam, P. Noordhoff, Groningen, and D. Van Nostrand, Princeton, New York and Toronto. Eighth reprint, Wolters-Noordhoff, Groningen North-Holland, Amsterdam, New York and Oxford, and Elsevier North-Holland, New York, 1980.

[**1959**] *Recursive functionals and quantifiers of finite types* I, Trans. Amer. Math. Soc. **91**, 1–52.

[**1963**] *Recursive functionals and quantifiers of finite types* II, Trans. Amer. Math. Soc. **108**, 106–142.

[**1978**] *Recursive functionals and quantifiers of finite types revisited* I. Generalized recursion theory II, Proceedings of the 1977 Oslo Symposium (J. E. Fenstad, R. O. Gandy and G. E. Sacks, eds.), North-Holland, Amsterdam, New York and Oxford, pp. 185–222.

[**1980**] *Recursive functionals and quantifiers of finite types revisited* II. The Kleene Symposium (J. Barwise, H. J. Keisler and K. Kunen, eds.), North-Holland, Amsterdam, New York and Oxford, pp. 1–29.

[**1982**] *Recursive functionals and quantifiers of finite types revisited* III. Patras Logic Symposion (G. Metakides, ed.), North-Holland, Amsterdam, New York and Oxford, pp. 1–40.

Turing, Alan Mathison

[**1939**] *Systems of logic based on ordinals*, Proc. London Math. Soc. Ser. 2 **45**, pp. 161–228.

Department of Mathematics, University of Wisconsin, Madison, Wisconsin 53706

Proceedings of Symposia in Pure Mathematics
Volume **42**, 1985

Canonical Forms and Hierarchies in Generalized Recursion Theory

PHOKION G. KOLAITIS[1]

The present paper is divided into two parts. The first is a contribution to the foundations of recursion in an arbitrary sequence of functionals $\Phi_1, \ldots, \Phi_s$. Kechris and Moschovakis [**1977**] showed how to develop this theory within the context of inductive definability. Functional recursion contains as important special cases both Kleene's theory of recursion in higher types and the theory of positive elementary induction. We introduce here a restricted class of explicitly defined functionals, which we call functionals in canonical form, and we show that iterating them once suffices to generate all partial functions which are recursive in $\Phi_1, \ldots, \Phi_s$. The functionals in canonical form provide useful substitutes for the Kleene master recursion when no coding apparatus is available.

In the second part of the paper we concentrate on recursion in functionals representing quantifiers over a structure. We relate this theory to second-order definability by establishing the existence of a hierarchy for the relations which are recursive in the type 2 objects and functionals associated with quantifiers. This hierarchy is obtained by iterating "lightface" Δ^1_1 definability with basis the previously constructed relations. As direct consequences we derive model-theoretic characterizations of recursion in quantifiers using "lightface" versions of the schemata of Δ^1_1 Comprehension and Σ^1_1 Collection.

The first section of each part contains the basic definitions and a minimal amount of the necesssary background material. We assume, however, some familiarity with Kechris and Mochovakis [**1977**] for Part I, while Part II depends to a certain extent on Kolaitis [**1979** and **1980**].

1980 *Mathematics Subject Classification*. Primary 03D65, 03D70; Secondary 03D75, 03D55.
[1]Research for this paper was partially supported by NSF Grant MCS 80-02763.

PART I. CANONICAL FORMS FOR FUNCTIONAL RECURSION

1. Introduction.

1.1. The theory of recursion in higher types was introduced by Kleene [**1959**a, **1963**]. The partial recursive functions of this theory are defined by means of nine schemata on the objects of finite type. They are the schemata S1–S8 for the primitive recursive functions on the finite types augmented with a powerful reflection schema S9, which "builds in" the enumeration property.

Recursion in higher types attracted many researchers, but at the same time it was considered a rather hard subject, because of the conceptual and technical difficulties surrounding it. Platek [**1966**] made a significant contribution to the foundations of this theory. He took as basic the fixed-point operation and succeeded in providing a rigorous approach to higher type recursion within the framework of the theory of induction. His work, however, has certain limitations, which are discussed at length in Feferman [**1977**] and also in Fenstad [**1980**]. Platek's account is technically quite difficult and, moreover, Kleene's theory is obtained from it in an involved way.

Moschovakis [**1977**] and Feferman [**1977**] introduced a new approach to recursion in higher types and generalized recursion theory within the context of inductive definability. The development of the theory is due to Moschovakis and is presented in detail in Kechris and Moschovakis [**1977**]. This is a coding-free approach which is both conceptually and technically simpler. It overcomes the limitations of Platek's account and contains in a natural way both the theory of recursion in higher types and the theory of positive elementary induction.

In Moschovakis' treatment the partial recursive functions are obtained by iterating once the operative functionals in the smallest class which contains a given list of functionals and satisfies certain minimal closure conditions (such as being closed under composition, definition by cases, substitution of projections, etc.). Therefore, in this approach the main idea is to specify first the kinds of inductions which are allowed (in terms of closure conditions) and then to take once the fixed points of all operative functionals.

In the light of the above work one can go back and view Kleene's schemata S1–S9 as the clauses of a single inductive definition. In other words, instead of specifying the kinds of the allowable inductions, Kleene introduces directly a single master operative functional and obtains the partial recursive functionals from its fixed point. It is, of course, the availability of the coding apparatus and the systematic use of indices that makes possible the simulation of the various inductions by a single functional.

1.2. In the first part of this paper we establish that the theory of functional recursion in Moschovakis [**1977**] and Kechris and Moschovakis [**1977**] can be derived by iterating once certain explicitly defined functionals which have the form of the Kleene master operator and which we call *functionals in canonical form*. More specifically, we show that even in the absence of a coding machinery we can generate all the partial recursive functions by considering only those functionals which are definitions by a finite number of cases and each case is an *evaluation functional* or a *composition functional* or an *initial functional*.

At this point we should mention the reasons for introducing this restricted class of functionals. Our main motivation is to provide a useful technical tool in the case where no coding apparatus is available. The functionals in canonical form are easier to handle and work with than the arbitrary operative functionals determined by the closure conditions. For example, when proving a result about the stages or the fixed point of an arbitrary operative functional Φ, one is often forced to use an induction on the subfunctionals of Φ. However, a subfunctional of such an operative functional is not necessarily operative and this fact usually requires devising an elaborate induction hypothesis. In contrast, this difficulty does not occur when dealing with operative functionals in canonical form. The above point will become more transparent in the second part of the paper, where we give concrete applications of these functionals on structures which do not possess definable coding functions.

2. Functional recursion and suitable classes. In this section we review briefly the definitions of the basic notions in functional recursion. In general, we follow the notation and terminology of Moschovakis **[1977]** and Kechris and Moschovakis **[1977]**, where most of these definitions were introduced first.

2.1. Let A be a set with two distinguished elements 0 and 1, let W be a set such that $\{0,1\} \subseteq W \subseteq A$ and for each $k \in \omega$ let $\mathcal{PF}_k$ be the set of all k-ary partial functions from A to W. A *functional* (on A with values in W) is a partial mapping

$$\Phi\colon A^n \times \mathcal{PF}_{k_1} \times \mathcal{PF}_{k_2} \times \cdots \times \mathcal{PF}_{k_m} \to W$$

which is *monotone*, i.e. if $p_1 \subseteq q_1, p_2 \subseteq q_2, \ldots, p_m \subseteq q_m$ and $\Phi(\bar{x}, p_1, \ldots, p_m) = w$, then $\Phi(\bar{x}, q_1, \ldots, q_m) = w$. The *signature* of such a functional is the tuple $(n, k_1, \ldots, k_m)$. In particular, the n-ary partial functions from A into W are identified with the functionals of signature (n). We say that a functional is *operative* in case its signature is of the form $(n, n, k_1, \ldots, k_m)$. If $\Phi(\bar{x}, p, \bar{q})$ is an operative functional, then we can iterate it and define by induction on the ordinals its stages $\{\Phi^\xi\}$ as follows:

$$\Phi^\xi(\bar{x}, \bar{q}) = \Phi(\bar{x}, \lambda\bar{y}\Phi^{<\xi}(\bar{y}, \bar{q}), \bar{q}),$$

where $\Phi^{<\xi}(\bar{y}, \bar{q}) = w \Leftrightarrow$ for some $\eta < \xi$, $\Phi^\eta(\bar{y}, \bar{q}) = w$. The union of all stages gives rise to the functional Φ^∞ *inductively defined by* Φ, i.e.

$$\Phi^\infty(\bar{x}, \bar{q}) = w \Leftrightarrow \text{for some } \xi, \quad \Phi^\xi(\bar{x}, \bar{q}) = w.$$

Notice that if the functional Φ is of signature (n, n), then Φ^∞ is actually an n-ary partial function from A into W. In this case the least ordinal ξ such that $\Phi^\xi = \Phi^{<\xi}$ is called *the closure ordinal of* Φ and is denoted by $\|\Phi\|$.

Let $\mathcal{F}$ be a class of functionals on A with values in W. We say that a functional $\Psi(\bar{x}, \bar{q})$ is *$\mathcal{F}$-recursive* if there is an operative functional $\Phi(\bar{u}, \bar{x}, p, \bar{q})$ in $\mathcal{F}$ and a sequence $\bar{a}$ of 0's and 1's such that

$$\Psi(\bar{x}, \bar{q}) = \Phi^{\infty}(\bar{a}, \bar{x}, \bar{q}).$$

We write $\mathscr{F}$-REC for the collection of all $\mathscr{F}$-recursive functionals on A with values in W.

2.2. Let $\mathfrak{A} = \langle A, R_1,\dots,R_n, f_1,\dots,f_m, c_1,\dots,c_k\rangle$ be a structure such that 0 and 1 are among the constants $c_1,\dots,c_k$ and let W be a set such that $\{0,1\} \subseteq W \subseteq A$.

DEFINITION. A class $\mathscr{F}$ of functionals on A with values in W is *suitable* if it contains the functions and functionals (i)–(iv) and is closed under the rules (v)–(x) below.

(i) *The constant functions* φ_0 *and* φ_1, where $\varphi_0(x) = 0$ and $\varphi_1(x) = 1$ for all $x \in A$.

(ii) *The characteristic function of the set* $2 = \{0,1\}$, i.e. the function

$$\chi_2 = \begin{cases} 0 & \text{if } x = 0 \text{ or } x = 1, \\ 1 & \text{if } x \neq 0 \text{ and } x \neq 1. \end{cases}$$

(iii) *The characteristic function* χ_{R_j} *of each relation* R_j, $j = 1,2,\dots,n$, *of the structure*, i.e. the function

$$\chi_{R_j}(\bar{x}) = \begin{cases} 0 & \text{if } \bar{x} \in R_j, \\ 1 & \text{if } \bar{x} \notin R_j. \end{cases}$$

(iv) *The evaluation functional* $\Phi(\bar{x}, p) = p(\bar{x})$.

(v) *Addition of variables.* If $\Phi(\bar{x}, \bar{p})$ is in $\mathscr{F}$, then so is $\Psi(\bar{y}, \bar{x}, \bar{z}, \bar{q}, \bar{p}, \bar{r}) = \Phi(\bar{x}, \bar{p})$.

(vi) *Composition.* If $\Phi(a, \bar{x}, \bar{p})$ and $X(\bar{x}, p)$ are in $\mathscr{F}$, then so is $\Psi(\bar{x}, \bar{p}) = \Phi(X(\bar{x}, \bar{p}), \bar{x}, \bar{p})$.

(vii) *Definition by cases.* If $\Phi(\bar{x}, \bar{p})$ and $X(\bar{x}, p)$ are in $\mathscr{F}$, then so is

$$\Psi(a, \bar{x}, \bar{p}) = \begin{cases} \Phi(\bar{x}, \bar{p}) & \text{if } a = 0, \\ X(\bar{x}, \bar{p}) & \text{if } a \neq 0. \end{cases}$$

(viii) *Substitution by the functions and the constants of the structures.* If $\Phi(a, \bar{x}, p)$ is in $\mathscr{F}$ and f is one of the functions $f_1,\dots,f_m$ of the structure or one of the constant functions $\bar{x} \mapsto c_1,\dots,\bar{x} \mapsto c_k$, then $\Psi(\bar{x}, p)$ is in $\mathscr{F}$, where $\Psi(\bar{x}, \bar{p}) = \Phi(f(\bar{x}), \bar{x}, \bar{p})$.

(ix) *Substitution by projections.* A projection is any mapping π of the form $\pi(x_1,\dots,x_l) = x_j$, where $1 \leqslant j \leqslant l$. Thus, if $\Phi(y_1,\dots,y_s, \bar{p})$ is in $\mathscr{F}$ and $\pi_1,\dots,\pi_s$ are projections, then $\Psi(\bar{x}, \bar{p})$ is in $\mathscr{F}$, where $\Psi(\bar{x}, \bar{p}) = \Phi(\pi_1(\bar{x}),\dots,\pi_s(\bar{x}), \bar{p})$.

(x) *Function substitution.* If $\Phi(\bar{x}, q_1,\dots,q_m)$ and $X_1,\dots,X_l$ are in $\mathscr{F}$, then so is

$$\Psi(\bar{x}, \bar{p}) = \Phi(\bar{x}, \lambda\bar{y}_1 X_1(\bar{y}_1, \bar{x}, \bar{p}),\dots,\lambda\bar{y}_m X_m(\bar{y}_m, \bar{x}, \bar{p})).$$

2.3. The notion of a suitable class of functionals plays a crucial role in developing higher type recursion within the framework of inductive definability. Moschovakis **[1977]** and Kechris and Moschovakis **[1977]** showed that suitable classes possess a smooth recursion theory. In fact, in these papers it is proved that if $\mathscr{F}$ is a suitable class of functionals on A with values in W, then the resulting collection $\mathscr{F}$-REC of all $\mathscr{F}$-recursive functionals is also a suitable class. Moreover, iteration of $\mathscr{F}$-recursive functionals does not lead outside the class $\mathscr{F}$-REC, and consequently

$$(\mathscr{F}\text{-REC})\text{-REC} = \mathscr{F}\text{-REC}.$$

Assume now that $\overline{\Phi} = (\Phi_1, \ldots, \Phi_s)$ is a finite sequence of functionals on A with values in W. We can associate a recursion theory with these functionals by defining the notion of a *recursive in* $\overline{\Phi}$ partial function from A into W. For this we let $\mathscr{F}[\overline{\Phi}]$ be the smallest suitable class of functionals which contains the functionals $\Phi_1, \ldots, \Phi_s$. We say then that a partial function $\varphi: A^k \to W$ is *recursive in* $\overline{\Phi}$ if it is $\mathscr{F}[\overline{\Phi}]$-recursive.

A relation R on A is *semirecursive in* $\overline{\Phi}$ if it is the domain of a recursive in $\overline{\Phi}$ partial function. We say that the relation R is *recursive in* $\overline{\Phi}$ if its characteristic function is recursive in $\overline{\Phi}$. The class ENV$[\overline{\Phi}]$ (the *Envelope of* $\overline{\Phi}$) is the collection of all semirecursive in $\overline{\Phi}$ relations, while the class SEC$[\overline{\Phi}]$ (the *Section of* $\overline{\Phi}$) is the collection of all recursive in $\overline{\Phi}$ relations. If $\bar{x}$ is a finite sequence from A, then all the above concepts relativize in a direct way. For example, a partial function φ is *recursive in* $\overline{\Phi}$ *from* $\bar{x}$ if there is a recursive in $\overline{\Phi}$ partial function ψ such that $\varphi(\bar{y}) = \psi(\bar{x}, \bar{y})$. We also put

$$\text{ENV}[\overline{\Phi}, \bar{x}] = \text{all semirecursive relations in } \overline{\Phi} \text{ from } \bar{x},$$
$$\text{SEC}[\overline{\Phi}, \bar{x}] = \text{all recursive relations in } \overline{\Phi} \text{ from } \bar{x}.$$

2.4. Most branches of generalized recursion theory can be developed within the above framework by considering specific lists $\overline{\Phi} = (\Phi_1, \ldots, \Phi_s)$ of functionals over appropriate structures. The following principal examples illustrate the applicability of this approach.

(i) If we take $W = A$ and we let $\Phi_1, \ldots, \Phi_s$ be the functions $f_1, \ldots, f_s$ of an abstract structure $\mathfrak{A} = \langle A, R_1, \ldots, R_n, f_1, \ldots, f_s, c_1, \ldots, c_k\rangle$, then we obtain the theory of *prime computability over* $\mathfrak{A}$, which was introduced by Moschovakis [**1969a**]. In particular, on the structure of arithmetic $\mathbf{N} = \langle \omega, =, s, 0, 1\rangle$ (where s is the successor function) the prime computable functions are precisely the ordinary recursive partial functions.

(ii) Let $\mathfrak{A} = \langle A, R_1, \ldots, R_n, c_1, \ldots, c_k\rangle$ be a structure such that the equality $=_A$ on A is among the relations $R_1, \ldots, R_n$, let $W = \{0, 1\}$ and let $\Phi_1 = \mathbf{E}^\#$ be the functional which embodies unrestricted quantification on A (i.e. the functional $\mathbf{E}^\#$ can tell if a partial function from A into W has a root or not). It is then well known that recursion in $\mathbf{E}^\#$ coincides with the theory of *positive elementary induction on* $\mathfrak{A}$, in the sense that a relation R on A is semirecursive in $\mathbf{E}^\#$ if and only if it is ("lightface") inductive on $\mathfrak{A}$. Thus recursion in $\mathbf{E}^\#$ on an abstract structure is a natural generalization of the hyperarithmetic theory on the integers.

(iii) Another interesting recursion theory arises if we consider a structure $\mathfrak{A} = \langle A, R_1,\ldots,R_n, f_1,\ldots,f_m, c_1,\ldots,c_k\rangle$ such that $=_A$, ω, $\leqslant_\omega$ are among the relations $R_1,\ldots,R_n$, we take $W = \omega$ and we let $\Phi_1 = \mathbf{E}$ be the the functional which expresses equality for subsets of the universe A of the structure (or, equivalently, we take the type 2 object on A which can tell if a total function from A into ω has a root or not). This theory is a direct generalization of Kleene's *recursion in normal higher type objects*, because recursion in $\mathbf{E}$ on the appropriate structure with universe $\mathrm{Tp}(n)$ (the set of objects of type n) is precisely Kleene recursion in the type $(n + 2)$ object ${}^{n+2}\mathbf{E}$.

Finally, if one takes the functional $\mathbf{E}$ uniformly on every element of the universe of sets and makes certain minor modifications, then one arrives at the theory of *set recursion* or **E**-*recursion*, which was introduced by Normann [**1978**] and is currently the subject of active investigation.

3. Functionals in canonical form. The blanket assumption throughout this section is that $\mathfrak{A} = \langle A, R_1,\ldots,R_n, f_1,\ldots,f_m, c_1,\ldots,c_k\rangle$ is a structure with 0 and 1 among its constants, W is a set such that $\{0,1\} \subseteq W \subseteq A$, and $\overline{\Phi} = (\Phi_1,\ldots,\Phi_s)$ is a sequence of functionals on A with values in W.

We introduce here the class of *functionals in canonical form* and we establish that all recursive in $\overline{\Phi}$ functionals or partial functions can be obtained by iterating only operative functionals in canonical form. These functionals are not determined by closure conditions, but they are instead defined explicitly and they constitute a proper subclass of $\mathscr{F}[\overline{\Phi}]$, the smallest suitable class of functionals containing $\Phi_1,\ldots,\Phi_s$.

3.1. DEFINITIONS. We give below a sequence of definitions which lead to the notion of a *functional in canonical form*.

(i) Let $\bar{x} = (x_1,\ldots,x_l)$ be a finite sequence of individual variables. We say that an expression t is *a basic term in* $\bar{x}$ if

(a) t is one of the variables $x_1,\ldots,x_l$, or

(b) t is one of the constants $c_1,\ldots,c_k$ of the structure $\mathfrak{A}$, or

(c) t is $f(v_1,\ldots,v_r)$, where f is one of the functions of the structure $\mathfrak{A}$ and $v_1,\ldots,v_r$ are among the variables $x_1,\ldots,x_l$.

Assume that $\Psi(x_1,\ldots,x_l, p_1,\ldots,p_r)$ is a functional on A with values in W of signature $(l, k_1,\ldots,k_r)$.

(ii) We say that Ψ is an *evaluation functional* if for some $j \leqslant r$

$$\Psi(x_1,\ldots,x_l, p_1,\ldots,p_r) = p_j\big(t_1,\ldots,t_{k_j}\big),$$

where $t_1,\ldots,t_{k_j}$ are all basic terms in $\bar{x}$. We single out now certain evaluation functionals which will play a special role later on. We say that Ψ is a *simple evaluation functional* if for some $j \leqslant r$

$$\Psi(x_1,\ldots,x_l, p_1,\ldots,p_r) = p_j\big(t_1,\ldots,t_{k_j}\big),$$

where $t_1,\ldots,t_{k_j}$ are among the constants $c_1,\ldots,c_k$ of the structure or they are *distinct* variables among $x_1,\ldots,x_l$.

EXAMPLE. $\Psi(x_1,\dots,x_7, p_1,\dots,p_5) = p_3(x_1, 0, c_3, x_5, 1, x_2)$ is a simple evaluation functional, while $X(x_1,\dots,x_7, p_1,\dots,p_5) = p_3(x_1, 0, c_3, x_1, 1, x_2)$ is an evaluation functional but not a simple one.

(iii) We say that Ψ is a *composition functional* if for some $j \leqslant r$

$$\Psi(x_1,\dots,x_l, p_1,\dots,p_r) = p_j\left(t_1,\dots,t_m, p_j\left(t'_1,\dots,t'_{k_j}\right), t_{m+1},\dots,t_{k_j-1}\right),$$

where $t_1,\dots,t_{k_j-1}, t'_1,\dots,t'_{k_j}$ are all among the constants $c_1,\dots,c_k$ of the structure or among the variables $x_1,\dots,x_l$.

EXAMPLE. $\Psi(x_1, x_2, p_1, p_2, p_3) = p_2(x_2, 0, x_1, x_2, p_2(x_2, x_1, x_1, x_1, 0, 1), c_3)$ is a typical composition functional. Notice however that

$$X(x_1, x_2, p_1, p_2, p_3) = p_2(x_2, 0, x_1, x_2, p_3(x_2, x_2, x_1, x_1, 0, 1), c_3)$$

is not a composition functional, because different function variables occur in the right-hand side.

(iv) We say that Ψ is an *initial functional* if

(a) $\Psi(x_1,\dots,x_l, p_1,\dots,p_r) = 0$, or

(b) $\Psi(x_1,\dots,x_l, p_1,\dots,p_r) = 1$, or

(c) $\Psi(x_1,\dots,x_l, p_1,\dots,p_r) = \chi_{R_i}(v_1,\dots,v_{n_i})$, where R_i is one of the relations of the structure and $v_1,\dots,v_{n_i}$ are among the variables $x_1,\dots,x_l$, or

(d) Ψ is obtained from one of the functionals $\Phi_1,\dots,\Phi_s$ by functional substitution of evaluation functionals.

(v) Finally, assume that $\Psi(u_1,\dots,u_m, x_1,\dots,x_l, p_1,\dots,p_r)$ is a functional on A with values in W of signature $(m + l, k_1,\dots,k_r)$. We say that Ψ is a *functional in canonical form* if

$$\Psi(u_1,\dots,u_m, x_1,\dots,x_l, p_1,\dots,p_r) = \begin{cases} \Psi_1(x_1,\dots,x_l, p_1,\dots,p_r) & \text{if } \bar{u} = \bar{a}_1, \\ \Psi_2(x_1,\dots,x_l, p_1,\dots,p_r) & \text{if } \bar{u} = \bar{a}_2, \\ \vdots & \\ \Psi_n(x_1,\dots,x_l, p_1,\dots,p_r) & \text{if } \bar{u} = \bar{a}_n, \\ 0 & \text{otherwise,} \end{cases}$$

where $\bar{a}, \bar{a}_2,\dots,\bar{a}_n$ are distinct sequences of 0's and l's of length m and each $\Psi_i(x_1,\dots,x_l, p_1,\dots,p_r)$, $1 \leqslant i \leqslant r$, is an evaluation functional, or a composition functional or an initial functional.

We write CF$[\bar{\Phi}]$ for the class of all functionals in canonical form.

3.2. It is clear from the preceding definitions that the collection of all functionals in canonical form is a proper subclass of the smallest class of functionals containing $\Phi_1,\dots,\Phi_s$, i.e. CF$[\bar{\Phi}] \subsetneqq \mathscr{F}[\bar{\Phi}]$. In what follows, however, we will show that CF$[\bar{\Phi}]$-REC $= \mathscr{F}[\bar{\Phi}]$-REC, establishing therefore that the functionals in canonical form suffice to generate the theory of functional recursion in $\bar{\Phi}$. The proof of this result involves two main steps. We will show first that the collection CF$[\bar{\Phi}]$-REC is a suitable class; this will imply immediately that $\mathscr{F}[\bar{\Phi}] \subseteq$ CF$[\bar{\Phi}]$-REC. The second step amounts to showing that an induction completeness theorem holds for the class CF[$\overline{\Phi}$]-REC, in the sense that (CF[$\overline{\Phi}$]-REC)-REC = CF[$\overline{\Phi}$]-REC. In proving that CF[$\overline{\Phi}$]-REC is a suitable class we have to establish that this collection of functionals satisfies all the closure conditions of Definition 2.2. This will be carried out in a series of lemmas and theorems, the nontrivial cases being the closure under functional substitution and composition. We begin with a preliminary lemma which describes a closure property of the class CF[$\overline{\Phi}$] and justifies in part why we introduced before the notion of a simple evaluation functional.

3.3. LEMMA. *The class* CF[$\overline{\Phi}$] *of functionals in canonical form is closed under functional substitutions of simple evaluation functionals.*

PROOF. It is enough to show that both the evaluation functionals and the composition functionals are closed under functional substitutions of simple evaluation functionals. For simplicity we will only give two examples, which, however, are general enough to illustrate what is involved.

(i) Assume that we have the evaluation functional

$$\Psi(x_1, x_2, x_3, p) = p(f_1(x_1, x_2, x_3), x_2, x_2, x_1, 0)$$

(where f_1 is one of the functions of the structure) of signature (3, 5) and we substitute for p the simple evaluation functional

$$X(x_1, x_2, x_3, y_1, y_2, y_3, y_4, y_5, q) = q(y_2, x_3, y_3, y_1, 1).$$

Then

$$\begin{aligned}\Psi'(x_1, x_2, x_3, q) &= \Psi(x_1, x_2, x_3, \lambda \bar{y} X(x_1, x_2, x_3, \bar{y}, q)) \\ &= \Psi(x_1, x_2, x_3, \lambda y_1 \cdots y_5 q(y_2, x_3, y_3, y_1, 1)) \\ &= q(x_2, x_3, x_2, f_1(x_1, x_2, x_3), 1),\end{aligned}$$

which is clearly an evaluation functional.

(ii) Assume that we have the composition functional

$$\Psi(x_1, x_2, x_3, p) = p(x_1, 0, x_1, x_2, p(x_2, x_2, x_1, x_1, 0, x_3), 1)$$

of signature (3, 6) and we substitute for p the simple evaluation functional

$$X(x_1, x_2, x_3, y_1, y_2, y_3, y_4, y_5, y_6, q) = q(x_2, y_3, y_2, y_5, y_4, x_1, 1).$$

Then

$$\begin{aligned}\Psi'(x_1, x_2, x_3, q) &= \Psi(x_1, x_2, x_3, \lambda \bar{y} X(x_1, x_2, x_3, \bar{y}, q)) \\ &= \Psi(x_1, x_2, x_3, \lambda y_1 \cdots y_6 q(x_2, y_3, y_2, y_5, y_4, x_1, 1)) \\ &= q(x_2, x_1, 0, q(x_2, x_1, x_2, 0, x_1, x_1, 1), x_2, x_1, 1),\end{aligned}$$

which is clearly a composition functional. □

3.4. REMARK. The above lemma is false if we allow arbitrary evaluation functionals in the substitutions. The problem actually arises when substituting evaluation functionals in the composition functional. For example, if in the composition functional

$$\Psi(x, p) = p(x, p(x, x, 1), 0)$$

we substitute for p the evaluation functional

$$X(x, y_1, y_2, y_3, q) = q(y_2, y_2, x, 0),$$

then the resulting functional is

$$\begin{aligned}\Psi'(x, q) &= \Psi(x, \lambda \bar{y} X(x, \bar{y}, q)) = \Psi(x, \lambda y_1 y_2 y_3 q(y_2, y_2, x, 0)) \\ &= q(q(x, x, x, 0), q(x, x, x, 0), x, 0),\end{aligned}$$

which is not one of the composition functionals allowed.

The next lemma gives another closure property of the class $\mathrm{CF}[\overline{\Phi}]$ of functionals in canonical form. Its proof follows easily from the definitions and is omitted.

3.5. LEMMA. *The class* $\mathrm{CF}[\overline{\Phi}]$ *of functionals in canonical form is closed under definitions by a finite number of cases.* □

We proceed now into establishing the main properties of the class $\mathrm{CF}[\overline{\Phi}]$-REC. The presentation follows closely the corresponding development of the properties of suitable classes in Kechris and Moschovakis **[1977]**. The proof of the next result, in particular, is almost identical to the proof of the Simultaneous Induction Lemma in that paper, but we have included it here in order to exhibit the use of the simple evaluation functionals.

3.6. SIMULTANEOUS INDUCTION LEMMA. *Let* $\Psi_1(\bar{x}_1, \bar{w}, p_1, \ldots, p_m, \bar{q}), \ldots, \Psi_m(\bar{x}_m, \bar{w}, p_1, \ldots, p_m, \bar{q})$ *be a finite sequence of functionals in canonical form and consider the simultaneous induction*

$$\begin{aligned}&X_1^{\xi}(\bar{x}_1, \bar{w}, \bar{q}) = \Psi_1(\bar{x}_1, \bar{w}, \lambda \bar{x}_1' X_1^{<\xi}(\bar{x}_1', \bar{w}, \bar{q}), \ldots, \lambda \bar{x}_m' X_m^{<\xi}(\bar{x}_m', \bar{w}, \bar{q})), \\ &\quad\vdots \\ &X_m^{\xi}(\bar{x}_m, \bar{w}, \bar{q}) = \Psi_m(\bar{x}_m, \bar{w}, \lambda \bar{x}_1' X_1^{<\xi}(\bar{x}_1', \bar{w}, \bar{q}), \ldots, \lambda \bar{x}_m' X_m^{<\xi}(\bar{x}_m', \bar{w}, \bar{q})).\end{aligned}$$

Then $X_1^{\infty}, \ldots, X_n^{\infty}$ *are all* $\mathrm{CF}[\overline{\Phi}]$*-recursive.*

PROOF. Take $m = 2$ for simplicity and assume without loss of generality that the sequences of 0's and 1's which determine the cases in the functionals Ψ_1, Ψ_2 have the same length. Put

$$X(a, b, \bar{x}_1, \bar{x}_2, \bar{w}, p, \bar{q})$$

$$= \begin{cases}\Psi_1(\bar{x}_1, \bar{w}, \lambda \bar{x}_1' p(0, 0, \bar{x}_1', \bar{0}, \bar{w}), \lambda \bar{x}_2' p(0, 1, \bar{0}, \bar{x}_2', \bar{w}), \bar{q}) & \text{if } a = 0, b = 0, \\ \Psi_2(\bar{x}_2, \bar{w}, \lambda \bar{x}_2' p(0, 0, \bar{x}_1', \bar{0}, \bar{w}), \lambda \bar{x}_2' p(0, 1, \bar{0}, \bar{x}_2', \bar{w}), \bar{q}) & \text{if } a = 0, b = 1, \\ 0 & \text{otherwise.}\end{cases}$$

The functional X is in canonical form, because it is obtained from the functionals Ψ_1, Ψ_2 in the class $\mathrm{CF}[\overline{\Phi}]$ using substitutions of *simple* evaluation functionals

(Lemma 3.3) and definition by cases (Lemma 3.5). Moreover, a trivial induction on ξ shows that

$$X_1^\xi(\bar{x}_1, \bar{w}, \bar{q}) = X^\xi(0, 0, \bar{x}_1, \bar{0}, \bar{w}, \bar{q}), \quad X_2^\xi(\bar{x}_2, \bar{x}, \bar{q}) = X^\xi(0, 1, \bar{0}, \bar{x}_2, \bar{w}, \bar{q}),$$

and therefore the functionals X_1^∞, X_2^∞ are CF[$\overline{\Phi}$]-recursive. □

The preceding Lemmas 3.3 and 3.6 will be used below in order to obtain the following basic result about the CF[$\overline{\Phi}$]-recursive functionals.

3.7. FUNCTIONAL SUBSTITUTION THEOREM. *The class* CF[$\overline{\Phi}$]-*REC of the* CF[$\overline{\Phi}$]-*recursive functionals is closed under full functional substitution, i.e. if* $X(\bar{x}, p_1, \ldots, p_m)$, $X_1, \ldots, X_m$ *are all* CF[$\overline{\Phi}$]-*recursive functionals, then so is*

$$\Psi(\bar{x}, \bar{q}) = X(\bar{x}, \lambda\bar{y}_1 X_1(\bar{y}_1, \bar{x}, \bar{q}), \ldots, \lambda\bar{y}_m X_m(\bar{y}_m, \bar{x}, \bar{q})).$$

PROOF. In order to simplify the already heavy notation, we will give the proof for the case $m = 1$ only. Assume then that

$$\Psi(\bar{x}, \bar{q}) = X(\bar{x}, \lambda\bar{y}_1 X_1(\bar{y}_1, \bar{x}, \bar{q})),$$
$$X_1(\bar{y}_1, \bar{x}, \bar{q}) = \Psi_1^\infty(\bar{a}_1, \bar{y}_1, \bar{x}, \bar{q}), \quad X(\bar{x}, p_1) = \Psi_2^\infty(\bar{a}_2, \bar{x}, p_1),$$

where $\Psi_1(\bar{u}_1, \bar{y}_1, \bar{x}, p, \bar{q})$, $\Psi_2(\bar{u}_2, \bar{x}, p', p_1)$ are functionals in canonical form and $\bar{a}_1$, $\bar{a}_2$ are sequences of 0's and 1's. Consider now the following simultaneous induction:

$$Z_1^\xi(\bar{u}_1, \bar{y}_1, \bar{x}, \bar{q}) = \Psi_1(\bar{u}_1, \bar{y}_1, \bar{x}, \lambda\bar{u}_1'\bar{y}_1'\bar{x}' Z_1^{<\xi}(\bar{u}_1', \bar{y}_1', \bar{x}', \bar{q}), \bar{q}),$$
$$Z_2^\xi(\bar{u}_2, \bar{x}, \bar{w}, \bar{q}) = \Psi_2(\bar{u}_2, \bar{x}, \lambda\bar{u}_2'\bar{x}' Z_2^{<\xi}(\bar{u}_2', \bar{x}', \bar{w}, \bar{q}), \lambda\bar{y}_1' Z_1^{<\xi}(\bar{a}_1, \bar{y}_1', \bar{w}, \bar{q})),$$
$$Z_3^\xi(\bar{u}_2, \bar{x}, \bar{q}) = Z_2^{<\xi}(\bar{u}_2, \bar{x}, \bar{x}, \bar{q}).$$

It is clear that $Z_1^\xi(\bar{u}_1, \bar{y}_1, \bar{x}, \bar{q}) = \Psi_1^\xi(\bar{u}_1, \bar{y}_1, \bar{x}, \bar{q})$ and hence $Z_1^\infty(\bar{a}_1, \bar{y}_1, \bar{x}, \bar{q}) = X_1(y_1, \bar{x}, \bar{q})$. A straightforward induction on ξ, using the monotonicity of the functionals, shows that

$$\lambda\bar{u}_2\overline{xw} Z_2^\xi(\bar{u}_2, \bar{x}, \bar{w}, \bar{q}) \subseteq \lambda\bar{u}_2\overline{xw}\Psi_2^\infty(\bar{u}_2, \bar{x}, \lambda\bar{y}_1 X_1(\bar{y}_1, \bar{w}, \bar{q}))$$

and consequently

$$\text{(1)} \qquad \lambda\bar{u}_2\overline{xw} Z_2^\infty(\bar{u}_2, \bar{x}, \bar{w}, \bar{q}) \subseteq \lambda\bar{u}_2\overline{xw}\Psi_2^\infty(\bar{u}_2, \bar{x}, \lambda\bar{y}_1 X_1(\bar{y}_1, \bar{w}, \bar{q})).$$

Using again the monotonicity of the functionals involved, an induction on ξ shows that

$$\lambda\bar{u}_2\overline{xw}\Psi_2^\xi(\bar{u}_2, \bar{x}, \lambda\bar{y}_1 X_1(\bar{y}_1, \bar{w}, \bar{q})) \subseteq \lambda\bar{u}_2\overline{xw} Z_2^\infty(\bar{u}_2, \bar{x}, \bar{w}, \bar{q})$$

and consequently

$$\text{(2)} \qquad \lambda\bar{u}_2\overline{xw}\Psi_2^\infty(\bar{u}_2, \bar{x}, \lambda\bar{y}_1 X_1(\bar{y}_1, \bar{w}, \bar{q})) \subseteq \lambda\bar{u}_2\overline{xw} Z_2^\infty(\bar{u}_2, \bar{x}, \bar{w}, \bar{q}).$$

From (1) and (2) we conclude that

$$Z_2^\infty(\bar{u}_2, \bar{x}, \bar{w}, \bar{q}) = \Psi_2^\infty(\bar{u}_2, \bar{x}, \lambda\bar{y}_1 X_1(\bar{y}_1, \bar{w}, \bar{q}))$$

for any $\bar{u}_2$, $\bar{x}$, $\bar{z}$, $\bar{q}$, so that finally

$$Z_3^\infty(\bar{u}_2, \bar{x}, \bar{q}) = \Psi_2^\infty(\bar{u}_2, \bar{x}, \lambda\bar{y}_1 X_1(\bar{y}_1, \bar{x}, \bar{q}))$$

and

(3) $Z_3^\infty(\bar{a}_2, \bar{x}, \bar{q}) = \Psi_2^\infty(\bar{a}_2, \bar{x}, \lambda\bar{y}_1 X_1(\bar{y}_1, \bar{x}, \bar{q})) = X(\bar{x}, \lambda\bar{y}_1 X_1(\bar{y}_1, \bar{x}, \bar{q}))$.

Notice now that the simultaneous induction Z_1, Z_2, Z_3 above involves functionals in canonical form, because Z_1 and Z_2 are obtained from functionals in canonical form by functional substitutions of simple evaluation functionals, while Z_3 amounts to the evaluation functional $\Psi_3(\bar{u}_3, \bar{x}, p_2) = p_2(\bar{u}_2, \bar{x}, \bar{x})$. Therefore from (3) above and the preceding Simultaneous Induction Lemma 3.6 we conclude that the functional

$$\Psi(\bar{x}, \bar{q}) = X(\bar{x}, \lambda\bar{y}_1 X_1(\bar{y}_1, \bar{x}, \bar{q})) = Z_3^\infty(\bar{a}_2, \bar{x}, \bar{q})$$

is CF[$\bar{\Phi}$]-recursive. □

3.8. REMARK. It has been pointed out by other people before that there is a mistake in the proof of the Functional Substitution Theorem 3.2 for suitable classes of functionals in Kechris and Moschovakis [**1977**]. The mistake arises from the fact that in their system of simultaneous induction there is no distinction between the variables on which the induction takes place and the variables which are carried as parameters. This difficulty is bypassed in the proof of the preceding Theorem 3.7 by introducing the new variables $\bar{w}$ as parameters in the functional Z_2 and this way keeping the variables $\bar{x}$ only for the induction. In addition, an extra clause is needed, given by the functional Z_3, which identifies $\bar{w}$ with $\bar{x}$ one step later.

One of the chief differences between the smallest suitable class of functionals $\mathscr{F}[\bar{\Phi}]$ containing $\Phi_1, \ldots, \Phi_s$ and the class CF[$\bar{\Phi}$] of functionals in canonical form is that the former is closed under composition, while the latter is not. This should be contrasted to the following

3.9. THEOREM. *The class* CF[$\bar{\Phi}$]-REC *of the* CF[$\bar{\Phi}$]-*recursive functionals is closed under composition, i.e. if the functionals* $X_1(y, \bar{x}, \bar{q})$, $X_2(\bar{x}, \bar{q})$ *are* CF[$\bar{\Phi}$]-*recursive, then so is* $X(\bar{x}, \bar{q}) = X_1(X_2(\bar{x}, \bar{q}), \bar{x}, \bar{q})$.

PROOF. Let $X_1(y, \bar{x}, \bar{q})$ and $X_2(\bar{x}, \bar{q})$ be CF[$\bar{\Phi}$]-recursive functionals. There are then operative functionals $\Psi_1(\bar{u}_1, y, \bar{x}, p', \bar{q})$, $\Psi_2(\bar{u}_2, \bar{x}, p'', \bar{q})$ in canonical form and sequences $\bar{a}_1$, $\bar{a}_2$ of 0's and 1's such that

$$X_1(y, \bar{x}, \bar{q}) = \Psi_1^\infty(\bar{a}_1, y, \bar{x}, \bar{q}), \qquad X_2(\bar{x}, q) = \Psi_2^\infty(\bar{a}_2, \bar{x}, q).$$

Consider the functional $\Psi_c(\bar{x}, p_1, p_2) = p_1(p_2(\bar{x}), \bar{x})$ and observe that the composition of X_1 with X_2 amounts to functional substitution of X_1, X_2 in Ψ_c, i.e.

$$X(\bar{x}, \bar{q}) = X_1(X_2(\bar{x}, \bar{q}), \bar{x}, \bar{q}) = \Psi_c(\bar{x}, \lambda y'\bar{x}' X_1(y', \bar{x}, \bar{q}), \lambda\bar{x}' X_2(\bar{x}', \bar{q})).$$

The functional $\Psi_c(\bar{x}, p_1, p_2) = p_1(p_2(\bar{x}), \bar{x})$ is *not* one of the composition functionals we allowed in Definition 3.1, and therefore it is *not* in the class CF[$\bar{\Phi}$] of functionals in canonical form. We claim, however, that the functional Ψ_c *is* CF[$\bar{\Phi}$]-recursive. For this take the operative functional $\Psi(a, b, y, \bar{x}, p, p_1, p_2)$ in canonical form, where

$$\Psi(a, b, y, \bar{x}, p, p_1, p_2) = \begin{cases} p_1(y, \bar{x}) & \text{if } a = 0, b = 0, \\ p_2(\bar{x}) & \text{if } a = 0, b = 1, \\ p(0, 0, p(0, 1, 0, \bar{x}), \bar{x}) & \text{if } a = 1, b = 0, \\ 0 & \text{otherwise,} \end{cases}$$

and observe that

$$\Psi_c(\bar{x}, p_1, p_2) = p_1(p_2(\bar{x}), \bar{x}) = \Psi^\infty(1, 0, 0, \bar{x}, p_1, p_2).$$

The preceding Functional Substitution Theorem 3.7 implies now immediately that the functional

$$X(\bar{x}, \bar{q}) = X_1(X_2(\bar{x}, \bar{q}), \bar{x}, \bar{q}) = \Psi_c(\bar{x}, \lambda y' \bar{x}' X_1(y', \bar{x}', q'), \lambda \bar{x}' X_2(\bar{x}', q))$$

is CF[$\bar{\Phi}$]-recursive. □

We establish next the remaining closure properties of the CF[$\bar{\Phi}$]-recursive functionals.

3.10. THEOREM. *The collection* CF[$\bar{\Phi}$]-REC *of the* CF[$\bar{\Phi}$]-*recursive functionals is a suitable class containing the functionals* $\Phi_1, \ldots, \Phi_s$.

PROOF. From the Definitions 2.2 and 3.1, the preceding Theorems 3.7 and 3.9, and the fact that CF[$\bar{\Phi}$] $\subseteq$ CF[$\bar{\Phi}$]-REC, it follows that it is enough to show that the collection of the CF[$\bar{\Phi}$]-recursive functionals is closed under definitions by cases, substitutions by projections and substitutions by the functions or the constants of the structure. These closure properties are now immediate consequences of the Functional Substitution Theorem 3.7 and the fact that CF[$\bar{\Phi}$] $\subseteq$ CF[$\bar{\Phi}$]-REC. Indeed, consider first the functional

$$\Psi(u, \bar{x}, p_1, p_2) = \begin{cases} p_1(\bar{x}) & \text{if } u = 0, \\ p_2(\bar{x}) & \text{if } u = 1, \\ 0 & \text{otherwise,} \end{cases}$$

which is in canonical form (and hence CF[$\bar{\Phi}$]-recursive) and observe that definition by cases is equivalent to functional substitution in Ψ. Similarly, if the mappings $g_1(\bar{x}), \ldots, g_l(\bar{x})$ are projections or they are among the functions and the constants of the structure, then take the evaluation functional

$$X(\bar{x}, p) = p(g_1(\bar{x}), \ldots, g_l(\bar{x}))$$

and notice that substitutions by $g_1, \ldots, g_l$ amount to functional substitution in X. □

The following result shows that the class CF[$\bar{\Phi}$]-REC is closed under the fixed-point operation.

3.11. INDUCTION COMPLETENESS THEOREM. *If* $X(\bar{x}, p, \bar{q})$ *is an operative functional which is* CF[$\bar{\Phi}$]-*recursive, then the functional* $X^\infty(\bar{x}, \bar{q})$ *is also* CF[$\bar{\Phi}$]-*recursive.*

PROOF. We only outline the proof here, because it is the same as the one of the corresponding result for suitable classes of functionals in Kechris and Moschovakis [**1977**].

If $X(\bar{x}, p, \bar{q})$ is an operative CF[$\bar{\Phi}$]-recursive functional, then there is a functional $\Psi(\bar{u}, \bar{x}, r, p, \bar{q})$ in canonical form and a sequence $\bar{a}$ of 0's and 1's such that

$$X(\bar{x}, p, \bar{q}) = \Psi^\infty(\bar{a}, \bar{x}, p, \bar{q}).$$

Consider now the functional

$$Z(\bar{u}, \bar{x}, r, \bar{q}) = \Psi(\bar{u}, \bar{x}, r, \lambda \bar{x}' r(\bar{a}, \bar{x}'), \bar{q})$$

which is in canonical form, because it is obtained from Ψ by functional substitution of a simple evaluation functional. As in Kechris and Moschovakis [**1977**] it can be shown that $Z^\infty(\bar{a}, \bar{x}, \bar{q}) = \Psi^\infty(\bar{x}, \bar{q})$ and consequently Ψ^∞ is CF[$\bar{\Phi}$]-recursive. □

Finally, by combining the preceding Theorems 3.10 and 3.11, we establish the main result in this part of the paper.

3.12. THEOREM. *Let* $\mathfrak{A} = \langle A, R_1, \ldots, R_n, f_1, \ldots, f_m, c_1, \ldots, c_k \rangle$ *be a structure with* 0 *and* 1 *among its constants, let* W *be a set such that* $\{0, 1\} \subseteq W \subseteq A$, *and let* $\bar{\Phi} = (\Phi_1, \ldots, \Phi_s)$ *be a sequence of functionals on* A *with values in* W. *Then any recursive in* $\bar{\Phi}$ *functional can be obtained by iterating a functional in canonical form and therefore*

$$\mathrm{CF}[\bar{\Phi}]\text{-REC} = \mathscr{F}[\bar{\Phi}]\text{-REC}.$$

PROOF. It follows from the definitions and the preceding Theorem 3.10 that

$$\mathrm{CF}[\bar{\Phi}] \subseteq \mathscr{F}[\bar{\Phi}] \subseteq \mathrm{CF}[\bar{\Phi}]\text{-REC}$$

and consequently

$$\mathrm{CF}[\bar{\Phi}]\text{-REC} \subseteq \mathscr{F}[\bar{\Phi}]\text{-REC} \subseteq (\mathrm{CF}[\bar{\Phi}]\text{-REC})\text{-REC}.$$

The Induction Completeness Theorem 3.11, however, implies that

$$(\mathrm{CF}[\bar{\Phi}]\text{-REC})\text{-REC} \subseteq \mathrm{CF}[\bar{\Phi}]\text{-REC},$$

which completes the proof. □

3.13. REMARK. The functionals in canonical form are explicitly defined and therefore they are easier to work with in general than the arbitrary functionals in the suitable class $\mathscr{F}[\bar{\Phi}]$. They can be viewed as providing substitutes for the Kleene master operator on structures which do not possess a coding apparatus. The preceding work also makes quite clear why in the presence of coding the approach to generalized recursion theory via functional recursion and suitable classes is equivalent to the one via Kleene schemata: what happens is that one can use the coding to simulate all functionals in canonical form by a single such functional, namely the Kleene master operator.

3.14. APPLICATIONS. We conclude this part of the paper by looking at the functionals in canonical form in two important cases.

As mentioned before in §2.4, ordinary recursion theory coincides with prime computability on the structure of arithmetic $\mathbf{N} = \langle \omega, =, \mathrm{s}, 0, 1\rangle$. Therefore any ordinary partial recursive function can be obtained from the fixed point of a functional which is a definition by cases and each case is one of the following: the constant function 0, the characteristic function of equality, the successor function, an evaluation functional, or a composition functional.

The theory of positive elementary induction on a structure $\mathfrak{A} = \langle A, R_1, \ldots, R_n, c_1, \ldots, c_k\rangle$ (where the equality on A is among the relations $R_1, \ldots, R_n$) is equivalent to recursion in the functional $\mathbf{E}^{\#}$. The conclusion in this case is that, even if the structure $\mathfrak{A}$ does not have a coding machinery, any inductive relation on $\mathfrak{A}$ is reducible to the fixed point of a functional which is a definition by cases and each case is one of the following: the constant functions 0 and 1, the characteristic functions of the relations $R_1, \ldots, R_n$, an evaluation functional, a composition functional, or a substitution of an evaluation functional in the functional $\mathbf{E}^{\#}$. One immediate consequence is that any inductive relation on $\mathfrak{A}$ can be generated by iterating a positive formula which is a finite disjunction of existential or universal formulas with one quantifier each. This is a well-known result in positive elementary induction and is used in Chapter 7 of Moschovakis [**EIAS**] in order to derive model-theoretic characterizations of the hyperelementary relations. In connection with our work here this result was one of the motivating factors for introducing and studying the functionals in canonical form.

PART II. HIERACHIES AND MODEL-THEORETIC CHARACTERIZATIONS IN GENERALIZED RECURSION THEORY

In the second part of this paper we concentrate on the study of recursion in functionals which represent quantifiers on a structure $\mathfrak{A}$. The main result is a hierarchy theorem for the recursive relations in the functionals $\mathbf{E}$, $\mathbf{F}_{Q_1}$, $\mathbf{F}^{\wedge}_{Q_2}$, $\mathbf{F}^{\#}_{Q_3}$, where Q_1, Q_2, Q_3 are arbitrary generalized quantifiers on the universe of the structure $\mathfrak{A}$. The hierarchy is obtained by iterating "lightface" Δ^1_1 definability with basis the previously defined relations and without second order parameters. As consequences of the main theorem, we establish model-theoretic and invariant definability characterizations of recursion in $\mathbf{E}$, $\mathbf{F}_{Q_1}$, $\mathbf{F}^{\wedge}_{Q_2}$, $\mathbf{F}^{\#}_{Q_3}$ in terms of "lightface" versions of the Δ^1_1 Comprehension Schema.

The results mentioned above contain as special cases the model-theoretic characterizations of the "boldface" hyperelementary relations of Moschovakis [**EIAS**] and the "boldface" Q-hyperelementary and recursive in $\mathbf{E}^{\#}$, $\mathbf{F}_Q$ relations of Harrington, Kirousis and Schlipf [**1978**]. They imply also the "lightface" characterizations of recursion in $\mathbf{E}$ and in $\mathbf{E}^{\#}$ which were presented without proofs in Kolaitis [**1981**]. There one can find in addition a summary of the classical results in hyperarithmetic theory which constitute the origins of and the motivation for the work reported here.

4. Recursion in quantifiers on a structure and the stage comparison theorems. In this section we present the definitions of the basic notions involved, as well as a minimal amount of the necessary background material.

4.1. From now on we will assume that $\mathfrak{A} = \langle A, R_1, \ldots, R_n, f_1, \ldots, f_m, c_1, \ldots, c_k \rangle$ is a structure such that both ω and the ordering relation $\leqslant_\omega$ are first order definable on $\mathfrak{A}$ (without parameters). In certain cases we will need the additional hypothesis that $\mathfrak{A}$ possesses a first order definable coding machinery. More precisely, we say that the structure $\mathfrak{A}$ is *acceptable* if there is a one-to-one coding function $\langle\ \rangle: A^{<\omega} \to A$ such that the associated notions of finite sequence, length of a finite sequence etc. are first order definable on $\mathfrak{A}$ (without parameters). Typical examples of acceptable structures are the structure of arithmetic $\mathbf{N} = \langle \omega, +, \cdot \rangle$, the rationals $\mathbf{Q} = \langle Q, +, \cdot \rangle$, the structure of analysis $\mathbf{R} = \langle \omega^\omega \cup \omega, \omega, +, \cdot, \mathrm{Ap} \rangle$ (where $\mathrm{Ap}(\alpha, n) = \alpha(n)$), and for each ordinal λ the initial segment of the universe $\mathbf{V}_\lambda = \langle \mathrm{V}_\lambda, \in \rangle$.

The ("*lightface*") *first order language* $\mathscr{L}^{\mathfrak{A}}$ of the structure $\mathfrak{A}$ has both individual variables $x, y, z, \ldots$ and relation variables $S, T, V, \ldots,$ has constant symbols $\mathbf{c}_1, \ldots, \mathbf{c}_k$, a constant symbol $\mathbf{l}$ for each $l \in \omega$, relation symbols $\mathbf{R}_1, \ldots, \mathbf{R}_n$ and function symbols $\mathbf{f}_1, \ldots, \mathbf{f}_m$. The "*boldface*" *first order language* $\mathbf{L}^{\mathfrak{A}}$ of the structure $\mathfrak{A}$ has, in addition, a constant symbol $\mathbf{a}$ for each $a \in A$. In both languages $\mathscr{L}^{\mathfrak{A}}$ and $\mathbf{L}^{\mathfrak{A}}$ the quantifiers $\forall$ and $\exists$ range *only* over the individual variables.

A (*generalized*) *quantifier on* $\mathfrak{A}$ is a collection Q of subsets of A such that $\varnothing \subsetneqq Q \subsetneqq \mathscr{P}(A)$ and such that if $X \subseteq Y$ and $X \in Q$, then $Y \in Q$. The *dual quantifier* $\breve{Q}$ of Q is defined by the condition

$$X \in \breve{Q} \Leftrightarrow (A - X) \notin Q.$$

Assume that $\overline{Q} = (Q_1, \ldots, Q_l)$ is a finite sequence of quantifiers on A. We expand the language $\mathscr{L}^{\mathfrak{A}}$ of the structure $\mathfrak{A}$ to the language $\mathscr{L}^{\mathfrak{A}}(\overline{Q})$ or $\mathscr{L}^{\mathfrak{A}}(Q_1, \ldots, Q_l)$ by adding the symbols $Q_1, \breve{Q}_1, Q_2, \breve{Q}_2, \ldots, Q_l, \breve{Q}_l$, so that if φ is a formula of $\mathscr{L}^{\mathfrak{A}}(Q)$, then the expressions

$$Q_i x \varphi \quad \text{and} \quad \breve{Q}_i x \varphi, \qquad 1 \leqslant i \leqslant l,$$

are also formulas of $\mathscr{L}^{\mathfrak{A}}(\overline{Q})$. The interpretation is the standard one, namely

$$\mathfrak{A} \models Qx\varphi(x) \quad \text{if and only if} \quad \{x : \mathfrak{A} \models \varphi(x)\} \in Q,$$

where Q is one of the quantifiers $Q_1, \breve{Q}_1, \ldots, Q_l, \breve{Q}_l$. The "boldface" language $\mathbf{L}^{\mathfrak{A}}(\overline{Q})$ is obtained from $\mathscr{L}^{\mathfrak{A}}(\overline{Q})$ by adding a constant symbol $\mathbf{a}$ for each element a of A.

4.2. In generalized recursion theory a quantifier Q on a structure $\mathfrak{A}$ is represented by one of the functionals $\mathbf{F}_Q, \mathbf{F}^{\wedge}_Q, \mathbf{F}^{\#}_Q$, each of signature $(0, 1)$, where

$$\mathbf{F}_Q(p) = \begin{cases} 0 & \text{if } p \text{ is total and } (Qx)(p(x) = 0), \\ 1 & \text{if } p \text{ is total and } (\breve{Q}x)(p(x) \neq 0), \\ \text{undefined} & \text{if } p \text{ is not total;} \end{cases}$$

$$\mathbf{F}^{\wedge}_Q(p) = \begin{cases} 0 & \text{if } (Qx)(p(x) = 0), \\ 1 & \text{if } p \text{ is total and } (\breve{Q}x)(p(x) \neq 0), \\ \text{undefined} & \text{otherwise;} \end{cases}$$

$$\mathbf{F}_Q^{\#}(p) = \begin{cases} 0 & \text{if } (Qx)(p(x) = 0), \\ 1 & \text{if } (\breve{Q}x)(p(x)\downarrow \neq 0), \\ \text{undefined} & \text{otherwise.} \end{cases}$$

If Q is the existential quantifier $\exists = \{X \subseteq A: X \neq \varnothing\}$, then we write $\mathbf{E}$ for the functional $\mathbf{F}_\exists$ and $\mathbf{E}^{\#}$ for the functional $\mathbf{F}_\exists^{\#}$, i.e.

$$\mathbf{E}(p) = \begin{cases} 0 & \text{if } p \text{ is total and } (\exists x)(p(x) = 0), \\ 1 & \text{if } p \text{ is total and } (\forall x)(p(x) \neq 0), \\ \text{undefined} & \text{if } p \text{ is not total;} \end{cases}$$

and

$$\mathbf{E}^{\#}(p) = \begin{cases} 0 & \text{if } (\exists x)(p(x) = 0), \\ 1 & \text{if } (\forall x)(p(x)\downarrow \neq 0), \\ \text{undefined} & \text{otherwise.} \end{cases}$$

Notice that $\mathbf{F}_\exists^{\hat{}} = \mathbf{E}^{\#}$ and that $F_\forall^{\hat{}}(p) = 1 \dot{-} \mathbf{E}(\lambda x(1 \dot{-} p(x)))$, where

$$1 \dot{-} a = \begin{cases} 1 & \text{if } a = 0, \\ 0 & \text{if } a \neq 0. \end{cases}$$

The functionals $\mathbf{E}$ and $\mathbf{F}_Q$ are type 2 objects on A in the sense of Kleene [**1959a**, **1963**]. Moreover, for recursion-theoretic purposes, the functional $\mathbf{E}$ can be identified with the equality predicate for subsets of A, i.e.

$$\mathbf{E}(X, Y) = \begin{cases} 0 & \text{if } X = Y, \\ 1 & \text{if } X \neq Y, \end{cases}$$

where X and Y range over the subsets of A. Recursion in the functionals $\mathbf{F}_Q$ and $\mathbf{F}_Q^{\#}$ (when $A = \omega$) was first studied by Hinman [**1969**] and Aczel [**1970**]. The functionals $\mathbf{F}_Q$ and $\mathbf{F}_Q^{\#}$ do not distinguish the quantifier Q from its dual $\breve{Q}$, since

$$\mathbf{F}_{\breve{Q}}(p) = 1 \dot{-} \mathbf{F}_Q(\lambda x(1 \dot{-} p(x))) \quad \text{and} \quad \mathbf{F}_{\breve{Q}}^{\#}(p) = 1 \dot{-} \mathbf{F}_Q^{\#}(\lambda x(1 \dot{-} p(x))).$$

The theory of recursion in $\mathbf{E}^{\#}$, $\mathbf{F}_Q^{\#}$ coincides with positive elementary induction in the quantifier Q. Finally the functional $\mathbf{F}_Q^{\hat{}}$ was introduced in Kolaitis [**1980**] and used in order to characterize certain nonmonotone inductive definitions in the quantifier Q. In many cases recursion in $\mathbf{F}_Q^{\hat{}}$ is different from recursion in $\mathbf{F}_{\breve{Q}}^{\hat{}}$.

Each of the functionals $\mathbf{F}_Q$, $\mathbf{F}_Q^{\hat{}}$ and $\mathbf{F}_Q^{\#}$ embodies in general different strength of Q or $\breve{Q}$ quantification. Recall that if $\overline{\Phi}$ is a sequence of functionals on A, then $\mathscr{F}[\overline{\Phi}]$ is the smallest suitable class of functionals containing the sequence $\overline{\Phi}$. It is easy now to see that always

$$\mathscr{F}[\mathbf{E}, \mathbf{F}_Q] \subseteq \mathscr{F}[\mathbf{E}, \mathbf{F}_Q^{\hat{}}] \subseteq \mathscr{F}[\mathbf{E}, \mathbf{F}_Q^{\#}]$$

and consequently we have the following inclusions for the corresponding classes of semirecursive relations:

$$\mathrm{ENV}[\mathbf{E}, \mathbf{F}_Q] \subseteq \mathrm{ENV}[\mathbf{E}, \mathbf{F}_Q^{\hat{}}] \subseteq \mathrm{ENV}[\mathbf{E}, \mathbf{F}_Q^{\#}].$$

Moreover, we have that $\mathscr{F}[\mathbf{E}, \mathbf{F}_Q^{\#}] = \mathscr{F}[\mathbf{E}, \mathbf{F}_Q^{\hat{}}, \mathbf{F}_{\breve{Q}}^{\hat{}}]$ and so

$$\mathrm{ENV}[\mathbf{E}, \mathbf{F}_Q^{\#}] = \mathrm{ENV}[\mathbf{E}, \mathbf{F}_Q^{\hat{}}, \mathbf{F}_{\breve{Q}}^{\hat{}}].$$

In what follows we will concentrate on the theory of recursion in a sequence of functionals $\mathbf{E}$, $\mathbf{F}_{Q_1},\ldots,\mathbf{F}_{Q_l}$, $\mathbf{F}^{\wedge}_{S_1},\ldots,\mathbf{F}^{\wedge}_{S_m}$, where $Q_1,\ldots,Q_l$, $S_1,\ldots,S_m$ are quantifiers on A. It is clear that this is the most general case of recursion in quantifiers. For example, the theory of recursion in $\mathbf{E}^{\#}$, $\mathbf{F}^{\#}_{Q}$ (i.e. positive elementary induction in Q) is the same as recursion in $\mathbf{E}, \mathbf{F}^{\wedge}_{\exists}$, $\mathbf{F}^{\wedge}_{Q}$, $\mathbf{F}^{\wedge}_{\check{Q}}$. From now on we will use the abbreviation $\mathbf{E}$, $\overline{\mathbf{F}}_Q$, $\overline{\mathbf{F}}^{\wedge}_S$ to denote the sequence of functionals $\mathbf{E}$, $\mathbf{F}_{Q_1},\ldots,\mathbf{F}_{Q_l}$, $\mathbf{F}^{\wedge}_{S_1},\ldots,\mathbf{F}^{\wedge}_{S_m}$.

4.3. The functionals $\mathbf{E}$, $\overline{\mathbf{F}}_Q$, $\overline{\mathbf{F}}^{\wedge}_S$ possess a smooth recursion theory with interesting closure and structural properties. Many of these properties are consequences of the fact that the functionals $\mathbf{E}$, $\overline{\mathbf{F}}_Q$, $\overline{\mathbf{F}}^{\wedge}_S$ are *normal* in the sense of Moschovakis [**1977**] and Kechris and Moschovakis [**1977**]. Normality is a technical condition which guarantees that one can compare the stages of operative functionals in a recursive way. More precisely, if Φ is an operative functional in a suitable class $\mathscr{F}$ and $\bar{x}$ is a sequence of elements from A for which $\Phi^{\infty}(\bar{x})$ is defined, then we write $|\bar{x}|$ for the smallest ordinal ξ such that $\Phi^{\xi}(\bar{x})$ is defined. We associate with Φ the relations $\leqslant^*_{\Phi}$ and $<^*_{\Phi}$, where

$$\bar{x} \leqslant^*_{\Phi} \bar{y} \Leftrightarrow (\Phi^{\infty}(\bar{x})\downarrow)\&\big((\Phi^{\infty}(\bar{y})\uparrow) \text{ or } (|\bar{x}| \leqslant |\bar{y}|)\big)$$

and

$$\bar{x} <^*_{\Phi} \bar{y} \Leftrightarrow (\Phi^{\infty}(\bar{x})\downarrow)\&\big((\Phi^{\infty}(\bar{y})\uparrow) \text{ or } (|\bar{x}| < |\bar{y}|)\big).$$

Here $\Phi^{\infty}(\bar{x})\downarrow$ and $\Phi^{\infty}(\bar{x})\uparrow$ abbreviate, respectively "$\Phi^{\infty}(\bar{x})$ is defined" and "$\Phi^{\infty}(\bar{x})$ is undefined".

In Kolaitis [**1980**] it is proved that the sequence $\mathbf{E}$, $\overline{\mathbf{F}}_Q$, $\overline{\mathbf{F}}^{\wedge}_S$ is normal. This fact combined with Theorem 6.3 in Kechris and Mochovakis [**1977**] implies immediately the following

4.4. Stage Comparison Theorem. *Let $\Phi(\bar{x}, g)$ be an operative functional in the suitable class $\mathscr{F}[\mathbf{E}, \overline{\mathbf{F}}_Q, \overline{\mathbf{F}}^{\wedge}_S]$. Then the stage comparison function $\chi(\bar{x}, \bar{y})$ of Φ is recursive in $\mathbf{E}$, $\overline{\mathbf{F}}_Q$, $\overline{\mathbf{F}}^{\wedge}_S$, where*

$$\chi(\bar{x}, \bar{y}) = \begin{cases} 0 & \textit{if } \bar{x} \leqslant^*_{\Phi} \bar{y}, \\ 1 & \textit{if } \bar{y} <^*_{\Phi} \bar{x}, \\ \uparrow & \textit{otherwise}. \end{cases} \quad \square$$

Assume that $\bar{x}$ is a sequence from A such that $\Phi^{\infty}(\bar{x})\downarrow$ and let $|\bar{x}| = \xi$. One of the direct consequences of the above Theorem 4.4 is that the ξth stage Φ^{ξ} of the operative functional Φ is recursive in $\mathbf{E}$, $\overline{\mathbf{F}}_Q$, $\overline{\mathbf{F}}^{\wedge}_S$ from $\bar{x}$. Moreover the *stage comparison function Ψ^{ξ} of Φ up to ξ* is also recursive in $\mathbf{E}$, $\overline{\mathbf{F}}_Q$, $\overline{\mathbf{F}}^{\wedge}_S$ from $\bar{x}$, where

$$\Psi^{\xi}(\bar{x}', \bar{y}') = \begin{cases} 0 & \text{if } \bar{x}' \leqslant_{\Phi} \bar{y}' \text{ and } |\bar{x}'| \leqslant \xi, \\ 1 & \text{if } \bar{y}' <^*_{\Phi} \bar{x}' \text{ and } |\bar{y}'| \leqslant \xi. \end{cases}$$

We put $\Psi^{<\xi} = \bigcup_{\eta<\xi}\Psi^{\eta}$, so that

$$\Psi^{<\xi}(\bar{x}', \bar{y}') = \begin{cases} 0 & \text{if } \bar{x}' \leqslant_{\Phi} \bar{y}' \text{ and } |\bar{x}'| < \xi, \\ 1 & \text{if } \bar{y}' <^*_{\Phi} \bar{x}' \text{ and } |\bar{y}'| < \xi. \end{cases}$$

Using again Theorem 4.4, we can easily see that the function $\Psi^{<\xi}$ is recursive in $\mathbf{E}, \bar{\mathbf{F}}_Q, \bar{\mathbf{F}}_S^{\hat{}}$ from $\bar{x}$.

In Kolaitis [**1979**] it was shown that if $\bar{\Phi}$ is a sequence of normal functionals, then a Second Stage Comparison Theorem holds for the suitable class $\mathscr{F}[\bar{\Phi}]$. This is a technical tool which relates the stages of operative functionals to second order definability. The normality of the functionals $\mathbf{E}, \bar{\mathbf{F}}_Q, \bar{\mathbf{F}}_S^{\hat{}}$, Lemma 4.4 and Theorem 4.5 in Kolaitis [**1979**] yield now the following

4.5. Second Stage Comparison Theorem. *Let $\Phi(\bar{x}, g)$ be an operative functional in the suitable class $\mathscr{F}[\mathbf{E}, \bar{\mathbf{F}}_Q, \bar{\mathbf{F}}_S^{\hat{}}]$. Then there is a formula $\varphi(Y_1, Y_2)$ of the language $\mathscr{L}^{\mathfrak{A}}(\bar{Q}, \bar{S})$ of the structure $\mathfrak{A}$ such that:*

(i) *If $\Phi^{\infty}(\bar{x})$ is defined and $|\bar{x}| = \xi$, then $\varphi(\text{graph}\,\Phi^{\xi}, \text{graph}\,\Psi^{\xi})$ holds.*

(ii) *Assume that f and g are partial functions satisfying $\varphi(\text{graph}\,f, \text{graph}\,g)$:*

(a) *if $\bar{x} \leqslant^*_{\Phi} \bar{y}$ and $f(\bar{x}, \bar{y})$ is defined, then $f(\bar{x}, \bar{y}) = 0$ and $g(\bar{x}) = \Phi^{|\bar{x}|}(\bar{x})$,*

(b) *if $\bar{y} <^*_{\Phi} \bar{x}$ and $f(\bar{x}, \bar{y})$ is defined, then $f(\bar{x}, \bar{y}) = 1$ and $g(\bar{y}) = \Phi^{|\bar{y}|}(\bar{y})$.*

(iii) *If $\Phi^{\infty}(\bar{x})$ is defined or $\Phi^{\infty}(\bar{y})$ is defined, then*

(a) $\bar{x} \leqslant^*_{\Phi} \bar{y} \Leftrightarrow (\exists f, g \in \text{SEC}[\mathbf{E}, \bar{\mathbf{F}}_Q, \bar{\mathbf{F}}_S^{\hat{}}, \bar{x}])(\varphi(\text{graph}\,f, \text{graph}\,g) \,\&\, f(\bar{x}, \bar{y}) = 0)$
$\Leftrightarrow (\exists f, g)(\varphi(\text{graph}\,f, \text{graph}\,g) \,\&\, f(\bar{x}, \bar{y}) = 0)$,

(b) $\bar{y} <^*_{\Phi} \bar{x} \Leftrightarrow (\exists f, g \in \text{SEC}[\mathbf{E}, \bar{\mathbf{F}}_Q, \bar{\mathbf{F}}_S^{\hat{}}, \bar{y}])(\varphi(\text{graph}\,f, \text{graph}\,g) \,\&\, f(\bar{x}, \bar{y}) = 1)$
$\Leftrightarrow (\exists f, g)(\varphi(\text{graph}\,f, \text{graph}\,g) \,\&\, f(\bar{x}, \bar{y}) = 1)$. □

Here $\text{SEC}[\mathbf{E}, \bar{\mathbf{F}}_Q, \bar{\mathbf{F}}_S^{\hat{}}, \bar{x}]$ is the collection of all relations which are recursive in $\mathbf{E}, \bar{\mathbf{F}}_Q, \bar{\mathbf{F}}_S^{\hat{}}$ from $\bar{x}$. The notation "$f \in \text{SEC}[\mathbf{E}, \bar{\mathbf{F}}_Q, \bar{\mathbf{F}}_S^{\hat{}}, \bar{x}]$" means that graph f is a relation recursive in $\mathbf{E}, \bar{\mathbf{F}}_Q, \bar{\mathbf{F}}_S^{\hat{}}$ from $\bar{x}$. The crucial formula $\varphi(Y_1, Y_2)$ in the above Theorem 4.5 is defined as follows:

$$\varphi(Y_1, Y_2) \Leftrightarrow (Y_1, Y_2 \text{ are graphs of partial functions}) \,\&\, \psi(Y_1, Y_2),$$

where $\psi(Y_1, Y_2)$ is the conjunction of the following four conditions:

(i) If $g(\bar{x}) = w$, then $\Phi(\bar{x}, g \upharpoonright \{\bar{x}' : f(\bar{x}, \bar{x}') = 1\}) = w$.

(ii) If $f(\bar{x}, \bar{y}) = 0$, then $g(\bar{x})\downarrow$ and $\lambda\bar{x}' f(\bar{x}', \bar{x}), \lambda\bar{y}' f(\bar{x}, \bar{y}')$ are total functions.

(iii) If $f(\bar{x}, \bar{y}) = 1$, then $g(\bar{y})\downarrow$ and $\lambda\bar{x}' f(\bar{x}', \bar{y}), \lambda\bar{y}' f(\bar{y}, \bar{y}')$ are total functions.

(iv) Range $f \subseteq \{0, 1\}$, and if $f(\bar{x}, \bar{y}) = w$, then $\Delta_{\Phi}(\bar{x}, g, \lambda\bar{x}'(1 \dot{-} f(\bar{y}, \bar{x}'))) = w$, where $\Delta_{\Phi}(\bar{x}, g, \delta)$ is a normalizing functional for Φ in the class $\mathscr{F}[\mathbf{E}, \bar{\mathbf{F}}_Q, \bar{\mathbf{F}}_S^{\hat{}}]$.

Intuitively, $\varphi(Y_1, Y_2)$ describes certain first order properties of the functions Φ^{ξ} and Ψ^{ξ}. The content of the Second Stage Comparison Theorem is that if f and g are partial functions satisfying $\varphi(\text{graph}\,f, \text{graph}\,g)$, then f and g are, respectively, approximations of Ψ^{ξ} and Φ^{ξ}.

5. A second order hierarchy for recursion in quantifiers. Assume that $\mathfrak{A} = \langle A, \bar{R}, \bar{f}, \bar{c}\rangle$ is a structure such that ω, $\leqslant_\omega$ are first order definable on $\mathfrak{A}$ and let $\bar{Q} = (Q_1, \ldots, Q_l)$, $\bar{S} = (S_1, \ldots, S_m)$ be two sequences of quantifiers on A. We will establish here the existence of a hierarchy for the family

$$\mathscr{R} = \left\{\mathrm{SEC}\left[\mathbf{E}, \bar{\mathbf{F}}_Q, \bar{\mathbf{F}}_S^{\wedge}, \bar{x}\right] : \bar{x} \in A^{<\omega}\right\}$$

of the classes of relations which are recursive in $\mathbf{E}$, $\bar{\mathbf{F}}_Q$, $\bar{\mathbf{F}}_S^{\wedge}$ from $\bar{x}$, $\bar{x} \in A^{<\omega}$. In order to describe this hierarchy we first need some definitions.

5.1. An *indexed family on* $\mathfrak{A}$ is a collection $\mathscr{G} = \{\Gamma(\bar{x}): \bar{x} \in A^{<\omega}\}$ of nonempty classes of relations on A with the following properties:

(i) If $\bar{x} \in A^{<\omega}$, $\bar{n} \in \omega^{<\omega}$ and $\bar{d}$ is a sequence from the constants $\bar{c}$ of $\mathfrak{A}$, then $\Gamma(\bar{x}, \bar{n}, \bar{d}) = \Gamma(\bar{x})$.

(ii) If $\bar{x} = (x_1, \ldots, x_k)$, $\bar{y} = (y_1, \ldots, y_s)$ are finite sequences from A and $\{y_j: j = 1, 2, \ldots, s\} \subseteq \{x_i: i = 1, 2, \ldots, k\}$, then $\Gamma(\bar{y}) \subseteq \Gamma(\bar{x})$.

(iii) If $\bar{x} = (x_1, \ldots, x_k)$, $\bar{y} = (y_1, \ldots, y_s)$ are finite sequences from A and $R \subseteq A^{s+k}$ is an element of $\Gamma(\bar{x})$, then the relation $R_{\bar{y}} = \{\bar{z}: R(\bar{y}, \bar{z})\}$ is an element of $\Gamma(\bar{x}, \bar{y})$.

The above definition abstracts the combinatorial properties of relativization to finite sequences. The collection $\mathscr{R} = \{\mathrm{SEC}[\mathbf{E}, \bar{\mathbf{F}}_Q, \bar{\mathbf{F}}_S^{\wedge}, \bar{x}]: \bar{x} \in A^{<\omega}\}$ is a typical example of an indexed family.

5.2. Assume that $\mathscr{G} = \{\Gamma(\bar{x}): \bar{x} \in A^{<\omega}\}$ is an indexed family on A, $\bar{x}$ is a finite sequence from A and that R is a relation on A. We say that the relation R is *simply* $\Sigma_1^1(\exists, Q_1, \ldots, Q_l, \hat{S_1}, \ldots, \hat{S_m})$ *definable from* $\bar{x}$ *with basis* $\mathscr{G}$ if there is a formula φ of the language $\mathscr{L}^{\mathfrak{A}}(Q_1, \ldots, Q_l, S_1, \ldots, S_m)$ such that

$$(\forall \bar{y})[R(\bar{y}) \Leftrightarrow (\exists Y \in \Gamma(\bar{x}, \bar{y}))\varphi(\bar{x}, \bar{y}, Y) \Leftrightarrow (\exists Y)\varphi(\bar{x}, \bar{y}, Y)]$$

or

$$(\forall \bar{y})[R(\bar{y}) \Leftrightarrow (Sz)(\exists Y \in \Gamma(\bar{x}, \bar{y}, z))\varphi(\bar{x}, \bar{y}, z, y)$$
$$\Leftrightarrow (Sz)(\exists Y)\varphi(\bar{x}, \bar{y}, z, Y)],$$

where S is one of the quantifiers $S_1, \ldots, S_m$.

Notice that the formula φ may involve some or all of the quantifiers Q_1, $\breve{Q}_1, \ldots, Q_l$, $\breve{Q}_l$, S_1, $\breve{S}_1, \ldots, S_m$, $\breve{S}_m$, but only one of the quantifiers $S_1, \ldots, S_m$ is allowed in front of the second order existential quantifier. In order to clarify further this definition we give a few examples.

(a) If $\varphi(\bar{u}, \bar{y}, Y)$ is a formula of $\mathscr{L}^{\mathfrak{A}}$ and

$$(\forall \bar{y})[R(\bar{y}) \Leftrightarrow (\exists Y \in \Gamma(\bar{x}, \bar{y}))\varphi(\bar{x}, \bar{y}, Y) \Leftrightarrow (\exists y)\varphi(\bar{x}, \bar{y}, Y)],$$

then R is $\Sigma_1^1(\exists)$ definable from $\bar{x}$ with basis $\mathscr{G}$.

(b) If $\varphi(\bar{u}, \bar{y}, z, Y)$ is a formula of $\mathscr{L}^{\mathfrak{A}}$ and

$$(\forall \bar{y})[R(\bar{y})$$
$$\Leftrightarrow (\exists z)(\exists Y \in \Gamma(\bar{x}, \bar{y}, z))\varphi(\bar{x}, \bar{y}, z, Y) \Leftrightarrow (\exists z)(\exists Y)\varphi(\bar{x}, \bar{y}, z, Y)],$$

then R is simply $\Sigma_1^1(\exists, \exists^{\wedge})$ definable from $\bar{x}$ with basis $\mathcal{G}$. On the other hand, if φ is a formula of $\mathcal{L}^{\mathfrak{A}}(Q_1)$ and R satisfies the above equivalence, then R is simply $\Sigma_1^1(\exists, Q_1, \exists^{\wedge})$ definable from $\bar{x}$ with basis $\mathcal{G}$.

(c) Finally, if $\varphi(\bar{u}, \bar{y}, z, Y)$ is a formula of $\mathcal{L}^{\mathfrak{A}}(Q_1, S_1)$ and

$$(\forall \bar{y})\big[R(\bar{y}) \Leftrightarrow (S_1 z)(\exists Y \in \Gamma(\bar{x}, \bar{y}, z))\varphi(\bar{x}, \bar{y}, z, Y) \Leftrightarrow (S_1 z)(\exists Y)\varphi(\bar{x}, \bar{y}, z, Y)\big],$$

then R is simply $\Sigma_1^1(\exists, Q_1, S_1^{\wedge})$ definable from $\bar{x}$ with basis $\mathcal{G}$.

If in the formula φ we allow second order parameters from $\Gamma(\bar{x})$, then we say that R is *simply* $\Sigma_1^1(\exists, \bar{Q}, \bar{S}^{\wedge})$ *definable from* $\bar{x}$ *with basis* $\mathcal{G}$ *and with parameters from* $\Gamma(\bar{x})$.

We say that a relation R is $\Sigma_1^1(\exists, \bar{Q}, \bar{S}^{\wedge})$ *definable from* $\bar{x}$ *with basis* $\mathcal{G}$ if it is a finite disjunction of simply $\Sigma_1^1(\exists, \bar{Q}, \bar{S}^{\wedge})$ definable relations from $\bar{x}$ with basis $\mathcal{G}$. A relation R is $\Delta_1^1(\exists, \bar{Q}, \bar{S}^{\wedge})$ *definable from* $\bar{x}$ *with basis* $\mathcal{G}$ if both R and $\neg R$ are $\Sigma_1^1(\exists, \bar{Q}, \bar{S}^{\wedge})$ definable from $\bar{x}$ with basis $\mathcal{G}$. In a similar way we define the concepts of $\Sigma_1^1(\exists, \bar{Q}, \bar{S}^{\wedge})$ *and* $\Delta_1^1(\exists, \bar{Q}, \bar{S}^{\wedge})$ *definable realtions from* $\bar{x}$ *with basis* $\mathcal{G}$ *and with parameters from* $\Gamma(\bar{x})$.

5.3. Let $\mathfrak{A} = \langle A, \bar{R}, \bar{f}, \bar{c}\rangle$ be a structure such that ω and $\leqslant_\omega$ are first order definable on $\mathfrak{A}$ and assume that $\bar{Q} = (Q_1, \ldots, Q_l)$, $\bar{S} = (S_1, \ldots, S_m)$ are two sequences of quantifiers on $\mathfrak{A}$. For any ordinal ξ and any finite sequence $\bar{x} \in A^{<\omega}$ we define by induction on ξ a class of relation $\Delta^\xi(\bar{x})$ as follows:

$\Delta^0(\bar{x})$ = all relations on A which are first order definable on $\mathfrak{A}$ from $\bar{x}$,

$\Delta^\xi(\bar{x})$ = all relations on A which are $\Delta_1^1(\exists, \bar{Q}, \bar{S}^{\wedge})$ definable from $\bar{x}$ with basis $\mathcal{D}^{<\xi} = \{\Delta^{<\xi}(\bar{w}): \bar{w} \in A^{<\omega}\}$, where

$$\Delta^{<\xi}(\bar{x}) = \bigcup_{\eta<\xi} \Delta^\eta(\bar{x}).$$

For each $\bar{x} \in A^{<\omega}$ we put $\Delta(\bar{x}) = \bigcup_\xi \Delta^\xi(\bar{x})$ and we let $\mathcal{D} = \{\Delta(\bar{x}): \bar{x} \in A^{<\omega}\}$.

The main theorem in this section is that the indexed family

$$\mathcal{R} = \left\{\mathrm{SEC}\left[\mathbf{E}, \bar{\mathbf{F}}_Q, \bar{\mathbf{F}}_S^{\wedge}, \bar{x}\right]: \bar{x} \in A^{<\omega}\right\}$$

of the recursive relations in $\mathbf{E}, \mathbf{F}_{Q_1}, \ldots, \mathbf{F}_{Q_l}, \mathbf{F}_{S_1}^{\wedge}, \ldots, \mathbf{F}_{S_m}^{\wedge}$ from $\bar{x}$ is contained in the indexed family $\mathcal{D} = \{\Delta(\bar{x}): \bar{x} \in A^{<\omega}\}$, in the sense that $\mathrm{SEC}[\mathbf{E}, \bar{\mathbf{F}}_Q, \bar{\mathbf{F}}_S^{\wedge}, \bar{x}] \subseteq \Delta(\bar{x})$ for any $\bar{x} \in A^{<\omega}$. Moreover, if the structure $\mathfrak{A}$ is acceptable, then equality holds for any $\bar{x} \in A^{<\omega}$.

Moschovakis [**1969b**, **EIAS**] and Harrington, Kirousis and Schlipf [**1978**] established hierarchy theorems for recursion in $\mathbf{E}^{\#}$ (positive elementary induction) and recursion in $\mathbf{E}^{\#}$, $\mathbf{F}_Q^{\#}$ (positive elementary induction in the quantifier Q). Their hierarchies are obtained by iterating similar notions of Δ_1^1 definability with basis. There are however two main differences between their approach and ours which should be pointed out. Moschovakis, and Harrington, Kirousis and Schlipf consider a single hierarchy in which arbitrary parameters from A are allowed. In other words they work with "boldface" notions and as a result they study the classes

$$\mathbf{SEC}[\mathbf{E}^{\#}] = \bigcup\{\mathrm{SEC}[\mathbf{E}^{\#}, \bar{x}] : \bar{x} \in A^{<\omega}\}$$

and

$$\mathbf{SEC}\left[\mathbf{E}^{\#}, \mathbf{F}_Q^{\#}\right] = \bigcup\{\mathrm{SEC}[\mathbf{E}^{\#}, \bar{x}] : \bar{x} \in A^{<\omega}\}$$

of the "boldface" recursive in $\mathbf{E}^{\#}$ and recursive in $\mathbf{E}^{\#}$, $\mathbf{F}_Q^{\#}$ relations respectively. In contrast here we study the indexed family

$$\mathscr{R} = \left\{\mathrm{SEC}\left[\mathbf{E}, \overline{\mathbf{F}}_Q, \overline{\mathbf{F}}_S^{\hat{}}, \bar{x}\right] : \bar{x} \in A^{<\omega}\right\}$$

by keeping track of the parameters involved and by working with an indexed family of hierarchies. The need of such finer "lightface" results is explained and justified in Kolaitis [**1981**]. Roughly speaking it is dictated by the fact that if the functional $\mathbf{E}^{\#}$ is not available, then the semirecursive relations are not closed under existential quantification, so that one has to introduce "lightface" notions and to account for the parameters from A. The second difference is that at the ξth level of the hierarchy we do not allow second order parameters from the previous levels, while the hierarchies of Moschovakis, and Harrington, Kirousis and Schlipf are built using relations from the lower stages as parameters. Our approach will cause some technical difficulties in the proof of the main theorem, but it yields stronger results and is closer to the spirit of Kleene's [**1959b**] work which relates the hyperarithmetic sets to the ramified analytic hierarchy.

5.4. LEMMA. (i) *For any finite sequence* $\bar{x} \in A^{<\omega}$ *and any ordinals* η, ξ, *if* $\eta < \xi$, *then* $\Delta^{\eta}(\bar{x}) \subseteq \Delta^{\xi}(\bar{x})$.

(ii) *For any ordinal* ξ, *the collections* $\mathscr{D}^{\xi} = \{\Delta^{\xi}(\bar{x}) \colon \bar{x} \in A^{<\omega}\}$ *and* $\mathscr{D}^{<\xi} = \{\Delta^{<\xi}(\bar{x}) \colon \bar{x} \in A^{<\omega}\}$ *are both indexed families on* $\mathfrak{A}$.

(iii) *Assume that* $t_i \colon A^{n_i} \to A$, $1 \leqslant i \leqslant m$, *are functions whose graphs are first order definable on* $\mathfrak{A}$. *Then for any sequences* $\bar{x}_1, \ldots, \bar{x}_m, \bar{y} \in A^{<\omega}$ *we have that* $\Delta^{\xi}(t_1(\bar{x}_1), \ldots, t_m(\bar{x}_m), \bar{x}_1, \ldots, \bar{x}_m, \bar{y}) \subseteq \Delta^{\xi}(\bar{x}_1, \ldots, \bar{x}_m, \bar{y})$ *for any ordinal* ξ. □

The above lemma can be proved by easy inductions on ξ, using the definitions and the monotonicity of the quantifiers involved. The next result is the key technical tool in establishing the hierarchy theorem for recursion in $\mathbf{E}$, $\overline{\mathbf{F}}_Q$, $\overline{\mathbf{F}}_S^{\hat{}}$. It is in this result that we will make essential use of the functionals in canonical form which were introduced in Part I of the paper.

5.5. THEOREM. *Let* $\mathfrak{A} = \langle A, \bar{R}, \bar{f}, \bar{c}\rangle$ *be a structure such that* $\omega, \leqslant_{\omega}$ *are elementary on* $\mathfrak{A}$, *let* $\bar{Q} = (Q_1, \ldots, Q_l)$, $\bar{S} = (S_1, \ldots, S_m)$ *be two sequences of quantifiers on* A *and let* $\Phi(\bar{x}, g)$ *be an operative functional in the class* $\mathscr{F}[\mathbf{E}, \overline{\mathbf{F}}_Q, \overline{\mathbf{F}}_S^{\hat{}}]$ *which is in canonical form. Assume that* $\bar{x}$ *is a finite sequence from* A *such that* $\Phi^{\infty}(\bar{x})$ *is defined and let* $|\bar{x}| = \lambda + m$, *with* λ *limit and* $m \in \omega$. *Then*

(i) (graph $\Phi^{\lambda+m}) \in \Delta^{\lambda+3m+2}(\bar{x})$, *where* $\Phi^{\lambda+m}$ *is the* $(\lambda + m)$th *stage of* Φ,

(ii) (graph $\Psi^{\lambda+m}) \in \Delta^{\lambda+3m+2}(\bar{x})$, *where* $\Psi^{\lambda+m}$ *is the stage comparison function of* Φ *up to* $\lambda + m$.

PROOF. We will prove (i) and (ii) simultaneously by induction on $|\bar{x}|$ in a series of steps. We will use repeatedly the Second Stage Comparison Theorem 4.5, as well as the hypothesis that the functional Φ is in canonical form.

Let $\bar{x} \in A^{<\omega}$ be a finite sequence from A such that $\Phi^\infty(\bar{x})$ is defined and let $|\bar{x}| = \lambda + m$. Assume that for all $\bar{y} \in A^{<\omega}$ and all ordinals $\lambda' + m' < \lambda + m$ if $|\bar{y}| = \lambda' + m'$, then both (graph $\Phi^{\lambda'+m'}$) and (graph $\Psi^{\lambda'+m'}$) are in the class $\Delta^{\lambda'+3m'+2}(\bar{y})$. Using this inductive hypothesis it is quite easy to show that if $\eta < \lambda + m$ and $|\bar{y}| = \eta$, then both (graph Φ^η) and (graph Ψ^η) are in $\Delta^{<\lambda+3m}(\bar{y})$. We present now the steps which will establish that the relations (graph $\Phi^{\lambda+m}$) and (graph $\Psi^{\lambda+m}$) are in the class $\Delta^{\lambda+3m+2}(\bar{x})$.

Step 1. Both (graph $\Phi^{<\lambda+m}$) and (graph $\Psi^{<\lambda+m}$) are $\Sigma_1^1(\exists, \overline{Q}, \overline{S}^{\,\hat{}}\,)$ definable from $\bar{x}$ with basis $\mathscr{D}^{<\lambda+3m}$.

We give the argument for (graph $\Psi^{<\lambda+m}$) only, since (graph $\Phi^{<\lambda+m}$) can be treated in a similar and actually somewhat simpler way.

$$
\begin{aligned}
&\Psi^{<\lambda+m}(\bar{x}', \bar{y}') = u \Rightarrow (\exists \eta < \lambda + m)(\Psi^\eta(\bar{x}', \bar{y}') = u)\\
&\quad\Rightarrow \big(|\bar{x}'| = \eta \,\&\, \Psi^\eta(\bar{x}', \bar{y}') = 0 \,\&\, \Psi^\eta(\bar{x}, \bar{x}') = 1\big)\\
&\qquad \vee \big(|\bar{y}'| = \eta \,\&\, \Psi^\eta(\bar{x}', \bar{y}') = 1 \,\&\, \Psi^\eta(\bar{x}, \bar{y}')\big) = 1\\
&\quad\Rightarrow \text{(by induction hypothesis and Theorem 4.5)}\\
&\quad\Rightarrow \big(\exists f, g \in \Delta^{<\lambda+3m}(\bar{x}', \bar{y}')\big)\big[(\varphi(\text{graph } f, \text{graph } g) \,\&\, u = 0 \,\&\, f(\bar{x}', \bar{y}') = 0\\
&\qquad\qquad \&\, f(\bar{x}, \bar{x}') = 1) \vee (\varphi(\text{graph } f, \text{graph } g)\\
&\qquad\qquad \&\, u = 1 \,\&\, f(\bar{x}', \bar{y}') = 1 \,\&\, f(\bar{x}, \bar{y}') = 1)\big] \Rightarrow (\exists f, g)\\
&\quad\big[(\varphi(\text{graph } f, \text{graph } g) \,\&\, u = 0 \,\&\, f(\bar{x}', \bar{y}') = 0 \,\&\, f(\bar{x}, \bar{x}') = 1)\\
&\qquad \vee (\varphi(\text{graph } f, \text{graph } g) \,\&\, u = 1 \,\&\, f(\bar{x}', \bar{y}') = 1 \,\&\, f(\bar{x}, \bar{y}') = 1)\big]\\
&\quad\Rightarrow \text{(by Theorem 4.5)}\\
&\quad\Rightarrow \big[\big(u = 0 \,\&\, \bar{x}' <_\Phi^* \bar{x} \,\&\, \bar{x}' \leqslant_\Phi^* \bar{y}'\big) \vee \big(u = 1 \,\&\, \bar{y}' <_\Phi^* \bar{x} \,\&\, \bar{y}' <_\Phi^* \bar{x}'\big)\big]\\
&\quad\Rightarrow \Psi^{<\lambda+m}(\bar{x}', \bar{y}') = u.
\end{aligned}
$$

The formula φ above is the same as the one in Theorem 4.5. We indicate below how the theorem is used in deriving the last two implications. Assume for example that

$$(\exists f, g)(\varphi(\text{graph } f, \text{graph } g) \,\&\, u = 0 \,\&\, f(\bar{x}', \bar{y}') = 0 \,\&\, f(\bar{x}, \bar{x}') = 1).$$

Since $\Phi^\infty(\bar{x})\downarrow$, we have that $\bar{x} \leqslant_\Phi^* \bar{x}'$ or $\bar{x}' <_\Phi^* \bar{x}$, but since $f(\bar{x}, \bar{x}') = 1$ we must have by Theorem 4.5(iii) that $\bar{x}' <_\Phi^* \bar{x}$. Therefore $\Phi^\infty(\bar{x}')\downarrow$ and hence $\bar{x}' \leqslant_\Phi^* \bar{y}'$ or $\bar{y}' <_\Phi^* \bar{x}'$, but $f(\bar{x}', \bar{y}') = 0$ and hence again by 4.5(iii) we get $\bar{x}' \leqslant_\Phi^* \bar{y}'$. Now we have $\bar{x}' \leqslant_\Phi^* \bar{y}'$ and $\bar{x}' <_\Phi^* \bar{x}$, hence $\Psi^{<\lambda+m}(\bar{x}', \bar{y}') = 0$.

Step 2. The relation $P_1(\bar{y}) \Leftrightarrow (\bar{y} <_\Phi^* \bar{x})$ is $\Sigma_1^1(\exists, \overline{Q}, \overline{S}^{\,\hat{}}\,)$ definable from $\bar{x}$ with basis $\mathscr{D}^{<\lambda+3m}$. Immediate from Step 1, since $P_1(\bar{y}) \Leftrightarrow \Psi^{<\lambda+m}(\bar{x}, \bar{y}) = 1$.

Step 3. The relation $P_2(\bar{y}) \Leftrightarrow (\bar{x} \leqslant_\Phi^* \bar{y})$ is $\Sigma_1^1(\exists, \overline{Q}, \overline{S}^{\,\hat{}}\,)$ definable from $\bar{x}$ with basis $\mathscr{D}^{\lambda+3m} = \mathscr{D}^{<\lambda+3m+1}$.

This is the key step in establishing the theorem. Since Φ belongs to the class $\mathscr{F}[\mathbf{E}, \overline{\mathbf{F}}_Q, \overline{\mathbf{F}}_S^{\,\hat{}}]$ and it is a functional in canonical form, we know that Φ is defined by cases where in each of the cases we have an evaluation functional or a composition functional or a substitution of an evaluation functional in one of the functionals $\mathbf{E}, \mathbf{F}_{Q_1}, \ldots, \mathbf{F}_{Q_l}, \mathbf{F}_{S_1}^{\hat{}}, \ldots, \mathbf{F}_{S_m}^{\hat{}}$. Therefore the analysis of the relation $P_2(\bar{y})$ breaks into several cases depending on the clause which makes $\lambda + m$ the smallest ordinal ξ such that $\Phi^{\xi}(\bar{x})$ is defined. We omit the cases of evaluation and composition, since they are straightforward using the induction hypothesis and Theorem 4.5. Actually in these cases the $\Sigma_1^1(\exists, \overline{Q}, \overline{S}\,\hat{})$ definition of P_2 is with basis $\mathscr{D}^{<\lambda+3m}(\bar{x})$. We consider now the remaining cases.

Case 1. $\Phi^{\lambda+m}(\bar{x}) = 0$, because $\mathbf{F}^{\hat{}}_{S}(\lambda z \Phi^{<\lambda+m}(\bar{t})) = 0$, where S is one of the quantifiers $S_1, \ldots, S_m$ and $\bar{t}$ is a sequence of terms in $\bar{x}$ and z. Since $|\bar{x}| = \lambda + m$, it follows that in this case $\lambda + m$ is the smallest ordinal ξ such that $(Sz)(\Phi^{<\xi}(\bar{t}) = 0)$. But then,

$$
\begin{aligned}
\bar{x} \leqslant_{\Phi}^{*} \bar{y} &\Rightarrow (Sz)\big(\bar{t} <_{\Phi}^{*} \bar{x} \,\&\, \bar{t} <_{\Phi}^{*} \bar{y} \,\&\, \Phi^{|\bar{t}|}(\bar{t}) = 0\big)\\
&\Rightarrow (Sz)(\exists \eta < \lambda + m)\big(|\bar{t}| = \eta \,\&\, \varphi(\operatorname{graph}\Psi^{\eta}, \operatorname{graph}\Phi^{\eta}) \,\&\, \Psi^{\eta}(\bar{x}, \bar{t}) = 1\\
&\qquad\qquad \&\Psi^{\eta}(\bar{y}, \bar{t}) = 1 \,\&\, \Phi^{\eta}(\bar{t}) = 0\big)\\
&\Rightarrow (Sz)\big(\exists f, g \in \Delta^{<\lambda+3m}(\bar{t})\big)\big(\varphi(\operatorname{graph} f, \operatorname{graph} g) \,\&\, f(\bar{x}, \bar{t}) = 1\\
&\qquad\qquad\qquad \& f(\bar{y}, \bar{t}) = 1 \,\&\, g(\bar{t}) = 0\big)\\
&\Rightarrow (Sz)(\exists f, g)\big(\varphi(\operatorname{graph} f, \operatorname{graph} g) \,\&\, f(\bar{x}, \bar{t}) = 1\\
&\qquad\qquad f(\bar{y}, \bar{t}) = 1 \,\&\, g(t) = 0\big)\\
&\Rightarrow r(\text{by Lemma 5.4})\\
&\Rightarrow (Sz)(\exists f, g)\big(\varphi(\operatorname{graph} f, \operatorname{graph} g) \,\&\, f(\bar{x}, \bar{t}) = 1 \,\&\, f(\bar{y}, \bar{t}) = 1 \,\&\, g(\bar{t}) = 0\big)\\
&\Rightarrow (\text{by Theorem 4.5})\\
&\Rightarrow (Sz)\big(\bar{t} <_{\Phi}^{*} \bar{x} \,\&\, \bar{t} <_{\Phi}^{*} \bar{y} \,\&\, \Phi^{|\bar{t}|}(\bar{t}) = 0\big)\\
&\Rightarrow \bar{x} \leqslant_{\Phi}^{*} \bar{y}
\end{aligned}
$$

(otherwise $\bar{y} <_{\Phi}^{*} \bar{x}$ and then $(Sz)(\Phi^{<|\bar{y}|}(\bar{t}) = 0)$ violating thus the minimality of $\lambda + m$).

Case 2. $\Phi^{\lambda+m}(\bar{x}) = 1$, because $\mathbf{E}(\lambda z \Phi^{<\lambda+m}(\bar{t})) = 1$, where $\bar{t}$ is a sequence of terms in $\bar{x}$ and z. In this case $\lambda + m$ is the smallest ordinal ξ such that $(\forall z)(\Phi^{<\xi}(\bar{t})\downarrow$ and $\Phi^{<\xi}(\bar{t}) \neq 0)$.

It turns out that this is the hardest case, especially because we do not allow relations from $\Delta^{<\xi}(\bar{x})$ to be used as parameters when we define the relations in $\Delta^{\xi}(\bar{x})$. We will introduce first three relations R_1, R_2, R_3 and we will show that they are $\Delta_1^1(\exists, \overline{Q}, \overline{S}\,\hat{})$ definable from $\bar{x}$ with basis $\mathscr{D}^{<\lambda+3m}$, and hence they are elements of $\Delta^{\lambda+3m}(\bar{x})$. Using these three relations and their first order properties, we will prove that in this case the relation $P_2(\bar{y}) \Leftrightarrow \bar{x} \leqslant_{\Phi}^{*} \bar{y}$ has a $\Sigma_1^1(\exists, \overline{Q}, \overline{S}\,\hat{})$ definition from $\bar{x}$ with basis $\mathscr{D}^{\lambda+3m} = \mathscr{D}^{<\lambda+3m+1}$.

We now put

$$R_1(z, \bar{y}, w) \Leftrightarrow \Phi^{<\lambda+m}(\bar{y}) = w \,\&\, \bar{y} \leqslant^*_\Phi \bar{t},$$
$$R_2(z, \bar{y}_1, \bar{y}_2) \Leftrightarrow (\bar{y}_1 \leqslant^*_\Phi \bar{y}_2 \,\&\, \bar{y}_2 \leqslant^*_\Phi \bar{t}) \vee (\bar{y}_1 \leqslant^*_\Phi \bar{t} \,\&\, \bar{t} <^*_\Phi \bar{y}_2),$$
$$R_3(z, \bar{y}_1, \bar{y}_2) \Leftrightarrow (\bar{y}_2 <^*_\Phi \bar{y}_1 \,\&\, \bar{y}_1 \leqslant^*_\Phi \bar{t}) \vee (\bar{y}_2 \leqslant^*_\Phi \bar{t} \,\&\, \bar{t} <^*_\Phi \bar{y}_1).$$

To motivate the relevance of the above relations we prove first that the functions $\Phi^{<\lambda+m}$, $\Psi^{<\lambda+m}$ are first order in R_1, R_2, R_3 according to the following

LEMMA 1. (i) $\Phi^{<\lambda+m}(\bar{y}) = w \Leftrightarrow (\exists z) R_1(z, \bar{y}, w)$,
(ii) $\Psi^{<\lambda+m}(\bar{y}_1, \bar{y}_2) = 0 \Leftrightarrow (\exists z) R_2(z, \bar{y}_1, \bar{y}_2)$,
(iii) $\Psi^{<\lambda+m}(\bar{y}_1, \bar{y}_2) = 1 \Leftrightarrow (\exists z) R_3(z, \bar{y}_1, \bar{y}_2)$.

PROOF OF LEMMA 1. We show that if $\Psi^{<\lambda+m}(\bar{y}_1, \bar{y}_2) = 0$, then $(\exists z) R_2(z, \bar{y}_1, \bar{y}_2)$. The other implications are easier or similar and we omit them. Since $\Psi^{<\lambda+m}(\bar{y}_1, \bar{y}_2) = 0$, we have that $\bar{y}_1 \leqslant^*_\Phi \bar{y}_2$ and $\bar{y}_1 \leqslant^*_\Phi \bar{x}$. We distinguish two cases.

(a) If $\bar{y}_2 <^*_\Phi \bar{x}$, then there is some w such that $\Phi^{<\lambda+m}(\bar{y}_2) = w$. We claim now that there is some z such that $\bar{y}_2 \leqslant^*_\Phi \bar{t}$ (where the terms in t are evaluated in $\bar{x}$ and this z). Otherwise we would have that $(\forall z)(\bar{t} <^*_\Phi \bar{y}_2)$ and hence $(\forall z)(\Phi^{<|\bar{y}_2|}(\bar{t})\downarrow)$ which violates the minimality of $\lambda + m = |\bar{x}| > |\bar{y}_2|$. We have found thus a z such that $\bar{y}_2 \leqslant^*_\Phi \bar{t}$, but also $\bar{y}_1 \leqslant^*_\Phi \bar{y}_2$, hence $R_2(z, \bar{y}_1, \bar{y}_2)$.

(b) If $\bar{x} \leqslant^*_\Phi \bar{y}_2$, then we pick a z so that $\bar{y}_1 \leqslant^*_\Phi \bar{t}$ (such a z exists since $\bar{y}_1 <^*_\Phi \bar{x}$) and we notice that also $\bar{t} <^*_\Phi \bar{y}_2$, since $\bar{t} <^*_\Phi x$. □

In order to prove that the relations R_1, R_2, R_3 are in the class $\Delta^{\lambda+3m}(\bar{x})$ we need first another lemma.

LEMMA 2. *The relation* $R(\bar{y}, z) \Leftrightarrow \bar{y} \leqslant^*_\Phi \bar{t}$ *is* $\Delta^1_1(\exists, \overline{Q}, \overline{S}^\wedge)$ *definable from* $\bar{x}$ *with basis* $\mathscr{D}^{<\lambda+3m}$.

PROOF OF LEMMA 2.

$$R(\bar{y}, z) \Rightarrow \bar{y} \leqslant^*_\Phi \bar{t} \Rightarrow (\exists \eta < \lambda + m)(\Psi^\eta(\bar{y}, \bar{t}) = 0 \,\&\, \Psi^\eta(\bar{x}, \bar{t}) = 1)$$
$$\Rightarrow (\exists f, g \in \Delta^{<\lambda+3m}(\bar{x}, \bar{y}, z))(\varphi(\text{graph } f, \text{graph } g)$$
$$\&f(\bar{x}, \bar{t}) = 1 \,\&\, f(\bar{y}, \bar{t}) = 0)$$
$$\Rightarrow (\exists f, g)(\varphi(\text{graph } f, \text{graph } g) \,\&\, f(\bar{x}, \bar{t}) = 1 \,\&\, f(\bar{y}, \bar{t}) = 0)$$
$$\Rightarrow \bar{y} \leqslant^*_\Phi t. \quad \text{(by Theorem 4.5)}$$

The $\Sigma^1_1(\exists, \overline{Q}, \overline{S}^\wedge)$ definition of the complement $\neg R$ is obtained in a similar way by noticing that since $(\forall z)(\Phi^{<\lambda+m}(t)\downarrow)$ we have that

$$\neg(\bar{y} \leqslant^*_\Phi \bar{t}) \Leftrightarrow (\bar{t} <^*_\Phi \bar{y}). \qquad \square$$

LEMMA 3. *The relations* R_1, R_2, R_3 *are* $\Delta^1_1(\exists, \overline{Q}, \overline{S}^\wedge)$ *definable from* $\bar{x}$ *with basis* $\mathscr{D}^{<\lambda+3m}$, *and hence they are in the class* $\Delta^{\lambda+3m}(\bar{x})$.

PROOF OF LEMMA 3. The argument uses the preceding Lemma 2 and, of course, Theorem 4.5. For example we have

$$R_1(z, \bar{y}, w) \Leftrightarrow \Phi^{<\lambda+m}(\bar{y}) = w \,\&\, \bar{y} \leqslant_{\Phi}^{*} \bar{t}$$

so R_1 is $\Sigma_1^1(\exists, \bar{Q}, \bar{S}^{\wedge})$ definable from $\bar{x}$ with basis $\mathscr{D}^{<\lambda+3m}$ by Step 1 and Lemma 2. Moreover,

$$\begin{aligned}\neg\, R_1(z, \bar{y}, w) &\Leftrightarrow \left(\bar{t} <_{\Phi}^{*} \bar{y}\right) \vee \left(\bar{y} \leqslant_{\Phi}^{*} \bar{t} \,\&\left(\Phi^{<\lambda+m}(\bar{y})\downarrow\; \neq w\right)\right)\\
&\Leftrightarrow \left(\exists f, g \in \Delta^{<\lambda+3m}(\bar{x}, \bar{y}, z)\right)\\
&\qquad \big[\left(\varphi(\operatorname{graph} f, \operatorname{graph} g) \,\&\, f(\bar{x}, \bar{t}) = 1 \,\&\, f(\bar{y}, \bar{t}) = 1\right)\\
&\qquad \vee \left(\varphi(\operatorname{graph} f, \operatorname{graph} g) \,\&\, f(\bar{x}, \bar{t}) = 1\right.\\
&\qquad \left.\&\, f(\bar{y}, \bar{t}) = 0 \,\&\, (\exists w')\left(w' \neq w \,\&\, g(\bar{t}) = w'\right)\right)\big]\\
&\Leftrightarrow (\exists f, g)\big[\left(\varphi(\operatorname{graph} f, \operatorname{graph} g) \,\&\, f(\bar{x}, \bar{t}) = 1 \,\&\, f(\bar{y}, \bar{t}) = 1\right)\\
&\qquad \vee\, \varphi(\operatorname{graph} f, \operatorname{graph} g) \,\&\, f(\bar{x}, \bar{t}) = 1\\
&\qquad \&\, f(\bar{y}, \bar{t}) = 0 \,\&\, (\exists w')\left(w' \neq w \,\&\, g(\bar{t}) = w'\right)\big].\end{aligned}$$

The analysis of $R_2, \neg R_2, R_3, \neg R_3$ is similar and we omit it. □

We complete now the proof for the Case 2 by exhibiting the following definition of the relation $P_2(\bar{y})$ in this case:

$$\begin{aligned}P_2(\bar{y}) &\Leftrightarrow \bar{x} \leqslant_{\Phi}^{*} \bar{y}\\
&\Leftrightarrow \left(\exists Y_1, Y_2, Y_3 \in \Delta^{<\lambda+3m+1}(\bar{x})\right)\left[\varphi(\operatorname{graph} f, \operatorname{graph} g) \,\&\, (\forall z) Y_2(z, \bar{t}, \bar{t})\right.\\
&\qquad \left.\&\, (\forall z)(\forall w)((z, \bar{y}, w) \notin Y_1)\right]\\
&\Leftrightarrow (\exists Y_1, Y_2, Y_3)\left[\varphi(\operatorname{graph} f, \operatorname{graph} g)\right.\\
&\qquad \left.\&\, (\forall z) Y_2(z, \bar{t}, \bar{t}) \,\&\, (\forall z)(\forall w)((z, \bar{y}, w) \notin Y_1)\right],\end{aligned}$$

where graph $f = \{(\bar{y}_1, \bar{y}_2, 0)\colon (\exists z) Y_2(z, \bar{y}_1, \bar{y}_2)\} \cup \{(\bar{y}_1, \bar{y}_2, 1)\colon (\exists z) Y_3(z, \bar{y}_1, \bar{y}_2)\}$ and graph $g = \{(\bar{y}, w)\colon (\exists z) Y_1(z, \bar{y}, w)\}$.

In order to prove the direction " $\Rightarrow$ " in the above equivalence, it is enough to take $Y_1 = R_1$, $Y_2 = R_2$, $Y_3 = R_3$ and to appeal to the preceding Lemmas 1 and 3. Towards the other direction assume that Y_1, Y_2, Y_3 are relations such that

$$\varphi(\operatorname{graph} f, \operatorname{graph} g) \,\&\, (\forall z)\left(Y_2(z, \bar{t}, \bar{t})\right) \,\&\, (\forall z)(\forall w)((z, \bar{y}, w) \notin Y_1),$$

where f and g are defined from Y_1, Y_2, Y_3 as above. We have then that $(\forall z)(f(\bar{t}, \bar{t}) = 0) \,\&\, (\forall w)((\bar{y}, w) \notin \operatorname{graph} g)$. We claim that these conditions imply that $\bar{x} \leqslant_{\Phi}^{*} \bar{y}$. Otherwise, we have $\bar{y} <_{\Phi}^{*} \bar{x}$, hence $\Phi^{<\lambda+m}(\bar{y})\downarrow$ and so there is some z with $\bar{y} \leqslant_{\Phi}^{*} \bar{t}$. Since $f(\bar{t}, \bar{t}) = 0$, the definition of φ in Theorem 4.5 implies that the function $\lambda \bar{y}' f(\bar{y}', \bar{t})$ is total and in particular $f(\bar{y}, \bar{t})\downarrow$. But $\bar{y} \leqslant_{\Phi}^{*} \bar{t}$ and so $f(\bar{y}, \bar{t}) = 0$ and $g(\bar{y})\downarrow\; = \Phi^{|\bar{y}|}(\bar{y})$, contradicting the fact that $(\forall w)((\bar{y}, w) \notin$ graph $g)$. This completes the argument for Case 2.

Case 3. $\Phi^{\lambda+m}(\bar{x}) = 0$, because $\mathbf{E}(\lambda z \Phi^{<\lambda+m}(\bar{t})) = 0$ or $\mathbf{F}_{Q_i}(\lambda z \Phi^{<\lambda+m}(\bar{t})) = 0$ for some i with $1 \leqslant i \leqslant l$, where $\bar{t}$ is a sequence of terms in $\bar{x}$, z. In this case

$\lambda + m$ is the smallest ordinal ξ such that $(\forall z)(\Phi^{<\xi}(\bar{t})\downarrow)\&(Qz)(\Phi^{<\xi}(\bar{t}) = 0)$, where Q is one of the quantifiers $\exists, Q_1, \ldots, Q_l$.

We claim that in this case $\lambda + m$ is also the smallest ordinal ξ such that $(\forall z)(\Phi^{<\xi}(\bar{t})\downarrow)$. Once this is proved, then we can show that the relation $P_2(\bar{y}) \Leftrightarrow \bar{x} \leqslant^*_\Phi \bar{y}$ is $\Sigma^1_1(\exists, \bar{Q}, \bar{S}^{\,\hat{}})$ definable from $\bar{x}$ with basis $\mathscr{D}^{<\lambda+3m+1}$ using the same method as in Case 2. Towards a contradiction assume that there is an ordinal $\eta < \lambda + m$ such that $(\forall z)(\Phi^{<\eta}(\bar{t})\downarrow)$. It follows then that $(Qz)(\Phi^{<\eta}(\bar{t}) = 0)$. Indeed, otherwise we would have $(\breve{Q}z)(\Phi^{<\eta}(\bar{t})\downarrow \neq 0)$ and by monotonicity $(\breve{Q}z)(\Phi^{<\lambda+m}(\bar{t})\downarrow \neq 0)$. However, since $(Qz)(\Phi^{<\lambda+m}(\bar{t}) = 0)$ we can find a z in the intersection, i.e. there is a z such that $\Phi^{<\lambda+m}(\bar{t})\downarrow\ = 0$ and $\Phi^{<\lambda+m}(\bar{t}) = 0$, which is absurd. This way we have showed that $(\forall z)(\Phi^{<\eta}(\bar{t})\downarrow)\&(Qz)(\Phi^{<\eta}(\bar{t}) = 0)$ with $\eta < \lambda + m$, violating thus the minimality of $\lambda + m$.

Case 4. $\Phi^{\lambda+m}(\bar{x}) = 1$, because $\mathbf{F}_{Q_i}(\lambda z\Phi^{<\lambda+m}(\bar{t})) = 1$ for some i with $1 \leqslant i \leqslant l$, or $\mathbf{F}^{\hat{}}_{S_j}(\lambda z\Phi^{<\lambda+m}(\bar{t})) = 1$ for some j with $1 \leqslant j \leqslant m$, where $\bar{t}$ is a sequence of terms in $\bar{x}$, z. In this case we have that $\lambda + m$ is the smallest ordinal ξ such that $(\forall z)(\Phi^{<\xi}(\bar{t})\downarrow)\,\&\,(\breve{Q}z)(\Phi^{<\eta}(\bar{t})\downarrow\ \neq 0)$, where Q is one of the quantifiers $Q_1, \ldots, Q_l, S_1, \ldots, S_m$. The same argument as in Case 3 shows that $\lambda + m$ is the smallest ordinal ξ such that $(\forall z)(\Phi^{<\xi}(\bar{t})\downarrow)$ and then we can proceed as in Case 2.

The proof of Step 3 is now complete and thus we have established that the relation $P_2(\bar{y}) \Leftrightarrow \bar{x} \leqslant^*_\Phi \bar{y}$ is $\Sigma^1_1(\exists, \bar{Q}, \bar{S}^{\,\hat{}})$ definable from $\bar{x}$ with basis $\mathscr{D}^{<\lambda+3m+1}$.

Step 4. Both (graph $\Phi^{<\lambda+m}$) and (graph $\Psi^{<\lambda+m}$) are $\Delta^1_1(\exists, \bar{Q}, \bar{S}^{\,\hat{}})$ definable from $\bar{x}$ with basis $\mathscr{D}^{<\lambda+3m+1}$, and hence they are in $\Delta^{\lambda+3m+1}(\bar{x})$.

The $\Sigma^1_1(\exists, \bar{Q}, \bar{S}^{\,\hat{}})$ definitions of these relations were obtained in Step 1. For their complements we have now the following equivalences:

$$\neg\big(\Phi^{<\lambda+m}(\bar{y}) = w\big) \Leftrightarrow \big(\Phi^{<\lambda+m}(\bar{y})\downarrow\ \neq w\big) \vee \big(\bar{x} \leqslant^*_\Phi \bar{y}\big),$$

$$\neg\big(\Psi^{<\lambda+m}(\bar{y}_1, \bar{y}_2) = 0\big) \Leftrightarrow \big(\Psi^{<\lambda+m}(\bar{y}_1, \bar{y}_2) = 1\big) \vee \big(\bar{x} \leqslant^*_\Phi \bar{y}_1 \,\&\, \bar{x} \leqslant^*_\Phi \bar{y}_2\big),$$

$$\neg\big(\Psi^{<\lambda+m}(\bar{y}_1, \bar{y}_2) = 1\big) \Leftrightarrow \big(\Psi^{<\lambda+m}(\bar{y}_1, \bar{y}_2) = 0\big) \vee \big(\bar{x} \leqslant^*_\Phi \bar{y}_1 \,\&\, \bar{x} \leqslant^*_\Phi \bar{y}_2\big).$$

The desired $\Sigma^1_1(\exists, \bar{Q}, \bar{S}^{\,\hat{}})$ definitions from $\bar{x}$ with basis $\mathscr{D}^{<\lambda+3m+1}$ follow now easily from the preceding Steps 1 and 3.

Step 5. The relation $P_3(\bar{y}) \Leftrightarrow \bar{x} <^*_\Phi \bar{y}$ is $\Sigma^1_1(\exists, \bar{Q}, \bar{S}^{\,\hat{}})$ definable from $\bar{x}$ with basis $\mathscr{D}^{\lambda+3m+1} = \mathscr{D}^{<\lambda+3m+2}$.

$$\begin{aligned}
\bar{x} <^*_\Phi \bar{y} \Leftrightarrow\ & (\exists f, g \in \Delta^{<\lambda+3m+2}(\bar{x}))\big[\varphi(\text{graph } f, \text{graph } g) \\
&\qquad \&(\exists w)\big(\Phi\big(\bar{x}, g \upharpoonright \{\bar{y}' : f(\bar{x}, \bar{y}') = 1\}\big) = w\big) \\
&\qquad \&(\forall w)\neg\big(\Phi\big(\bar{y}, g \upharpoonright \{\bar{y}' : f(\bar{x}, \bar{y}') = 1\}\big) = w\big)\big] \\
\Leftrightarrow\ & (\exists f, g)\big[\varphi(\text{graph } f, \text{graph } g) \\
&\qquad \&(\exists w)\big(\Phi\big(\bar{x}, g \upharpoonright \{\bar{y}' : f(\bar{x}, \bar{y}') = 1\}\big) = w\big) \\
&\qquad \&(\forall w)\neg\big(\Phi\big(\bar{y}, g \upharpoonright \{\bar{y}' : f(\bar{x}, \bar{y}') = 1\}\big) = w\big)\big].
\end{aligned}$$

For the direction "$\Rightarrow$" we take $f = \Psi^{<\lambda+m}$, $g = \Phi^{<\lambda+m}$ whose graphs are in $\Delta^{<\lambda+3m+2}(\bar{x})$ by Step 4. The other direction is proved using Theorem 4.5 and the properties of the formula φ as in Case 2.

Step 6. The relation graph $\Phi^{\lambda+m}$ is $\Delta_1^1(\exists, \bar{Q}, \bar{S}^{\wedge})$ definable from $\bar{x}$ with basis $\mathscr{D}^{<\lambda+3m+2}$, and hence it is in the class $\Delta^{\lambda+3m+2}(\bar{x})$.

The $\Sigma_1^1(\exists, \bar{Q}, \bar{S}^{\wedge})$ definition is provided by the equivalences

$$\begin{aligned}\Phi^{\lambda+m}(\bar{y}) = w \Leftrightarrow{} & \Phi(\bar{y}, \Phi^{<\lambda+m}) = w \\ \Leftrightarrow{} & (\exists f, g \in \Delta^{<\lambda+3m+2}(\bar{x}))(\varphi(\text{graph } f, \text{graph } g) \\ & \qquad \& \Phi(\bar{y}, g \restriction \{\bar{y}' : f(\bar{x}, \bar{y}') = 1\}) = w) \\ \Leftrightarrow{} & (\exists f, g)(\varphi(\text{graph } f, \text{graph } g) \& \Phi(\bar{y}, g \restriction \{\bar{y}' : f(\bar{x}, \bar{y}') = 1\}) = w).\end{aligned}$$

In the above equivalences the direction " $\Rightarrow$ " is proved by taking $f = \Psi^{<\lambda+m}$, $g = \Phi^{<\lambda+m}$ which are in $\Delta^{<\lambda+3m+2}(\bar{x}) = \Delta^{\lambda+3m+1}(\bar{x})$ by Step 4. The proof of the other direction uses once more Theorem 4.5. For the complement of graph $\Phi^{\lambda+m}$ we have that

$$\begin{aligned}\neg(\Phi^{\lambda+m}(\bar{y}) = w) \Leftrightarrow{} & (\Phi^{\lambda+m}(\bar{y})\downarrow \neq w) \vee (\Phi^{\lambda+m}(\bar{y})\uparrow) \\ \Leftrightarrow{} & (\Phi^{\lambda+m}(\bar{y})\downarrow \neq w) \vee (\bar{x} <^*_\Phi \bar{y})\end{aligned}$$

and therefore the $\Sigma_1^1(\exists, \bar{Q}, \bar{S}^{\wedge})$ definition of $\neg(\text{graph } \Phi^{\lambda+m})$ is obtained from the $\Sigma_1^1(\exists, \bar{Q}, \bar{S}^{\wedge})$ definition of graph $\Phi^{\lambda+m}$ and the preceding Step 5.

We conclude the proof of the theorem by establishing the remaining

Step 7. The relation graph $\Psi^{\lambda+m}$ is $\Delta_1^1(\exists, \bar{Q}, \bar{S}^{\wedge})$ definable from $\bar{x}$ with basis $\mathscr{D}^{<\lambda+3m+2}$, and hence it is in the class $\Delta^{\lambda+3m+2}(\bar{x})$.

Notice first that the relation $P_4(\bar{y}) \Leftrightarrow \bar{y} \leqslant^*_\Phi \bar{x}$ is $\Sigma_1^1(\exists, \bar{Q}, \bar{S}^{\wedge})$ definable from $\bar{x}$ with basis $\mathscr{D}^{<\lambda+3m+2}$. Indeed,

$$\bar{y} \leqslant^*_\Phi \bar{x} \Leftrightarrow (\exists w)(\Phi^{\lambda+m}(\bar{y}) = w)$$

and so the $\Sigma_1^1(\exists, \bar{Q}, \bar{S}^{\wedge})$ definition can be obtained from Step 6. Now we have

$$\begin{aligned}\Psi^{\lambda+m}(\bar{y}_1, \bar{y}_2) = 0 \Leftrightarrow{} & \bar{y}_1 \leqslant^*_\Phi \bar{y}_2 \& \bar{y}_1 \leqslant^*_\Phi \bar{x} \\ \Leftrightarrow{} & (\Psi^{<\lambda+m}(\bar{y}_1, \bar{y}_2) = 0) \vee (\bar{y}_1 \leqslant^*_\Phi \bar{x} \& \bar{x} \leqslant^*_\Phi \bar{y}_1 \& \bar{x} \leqslant^*_\Phi \bar{y}_2)\end{aligned}$$

and

$$\begin{aligned}\Psi^{\lambda+m}(\bar{y}_1, \bar{y}_2) = 1 \Leftrightarrow{} & \bar{y}_2 <^*_\Phi \bar{y}_1 \& \bar{y}_2 \leqslant^*_\Phi \bar{x} \\ \Leftrightarrow{} & (\Psi^{<\lambda+m}(\bar{y}_1, \bar{y}_2) = 1) \vee (\bar{y}_2 \leqslant^*_\Phi \bar{x} \& \bar{x} \leqslant^*_\Phi \bar{y}_2 \& \bar{x} <^*_\Phi \bar{y}_1).\end{aligned}$$

The preceding Steps 1, 3, 5 and the observation above show that graph $\Psi^{\lambda+m}$ is $\Sigma_1^1(\exists, \bar{Q}, \bar{S}^{\wedge})$ definable from $\bar{x}$ with basis $\mathscr{D}^{<\lambda+3m+2}$. Finally, for the complement of graph $\Psi^{\lambda+m}$ we have that

$$\neg(\Psi^{\lambda+m}(\bar{y}_1, \bar{y}_2) = 0) \Leftrightarrow (\Psi^{\lambda+m}(\bar{y}_1, \bar{y}_2) = 1) \vee (\bar{x} <^*_\Phi \bar{y}_1 \& \bar{x} <^*_\Phi \bar{y}_2),$$

$$\neg(\Psi^{\lambda+m}(\bar{y}_1, \bar{y}_2) = 1) \Leftrightarrow (\Psi^{\lambda+m}(\bar{y}_1, \bar{y}_2) = 0) \vee (\bar{x} <^*_\Phi \bar{y}_1 \& \bar{x} <^*_\Phi \bar{y}_2),$$

and therefore the $\Sigma_1^1(\exists, \bar{Q}, \bar{S}^{\wedge})$ definition of $\neg(\text{graph } \Psi^{\lambda+m})$ is obtained from the $\Sigma_1^1(\exists, \bar{Q}, \bar{S}^{\wedge})$ definition of graph $\Psi^{\lambda+m}$ and the preceding Step 5. □

5.6. HIERARCHY THEOREM. *Let $\mathfrak{A} = \langle A, \bar{R}, \bar{f}, \bar{c}\rangle$ be a structure such that ω, $\leqslant_\omega$ are elementary on $\mathfrak{A}$ and assume that $\bar{Q} = (Q_1, \ldots, Q_l)$, $\bar{S} = (S_1, \ldots, S_m)$ are two sequences of quantifiers on $\mathfrak{A}$. For each $\bar{x} \in A^{<\omega}$ and each ordinal ξ let $\Delta^\xi(\bar{x})$ be the* *class of relations defined by the induction*

$\Delta^0(\bar{x})$ = *the class of first order definable relations on* $\mathfrak{A}$ *from* $\bar{x}$,

$\Delta^\xi(\bar{x})$ = *the class of relations which are* $\Delta^1_1(\exists, \bar{Q}, \bar{S}^\frown)$ *definable from* $\bar{x}$ *with basis* $\mathscr{D}^{<\xi} = \{\bigcup_{\eta<\xi}\Delta^\eta(\bar{w})\colon \bar{w} \in A^{<\omega}\}$.

Then

(i) *The indexed family* $\mathscr{R} = \{\mathrm{SEC}[\mathbf{E}, \bar{\mathbf{F}}_Q, \bar{\mathbf{F}}^\frown_S, \bar{x}]\colon \bar{x} \in A^{<\omega}\}$ *of the recursive relations in* $\mathbf{E}, \mathbf{F}_{Q_1}, \ldots, \mathbf{F}_{Q_l}, \mathbf{F}^\frown{}_{S_1}, \ldots, \mathbf{F}^\frown{}_{S_m}$ *from* $\bar{x}$ *is contained in the indexed family* $\mathscr{D} = \{\bigcup_\xi \Delta^\xi(\bar{x})\colon \bar{x} \in A^{<\omega}\}$, *i.e.*

$$\mathrm{SEC}\left[\mathbf{E}, \bar{\mathbf{F}}_Q, \bar{\mathbf{F}}^\frown_S, \bar{x}\right] \subseteq \bigcup_\xi \Delta^\xi(\bar{x}) \quad \textit{for every } \bar{x} \in A^{<\omega}.$$

(ii) *If, in addition, the structure* $\mathfrak{A}$ *is acceptable, then*

$$\mathrm{SEC}\left[\mathbf{E}, \bar{\mathbf{F}}_Q, \bar{\mathbf{F}}^\frown_S, \bar{x}\right] = \bigcup_\xi \Delta^\xi(\bar{x}) \quad \textit{for every } \bar{x} \in A^{<\omega}.$$

PROOF. Part (i) follows almost immediately from the preceding Theorems 3.12 and 5.5. If $\mathfrak{A}$ is an acceptable structure, then for every $\bar{x} \in A^{<\omega}$ the collection $\mathrm{ENV}[\mathbf{E}, \bar{\mathbf{F}}_Q, \bar{\mathbf{F}}^\frown_S, \bar{x}]$ of the semirecursive relations in $\mathbf{E}, \bar{\mathbf{F}}_Q, \bar{\mathbf{F}}^\frown_S$ from $\bar{x}$ is a semi-Spector class which is closed under the quantifiers $S_1, \ldots, S_m$ (for background on semi-Spector classes see Kolaitis [**1978**]). Using the good parametrization theorem for semi-Spector classes, and the closure properties of $\mathrm{ENV}[\mathbf{E}, \bar{\mathbf{F}}_Q, \bar{\mathbf{F}}\hat{s}, \bar{x}]$ it is not hard to verify that in this case

$$\bigcup_\xi \Delta^\xi(\bar{x}) \subseteq \mathrm{SEC}\left[\mathbf{E}, \bar{\mathbf{F}}_Q, \bar{\mathbf{F}}^\frown_S, \bar{x}\right]$$

for every $\bar{x} \in A^{<\omega}$ and so by part (i)

$$\mathrm{SEC}\left[\mathbf{E}, \bar{\mathbf{F}}_Q, \bar{\mathbf{F}}^\frown_S, \bar{x}\right] = \bigcup_\xi \Delta^\xi(\bar{x})$$

for every $\bar{x} \in A^{<\omega}$. □

5.7. REMARKS. (i) The *closure ordinal of the structure* $\mathfrak{A}$ *for recursion in* $\mathbf{E}, \bar{\mathbf{F}}_Q, \bar{\mathbf{F}}^\frown_S$ is the ordinal

$$\kappa^{\mathfrak{A}}\left(\mathbf{E}, \bar{\mathbf{F}}_Q, \bar{\mathbf{F}}^\frown_S\right) = \sup\{\mathrm{rank}(<)\colon < \text{ is a prewellordering which is "boldface" recursive in } \mathbf{E}, \bar{\mathbf{F}}_Q, \bar{\mathbf{F}}^\frown_S\}.$$

The proof of Theorem 5.5 actually shows that for any $\bar{x} \in A^{<\omega}$

$$\mathrm{SEC}\left[\mathbf{E}, \bar{\mathbf{F}}_Q, \bar{\mathbf{F}}^\frown_S, \bar{x}\right] \subseteq \Delta^{\kappa^{\mathfrak{A}}(\mathbf{E}, \bar{\mathbf{F}}_Q, \bar{\mathbf{F}}^\frown_S)}(\bar{x}).$$

If, in addition, the structure $\mathfrak{A}$ is acceptable, then it is easy to prove that the hierarchy $\{\Delta^\xi(\bar{x})\colon \xi \text{ is an ordinal}\}$ closes off at $\kappa^{\mathfrak{A}}(\mathbf{E}, \bar{\mathbf{F}}_Q, \bar{\mathbf{F}}^\frown_S)$. It follows that if $\mathfrak{A}$ is an acceptable structure, then

$$\mathrm{SEC}\left[\mathbf{E}, \bar{\mathbf{F}}_Q, \bar{\mathbf{F}}^\frown_S, \bar{x}\right] = \Delta^{\kappa^{\mathfrak{A}}(\mathbf{E}, \bar{\mathbf{F}}_Q, \bar{\mathbf{F}}^\frown_S)}(\bar{x}) \quad \text{for every } \bar{x} \in A^{<\omega}.$$

(ii) As we mentioned earlier in 5.3, Moschovakis [**EIAS**] obtained a hierarchy for the "boldface" hyperelementary relations on an acceptable structure $\mathfrak{A}$. The "boldface" hyperelementary relations on $\mathfrak{A}$ are exactly the relations which are "boldface" recursive in the functional $\mathbf{E}^{\#}$, or equivalently "boldface" recursive in $\mathbf{E}, \mathbf{F}^{\wedge}_{\exists}$. Moschovakis' theorem follows now easily by applying first the preceding hierarchy Theorem 5.6 to recursion in $\mathbf{E}^{\#}$ and showing then that for any limit ordinal ξ

$$\bigcup\{\Delta^{\xi}(\bar{x}) : \bar{x} \in A^{<\omega}\} \subseteq \mathbf{D}^{\xi},$$

where $\mathbf{D}^{\xi}$ is the ξth stage of the hierarchy in 7.E of Moschovakis [**EIAS**]. In a similar way one can derive the hierarchy theorems for the "boldface" recursive relations in $\mathbf{E}^{\#}, \mathbf{F}^{\#}_{Q}$ and $\mathbf{E}^{\#}, \mathbf{F}_{Q}$ which were proved first by Harrington, Kirousis and Schlipf [**1978**] and Kirousis [**1978**].

6. Model-theoretic characterizations for recursion in quantifiers. Our aim in this section is to present model-theoretic characterizations of the recursive relations in $\mathbf{E}, \bar{\mathbf{F}}_{Q}, \bar{\mathbf{F}}^{\wedge}_{S}$ using "lightface" versions of the schemata of Δ^1_1 Comprehension and Σ^1_1 Collection. We restrict ourselves here into explaining the definitions and stating the results, since the proofs involve direct applications of the Hierarchy Theorem 5.6 together with the parametrization theorems for recursion in $\mathbf{E}, \bar{\mathbf{F}}_{Q}$, $\bar{\mathbf{F}}^{\wedge}_{S}$.

6.1. DEFINITIONS. Assume that $\mathcal{G} = \{\Gamma(\bar{z}) : \bar{z} \in A^{<\omega}\}$ is an indexed family on a structure $\mathfrak{A}$, $\bar{Q} = (Q_1, \ldots, Q_l)$, $\bar{S} = (S_1, \ldots, S_m)$ are two sequences of quantifiers on A and $\bar{x}$ is a finite sequence from A. We say that a relation R on A is *simply* $\Sigma^1_1(\exists, \bar{Q}, \bar{S}^{\wedge})$ *definable from* $\bar{x}$ *with range* $\mathcal{G}$ if there is a formula φ of the language $\mathcal{L}^{\mathfrak{A}}(\bar{Q}, \bar{S})$ such that

$$(\forall \bar{y})[R(\bar{y}) \Leftrightarrow (\exists Y \in \Gamma(\bar{x}, \bar{y}))\varphi(\bar{x}, \bar{y}, Y)]$$

or

$$(\forall \bar{y})[R(\bar{y}) \Leftrightarrow (Sz)(\exists Y \in \Gamma(\bar{x}, \bar{y}, z))\varphi(\bar{x}, \bar{y}, z, Y)],$$

where S is one of the quantifiers $S_1, \ldots, S_m$.

A relation R is $\Sigma^1_1(\exists, \bar{Q}, \bar{S}^{\wedge})$ *definable from* $\bar{x}$ *with range* $\mathcal{G}$ if it is the finite disjunction of simply $\Sigma^1_1(\exists, \bar{Q}, \bar{S}^{\wedge})$ definable relations. If we allow second order parameters from $\Gamma(\bar{x})$ in the defining formulas φ, then we have the concept of a $\Sigma^1_1(\exists, \bar{Q}, \bar{S}^{\wedge})$ *definable relation from* $\bar{x}$ *with range* $\mathcal{G}$ *and with parameters from* $\Gamma(\bar{x})$. A relation $R \subseteq A^n$ is $\Delta^1_1(\exists, \bar{Q}, \bar{S}^{\wedge})$ *definable from* $\bar{x}$ *with range* $\mathcal{G}$ (*and with parameters from* $\Gamma(\bar{x})$) if both R and $\neg R = A^n - R$ are $\Sigma^1_1(\exists, \bar{Q}, \bar{S}^{\wedge})$ definable from $\bar{x}$ with range $\mathcal{G}$ (and with parameters from $\Gamma(\bar{x})$). Notice that in the above definitions only one of the quantifiers $S_1, \ldots, S_m$ is allowed in front of the second order existential quantifier, while one or more of the quantifiers $Q_1, \breve{Q}_1, \ldots, Q_l, \breve{Q}_l, S_1, \breve{S}_1, \ldots, S_m, \breve{S}_m$ can appear in the defining formula φ. We say that the indexed family $\mathcal{G} = \{\Gamma(\bar{z}) : \bar{z} \in A^{<\omega}\}$ satisfies *the schema of* $\Delta^1_1(\exists, \bar{Q}, \bar{S}^{\wedge})$ *comprehension* if for any $\bar{x} \in A^{<\omega}$ and any relation R which is $\Delta^1_1(\exists, \bar{Q}, \bar{S}^{\wedge})$ definable from $\bar{x}$ with range $\mathcal{G}$, we have that R is in the class $\Gamma(\bar{x})$. In an analogous way we define what it means for an indexed family $\mathscr{G}$ to satisfy the schema of $\Delta_1^1(\exists, \overline{Q}, \overline{S}^\frown)$ *comprehension with parameters*. It is obvious that if $\mathscr{G}$ satisfies the schema of $\Delta_1^1(\exists, \overline{Q}, \overline{S}^\frown)$ comprehension with parameters, then it is also a model of the $\Delta_1^1(\exists, \overline{Q}, \overline{S}^\frown)$ comprehension schema (without parameters), although it can be shown that the converse is not true in general. The content of the text theorem is that on acceptable structures the indexed family $\mathscr{R} = \{\mathrm{SEC}[\mathbf{E}, \overline{\mathbf{F}}_Q, \overline{\mathbf{F}}_S^\frown, \overline{x}]: \overline{x} \in A^{<\omega}\}$ is the smallest model of these schemata.

6.2. THEOREM. *Let $\mathfrak{A} = \langle A, \overline{R}, \overline{f}, \overline{c}\rangle$ be a structure such that ω, $\leqslant_\omega$ are elementary on $\mathfrak{A}$ and assume that $\overline{Q} = (Q_1, \ldots, Q_l)$, $\overline{S} = (S_1, \ldots, S_m)$ are two sequences of quantifiers on $\mathfrak{A}$.*

(i) *The indexed family $\mathscr{R} = \{\mathrm{SEC}[\mathbf{E}, \overline{\mathbf{F}}_Q, \overline{\mathbf{F}}_S^\frown, \overline{x}]: \overline{x} \in A^{<\omega}\}$ of the recursive relations in $\mathbf{E}, \overline{\mathbf{F}}_Q, \overline{\mathbf{F}}_S^\frown$ is contained in any indexed family $\mathscr{G} = \{\Gamma(\overline{x}): \overline{x} \in A^{<\omega}\}$ which satisfies the schema of $\Delta_1^1(\exists, \overline{Q}, \overline{S}^\frown)$ comprehension, i.e. if $\mathscr{G}$ is any such family, then*

$$\mathrm{SEC}\left[\mathbf{E}, \overline{\mathbf{F}}_Q, \overline{\mathbf{F}}_S^\frown, \overline{x}\right] \subseteq \Gamma(\overline{x}) \quad \textit{for every } \overline{x} \in A^{<\omega}.$$

(ii) *If, in addition, the structure $\mathfrak{A}$ is acceptable, then the indexed family $\mathscr{R} = \{\mathrm{SEC}[\mathbf{E}, \overline{\mathbf{F}}_Q, \overline{\mathbf{F}}_S^\frown, \overline{x}]: \overline{x} \in A^{<\omega}\}$ is the smallest model of both the schema of $\Delta_1^1(\exists, \overline{Q}, \overline{S}^\frown)$ comprehension and the schema of $\Delta_1^1(\exists, \overline{Q}, \overline{S}^\frown)$ comprehension with parameters.* $\square$

The preceding Theorem 6.2 has numerous consequences which shed light into the differences between the functionals $\mathbf{E}, \mathbf{E}^\#, \mathbf{F}_Q$ and $\mathbf{F}_Q^\#$, where Q is a quantifier on A. We state just two results about recursion in $\mathbf{E}, \mathbf{F}_Q$ and recursion in $\mathbf{E}^\#, \mathbf{F}_Q^\#$, i.e. about recursion in type 2 objects vs. positive elementary induction.

6.3. COROLLARY. *Let $\mathfrak{A} = \langle A, \overline{R}, \overline{f}, \overline{c}\rangle$ be an acceptable structure and let Q be a quantifier on A.*

(i) *The family $\mathscr{R}(\mathbf{E}, \mathbf{F}_Q) = \{\mathrm{SEC}[\mathbf{E}, \mathbf{F}_Q, \overline{x}]: \overline{x} \in A^{<\omega}\}$ of the recursive relations in the type two objects $\mathbf{E}, \mathbf{F}_Q$ is the smallest indexed family on $\mathfrak{A}$ which satisfies the schema of $\Delta_1^1(\exists, Q)$ comprehension.*

(ii) *The family $\mathscr{R}(\mathbf{E}^\#, \mathbf{F}_Q^\#) = \{\mathrm{SEC}[\mathbf{E}^\#, \mathbf{F}_Q^\#, \overline{x}]: \overline{x} \in A^{<\omega}\}$ of the recursive relations in the functionals $\mathbf{E}^\#, \mathbf{F}_Q^\#$ is the smallest indexed family on $\mathfrak{A}$ which satisfies the schema of $\Delta_1^1(\exists, \exists^\frown, Q^\frown, \breve{Q}^\frown)$ comprehension.* $\square$

The difference beween the schemata of $\Delta_1^1(\exists, Q)$ and $\Delta_1^1(\exists, \exists^\frown, Q^\frown, \breve{Q}^\frown)$ comprehension is that in the latter the first order quantifiers $\exists$, Q and $\breve{Q}$ are allowed in front of second order existential quantifiers. These schemata provide therefore a way to capture the different strength of existential, Q and $\breve{Q}$ quantification embodied by the functionals $\mathbf{E}, \mathbf{F}_Q$ and $\mathbf{E}^\#, \mathbf{F}_Q^\#$, using second order definability notions. Part (ii) of the above corollary yields easily the characterizations of the "boldface" hyperelementary and "boldface" Q-hyperelementary relations in Moschovakis [**EIAS**] and Harrington, Kirousis and Schlipf [**1978**].

6.4. DEFINITIONS. Assume that $\mathscr{G} = \{\Gamma(\bar{x}): \bar{x} \in A^{<\omega}\}$ is an indexed family on a structure $\mathfrak{A}$ and $\bar{Q} = (Q_1, \ldots, Q_l)$, $\bar{S} = (S_1, \ldots, S_m)$ are two sequences of quantifiers on A. We say that $\mathscr{G}$ satisfies the *schema of* $\Delta^0_\infty(\bar{Q}, \bar{S})$ *comprehension* if for any $\bar{x} \in A^{<\omega}$, any formula $\varphi(\bar{u}, \bar{y}, Y)$ of the language $\mathscr{L}^{\mathfrak{A}}(\bar{Q}, \bar{S})$ and any relation P in $\Gamma(\bar{x})$ there is a relation R in $\Gamma(\bar{x})$ such that

$$R(\bar{y}) \Leftrightarrow \varphi(\bar{x}, \bar{y}, P).$$

We say that the indexed family $\mathscr{G}$ satisfies the schema of $\Sigma^1_1(\exists, \bar{Q}, \bar{S}\,\hat{}\,)$ *collection* if for any $\bar{x} \in A^{<\omega}$ and any formula $\varphi(\bar{u}, z, Y)$ of the language $\mathscr{L}^{\mathfrak{A}}(\bar{Q}, \bar{S})$ we have that

$$(\forall z)(\exists Y \in \Gamma(\bar{x}, z))\varphi(\bar{x}, z, Y) \Leftrightarrow (\exists W \in \Gamma(\bar{x}))(\forall z)(\exists n \in \omega)\varphi(\bar{x}, z, W_{z,n})$$

and

$$(Sz)(\exists Y \in \Gamma(\bar{x}, z))\varphi(\bar{x}, z, Y) \Leftrightarrow (\exists W \in \Gamma(\bar{x}))(Sz)(\exists n \in \omega)\varphi(\bar{x}, z, W_{z,n}),$$

where S is one of the quantifiers $S_1, \ldots, S_m$ and $W_{z,n} = \{\bar{w}: (z, n, \bar{w}) \in W\}$.

If in the formula φ we can substitute also relations from $\Gamma(\bar{x})$, then we say that $\mathscr{G}$ satisfies the schema of $\Sigma^1_1(\exists, \bar{Q}, \bar{S}\,\hat{}\,)$ *collection with parameters*.

The last result of this paper is a minimality characterization of recursion in $\mathbf{E}, \bar{\mathbf{F}}_Q, \bar{\mathbf{F}}\hat{}_S$ in terms of $\Delta^0_\infty(\bar{Q}, \bar{S})$ comprehension and $\Sigma^1_1(\exists, \bar{Q}, \bar{S}\,\hat{}\,)$ collection.

6.5. THEOREM. *Let* $\mathfrak{A} = \langle A, \bar{R}, \bar{f}, \bar{c}\rangle$ *be a structure such that* ω, $\leqslant_\omega$ *are elementary on* A *and assume that* $\bar{Q} = (Q_1, \ldots, Q_l)$, $\bar{S} = (S_1, \ldots, S_m)$ *are two sequences of quantifiers on* A.

(i) *The family* $\mathscr{R} = \{\mathrm{SEC}[\mathbf{E}, \bar{\mathbf{F}}_Q, \bar{\mathbf{F}}\hat{}_S, \bar{x}]: \bar{x} \in A^{<\omega}\}$ *of recursive relations in* $\mathbf{E}, \bar{\mathbf{F}}_Q, \bar{\mathbf{F}}\hat{}_S$ *is contained in any indexed family* $\mathscr{G} = \{\Gamma(\bar{x}): \bar{x} \in A^{<\omega}\}$ *which satisfies the schemata of* $\Delta^0_\infty(\bar{Q}, \bar{S})$ *comprehension and* $\Sigma^1_1(\exists, \bar{Q}, \bar{S}\,\hat{}\,)$ *collection.*

(ii) *If, in addition, the structure* $\mathfrak{A}$ *is acceptable, then the indexed family* $\mathscr{R} = \{\mathrm{SEC}[\mathbf{E}, \bar{\mathbf{F}}_Q, \bar{\mathbf{F}}\hat{}_S, \bar{x}]: \bar{x} \in A^{<\omega}\}$ *is the smallest model of the schemata of* $\Delta^0_\infty(\bar{Q}, \bar{S})$ *comprehension and* $\Sigma^1_1(\exists, \bar{Q}, \bar{S}\,\hat{}\,)$ *collection (with or without parameters).*

HINT OF PROOF. In order to establish part (i) of the theorem one has to show first that if $\mathscr{G}$ is an indexed family which satisfies both $\Delta^0_\infty(\bar{Q}, \bar{S})$ comprehension and $\Sigma^1_1(\exists, \bar{Q}, \bar{S}\,\hat{}\,)$ collection, then $\mathscr{G}$ satisfies also the schema of $\Delta^1_1(\exists, \bar{Q}, \bar{S}\,\hat{}\,)$ comprehension. The result follows then from the preceding Theorem 6.2.

Part (ii) amounts to showing that if $\mathfrak{A}$ is an acceptable structure, then the indexed family $\mathscr{R} = \{\mathrm{SEC}[\mathbf{E}, \bar{\mathbf{F}}_Q, \bar{\mathbf{F}}\hat{}_S, \bar{x}]: \bar{x} \in A^{<\omega}\}$ satisfies the schema of $\Sigma^1_1(\exists, \bar{Q}, \bar{S}\,\hat{}\,)$ collection. This in turn involves a "lightface" boundedness argument using the good parametrization theorem for the recursive relations in $\mathbf{E}, \bar{\mathbf{F}}_Q, \bar{\mathbf{F}}\hat{}_S$. □

REFERENCES

P. Aczel [**1970**], *Representability in some systems of second order arithmetic*, Israel J. Math. **8** (1970), 309–328.

J. E. Fenstad [**1980**], *General recursion theory, an axiomatic approach*, Springer-Verlag, Berlin and New York, 1980.

S. Feferman [**1977**], *Inductive schemata and recursively continuous functionals*, Logic Colloquium '76 (R. Gandy and M. Hyland, eds.), North-Holland, Amsterdam, 1977, pp. 373–392.

L. A. Harrington, L. M. Kirousis and J. S. Schlipf [**1978**], *A generalized Kleene-Moschovakis theorem*, Proc. Amer. Math. Soc. **68** (1978), 209–213.

P. G. Hinman [**1969**], *Hierarchies of effective descriptive set theory*, Trans. Amer. Math. Soc. **142** (169), 265–299.

A. S. Kechris and Y. N. Moschovakis [**1977**], *Recursion in higher types*, Handbook of Mathematical Logic (J. Barwise, ed.), North-Holland, Amsterdam, 1977, pp. 681–737.

L. M. Kirousis [**1978**], *On abstract recursion theory and recursion in the universe of sets*, Ph.D. Dissertation, University of California, Los Angeles, 1978.

S. C. Kleene [**1959a**], *Recursive functionals and quantifiers of finite type*. I, Trans. Amer. Math. Soc. **91** (1959), 1–52.

______ [**1959b**], *Quantification of number theoretic functions*, Compositio Math. **14** (1959), 23–40.

______ [**1963**], *Recursive functionals and quantifiers of finite type*. II, Trans. Amer. Math. Soc. **108** (1963), 106–142.

Ph. G. Kolaitis [**1978**], *On recursion in* **E** *and semi-Spector classes*, Cabal Seminar '76–'77 (A. Kechris and V. Moschovakis, eds.), Lecture Notes in Math., vol. 689, Springer-Verlag, Berlin and New York, 1978, pp. 209–243.

______ [**1979**], *Recursion in a quantifier vs. elementary induction*, J. Symbolic Logic **44** (1979), 235–259.

______ [**1980**], *Recursion and nonmonotone induction in a quantifier*, The Kleene Symposium (J. Barwise, T. Keisler and K. Kunen, eds.), North-Holland, Amsterdam, 1980, pp. 367–389.

______ [**1981**], *Model theoretic characterizations in generalized recursion theory*, Logic Year 1979–80 (The University of Connecticut) (M. Lerman, J. Schmerl and R. Soare, eds.), Lecture Notes in Math., vol. 859, Springer-Verlag, Berlin and New York, 1981, pp. 104–119.

Y. N. Moschovakis [**1969a**], *Abtract first order computability*. I, Trans. Amer. Math. Soc. **138** (1969), 427–464.

______ [**1969b**], *Abstract computability and invariant definability*, J. Symbolic Logic **34** (1969), 605–633.

______ [**EIAS**], *Elementary induction on abstract structures*, North-Holland, Amsterdam, 1974.

______ [**1977**], *On the basic notions in the theory of induction*, Logic, Foundations of Mathematics and Computability (R. Butts and T. Hintikka, eds.), Reidel, Dordrecht, 1977, pp. 207–236.

D. Normann [**1978**], *Set recursion*, Generalized Recursion Theory. II (T. Fenstad, R. Gandy and G. Sacks, eds.), North-Holland, Amsterdam, 1978, pp. 303–320.

R. A. Platek [**1966**], *Foundations of recursion theory*, Ph.D. Thesis, Stanford University, 1966.

DEPARTMENT OF MATHEMATICS, OCCIDENTAL COLLEGE, LOS ANGELES, CALIFORNIA 90041

Proceedings of Symposia in Pure Mathematics
Volume **42**, 1985

Aspects of the Continuous Functionals

DAG NORMANN

The continuous functionals were first defined by Kreisel [**5**], and a reasonably equivalent family, the countable functionals, were independently defined by Kleene [**4**]). The continuous functionals can informally be described as the richest hierarchy of total functionals of pure type where each element ψ can be coded by *associates* α: $\mathbf{N} \to \mathbf{N}$ in such a way that application $\psi(\phi)$ is uniformly recursive in corresponding associates. That ψ of type $k + 1$ is total means that $\psi(\phi)$ is defined for all continuous functionals ϕ of type k.

Here we will give a definition close to the one used in Kleene [**4**].

DEFINITION 1. (a) $\mathrm{Ct}(0) = \mathbf{N}$.

(b) $\mathrm{Ct}(1) = \mathbf{N}^{\mathbf{N}}$, all functions are their own associates.

(c) $\mathrm{Ct}(k + 1) = \{\psi\colon C(k) \to \mathbf{N}, \psi \text{ has an associate}\}$, where β is an associate for ψ ($\beta \in \mathrm{As}(\psi)$) if

$$\forall \phi \in \mathrm{Ct}(k)\, \forall \alpha \in \mathrm{As}(\phi)\, \exists n\, \forall m\, (m < n \to \beta(\bar{\alpha}(m)) = 0$$
$$\wedge\, m \geqslant n \to \beta(\bar{\alpha}(m)) = \psi(\phi) + 1).$$

We use the standard notion $\bar{\alpha}(m) = \langle \alpha(0), \ldots, \alpha(m - 1) \rangle$.

We let $\mathrm{As}(k)$ be the set of associates for type-k functionals. This set will have a natural topology inherited from $\mathbf{N}^{\mathbf{N}}$. We use this to define a topology on $\mathrm{Ct}(k)$ as follows:

Let ρ_k: $\mathrm{As}(k) \to \mathrm{Ct}(k)$ be the associating function. $O \subseteq \mathrm{Ct}(k)$ is *open* if $\rho_k^{-1}O$ is open in $\mathrm{As}(k)$. Then $\mathrm{Ct}(k + 1)$ is exactly the set of continuous operators w.r.t. this topology.

Various characterizations of these topologies due to Martin Hyland show that there is nothing artificial about our hierarchy of functionals, see Hyland [**2**]. $\mathrm{Ct}(k)$ turns out to be the natural completion of the set of finite operators of type k.

1980 *Mathematics Subject Classification*. Primary 03D65.

Moreover Bergstra [1] showed that the continuous functionals constitute the richest typestructure with a decent recursion theory, but without the functional ${}^2E$.

There are two natural notions of algorithms on the continuous functionals, Kleene's S1–S9-computations from Kleene [3] and countable recursion. S1–S9 are nine schemes defining the set of computations

$$\{e\}(\vec{\psi}) \simeq t$$

inductively. With this notion we use "computable" and "computable in" ($<_K$).

Countable recursion is defined by uniform algorithms on the associates:

DEFINITION 2. $[e](\psi_1,\ldots,\psi_k) \cong t$ if for all associates $\alpha_1,\ldots,\alpha_k$, for $\psi_1,\ldots,\psi_k$ respectively, we have $T_e(\alpha_1,\ldots,\alpha_k) = t$, where T_e is the eth Turing algorithm.

With this notion we use "recursive" and "recursive in" ($<_c$). Tait (unpublished) showed that the fan-functional defined by

$$\Phi(F) = \mu n \forall f, g\colon N \to \{0,1\}(\bar{f}(n) = \bar{g}(n) \to F(f) = F(g))$$

is recursive but not computable, so the theories are not equivalent. The result has later been improved, and so far the best result along these lines seems to be

THEOREM 1 (NORMANN [7]). *Let* $\Phi \in \mathrm{Ct}(k)$, $k > 2$. *There is a* $\Psi \in \mathrm{Ct}(k)$ *uniformly recursive in* Φ *such that* Ψ *is not computable in* Φ *and any* $\phi \in \mathrm{Ct}(k-1)$.

There is a limit to how far we can go along these lines, as the following two results show:

THEOREM 2 (NORMANN [7]). (a) *Let* $k > 2$ *be fixed. Then there is a recursive functional* $R \in \mathrm{Ct}(k+1)$ *such that each* $\Phi \in \mathrm{Ct}(k)$ *is uniformly computable from* R *and any associate for* Φ.

(b) *Let* k *be fixed. There is a recursive functional* $K \in \mathrm{Ct}(k+2)$ *such that countable recursion over* $\mathrm{Ct}(k)$ *is equivalent to Kleene-computability over* $\mathrm{Ct}(k)$ *modulo* K.

(*There are primitive recursive functions* s, t *such that*

$$[e](\psi_1,\ldots,\psi_n) \simeq \{s(e)\}(\psi_1,\ldots,\psi_n K),$$
$$\{e\}(\psi_1,\ldots,\psi_n, K) \simeq [t(e)](\psi_1,\ldots,\psi_n).)$$

One question remains unsolved: Is there a natural notion of some internal system of algorithms on $\mathrm{Ct}(k)$ (or just $\mathrm{Ct}(3)$) equivalent to countable recursion?

So far we have not discussed any of the methods used within this branch of recursion theory, and one reason is that some of the basic results require quite lengthy proofs. These results can be found more or less explictly in the original paper of Kleene [4] and Kreisel [5]. Moreover, Normann [6] gives a coherent introduction to the recursion theory of the continuous functionals and contains detailed proofs.

We will now give some results without the lengthy proofs.

THEOREM 3. (a) *Uniformly in k there is a primitive recursive family* $\{\xi_i^k\}_{i\in\mathbf{N}}$ *of finitary functionals dense in* Ct(k) $(\xi_i^0 = i)$.

(b) *Let the* trace $h_\phi(i) = \phi(\xi_i^k)$ *when* $\phi \in \mathrm{Ct}(k+1)$. *Uniformly in* h_ϕ *we can compute a function* $f\colon \mathbf{N} \to \mathbf{N}$ *and a functional* $\psi \in \mathrm{Ct}(k+1)$ *such that*

$$\phi = \lim \xi_{f(i)}^{k+1} \quad \text{and} \quad \forall i \geqslant \psi(\eta)\big(\xi_{f(i)}^{k+1}(\eta) = \phi(\eta)\big)$$

for all $\eta \in \mathrm{Ct}(k)$. ψ *is called a* modulus *for the sequence.*

(c) *Let* $A \subseteq \mathbf{N}^{\mathbf{N}}$ *be* Π_k^1. *Then there is a recursive relation R such that*

$$f \in A \leftrightarrow \forall\phi \in \mathrm{Ct}(k)\,\exists n\, R(f, h_\phi, n).$$

(d) *Let* $B \subseteq \mathbf{N}^{\mathbf{N}}$ *be* Σ_{k+1}^1. *Then there is a recursive relation S such that*

$$g \in B \leftrightarrow \forall\psi \in \mathrm{Ct}(k+1)\,\exists n\, S(g, h_\psi, n)$$

where we uniformly computable in $g \notin B$ *may find* ψ *such that* $\forall n \neg S(g, h_\psi, n)$.

Though the proofs of (c) and (d) are fairly simple, they seem to have a tremendous effect on the theory of the continuous functionals. Theorems 1 and 2 are based on these results together with an estimate of the complexity of computations and recursions. They can also be used to give a complete description of the sections and envelopes of the continuous functionals, and they can even be used to give a characterization of truth in Π_k^1-absolute models of second order arithmetic in terms of continuous proofs. We will not go into any of these applications here, but rather dwell at a fairly new concept, the continuous r.e. degrees.

In classical recursion theory the concepts of recursive enumerability and semirecursiveness (domain of partial algorithm) coincide, and this is also the case for many generalisations where the concept of "finite" is generalised as well. But for the continuous funtionals semirecursion is an absolutely inadequate concept. Thus we give a direct generalisation of recursive enumerability.

In Normann [**7** and **8**] we introduce the degree theory based on Kleene computations. The results for countable recursion follow as corollaries. Here we will indicate a direct approach based on countable recursion.

DEFINITION 3. (a) Let $A \subseteq \mathrm{Ct}(k)$. A is *recursively enumerable* (r.e.) if there is a computable family $\{A_n\}_{n\in\mathbf{N}}$ of subsets of Ct(k) such that $A = \bigcup_{n\in\mathbf{N}} A_n$.

(b) If A is r.e. and closed in the topology on Ct(k) we call A continuously r.e.

We have the following characterizations:

THEOREM 4. *Let* $\phi \in \mathrm{Ct}(k)$. *Then the following are equivalent*:

(i) ϕ *is of r.e. degree.*

(ii) *There is a recursive sequence* $\phi_i \to \phi$ *with a modulus recursive in* ϕ.

(iii) *There is a* ψ *equivalent to* ϕ *such that* h_ψ *is recursive.*

The hard part in the proof of Theorem 4 is to show the following

LEMMA. *If* $\psi_i \to \psi$ *with a modulus recursive in* ψ *and if* $\phi \leqslant_c \psi$, *then we can compute* ϕ_i *from* ψ_i *such that* $\phi_i \to \phi$ *with a modulus recursive in* ψ.

The idea behind the proof of the Lemma is that recursions are continuous, and that we just apply the reduction procedure on ψ_i. The combinatorial difficulty is that the reduction procedure may not work on ψ_i, a difficulty that is overcome by dealing with approximations to the procedure that will work.

By the Lemma it is sufficient to prove (iii) → (ii) when h_ψ is recursive; but this follows from Theorem 3(b).

(ii) → (i). Let $\phi_i \to \phi$. Let

$$(m, \xi) \in A_n \leftrightarrow \exists i(m \leqslant i \leqslant n \,\&\, \phi_i(\xi) \neq \phi_m(\xi)).$$

Let $A = \bigcup_{n \in \mathbf{N}} A_n$. Then A is equivalent to the modulus for $\{\phi_i\}_{i \in \mathbf{N}}$, and thus equivalent to ϕ, so ϕ is of r.e. degree.

(i) → (iii). Let f be recursive but not primitive recursive. Let $A = \bigcup_{n \in \mathbf{N}} A_n$ be r.e. Let

$$\phi(\xi, g) = \begin{cases} 0 & \text{if } \exists n(\bar{g}(n) = \bar{f}(n) \wedge \xi \in A_n), \\ 1 & \text{if } \forall n(\bar{g}(n) = 0(n) \to \xi \notin A_n). \end{cases}$$

Then $\phi \equiv_c A$ and h_ϕ is recursive since ϕ is recursive when restricted to primitive recursive functions. (Then there are finitely many n such that $\bar{g}(n) = \bar{f}(n)$.)

The ordinary or classical r.e. degrees is essentially a subclass of the higher type r.e. degrees; we do not introduce more reduction procedures between the functions of type 1. But they form by no means an elementary substructure, as the following result shows.

THEOREM 5. *If $A = \bigcup_{n \in \mathbf{N}} A_n$, $B = \bigcup_{n \in \mathbf{N}} B_n$ are nonrecursive r.e. subsets of* $\mathbf{N}$ *then there is a type-3 functional Φ such that $\Phi <_c A$ and $\Phi <_c B$.*

PROOF. Let f, g, f_n, g_n be the characteristic functions of A, B, A_n and B_n respectively. Let

$$\Phi(F) = \mu n \forall m \geqslant n(F(f_m) = F(f_{m+1}) \vee F(g_m) = F(g_{m+1})).$$

Then

(i) $\Phi < f$.

PROOF. Let F be given and let α be an associate for F. Find n such that for some t we have for all $m \geqslant n$

$$\alpha(\bar{f}(t)) > 0 \vee \bar{f}_m(t) = \bar{f}(t).$$

This can be done recursively in f and α. This n is an upper bound for $\Phi(F)$ from which we can compute $\Phi(F)$.

(ii) $\Phi < g$ follows by the same argument.

(iii) Φ is not recursive.

PROOF. Let β be an associate for Φ.

Claim 1. If $\beta(\delta) > 0$ then there is some t such that $\delta(\bar{f}(t)) > 0$ or $\delta(\bar{g}(t)) > 0$.

PROOF. Assume not. Let i_0 be such that for all $i \geqslant i_0$ and all t we do not have

$$\delta(\bar{f}_i(t)) > 0 \quad \text{or} \quad \delta(\bar{g}_i(t)) > 0.$$

Choose $i > \max\{\beta(\delta), i_0\}$. We may extend δ to some δ' such that for any functional F' with an associate extending δ'

$$F(f_i) \neq F(f_{i+1}), \qquad F(g_i) \neq F(g_{i+1}).$$

But then $\Phi(F) \geqslant i$ contradicting the value of $\beta(\delta') = \beta(\delta)$.

Claim 2. $\{\delta; \exists t\, \delta(\bar{f}(t)) > 0\}$ is Π^0_1.

PROOF. The quantifier $\exists t$ is bounded since δ is a finite sequence. Thus we are reduced to showing that the relation $\tau = \bar{f}(n)$ is Π^0_1, but that is trivial.

The proof of (iii) is now divided into three cases:

(1) For arbitrary large t there is a δ such that $\beta(\delta) > 0$, for all τ, $\delta(\tau) > 0 \to \mathrm{lh}(\tau) \geqslant \tau$ and δ does not secure g (i.e. $\delta(\bar{g}(t)) = 0$ whenever defined).

(2) Same as above except that δ now does not secure f.

(3) For some t we have

For all δ with $\beta(\delta) > 0$,

$$\text{if } \delta(\tau) \to \mathrm{lh}(\tau) > \tau \text{ then } \delta \text{ secures both } f \text{ and } g.$$

In case 1 we compute f from β; in case 2 we compute g from β; and in case 3 we compute both f and g from β. This ends the proof of the theorem.

A similar construction can be used for more striking results.

DEFINITION 4. Let ϕ, ψ be continuous functionals of r.e. degree such that h_ϕ, h_ψ are recursive. Let ϕ_i, ψ_i be recursive approximations. Let

$$I(\Phi) = \mu n \forall m \geqslant n(\Phi(\phi_m) = \Phi(\phi_{m+1}) \vee \Phi(\psi_m) = \Phi(\psi_{m+1})).$$

I is of r.e. degree and we have

THEOREM 6. *Assume that* $\psi \leqslant_c \phi, \Phi, \Psi \leqslant_c \psi, \Phi$. *Then* $\Psi \leqslant_c I, \Phi$.

INDICATION OF PROOF. W.l.o.g. we may assume that we use the same index e in both reductions. Let α, β be associates for Φ and I respectively. We want to compute $\Psi(\xi)$ uniformly in α, β and an associate γ for ξ.

$\Psi(\xi) = T_e(\gamma, f, \alpha) = T_e(\gamma, g, \alpha)$ where f, g are associates for ϕ, ψ respectively. Find σ, t such that $\beta(\sigma) > 0$ and such that for all δ $(\sigma(\delta) > 0 \to \delta$ is not extendable to an associate for ϕ or ψ or $T_e(\gamma, \delta, \alpha) = t)$. (Here σ and τ are finite sequences of numbers. The relation "δ is not extendable..." is r.e. so we may effectively search for one such pair σ, t.) Then $t = \Psi(\xi)$.

COROLLARY. *The continuous r.e. degrees form a distributive lattice.*

There are many open problems concerning these r.e. degrees, for example: Is the ordering dense? Do they contain minimal pairs?

The corresponding Kleene degrees are also a distributive lattice: they are densly ordered and without minimal pairs. The main problem for them is whether the theory of these degrees is decidable, and if not, how strong is the theory.

REFERENCES

1. J. A. Bergstra, *The continuous functionals and* ^2E, Generalized Recursion Theory. II (J. E. Fenstad, R. O. Gandy and G. E. Sacks, eds.), North-Holland, Amsterdam, 1978, pp. 39–54.

2. J. M. E. Hyland, *Filter spaces and continuous functionals*, Ann. Math. Logic **16** (1979), 101–143.

3. S. C. Kleene, *Recursive functionals and quantifiers of finite types*. I, Trans. Amer. Math. Soc. **91** (1959) 1–52.

4. ______, *Countable functionals*, Constructivity in Mathematics (A. Heyting, ed.), North-Holland, Amsterdam, 1959, pp. 81–100.

5. G. Kreisel, *Interpretation of analysis by means of functionals of finite type*, Constructivity in Mathematics (A. Heyting, ed.), North-Holland, Amsterdam, 1959, pp. 101–128.

6. D. Normann, *Recursion on the countable functionals*, Lecture Notes in Math., vol. 811, Springer-Verlag, Berlin and New York, 1980.

7. ______, *The continuous functionals: computations, recursions and degrees*, Ann. Math. Logic **21** (1981), 1–26.

8. ______, *R.e. degrees of continuous functionals*, Arch. Math. Logik Grundlag. **23** (1983), 79–98.

INSTITUTE OF MATHEMATICS, UNIVERSITY OF OSLO, P. O. BOX 1053, BLINDERN, OSLO 3, NORWAY

Proceedings of Symposia in Pure Mathematics
Volume **42**, 1985

Post's Problem in E-Recursion

GERALD E. SACKS[1]

Abstract. In the setting of E-recursion a positive solution to Post's problem is given for a class of structures that includes all E-closed, initial segments of L. The solution, based largely on fine structure and reflection properties, does not give rise to injuries when $L(\kappa)$ is Σ_1 inadmissible; i.e., each Friedberg-Muchnik requirement is attempted at most once.

1. Introduction. The partial E-recursive functions were discovered by Normann [**1**], and independently by Moschovakis. They constitute a generalization of Kleene's theory of higher types recursion from normal objects of finite type to all sets. Thus they give a meaning to $\{e\}(x)$ for all x based on computations. The fundamentals of E-recursion are reviewed formally in §2. They are dealt with intuitively in the present section in order to provide a rough idea of the proof of the main result of the paper.

By a positive solution to Post's problem is usually meant the construction of two incomparable r.e. sets A and B. The meanings of "incomparable" and "r.e." depend on the choice of underlying structure $\mathscr{E}$, of which A and B are subsets, and the choice of recursion theory on $\mathscr{E}$. In this paper $\mathscr{E}$ is always E-closed; that is, $f(x) \in \mathscr{E}$ for all $x \in \mathscr{E}$ and every partial E-recursive f.

Let $E(x)$ be the least transitive E-closed set with x as an element. This paper was initiated by a striking result of Normann [**2**] concerning Post's problem for Kleene recursion in ${}^3E$. His result, restated in the language of E-recursion, is:

(1) If $E(2^\omega) \models [2^\omega$ is well ordered and of regular cardinality], then there exist two subsets of $E(2^\omega)$, each E-recursively enumerable on $E(2^\omega)$, such that neither is E-reducible to the other on $E(2^\omega)$.

1980 *Mathematics Subject Classification*. Primary 03D65; Secondary 03E45.

[1] The preparation of this paper was supported in part by National Science Foundation Grant MCS-81-04266. The author is grateful to Theodore Slaman for his generous assistance.

The proof of (1) bears only superficial resemblance to the positive solution of Post's problem given in classical recursion theory [**3**, **4**] or in α-recursion theory [**5**]. The results of the present paper originated in the desire to circumvent the assumption of "regular cardinality" in (1).

In order to clarify the consequent of (1), consider subsets A and B of some E-closed transitive set $\mathscr{E}$. A is E-r.e. on $\mathscr{E}$ if there are a partial E-rec. function $f(x, y)$ and a $p \in \mathscr{E}$ such that

(2) $$A = \{x \mid x \in \mathscr{E} \,\&\, f(x, p) \text{ converges}\}.$$

The schemes for computing E-recursive functions can be relativized to an arbitrary class C, thereby giving a meaning to $\{e\}^C(x)$, by adding the scheme

(3) $$g(x) = x \cap C.$$

A is E-reducible to B if there are a partial function f, E-rec. relative to B, and a $p \in \mathscr{E}$ such that $f(x, p)$ converges for all $x \in \mathscr{E}$, and $A = \{x \mid x \in \mathscr{E} \,\&\, f(x, p) = 0\}$.

Definition (2) is based on the idea of an r.e. set as the domain, rather than range, of a partial recursion function. Thus in (2) x lands in A iff some convergent computation c puts x into A. c is generated by starting with x and following the rules of computation. The idea of an r.e. set as the range of a partial recursive function does not, in general, yield an effective method of working back from an element y of an r.e. set B to a computation that puts y into B. In particular, there is a partial E-recursive function f whose range is not E-recursively enumerable. The trouble with f is that there is no partial E-recursive g such that for y in the range of f, $g(y)$ converges and $y = f(g(y))$.

One advantage of the idea of an r.e. set as the range of a partial recursive function is that it leads directly to the view of an r.e. set as an enumerable set when the underlying structure can be effectively well ordered. This dynamic view of an r.e. set has been essential in r.e. set constructions since the time of Post [**6**]. A new r.e. set is enumerated by simultaneously enumerating all r.e. sets and putting elements in the new set whenever dictated by information so far developed. Definition (2) leads indirectly to a sort of dynamic enumeration of A. Suppose $x \in A$. Then some computation c puts x into A. Since $\mathscr{E}$ is E-closed, both c and $|c|$, the ordinal length of c, belong to $\mathscr{E}$. Define

$$A^\sigma = \{x \mid x \in A \text{ by virtue of some computation } c \text{ such that } |c| \leqslant \sigma\}$$

for each $\sigma \in \mathscr{E}$. Then $A = \bigcup\{A^\alpha \mid \sigma < \kappa\}$, where κ is the least ordinal not in $\mathscr{E}$. Thus, stage σ of the dynamic enumeration of A consists of putting into A all those elements of A that belong to A by virtue of computations of length σ. If A is a subset of some $b \in \mathscr{E}$, then each $A^\sigma \in \mathscr{E}$, and $\lambda\sigma \mid A^\sigma$ is partial E-recursive on $\mathscr{E}$.

$\mathscr{E}$ is effectively well orderable if there exists a partial E-recursive-on-$\mathscr{E}$ map from κ, the least ordinal not in $\mathscr{E}$, onto $\mathscr{E}$. $\mathscr{E}$ is so orderable if $\mathscr{E}$ is $L(\kappa)$ or is $E(x)$ for some set x of ordinals. Note that the hypothesis of (1) implies $E(2^\omega)$ has the latter property. The assumption that $\mathscr{E}$ be effectively well orderable is essential to all existing positive solutions of Post's problem in the setting of E-recursion theory.

The central positive result of the paper is:

(4) If $L(\kappa)$ is E-closed, then there exist two subsets of $L(\kappa)$, each E-recursively enumerable on $L(\kappa)$, such that neither is E-reducible to the other on $L(\kappa)$.

A curious feature of the proof of (4) is the presence of a Friedberg-Muchnik scheme of requirements without injuries when $L(\kappa)$ is not Σ_1 admissible. At most, one positive attempt is made to satisfy each requirement, and no negative requirement is injured. Further results, such as Slaman's density theorem [7], do necessitate Friedberg-Muchnik injury arguments. The reasons that r.e. degree theory on E-closed structures is less troubled by injuries than its counterparts in classical or α-recursion theory have to do with reflection phenomena. There is no hope in any recursion theory worthy of the name of effectively enumerating all divergent computations. But something very close to such an enumeration can be realized in many E-closed structures.

To define the concept of reflection in E-recursion it is helpful to make a connection between $E(x)$ and $L(x)$. Say y is E-recursive in x if $y = \{e\}(x)$ for some e. Define

$$\kappa_0^x = \sup\{\gamma \mid \gamma \text{ is an ordinal } E\text{-recursive in } x\}$$

and

$$\kappa^x = \sup\{\kappa_0^{x,a} \mid a \in \mathrm{TC}(x)\}.$$

($\mathrm{TC}(x)$ is the transitive closure of x.) Then

$$E(x) = L(\kappa^x, \mathrm{TC}(\{x\})).$$

γ is said to be x-reflecting if

$$[L(\delta, \mathrm{TC}(\{x\})) \models \mathscr{F}] \to [L(\kappa_0^x, \mathrm{TC}(\{x\})) \models \mathscr{F}]$$

for every $\mathscr{F}$ a Σ_1^{ZF} sentence with x as parameter. Define

$$\kappa_r^x = \text{sup of all } x\text{-reflecting ordinals}.$$

Clearly, κ_r^x is the greatest x-reflecting ordinal. It can be shown that $\kappa_r^x \leqslant \kappa^x$. More important is the fact that:

(5) If $L(\kappa)$ is not Σ_1 admissible, then $\kappa_r^x < \kappa$ for all $x \in L(\kappa)$.

If x is a set of ordinals, then κ_r^x turns out to be large enough to decide whether or not $\{e\}(x)$ converges for all e. If $\{e\}(x)$ converges, the computation of $\{e\}(x)$ is a well-founded tree whose height is an ordinal recursive in x, hence less than κ_0^x. If $\{e\}(x)$ diverges, then any infinite, descending path through its computation tree is called a Moschovakis witness to the divergence of $\{e\}(x)$. The connection between reflection and divergence needed for the solution to Post's problem is:

(6) If x is a set of ordinals and $\{e\}(x)$ diverges, then a Moschovakis witness to the divergence is first order definable over $L(\kappa_r^x, x)$.

It follows from (5) and (6) that:

(7) If $L(\kappa)$ is not Σ_1 admissible, $x \in L(\kappa)$ and $\{e\}(x)$ diverges, then some Moschovakis witness to the divergence belongs to $L(\kappa)$.

It is (7) that makes possible a positive solution to Post's problem by a Friedberg-Muchnik scheme without injuries. A requirement is satisfied in §4 below by waiting for a certain computation to converge or diverge. If it converges, then the usual steps are taken to create and preserve an inequality. If it diverges, then (7) implies there is a stage of the construction by which the divergence is apparent. At that stage a witness to the divergence is chosen and its role as a witness is preserved forever.

If $L(\kappa) = E(\omega_1)$, then the requirements can be indexed by countable ordinals, and each can be satisfied in its turn as in the previous paragraph. For an arbitrary Σ_1 inadmissible $L(\kappa)$, an indexing of requirments has to be chosen short enough to allow all requirements to be met before time runs out. In the case of $E(\omega_1) = L(\kappa)$, any countable set of requirements can be satisfied by stage σ for some $\sigma < \kappa$. In general, the requirements are indexed by ordinals less than ρ, the greater r.e. projectum of $L(\kappa)$.

$$\rho = \mu_{\gamma \leqslant \kappa}(Ef)[f \text{ is a partial function, } E\text{-rec. on } L(\kappa), \text{ from } \gamma \text{ onto } L(\kappa)].$$

It is shown in §3 that:

(8) If $x < \rho$ and $p \in L(\kappa)$, then $\sup\{\kappa_r^{p,y} | y < x\} < \kappa$.

It follows from (8) that any set of requirements bounded below ρ can be satisfied by stage σ for some $\sigma < \rho$. (There is an additional complication having to do with the r.e. cofinality of ρ and a technique similar to Shore blocking, but it can wait until §§3 and 4.) The proof of (8) has two parts. Define the lesser r.e. projection

$$\eta = \mu\gamma_{\gamma \leqslant \kappa}(EA)[A \in 2^\gamma - L(\kappa) \,\&\, A \text{ is } E\text{-r.e. on } L(\kappa)].$$

A condensation argument given in §3 shows:

(9) If $L(\kappa)$ is E-closed, then $\eta = \rho$.

The proof of (8) consists of proving (8) with η in place of ρ and then invoking (9).

The problem of extending the solution of Post's problem from initial segments of L to other effectively well-orderable E-closed structures is discussed in §5. The only substantial difficulty resides in the proof of (9), a Gödel type condensation argument that appears to be useless when $L(\kappa)$ is replaced by $L[\kappa, A]$ for some $A \subseteq L(\kappa)$. An important special case is $E(2^\omega)$. Normann's result (1) can be redone as follows. In the first half of his proof he shows:

(10) If $E(2^\omega) \models [2^\omega$ is well ordered and of regular cardinality], then $\eta = \rho$ for $E(2^\omega)$.

Then the second half of his argument can be replaced by the proof given in §4. Thus Normann's assumption of "regularity" in (1) was needed only to show $\eta = \rho$. Slaman [**8**] has shown:

(11) There exists a model M of ZFC in which $E(2^\omega) \models [2^\omega$ is well ordered and of singular cardinality] and $\eta = \rho$ for $E(2^\omega)$.

It follows from the argument of §4 that in Slaman's model M, $E(2^\omega)$ admits a positive solution to Post's problem. Thus the study of Post's problem for $E(2^\omega)$, when 2^ω is well orderable in $E(2^\omega)$, appears to turn more on the equality of η and ρ than on the regularity of 2^ω. The existing technology of forcing over E-closed structures [**8**–**10**] gives no hint of how to create an $E(2^\omega)$ in which 2^ω is well ordered and for which $\eta \neq \rho$.

2. Review of E-recursion. The partial E-recursive functions are defined via Normann's schemes [**1**]. $\{e\}$ is the partial E-recursive function with index e. $x_1,\ldots,x_n$ are arbitrary sets.

(1) $$\{e\}(x_1,\ldots,x_n) = x_i \quad \text{if } e = \langle 1, n, i\rangle.$$

(2) $$\{e\}(x_1,\ldots,x_n) = x_i - x_j \quad \text{if } e = \langle 2, n, i, j\rangle.$$

(3) $$\{e\}(x_1,\ldots,x_n) = \{x_i, x_j\} \quad \text{if } e = \langle 3, n, i, j\rangle.$$

(4) $$\{e\}(x_1,\ldots,x_n) \simeq \bigcup\{\{c\}(y, x_2,\ldots,x_n) \mid y \in x_1\} \quad \text{if } e = \langle 4, n, c\rangle.$$

(5) $$\{e\}(x_1,\ldots,x_n) \simeq \{c\}\{\{d_1\}(x_1,\ldots,x_n),\ldots,\{d_m\}(x_1,\ldots,x_n)\} \quad \text{if } e = \langle 5, n, m, c, d_1,\ldots,d_m\rangle.$$

(6) $$\{e\}(c, x_1,\ldots,x_n, y_1,\ldots,y_m) \simeq \{c\}(x_1,\ldots,x_n) \quad \text{if } e = \langle 7, n, m\rangle.$$

$\{e\}(x)\downarrow$ means $\{e\}(x)$ is defined or converges, and $\{e\}(x)\uparrow$ means $\{e\}(x)$ is undefined or diverges. In scheme (4), $\{e\}(x_1,\ldots,x_n)\downarrow$ iff $\{c\}(y, x_2,\ldots,x_n)\downarrow$ for all $y \in x_1$. An important derived scheme, called *bounding*, is

(7) $$\{2^m \cdot 3^n\}(x) \simeq \{\{n\}(y) \mid y \in \{m\}(x)\}.$$

Implementation of the schemes generates the universal computation tree $>_U$. Each node of the tree is of the form $\langle e, x_1,\ldots,x_n\rangle$, or more simply $\langle e, x\rangle$, and is called a computation instruction. $a >_U b$ is read: b is a subcomputation instruction of a, and is defined by: There exists a finite sequence $b_0,\ldots,b_n$ such that $a = b_0$, $b_n = b$, and b_{i+1} is an immediate subcomputation instruction of b_i $(i < n)$. The inductive definition of immediate subcomputation instruction has a clause corresponding to each Normann scheme. It is reasonable to regard (7) as typical, because it combines (4) and (5). The corresponding clauses are:

(8) $\langle m, x\rangle$ is an immed. subcomp. instruc. of $\langle 2^m \cdot 3^n, x\rangle$.

(9) If $\{m\}(x)\downarrow$ and $y \in \{m\}(x)$, then $\langle n, y\rangle$ is an immed. subcomp. instruc. of $\langle 2^m \cdot 3^n, x\rangle$.

In addition:

(10) If c is not the index of a Normann scheme, then $\langle c, x\rangle$ is an immed. subcomp. instruc. of $\langle c, x\rangle$.

The tree-like object consisting of $\langle e, x\rangle$ and that part of $>_U$ below $\langle e, x\rangle$ is called the *computation* of $\{e\}(x)$. Note that $\{e\}(x)\downarrow$ iff the computation of $\{e\}(x)$ is well founded.

Let A be a class. A is *E-recursively enumerable* in x if, for some e,

$$A = \{y|\{e\}(x, y)\downarrow\}.$$

If $x = 0$, then A is E-recursively enumerable. B is said to be E-recursively enumerable in x on C if $B = A \cap C$ for some A E-recursively in x.

The relation $\{e\}(x)\downarrow$ is E-recursively enumerable, and consequently the relation "b is an immediate subcomputation instruction of c" is E-recursively enumerable. *Note well*: The subcomputation relation $>_U$ is not E-recursively enumerable.

If $\{e\}(x)\downarrow$, let $|\{e\}(x)|$ be the ordinal height of the computation of $\{e\}(x)$. $|\{e\}(x)|$ is a partial E-recursive function of e and x.

A set y is said to be E-recursive in a set x if $y = \{e\}(x)$ for some e; in symbols, $y \leqslant_E x$. Thus $|\langle e, x\rangle| \leqslant_E x$ when $\{e\}(x)\downarrow$. Let $\alpha, \beta, \gamma, \ldots$ denote ordinal numbers. Define

$$\kappa_0^x = \bigcup\{\gamma|\gamma \leqslant_E x\} \quad \text{and} \quad \kappa^x = \bigcup\{\kappa_0^{x,a}|a \in \mathrm{TC}(x)\}.$$

$\mathrm{TC}(x)$ is the transitive closure of x.

A set z is E-closed if

$$x \in z \,\&\, \{e\}(x)\downarrow \;\rightarrow\; \{e\}(x) \in z$$

for all $e < \omega$. Define $E(x)$, the E-closure of x, to be the least E-closed z such that $z \supseteq \mathrm{TC}(\{x\})$. Then

$$E(x) = L(\kappa^x, \mathrm{TC}(\{x\})).$$

δ is said to be x-reflecting if

$$[L(\delta, \mathrm{TC}(\{x\})) \models \mathscr{F}] \rightarrow [L(\kappa_0^x, \mathrm{TC}(\{x\})) \models \mathscr{F}]$$

for every Σ_1 sentence $\mathscr{F}$ of ZF with parameter x. Define κ_r^x to be the supremum of all x-reflecting ordinals. Thus κ_r^x is the greatest x-reflecting ordinal. Note that $\kappa_r^x \leqslant \kappa^x$.

κ_r originated in an application of forcing [**11**] to the theory of Kleene recursion in finite types, where it was necessary to know that

$$\kappa_r^{2^{2^\omega}} > \kappa_0^{2^{2^\omega}}.$$

Later Harrington [**12**] showed $\kappa_r^{2^\omega} > \kappa_0^{2^\omega}$, and more importantly, gave a characterization of $\kappa_r^{2^\omega}$ that inspired Lemma 2.1.

If $\{e\}(x)\uparrow$, then there exists an infinite descending path below $\langle e, x\rangle$ in $>_U$. Any such path is termed a Moschovakis witness to the divergence of $\{e\}(x)$. They were introduced by Moschovakis [**13**] to show $E(2^\omega)$ is not Σ_1 admissible. Observe that the relation "z is an M-witness to $\{e\}(x)\uparrow$" is E-recursively enumerable. Let $\mathscr{E}$ be E-closed. $\mathscr{E}$ is said to *admit M-witnesses* if for all $e < \omega$ and $x \in \mathscr{E}$, if $\{e\}(x)\uparrow$, then some M-witness to $\{e\}(x)\uparrow$ is a member of M. $E(2^\omega)$ admits M-witnesses, but $E(\omega)$ does not.

LEMMA 2.1 [9]. *Suppose some well-ordering of* TC(x) *is E-recursive in x. If* $\{e\}(x)\uparrow$, *then some Moschovakis witness to* $\{e\}(x)\uparrow$ *is first-order definable over* $L(\kappa_r^x, \mathrm{TC}(\{x\}))$.

Harrington [12] showed $\kappa_r^{2^\omega,a,b} \geqslant \kappa_r^{2^\omega,a}$ for all $a, b \in 2^\omega$. But Slaman [14] found sets x and y such that $\kappa_r^{x,y} < \kappa_r^x$. It follows from Lemma 2.1 that the function $\lambda x | \kappa_r^x$, restricted to $x \subseteq On$, is E-r.e.

LEMMA 2.2 [9]. *Suppose some well-ordering of* TC(x) *is E-recursive in x. Then* $\kappa_r^{x,y} \geqslant \kappa_r^x$ *for all* y, *and* $\kappa_r^z \leqslant \kappa_r^x$ *for all* $z \leqslant_E x$.

LEMMA 2.3 [9]. *Suppose x is a set of ordinals, and $L(\kappa, x)$ is E-closed and Σ_1 inadmissible. Then* $\kappa_r^{x,y} < \kappa$ *for all* $y \in L(\kappa, x)$.

THEOREM 2.4 [9]. *Let x be a set of ordinals. Then the following are equivalent.*

(i) $\kappa_r^{x,y} \in E(x)$ *for all* $y \in E(x)$.

(ii) $E(x)$ *admits Moschovakis witnesses.*

(iii) $E(x)$ *is* Σ_1 *inadmissible.*

THEOREM 2.5 [9]. *Suppose x is a set of ordinals and $L(\kappa, x)$ is E-closed. If $L(\kappa, x)$ is not Σ_1 admissible, then $L(\kappa, x)$ admits Moschovakis witnesses.*

Slaman [14] has devised an E-closed, Σ_1 inadmissible set that does not admit Moschovakis witnesses.

Let $\mathscr{E}$ be an E-closed structure. The notion of Turing degree generalizes from r.e. subsets of ω to r.e. subsets of $\mathscr{E}$ as follows. The principal step consists of relativizing the concept of E-recursiveness. Let A be a class. $\{e\}^A$ is the eth partial E-recursive function relative to A. $\{e\}^A$ is defined by Normann schemes (1)–(5), with $\{\ \}^A$ in place of $\{\ \}$ throughout, and one additional Normann scheme:

$$(11) \qquad \{e\}^A(x_1,\dots,x_m) = A \cap x_i \quad \text{if } e = \langle 6, n, i \rangle.$$

B is said to be E-reducible to A if for some e, $B(x) = \{e\}^A(x)$ for all x.

Note well the difference between (i) x is E-recursive in y, and (ii) x is E-reducible to y. In (i) y occurs as an argument or parameter, and in (ii) as an additional predicate. For example, suppose x and y are infinite subsets of ω. Then (i) means x is hyperarithmetic in y, and (ii) says x is Turing reducible to y.

Now let A and B be subsets of $\mathscr{E}$. A is E-recursive on $\mathscr{E}$ if there exist e and $p \in \mathscr{E}$ such that

$$\{e\}(x, p)\downarrow \text{ for all } x \in \mathscr{E} \quad \text{and} \quad A = \{x | x \in \mathscr{E} \,\&\, \{e\}(x, p) = 0\}.$$

A is E-recursively enumerable on $\mathscr{E}$ if there exist e and $p \in \mathscr{E}$ such that

$$A = \{x | x \in \mathscr{E} \,\&\, \{e\}(x, p)\downarrow\}.$$

Finally, A is E-reducible to B on $\mathscr{E}$ (in symbols, $A \leqslant_{\mathscr{E}} B$) if there exist e and $p \in \mathscr{E}$ such that

$$\{e\}^B(x, p)\downarrow \text{ for all } x \in \mathscr{E} \quad \text{and} \quad A = \{x | x \in \mathscr{E} \,\&\, \{e\}^B(x, p) = 0\}.$$

The notion of E-reducibility is not, in general, transitive. Call $A \subseteq \mathscr{E}$ subgeneric if for all e and $x \in \mathscr{E}$, if $\{e\}^A(x)\downarrow$, then the computation of $\{e\}^A(x)$ belongs to $\mathscr{E}$. For such an A, $\mathscr{E}$ may be said to be $\mathscr{E}$-closed relative to A, or in other words the structure $\langle \mathscr{E}, A\rangle$ is E-closed.

PROPOSITION 2.6. *Suppose $\mathscr{E}$ is E-closed and $A, B, C \subseteq \mathscr{E}$.*

(i) *If $B \leqslant_{\mathscr{E}} C$ and C is subgeneric, then B is subgeneric.*

(ii) *If $A \leqslant_{\mathscr{E}} B$, $B \leqslant_{\mathscr{E}} C$ and C is subgeneric, then $A \leqslant_{\mathscr{E}} C$.*

(iii) *If A is $\mathscr{E}$-r.e. and not subgeneric and B is E-recursively enumerable on $\mathscr{E}$, then $B \leqslant_{\mathscr{E}} A$.*

(iv) *If A, B and C are E-recursively enumerable on $\mathscr{E}$, $A \leqslant_{\mathscr{E}} B$ and $B \leqslant_{\mathscr{E}} C$, then $A \leqslant_{\mathscr{E}} C$.*

PROOF. (i) is immediate. (ii) follows from (i) and effective bounding. Similarly, (iv) follows from (iii). To prove (iii), observe that the failure of A to be subgeneric implies there is a function f, E-recursive in A on $\mathscr{E}$, such that

$$(\operatorname{dom} f) = d \in \mathscr{E} \quad \text{and} \quad \sup\{|f(x)| | z \in d\} = \kappa,$$

where κ is the least ordinal not in $\mathscr{E}$. Suppose B is $\mathscr{E} \cap \operatorname{dom}(\lambda x|\{e\}(x, p))$. Then, for all $x \in \mathscr{E}$,

$$x \in B \leftrightarrow (Ez)\big[z \in d \,\&\, |\{e\}(x, p)| \leqslant |f(z)|\big]. \quad \square$$

The proof of (iii) is essentially the same as Spector's proof [**15**] that every nonhyperarithmetic Π_1^1 subset of ω has the same hyperdegree as Kleene's $\mathscr{O}$.

Let A and B be E-recursively enumerable on $\mathscr{E}$. In the light of Proposition 2.6 it makes sense to say A and B have the same *degree* if each is E-reducible to the other on $\mathscr{E}$. The proof of Theorem 2.4, in particular the equivalence of 2.4(i) and (iii), yields a connection between E-recursion and recursion on an admissible set.

LEMMA 2.7. *Suppose x is a set of ordinals, $E(x)$ is Σ_1 admissible, and A, $B \subseteq E(x)$.*

(i) *A is E-recursively enumerable on $E(x)$ iff A is Σ_1^A.*

(ii) *If A is E-reducible to B on $E(x)$, then A is $E(x)$-recursive in B (in the sense of admissibility theory).*

Another notion of some use in the study of recursion on an E-closed set $\mathscr{E}$ is that of regularity. A set $A \subseteq \mathscr{E}$ is said to be *regular* if $(A \cap x) \in \mathscr{E}$ for all $x \in \mathscr{E}$.

LEMMA 2.8. *Suppose $\mathscr{E}$ is E-closed and $A, B \subseteq \mathscr{E}$ are E-recursively enumerable on $\mathscr{E}$.*

(i) *If A is not regular, then $B \leqslant_{\mathscr{E}} A$.*

(ii) *There exists a regular $C \subseteq \mathscr{E}$ such that C is E-recursively enumerable on $\mathscr{E}$ and has the same degree as A.*

PROOF. (i) Suppose $(A \cap x) \notin \mathscr{E}$. Then the lengths of computations needed to enumerate $A \cap x$ are unbounded in $\mathscr{E}$, and so A is not subgeneric. Now apply Proposition 2.6(iii).

(ii) By (i) it is safe to assume A is "complete". Thus it suffices to find a regular, r.e. C such that $A \leqslant_{\mathscr{E}} C$. The original construction of C is a transfinite recursion that applies some of the thinking behind the regular sets theorem of α-recursion theory. A much shorter proof was subsequently found by Slaman [7]. (2.7 is not needed in this paper, but is used in the proof of Slaman's density theorem [7].) □

Let A be a class. All of the definitions and results of this section through Theorem 2.4 relativitize to A. $\{e\}^A(x)$ has already been defined. The computation tree for $\{e\}^A$ differs from that of $\{e\}$ only by nodes corresponding to scheme (11). A set y is E-recursive, relative to A, in a set x if $y = \{e\}^A(x)$ for some e; in symbols, $y \leqslant_E^A x$. Define

$$\kappa_0^{A;x} = \bigcup\{\gamma \mid \gamma \leqslant_E^A x\} \quad \text{and} \quad \kappa^{A;x} = \bigcup\{\kappa_0^{A;x,a} \mid a \in \mathrm{TC}(x)\}.$$

A set z is E-closed relative to A if, for all e,

$$x \in z \,\&\, \{e\}^A(x)\downarrow \;\rightarrow\; \{e\}^A(x) \in Z.$$

Define $E^A(x)$, the E-closure of x relative to A, to be the least $z \supseteq \mathrm{TC}(\{x\})$ such that z is E-closed relative to A. Then

$$E^A(x) = L[\kappa^{A;x}, A; \mathrm{TC}(\{x\})]. \tag{12}$$

The right side of (12) is the set of all sets constructible via ordinals less than $\kappa^{A;x}$ with $\mathrm{TC}(\{x\})$ as the starting set and A as an additional predicate. (In addition to the usual Gödel operations, $A \cap x$ is allowed.)

δ is x-reflecting relative to a if

$$[L[\delta, A; \mathrm{TC}(\{x\})] \models \mathscr{F}] \rightarrow [L[\kappa_0^{A;x}, A; \mathrm{TC}(\{x\})] \models \mathscr{F}]$$

for every Σ_1 sentence $\mathscr{F}$ of ZF with parameter x and additional predicate $y \in A$. Let $\kappa_r^{A;x}$ be the greatest x-reflecting ordinal relative to A.

The treatment of κ_r and Moschovakis witnesses in 2.1–2.5 relativizes to A without any surprises. For example, Theorem 2.5 becomes

THEOREM 2.9. *Suppose x is a set of ordinals and $\langle L[\kappa, A; x], A\rangle$ is E-closed, but not Σ_1 admissible. Then $L[\kappa, A; x]$ admits Moschovakis witnesses.*

3. Fine structure. Assume $L(\kappa)$ is E-closed. This section concentrates on several structural properties of $L(\kappa)$ needed in the next section to establish a suitable indexing of Friedberg-Muchnik requirements. Define

$$\rho = \mu\gamma_{\gamma \leqslant \kappa}(Ef)\,[f \text{ is a partial } E\text{-rec. (on } L(\kappa)) \text{ map of } \gamma \text{ onto } L(\kappa)]$$

and

$$\eta = \mu\gamma_{\gamma \leqslant \kappa}(EA)[A \in 2^\gamma - L(\kappa) \,\&\, A \text{ is } E\text{-r.e. on } L(\kappa)].$$

LEMMA 3.1. $\eta = \rho$.

PROOF. First suppose that

$$L(\kappa) \text{ is } \Sigma_1 \text{ admissible and equal to } E(\beta) \tag{1}$$

for some $\beta < \kappa$. By Proposition 2.7, E-r.e. coincides with $\Sigma_1(L(\kappa))$, and ρ and η are equivalent definitions of the $\Sigma_1(L(\kappa))$ projectum of $L(\kappa)$.

$$(2) \qquad \text{Now suppose (1) is false.}$$

To see $\eta \leqslant \rho$, let f be a partial E-rec. map of ρ onto $L(\kappa)$. If $\rho < \eta$, then $(\operatorname{dom} f) \in L(\kappa)$, and consequently $(\operatorname{range} f) \in L(\kappa)$.

To see $\eta = \rho$, fix $\gamma < \rho$ and let $A \subseteq \gamma$ be E-r.e. in some $p \in L(\kappa)$ via Gödel number e with the intention of showing $A \in L(\kappa)$. It is safe to assume p is an ordinal, since each member of $L(\kappa)$ is E-recursive in some ordinal less than κ. Let g be a universal, partial E-recursive function. Define

$$H = g[\gamma \cup \{p\}].$$

(H is said to be the partial E-recursive hull of $\gamma \cup \{p\}$.)

Suppose H is bounded below κ, i.e. $H \subseteq L(\delta)$ for some $\delta < \kappa$. Then for $y < \gamma$,

$$y \in A \leftrightarrow L(\delta) \models \{e\}(y, p)\downarrow,$$

and so $A \in L(\kappa)$.

Suppose H is unbounded in $L(\kappa)$. Observe that

$$(3) \qquad z \in H \to \mathcal{O}(z) \in H,$$

where $\mathcal{O}$ is Gödel's order of constructibility function. If $z \in H$, then $z = g(x)$ for some $x \in \gamma \cup \{p\}$, and $g(x)$ is first-order definable over $L(x + |g(x)| + \omega)$. Since H is E-closed, $|g(x)| \in H$.

The unboundedness of H implies H admits M-witnesses. Fix $x \in H$. It follows from (2) and Lemma 2.3 that $\kappa_r^x < \kappa$. By (3) the supremum of all ordinals in H is κ. Choose $\delta \in H$ so that $\kappa_r^x \leqslant \delta$. Then $\kappa_r^x \leqslant_E \delta, x$, and so $\kappa_r^x \in H$. Consequently, H admits M-witnesses by Lemma 2.1.

It follows from (3) that every element of H is well ordered in H. Thus H is extensional, hence isomorphic, to a transitive set H_0 via the collapsing map t. (3) also implies $H_0 = L(\beta)$ for some $\beta \leqslant \kappa$. The partial E-recursive function g maps $\gamma \cup \{t(p)\}$ onto $L(\beta)$. Hence $\beta < \kappa$, since $\gamma < \kappa$.

Since H admits M-witnesses, the map t preserves divergence, as well as convergence, facts. Hence A is first-order definable over $L(\beta)$. □

PROPOSITION 3.2. *Let p be an ordinal and A a set of ordinals. Suppose the function $\lambda x|\kappa_r^{p,x}$ does not attain a maximum on A. Then the range of $\lambda x|\kappa_r^{p,x}$ is bounded on A by* $\sup\{\kappa_0^{p,y,z}|y, z \in A\}$.

PROOF. Suppose not. Then there exist $c, d \in A$ such that

$$(4) \qquad \kappa_r^{p,c} > \kappa_r^{p,d} \geqslant \sup\{\kappa_0^{p,y,z}|y, z \in A\}.$$

It follows from Lemma 2.2 that $\kappa_r^{p,c,d} > \kappa_r^{p,d}$. The desired contradiction is: $\kappa_r^{p,c,d}$ is a $\langle p, d\rangle$-reflecting ordinal. Suppose some Σ_1 sentence $\mathcal{F}$ about $\langle p, d\rangle$ is true below $\kappa_r^{p,c,d}$. Then $\mathcal{F}$ is true below $\kappa_0^{p,c,d}$ by definition of κ_r. But then by (4), $\mathcal{F}$ is true below $\kappa_r^{p,d}$. □

THEOREM 3.3. *Assume $L(\kappa)$ is E-closed and $p \in L(\kappa)$. Further assume $L(\kappa)$ is either Σ_1 inadmissible or not of the form $E(\beta)$ for any $\beta < \kappa$. If $\gamma < \rho$, then*

$$\sup\{\kappa_r^{p,\delta} | \delta < \gamma\} < \kappa.$$

PROOF. As in the proof of 3.1, p can be assumed to be an ordinal. First observe that

$$(5) \qquad \sup\{\kappa_0^{p,\delta} | \delta < \gamma\} < \kappa.$$

Define $f(e, \delta) \simeq |\{e\}(p, \delta)|$. f is a partial E-rec. map from $\omega \times \gamma$ into κ. Since ρ is either κ or an $L(\kappa)$-cardinal, it is safe to assume γ is closed under pairing in the following sense:

$$(6) \qquad x, y < \gamma \to (E\delta)(\delta < \gamma \,\&\, x \leqslant_E \delta \,\&\, y \leqslant_E \delta).$$

Consequently, the domain of f can be construed as a subset of γ. It follows from Lemma 3.1 that (dom f) $\in L(\kappa)$, hence (range f) $\in L(\kappa)$, and so (5) is proved.

Suppose $\lambda\delta|\kappa_r^{p,\delta}$ does not attain a maximum on γ. Then Proposition 3.2 yields

$$(7) \qquad \sup\{\kappa_r^{p,\delta} | \delta < \gamma\} \leqslant \sup\{\kappa_0^{p,x,y} | x, y \in \gamma\}.$$

The theorem now follows from (5) and (6).

If $\lambda\delta|\kappa_r^{p,\delta}$ does attain a maximum on γ, then that maximum is less than κ by Lemma 2.3. □

Cofinality considerations figure prominently in the solution to Post's problem. Let $\lambda \leqslant \kappa$. Define r.e. cf(λ), the r.e.-on-$L(\kappa)$ cofinality of λ, to be

$$\mu\gamma(EA)[\sup A = \lambda \,\&\, A \text{ is } E\text{-r.e. on } L(\kappa) \,\&\, \text{ordertype of } A \text{ is } \gamma].$$

LEMMA 3.4. *r.e.* cf(κ) = *r.e.* cf(ρ).

PROOF. Assume $L(\kappa)$ is either Σ_1 inadmissible or not of the form $E(\beta)$ for any $\beta < \kappa$. To show cf(κ) $\leqslant$ cf(ρ), let A be an unbounded subset of ρ and E-r.e. on $L(\kappa)$. Let p be the parameter needed to define a partial E-rec. f from ρ onto $L(\kappa)$. As in the proof of 3.1, p can be assumed to be an ordinal. Define

$$x^r = \sup\{\kappa_r^{p,y} | y \leqslant x\} \quad \text{and} \quad A^r = \{x^r | x \in A\}.$$

Theorem 3.3 implies $A^r \subseteq \kappa$. A^r is unbounded in κ, since $\kappa_r^{f(y)} \leqslant \kappa_r^{p,y}$ by Lemma 2.2 if $f(y)\downarrow$. $|A^r|$, the ordertype of A^r, is at most $|A|$. The predicate $v = \kappa_r^w$, restricted to $w \in \mathcal{O}n$, is E-rec. by Lemma 2.1. It follows that A^r is E-rec. on $L(\kappa)$.

To show cf(ρ) $\leqslant$ cf(κ) let B be an unbounded subset of κ, E-r.e. on $L(\kappa)$, and of ordertype cf(κ). B can be replaced by an E-rec.-on-$L(\kappa)$ set with the same ordertype as B. Assume (a) no such set exists. Define B^0 to be the set of all σ such that: σ is the length of a computation that put some x into B greater than any y put into B by any computation of length less than σ. Clearly B^0 is E-rec. on $L(\kappa)$. $|B^0| \leqslant |B|$ since B^0 is "strictly increasing". If B^0 is bounded below κ, then there exists a σ_0 such that computations of length at most σ_0 put an unbounded set of ordinals into B; but that contradicts assumption (a). Thus it is safe to assume B^0 is the set that replaces B.

B^0 divides κ into blocks. Let $\delta_0 < \delta_1 < \cdots < \delta_\tau < \cdots$ ($\tau < |B^0|$) be a listing of B^0. Define block τ to be $[\delta_\tau, \delta_{\tau+1})$. The blocks are needed to describe the enumeration of A_0, a subset of dom f. A_0 will be unbounded in ρ, and its ordertype will be at most $|B^0|$.

A^0 is the range of a partial function g defined by a partial recursion on $|B^0|$.

$g(\tau) \simeq ux$ [x is enumerated in dom f via a computation of length in block τ and $(\beta)_{\beta<\tau}(g(\beta)$ defined $\to g(\beta) < x)$].

$|A_0| \leqslant |B^0|$ because g is strictly increasing.

To check that A_0 is unbounded in ρ, suppose otherwise. Thus $A_0 \subseteq (\text{dom } f) \cap z$ for some $z < \rho$. By Lemma 3.1, $((\text{dom } f) \cap z) \in L(\kappa)$, hence the computations needed to enumerate $(\text{dom } f) \cap z$ all lie below some δ_{τ_0}. It follows that g *is not* defined for any $\tau \geqslant \tau_0$. But $((\text{dom } f) - z) \notin L(\kappa)$, so some $x \in (\text{dom } f) - z$ must be enumerated by a computation of length greater than δ_{τ_0}, and consequently g *is* defined for some $\tau \geqslant \tau_0$.

x is recursively enumerated in A_0 as follows. First enumerate x in dom f. Suppose the length of computation used to put x into dom f belongs to block τ. Compute $g(\beta)$ for all $\beta \leqslant \tau$. Enumerate x in A_0 if $g(\tau) = x$.

The above instructions are effective largely because B^0 is E-recursive on $L(\kappa)$. □

PROPOSITION 3.5. *Suppose* γ = *r.e.* cf(λ) *for some* $\lambda \leqslant \kappa$. *then there exists a* g: $\gamma \to \lambda$ *such that*:

(i) *the graph of* g *is E-r.e. on* $L(\kappa)$,

(ii) sup(range g) = λ,

(iii) g *is strictly increasing, and*

(iv) $\tau_1 < \tau_2 < \lambda \to |\langle\tau_1, g(\tau_1)\rangle| < |\langle\tau_2, g(\tau_2)\rangle|$.

PROOF. The proof is similar to the second half of the proof of 3.4. □

4. Solution to Post's problem.

THEOREM 4.1. *Suppose* $L(\kappa)$ *is E-closed. Then there exist two subsets of* $L(\kappa)$, *each E-recursively enumerable on* $L(\kappa)$, *such that neither is E-reducible to the other on* $L(\kappa)$.

Assume $L(\kappa)$ is either Σ_1 inadmissible or not equal to $E(\beta)$ for any $\beta < \kappa$. Otherwise Proposition 2.7 implies that the positive solution of Post's problem for $L(\kappa)$ provided by α-recursion theory [5] is also a solution in the sense of E-recursion theory. Thus $L(\kappa)$ has certain structural properties described in §§2 and 3. In particular, $\kappa_r^x < \kappa$ for every $x \in L(\kappa)$.

The subsets of $L(\kappa)$ to be constructed are A and B. All requirements are of the form $A \neq \{y\}^B$ or $B \neq \{y\}^A$, and are indexed by ordinals less than ρ. Let f be a partial function, E-rec. on $L(\kappa)$, from ρ onto $L(\kappa)$. Then

$$\text{requirement } 2x \text{ is } A \neq \{f(x)\}^B \quad \text{and} \quad \text{requirement } 2x+1 \text{ is } B \neq \{f(x)\}^A.$$

Each requirement belongs to a block labeled by an ordinal less than γ, the r.e.-on-$L(\kappa)$ cofinality of ρ. By Proposition 3.5, with ρ in place of λ, there exists a function g suitable for dividing ρ into blocks. Assume $g(0) = 0$. Define

$$(\text{req. } 2x) \in \text{block } 2\tau \quad \text{and} \quad (\text{req. } 2x + 1) \in \text{block } 2\tau + 1,$$

if $g(\tau) \leqslant x < g(\tau + 1)$.

The construction has the form of an effective transfinite recursion on κ. At stage σ ($\sigma < \kappa$) all decisions are made by examining the set of all computations of length at most σ. (It will be convenient to assume that the "length" of a computation exceeds all ordinals mentioned in the computation.) Let

$$A^{<\sigma} = \text{set of all elements put in } A \text{ before stage } \sigma$$

and

$$A^{\sigma} = A^{<\sigma} \cup \{\text{elements put in } A \text{ at stage } \sigma\}.$$

Each requirement is acted on at most once.

Requirement $2x$ is *acted on* at stage σ if (i)–(iv) hold.

(i) req. $2x$ has not been acted on prior to stage σ.

(ii) $\langle \delta, g(\delta) \rangle$ has been computed for all $\delta \leqslant \tau$ prior to stage σ. (Req. $2x$) $\in$ block 2τ. $f(x)$ has been computed prior to stage σ.

(iii) Every requirement in block y, for all $y < 2\tau$, has been acted on prior to stage σ.

(iv) Let $\sigma(2\tau)$ be the first stage after all the activity of (ii) and (iii) has occurred. Then either (iv)(a) or (iv)(b) holds.

(iv)(a) $\{f(x)\}^{B<\sigma(2\tau)}(\langle \sigma(2\tau), f(x) \rangle)$ converges via a computation of length at most σ. (Let $\langle \sigma(2\tau), f(x) \rangle$ be an ordinal that encodes the ordered pair $(\sigma(2\tau), f(x))$ and exceeds both of its elements.)

(b) A Moschovakis witness to the divergence of $\{f(x)\}^{B^{<\sigma(2\tau)}}(\langle \sigma(2\tau), f(x) \rangle)$ is first-order definable over $L(\sigma)$.

Suppose req. $2x$ is acted on at stage σ. If (iv)(a) holds and

$$\{f(x)\}^{B^{<\sigma(2\tau)}}(\langle \sigma(2\tau), f(x) \rangle) = 1.$$

then $\langle \sigma(2\tau), f(x) \rangle$ is put in A at stage σ.

Requirement $2x + 1$ is handled similarly. Simply replace $2x$ by $2x + 1$, 2τ by $2\tau + 1$, and exchange A and B in the above instructions.

PROPOSITION 4.2. *Suppose every requirement in block z is acted on. Then there exists a σ such that every requirement in block z is acted on prior to stage σ.*

PROOF. Suppose $z = 2\tau$ and (req. $2x$) $\in$ block 2τ. Let σ_{2x} be the unique σ such that req. $2x$ is acted on at stage σ. Let p be an ordinal that encodes $\langle B^{<\sigma(2\tau)}, \sigma(2\tau), q \rangle$, where q is the parameter needed to compute f. Then

$$\sigma_{2x} \leqslant p + x + \kappa_r^{p,x}$$

by Lemma 2.1. And $\sup\{\kappa_r^{p,x} | 2x \in \textit{block } 2\tau\} < \kappa$ by Theorem 3.3. $\square$

PROPOSITION 4.3. *Assume* $z <$ *r.e.* $\mathrm{cf}(\rho)$. *Suppose every requirement in block* y, *for all* $y < z$, *is acted on. Then there exists a* σ *such that every requirement in block* y, *for all* $y < z$, *is acted on prior to stage* σ.

PROOF. If $u < v$, then no requirement in block v is acted on until every requirement in block u has been acted on. Consequently, if z is a successor ordinal, then the desired σ exists by Proposition 4.2. Assume z is a limit. For each $y < z$, let $\tau(y)$ be the least σ such that every requirement in block y is acted on prior to stage σ. $\tau(y) < \kappa$ by Proposition 4.2. Since

$$\tau(y) \leqslant \sigma(y+1) \leqslant \tau(y+1)$$

for all $y < z$, it remains only to show the supremum of

$$\mathcal{S} = \{\sigma(y) \mid y < z\}$$

is less than κ. $\mathcal{S}$ is E-r.e. on $L(\kappa)$, since the graph of $\lambda x|\kappa_r^x$ is E-r.e., as was noted after Lemma 2.1. The ordertype of $\mathcal{S}$ is at most z, since $\lambda y|\sigma(y)$ is nondecreasing. By Lemma 3.4, $z <$ r.e. $\mathrm{cf}(\kappa)$, hence $\sup \mathcal{S} < \kappa$. $\square$

Now the proof of Theorem 4.1 is easily completed. Fix $x < \rho$ in order to see req. $2x$ is acted on. Suppose (req. $2x$) $\in$ block 2τ. Assume for the sake of an induction that every requirement in block y, for all $y < 2\tau$, has been acted on. By Proposition 4.3, $\sigma(2\tau)$ is well defined. Suppose

$$\{f(x)\}^{B^{<\sigma(2\tau)}}(\langle\sigma(2\tau), f(x)\rangle) \tag{1}$$

converges. Then req. $2x$ is acted on at some stage $\sigma \geqslant \sigma(2\tau)$, clause (iv)(a) holds, and $A^\sigma(\langle\sigma(2\tau), f(x)\rangle)$ disagrees with (1).

The construction insures that

$$A^\sigma(\langle\sigma(2\tau), f(x)\rangle) = A(\langle\sigma(2\tau), f(x)\rangle);$$

and

$$B^{\sigma(2\tau)} = B^\sigma = B \cap \sigma. \tag{2}$$

Consequently $A \neq \{f(x)\}^B$.

Suppose (1) diverges. Then Theorem 2.5 implies req. $2x$ is acted on at some stage $\sigma \geqslant \sigma(2\tau)$ and clause (iv)(b) holds. Thus there is a Moschovakis witness w to the divergence of (1) first-order definable over $L(\sigma)$. It follows from (2) that w also witnesses the divergence of

$$\{f(x)\}^B(\langle\sigma(2\tau), f(x)\rangle).$$

Thus again $A \neq \{f(x)\}^B$.

To complete the induction observe that the above argument shows every requirement in block 2τ is acted on.

It remains only to show A and B are E-r.e. on $L(\kappa)$. Suppose z, a typical member of A, is put in A at stage σ. The decision is based on the set K_σ, the set of all computations of length at most σ. Since $K_\sigma \leqslant_E \sigma$ (uniformly in σ), it suffices to show $\sigma \leqslant_E z$.

Let z be $\langle\sigma(2\tau), f(x)\rangle$ and assume clause (iv)(a) of the definition of *acted on* holds. Then $\sigma \leqslant_E z$, $B^{<\sigma(2\tau)}$, and $B^{<\sigma(2\tau)} \leqslant_E K_{\sigma(2\tau)} \leqslant_E z$. $\square$

5. Further results and open questions. As in §1, a positive solution to Post's problem consists of two "incomparable" r.e. sets. The method of §4 sheds a little light on Post's problem for structures other than initial segments of L. Assume $A \subseteq \kappa$. Let $L[\kappa, A]$ as usual be the set of all sets constructible via ordinals less than κ with $x \in A$ as an additional predicate. (Simply add $A \cap y$ to the nine Gödel operations that define constructibility.) The structure $\langle L[\kappa, A], A\rangle$ is said to be E-closed if $\{e\}^A(x) \in L[\kappa, A]$ when $x \in L[\kappa, A]$.

As described in §2, the definitions of κ_0 and κ_r relativize straightforwardly to A; $\{e\}$ is replaced by $\{e\}^A$ throughout. Similarly η, ρ and r.e. cofinality relativize to $\langle L[\kappa, A], A\rangle$. The proof of Theorem 4.1 shows:

(1) Assume $\langle L[\kappa, A], A\rangle$ is E-closed. If $\eta = \rho$ and $\kappa_r^{A;x} < \kappa$ for all $x \in L[\kappa, A]$, then $\langle L[\kappa, A], A\rangle$ admits a positive solution to Post's problem.

It seems unlikely that $\eta = \rho$ in general. An immediate corollary of a selection theorem of Griffor and Normann [**16**] states:

(2) If $L[\kappa, A] \models$ [there exists a greatest cardinal and it is regular], then $\eta = \rho$ for $\langle L[\kappa, A], A\rangle$.

The assumption in (1) on κ_r can be related to inadmissibility as in Theorem 2.9.

(3) If $\langle L[\kappa, A], A\rangle$ is E-closed, but not Σ_1 admissible, then $\kappa_r^{A;x} < \kappa$ for all $x \in L[\kappa, A]$.

The proof of Theorem 4.1 also applies to an arbitrary E-closed structure $\mathscr{E}$. Let κ be the least ordinal not in $\mathscr{E}$. $\mathscr{E}$ is said to be *effectively well orderable* if there exists a partial, E-recursive-on-$\mathscr{E}$ map from κ onto $\mathscr{E}$.

(4) Assume $\mathscr{E}$ is E-closed and effectively well orderable. If $\eta = \rho$ for $\mathscr{E}$, and $\mathscr{E}$ admits Moschovakis witnesses, then $\mathscr{E}$ admits a positive solution to Post's problem.

The proof of Theorem 4.1 yields a property stronger than subgenericity for those κ where blocking is unnecessary, that is, when r.e. $\mathrm{cf}(\rho) = \rho$. The simplest example is:

(5) $E(\omega_1)$ admits a positive solution (A, B) to Post's problem such that for all $\delta < \omega_1$,

(6) $\kappa_0^{A;\omega_1,\delta} = \kappa_0^{B;\omega_1,\delta} = \kappa_0^{\omega_1,\delta}$.

A set $A \subseteq E(\omega_1)$ with property (6) is said to preserve the κ_0-spectrum of $E(\omega_1)$. It is not known if the κ_0-spectrum can be preserved when ρ fails to be r.e. regular.

A consequence of (4) and Theorem 2.5 is:

(7) Suppose x is a set of ordinals, and $L(\kappa, x)$ is E-closed but not Σ_1 admissible. If $\eta = \rho$ for $L(\kappa, x)$, then $L(\kappa, x)$ admits a positive solution to Post's problem.

At this writing virtually nothing is known about situations where Post's problem has a negative solution, or where $\eta \neq \rho$.

(8) Does there exist a set x of ordinals such that $L(\kappa, x)$ is E-closed and Σ_1 inadmissible, and does not admit a positive solution to Post's problem for E-recursion theory?

(9) Does there exist a set x of ordinals such that $L(\kappa, x)$ is E-closed and Σ_1 inadmissible, and $\eta \neq \rho$ for $L(\kappa, x)$?

According to (7) a negative answer for (8) implies a negative answer for (9). It might be easier, however, to deal with (9) directly. Both questions are good candidates for forcing arguments. Forcing over E-closed structures has been extensively studied in [**9**, **10** and **14**], but no technique so far developed appears to be of any use.

Normann's result on Post's problem for $E(2^\omega)$, quoted in §1, leads to the following terminal question:

Does there exist a model of ZFC in which

(10) $$E(2^\omega) \models [2^\omega \text{ is well ordered}]$$

and $E(2^\omega)$ does not admit a positive solution to Post's problem?

References

1. D. Normann, *Set recursion*, Generalized Recursion Theory. II, North-Holland, Amsterdam 1978, pp. 303–320.

2. ______, *Degrees of functionals*, Preprint Series in Math., No. 22, Univ. of Oslo, Oslo, 1975.

3. R. Friedberg, *Two recursively enumerable sets of incomparable degrees of unsolvability*, Proc. Nat. Acad. Sci. U.S.A. **43** (1957), 236–238.

4. A Muchnik, *On the unsolvability of the problem of reducibility in the theory of algorithms*, Dokl. Akad. Nauk SSSR **108** (1956), 194–197.

5. G. Sacks and S. Simpson, *The α-finite injury method*, Ann. Math. Logic **4** (1972), 323–367.

6. E. Post, *Recursively enumerable sets of positive integers and their decision problem*, Bull. Amer. Math. Soc. **50** (1944), 284–316.

7. T. Slaman, *The E-recursively enumerable degrees are dense*, these PROCEEDINGS.

8, ______, *The extended plus-one hypotheses—a relative consistency result*, Nagoya Math. J. **92** (1983) (to appear).

9. G. Sacks, *The limits of E-recursive enumerability*, Nagoya Math. J. **92** (1983), 107–120.

10. G. Sacks and T. Slaman, *Inadmissible forcing*, Ann. Pure Appl. Logic (to appear).

11. G. Sacks, *The k-section of a type n object*, Amer. J. Math., **99** (1977), 901–917.

12. L. Harrington, *Contributions to recursion theory on higher types*, Ph.D. Thesis, M.I.T., Cambridge, Mass., 1973.

13. Y. Moschovakis, *Hyperanalytic predicates*, Trans. Amer. Math. Soc. **138** (1967), 249–282.

14. T. Slaman, *Forcing and reflection in E-recursion*, Ann. Pure and Appl. Logic (to appear).
15. C. Spector, *Recursive wellorderings*, J. Symbolic Logic **20** (1955), 151–163.
16. E. Griffor and D. Normann, *Effective cofinalities and admissibility in E-recursion*, preprint, University of Oslo, 1982.
17. R. Shore, *The recursively enumerable α-degrees are dense*, Ann. Math. Logic **9** (1976), 123–155.

DEPARTMENT OF MATHEMATICS, HARVARD UNIVERSITY, CAMBRIDGE, MASSACHUSETTS 02138

MASSACHUSETTS INSTITUTE OF TECHNOLOGY, CAMBRIDGE, MASSACHUSETTS 02139

Proceedings of Symposia in Pure Mathematics
Volume **42**, 1985

The E-Recursively Enumerable Degrees are Dense

THEODORE A. SLAMAN[1]

Abstract. In the context of E-recursion on initial segments of L it is shown that every E-recursively enumerable degree splits over any strictly smaller one. The proof proceeds by means of a finite injury construction using the tame Σ_1 projectum to assign priority. The injury is made finite by means of exploitation of a phenomenon unique to E-recursion: witnesses to divergence.

E-recursion generalizes Kleene's definition of recursion in a normal object of higher type. The notions associated with E-recursion were introduced by Normann [5] and, independently, by Moschovakis. The intention is to describe algorithms which can be used to compute one set from another. This approach isolates the inductive nature of computation, as distinguished from general Σ_1 definability.

§1 reviews the underpinnings of the subject. Briefly, E-recursive computations from a predicate A are generated from schemes for the rudimentary set-theoretic functions, a universal machine scheme, and a scheme for intersection of a set with A; a function produced by such computations is called E-recursive in A or E^A-recursive. A set is E^A-closed if it is closed under application of the E^A-recursive functions. A predicate is E-recursively enumerable if it is the domain of a partial E-recursive function.

Over a fixed E-closed structure $\mathfrak{A}$, predicates on $\mathfrak{A}$ are partially ordered by relative computability: roughly, A is E-reducible to B on $\mathfrak{A}$ ($A \leqslant_{\mathfrak{A}} B$) if there are

1980 *Mathematics Subject Classification.* Primary 03D25.

[1] The preparation of this paper was supported by the National Science Foundation Postdoctoral Fellowship MCS81-14165. The results herein form a part of the author's doctoral thesis, Harvard University, 1981. The author wishes to thank the University of Chicago, the University of California at Los Angeles, and Harvard University for their hospitality during the academic years 1981–1983. Also, the author is grateful for the tutelage of Gerald E. Sacks; "those having torches will pass them onto others"—Plato.

a parameter p and an E-reducibility $\{e\}$ so that for every y in $\mathfrak{A}$, $\{e\}^B(y, p) = A(y)$. This paper is concerned with studying this partial order over E-closed initial segments of L. In this context, $\mathfrak{A} = L_\kappa$, $\leqslant_{\mathfrak{A}}$ is denoted $\leqslant_\kappa$, and a predicate X is called E-r.e.(κ) if $X \subseteq \kappa$ and X is E-recursively enumerable on L_κ.

The computation of the E^A-recursive function $\{e\}^A$ at the argument y can be analyzed by examining the associated tree of subcomputations $T^A_{\langle e,y\rangle}$, which traces the process of computing $\{e\}^A$ at y. This tree is well founded exactly when $\{e\}^A(y)$ has a value (the computation converges). Otherwise, any infinite descending sequence in $T^A_{\langle e,y\rangle}$ constitutes a witness to the divergence of $\{e\}^A$ at y and is called a Moschovakis witness. An E-closed set M exhibits the Moschovakis phenomenon (MP) if, whenever $\{e\}$ diverges at an element of M, there is a witness to that fact in M. Sacks [**9**], generalizing work of Harrington [**2**] showed that in E-closed L_κ exactly one of two situations is obtained: either L_κ is admissible and the notions of E-recursion and admissible recursion theory agree on L_κ, or L_κ exhibits MP. Since the first case has been extensively studied under the heading of α-recursion theory, in what follows MP will be assumed.

§2 develops some of the basic facts concerning the E-r.e.(κ) prerdicates. For example, a predicate D is called regular on L_κ if, for every y in L_κ, $y \cap D \in L_\kappa$; if D is E-r.e.(κ) and not regular, then D is $\leqslant_\kappa$-complete for the E-r.e.(κ) predicates. Sacks has shown that there is a regular complete E-r.e.(κ) predicate.

Historically, the study of the $\leqslant_\kappa$-degrees of E-r.e.(κ) predicates begins with Normann [**6**]. Normann uses the finite injury method to construct two $\leqslant_{\kappa_1}$-incomparable E-r.e.(κ_1) predicates, where κ_1 is the least ordinal greater than ω_1^L which is E-closed (i.e., L_{κ_1} is E-closed). Sacks [**9**] extends this result, showing by noninjury methods that on any E-closed L_κ exhibiting MP there are $\leqslant_\kappa$-incomparable E-r.e.(κ) predicates. The Sacks construction uses a priority ranking of requirements in order type ρ_κ, where ρ_κ is the E-recursively enumerable projection of κ, defined analogously to the Σ_1 projection. Normann [**7**] also showed that the splitting theorem is true for a certain class of well-behaved E-r.e.(κ_1) predicates. Sacks improved this result to prove the splitting theorem for L_{κ_1} and many other κ as well. Griffor [**1**] extended Sacks' techniques to prove that the E-r.e.(κ) degrees are dense provided that ρ_κ is sufficiently regular. The obstacle to applying these methods in general was the existence (see Slaman [**12**]) of non-E-recursive E-r.e.(κ) predicates which are scattered (of order type less than ρ_κ).

The new ingredient needed to prove the full splitting and density theorems, which is supplied in §3, is the closure of L_κ with regard to a certain Σ_1 function on any domain which is smaller than $t\sigma 1p(\kappa)$, the tame Σ_1-projection of κ. Tameness merely asserts that the projection of L_κ into $t\sigma 1p(\kappa)$ can be expressed as the limit of an E-recursive sequence of E-recursive approximations which converges on initial segments of $t\sigma 1p(\kappa)$. It is shown in §3 that if E-r.e.(κ) predicates, $D <_\kappa C$, and a size less than $t\sigma 1p(\kappa)$ set of indices for E-reducibilities on L_κ, $R \in L_\kappa$, are given, then there is an ordinal $\beta < \kappa$ such that whenever f is indexed by an element of R, $f^D \neq C$ is witnessed in L_β. These many include witnesses to divergence as well as to convergence.

In §4, the construction used to prove the combined splitting and density theorem is provided; it is basically an amalgamation of the Sacks splitting strategy from classical recursion theory [**8**] and the Shore blocking method from α-recursion theory [**10**]. The Sacks strategy is altered to include the possibility of witnessing inequalities via divergence; the Shore blocking method is used to arrange the global construction using the sequence of approximations to the tame Σ_1 function mapping L_κ into $t\sigma 1 p(\kappa)$.

§5 mentions two open problems.

1. Groundwork.

DEFINITION. 1.1. A partial function is E-recursive relative to the prediate A (E^A-recursive) if it can be defined from schemes for the primitive recursive functions, intersection of a set with A and a universal machine ($u(e, x) = \{e\}^A(x)$). The eth E^A-recursive function in the parameter p is obtained from the eth E^A-recursive function by fixing one of its arguments to be p; it is denoted by $\{e\}^A(-, p)$.

The universal machine or enumeration scheme allows the proofs of the Kleene recursion theorems to apply to the E^A-recursive functions. In particular, a function which is defined by effective transfinite recursion is E-recursive. For example, the function mapping each ordinal δ to $L_\delta(A)$ is E^A-recursive. This method of defining E^A-recursive functions will be used freely.

One set y is E^A-recursive in another x if there is an e so that $\{e\}^A(x) = y$. A predicate B is E-reducible to A if there is a set p and an e so that $\{e\}^A(-, p)$ is the characteristic of B. This will, ambiguously, be written as $\{e\}^A(-, p) = B$. B is E^A-recursively enumerable if there is an e and a p so that B is the domain of $\{e\}^A(-, p)$.

Easily derived from the inductive defnition of $\{e\}^A(x) = y$ are the associated definitions of the recursively enumerable computation tree $T^A_{\langle e,x\rangle}$ and norm $\|\langle e, x\rangle\|^A$. Write $\{e\}^A(x)\downarrow$if there is a y so that $\{e\}^A(x) = y$ and $\{e\}^A(x)\uparrow$ otherwise. $\{e\}^A(x)\downarrow$ if and only if $T^A_{\langle e,x\rangle}$ is well founded. $T^A_{\langle e,x\rangle}$ will occasionally be referred to as the computation of $\{e\}^A$ at x.

The E-recursive functions can be examined on restricted domains. A set E-closed if it is closed under application of the E-recursive functions. The E-closure $E(x)$ of x is the smallest transitive E-closed set M with $x \in M$. In what follows, the structures of concern are the E-closed initial segments of L, Gödel's hierarchy of constructible sets. The predicates of interest will all be defined on the ordinal numbers.

1.2. L: As might be anticipated, there are advantages to working in L. In particular, the absoluteness of the E-recursive functions and the absoluteness of the generation of L blend together to allow the formation of Skolem hulls with the property that the E-recursive functions commute with the collapsing map.

DEFINITION 1.3. Let κ be an ordinal. Define ρ_κ to be the least ordinal γ so that there is a parameter p from L_κ such that every set in L_κ is E-recursive in p and an element of γ.

ρ_κ, the E-recursive projectum of L_κ, is the least ordinal γ so that there is a partial function which is E-recursive in a parameter from L_κ and maps a subset of γ onto L_κ. It is the E-recursion-theoretic analog of the Σ_1 projectum of κ.

PROPOSITION 1.4 (SACKS [9]). *Suppose that p is an element of L_κ and β is less than ρ_κ. Given any W, if W is an E-recursively in p enumerable subset of β, then W is an element of L_κ.*

Proposition 1.4 has a host of corollaries. In general, it says that there is an E-recursive indexing of L_κ via a subset of ρ_κ, any proper initial segment of which is an element of L_κ. Sacks used this indexing in order to solve Post's problem for L_κ in E-recursion. Another corollary of some use is that if p is an element of L_κ and γ is less than ρ_κ, then the set of sets which are E-recursive in p and some element of γ is an element of L_κ.

1.5. *Moschovakis witnesses.* The tree of a computation is well founded exactly when the computation is convergent.

DEFINITION 1.6. (i) A Moschovakis witness to the divergence of $\{e\}$ at x is an infinite descending path in $T_{\langle e,x\rangle}$.

(ii) An E-closed set M exhibits the Moschovakis phenomenon (MP) if, whenever $x \in M$ and $\{e\}(x)\uparrow$, there is a Moschovakis witness in M.

When M exhibits MP, the complete E-recursively enumerable subset of M is Δ_1 over M. When M is an initial segment of L the converse is also true as will be seen.

1.7. *Reflection.*

DEFINITION 1.8. Let x be a set.

(i) κ_0^x is the supremem of the ordinals which are E-recursive in x.

(ii) An ordinal α is x-reflecting if, for every $\Sigma_1(x)$ sentence φ,

$$L_\alpha[x] \models \varphi \Rightarrow L_{\kappa_0^x}[x] \models \varphi.$$

κ_r^x is the largest x-reflecting ordinal.

κ_r^x measures the amount of Σ_1-definability in $L[x]$ which is E-recursively enumerable. It also has other uses.

THEOREM 1.9 (KECHRIS BASIS THEOREM; SEE [3]). *Suppose B is a nonempty corecursively in p enumerable subset of x. There is a y in B so that $\kappa_r^{\langle x,p,y\rangle} \leqslant \kappa_r^{\langle p,x\rangle}$.*

1.10. *MP in L.* When x is an ordinal, κ_r^x can be used to define an element of a nonempty corecursively enumerable subset of x. This leads to a characterization of κ_r^x in terms of Moschovakis witnesses. This characterization is inspired by the analogous one due to Harrington in the context of recursion in objects of finite type.

THEOREM 1.11 (SACKS [9]). *Let x be a set of ordinals. If $\{e\}(x)\uparrow$, then there is a Moschovakis witness w in $L_{\kappa_r^x+1}[x]$ to this divergence. Moreover, w can be pointwise enumerated in $T_{\langle e,x\rangle}$ via computations of norm less than κ_r^x.*

COROLLARY 1.12. *Let L_κ be E-closed. Either L_κ exhibits MP or there is a γ below κ so that $\kappa_r^\gamma = \kappa$, L_κ is admissible, and the complete Σ_1 predicate on L_κ is E-recursively enumerable in γ.*

Corollary 1.12 can be interpreted as saying that it is always safe to assume that L_κ exhibits MP. Otherwise, E-recursive theory on L_κ is identical with admissible set theory on L_κ.

1.13. *The function $x \mapsto \kappa_r^x$.* By Proposition 1.4, if L_κ is E-closed, then it satisfies replacement for partial E-recursive functions evaluated on domains of size smaller than ρ_κ. In fact, if L_κ exhibits MP then it is closed under the function $x \to \kappa_r^x$ on small domains.

PROPOSITION 1.14 (SACKS [9]). *Suppose L_κ exhibits MP. Let β be less than ρ_κ and let p be a parameter from L_κ. The supremum of $\{\kappa_r^{\langle\gamma,p\rangle} | \gamma < \beta\}$ is strictly less than κ.*

The use of Proposition 1.14 in constructions is that not only can L_κ be approximated using ρ_κ but the complete E-recursively enumerable subset of L_κ can also be approximated in a way that is eventually verified as correct on initial segments (of the indexing via ρ_κ). With regard to a dynamic construction, it is acceptable to wait for fewer than ρ_κ computations to either converge or diverge before going on to the next phase of the construction.

2. *E*-recursively enumerable predicates on κ.

2.1. *E-reducibilities on L_κ.*

DEFINITION 2.2. A predicate $A \subseteq \kappa$ is E-recursive on L_κ (E-recursively enumerable on L_κ; abbreviated by E-r.e.(κ)) if there is a predicate A^* and a parameter p from L_κ so that A^* is E-recursive in p (E-recursively in p enumerable) and A is the restriction of A^* to L_κ.

Let L_κ be E-closed and let A be E-r.e.(κ). There is an immediate difficulty in using L_κ as a reservoir for E^A-computations: L_κ need not be E^A-closed. There is some flexibility in choosing how to interpret E-recursive reductions between predicates on L_κ. The definition given below is the most restrictive one possible; however, in the case of predicates which are E-r.e.(κ) the different possibilities all result in the same notion of degree.

DEFINITION 2.3. Let A and B be subsets of an E-closed ordinal κ. Say that $A \leqslant_\kappa B$ if there is a parameter p in L_κ and an e so that

(i) $B = \{e\}^A(-, p)$, and,

(ii) for all β below κ, $T^A_{\langle e,\beta,p\rangle}$ is an element of L_κ.

2.4. *Completeness.* In Definition 2.3 the only acceptable computations are those in L_κ. In 2.4, it will be shown that if L_κ exhibits MP and is not E^A-closed and A is E-r.e.(κ), then A is $\leqslant_\kappa$-complete. That is, every E-r.e.(κ) predicate can be computed from A using computations in L_κ. To start, simply intersecting with A may lead out of L_κ.

DEFINITION 2.5. A subset A of κ is regular if for all β below κ, $A \cap \beta$ is an element of L_κ.

LEMMA 2.6. *Let L_κ be E-closed. Suppose that A is E-r.e.(κ) and not regular. There is an ordinal γ below κ, $e \in \omega$ and a parameter p from L_κ so that*

$$\{e\}^A(-, p): \gamma \xrightarrow{\text{cofinally}} \kappa.$$

Moreover, for all β less than γ the set $\{T^A_{\langle e,\alpha,p\rangle} | \alpha < \beta\}$ is an element of L_κ.

PROOF. Since A is E-r.e.(κ), fix p in L_κ and a in ω so that A is the domain of $\{a\}(-, p)$ on L_κ. Let γ be the least ordinal δ so that $A \cap \delta$ is not an element of L_κ. Define the E^A-recursive in p function f by

$$f(\alpha) = \begin{cases} \|\langle a, \alpha, p\rangle\| & \text{if } \alpha \in A, \\ 0 & \text{otherwise.} \end{cases}$$

The range of f on γ must be unbounded in κ, or $A \cap \gamma$ would be an element of L_κ. f restricted to some β below γ is E-recursive in $A \cap \beta$ and p and so can be completely computed in L_κ.

PROPOSITION 2.7. *Suppose that L_κ is E-closed and exhibits MP. Suppose also that A is E-r.e.(κ) and L_κ is not E^A-closed. Then for any E-r.e.(κ) predicate B, $A \geqslant_\kappa B$.*

PROOF. If A is not regular, then Lemma 2.6 guarantees the existence of a γ below κ and a cofinal function f mapping γ into κ so that f is E^A-recursive in some parameter from L_κ. Moreover, the computation of the restriction of f to some β below γ from A lies in L_κ. If A happens to be regular, any A-computation which has norm κ can be converted into a function f with the above properties. Fix f and γ as above.

Suppose $B \subseteq \kappa$ is E-r.e.(κ). Let b be an index and q a parameter from L_κ so that B is the domain of $\{b\}(-, q)$ on L_κ. Given β below κ, A can compute to decide if $\beta \in B$ as follows. The method of computation is by means of an effective transfinite recursion of length γ. An auxiliary function g is defined with domain γ:

Step 1. Compute $f(1)$. Let $g(1) = f(1)$.

Step α. Let $g^{<\alpha} = \sup\{g(\alpha') | \alpha' < \alpha\}$. If $\|\langle b, \beta, q\rangle\| < g^{<\alpha}$ or there is a Moschovakis witness to $\{b'\}(\beta, q)\uparrow$ in $L_{g^{<\alpha}}$, all of whose points can be enumeratd onto $T_{\langle b,\beta,q\rangle}$ with norm below $g^{<\alpha}$, then let $g(\alpha) = g^{<\alpha}$. Otherwise, let $g(\alpha) = f(\alpha)$.

Let $g^{<\gamma}$ be equal to the supremum of $\{g(\alpha) | \alpha < \gamma\}$. There is an $\alpha < \gamma$ so that $g(\alpha') = g(\alpha)$ for all $\alpha' \geqslant \alpha$ as L_κ exhibits MP. The result of $\{b\}(\beta, q)$ is apparent from the vantage point $g^{<\gamma}$ by a witness one way or the other. The point is that B is Δ_1 over L_κ and A can E^A-recursively perform an unbounded search through L_κ;

any atomic fact about B can be computed from A. The computation from A lies in L_κ since it only uses some proper initial segment of f.

COROLLARY 2.8. *Suppose that L_κ is the E-closure of an ordinal δ and exhibits MP. If A is E-r.e.(κ), then A is complete if and only if L_κ is not E^A-closed.*

PROOF. Proposition 2.7 provides one direction of the corollary. For the other half of the equivalence, suppose that A is complete. Then A uniformly computes the universal E-recursively enumerable predicate U on δ using a parameter p from L_κ. Then $U \cap \delta$ is E^A-recursive in δ and p as a set. Since $U \cap \delta$ is not an element of L_κ, L_κ is not E^A-closed.

For those E-r.e.(κ) A which do not violate the E-closure of L_κ, all of the machinery of §1 which did not depend upon the absoluteness of L is available. In particular, if A is an incomplete E-r.e.(κ) predicate, then either $L_\kappa[A]$ exhibits MP or it is A-admissible and the E^A-recursively enumerable predicates on $L_\kappa[A]$ are closed under $\exists x \in L_\kappa[A]$. Surprisingly, much of the fine structure of L comes over to $L_K[A]$ when A is E-r.e.(κ) as will be shown in §3.

The next proposition shows that $\leqslant_\kappa$ induces a notion of degree on the E-r.e.(κ) predicates when L_κ exhibits MP.

PROPOSITION 2.9. *Suppose that L_κ is E-closed, exhibits MP and that A, B and C are E-r.e.(κ). If $A \geqslant_\kappa B$ and $B \geqslant_\kappa C$, then $A \geqslant_\kappa C$.*

PROOF. Suppose $A \geqslant_\kappa B$ and $B \geqslant_\kappa C$. If A is complete, then $A \geqslant_\kappa C$. Assume then that A is incomplete. By Proposition 2.7, L_κ is E^A-closed. So $B \restriction \gamma$ is uniformly E^A-recursive in γ using computations which come from L_κ. This condition is sufficient to compose the E-reductions from B to A and C to B and produce one from C to A.

2.10. $\equiv_\kappa$.

DEFINITION 2.11. Suppose that A and B are subsets of κ. $A \equiv_\kappa B$ if $A \leqslant_\kappa B$ and $A \geqslant_\kappa B$.

By the preceding results, $\equiv_\kappa$ is an equivalence relation on the E-r.e.(κ) predicates. The equivalence class of an E-r.e.(κ) predicate is called its degree.

By the results of 2.4 every representative of an incomplete degree is regular. In some cases, the natural representative of the complete degree is not regular. Even in these cases, there are regular members of the complete degree.

THEOREM 2.12. (SACKS; SEE [**12**]). *Suppose that L_κ is E-closed. There is a regular subset of κ which is E-r.e.(κ) and of complete degree.*

Regularity plays an important role in any dynamic construction in recursion theory. During a typical construction there are steps which cannot be executed until some given enumeration exhibits certain properties and other steps which must be executed whenever the enumeration exhibits other properties. If the enumeration is not regular then the first sort of step may never be attempted and the second sort may run amok.

3. Relativzation to an incomplete E-r.e.(κ) predicate.

3.1. $L_\kappa[D]$: The main theorem of this paper, proven in §4, is that any E-r.e.(κ) degree C splits into two E-r.e.(κ) degrees over any smaller one D. Much of the work has to do with the analysis of the functions which are E^D-recursive. The global and combined properties of these functions need be well understood in order to first design the strategy to satisfy a single requirement in the construction and second to combine the strategies to satisfy all of the requirements.

As mentioned at the conclusion of 2.4, some of the use of Skolem hulls in L_κ can be adapted to $L_\kappa[D]$ when D is E-r.e.(κ). Let ρ_κ^D be the E^D-recursive projectum of L_κ as in Definition 1.3. The proof of the next proposition closely follows an argument of Shore [**10**] in α-recursion theory.

PROPOSITION 3.2 (SACKS). *Suppose that L_κ is E-closed and D is E-r.e.(κ) and regular. If γ is smaller than ρ_κ^D and $W \subset \gamma$ is E^D-recursively enumerable on L_κ, then W is an element of L_κ.*

PROOF. Fix γ, e and p from L_κ so that $W = \{\beta| < \gamma \text{ and } \{e\}^D(\beta, p)\downarrow\}$ is not an element of L_κ. Also, fix d an index and q from L_κ so that D is the domain of $\{d\}(-, q)$ on L_κ. It remains to be seen that there is an E^D-recursive partial function from γ onto L_κ.

For simplicity, assume that γ is a cardinal in L_κ; if not, then there is a smaller ordinal γ', the L_κ-cardinality of γ, which can be used in its place. This assumption is used solely to insure that γ is closed under the Gödel (E-recursive) pairing function. It is also safe to assume that p and q are ordinals since any element of L_κ is E-recursive in some ordinal.

Let $\mathfrak{H} = \{y\,|(\exists e' \in \omega)(\exists \beta < \gamma)\{e'\}^D(\beta, p, q, \gamma) = y\}$. $\mathfrak{H}$ is the Skolem hull of $\gamma \cup \{p, q, \gamma\}$ formed using the universal partial E^D-recursive Skolm function. Since every element in $\mathfrak{H}$ is hereditarily constructed from ordinals in an E^D-recursive way, every element of $\mathfrak{H}$ is well ordered in $\mathfrak{H}$. This coupled with the fact that $\mathfrak{H}$ is E-closed implies that $\mathfrak{H}$ is extensional. Let $\overline{\mathfrak{H}}$ be the transitive collapse of $\mathfrak{H}$; let $\bar{\kappa}$ be the ordinal height of $\overline{\mathfrak{H}}$; let π: $\mathfrak{H} \to \overline{\mathfrak{H}}$ be the collapsing map. A straightforward transfinite recursion shows that π commutes with any E-recursive function. That is, if $h \in \mathfrak{H}$, then $\{e'\}(h)\downarrow$ if and only if $\|\langle e', \pi(h)\rangle\| < \bar{\kappa}$. In particular, $\overline{D}$, the pointwise image of D under π, is equal to the domain of $\{d\}(-, \pi(q))$ on $\overline{\mathfrak{H}}$. A similar transfinite recursion shows that, for any e' and h, $\{e'\}^D(h)\downarrow$ if and only if $\|\langle e', \pi(h)\rangle\|^{\overline{D}} < \bar{\kappa}$. Since γ is a subset of $\mathfrak{H}$, π is the identity on all elements of γ; if $\beta < \gamma$, then $\beta \in W$ if and only if $\|\langle e, \beta, \pi(p)\rangle\|^{\overline{D}} < \bar{\kappa}$. If $\bar{\kappa} < \kappa$, then $\overline{D}$ and hence W are elements of L_κ, an impossibility. Thus, $\bar{\kappa} = \kappa$.

In $\overline{\mathfrak{H}}$ there is an $E^{\overline{D}}$-partial recursive function mapping γ onto $\overline{\mathfrak{H}}$: the universal one (as used to form $\mathfrak{H}$ E^D-recursively from γ) using parameters γ, $\pi(p)$ and $\pi(q)$. Since $\overline{\mathfrak{H}}$ has height κ the range of this function includes all of L_κ. It need only be shown that D can approximate the $\overline{D}$ surjection with only γ many steps.

First, it will be shown that $\overline{D}$ is regular and hence $\overline{\mathfrak{H}} = L_\kappa$. Let β be an ordinal less than κ. Since $\kappa = \bar{\kappa}$, there is a β' so that $\beta = \pi(\beta')$. D is regular so

$D \cap \beta' \in L_\kappa$. Let α_1 be the least ordinal so that $D \cap \beta' \in L_{\alpha_1}$ and let α_2 be the location of $D \cap \beta'$ in the canonical well-ordering of L_{α_1}. As L_κ is E-closed, α_2 is an element of L_κ; as $\mathfrak{H}$ is E-closed and cofinal in L_κ, both α_1 and α_2 are elements of $\mathfrak{H}$. $\mathfrak{H}$ satisfies the sentence that $(\forall\alpha < \beta')[\{d\}(\alpha, q)\downarrow$ if and only if α is an element of the α_2th member of $L_{\alpha_1}]$. In $\overline{\mathfrak{H}}$, $(\forall\alpha < \beta)[\{d\}(\alpha, \pi(q))\downarrow$ if and only if α is an element of the $\pi(\alpha_2)$th member of $L_{\pi(\alpha_1)}]$. Thus, $\overline{D} \cap \beta \in L_\kappa$ since $\overline{D}$ is the domain of $\{d\}(\alpha, \pi(q))$.

For each δ below κ, define $\overline{D}_\delta$, the stage δ approximation of $\overline{D}$, by

$$\beta \in \overline{D}_\delta \quad \text{if and only if} \quad \beta < \delta \text{ and } \|\langle d, \beta, \pi(q)\rangle\| < \delta.$$

Let u be an index for the universal $E^{\overline{D}}$-partial recursive function. Define the function $g\colon \gamma \times \gamma \xrightarrow{\text{onto}} L_\kappa$ as follows:

$$g(\alpha, \beta) \cong \begin{cases} \{u\}^{\overline{D}_\delta}(\beta, \gamma, \pi(p), \pi(q)) & \text{if } \|\langle e, \alpha, p\rangle\|^D = \delta \text{ and} \\ & \{u\}^{\overline{D}_\delta}(\beta, \gamma, \pi(p), \pi(q))\downarrow, \\ \text{undefined otherwise.} & \end{cases}$$

g is E^D-recursive on L_κ by way of its definition. Suppose that ν is some ordinal less than κ. Since the universal $E^{\overline{D}}$-recursive function in γ, $\pi(p)$, $\pi(q)$, maps γ onto L_κ in L_κ, fix β below γ so that

$$\{u\}^{\overline{D}}(\beta, \gamma, \pi(p), \pi(q)) = \nu \quad \text{with } \|\langle u, \beta, \gamma, \pi(p), \pi(q)\rangle\|^{\overline{D}} < \kappa.$$

Since $\overline{D}$ is regular and L_κ is E-closed, fix σ below κ so that, for all $\sigma' \geqslant \sigma$, $\overline{D}$ and $\overline{D}_{\sigma'}$, agree below $\|\langle u, \beta, \gamma, \pi(p), \pi(q)\rangle\|^{\overline{D}}$. As W is not an element of L_κ, fix below γ so that $\|\langle e, \alpha, p\rangle\|^D > \sigma$. Then, $g(\alpha, \beta) = \nu$.

The existence of g implies that $\gamma \geqslant \rho_\kappa^D$, the desired result.

3.3. $t\sigma 1p^D(\kappa)$. The other definitions of §1 (Moschovakis witness, MP,...) lead directly to sensible definitions relative to D. When D is incomplete and E-r.e.(κ), the remaining propositions of §1 are true of D as well.

Unfortunately, in the forthcoming construction, not all of the functions considered are E^D-recursive. The Σ_1^D definable over L_κ functions come into play.

Definition 3.4. (i) The Σ_1^D projectum of κ, $\sigma 1p^D(\kappa)$, is the least γ so that there is a function mapping a subset of γ onto $L_\kappa^{[D]}$ which is Σ_1^D definable over $L_\kappa[D]$.

(ii) Suppose that f is a Σ_1^D function with domain contained in γ defined by

$$f(x) = y \quad \text{if and only if } L_\kappa[D] \models \varphi(x, y),$$

where φ is Σ_1^D. f is tamely defined on γ if for every δ less than γ there is a β less than κ so that

$$(\forall x < \delta)\{(\exists y)[f(x) = y] \text{ if and only if } (\exists y \in L_\beta[D])[L_\beta[D] \models \varphi(x, y)]\}.$$

(iii) The tame Σ_1^D projection of κ, $t\sigma 1p^D(\kappa)$, is the least γ so that there is a function mapping a subset of γ onto $L_\kappa[D]$ which is Σ_1^D definable over $L_\kappa[D]$ and tamely defined on γ.

The advantage to working with a tamely defined function is that it can be approximated by a recursive sequence of recursive functions which are eventually constant on any initial segment of the final domain. The use of Σ_1 and Σ_2 projecta and of tameness goes back to α-recursion theory; the particular example of $t\sigma 1p(\kappa)$ was introduced by Maass [**4**].

3.5. *Sacks preservation*. The salient feature of the proof of splitting and density in E-recursion which transcends the contruction of two incomparable degrees is the use of the Sacks preservation strategy from classical recursion theory [**8**]. In its purest form the strategy is designed to insure that a nonrecursive set A under construction does not compute a given recursively enumerable set C. The strategy to insure $\{e\}^A \neq C$ works as follows. While enumerating C, via a given procedure, build A to be nonrecursive. However, globally control the construction of A so that for each n, if $\{e\}^A \upharpoonright n = C \upharpoonright n$ appears during any phase of the construction, then for every later stage the construction is not allowed to change A at any number needed to compute $\{e\}^A \upharpoonright n$. If $\{e\}^A = C$, then C can be computed recursively; $C \upharpoonright n$ is equal to the first common value of $\{e\}^A \upharpoonright n$ and $C \upharpoonright n$ seen during the construction. So, eventually some n is found so that $\{e\}^A(n)\uparrow$ or $\{e\}^A(n) \neq C(n)$. The new compliction, beyond the Friedberg-Muchnik construction, is that n depends upon C.

3.6. *Inequalities*. In E-recursion theory the Moschovakis phenomenon can be used to make the Sacks strategy always have an existential resolution. This simplifies the definition of the function, taking a strategy for $\{e\}^A \neq C$ to an argument n, where the strategy succeeds in making $\{e\}^A(n) \not\simeq C(n)$. This function becomes Σ_1 instead of recursive in $0'$.

DEFINITION 3.7. Suppose L_κ is E-closed and exhibits MP. If C is E-r.e.(κ) and equal to the domain of $\{c\}(-, p)$, and A is a regular predicate on κ, then a witness to $\{e\}^A(-, q) \neq C$ is any one of the following:

(i) a Moschovakis witness w to the divergence of $\{e\}^A(\beta, q)$ for some β below κ and an ordinal γ so that γ is an upper bound on the norms of the computations enumerating w pointwise into $T^A_{\langle e,\beta,p\rangle}$ ($\langle \omega, \gamma\rangle$ witnesses that $\{e\}^A(-, q)$ is not total);

(ii) for some β below κ, a pair of computation trees $T_{\langle c,\beta,p\rangle}$ and $T^A_{\langle e,\beta,q\rangle}$ and an ordinal γ which is greater than their heights so that $\{e\}^A(\beta, q) = 0$. ($\langle T_{\langle c,\beta,p\rangle}, T^A_{\langle e,\beta,q\rangle}, \gamma\rangle$) witnesses that $\beta \in C$ but $\{e\}^A(\beta, q) = 0$ saying that $\beta \notin C$);

(iii) for some β below κ, a Moschovakis witness w to $\{c\}(\beta, p)\uparrow$, a well-founded computation tree $T^A_{\langle e,\beta,q\rangle}$ so that $\{e\}^A(\beta, q) = 1$ and an ordinal γ bounding the height of $T^A_{\langle e,\beta,q\rangle}$ and the norms of the computations enumerating w pointwise into $T_{\langle c,\beta,p\rangle}$ ($\langle w, T^A_{\langle e,\beta,q\rangle}, \gamma\rangle$ witnesses that $\beta \notin C$ but $\{e\}^A(\beta, q) = 1$, saying that $\beta \in C$).

The ordinal γ, called the ordinal of a witness, is large enough to decide whether a purported witness is one in truth. Thus, the property of being a witness to $\{e\}^A(-, q) \neq C$ is E^A-recursive on $L_\kappa[A]$. If $L_\kappa[A]$ exhibits MP, C is E-r.e.(κ), and $\{e\}^A(-, q) \neq C$, then there is a witness to the inequality in $L_\kappa[A]$.

Using inequality witnesses, satisfying one requirement $\{e\}^A(-, q) \neq C$ during the construction of A is even more direct than in the classical setting. Simply restrict any new members from entering A until a witness is found to $\{e\}^A(-, q) \neq C$. If A is preserved below the oridnal of this witness, then the requirement is satisfied.

3.8. δ_C^D. The final problem, solved by using $t\sigma 1p^D(\kappa)$ to index requirements, is how to arrange a priority ranking of strategies so that every initial segment of it finds a complete set of witnesses at some bounded point in the construction.

DEFINITION 3.9. Let L_κ be E-closed and exhibit MP. Let C be E-r.e.(κ) and regular. If D is contained in κ let δ_C^D be defined by

$$\begin{aligned}\delta_C^D = {} & \text{the least } \delta \leqslant \kappa \text{ so that there is a } p \text{ in } L_\kappa \text{ and an index } e \\ & \text{such that the set } \left\{\beta \mid (\exists\gamma < \delta)(\forall\beta' < \beta)\left[C(\beta') = \{e\}^D(\beta', \gamma, p)\right]\right\} \\ & \text{is cofinal in } \kappa.\end{aligned}$$

δ_C^D is the longest order type of requirements so that the preservation strategy can work to build a predicate E^D-recursively enumerable on $L_\kappa[D]$ and above D which meets the requirements. Notice that $\delta_C^D \leqslant \rho_\kappa^D$ by the reqularity of C. If $D \geqslant_\kappa C$, then $\delta_C^D = 1$. The main result of this section is that if D is E-recursively enumerable on L_κ and $D \ngeqslant_\kappa C$, then $\delta_C^D \geqslant t\sigma 1p^D(\kappa)$. Recall a result from α-recursion theory (adapted to the present context).

PROPOSITION 3.10 (SHORE [**10**]). *Let L be E-closed and exhibit MP. Let $D \subseteq \kappa$ be regular and E-recursively enumerable on L_κ. If γ is less than $\sigma 1p^D(\kappa)$ and W is a Σ_1^D over $L_\kappa[D]$ subset of γ, then $W \in L_\kappa$.*

The proof of Proposition 3.10 involves forming Σ_1^D Skolem hulls in L_κ. The construction is essentially the same as in the proof of Proposition 3.2 with the universal partial E^D-recursive function replaced by the universal Σ_1^D one.

The next lemma says that, under the appropriate hypotheses, if a set I (of indices for E-reducibilities on L_κ) has size smaller than δ_C^D, then there is a set in L_κ including a witness for each element of I that it does not reduce C to D. The important application is that during a construction all of these witnesses appear at some ordinal strictly smaller than κ.

LEMMA 3.11. *let L_κ be E-closed and exhibit MP. Let C and D be regular and E-r.e.(κ). Suppose $D \ngeqslant_\kappa C$. Then, for any p in L_κ, index e, and γ below δ_C^D, there is a β below κ so that for all γ' less than γ there is a w in $L_\beta[D]$ which is a witness to $\{e\}^D(-, \gamma', p) \neq C$.*

PROOF. Let p and e be fixed and let $\{c\}(-, q)$ enumerate C. Since γ is strictly less than δ_C^D, there is a β_1 below κ so that for all γ' less than γ there is an x below β_1 so that $\{e\}^D(x, \gamma', p) \neq C(x)$. This is by appeal to the definition of δ_C^D. Since C is regular and L_κ is E-closed, fix a β_2 below κ so that, for all x below β_1, if $x \in C$ then $\|\langle c, x, q\rangle\| < \beta_2$. Let β_3 be the supremum of $\{\kappa_0^{\gamma', p, q, p, \beta_1, \beta_2; D} \mid \gamma' < \gamma\}$. β_3 is strictly smaller than κ as $\gamma < \delta_C^D \leqslant \rho_\kappa^D$ and 1.2 is relevant. Let β_4 be the supremum of $\{\kappa_r^{\gamma',p,q,\beta_1,\beta_2,\beta_3;D} | \gamma' < \gamma\}$; β_4 is smaller than κ by Proposition 1.14 relativized to D using Proposition 3.2. $\beta_4 + \omega$ is sufficiently large to construct the necessarry witnesses. Let γ' be a fixed ordinal which is less than γ. There are three cases corresponding to the three possible ways in which $\{e\}(-, \gamma', p) \neq C$.

Case 1. There is an x less than β_1 so that $\{e\}^D(x, \gamma', p)\uparrow$. By the Kechris Basis Theorem 1.9, there is such an x so that $\kappa_r^{x,\beta_1,\gamma',p;D} \leqslant \kappa_r^{\beta_1,\gamma',p;D} \leqslant \beta_4$. A Moschovakis witness to this divergence is constructed in $L_{\beta_4+1}[D]$ and the ordinal β_4 is sufficient to verify the relevant E^D-recursively enumerable facts by Theorem 1.11. Thus, there is a clause (1) witness to $\{e\}^D(-, \gamma', p) \neq C$ in $L_{\beta_4+\omega}[D]$.

Case 2. $\{e\}^D(-, \gamma', p)$ is total on β_1 but, for some x below β_1 which is an element of C, $\{e\}^D(x, \gamma', p) = 0$. Fix x_0 to be such an x. Since $\{e\}^D(-, \gamma', p)$ is total on β_1, the supremum of $\{\|\langle e, x, \gamma', p\rangle\|^D | x < \beta_1\}$ is E^D-recursive in γ', p and β_1, and so is less than $\kappa_0^{\beta_1,\gamma',p;D} < \beta_4$. By the choice of β_2, $\|\langle c, x_0, q\rangle\| < \beta_4$. The two conflicting computations are present in $L_{\beta_4}[D]$ and together with the ordinal β_4 constitute a clause (2) witness to $\{e\}^D(-, \gamma', p) \neq C$.

Case 3. $\{e\}^D(-, \gamma', p)$ is total on β_1 but there is an x below β_1 so that x is not an element of C and $\{e\}^D(x, \gamma', p) \neq 0$. Let G be the subset of β_1: $\{x | x < \beta_1$ and $\{e\}^D(x, \gamma', p) \neq 0\}$. G is recursive in β_1, γ' and p. Let $\mathfrak{H}$ be the subset of G which lies in the complement of C. $\mathfrak{H}$ is co-E^D-recursively enumerable in the parameters β_1, γ', p and q and is not empty by assumption. The Kechris Basis Theorem 1.9 implies that $\mathfrak{H}$ has a member, call it x_0, so that

$$\kappa_r^{x_0,\gamma',p,q,\beta_1;D} \leqslant \kappa_r^{\gamma',p,q,\beta_1;D} \leqslant \beta_4.$$

By Theorem 1.11, there is a Moschovakis witness w to $\{c\}(x_0, q)\uparrow$ in $L_{\beta_4+1}[D]$ which can be pointwise verified to lie in $T^D_{\langle c,x_0,q\rangle}$ using β_4. $T^D_{\langle e,x_0,\gamma',p\rangle}$, w and β_4 constitute a clause (3) witness to $\{e\}^D(-, \gamma', p) \neq C$.

3.12. $\delta_C^D \leqslant \tau\sigma 1 p^D(\kappa)$.

LEMMA 3.13. *With the same hypotheses as in the previous lemma, δ_C^D is greater than or equal to $\sigma 1 p^D(\kappa)$.*

PROOF. Suppose that $\delta_C^D < \sigma 1 p^D(\kappa)$. Let e be the index and p the parameter which appear in the definition of δ_C^D. Define a Σ_1^D function f: $\delta_C^D \xrightarrow{\text{cofinally}} \kappa$ using p as follows:

$f(\gamma) = \beta$ if and only if for all γ' below γ there is a witness in $L_\beta[D]$ to $\{e\}^D(-, \gamma', p) \neq C$ and also for all β' below β there is a γ' below γ so that $\{e\}^D(-, \gamma', p) \neq C$ is not witnessed by any element of $L_{\beta'}[D]$.

f is Σ_1^D in p and q and, by preceding lemma, defined for all elements γ of δ_C^D. (q is used to define witness to inequality.) f applied to δ_C^D is cofinal in κ as the length of equality between $\{e\}^D(-, \gamma, p)$, and C is cofinal in κ as γ ranges over δ_C^D.

Define a set G as follows:

$$G = \left\{ \begin{array}{l|l} \langle \gamma, \gamma_1 \rangle & \text{There is an } x \text{ below } f(\gamma_1) \text{ so that} \\ & \{e\}^D(x, \gamma, p) \neq C(x) \text{ is witnessed in} \\ & L_\kappa[D];\ \gamma, \gamma_1 \leqslant \delta_C^D \end{array} \right\}$$

G is Σ_1^D in the parameters δ_C^D, p and q. Since $\delta_C^D < \sigma 1 p^D(\kappa)$, Shore's Proposition 3.10 implies that G is an element of L_κ. However, G can be used to E-reduce C to D.

To compute if $x \in C$, first compute the least γ_1 so that $f(\gamma_1) > x$. The value of γ_1 can be read off from $L_x[D]$ which is E^D-recursive in x. From G, let γ be the least ordinal so that $\langle \gamma, \gamma_1 \rangle \notin G$. There must be such a γ by the definition of δ_C^D and it must be the case that $\{e\}^D(x, \gamma, p) = C(x)$. The crucial fact is that G can be used to select a γ so that $\{e\}(-, \gamma, p)$ agrees with C on a large enough domain to compute $C(x)$.

By assumption, $D \not\geq_\kappa C$ so it must be the case that $\delta_C^D \geqslant \sigma 1 p^D(\kappa)$.

PROPOSITION 3.14. *With the same assumptions as in the previous two lemmas,* δ_C^D *is greater than or equal to* $t\sigma 1 p^D(\kappa)$.

PROOF. If δ_C^D is equal to κ, then the proposition is true. Otherwise, δ_C^D is a cardinal in the sense of L_κ and so is closed under the usual Gödel pairing function. Recall that this function, $\langle\, , \rangle$, has the property that, for all δ_1 and δ_2, $\langle \delta_1, \delta_2 \rangle$ is greater than either δ_1 or δ_2.

A tame Σ_1^D function from δ_C^D onto L_κ can be defined as follows. By the preceding proposition, let g be a partial Σ_1^D function mapping δ_C^D onto κ. Also, let p be the parameter and e the index in the definition of δ_C^D. Define f by

$$f(\delta) = \begin{cases} g(\delta_1) & \text{if } \delta = \langle \delta_1, \delta_2 \rangle \text{ and there is a } \delta \leqslant \delta_2 \text{ so that} \\ & \text{there is no witness to } \{e\}^D(-, \delta, p) \neq C \text{ before,} \\ & \text{the least witness to } g(\delta_1) = y \text{ for some } y, \\ 0 & \text{otherwise.} \end{cases}$$

f is Σ_1^D since g is. f is tame since the restrictions of f to γ below δ_C^D is determined at the least ordinal β, where witnesses are available for all γ' below γ to the effect that $\{e\}^D(-, \gamma', p) \neq C$. By Lemma 3.11, β is strictly smaller than κ. Finally, fixing α below κ, let δ_1 be an ordinal below δ_C^D so that $g(\delta_1) = \alpha$. By the definition of δ_C^D, there is a δ_2 below δ_C^D so that the least witness to $\{e\}^D(-, \delta_2, p) \neq C$ has ordinal greater than the ordinal needed to construct a witness to $g(\delta_1) = \alpha$. Thus, $f(\langle \delta_1, \delta_2 \rangle) = \alpha$.

This final proposition concludes the development of the technical underpinnings needed to organize a successful priority construction using the Sacks preservation strategy.

4. Splitting and density.

THEOREM 4.1. *Let* L_κ *be E-closed and exhibit MP. For any regular subsets of* κ, C *and* D, *which are E-r.e.*(κ) *if* $C >_\kappa D$, *there are E-r.e.*(κ) *sets* A *and* B *so that*

(i) $A \cup B = C$, $A \cap B = \varnothing$;

(ii) $A \oplus D <_\kappa C$, $B \oplus D <_\kappa C$

$$\big(A \oplus D = \{\langle 0, x \rangle | x \in A\} \cup \{\langle 1, x \rangle | x \in D\}\big).$$

Let κ, C and D be fixed for the remainder of §4 as in the hypothesis of the theorem; the sets A and B are defined by means of a finite injury priority construction. The requirements are indexed and assigned priority by $t\sigma 1 p^D(\kappa)$. The tameness of the projection of κ into $t\sigma 1 p^D(\kappa)$ and the regularity of D together insure that the assignment of priority can be E-recursively approximated. Lemma 3.11 will be used to show that the activity for the sake of any initial segment of the priority listing of requirements will be finished by some bounded point of the construction.

4.2. *The requirements and atomic strategies.* Adopt the notational convention that script letters denote requirements and Roman letters the associated strategies. The requirements in constructing A and B are two sorts:

$\mathfrak{P}$: For all x, if $x \in C$ then $x \in A$ or $x \in B$ but not both.

$\mathfrak{N}^A_{\langle e,p\rangle}, \mathfrak{N}^B_{\langle e,p\rangle}$: $\{e\}^{A\oplus D}(-, p) \neq C$; $\{e\}^{B\oplus D}(-, p) \neq C$.

The requirement $\mathfrak{P}$ will be satisfied in the usual way. During a stage σ of the construction, some ordinal x may be enumerated in C (via $\|\langle c, x, q\rangle\| = \sigma$ if C is the domain of $\{c\}(-, q)$). Then x will be enumerated into exactly one of A or B during stage σ. The choice as to which of these two predicates x enters will depend upon the other (negative) strategies of the construction.

The atomic strategy $N^A_{\langle e,p\rangle}$ to satisfy one of the negative requirements, $\mathfrak{N}^A_{\langle e,p\rangle}$, is the one discussed in 3.6. $N^A_{\langle e,p\rangle}$ keeps all ordinals out of A until a stage σ when a witness to $\{e\}^{A_\sigma\oplus D_\sigma}(-, p) \neq C$ is found; A_σ and D_σ are the stage σ approximations of A and D, namely, the set of ordinals less than σ enumerated in A or D via computations of norm less than σ. This witness is preserved by not allowing any ordinals less than β, the ordinal of the witness, to enter A. Some ordinal less than β may enter D in which case $N^A_{\langle e,p\rangle}$ again restrains A. If no permanent witness were ever found so $C \leqslant_\kappa A \oplus D$, then $A \leqslant_\kappa D$ (since D must permit ordinals to enter A) and so $C \leqslant_\kappa D$, an absurdity. Thus some real witness is eventually discovered and $N^A_{\langle e,p\rangle}$ has only a bounded effect on the construction.

4.3. *Combining strategies via blocks.* Given an index e and a parameter p from L_κ, the least witness to $\{e\}^D(-, p) \neq C$ was calculated in §3 to be constructed reasonably close to the least argument x so that $\{e\}^D(x, p) \neq C(x)$. However, there is no a priori bound on the least x which works for e and p. For example, there is no a priori bound on the least element of C.

This poses an impediment to combining the strategies N^A_- and N^B_- by ordering them sequentially. Namely, an order type less than δ^D_C sequence $N^A_{\langle e_1,p_1\rangle}$, $N^B_{\langle e_1,p_1\rangle}$, $N^A_{\langle e_2,p_2\rangle}$, $N^B_{\langle e_2,p_2\rangle}, \ldots$ could conspire to restrain all the ordinals below κ from entering A or B. The defining feature of δ^D_C was, in effect, that if $\gamma < \delta^D_C$ then a set of strategies $\{N^A_{\langle e,p_\delta\rangle} | \gamma\}$ would simultaneously stop acting below κ. The solution then is to arrange the strategies in blocks as in Shore's infinite injury method in α-recursion theory (Shore **[10]**). The sequence of blocks will be short enough to insure that all the strategies in any initial segment of blocks eventually stop acting; the blocks will be large enough that every requirement has a strategy in all but boundedly many of the blocks.

A typical scenario is as follows: At at stage β a block B^A of negative A strategies is given highest priority for all ordinals greater than or equal to β. No element greater than β is allowed to enter A (they all enter B) until a stage β' greater than β when every strategy $N^A_{\langle e,p\rangle}$ in B^A has established a witness to the inequality $\{e\}^{A\oplus D}(-, p) \neq C$. There will be such a stage if B^A has less than δ^D_C many elements. If no injury occurred, then each requirement $\mathfrak{N}^A_{\langle e,p\rangle}$ mentioned by a strategy in B^A would be satisfied by way of one of the witnesses. Moreover, above β' another block of strategies B^B could be put into effect causing elements greater than β' to enter A instead of B.

Continuing the scenario, injury can occur in one of two ways. The less harmful to B^A is if an ordinal between β and β' is enumerated in D and eliminates an inequality witness. The strategies in B^A take over again, keeping elements from entering A until a new witness is discovered. More dramatic injury occurs to B^A when an ordinal less than β enters C. It must enter either A or B and will effect some block which has higher priority than B^A, perhaps even causing B^A to be cancelled (as B^B was when B^A took over above). By defining the blocks simultaneously with constructing A and B, the blocks also interact as in a finite injury construction.

4.4. *Nomenclature.* A negative block B^A_β or B^B_β will be a set of strategies $N^A_{\langle e,p\rangle}$ or $N^B_{\langle e,p\rangle}$ respectively. The block B^A_β acts so as to keep all elements out of A above β. B^A_β is satisfied at β' during stage σ if for each $N^A_{\langle e,p\rangle}$ in B^A_β there is a witness to $\{e\}^{A_\sigma\oplus D_\sigma}(-, p) \neq C$ with ordinal less than β'. If B^A_β was satisfied at β' during stage σ, then B^A_β is injured at stage $\sigma+1$ if $(A_{\sigma+1} \oplus D_{\sigma+1}) \cap \beta' \neq (A_\sigma \oplus D_\sigma) \cap \beta'$. The B versions are analogous. Also, let q_1 and q_2 be parameters from L_κ so that $\{c\}(-, q_1)$ and $\{d\}(-, q_2)$ enumerate C and D, respectively.

4.5. *The construction.* During the course of the construction, various blocks will be activated. An A-block B^A_β, activated during stage β, will constitute a negative condition on A in $[\beta, \beta')$ where β' is the least ordinal at which B^A_β is satisfied. As long as B^A_β remains in effect all of the ordinals from $[\beta, \beta')$ which enter C are enumerated into B. At stage α, the active blocks will continuously divide α into disjoint intervals; each interval will be controlled by exactly one block of either the A or B persuasion.

For convenience, assume for each β below κ that there is at most one x so that $\|\langle c, x, q_0\rangle\| = \beta$ or $\|\langle d, x, q_1\rangle\| = \beta$. Also, assume that, for all x below κ and all β, if $\|\langle c, x, q_0\rangle\| = \beta$ or $\|\langle d, x, q_1\rangle\| = \beta$ then β is a successor ordinal greater than x. there is no loss in generality since any enumerations of C and D can be mildly perturbed so as to have the desired properties on L_κ. (Note that this fact uses the E-closure of L_κ.)

Let f be a tame Σ^D_1 function mapping a subset of $t\sigma 1p(\kappa)$ onto L_κ. Suppose $f(x) = y$ is defined by $f(x) = y$ if and only if $\exists\beta\varphi(x, y, \beta, D \cap \beta)$, where φ is Δ_0 on L_κ. For each β below κ define $P^\beta(x) = y$ if and only if there is a $\beta' < \beta$ and y in $L_\beta[D_\beta]$ so that $\varphi(x, y, \beta', D_\beta \cap \beta')$. P^β is the stage β approximation of f. Recall D_β is the stage β approximation of D. Fix p to be the parameter mentioned in φ.

The predicates A and B are defined in stages, one for each ordinal below κ. A_β and B_β are the subsets of A and B which have been enumerated by the end of stage β. The activity during a stage has two phases: first, define the sequence of active blocks for stage β; second, enumerate any element of C which entered C via a computation of norm β into one of A or B.

Stage 0. Activate an A-block $B^A = \{N^A_{\langle 0,0\rangle}\}$.

Stage $\alpha + 1$. Let $\langle B_{\beta_{\gamma'}} | \gamma' < \gamma\rangle$ be the sequence of active blocks in increasing order of subscript at the end of stage α. Inductively, $\langle \beta_{\gamma'} | \gamma' < \gamma\rangle$ is a continuous sequence of ordinals and γ is a successor ordinal. If $\gamma' < \gamma$, then $B_{\beta_{\gamma'}}$ is an A-block if γ' is even (i.e., a limit or an even successor of a limit) and $B_{\beta_{\gamma'}}$ is a B-block of γ' is odd. Moreover, if $\gamma'' < \gamma' < \gamma$, then $B_{\beta_{\gamma''}}$ was satisfied at $\beta_{\gamma''+1}$ during stage α. The stage $\alpha + 1$ of the construction takes place in several steps.

Step 1. If there is an element ν_0 of $D_{\alpha+1} - D_\alpha$ let γ' be the least ordinal less than γ so that $\nu_0 \geqslant \beta_{\gamma'}$. Cancel any block $B_{\beta_{\gamma''}}$, where $\gamma'' > \gamma'$ if any such exists in the sequence. Go on to Step 2.

Step 2. If there is an element ν_1 of $C_{\alpha+1} - C_\alpha$ let γ' be the greatest ordinal less than γ so that $\beta_{\gamma'} \leqslant \nu_1$ and $B_{\beta_{\gamma'}}$ is still active after Step 1. If $B_{\beta_{\gamma'}}$ is an A-block let $A_{\alpha+1} = A_\alpha$ and $B_{\alpha+1} = B_\alpha \cup \{\nu_1\}$. If $B_{\beta_{\gamma'}}$ is a B-block let $A_{\alpha+1} = A_\alpha \cup \{\nu_1\}$ and $B_{\alpha+1} = B_\alpha$. This action will injure any blocks of lower priority than $B_{\beta_{\gamma'}}$; all but the immediate next block are cancelled. $B_{\beta_{\gamma'}}$ is not injured since it managed to preserve its associated set A or B. Cancel all blocks $B_{\beta_{\gamma''}}$ if $\gamma'' > \gamma' + 1$. Continue with Step 3.

Step 3. If no such ν_1 exists let $A_{\alpha+1} = A_\alpha$ and let $B_{\alpha+1} = B_\alpha$.

Step 3 concludes Stage $\alpha + 1$.

Stage λ (*a limit*). The limit stages of the construction are used solely for the activation of blocks. Let $\langle B_{\beta_{\gamma'}} | \gamma' < \gamma\rangle$ be the sequence of blocks which were activated at stages $\beta_{\gamma'} < \lambda$ and were not cancelled during any stage below λ.

Step 1. Let $A_\lambda = \bigcup_{\alpha<\lambda} A_\alpha$ and $B_\lambda = \bigcup_{\alpha<\lambda} B_\alpha$.

Step 2. If each $B_{\beta_{\gamma'}}$ with $\gamma' < \gamma$ is satisfied at an ordinal less than λ (in fact, at $\beta_{\gamma'+1}$) during Stage λ, then activate a block B_λ as follows.

Let β be defined by

$$\beta = \sup\Big\{\beta' < t\sigma 1 p(\kappa) \,|\, (\exists\gamma' < \gamma)(\forall\gamma'' < \gamma) \cdot \left[\begin{matrix}(\gamma'' \geqslant \gamma' \wedge \gamma'' \text{ even} \Leftrightarrow \gamma \text{ even}) \Rightarrow \\ (\forall\beta'' < \beta')(P^{\beta_{\gamma''}}(\beta'') \cong P^\lambda(\beta''))\end{matrix}\right]\Big\}.$$

(a) If $\beta = t\sigma 1 p^D(\kappa)$ let

$$B_\lambda = \begin{cases} \{N^A_{\langle 0,0\rangle}\} & \text{if } \gamma \text{ is even,} \\ \{N^B_{\langle 0,0\rangle}\} & \text{if } \gamma \text{ is odd.}\end{cases}$$

(b) If $\beta < t\sigma 1 p^D(\kappa)$ let

$$B_\lambda = \begin{cases} \{N^A_{\langle e,p\rangle} | (\exists \beta' < \beta)[P^\lambda(\beta') \cong \langle e, p\rangle]\} & \text{if } \gamma \text{ is even,} \\ \{N^B_{\langle e,p\rangle} | (\exists \beta' < \beta)[P^\lambda(\beta') \cong \langle e, p\rangle]\} & \text{if } \gamma \text{ is odd.} \end{cases}$$

In the nontrivial case, there is a previous block $B_{\beta_{\gamma'}}$ of the same sort as the one being defined and the approximation P to the projection f has changed between $\beta_{\gamma'}$ and λ. Roughly, β is the least changed point. This ends the description of the construction.

4.6. *Verification.* The sets A and B are shown to have the required properties via a sequence of lemmas.

LEMMA 4.7. *$A \cup B = C$; $A \cap B = \varnothing$; A and B are E-r.e.(κ); $A, B \leqslant C$.*

PROOF. The first two are immediate from the construction; when x is enumerated into C it is also enumerated into exactly one of A or B.

A and B are E-r.e.(κ) since x is enumerated into A or B only if $x \in C$ and then at stage $\|\langle c, x, q_0\rangle\|$. The decision as to which of A or B that x enters is E-recursive in $\|\langle c, x, q_0\rangle\|$ and the parameters of the construction: q_0, q_1, $t\sigma 1 p^D(\kappa)$ and p. Recall q_0 is used to enumerate C, q_1 to enumerate D, and p to define P.

A and B are E-reducible to C on L_κ since $x \in A$ (or B) if and only if $x \in C$ and x is put into A (B) at stage $\|\langle c, x, q_0\rangle\|$.

LEMMA 4.8. *If $\delta < \kappa$ and $\langle B_{\beta_{\gamma'}} | \gamma' < \gamma\rangle$ is the sequence of active blocks at the end of stage δ, then there are only finitely many stages after stage δ, where some $B_{\beta_{\gamma'}}$ with $\gamma' < \gamma$ is cancelled.*

PROOF. A block B_β can only be activated at stage β, hence no block is ever reactivated after being cancelled. When B_β is cancelled at Stage α so are all the $B_{\beta'}$ where $\beta' > \beta$ which were active at the beginning of that stage. At a letter stage, in order to cancel another block with index below γ, a block B_{β^*} with $\beta^* < \beta$ must be cancelled. If blocks from the sequence $\langle B_{\beta_{\gamma'}} | \gamma' < \gamma\rangle$ were cancelled at infinitely many stages, there would be an infinite descending sequence of ordinals, an impossibility.

LEMMA 4.9. *Suppose that B_β is activated and never cancelled. Then there is an ordinal $\delta < \kappa$ so that B_β is satisfied at δ during every stage $\sigma > \delta$.*

PROOF. Suppose not. Then either B_β is never satisfied or any block $B_{\beta'}$ with $\beta' > \beta$ is eventually cancelled. Assume B_β $(= B^A_\beta)$ is an A-block. It must first be verified that A is E-reducible to D on L_κ.

Notice that since B_β is never cancelled, $D \cap \beta = D_\beta \cap \beta$. Let γ_1 be an ordinal below κ so that $C_{\gamma_1} \cap \beta = C \cap \beta$. There is such a γ_1 as C is a regular predicate on κ. To tell if x is an element of A compute from D as follows.

First, if x is below β, then x is an element A if and only if x is an element of A_{γ_1}.

If x is greater than or equal to β and, at stage $\sup\{x, \gamma_1\}$, β is the largest index below x of an active block, then $x \notin A$. The reason is that if x is later enumerated in C then x is put into B.

If there is an active block $B_{\beta'}^B$ which is active during stage $\sup\{x, \gamma_1\}$ so that $\beta' > \beta$, then x may later enter A. Let β_1 be the least index greater than β so that there is a necessarily B-block $B_{\beta_1}^B$ which is active during this stage. $B_{\beta_1}^B$ is eventually cancelled; this could only happen when some ν in $[\beta, \beta_1)$ is enumerated in D. Notice that when such a ν enters C, B_β^A is not injured as ν enters B, and $B_{\beta_1}^B$ is not cancelled since it is the immediate next block. Let γ_2 be the supremum of $\{\|\langle d, \nu, q_1\rangle\| \mid \nu \in D \cap [\beta, \beta_1)\}$. γ_2 is E^D-recursive in x, β, γ_1 and the fixed parameters of the construction. Since $B_{\beta_1}^B$ is cancelled it must be cancelled at or before stage γ_2. After stage γ_2, x is prohibited from entering A by B_β^A. Thus, $x \in A$ if and only if $x \in A_{\gamma_2}$ and A has been reduced to D.

Since $A \leqslant_\kappa D$, any computation of the form $\{e\}^{A\oplus D}(x)$ can uniformly be duplicated via $\{e'\}^D(e, x, p, r)$, where r is a parameter from L_κ. B_β^A has order type less than $t\sigma 1p(\kappa)$. By 3.11 and 3.14 there is an ordinal δ below κ so that for all $N_{\langle e,p\rangle}^A$ in B_β^A there is a witness in $L_\delta[D]$ to $\{e'\}^D(e, -, p, r) \neq C$. By the regularity of D let δ' be large enough so that $D_{\delta'} \cap \delta = D \cap \delta$. During stage $\sup\{\delta, \delta', \gamma_1\}$, B_β^A is satisfied at δ via computations which are correct in any information used about D. This is impossible since B_β^A purported to never to be permanently satisfied. Hence, each B_β^A is eventually satisfied permanently or cancelled.

LEMMA 4.10. *For any $\beta' < \kappa$ there is a $\beta > \beta'$ so that B_β is activated and never cancelled.*

PROOF. Let δ be the greatest ordinal less than or equal to β' so that B_δ is activated and never cancelled. By Lemma 4.9 there is a γ so that B_δ is permanently satisfied at γ during every sufficiently large stage; let γ_0 be the least such. There will be a block B_β activated and never cancelled at the least stage γ after γ_0 which is a fixed point in the enumeration of D (i.e., $D_\gamma \cap \gamma = D \cap \gamma$). Such a fixed point exists since D is incomplete and $\kappa > \omega$.

LEMMA 4.11. *For any $\langle e, p\rangle$ in L_κ, $\{e\}^{A\oplus D}(-, p) \neq C$ and $\{e\}^{B\oplus D}(-, p) \neq C$.*

PROOF. Let $\langle e, p\rangle$ be fixed. For A it is sufficient to show that there is a block B_β^A which is activated, never cancelled, and has $N_{\langle e,p\rangle}^A$ as an element (by Lemma 4.9).

Fix $\gamma < t\sigma 1p^D(\kappa)$ so that $f(\gamma) = \langle e, p\rangle$. Choose $\beta_1 \geqslant \gamma$ so that for all $\gamma_1 \leqslant \gamma$ there is a y in $L_{\beta_1}[D]$ and β' below β_1 so that $\varphi(\gamma_1, y, \beta' \cap D)$ if and only if $\exists y \exists \beta\, \varphi(\gamma_1, y, \beta \cap D)$. Choose β_2 so that $D_{\beta_2} \cap \beta_1 = D \cap \beta_1$. β_1 exists by the tameness of f and the regularity of D. For all β above β_2, $P^\beta \restriction \gamma \cong f \restriction \gamma$.

Let ν_0 be greater than β_2 so that B_{ν_0} is an A-block which is never cancelled. ν_0 exists by Lemmas 4.9 and 4.10. By examining the limit case of the construction it suffices to show that there is an A-block B_{ν_1} which is never cancelled, and has

$\nu_1 > \nu_0$ so that $P^{\nu_1} \not\equiv P^{\nu_2}$. The first such block would be defined in the nontrivial case. (To see this let $\beta_{\gamma'} = \nu_0$ in the definition. If $P^{\nu_1} \neq P^{\nu_0}$ and for all intervening A-blocks, B_{ν^*}, $P^{\nu^*} = P^{\nu_0}$, then the supremum of the argument between P^{ν^*} and P^{ν_1} is bounded.) $P^{\nu_0} \neq f$ since f has range κ and P^{ν_0} has range contained in ν_0. Let β_3 be the least ordinal so that $P^{\nu_0} \neq P^{\beta'}$ for all $\beta' > \beta_3$. β_3 has enough information to see that one of the P^{ν_0}'s values is incorrect. There is an A-block $B^A_{\nu_1}$ which is never cancelled so that $\nu_1 > \beta_3$ and so $P^{\nu_1} \neq P^{\nu_0}$.

The lemma is proven for B mutatis mutandis.

This lemma concludes the proof of the theorem.

5. Open questions. Using the techniques of this paper and of **[12]** it was shown in **[11]** that given L_κ exhibiting MP, and C and D E-r.e.(κ) predicates so that $C >_\kappa D$, there are E-r.e.(κ) sets A and B below C with infimum D. It is unknown if this result can be combined with the splitting of C over D to show that the diamond lattice can be embedded in the E-recursively enumerable degrees preserving meet and join and mapping 1 to C and 0 to D.

The presence of MP is exploited in constructing E-recursively enumerable sets in that divergences relative to an incomplete E-recursively enumerable on L_κ subset of κ become apparent at a bounded point in the construction. In classical recursion theory, divergences relative to a given recursively enumerable set are not final until the construction is completely over; in fact, the infinite injury method was developed to overcome this problem. In E-recursion on L_κ, MP allowed the density theorem to be proven via a finite rather than infinite injury construction; is there an analogue of the infinite injury method in E-recursion on L_κ? Is there a priority construction in E-recursion which transcends the level of Σ_1 phenomena?

References

1. E. Griffor, *E-recursively enumerable degrees*, Doctoral Dissertation, M.I.T., Cambridge, Mass. 1980.

2. L. A. Harrington, *Contributions to recursion theory on higher types*, Doctoral Dissertation, M.I.T. Cambridge, Mass. 1973.

3. A. S. Kechris and Y. N. Mochovakis, *Recursion in higher types*, Handbook of Math. Logic (J. Barwise, ed.), North-Holland, Amsterdam, 1976, pp. 681–737.

4. W. Maass, *Inadmissibility, tame r.e. sets and the admissible collapse*, Ann. of Math. Logic **13** (1978), 149–170.

5. D. Normann, *Set recursion*, Generalized Recursion Theory. II (J. E. Fenstad, R. O. Gandy, G. E. Sacks, eds.), North-Holland, Amsterdam, 1978.

6. ______, *Degrees of functionals*, Ann. of Math. Logic **16** (1979), 269–304.

7. ______, *Recursion in ^3E and a splitting theorem*, preprint.

8. G. E. Sacks, *Degrees of unsolvability*, Ann. of Math. Studies, No. 55, Princeton Univ. Press, Princeton, N. J., 1963.

9. ______, *Post's problem in E-recursion*, this PROCEEDINGS.

10. R. A. Shore, *The recursively enumerable α-degrees are dense*, Ann. of Math. Logic **9** (1976), 123–155.

11. T. A. Slaman, *Aspects of E-recursion*, Doctoral Dissertation, Harvard Univ., Cambridge, Mass., 1981.

12. ______, *Reflection and the priorithy method in E-recursion* (to appear).

Department of Mathematics, University of Chicago, Chicago, Illinois 60637

III. FINE STRUCTURE AND DESCRIPTIVE SET THEORY

Proceedings of Symposia in Pure Mathematics
Volume **42**, 1985

Uncountable ZF-Ordinals

RENÉ DAVID AND SY D. FRIEDMAN

Let T be a theory such as ZF, KP, KP_n ($= \Sigma_n$-admissibility). Say that α is a *T-ordinal* if L_α is a model of T. For a subset x of some cardinal κ, let $\alpha_T(x)$ be the least ordinal $\alpha > \kappa$ such that $L_\alpha(x)$ is a model of T.

Assume $V = L$. In [**3**, **4**] the second author gave a characterization of the ordinals $\alpha_{\mathrm{KP}_n}(x)$ ($n \geqslant 1$, $x \subset \kappa$) for every cardinal κ. This is a generalization of a theorem of Sacks which says that every countable KP-ordinal is an $\alpha_{\mathrm{KP}}(x)$ for some $x \subset \omega$.

In [**2**] the first author showed that every countable ZF-ordinal is an $\alpha_{\mathrm{ZF}}(x)$ for some $x \subset \omega$. This result has been proved independently by A. Beller (see [**1**]).

In this paper we give a characterization of the ordinals $\alpha_{\mathrm{ZF}}(x)$ ($x \subset \kappa$) for every cardinal κ.

We use both the techniques of [**3**, **4** and **2**]. The situation for ZF is very different from that of KP. For the latter the ordinals have cofinality equal to the cofinality of κ whereas in the present case they have cofinality ω.

Let us mention that to prove this characterization much of the work of R. Jensen on the fine structure of L is used: the usual tools for fine structure but also the coding theorem and even the covering theorem, although we are working inside L.

THEOREM. *Assume $V = L$. Let α be a* ZF-*ordinal of cardinality κ, $\alpha > \kappa$. Then α is an $\alpha_{\mathrm{ZF}}(x)$ for some $x \subset \kappa$ if and only if one of the following holds*:

(1) *$L_\alpha \models \kappa$ is singular and α is a successor* ZF *ordinal and $L_\alpha \models$ the* sup *of the ZF-ordinals has cardinality κ.*

(2) *κ is regular and there is a $\beta < \alpha$ and a sequence $(X_n | n < \omega)$ such that*

(i) $\forall \gamma < \kappa \forall f \colon \gamma \to \beta$ (*f bounded* $\to f \in L_\alpha$);

(ii) *$X_n \in L_\alpha$ and L_α* $-$ card$(X_n) < \beta$ *for $n \in \omega$, $L_\alpha = \bigcup_n X_n$*;

(iii) *β is a regular cardinal in L_α.*

1980 *Mathematics Subject Classification*. Primary 03E35, 03D25.

(3) *κ is singular but $L_\alpha \models \kappa$ is regular, and there is a $\beta < \alpha$ and a sequence $(X_n | n < \omega)$ such that* (ii) *and* (iii) *of* (2) *hold and*

(i) $\forall \lambda < \beta \exists f\colon L_\lambda \to \kappa$, *$f$ one-one and tame, i.e.,*

$$\forall \gamma < \kappa f^{-1}[\gamma] \in L_\alpha.$$

(*Note*. (3(i)) can be replaced by (3(i′)): $\exists f\colon L_\alpha \to \kappa$, f one-one and tame, when $\operatorname{cof} \kappa = \omega$. (2(i)) $\leftrightarrow$ (3(i)) when κ is regular.)

We shall deal with the three cases separately.

I. The following lemma will be often used:

LEMMA I.1. *Let κ be a cardinal $\geqslant \omega_2$, $x \subset \kappa$ such that $x \in L$ and $L_\alpha(x) \models$ ZF. Let $f \in L_\alpha(x)$ $f\colon \gamma < \kappa \to \alpha$.*

Then $f \in L_\alpha$.

PROOF. Let $A = \{\langle i, f(i)\rangle | i < \gamma\}$ where $\langle \cdot , \cdot \rangle$ is the Gödel pairing function. By Jensen's covering theorem there is a B such that: $B \in L_\alpha$, $B \supset A$ and $L_\alpha(x) \models \overline{\overline{B}} = \operatorname{Max}(\omega_1, \overline{\overline{A}}) < \kappa$. Since $x \in L$, $L_\alpha \models \mu = \overline{\overline{B}} < \kappa$. Let $g \in L_\alpha$ $g\colon \mu \to B$ bijective, and $c = g^{-1}[A]$; then c is a subset of μ and so $c \in L_\alpha$. It follows that A and $f \in L_\alpha$. $\square$

LEMMA I.2 *Let κ be a cardinal, $x \subset \kappa$, $x \in L$ such that $L_\alpha(x) \models$ ZF + κ singular. Then $x \in L_\alpha$.*

PROOF. By Jensen's covering theorem, $L_\alpha \models \kappa$ is singular and (since $x \in L$) $L_\alpha - \operatorname{cof}(\kappa) = L_\alpha(\mathrm{x}) - \operatorname{cof}(\kappa)$. Let $(\kappa_i | i < \lambda) \in L_\alpha$ be a normal sequence coverging to κ where $\lambda = L_\alpha - \operatorname{cof}(\kappa)$. Define $f\colon \lambda \to L_\alpha$ by $f(i) =$ the L-code for $x \cap \kappa_i$. By Lemma I.1, $f \in L_\alpha$ and so $x \in L_\alpha$. $\square$

Case (1) of the Theorem is now clear: If $L_\alpha \models \kappa$ singular by Lemma I.2, $x \in L_\alpha$. Let $\beta < \alpha$ be least such that $x \in L_\beta$. Then clearly α is the least ZF-ordinal greater than β and (since $x \subset \kappa$) $L_\alpha \models \beta < \kappa^+$.

The opposite is trivial: it is enough to take for x a code for an ordinal greater than the ZF-ordinals below α.

II.

LEMMA II.1. *Let κ be a cardinal and α be $\alpha_{\mathrm{ZF}}(x)$ for some $x \subset \kappa$. Then there is a $\beta < \alpha$ and a sequence $(X_n | n < \omega)$ such that* (2(ii)) *and* (2(iii)) *of the Theorem hold.*

PROOF. Let β be such that $L_\alpha(x) \models \beta = \kappa^+$ and set $y_n = \{t \in L_\alpha(x) | \ t$ is Σ_n-definable in $L_\alpha(x)$ with parameters from $\kappa \cup \{x\}\}$. Then clearly $y_n \in L_\alpha(x)$; $y_n \in y_{n+1}$ and $L_\alpha(x) - \operatorname{card}(y_n) = \kappa$.

Set $y = \bigcup_n y_n$. Clearly $y \prec L_\alpha(x)$.

Set $\pi\colon y \to^{\cong} L_\gamma(x)$. Then $\gamma = \alpha$ since $L_\alpha(x) \models$ ZF and $L_\alpha(x) \models \forall \delta\, L_\delta(x) \not\models$ ZF. So $y = L_\alpha(x)$ since every element of y is y-definable from $\kappa \cup \{x\}$.

Now let $x_n \in L_\alpha$ be such that $x_n \supset y_n \cap L_\alpha$ and $L_\alpha(x) \models \operatorname{card}(x_n) = \kappa$. Then clearly $L_\alpha = \bigcup x_n$ and $L_\alpha - \operatorname{card}(x_n) < \beta$. $\square$

To prove that (2(i)) is true in the case κ is regular, we shall first assume $\kappa \geqslant \omega_2$. This proof does not work for $\kappa = \omega_1$ since it uses Lemma I.1 which is not true for ω_1 by a theorem of Bukovsky. The proof we shall give for ω_1 works for every regular cardinal κ, but since it is a bit more complicated, it seems useful to give first the simplest one.

LEMMA II.2. *Let κ be a regular cardinal and α be $\alpha_{ZF}(x)$ for some $x \subset \kappa$. Then* (2) *of the Theorem holds.*

PROOF. It remains to show (2(i)); let $L_\alpha(x) \vDash \beta = \kappa^+$.

(*) Assume first $\kappa \geqslant \omega_2$: Let $f\colon \gamma < \kappa \to \mu < \beta$ and let $g \in L_\alpha(x)$ $g\colon \mu \to \kappa$ bijective, and $h = g \circ f\colon \gamma \to \kappa$. Since κ is regular, h is bounded and so $h \in L_k$ and $f \in L_\alpha(x)$. Now using Lemma I.1, $f \in L_\alpha$. Note that we have used here not only the fact that $L_\alpha(x) \vDash \kappa$ is regular, but also the fact that κ is regular.

(**) Assume now $\kappa = \omega_1$: the proof uses the second author's notion of critical projecta defined in [**3**]. We prove exactly as in [**3**, Lemmas 9–11] that the ρ_i, ρ_i' have cofinality ω_1, where the ρ_i, ρ_i' are the critical projecta of β and then (this is Theorem 13 in [**3**]) that (2(i)) holds. □

We now have to prove the converse part. So assume from now on that (2) of the Theorem holds. We have to find $x \subset \kappa$ such that $\alpha = \alpha_{ZF}(x)$. We shall build x by a 3-step forcing iteration over L_α. The main problems are to show that we can find *in* L the generics we need.

We first find an $x_0 \subset \beta$ such that

$$L_\alpha(x_0) \vDash \mathrm{ZF} + \beta = \kappa^+.$$

Since β is regular in L_α it is either a successor cardinal or an inaccessible one.

(*) Assume first $L_\alpha \vDash \beta = \theta^+$ for some $\theta < \beta$ let $\mathbf{P}$ be the usual poset to collapse θ on κ. By (2(i)), $\mathbf{P}$ is $< \kappa$-closed (that means: if $(p_i|\ i < \gamma < \kappa)$ is a decreasing sequence of conditions *in or out of* L_α, then there is a $p \in \mathbf{P}$ such that $p \leqslant p_i \forall i < \gamma$). Since $\overline{\overline{L_\alpha}} = \kappa$ it is easy to find in L a $\mathbf{P}$ generic over L_α and from that an x_0 such that $L_\alpha(x_0) \vDash \mathrm{ZF} + x_0 \subset \kappa + \beta = \kappa^+$.

(**) Assume next $L_\alpha \vDash \beta$ is inaccessible. Let $\mathbf{P}$ be the usual poset to collapse all the cardinals between κ and β: more precisely let $I = L_\alpha\text{-card} \cap]\kappa, \beta[$ and

$$\mathbf{P} = \Big\{ p = (p_j)_{j \in J} | J \subset I\, \overline{\overline{J}} < \kappa, \mathrm{dom}(p_j) \subset \kappa, \mathrm{card}(\mathrm{dom}\ p_j) < \kappa$$
$$p_j\colon \mathrm{dom}\ p_j \to j \Big\}.$$

Note that we do not ask $J \in L_\alpha$. Also note that (by (2(i))) $\forall j \in I p_j \in L_\alpha$.
Set $\tilde{\mathbf{P}} = \mathbf{P} \cap L_\alpha$. $\tilde{\mathbf{P}}$ is, in L_α, the usual poset to make $\beta = \kappa^+$.

LEMMA II.3. *Let $D \subset \tilde{\mathbf{P}}$, $D \in L_\alpha$ be dense in $\tilde{\mathbf{P}}$. Then D is predense in* $\mathbf{P}$.

PROOF. Let $A \subset D$ be a maximal antichain in $\tilde{\mathbf{P}}$, $A \in L_\alpha$. Since it is well known that $\tilde{\mathbf{P}}$ has, in L_α, the $< \beta$ chain condition there is a $\theta < \beta$ such that for $p \in A$ $J_p \subset \theta$.

Now let $q \in \mathbf{P}$ and define $\tilde{q}$ by $J_{\tilde{q}} = J_q \cap \theta$ and for $j \in J_{\tilde{q}}\ \tilde{q}_j = q_j$; then, by (2(i)), $\tilde{q} \in \tilde{\mathbf{P}}$. Since A is a maximal antichain $\tilde{q}$ is compatible with some $r \in A$ but since $J_r \subset \theta$, q also is compatible with r. □

Using this lemma it is not difficult to find a $\tilde{\mathbf{P}}$ generic over L_α. Let $(D_i|\ i < \kappa)$ be an enumeration of the open dense subsets of $\tilde{\mathbf{P}}$ in L_α. Define a decreasing sequence $(p_i|\ i < \kappa)$ of elements of $\mathbf{P}$ such that: $\forall i\ p_i \in D_i$ as follows: $p_0 = \varnothing$. Assume $(p_j|\ j < i < \kappa)$ has been defined. Set $p = \bigcup_{j<i} p_j$. Then $p \in \mathbf{P}$. By Lemma II.3 let p_i be the least q such that $q \leqslant p$ and $q \in D_i$.

Set $p_\kappa = \bigcup_{i<\kappa} p_i$. Then $G = \{q \in \tilde{\mathbf{P}}|\ \forall j \in J_q\ q_j = (p_\kappa)_j \upharpoonright \operatorname{dom} q_j\}$. It is clear that G is $\tilde{\mathbf{P}}$ generic over L_α.

So we have proved

LEMMA II.4. *There is a subset x_0 of β such that*
(1) $L_\alpha(x_0) \models \mathrm{ZF} + \beta = \kappa^+$;
(2) $L_\alpha(x_0) = \bigcup X_n$ *where* $X_n \in L_\alpha(x_0)$ *and* $L_\alpha(x_0) - \operatorname{card}(X_n) = \kappa$.

In the second step we use the results of [2] to find a subset x_1 of β to kill all the ZF-ordinals. Let $\mathbf{P} = \mathbf{P}_\kappa$ with the notations of [2]. $\mathbf{P}$ is a class in $L_\alpha(x_0)$. It is shown in [2] that in a $\mathbf{P}$ generic extension of $L_\alpha(x_0)$ all the ZF-ordinals are killed and that this extension satisfies $V = L_\alpha(x_1)$ for some $x_1 \subset \beta$. Moreover, $\mathbf{P}$ is κ-distributive in $L_\alpha(x_0)$.

For $n < \omega$ let $(\Delta_i^n|\ i < \kappa)$ be an enumeration of the open dense subclasses of $\mathbf{P}$ definable by a Σ_n-formula with parameters from X_n. By the distributivity of $\mathbf{P}$, $D_n = \bigcap_{i<\kappa} \Delta_i^n$ is an open dense subclass of $\mathbf{P}$.

Define a sequence $(p_n|n < \omega)$ of elements of $\mathbf{P}$ by $p_0 = \varnothing$, $p_{n+1} =$ some $p \leqslant p_n$ such that $p \in D_n$. Then clearly $\bigcup p_n$ is $\mathbf{P}$ generic over $L_\alpha(x_0)$.

It remains now to code x_1 by a subset of κ. So it is enough to show

LEMMA II.5. *Let κ be a regular cardinal, α, β be ordinals of cardinality κ, and x a subset of β such that*

$$L_\alpha(x) \models \mathrm{ZF} + \beta = \kappa^+.$$

Then there is, in L, a subset y of κ such that

$$L_\alpha(y) \models \mathrm{ZF} + \beta = \kappa^+ + x = \{\xi < \beta|\ S_\xi \cap y \text{ is bounded}\},$$

where $(S_\xi|\ \xi < \beta)$ is some nice sequence of almost disjoint subsets of κ; i.e., S_ξ is uniformly $L_\alpha(x \cap \xi)$-definable.

(*Note*. If β had (true) cofinality κ, there would be no problems since then the forcing that gives y would be $< \kappa$-closed. But here β has cofinality ω!)

PROOF. We use Solovay's trick (see [1, p. 12]); the S_ξ are $S(b_\xi)$ where the b_ξ are mutually generic. Let $\mathbf{P}$ be the poset of conditions (not necessarily in $L_\alpha(x)$) to code x by a subset of κ. Let $\tilde{\mathbf{P}} = \mathbf{P} \cap L_\alpha(x)$. The lemma similar to Lemma II.3 with the new forcing is proved in [1, Lemma 1.3, p. 13]. From that it is easy to find the generic we need: Do as after Lemma II.3. □

The proof of the second case is now complete.

III. We assume now that κ is a singular cardinal but $L_\alpha \vDash \kappa$ is regular.

For the "only if" part of the Theorem we have to prove (3(i)). The proof of that is exactly as in [**4**, Theorem 9], using the critical projecta of β.

To prove (3(i′)) in the case cof $\kappa = \omega$ we use the following, which is proved in [**4**] (see Theorem 3).

Claim. Assume $\mathrm{cof}(\alpha) = \omega$ and for all $y \in L_\alpha$ there is a $\lambda < \alpha$ and $(y_n | n < \omega)$ such that $y = \bigcup y_n$ and $\forall n < \omega\ y_n \in L_\lambda$ and $\mathrm{card}(y_n) < \kappa$. Then there is a tame injection from L_α into κ.

So it is enough to prove the hypothesis of that claim. By Lemma II.1 and by (3(i)), if $y \in L_\alpha$ we can write $y = \bigcup y_n$ with $y_n \in L_\alpha$ and $L_\alpha - \mathrm{card}(y_n) < \kappa$. Now since $y \subset L_\mu$ for some $\mu < \alpha$, $\forall n\ y_n \in L_{\mu^+}$.

To prove the converse part of the Theorem, it is enough to show that we can get a generic for the first and third steps of the iteration given in §II. (The second one is exactly as in §II.) This is done as follows: In each case we have to meet the open dense subsets of some poset **P** which is (inside L_α) $< \kappa$-closed (since in L_α, κ is regular).

It is enough to prove

LEMMA III.1. *Let* $\lambda = \mathrm{cof}(\kappa) < \kappa$ *and* $f: \gamma < \lambda \to \mu < \beta$. *Then* $f \in L_\alpha$ (*note that —in fact—in the "only if" part of the Theorem this is proved before proving* (3(*i*)), *but it turns out that it is a consequence of it*).

LEMMA III.2. *There is a sequence* $(D_i | i < \lambda)$ *of open dense subsets of* **P** *such that every subset of* **P** *that meets all the* D_i *is* **P** *generic over* L_α.

From these lemmas we can find—by the same techniques as in §II—the generics we need.

PROOF OF LEMMA III.1. Let $f: \gamma < \lambda \to \mu < \beta$. By (3(i)), there is a $g: \mu \to \kappa$ one-one and tame. Let $h = g \circ f: \gamma \to \kappa$. Then h is bounded in κ and $h \in L_\alpha$.

But then $f = (g^{-1} \upharpoonright \rho) \circ h$ for some $\rho < \kappa$ and so $f \in L_\alpha$ since g is tame. □

Note that we have used here that $g^{-1} \upharpoonright \rho \in L_\alpha$ for $\rho < \kappa$ and not only $g^{-1}[\rho] \in L_\alpha$. This comes from the fact that $g^{-1} \upharpoonright \rho = g_0 \circ g'$ where $g': \rho \to \rho' \in L_\kappa$, $\rho' = \mathrm{ordertype}(g^{-1}[\rho]) < \kappa$ and $g_0: \rho' \to g^{-1}[\rho]$ lists $g^{-1}[\rho]$ in increasing order.

PROOF OF LEMMA III.2. By (2(ii)) there is a sequence $(\Delta_n | n < \omega)$ such that $\Delta_n \in L_\alpha$, $L_\alpha - \mathrm{card}(\Delta_n) < \beta$, and $\bigcup_n \Delta_n$ is the set of the open dense subsets of P. Now by (3(i)) there is an enumeration $(\Delta^n_\xi | \xi < \kappa)$ of Δ_n for each n such that $(\Delta^n_\xi | \xi < \nu) \in L_\alpha$ for each $n < \omega$ and $\nu < \kappa$.

Let $(\kappa_i | i < \lambda)$ be a normal sequence converging to κ. Set $D_{n,i} = \bigcap_{\xi < \kappa_i} \Delta^n_\xi$. Then $D_{n,i} \in L_\alpha$ for $n < \omega$ and $i < \lambda$ and since **P** is, in L_α, $< \kappa$-closed: $D_{n,i}$ is open dense. It is then enough to rearrange the $\mathrm{D}_{n,i}$'s into a λ-sequence. □

This achieves the proof of the Theorem.

IV. Some final comments.

(1) The Theorem can be easily generalized to sequences of ZF ordinals: following [**2**] we can give *sufficient* conditions for a sequence of length $< \kappa^+$ of ZF

ordinals of cardinality κ to be an initial segment of the α, such that $L_\alpha(x)$ is a model of ZF, for some $x \subset \kappa$. As in [2] the essential fact is to assume $\operatorname{Sup} Q \cap \alpha < \alpha$ for $\alpha \in Q$, where Q is the given sequence.

(2) It would be interesting to find some classes A (or for which classes?) for which there is a subset x of ω such that A is exactly the class of the α such that $L_\alpha(x)$ is a model of ZF. This is done in [5] for KP instead of ZF.

(3) Finally note that in the Theorem (2(ii)) cannot be replaced by a simple condition on the cofinality of α and β; for example $\operatorname{cof}(\alpha) = \operatorname{cof}(\beta) = \omega$. To see that, assume that there is a $\bar{\beta}$ such that

$$\omega_1 < \bar{\beta} < \omega_2 \quad \text{and} \quad L_{\bar{\beta}+3} \vDash \bar{\beta} \text{ is inaccessible.}$$

We shall find α of cofinality ω for which there is no sequence $(X_n | \, n < \omega)$ satisfying (2(ii)): we first find a γ such that

$(*)$ ($L_{\gamma+2} \vDash \gamma$ is inaccessable) and ($\operatorname{cof} \gamma = \omega_1$) and (for $\delta < \gamma$ if $L_\gamma \vDash \delta$ regular then $\operatorname{cof}(\delta) = \omega_1$). (Define $x_0 =$ the Skolem Hull of ω_1 in $L_{\bar{\beta}+2}$;

$$x_{i+1} = \operatorname{SH}(x_i \cup \{x_i\}, L_{\bar{\beta}+2}); \qquad x_i = \bigcup_{j<i} x_j \quad \text{for limit } i.$$

Let $\pi\colon x_{\omega_1} \to^{\cong} L_{\gamma+2}$. It is easy to see that γ has the desired properties.)

Now define the sequence $(\alpha_n)_{n<\omega}$ as follows: $\alpha_0 = \omega_1$; $\alpha_{2n+1} = \alpha_{2n}^+$ in the sense of L_γ; $\alpha_{2n+2} =$ the least $\alpha > \alpha_{2n+1}$ such that $L_\alpha \prec L_\gamma$ (such an α exists since $L_{\gamma+2} \vDash \gamma$ is inaccessible). Set $\alpha = \bigcup \alpha_n$. Let $L_\alpha \prec L_\gamma$ so α is a ZF ordinal and $\operatorname{cof}(\alpha) = \omega$. Assume there is a $\beta < \alpha$ and a sequence $(x_n | \, n < \omega)$ such that (2(ii)) holds. Choose $\mu = \alpha_{2n+1} > \beta$; then, by $(*)$ $\operatorname{cof}(\mu) = \omega_1$ but $\mu = \bigcup_n (x_n \cap \mu)$, and since $L_\alpha - \operatorname{card}(x_n \cap \mu) < \mu$, $\operatorname{cof}(\mu) = \omega$, a contradiction.

References

1. A Beller, R. B. Jensen and P. Welch, *Coding the universe*, Cambridge Univ. Press, London and New York, 1981.

2. R. David, *Some applications of Jensen's coding theorem*, Ann. Math. Logic **22** (1982), 177–196.

3. Sy D. Friedman, *Uncountable admissibles*. I, *Forcing*, Trans. Amer. Math. Soc. **270** (1982), 61–73.

4. ______, *Uncountable admissibles*. II. *Compactness*, Israel J. Math. **40** (1981), 129–149.

5. ______, *Strong coding* (to appear).

6. R. B. Jensen, *Coding the universe by a real*, manuscript, 1975.

Department of Mathematics, Université Toulouse le Mirail, 109 rue Vauquelin 31058 Toulouse Cedex, France

Massachusetts Institute of Technology, Department of Mathematics, Cambridge, Massachusetts 02139

Proceedings of Symposia in Pure Mathematics
Volume **42**, 1985

Another Look at Gap-1 Morasses

HANS-DIETER DONDER

Abstract. Recently, Velleman considerably simplified the theory of gap-1 morasses by introducing much simpler structures which have the same strength. In this context he also defined stronger structures which he called "simplified $(\kappa, 1)$-morasses with linear limits". The main aim of this paper is to show that these structures exist in L for all "possible" κ; namely, κ has to be regular but not weakly compact. We assume familiarity with the fine structure theory of L and the notion of a gap-1 morass (see [5]). In analysing parameters we use the following well-ordering on $[ON]^{<\omega}$:

$$p <_* q \leftrightarrow \exists\alpha(p - \alpha = q - \alpha \text{ and } q \cap \alpha \neq \varnothing \text{ and}$$
$$(p \cap \alpha = \varnothing \text{ or } \max(p \cap \alpha) < \max(q \cap \alpha))).$$

1. Simplified $(\kappa, 1)$-morasses. We first recall Velleman's definition of a simplified $(\kappa, 1)$-morass (see [7]). Actually, we present the equivalent "expanded" version.

DEFINITION. Let $f, g\colon \alpha \to \gamma$ be order preserving s.t., for some $\tau < \alpha$, $(f \restriction \tau = g \restriction \tau$ and $\forall\beta < \alpha\, f(\beta) < g(\tau))$. Then $\{f, g\}$ is called an *amalgamation pair* for α.

DEFINITION. Let $\langle\theta_\alpha | \alpha \leqslant \kappa\rangle$ be a sequence of ordinals s.t. $\alpha < \kappa \to 0 < \theta_\alpha < \kappa$ and $\theta_\kappa = \kappa^+$. In addition, for $\alpha < \beta \leqslant \kappa$ let $\mathscr{F}_{\alpha\beta}$ be a set of order preserving functions $f\colon \theta_\alpha \to \theta_\beta$. $\langle\langle\theta_\alpha | \alpha \leqslant \kappa\rangle, \langle\mathscr{F}_{\alpha\beta} | \alpha < \beta \leqslant \kappa\rangle\rangle$ is a *simplified* $(\kappa, 1)$-*morass* if the following properties are satisfied:

(P1) $\forall\alpha < \beta < \kappa\, |\mathscr{F}_{\alpha\beta}| < \kappa$.

(P2) $\forall\alpha < \beta < \gamma \leqslant \kappa\, \mathscr{F}_{\alpha\gamma} = \{f \circ g | f \in \mathscr{F}_{\beta\gamma}, g \in \mathscr{F}_{\alpha\beta}\}$.

(P3) $\forall\alpha < \kappa\, \mathscr{F}_{\alpha,\alpha+1}$ is a singleton or an amalgamation pair for θ_α.

(P4) $\forall$ limit $\alpha \leqslant \kappa\, \forall\beta, \bar\beta < \alpha\, \forall f_1 \in \mathscr{F}_{\beta\alpha}\, \forall f_2 \in \mathscr{F}_{\bar\beta\alpha}\, \exists\gamma(\beta, \bar\beta < \gamma < \alpha$ and $\exists f_1' \in \mathscr{F}_{\beta\gamma}\, \exists f_2' \in \mathscr{F}_{\bar\beta\gamma}\, \exists g \in \mathscr{F}_{\gamma\alpha}(f_1 = g \circ f_1'$ and $f_2 = g \circ f_2'))$.

(P5) $\kappa^+ = \bigcup\{f''\theta_\alpha | \alpha < \kappa, f \in \mathscr{F}_{\alpha\kappa}\}$.

1980 *Mathematics Subject Classification*. Primary 03E45, 03E05.

Velleman has already shown (by a rather indirect argument) that for regular $\kappa > \omega$:

There exists a $(\kappa, 1)$-morass $\leftrightarrow$ there exists a simplified $(\kappa, 1)$-morass.

For later use we now give a direct proof of " $\rightarrow$ ".

So let $\mathcal{M} = \langle \mathfrak{S}, \ldots \rangle$ be a $(\kappa, 1)$-morass. We introduce additional notation:

$\bar{\nu} \prec_* \nu$ iff ν immediately succeeds $\bar{\nu}$ in $\prec$,

$\mu \vdash \nu$ iff there are $\bar{\mu}, \bar{\nu}$ such that $\bar{\mu} \in S_{\alpha_{\bar{\nu}}} \cap \bar{\nu}$,

$$\bar{\nu} \prec_* \nu \quad \text{and} \quad \bar{\mu} \prec_* \mu \prec \pi_{\bar{\nu}\nu}(\bar{\mu}).$$

We may also assume w.l.o.g. that we have

(i) $\alpha \in S^0 \cap \kappa \rightarrow S_\alpha$ is closed,

(ii) ν minimal in S_α, $\bar{\nu}$ minimal in $S_{\bar{\alpha}}$, $\bar{\alpha} < \alpha \rightarrow \bar{\nu} \prec \nu$.

For $\alpha \in S^0 \cap \kappa$ let $\nu_\alpha = \max S_\alpha$. Now let

$$A = \{\alpha \in S^0 \cap \kappa | \nu_\alpha \text{ is a successor in } \prec\},$$
$$A_0 = \{\alpha \in A | \nu_\alpha \text{ is a successor in } S_\alpha\},$$
$$A_1 = \{\alpha \in A | \nu_\alpha \text{ is a limit in } S_\alpha, \text{ but } \pi_{\bar{\nu}\nu_\alpha} \text{ is not cofinal in } \nu_\alpha, \text{ where } \bar{\nu} \prec_* \nu_\alpha\}.$$

We now add some more points to the morass. For $\alpha \in A$ set

$$S'_\alpha = \{\nu_\alpha + \tau | \tau \in S_\alpha \cap \nu_\alpha\}.$$

Extend the relation $\prec$ to $\prec'$ by setting (in addition), for $\nu_\alpha + \tau \in S'_\alpha$,

$$\bar{\tau} \prec' \nu_\alpha + \tau \quad \text{iff } \bar{\tau} \prec \tau.$$

Now let $\bar{S} = S^0 \cup \{\alpha + 1 | \alpha \in A_0 \cup A_1\}$. For $\alpha \in \bar{S}$ let

$$\bar{S}_\alpha = \begin{cases} S_\alpha & \text{if } \alpha \in S^0 - A, \\ S_\alpha \cup S'_\alpha & \text{if } \alpha \in A - (A_0 \cup A_1), \\ S_\alpha - \{\nu_\alpha\} & \text{if } \alpha \in A_0 \cup A_1, \\ S_{\bar{\alpha}} \cup S'_{\bar{\alpha}} & \text{if } \alpha = \bar{\alpha} + 1, \bar{\alpha} \in A_0 \cup A_1. \end{cases}$$

For $\nu \in \bigcup_{\alpha \in \bar{S}} \bar{S}_\alpha$ let

$$\alpha'_\nu = \text{the largest } \alpha \text{ s.t. } \nu \in \bar{S}_\alpha, \qquad \alpha''_\nu = \text{the smallest } \alpha \text{ s.t. } \nu \in \bar{S}_\alpha.$$

Note that there are at most two α s.t. $\nu \in \bar{S}_\alpha$.

We now construct a simplified $(\kappa, 1)$-morass. Let $\langle \gamma_\delta | \delta \leqslant \kappa \rangle$ be the monotone enumeration of $\bar{S}$. For $\delta \leqslant \kappa$ let W_δ be the set of finite sequences $\langle \eta_0, \ldots, \eta_n \rangle$ s.t. n is odd and

(i) $\eta_0 \in \bar{S}_{\gamma_\delta}$,

(ii) $\eta_{2k+1} \prec' \eta_{2k}$,

(iii) $\eta_{2k+2} \in S_{\alpha'_{\eta_{2k+1}}}$, $\eta_{2k+2} > \eta_{2k+1}$.

Let $<_\delta$ be the lexicographical ordering on W_δ. Clearly $\langle W_\delta, <_\delta \rangle$ is well ordered. So let $\theta_\delta = \text{otp}(\langle W_\delta, <_\delta \rangle)$. In the following we "identify" θ_δ and $\langle W_\delta, <_\delta \rangle$. Clearly, we have $0 < \theta_\delta < \kappa$ for $\delta < \kappa$ and $\theta_\kappa = \kappa^+$. We now have to define the $\mathscr{F}_{\alpha\beta}$. We first define some special maps.

(A) Let $\bar{\nu} \prec \nu$, $\alpha'_{\bar{\nu}} = \gamma_\alpha$, $\alpha''_\nu = \gamma_\beta$. We define $\tilde{\pi}_{\bar{\nu}\nu}$: $W_\alpha \to W_\beta$ by

$$\tilde{\pi}_{\bar{\nu}\nu}(\langle \eta_0, \ldots, \eta_n \rangle) = \begin{cases} \langle \pi_{\bar{\nu}\nu}(\eta_0), \eta_1, \ldots, \eta_n \rangle & \text{if } \eta_0 \leqslant \bar{\nu}, \\ \langle \nu, \bar{\nu}, \eta_0, \ldots, \eta_n \rangle & \text{if } \eta_0 > \bar{\nu}. \end{cases}$$

(B) Let $\eta \vdash \nu$. So let $\bar{\eta}$, $\bar{\nu}$ s.t. $\bar{\nu} \prec_* \nu$, $\bar{\eta} \prec_* \eta \prec \pi_{\bar{\nu}\nu}(\bar{\eta}) =: \eta'$.
Let $\alpha'_\eta = \gamma_\alpha$, $\alpha''_\nu = \gamma_\beta$, $\rho = \alpha'_{\bar{\nu}}$. We define $\sigma_{\eta\nu}$: $W_\alpha \to W_\beta$ by

$$\sigma_{\eta\nu}(\langle \eta_0, \ldots, \eta_n \rangle) = \begin{cases} \langle \pi_{\eta\eta'}(\eta_0), \eta_1, \ldots, \eta_n \rangle & \text{if } \eta_0 < \eta \text{ or } (\eta_0 = \eta \text{ and } \eta_1 \neq \bar{\eta}), \\ \langle \nu + \pi_{\eta\eta'}(\tau), \eta_1, \ldots, \eta_n \rangle & \text{if } \eta_0 = \eta + \tau, \tau > 0, \\ \tilde{\pi}_{\bar{\nu}\nu}(\langle \eta_2, \ldots, \eta_n \rangle) & \text{if } \eta_0 = \eta \text{ and } \eta_2 \in \bar{S}_\rho, \\ \langle \nu, \bar{\nu} \rangle & \text{if } \langle \eta_0, \ldots, \eta_n \rangle = \langle \eta, \bar{\eta} \rangle. \end{cases}$$

(C) Let $\alpha \in A_0$. So let $\bar{\nu} \prec_* \nu$ and let ν immediately succeed ρ in S_α. Let $\pi_{\bar{\nu}\nu}(\bar{\rho}) = \rho$ and $\alpha = \gamma_\delta$. Then define g_0^δ, g_1^δ: $W_\delta \to W_{\delta+1}$ by $g_0^\delta = \mathrm{id} \restriction W_\delta$ and

$$g_1^\delta(\langle \eta_0, \ldots, \eta_n \rangle) = \begin{cases} \langle \eta_0, \ldots, \eta_n \rangle & \text{if } \eta_0 < \rho \text{ or } (\eta_0 = \rho \text{ and } \eta_1 < \bar{\rho}), \\ \langle \nu, \bar{\nu}, \eta_2, \ldots, \eta_n \rangle & \text{if } \eta_0 = \rho, \eta_1 = \bar{\rho}, \eta_2 \neq \bar{\nu}, \\ \langle \nu, \eta_3, \ldots, \eta_n \rangle & \text{if } \eta_0 = \rho, \eta_1 = \bar{\rho}, \eta_2 = \bar{\nu}, \\ \langle \nu + \rho, \eta_1, \ldots, \eta_n \rangle & \text{if } \eta_0 = \rho \text{ and } \eta_1 > \bar{\rho}. \end{cases}$$

(D) Let $\alpha \in A_1$. So let $\bar{\nu} \prec_* \nu$ and $\lambda = \sup \pi''_{\bar{\nu}\nu}\bar{\nu}$. Hence we have $\bar{\nu} \prec \lambda$. Let $\alpha = \gamma_\delta$. Then define g_0^δ, g_1^δ: $W_\delta \to W_{\delta+1}$ by $g_0^\delta = \mathrm{id} \restriction W_\delta$ and

$$g_1^\delta(\langle \eta_0, \ldots, \eta_n \rangle) = \begin{cases} \langle \eta_0, \ldots, \eta_n \rangle & \text{if } \eta_0 < \lambda \text{ or } (\eta_0 = \lambda \text{ and } \eta_1 < \bar{\nu}), \\ \langle \nu, \eta_1, \ldots, \eta_n \rangle & \text{if } \eta_0 = \lambda \text{ and } \eta_1 = \bar{\nu}, \\ \langle \nu + \eta_0, \eta_1, \ldots, \eta_n \rangle & \text{if } \eta_0 > \lambda \text{ or } (\eta_0 = \lambda \text{ and } \eta_1 > \bar{\nu}). \end{cases}$$

The following properties are easily checked:

(1) $\tilde{\pi}_{\bar{\nu}\nu}$, $\sigma_{\eta\nu}$, g_i^δ are order preserving.

(2) $\{g_0^\delta, g_1^\delta\}$ is an amalgamation pair.

(3) $\bar{\nu} \prec \nu' \prec \nu \to \tilde{\pi}_{\bar{\nu}\nu} = \tilde{\pi}_{\nu'\nu} \circ \tilde{\pi}_{\bar{\nu}\nu'}$.

(4) $\eta \vdash \rho \vdash \nu \to \sigma_{\eta\nu} = \sigma_{\rho\nu} \circ \sigma_{\eta\rho}$.

(5) Let $\eta \vdash \nu$, where $\bar{\nu} \prec_* \nu$ and $\bar{\eta} \prec_* \eta$. Then $\tilde{\pi}_{\bar{\nu}\nu} = \sigma_{\eta\nu} \circ \tilde{\pi}_{\bar{\eta}\eta}$.

(6) Let $\bar{\nu} \prec_* \nu$, where $\nu = \nu_\alpha$ and $\alpha \in A_0 \cup A_1$, $\alpha = \gamma_\delta$. Let $\lambda = \sup \pi''_{\bar{\nu}\nu}(S_{\alpha_{\bar{\nu}}} \cap \bar{\nu})$ and $\bar{\lambda} \prec_* \lambda^* \prec \lambda$ s.t. $\alpha_{\bar{\lambda}} = \alpha_{\bar{\nu}}$.

(So $\bar{\nu} = \bar{\lambda}$ or $\bar{\nu}$ immediately succeeds $\bar{\lambda}$ in $S_{\alpha_{\bar{\nu}}}$.) Then $\tilde{\pi}_{\bar{\nu}\nu} = g_1^\delta \circ \tilde{\pi}_{\bar{\lambda}\lambda}$ and for all $\eta \vdash \nu$ we have $\sigma_{\eta\nu} = g_1^\delta \circ \tilde{\pi}_{\lambda\lambda^*} \circ \sigma_{\eta\lambda^*}$ or ($\eta = \lambda^*$ and $\sigma_{\eta\nu} = g_1^\delta \circ \tilde{\pi}_{\eta\lambda}$).

(7) Let $\tau \in S_\alpha \cap \nu_\alpha$, where $\alpha \in A_0 \cup A_1$, $\alpha = \gamma_\delta$. Then

$$\bar{\tau} \prec \tau \to \tilde{\pi}_{\bar{\tau}\tau} = g_0^\delta \circ \tilde{\pi}_{\bar{\tau}\tau}, \qquad \eta \vdash \nu \to \sigma_{\eta\nu} = g_0^\delta \circ \sigma_{\eta\nu}.$$

Now let $\bar{\mathscr{F}}$ be the set of $\tilde{\pi}_{\bar{\nu}\nu}$, $\sigma_{\eta\nu}$, g_i^δ. Let $\mathscr{F}$ be the closure of $\bar{\mathscr{F}}$ under finite compositions. For $\alpha < \beta \leqslant \kappa$ set

$$\mathscr{F}_{\alpha\beta} = \{ f \in \mathscr{F} \mid \mathrm{dom}\, f = W_\alpha \text{ and } \mathrm{rng}(f) \subseteq W_\beta \}.$$

We now check that $\langle\langle \theta_\alpha | \alpha \leqslant \kappa \rangle, \langle \mathscr{F}_{\alpha\beta} | \alpha < \beta \leqslant \kappa \rangle\rangle$ is a simplified $(\kappa, 1)$-morass. By previous remarks we only have to show that (P1)–(P5) are satisfied.

(P1) is clear.

(P2) " $\supseteq$ " is clear by definition. We prove " $\subseteq$ " by induction on γ. There are, of course, a lot of different cases. As an example we treat the case $\bar{\nu} \prec_* \nu$, $\bar{\nu}$ a limit point in $S_{\alpha_{\bar{\gamma}}}$, $\pi_{\bar{\nu}\nu}$ is cofinal. Then γ is a limit ordinal. So we only have to show (using the induction hypothesis)

(*) For cofinally many $\beta < \gamma \, \exists f \in \mathscr{F}_{\beta\gamma} \exists g \in \mathscr{F}_{\alpha\beta} \, \tilde{\pi}_{\bar{\nu}\nu} = f \circ g$ (where $\operatorname{dom} \tilde{\pi}_{\bar{\nu}\nu} = W_\alpha$).

Now by (5) we have

$$\tilde{\pi}_{\bar{\nu}\nu} = \sigma_{\eta\nu} \circ \tilde{\pi}_{\bar{\eta}\eta} \quad \text{for } \eta \vdash \nu.$$

By morass properties we have

$$\alpha_\nu = \sup\{\, \alpha_\eta | \eta \vdash \nu \,\}.$$

So (*) follows. For the other cases use (3), (4), (6), (7).

(P3) Let $\alpha < \kappa$. If $\gamma_\alpha \in A_0 \cup A_1$, then $\mathscr{F}_{\alpha,\alpha+1}$ is an amalgamation pair by (2). If $\gamma_\alpha \notin A_0 \cup A_1$, then $\bar{\gamma} = \gamma_{\alpha+1}$ is a successor point of S^0. So by morass properties $S_{\bar{\gamma}} = \{\nu\}$ for some ν. By the additional assumption (ii) of this section we have that $\bar{\nu} \prec_* \nu$ for some $\bar{\nu}$. So we have $\mathscr{F}_{\alpha,\alpha+1} = \{\, \tilde{\pi}_{\bar{\nu}\nu} \}$.

(P4) We only define a subset $\mathfrak{G}_\alpha \subseteq \bigcup_{\gamma<\alpha} \mathscr{F}_{\gamma\alpha}(\lim(\alpha))$. We leave it to the reader to check that one can always find a $g \in \mathfrak{G}_\alpha$ verifying (P4). So let $\alpha \leqslant \kappa$ be a limit ordinal. Set $\gamma = \gamma_\alpha$ and, if $\alpha < \kappa$, let $\nu = \nu_\gamma$.

Case 1. $\gamma \in A_1$ or $\alpha = \kappa$. Set $\mathfrak{G}_\alpha = \{\, \tilde{\pi}_{\bar{\tau}\tau} | \bar{\tau} \prec \tau, \tau \in S_\gamma, \tau < \sup S_\gamma \}$.

Case 2. $\gamma \in A_0$. So let ν immediately succeed τ in S_{γ_α}. Set $\mathfrak{G}_\alpha = \{\, \tilde{\pi}_{\bar{\tau}\tau} | \bar{\tau} \prec \tau \}$.

Case 3. $\gamma \in (S^0 \cap \kappa) - A$. Set $\mathfrak{G}_\alpha = \{\, \tilde{\pi}_{\bar{\nu}\nu} | \bar{\nu} \prec \nu \}$.

Case 4. $\gamma \in A - (A_0 \cup A_1)$. Set $\mathfrak{G}_\alpha = \{\, \sigma_{\eta\nu} | \eta \vdash \nu \}$.

(P5) is obvious.

2. Simplified morasses with linear limits in L. The following definition is due to Velleman again (see [7]).

DEFINITION. Let $\mathscr{M} = \langle \langle \theta_\alpha | \alpha \leqslant \kappa \rangle, \langle \mathscr{F}_{\alpha\beta} | \alpha < \beta \leqslant \kappa \rangle \rangle$ be a simplified $(\kappa, 1)$-morass. For limit ordinals $\alpha < \kappa$ let $X_\alpha = \{ \langle \beta, f \rangle | \beta < \alpha, f \in \mathscr{F}_{\beta\alpha} \}$. Order X_α as follows:

$$\langle \beta_1, f_1 \rangle < \langle \beta_2, f_2 \rangle \leftrightarrow \beta_2 < \beta_1 \quad \text{and} \quad \exists g \in \mathscr{F}_{\beta_2\beta_1} f_1 \circ g = f_2.$$

$\mathscr{M}$ is a *simplified* $(\kappa, 1)$*-morass with linear limits* if there are sequences $\langle \langle \beta_\delta^\alpha, f_\delta^\alpha \rangle | \delta < \tau \rangle$ for every limit ordinal $\alpha < \kappa$ s.t. $\langle \beta_\delta^\alpha, f_\delta^\alpha \rangle \in X_\alpha$ and

(P6)(a) $\delta < \bar{\delta} < \tau_\alpha \to \langle \beta_\delta^\alpha, \mathrm{f}_\delta^\alpha \rangle > \langle \beta_{\bar{\delta}}^\alpha, \mathrm{f}_{\bar{\delta}}^\alpha \rangle$.

(b)$\forall \langle \beta, g \rangle \in X_\alpha \exists \delta < \tau_\alpha \langle \beta_\delta^\alpha, f_\delta^\alpha \rangle < \langle \beta, g \rangle$.

(c) Let $\gamma < \tau_\alpha$, $\lim(\gamma)$, and set $\bar{\alpha} = \beta_\gamma^\alpha$.

Then $\tau_{\bar{\alpha}} = \gamma$ and $\forall \delta < \gamma (\beta_\delta^{\bar{\alpha}} = \beta_\delta^\alpha$ and $f_\gamma^\alpha \circ f_\delta^{\bar{\alpha}} = f_\delta^\alpha)$.

We first prove a simple result.

LEMMA 1. *Assume there exists a simplified* $(\kappa, 1)$*-morass with linear limits. Then* κ *is not weakly compact.*

PROOF. Let $\mathscr{M} = \langle\langle\theta_\alpha\rangle, \langle\mathscr{F}_{\beta\alpha}\rangle\rangle$ be such a structure and let the linear limits be given by $\langle\langle\beta^\alpha_\delta, f^\alpha_\delta\rangle|\delta < \tau_\alpha\rangle$. For limit ordinals $\alpha < \kappa$ set $C_\alpha = \{\beta^\alpha_\delta|\delta < \tau_\alpha\}$. It is easy to see that

(i) $C_\alpha \subseteq \alpha$ is club in α,

(ii) $\bar{\alpha}$ a limit point of $C_\alpha \to C_{\bar{\alpha}} = C_\alpha \cap \bar{\alpha}$.

We may assume that $\kappa > \omega$ is regular. We then show

(iii) there is no club $C \subseteq \kappa$ s.t.

$$\alpha \text{ a limit point of } C \to C_\alpha = C \cap \alpha.$$

Clearly, (i)–(iii) shows that κ is not weakly compact. So it remains to prove (iii). So assume that there is a counterexample C for (iii). We shall derive a contradiction by showing that $|\theta_\kappa| \leqslant \kappa$. For this we define $h\colon \theta_\kappa \to \kappa^2$ as follows. Let $\nu < \theta_\kappa$ be given. First let

$$B_\nu = \{\alpha < \kappa \,|\, \exists g \in \mathscr{F}_{\alpha\kappa}\, \exists \bar{\nu} < \theta_\alpha\, g(\bar{\nu}) = \nu\}.$$

Then B_ν is a nonempty final segment of κ by (P2), (P5). For each $\alpha \in B_\nu$ choose $\nu_\alpha < \theta_\alpha$ s.t. $g(\nu_\alpha) = \nu$ for some $g \in \mathscr{F}_{\alpha\kappa}$. By (P4), ν_α is uniquely determined by ν, α (and κ). Let C^* be the set of limit points of C. By Fodor there are δ, τ s.t.

$$E = \{\alpha \in C^* \cap B_\nu | f^\alpha_\delta(\tau) = \nu_\alpha\} \text{ is stationary in } \kappa.$$

We set $h(\nu) = \langle\delta, \tau\rangle$. By (P6)(c) and our assumption on C it follows that E is actually a final segment of C^*. But then it is easy to see that h is injective. Q.E.D.

We now prove a converse to Lemma 1 under the assumption $V = L$.

THEOREM 2. *Assume $V = L$. Let $\kappa > \omega$ be regular but not weakly compact. Then there is a simplified $(\kappa, 1)$-morass with linear limits.*

The rest of this section is devoted to the proof of Theorem 2. So assume $V = L$ from now on. Until further notice let $\kappa > \omega$ be an *arbitrary* regular $\kappa > \omega$. We shall need the "natural" $(\kappa, 1)$-morass in L and the "natural" $\square$-sequences. So we first recall the relevant definitions. Set

$$S = \{\nu|\nu > \omega, \nu \text{ p.r. closed}, \nu \text{ singular}\},$$

and

$$S^+ = \{\nu \in S|\alpha > \omega L_\nu \models \nu = \alpha^+ \wedge \alpha \text{ is regular}\}.$$

For $\nu \in S$ set

$\beta(\nu) =$ the least $\beta \geqslant \nu$ s.t. ν is Σ_ω-singular in J_β,

$n(\nu) =$ the least $n \geqslant 1$ s.t. ν is Σ_n-singular in $J_{\beta(\nu)}$,

$\rho(\nu) = \rho^{n(\nu)-1}_{\beta(\nu)}$,

$A(\nu) = A^{n(\nu)-1}_{\beta(\nu)}$ (= the appropriate master code),

$\mathfrak{A}(\nu) = \langle J_{\rho(\nu)}, A(\nu)\rangle$.

For $\nu \in S^+$ let

$\alpha_\nu =$ the unique α s.t. $L_\nu \models \nu = \alpha^+$,

$p(\nu) =$ the $<_*$-least $p \in J_{\rho(\nu)}$ s.t. $h(\alpha_\nu \cup p(\nu)) = J_{\rho(\nu)}$

where h is the canonical Σ_1-Skolem function for $\mathfrak{A}(\nu)$. A standard argument shows that:

(1) Let $\nu \in S$. Then $\mathrm{cf}(\nu) = \mathrm{cf}(\omega\rho(\nu))$.

We also set

$$S_\alpha = \{\nu \in S^+ | \alpha_\nu = \alpha\}.$$

For $\nu \in S_\alpha$ let

$$q(\nu) = \begin{cases} \langle \alpha, \tau, p(\nu)\rangle & \text{if } \rho(\nu) = \nu \text{ and immediately succeeds } \tau \text{ in } S_\alpha, \\ \langle \alpha, p(\nu)\rangle & \text{if } \rho(\nu) = \nu \text{ and } \nu \text{ is not a successor in } S_\alpha, \\ \langle \alpha, \nu, p(\nu)\rangle & \text{if } \nu < \rho(\nu). \end{cases}$$

We then define a relation $\prec$ on S^+ by:

Let $\bar{\nu}, \nu \in S^+$. Then

$$\bar{\nu} \prec \nu \text{ iff there exists } f\colon \mathfrak{A}(\bar{\nu}) \to_{\Sigma_1} \mathfrak{A}(\nu) \text{ s.t.}$$

$$f \restriction \alpha_{\bar{\nu}} = \mathrm{id} \restriction \alpha_{\bar{\nu}} \text{ and } q(\nu) \in \mathrm{rng}(f),$$

f is uniquely determined by $\bar{\nu}, \nu$.

We set $\pi_{\bar{\nu}\nu} = f \restriction \bar{\nu} \cup \langle \bar{\nu}, \nu\rangle$ and $\bar{\pi}_{\bar{\nu}\nu} = f$. Let $\mathfrak{S} = \{\langle \alpha, \nu\rangle | \nu \in S^+,\ \alpha = \alpha_\nu,$ $\alpha \leqslant \kappa$ and, if $\kappa = \lambda^+, \alpha > \lambda\}$. Jensen has shown

(2) $$\langle \mathfrak{S}, \prec, \langle \pi_{\bar{\nu}\nu}\rangle_{\bar{\nu} \prec \nu}\rangle \text{ is a } (\kappa, 1)\text{-morass.}$$

(Note that our definition of $q(\nu)$ guarantees that we do not have to consider Q-embeddings.) Now let

$$S^0 = \{\alpha | S_\alpha \neq \varnothing\}.$$

For $\alpha \in S^0 \cap S$ set $\nu_\alpha = \max S_\alpha$. We then set

$A' = \{\alpha \in S^0 \cap S | \nu_\alpha$ is not a limit point in $\prec\}$,

$A = \{\alpha \in S^0 \cap S | \mathfrak{A}(\nu_\alpha) = \mathfrak{A}(\alpha)\}$.

Clearly, we have $A \subseteq A'$ and by (1)

(3) $$\alpha \in A \to \mathrm{cf}(\alpha) = \mathrm{cf}(\nu_\alpha).$$

We also need

(4) $$\alpha \in A' - A \to \mathrm{cf}(\alpha) = \mathrm{cf}(\nu_\alpha) = \omega.$$

PROOF. Let $\alpha \in A' - A$, $\nu = \nu_\alpha$, $\mathfrak{A} = \mathfrak{A}(\nu)$, $\rho = \rho(\nu)$, $q = q(\nu)$ and $\delta = \sup\{\alpha_{\bar{\nu}} | \bar{\nu} \prec \nu\}$. So we have $\delta < \alpha$ and

(*) α is Σ_1-regular in $\mathfrak{A}$.

By fine structure theory we also have

(**) Let $X \prec_{\Sigma_1} \mathfrak{A}$ s.t. $\delta, q \in X$ and $X \cap \alpha$ is transitive. Then $X = \mathfrak{A}$.

Let h be the canonical Σ_1-Skolem function for $\mathfrak{A}$. Define a sequence $\langle \delta_n | n < \omega\rangle$ by

$$\delta_0 = \delta + 1, \qquad \delta_{n+1} = \sup(\alpha \cap h(\delta_n \cup \{q\})).$$

Then by (*), (**) we clearly have $\delta_n < \delta_{n+1} < \alpha$ and $\alpha = \sup_n \delta_n$. So we already have $\mathrm{cf}(\alpha) = \omega$. We now show $\mathrm{cf}(\omega\rho) = \omega$ which suffices by (1). This is clear if ρ is a successor ordinal. So let ρ be a limit ordinal.

Let $\mu_n = \sup((ON \cap h(\delta_n \cup \{q\})))$. Clearly, $\omega\rho = \sup_n \mu_n$. So it suffices to show that $\mu_n < \omega\rho$ for all $n < \omega$. Assume that $\mu_n = \omega\rho$. We shall derive a contradiction by showing $\delta_{n+1} = \delta_{n+2}$. For $\lambda < \rho$ set

$$\mathfrak{A} \upharpoonright \lambda = \langle J_\lambda^{A(\nu)}, A(\nu) \cap J_\lambda^{A(\nu)} \rangle$$

and let h_λ be the canonical Σ_1-Skolem function for $\mathfrak{A} \upharpoonright \lambda$. Then $h = \bigcup_{\lambda<\rho} h_\lambda$. For $\max q < \lambda < \rho$ and $\gamma < \alpha$ set

$$\tau_{\lambda,\gamma} = \sup(\alpha \cap h_\lambda(\gamma \cup \{q\})).$$

Then $\{\tau_{\lambda,\gamma}\}$ is Σ_1-definable in $\mathfrak{A}$ with the parameters λ, γ, q. (Note that α occurs in q.) Hence

$$\delta_{n+2} = \sup\{\tau_{\lambda,\gamma} | \lambda, \gamma \in h(\delta_n \cup \{q\})\} \leqslant \delta_{n+1}. \quad \text{Q.E.D.(4)}$$

We also have

(5) Let ν be a successor point in S_α. Then $\mathrm{cf}(\nu) = \omega$.

PROOF. Let τ be the predecessor of ν. Then ν = the least $\mu > \beta(\tau)$ s.t. μ is pr. rec. closed. Q.E.D. (5)

We also need the natural $\square_\infty$-sequence for L. Let $\alpha \in S$ and let h be the canonical Σ_1-Skolem function for $\mathfrak{A}(\alpha)$. We set

$$p^*(\alpha) = \text{the } <_*\text{-least } p \in J_{\rho(\alpha)} \text{ s.t. } h(\alpha \cup \{p\}) = J_{\rho(\alpha)},$$

$$\alpha_\alpha^* = \text{the maximal } \delta < \alpha \text{ s.t. } h(J_\delta \times \{p(\alpha)\}) \cap \alpha \subseteq \omega\delta,$$

$$q^*(\alpha) = \begin{cases} \langle \alpha_\alpha^*, p^*(\alpha), \alpha \rangle & \text{if } \alpha < \rho(\alpha), \\ \langle \alpha_\alpha^*, p^*(\alpha) \rangle & \text{if } \alpha = \rho(\alpha). \end{cases}$$

$\mathbf{F}^*(\alpha) = \{f | f\colon \overline{\mathfrak{A}} \to_{\Sigma_1} \mathfrak{A}(\alpha),\ \overline{\mathfrak{A}} \text{ transitive},\ q^*(\alpha) \in \mathrm{rng}(f)\}$. For $f \in \mathbf{F}^*(\alpha)$ set $\lambda(f) = \sup(\alpha \cap \mathrm{rng}(f))$. It is proved in [**1**] that for all $f \in \mathbf{F}^*(\alpha)$ there is a unique $f_0 \in \mathbf{F}^*(\lambda(f))$ s.t. $\alpha \cap \mathrm{rng}(f) = \alpha \cap \mathrm{rng}(f_0)$. We set $f_0 = \mathrm{red}(f)$. Then set $C_\alpha = \{\lambda(f) | f \in \mathbf{F}^*(\alpha)\} - \{\alpha\}$. The following properties are proved in [**1**].

(6)(a) $C_\alpha \subseteq S \cap \alpha$ is closed in α,

(b) $\mathrm{otp}(C_\alpha) < \alpha$,

(c) $\beta \in C_\alpha \to C_\beta = C_\alpha \cap \beta$,

(d) $\mathrm{cf}(\alpha) > \omega \to \sup C_\alpha = \alpha$.

(7)(a) Let $\beta \in C_\alpha$; then there is a unique $k = k_{\beta\alpha}$ s.t. $k\colon \mathfrak{A}(\beta) \to_{\Sigma_0} \mathfrak{A}(\alpha)$, $k \upharpoonright \beta = \mathrm{id} \upharpoonright \beta$ and $k(p^*(\beta)) = p^*(\alpha)$,

(b) let $\beta \in C_\alpha$, $k = k_{\beta\alpha}$, $f \in \mathbf{F}^*(\alpha)$, $\beta = \lambda(f)$ and $f_0 = \mathrm{red}(f)$; then $f = k \circ f_0$ (moreover, k is not cofinal and $k(q^*(\beta)) = q^*(\alpha)$),

(c) $\sup C_\alpha = \alpha \to \mathfrak{A}(\alpha) = \bigcup\{\mathrm{rng}\, k_{\beta\alpha} | \beta \in C_\alpha\}$.

Now let $\alpha \in A$. For $\beta \in C_\alpha$ set

$$\lambda_{\beta\alpha} = \sup(\mathrm{rng}\, k_{\beta\alpha} \cap \nu_\alpha).$$

Note that $\lambda_{\beta\alpha} < \nu_\alpha$. We have:

(8) Let $\alpha \in A$ and $\beta \in C_\alpha$. Then $\beta \in A$, $\nu_\beta \prec \lambda_{\beta\alpha}$ and $\pi_{\nu_\beta}, \lambda_{\beta\alpha} \upharpoonright \nu_\beta = k_{\beta\alpha} \upharpoonright \nu_\beta$.

PROOF. Let $k = k_{\beta\alpha}$, $\lambda = \lambda_{\beta\alpha}$,

$$\omega\rho' = \sup(\mathrm{rng}\, k \cap ON) \quad \text{and} \quad \mathfrak{A}' = \langle J_{\rho'}, A(\alpha) \cap J_{\rho'} \rangle.$$

Then $k\colon \mathfrak{A}(\beta) \to_{\Sigma_1} \mathfrak{A}'$ cofinally. So $\mathfrak{A}'$ is amenable. Let h' be the canonical Σ_1-Skolem function for $\mathfrak{A}'$, and let $\sigma\colon \langle J_{\bar\rho}, \bar A\rangle \leftrightarrow h'(\lambda \cup p(\alpha))$. Then $\langle J_{\bar\rho}, \bar A\rangle$ is amenable and $\sigma\colon \langle J_{\bar\rho}, \bar A\rangle \to_{\Sigma_0} \mathfrak{A}(\alpha)$. So by the downward extension of embeddings lemma there is some $\bar\beta$ s.t. $\langle J_{\bar\rho}, \bar A\rangle = \langle J_{\rho_{\bar\beta}^{n-1}}, A_{\bar\beta}^{n-1}\rangle$ where $n = n(\alpha)$. Clearly, λ is Σ_1-singular in $\langle J_{\bar\rho}, \bar A\rangle$. But then $\langle J_{\bar\rho}, \bar A\rangle = \mathfrak{A}(\lambda)$. It is also easy to see that $\lambda \in S_\alpha$, and a standard argument shows that $\sigma(p(\lambda)) = p^*(\alpha) = p(\nu_\alpha)$. Now set $f = \sigma^{-1} \circ k$. It is easy to check that $f\colon \mathfrak{A}(\nu_\beta) \to_{\Sigma_1} \mathfrak{A}(\lambda)$, $f \upharpoonright \beta = \mathrm{id} \upharpoonright \beta$, $f(q(\nu_\beta)) = q(\lambda)$. Moreover, we have $f \upharpoonright \nu_\beta = k \upharpoonright \nu_\beta$. Q.E.D.

(9) Let $\alpha \in A$, $\beta, \gamma \in C_\alpha$, $\gamma < \beta$. Then $\pi_{\nu_\beta, \lambda_{\beta\alpha}}(\lambda_{\gamma\beta}) = \lambda_{\gamma\alpha}$.

PROOF. Note that $k_{\gamma\alpha} = k_{\beta\alpha} \circ k_{\gamma\beta}$. So it suffices to show that $k_{\beta\alpha}(\lambda_{\gamma\beta}) = \sup\{k_{\beta\alpha}(\delta) | \delta < \lambda_{\gamma\beta}\}$. But this follows from $\mathfrak{A}(\beta) \models \mathrm{cf}(\lambda_{\gamma\beta}) \leqslant \gamma < \beta$ and $k_{\beta\alpha} \upharpoonright \beta = \mathrm{id} \upharpoonright \beta$. Q.E.D.

Now set

$\bar A = \{\alpha \in A | \nu_\alpha \text{ is a limit point of } S_\alpha\}$,

$\bar A_0 = \{\alpha \in \bar A \mid \nu_\alpha \text{ is minimal in } \prec\}$,

$\bar A_1 = \{\alpha \in \bar A \mid \exists \bar\nu(\bar\nu \prec_* \nu_\alpha \text{ and } \pi_{\bar\nu\nu_\alpha} \text{ is not cofinal }\}$,

$\bar A_2 = \{\alpha \in \bar A \mid \exists \bar\nu(\bar\nu \prec_* \nu_\alpha \text{ and } \pi_{\bar\nu\nu_\alpha} \text{ is cofinal }\}$.

So $\bar A$ is the disjoint union of $\bar A_0, \bar A_1, \bar A_2$.

(10)(a) $\alpha \in \bar A_0 \to C_\alpha \subseteq \bar A_0$,

(b) Let $\alpha \in \bar A_1$, $\bar\nu \prec_* \nu_\alpha$, $\lambda = \sup(\mathrm{rng}\, \pi_{\bar\nu\nu_\alpha} \cap \nu_\alpha)$, let $\beta \in C_\alpha$ and let $k_{\beta\alpha}(\bar\lambda) = \lambda$. Then we have

$$\beta \in \bar A_1, \bar\nu \prec_* \nu_\beta \quad \text{and} \quad \bar\lambda = \sup\left(\mathrm{rng}\, \pi_{\bar\nu\nu_\beta} \cap \nu_\beta\right),$$

(c) let $\alpha \in \bar A_2$ and $\bar\nu \prec_* \nu_\alpha$; then

$$\beta \in C_\alpha \to \left(\beta \in \bar A_2 \text{ and } \nu_\beta \vdash \nu_\alpha\right).$$

PROOF. Let $\alpha \in \bar A$, $\beta \in C_\alpha$, $k = k_{\beta\alpha}$. We know that $\beta \in A$ and $k | J_{\nu_\beta}\colon J_{\nu_\beta} \to_{\Sigma_1} J_{\nu_\alpha}$, $k \upharpoonright \beta = \mathrm{id} \upharpoonright \beta$, $k(\beta) = \alpha$. So we have $\beta \in \bar A$. Set $r(\alpha) = \alpha$, if $\rho(\alpha) = \nu_\alpha$, and $r(\alpha) = \langle \alpha, \nu\rangle$, if $\nu_\alpha < \rho(\alpha)$. Let $\alpha^* = \alpha^*_\alpha$. Let h be the canonical Σ_1-Skolem function of $\mathfrak{A}(\alpha) = \mathfrak{A}(\nu_\alpha)$. By the usual fine structure arguments contained in the proof of (2) we have

$(*)$ $\qquad \nu_\alpha$ is a successor in $\prec \leftrightarrow \left(\alpha^* > 0 \text{ and } r(\alpha) \in h(\alpha^* \cup p(\nu))\right)$.

If $\bar\nu \prec_* \nu_\alpha$, then $\mathfrak{A}(\bar\nu) \cong h(\alpha^* \cup p(\nu))$ and the unique isomorphism extends $\pi_{\bar\nu\nu_\alpha}$.

But then (a) follows from the facts that $\alpha^*_\beta = \alpha^*_\alpha$, $k(r(\beta)) = r(\alpha)$ and $k \upharpoonright \beta = \mathrm{id} \upharpoonright \beta$. So let us assume now that $\alpha \in \bar A_1 \cup \bar A_2$. Choose some $f \in \mathfrak{A}^*(\alpha)$ s.t. $\lambda(f) = \beta$ and let $f_0 = \mathrm{red}(f)$. By (7(b)) we have $f = k \circ f_0$. So letting $f(\bar r) = r(\alpha)$ it is easy to see that $f_0(\bar r) = r(\beta)$. It follows that $\nu_\beta \in \bar A_1 \cup \bar A_2$ and letting $\bar\tau \prec_* \nu_\beta$ we have $\bar\tau \in S_{\alpha^*} \cap (\bar\nu + 1)$. Set $\tilde\lambda = \lambda_{\beta\alpha}$. We have

$$\begin{aligned} \tilde\lambda \cap \mathrm{rng}\, \pi_{\bar\tau\tilde\lambda} &= k''\left(\nu_\beta \cap h_{\mathfrak{A}(\beta)}(\alpha^* \cup p^*(\beta))\right) \\ &\subseteq h_{\mathfrak{A}(\alpha)}\left(\alpha^* \cup p^*(\alpha)\right) = \nu_\alpha \cap \mathrm{rng}\, \pi_{\bar\nu\nu_\alpha}. \end{aligned}$$

So we get

$(**)$ $$\pi_{\bar\tau\tilde\lambda} \upharpoonright \bar\tau = \pi_{\bar\nu\nu_\alpha} \upharpoonright \bar\tau.$$

We now prove (b). It is easy to see that we only have to show that in this case $\bar{\tau} = \bar{\nu}$. So it suffices to show that $h_{\mathfrak{A}}(\alpha^* \cup p^*(\alpha)) \subseteq h_{\mathfrak{A} \restriction \eta}(\alpha^* \cup p^*(\alpha))$ where

$$\mathfrak{A} = \mathfrak{A}(\alpha), \quad \omega\eta = \sup(ON \cap \operatorname{rng} k) \quad \text{and} \quad \mathfrak{A} \restriction \eta = \langle J_\eta, A(\alpha) \cap J_\eta \rangle.$$

Now it is easy to see that

$$h_{\mathfrak{A}}(\alpha^* \cup p^*(\alpha)) \subseteq h_{\mathfrak{A} \restriction \bar{\eta}}(\alpha^* \cup p^*(\alpha)) \quad \text{where } \omega\bar{\eta} = \sup(On \cap \operatorname{rng} \pi_{\bar{\nu}\nu_\alpha}).$$

So it suffices to show that $\bar{\eta} \leqslant \eta$. But it is easy to check that $\lambda \in h_{\mathfrak{A} \restriction \eta}(\alpha \cup p^*(\alpha))$ (since $\lambda \in \operatorname{rng} k$) and $\lambda \notin h_{\mathfrak{A} \restriction \bar{\eta}}(\alpha \cup p^*(\alpha))$. This proves (b).

We now prove (c). It is sufficient to show that in this case

$$(\dagger) \qquad \pi_{\bar{\nu}\nu_\alpha}(\bar{\tau}) = \tilde{\lambda} \quad \text{and} \quad \tilde{\lambda} = \sup(\tilde{\lambda} \cap \operatorname{rng} \pi_{\bar{\nu}\nu_\alpha}).$$

For this we need more information from [**1**]. For $\eta \in S^+$ set

$\mathbf{F}(\eta) = \{ f | f \colon \bar{\mathfrak{A}} \to_{\Sigma_1} \mathfrak{A}(\nu), \bar{\mathfrak{A}} \text{ transitive}, q(\eta) \in \operatorname{rng}(f)\}$,

$C'_\eta = \{\sup(\eta \cap \operatorname{rng}(f)) | f \in \mathbf{F}(\eta)\} - \{\eta\}$.

We have:

(i) Let δ be pr. rec. closed s.t. $\eta = \sup(\eta \cap h_{\mathfrak{A}(\eta)}(\delta \cup q(\eta)))$, then $\operatorname{otp}(C'_\eta) \leqslant \delta$ (cf. the proof of Lemma 6.35 in [**1**, p. 224]),

(ii) $\bar{\eta} \in C'_\eta \to C'_{\bar{\eta}} = \bar{\eta} \cap C'_\eta$,

(iii) $\langle J_\eta, C'_\eta \rangle$ is amenable,

(iv) Let $\bar{\eta} \prec \eta$. Then $\pi_{\bar{\eta}\eta} \colon \langle J_{\bar{\eta}}, C'_{\bar{\eta}} \rangle \to_{\Sigma_0} \langle J_\eta, C'_\eta \rangle$.

Applying this to our special pair $\bar{\nu}$, ν_α we get that $\operatorname{otp}(C'_{\nu_\alpha}) \leqslant \alpha^*$, $\pi''_{\bar{\nu}\nu_\alpha} C'_{\bar{\nu}} = C'_{\nu_\alpha}$ and $\eta = \sup(\eta \cap \operatorname{rng} \pi_{\bar{\nu}\nu_\alpha})$ for every $\eta \in C'_{\nu_\alpha}$. Hence (†) will follow from

$$(\dagger\dagger) \qquad \tilde{\lambda} \in C'_{\nu_\alpha}.$$

To see this just note that if $\pi_{\bar{\nu}\nu_\alpha}(\tau') = \tilde{\lambda}$ then $\tau' \prec \tilde{\lambda}$, hence $\tau' = \bar{\tau}$ as $\bar{\tau} \prec \tilde{\lambda}$, $\alpha_{\tau'} = \alpha_{\bar{\tau}}$. So it remains to prove (††). Choose some $f \in \mathbf{F}^*(\alpha)$ s.t. $\lambda(f) = \beta$. A little argument (which we leave to the reader again) shows that $q(\nu_\alpha) \in \operatorname{rng}(f)$. Hence $f \in \mathbf{F}(\nu)$. Now let $f_0 = \operatorname{red}(f)$. By (7(b)) we have $f = k \circ f_0$. The argument from Lemma 6.19 in [**1**, p. 208] shows that $\nu_\beta = \sup(\nu_\beta \cap \operatorname{rng} f_0)$. But then $\sup(\nu_\alpha \cap \operatorname{rng} f) = \sup(\nu_\alpha \cap \operatorname{rng} k) = \tilde{\lambda}$. Hence $\tilde{\lambda} \in C'_{\nu_\alpha}$, since $f \in \mathbf{F}(\nu)$. Q.E.D.

We now use our assumption that κ is not weakly compact. So choose some $B \subseteq \kappa$ and a first order formula ϕ s.t.

$$\forall D \subseteq \kappa \, J_\kappa \vDash \phi(D, B),$$

but $\forall \beta < \kappa \exists D \subseteq \beta \, J_\beta \vDash \neg\phi(D, B \cap \beta)$. Using this we thin out the given $(\kappa, 1)$-morass. For $\alpha \leqslant \kappa$ set

$$\tilde{S}_\alpha = \{\nu \in S_\alpha | B \cap \alpha \in J_\nu \text{ and } \forall D \in J_\nu \cap \mathfrak{P}(\beta) \, J_\alpha \vDash \phi(D, B \cap \alpha)\},$$

$$\tilde{S}^+ = \bigcup_{\alpha \leqslant \kappa} \tilde{S}_\alpha, \quad \tilde{\mathfrak{S}} = \{\langle \alpha, \nu \rangle | \nu \in \tilde{S}^+, \alpha = \alpha_\nu\}, \quad \tilde{S}^0 = \{\alpha | \tilde{S}_\alpha \neq \varnothing\}.$$

For $\bar{\nu}, \nu \in \tilde{S}^+$ let

$$\bar{\nu} \prec' \nu \quad \text{iff } (\bar{\nu} \prec \nu \text{ and } B \cap \alpha_\nu \in \operatorname{rng} \pi_{\bar{\nu}\nu}).$$

We clearly have

(11)(a) $\langle \tilde{\mathfrak{S}}, \prec', \langle \pi_{\bar{\nu}\nu} \rangle_{\bar{\nu} \prec' \nu} \rangle$ is a $(\kappa, 1)$-morass,

(b) $\alpha \in \tilde{S}^0 \cap \kappa \to \max(\tilde{S}_\alpha)$ exists.

We need the $\square$-type sequence $\langle \tilde{C}_\alpha | \alpha \in \tilde{S}^0 \cap \kappa \rangle$ constructed in Theorem 6.1 of [**3**]. For $\alpha \in \tilde{S}^0 \cap \kappa$ set

$$\bar{\beta}(\alpha) = \text{the least } \beta > \alpha \text{ s.t. } B \cap \alpha \in J_\beta$$
$$\text{and } \exists D \in \mathfrak{P}(\alpha) \cap J_\beta J_\alpha \vDash \neg\phi(D, B \cap \alpha).$$

We then define $\tilde{C}_\alpha$ as follows:

Case 1. α is regular or $\bar{\beta}(\alpha) < \beta(\alpha)$. Let $\bar{\beta} = \bar{\beta}(\alpha)$. For $\delta < \alpha$ let

$$X_\delta = \text{the Skolem hull of } \delta \cup \{\alpha, B \cap \alpha\} \text{ in } J_{\bar{\beta}}.$$

Then set

$$\tilde{C}_\alpha = \{\delta < \alpha | \delta = X_\delta \cap \alpha\}.$$

Case 2. α is singular and $\beta(\alpha) \leqslant \bar{\beta}(\alpha)$. For $\gamma \in C_\alpha$ let $\tilde{k}_{\gamma\alpha}\colon J_{\beta(\gamma)} \to_{\Sigma_{n-1}} J_{\beta(\alpha)}$ be the canonical extension of $k_{\gamma\alpha} (n = n(\alpha))$. Then set

$$\tilde{C}_\alpha = \{\gamma \in C_\alpha | \bar{q}(\nu_\alpha), B \cap \alpha \in \operatorname{rng} \tilde{k}_{\gamma\alpha}\}$$
$$\text{where } \bar{q}(\nu_\alpha) \text{ is defined by } q(\nu_\alpha) = \bar{q}(\nu_\alpha) * p(\nu_\alpha).$$

The following properties are implicitly proved in [**3**]:

(12)(a) $\tilde{C}_\alpha \subseteq \tilde{S}^0 \cap \alpha$ is closed in α,

(b) $\operatorname{cf}(\alpha) > \omega \to \sup \tilde{C}_\alpha = \alpha$,

(c) $\gamma \in \tilde{C}_\alpha \to \tilde{C}_\gamma = \tilde{C}_\alpha \cap \gamma$,

(d) $E = \{\alpha \in \tilde{S}^0 \cap \kappa | \sup \tilde{C}_\alpha < \alpha\}$ is stationary in κ.

For $\alpha \in \tilde{S}^0 \cap \kappa$ set

$$\tilde{\nu}_\alpha = \max \tilde{S}_\alpha.$$

Set $P = \{\alpha \in \tilde{S}^0 \cap \kappa | \tilde{\nu}_\alpha \text{ is a limit point in } \prec'\}$.

(13) Let $\alpha \in P$. Then

$$\gamma \in \tilde{C}_\alpha \to \tilde{\nu}_\gamma \prec' \tilde{\nu}_\alpha.$$

PROOF. *Case* 1. α is regular or $\bar{\beta}(\alpha) < \beta(\alpha)$. So let X_γ be defined as above and let $\sigma\colon J_\eta \xrightarrow{\sim} X_\gamma \prec J_{\bar{\beta}(\alpha)}$. It is easy to check that $\sigma(\tilde{\nu}_\beta) = \tilde{\nu}_\alpha$ and $\sigma(\beta(\tilde{\nu}_\beta)) = \beta(\tilde{\nu}_\alpha)$. It follows easily that $\tilde{\nu}_\gamma \prec' \tilde{\nu}_\alpha$.

Case 2. α is singular and $\beta(\alpha) \leqslant \bar{\beta}(\alpha)$. Note that $\nu_\alpha = \tilde{\nu}_\alpha$. Let $k = k_{\gamma\alpha}$ and let $\tilde{k}$ be the canonical extension of k (by fine structure theory). If $\beta(\tilde{\nu}_\alpha) < \beta(\alpha)$, then it is easy to see that $\tilde{k}(\beta(\tilde{\nu}_\gamma)) = \beta(\nu_\alpha)$ and the claim follows as in Case 1. So let $\beta(\alpha) = \beta(\tilde{\nu}_\alpha)$. Since $\tilde{\nu}_\alpha$ is a limit point in $\prec'$, we have $n(\alpha) < n(\tilde{\nu}_\alpha)$. So we have

$$\tilde{k} \restriction J_{\rho(\tilde{\nu}_\gamma)}\colon \mathfrak{A}(\tilde{\nu}_\gamma) \to_{\Sigma_1} \mathfrak{A}(\tilde{\nu}_\alpha) \quad \text{and} \quad \tilde{k}\left(p^n_{\beta(\tilde{\nu}_\gamma)}\right) = p^n_{\beta(\tilde{\nu}_\alpha)} \quad \text{where } n = n(\tilde{\nu}_\alpha).$$

It follows easily that $\tilde{k}(p(\tilde{\nu}_\alpha)) = p(\tilde{\nu}_\alpha)$. But then by definition of $\tilde{C}_\alpha$ we have $q(\tilde{\nu}_\alpha)$, $B \cap \alpha \in \operatorname{rng} \tilde{k}$. So $\tilde{\nu}_\gamma \prec' \tilde{\nu}_\alpha$. Q.E.D.

We are now ready to finish the proof of Theorem 2. We may assume w.l.o.g. that our morass $\langle \tilde{\mathfrak{S}}, \dots \rangle$ satisfies the condition (ii) in §1. So let $\mathscr{M} = \langle \langle \theta_\alpha \rangle, \langle \mathscr{F}_{\alpha\beta} \rangle \rangle$ be the simplified $(\kappa, 1)$-morass constructed from $\langle \tilde{\mathfrak{S}}, \dots \rangle$ as in §1. We shall show that $\mathscr{M}$ has linear limits. We first set

$\tilde{A} = \{\alpha \in \tilde{S}^0 \cap \kappa | \tilde{\nu}_\alpha$ is not a limit point in $\prec'$, but $\tilde{\nu}_\alpha$ is a limit point in $\tilde{S}_\alpha\}$,

$\tilde{A}_0 = \{\alpha \in \tilde{A} | \tilde{\nu}_\alpha$ is minimal in $\prec'\}$,

$\tilde{A}_1 = \{\alpha \in \tilde{A} | \exists \bar{\nu} (\bar{\nu} \prec'_* \tilde{\nu}_\alpha$ and $\pi_{\bar{\nu}\tilde{\nu}_\alpha}$ is not cofinal $\}$,

$\tilde{A}_2 = \{\alpha \in \tilde{A} | \exists \bar{\nu} (\bar{\nu} \prec'_* \tilde{\nu}_\alpha$ and $\pi_{\bar{\nu}\tilde{\nu}_\alpha}$ is cofinal $\}$,

$\tilde{B} = \{\alpha \in \tilde{S}^0 \cap \kappa | \tilde{\nu}_\alpha$ is not a limit point in $\prec'$, $\tilde{\nu}_\alpha$ is a successor point in $\tilde{S}_\alpha\}$.

The following properties are easily verified:

(14)(a) $\alpha \in \tilde{A} \cup \tilde{B} \to \alpha \in S$ and $\tilde{\nu}_\alpha = \nu_\alpha$,

(b) the properties (8)–(10) are true if we replace $\prec$, C_α, $\overline{A_i}$ by $\prec'$, $\tilde{C}_\alpha$, $\tilde{A}_i \cap A$, we denote these new properties by $(\tilde{8})$, $(\tilde{9})$, $(\widetilde{10})$.

(Of course, we may have that $\alpha \in \tilde{A}_0$, but $\alpha \in \overline{A}_1$.)

Let $\bar{\tilde{S}}$ be defined from $\tilde{S}^0$ as in §1, and let g: $\bar{\tilde{S}} \to \kappa + 1$ be the order isomorphism. It is easy to see that $\{\alpha < \kappa | \lim(\alpha)\}$ is the disjoint union of $\tilde{A}_0$, $\tilde{A}_1$, $\tilde{A}_2$, $\tilde{B}$, P. We shall now define the linear limits distinguishing several cases. In each case the verification of (P6(a)), (P6(b)) is straightforward and is left to the reader. We shall check only (P6(c)). Note that (P6(c)) has only to be checked when $\tau_\alpha > \omega$. We use the notations from §1. In addition, we introduce the following notation:

Let $\bar{\nu} \prec' \nu$, $g(\alpha_{\bar{\nu}}) = \delta$. Then set

$$\tilde{\tilde{\pi}}_{\bar{\nu}\nu} = \begin{cases} \tilde{\pi}_{\bar{\nu}\nu} \circ g_\delta^0 & \text{if } g_\delta^0 \text{ is defined,} \\ \tilde{\pi}_{\bar{\nu}\nu} & \text{otherwise.} \end{cases}$$

Let $\langle \gamma_\delta^\alpha | \delta < \bar{\tau}_\alpha \rangle$ be the increasing enumeration of $\tilde{C}_\alpha$. Now let $\alpha < \kappa$, $\lim(\alpha)$, be given and let $\alpha = g(\bar{\alpha})$. We set $\tilde{\nu} = \tilde{\nu}_{\bar{\alpha}}$.

Case 1. $\bar{\alpha} \in P$, $\sup \tilde{C}_{\bar{\alpha}} < \bar{\alpha}$. Then $\mathrm{cf}(\bar{\alpha}) = \omega$ by (12(a)). Choose $\langle \nu_n | n < \omega \rangle$ s.t. $\nu_n \prec' \nu_{n+1} \prec' \tilde{\nu}$ and $\bar{\alpha} = \sup_n \alpha_{\nu_n}$. Then set $\tau_\alpha = \omega$ and

$$\langle \beta_n^\alpha, f_n^\alpha \rangle = \langle g(\alpha_{\nu_n}), \tilde{\tilde{\pi}}_{\nu_n\tilde{\nu}} \rangle.$$

Case 2. $\bar{\alpha} \in P$, $\sup \tilde{C}_{\bar{\alpha}} = \bar{\alpha}$. For $\gamma = \gamma_\delta^{\bar{\alpha}} \in \tilde{C}_\alpha$ set $\eta_\delta = \tilde{\nu}_\gamma$. Then $\eta_\delta \prec' \tilde{\nu}$ by (13). We set $\tau_\alpha = \bar{\tau}_{\bar{\alpha}}$ and $\langle \beta_\delta^\alpha, f_\delta^\alpha \rangle = \langle g(\gamma_\delta^{\bar{\alpha}}), \tilde{\tilde{\pi}}_{\eta_\delta\tilde{\nu}} \rangle$. (P6) follows from (12(c)).

Case 3. $\bar{\alpha} \in \tilde{B}$. Then $\mathrm{cf}(\bar{\alpha}) = \omega$ by (3), (4), and (14(a)). Let $\tilde{\nu}$ immediately succeed $\tilde{\tau}$ in $\tilde{S}_{\bar{\alpha}}$. Choose $\langle \tau_n | n < \omega \rangle$ s.t. $\tilde{\tau}_n \prec' \tilde{\tau}_{n+1} \prec' \tilde{\tau}$ and $\bar{\alpha} = \sup_n \alpha_{\tilde{\tau}_n}$. Then set $\tau_\alpha = \omega$ and $\langle \beta_n^\alpha, f_n^\alpha \rangle = \langle g(\alpha_{\tilde{\tau}_n}), \tilde{\tilde{\pi}}_{\tilde{\tau}_n,\tilde{\nu}} \rangle$.

Case 4. $\bar{\alpha} \in \tilde{A}_0 \cup \tilde{A}_1$ and ($\bar{\alpha} \notin A$ or $\sup \tilde{C}_{\bar{\alpha}} < \bar{\alpha}$). Then $\mathrm{cf}(\bar{\alpha}) = \mathrm{cf}(\tilde{\nu}) = \omega$ by (3), (4), (12(b)), and (14(a)). So we can choose sequences $\langle \nu_n | n < \omega \rangle$, $\langle \bar{\nu}_n | n < \omega \rangle$ s.t.

(i) $\nu_n < \nu_{n+1} < \nu$, $\nu_n \in \tilde{S}_\alpha$,

(ii) $\bar{\nu}_n \prec' \nu_n$,

(iii) $\nu_n \in \mathrm{rng}\, \pi_{\bar{\nu}_{n+1},\nu_{n+1}}$.

Then set $\tau_\alpha = \omega$ and $\langle \beta_n^\alpha, f_n^\alpha \rangle = \langle g(\alpha_{\bar{\nu}_n}), \tilde{\tilde{\pi}}_{\bar{\nu}_n,\nu_n} \rangle$.

Case 5. $\bar{\alpha} \in \tilde{A}_2$ and ($\bar{\alpha} \notin A$ or $\sup \tilde{C}_{\bar{\alpha}} < \bar{\alpha}$). Then we have again $\mathrm{cf}(\bar{\alpha}) = \mathrm{cf}(\tilde{\nu}) = \omega$. So we can choose $\langle \eta_n | n < \omega \rangle$ s.t. $\eta_n \vdash \eta_{n+1} \vdash \tilde{\nu}$ and $\bar{\alpha} = \sup_n \alpha_{\eta_n}$. Then set $\tau_\alpha = \omega$ and $\langle \beta_n^\alpha, f_n^\alpha \rangle = \langle g(\alpha_{\eta_n}), \sigma_{\eta_n \tilde{\nu}} \rangle$.

Case 6. $\bar{\alpha} \in \tilde{A}_0 \cap A$ and $\sup \tilde{C}_{\bar{\alpha}} = \bar{\alpha}$. For $\gamma = \gamma_\delta^{\bar{\alpha}} \in \tilde{C}_{\bar{\alpha}}$ set $\eta_\delta = \tilde{\nu}_\gamma$ and $\lambda_\delta = \lambda_{\gamma\bar{\alpha}}$. Then $\eta_\delta \prec' \lambda_\delta$ by ($\tilde{8}$). We set $\tau_\alpha = \bar{\tau}_{\bar{\alpha}}$ and $\langle \beta_\delta^\alpha, f_\delta^\alpha \rangle = \langle g(\gamma_\delta^{\bar{\alpha}}), \bar{\tilde{\pi}}_{\eta_\delta \lambda_\delta} \rangle$ for $\delta < \tau_\alpha$. Then (P6(c)) follows from (9), (10(a)),....

Case 7. $\bar{\alpha} \in \tilde{A}_1 \cap A$ and $\sup \tilde{C}_{\bar{\alpha}} = \bar{\alpha}$. For $\gamma = \gamma_\delta^{\bar{\alpha}} \in \tilde{C}_{\bar{\alpha}}$ set $\eta_\delta = \tilde{\nu}_\gamma$. Then $\eta_\delta \vdash \tilde{\nu}$ by $(\widetilde{10}(c))$. We set $\tau_\alpha = \bar{\tau}_{\bar{\alpha}}$ and $\langle \beta_\delta^\alpha, f_\alpha^\delta \rangle = \langle g(\gamma_\delta^{\bar{\alpha}}), \sigma_{\eta_\delta \tilde{\nu}} \rangle$. Then (P6(c)) follows from (10(c)),....

Case 8. $\bar{\alpha} \in \tilde{A}_2 \cap A$ and $\sup \tilde{C}_{\bar{\alpha}} = \bar{\alpha}$. Let $\tilde{\nu} \prec'_* \tilde{\nu}$ and $\lambda = \sup(\tilde{\nu} \cap \mathrm{rng}\, \pi_{\tilde{\nu}\tilde{\nu}})$. Set $C = \{\gamma \in \tilde{C}_{\bar{\alpha}} | \lambda \in \mathrm{rng}\, k_{\gamma\bar{\alpha}}\}$. Let $\langle \bar{\gamma}_\delta | \delta < \tau \rangle$ be the monotone enumeration of C. For $\gamma = \bar{\gamma}_\delta \in C$ set $\eta_\delta = \nu_\gamma$ and $\lambda_\delta = \lambda_{\gamma\bar{\alpha}}$. Then $\eta_\delta \prec' \lambda_\delta$ by ($\tilde{8}$). We set $\tau_\alpha = \tau$ and $\langle \beta_\delta^\alpha, f_\delta^\alpha \rangle = \langle g(\bar{\gamma}_\delta), \bar{\tilde{\pi}}_{\eta_\delta \nu_\delta} \rangle$ for $\delta < \tau$. (P6(c)) follows from ($\tilde{9}$), ($\widetilde{10}$(b)), and (12(c)).

3. Built-in $\diamondsuit$. For some applications (for example, to questions treated in [**4**]) one seems to need a special built-in $\diamondsuit$-principle. The relevant definition is due to Velleman again (see [7]). We present the obvious translation to the "expanded" version. Let $\mathscr{L}$ be a (recursive) language with countably many symbols of all types.

DEFINITION. Let $\mathscr{M} = \langle \langle \theta_\alpha | \alpha \leqslant \kappa \rangle, \langle \mathscr{F}_{\alpha\beta} | \alpha < \beta \leqslant \kappa \rangle \rangle$ be a simplified $(\kappa, 1)$-morass with linear limits $\langle \langle \beta_\delta^\alpha, f_\delta^\alpha \rangle | \delta < \tau_\alpha \rangle$ ($\alpha < \kappa$, $\lim(\alpha)$). Then $\langle \mathfrak{A}_\alpha | \alpha \in E \rangle$ is an *adequate $\diamondsuit$-sequence for $\mathscr{M}$* if the following conditions are satisfied:

(a) $E \subseteq \kappa$; for all limit ordinals $\alpha < \kappa$, $\delta < \tau_\alpha$: $\beta_\delta^\alpha \notin E$.

(b) $\mathfrak{A}_\alpha$ is an $\mathscr{L}$-structure with universe θ_α.

(c) Let $\mathfrak{A}$ be an $\mathscr{L}$-structure with universe κ^+. Then there are $\alpha \in E$ and $f \in \mathscr{F}_{\alpha\kappa}$ s.t. $f\colon \mathfrak{A}_\alpha \to_{\Sigma_\omega} \mathfrak{A}$.

We now strengthen Theorem 2.

THEOREM 3. *Assume $V = L$. Let $\kappa > \omega$ be regular but not weakly compact. Then there is a simplified* $(\kappa, 1)$*-morass with linear limits and an adequate $\diamondsuit$-sequence.*

PROOF. Let $\mathscr{M}$ be the simplified $(\kappa, 1)$-morass with linear limits constructed in the proof of Theorem 2. We use the notations of that proof. For simplicity, we assume that defining $\mathscr{M}$ we used the $<_L$-least counterexample B to weak compactness. Then $\mathscr{M}$ is definable in $J_{\kappa^{++}}$ (without parameters). We have to define the adequate $\diamondsuit$-sequence. First set

$$E = \left\{ \alpha \in \tilde{S}^0 \cap \kappa \,|\, |\tilde{S}_\alpha| \geqslant 2, \sup \tilde{C}_\alpha < \alpha \right\}.$$

Note that $g \restriction E = \mathrm{id} \restriction E$ where $g\colon \bar{\tilde{S}} \to \kappa + 1$ is the order isomorphism used in §2. So by construction we have:

(15) For all limit ordinals $\alpha < \kappa$, $\delta < \tau_\alpha$: $\beta_\delta^\alpha \notin E$.

We now define the adequate $\diamondsuit$-sequence $\langle \mathfrak{A}_\alpha | \alpha \in E \rangle$ by recursion as follows: Let $\alpha \in E$. $\mathfrak{A}_\alpha$ = the $<_L$-least $\bar{\mathfrak{A}}$ which satisfies

(i) $\bar{\mathfrak{A}}$ is an $\mathscr{L}$-structure with universe θ_α,

(ii) there is no $f \in \mathscr{F}_{\beta\alpha}$, $\beta \in E \cap \alpha$, s.t. $f\colon \mathfrak{A}_\beta \to_{\Sigma_\omega} \overline{\mathfrak{A}}$.

If there is no such $\overline{\mathfrak{A}}$, let $\overline{\mathfrak{A}}$ be the $<_L$-least $\overline{\mathfrak{A}}$ satisfying (i).

We only have to show that $\langle \mathfrak{A}_\alpha | \alpha \in E \rangle$ satisfies condition (c) of the relevant definition. For this, we shall need the following technical result. For notational reasons we again identify $\langle W_\alpha, <_\alpha \rangle$ and $\langle \theta_\alpha, < \rangle$.

(16) Let $\nu \in \tilde{S}_\kappa$ s.t. $\nu = \operatorname{otp}(\tilde{S}_\kappa \cap \nu)$ and let ν' be the minimal $\prec'$-predecessor of ν. Then there is a club $C \subseteq \kappa$ s.t.

$$\bar{\nu} \prec' \nu \quad \text{and} \quad \alpha_{\bar{\nu}} \in C \to \pi_{\bar{\nu}\nu} \upharpoonright \bar{\nu} = \tilde{\pi}_{\bar{\nu}\nu} \upharpoonright \langle \bar{\nu}, \nu' \rangle .$$

PROOF. It is easy to check that there is a club $\bar{C} \subseteq \kappa$ s.t.

$$\nu = \bigcup \{ \operatorname{rng}(\tilde{\pi}_{\bar{\nu}\nu} \upharpoonright \langle \bar{\nu}, \nu' \rangle) | \bar{\nu} \prec' \nu, \alpha_{\bar{\nu}} \in \bar{C} \} .$$

On the other hand, we also have

$$\nu = \bigcup \{ \operatorname{rng}(\pi_{\bar{\nu}\nu} \upharpoonright \bar{\nu}) | \bar{\nu} \prec' \nu \} .$$

So we have two representations of ν as the union of an increasing continuous union of sets of cardinality less than κ. Hence a familiar argument using the regularity of κ gives the conclusion. Q.E.D. (16)

Now let X be the smallest elementary submodel of $J_{\kappa^{++}}$ s.t. $X \cap \kappa$ is transitive. Let $\sigma\colon J_\gamma \xrightarrow{\sim} X$ and let $\sigma(\bar{\alpha}) = \kappa$, $\sigma(\bar{\nu}) = \kappa^+$. Finally, we show

(17)(a) $\bar{\alpha} \in E$,

(b) $\sigma(\langle\langle \theta_\alpha | \alpha \leqslant \bar{\alpha} \rangle, \langle \mathscr{F}_{\alpha\beta} | \alpha < \beta \leqslant \bar{\alpha} \rangle\rangle) = \mathscr{M}$,

(c) $\sigma(\langle \mathfrak{A}_\alpha | \alpha \in E \cap \bar{\alpha} \rangle) = \langle \mathfrak{A}_\alpha | \alpha \in E \rangle$,

(d) $\sigma \upharpoonright \bar{\nu} \in \mathscr{F}_{\bar{\alpha}\kappa}$.

We postpone the proof of (17) for a moment to show how to finish the proof of the theorem. We argue by contradiction. So let $\mathfrak{A}$ be the $<_L$-least counterexample to condition (c). Then $\mathfrak{A}$ is definable in $J_{\kappa^{++}}$. So let $\sigma(\overline{\mathfrak{A}}) = \mathfrak{A}$. then (17(a)–(c)) shows that $\overline{\mathfrak{A}} = \mathfrak{A}_{\bar{\alpha}}$. But then $f\colon \mathfrak{A}_{\bar{\alpha}} \to_{\Sigma_\omega} \mathfrak{A}$, where $f = \sigma \upharpoonright \bar{\nu}$, and $f \in \mathscr{F}_{\bar{\alpha}\kappa}$ (by (17(d))), $\bar{\alpha} \in E$. This contradicts the definition of $\mathfrak{A}$.

So we only have to prove (17).

PROOF OF (17). The argument from the proof of Lemma 5.2 in [**3**] shows that $\beta(\bar{\alpha}) = \gamma + 1$ and $n(\bar{\alpha}) = 1$. It follows easily that $\tilde{C}_{\bar{\alpha}} = \varnothing$. So we already get that $\bar{\alpha} \in E$. We also get

$$(*) \qquad \beta(\bar{\nu}) = \gamma + 1, \qquad n(\bar{\nu}) = 1, \qquad \bar{\nu} = \nu_\alpha = \tilde{\nu}_\alpha, \quad \bar{\nu} \text{ is minimal in } \prec'.$$

But then it is easy to see that $\bar{\nu} = \theta_{\bar{\alpha}}$. So the only nontrivial point concerning (b) is to show that $\sigma(\mathscr{F}_{\alpha\bar{\alpha}}) = \mathscr{F}_{\alpha\kappa}$ for all $\alpha < \bar{\alpha}$. The verification is left to the reader. (Note that $\tilde{\nu}_\alpha = \bar{\nu}$ is minimal in $\prec'$.) But then (c) is obvious. We now turn to (d). Let $\bar{X}$ be the smallest elementary substructure of $J_{\kappa^{++}}$ s.t. $\kappa \subseteq \bar{X}$ and let $\bar{\sigma}\colon \bar{X} \xrightarrow{\sim} J_\delta$, $\bar{\sigma}(\nu) = \kappa^+$. Clearly, $\nu \in \tilde{S}_\kappa$. The same argument as above shows that

$$(**) \qquad \beta(\nu) = \delta + 1, \qquad n(\nu) = 1.$$

Another easy argument shows $p(\nu) = \delta$. Now let $\pi = \bar{\sigma}^{-1} \circ \sigma$, hence $\pi\colon J_\gamma \to_{\Sigma_\omega} J_\delta$. Let $\bar{\pi} \supseteq \pi$ be the unique extension of π s.t. $\bar{\pi}\colon J_{\gamma+1} \to_{\Sigma_1} J_{\delta+1}$ and $\bar{\pi}(\gamma) = \delta$. Then $\bar{\pi}$ shows that $\bar{\nu} \prec' \nu$ and $\pi_{\bar{\nu}\nu} \upharpoonright \bar{\nu} = \bar{\pi} \upharpoonright \bar{\nu} = \sigma \upharpoonright \bar{\nu}$. Applying (16) we see that there is an unbounded $H \subseteq \tilde{S}_{\bar{\alpha}} \cap \bar{\nu}$ s.t., for all $\bar{\tau} \in H$, $\pi_{\bar{\tau}\tau} \upharpoonright \bar{\tau} = \tilde{\pi}_{\bar{\tau}\tau} \upharpoonright \langle \bar{\tau}, \tau' \rangle$, where $\tau = \sigma(\bar{\tau})$, $\tau' =$ the minimal $\prec'$ predecessor of $\bar{\tau}$. So finally we get

$$\sigma \upharpoonright \bar{\nu} = \pi_{\bar{\nu}\nu} \upharpoonright \bar{\nu} = \bigcup_{\bar{\tau} \in H} \pi_{\bar{\tau}\tau} \upharpoonright \bar{\tau} = \bigcup_{\bar{\tau} \in H} \tilde{\pi}_{\bar{\tau}\tau} \upharpoonright \langle \bar{\tau}, \tau' \rangle = \tilde{\pi}_{\bar{\nu}\nu} \in \mathscr{F}_{\bar{\alpha}\kappa}.$$

This finishes the proof of Theorem 3.

REMARK. Note that $\mathrm{cf}(\alpha) = \omega$ for all $\alpha \in E$.

A careful reader will have noticed that the arguments in §2 contain a lot of information for *arbitrary* regular $\kappa > \omega$ in L. Namely, we implicitly proved the following theorem. All details of the proof are left to the reader.

THEOREM 4. *Assume $V = L$. Let $\kappa > \omega$ be regular. Then there is a simplified $(\kappa, 1)$-morass such that for all singular limit ordinals $\alpha < \kappa$ there are sequences satisfying the conditions* (P6(a)-(c)).

If κ is not ineffable in L one can produce a slightly better version which is sufficient to construct a "Todorcevic tree" (see [7]). But we do not give details since we already constructed such trees in L by different methods (see [2]).

REFERENCES

1. A. Beller, R. Jensen and P. Welch, *Coding the universe*, London Math. Soc. Lecture Note Ser., vol. 47, Cambridge Univ. Press, London and New York, 1982.

2. H.-D. Donder, *Coarse morasses in L*, Set Theory and Model Theory (R. Jensen and A. Prestel, eds.), Lecture Notes in Math. vol. 872, Springer-Verlag, Berlin and New York, 1981, pp. 37–54.

3. R. B. Jensen, *The fine structure of the constructible hierarchy*, Ann. Math. Logic **4** (1972), 229–308.

4. A. Kanamori, *On Silver's and related principles*, Logic Colloquium '80 (D. van Dahlen, D. Lascar and J. Smiley, eds.), North-Holland, Amsterdam, 1982.

5. L. Stanley, *A short course on gap-1 morasses*, Proc. Cambridge Summer School in Set Theory, 1978.

6. D. Velleman, *Simplified morasses*, J. Symbolic Logic **49** (1984), 257–271.

7. ______, *Simplified morasses with linear limits* (to appear).

MATHEMATISCHES INSTITUT, UNIVERSITÄT BONN, BERINGSTR. 6, 5300 BONN, WEST GERMANY

Current address: Institut für Mathematik II, Freie Universität Berlin, Arnimallee 3, 1000 Berlin 33, West Germany

Proceedings of Symposia in Pure Mathematics
Volume **42**, 1985

Condensation-Coherent Global Square Systems

H. D. DONDER, R. B. JENSEN AND L. J. STANLEY

0. Introduction. Various refinements of Jensen's original construction [**5**] of square systems have been made (see, e.g., [**1**, 7]). Later, Jensen found another, superficially very different, construction [**6**]. In this paper we present a further refinement which has enabled us to construct a global system of condensation-coherent square sequences. If $\kappa < \mu$ are uncountable cardinals, the collapses to level κ of segments of the square-sequence based at μ cohere, in a sense to be made precise below (see §4), with the square-sequence based at κ. We are especially interested in singular cardinals.

Stanley, in [**9**], first isolated coherence properties of the square-sequences constructed in [**5**] and used these coherence properties as an approach to higher-gap morasses as well as to directly derive two cardinal versions of the $\diamondsuit^+$ principles. This was carried much further by Donder in 1979. He formulated versions of the coherence properties appropriate for singular cardinals and showed them to hold for the system of square-sequences which Jensen constructed using his second approach. He exploited these properties to obtain results on the regularity and indecomposability of uniform ultrafilters in different L-like situations. This work, as yet unpublished, will appear elsewhere [**4**].

Upon seeing Donder's proof, Jensen introduced the main technical innovation of this paper which allowed him to "globalize" Donder's coherence properties. Prior to this, the closed unbounded (club) subset of a pseudo-successor cardinal ν assigned to ν by a square-sequence was constructed simultaneously with a club subset of $\omega\rho$, the ordinal of the Σ_1-collapsing structure of ν. The construction involved α_ν, the largest cardinal in the sense of J_ν, in an essential way. This was the obstacle to a global coherence property. Jensen's solution was, essentially, to directly consider the class of Σ_1-collapsing structures (J_ρ, A, p) independently of the pseudo-successor cardinals ν of which they are the Σ_1-collapsing structures

1980 *Mathematics Subject Classification*. Primary 03E45, 03E05.

1 Research partially supported by NSF grant MCS 830142.

and to define a preliminary, global, totally condensation-coherent square-sequence assigning to each $\mathfrak{A}^+ = (J_\rho, A, p)$ a club subset $\tilde{C}_{\mathfrak{A}^+}$ of $\omega\rho$. The desired system of square sequences is then constructed as follows: Given a pseudo-successor cardinal ν, letting $\mathfrak{A}^+$ be the Σ_1-collapsing structure for ν, $\tilde{C}_{\mathfrak{A}^+}$ is thinned and projected onto a club subset $C_\nu \subseteq \nu$ which is assigned to ν by the appropriate square-sequence. α_ν still figures in the "ν-thinning" of $\tilde{C}_{\mathfrak{A}^+}$ so the coherence is not completely perfect (this seems unavoidable and was already the case in Donder's work) but the role of α_ν is much less important.

Jensen's work was not widely circulated when, in early 1982, Avraham and Shelah formulated a global combinatorial principle which they called "Squared Scales" and whose statement evolved into the principle of §4. They sought such a principle in order to extend Shelah's work on "strong covering", viz. [**8**, Chapter XIII], and suggested to Stanley that he prove that the principle holds in L. Stanley quickly saw the close connection to Donder's work and isolated the obstacle to globalizing the coherence properties, but was ignorant of Jensen's work. He had hit upon essentially the same approach as Jensen to globalizing the coherence properties (but without having fully worked out the details) when Donder informed him of Jensen's work.

1. Preliminaries. Our notation and terminology, where unexplicated, are intended to be standard or have a clear meaning, e.g., card x for the cardinality of x, o.t. (x) for the order type of x. We assume the reader is familiar with the *basics* of the fine structure theory of L (as developed, e.g., in [**2**, **5**, or **10**]) up to and including the definitions and basic properties of the J-hierarchy, the projecta, standard codes and standard parameters, and the downward-extension-of-embeddings lemma.

For ordinals α, $\breve{\alpha}$ denotes $\omega\alpha$.

To simplify the statements of definitions, lemmas and theorems, we assume, for the remainder of this paper, that $V = L$.

(1.1) DEFINITION. $\nu \in S \Leftrightarrow J_\nu \models$ "there is a largest cardinal, which is uncountable". If $\nu \in S$, α_ν is the largest cardinal in the sense of J_ν.

(1.2) DEFINITION. If $\breve{\nu}$ is not a cardinal ($\nu \notin$ CARD), $\beta(\nu)$ is the collapsing ordinal of $\breve{\nu}$, i.e. the least β such that

$$J_{\beta+1} \models \text{“}\breve{\nu} \textit{ is not a cardinal}\text{”}.$$

(1.3) Now suppose $\nu \in S \setminus$ CARD.

DEFINITION. $m(\nu) =$ the least m such that there is a $\Sigma_{m+1}(J_{\beta(\nu)})$-function from a subset of a smaller ordinal onto $\breve{\nu}$. Let $n(\nu) = 0$ if $m(\nu) = 0$; otherwise, let $n(\nu) = m(\nu) - 1$.

In other words $(\beta(\nu), n(\nu))$ is the lexicographically-least pair (β, n) such that $\breve{\nu}$ is not a Σ_{n+1}-cardinal in J_β.

(1.4) DEFINITION. If $\nu \in S \setminus$ CARD, let $\beta = \beta(\nu)$, $n = n(\nu)$. Then

$$\rho(\nu) = \rho^n_\beta; \quad A(\nu) = A^n_\beta; \quad \mathfrak{A}(\nu) = \left(J_{\rho(\nu)}, A(\nu)\right).$$

$\mathfrak{A}(\nu)$ is the Σ_1-collapsing structure for $\breve{\nu}$.

(1.5) This terminology is justified by the following

PROPOSITION (a) $\rho(\nu) \geqslant \nu$,
(b) *there is a* $\Sigma_1(J_{\rho(\nu)}, A(\nu))$*-function from a subset of* α_ν *onto* $\check{\nu}$,
(c) $\omega\rho_{\beta(\nu)}^{n(\nu)+1} \leqslant \alpha_\nu$.

PROOF. Let $\rho = \rho(\nu)$, $n = n(\nu)$, $\beta = \beta(\nu)$, $A = A(\nu)$. For $m \leqslant n$, we show by induction on m that $\rho_m \geqslant \nu$, where $\rho_m = \rho_\beta^m$, which gives (a). For $m = 0$, this is just the trivial remark that $\beta \geqslant \nu$. Now let $m < n$ be such that $\rho_m > \nu$. If $\rho_{m+1} < \nu$, then there is a $\Sigma_1(J_{\rho_m}, A_\beta^m)$-function g from a subset of $\check{\rho}_{m+1}$ onto J_{ρ_m}. $g|g^{-1}[\check{\nu}]$ is also $\Sigma_1(J_{\rho_m}, A_\beta^m)$ (in parameter $\check{\nu}$ if $\nu < \rho_m$ and the other parameters involved in the $\Sigma_1(J_{\rho_m}, A_\beta^m)$-definition of g) and maps a subset of $\check{\rho}_{m+1} < \check{\nu}$ onto $\check{\nu}$. By fine structure $g|g^{-1}[\check{\nu}]$ is $\Sigma_{m+1}(J_\beta)$. But $m < n$, so $m + 1 < n + 1$, contradicting the defining property of n, since $n > m \geqslant 0$.

(b) is now clear, since there is a function g from a subset of some $\tau < \check{\nu}$ onto $\check{\nu}$ which is $\Sigma_{n+1}(J_\beta)$. But by (a) and fine structure, g is $\Sigma_1(J_\rho, A)$. Now if $\tau > \alpha_\nu$, there is an $f \in J_\nu \subseteq J_\rho$ such that $f\colon \alpha_\nu \to_{\text{onto}} \tau$. Then $g \circ f$ is $\Sigma_1(J_\rho, A)$ (in parameter f and the parameters involved in the $\Sigma_1(J_\rho, A)$-definition of g) and maps a subset of α_ν onto $\check{\nu}$.

For (c), we first show that $\rho_{n+1} \leqslant \nu$. This is immediate from the fact that if $\nu < \rho_{n+1}$, then any g as in the proof of (b) must lie in J_ρ, since (J_ρ, g) is amenable, by fine structure. Now suppose $\rho_{n+1} = \nu$. Then, by (b), ρ_{n+1} fails to be a $\Sigma_1(J_\rho, A)$-cardinal, contradicting the minimality of ρ_{n+1} for a $\Sigma_1(J_\rho, A)$-mapping onto $\check{\rho}$. So $\rho_{n+1} < \nu$. If $\alpha_\nu < \check{\rho}_{n+1} < \check{\nu}$ then $\check{\rho}_{n+1}$ is not a cardinal in $J_\nu \subseteq J_\rho$, which again would contradict the minimality for a $\Sigma_1(J_\rho, A)$-mapping onto $\check{\rho}$ property.

(1.6) A careful analysis of the proof of (2.10) of [5] will show that for all ξ:

(1) There is a finite $a \subseteq \check{\xi}$ and $f\colon \check{\xi} \to_{\text{onto}} J_\xi$ such that f is $\Sigma_1(J_\xi)$ in parameter a.

This easily yields, as a corollary to (1.5)(b):

(2) There is a finite $p \subseteq \check{\rho}(\nu)$ such that $h''(\omega \times (\alpha_\nu \cup p)) = J_{\rho(\nu)}$, where h is the canonical Σ_1-Skolem function for $\mathfrak{A}(\nu)$.

DEFINITION. (a) If a, b are finite sets of ordinals, let $a <_* b$ iff there is an η such that $a \setminus \eta = b \setminus \eta$ and $(\max(a \cap \eta) < \max(b \cap \eta)$ or $a \cap \eta = \varnothing$ and $b \cap \eta \neq \varnothing)$.

(b) If $\nu \in S \setminus \mathrm{CARD}$, $\mathfrak{A} = \mathfrak{A}(\nu)$ and $h =$ the canonical Σ_1-Skolem function for $\mathfrak{A}(\nu)$, let $p(\nu) =$ the $<_*$-least finite $p \subset \omega \cdot \rho(\nu)$ such that $h''(\omega \times (\alpha_\nu \cup p)) = J_{\rho(\nu)}$.

(c) Let $\mathfrak{A}^+(\nu) = (\mathfrak{A}(\nu), p(\nu))$.

(1.7) The following result and its corollary, (1.8), will be very useful in §2. Though it is part of the folklore of the subject, to our knowledge it has not appeared in print prior to now. This reuslt is a uniform version of a theorem of Lévy.

LEMMA. $(\forall\beta)(\forall\gamma < \beta)(J_\beta \models$ "*γ is an uncountable cardinal*" $\Rightarrow J_\gamma \prec_{\Sigma_1} J_\beta)$.

PROOF. Clearly $\check{\gamma} = \gamma$. We shall assume that the necessary number of iterations of the $\cup$ function have been added to the finite basis for the rudimentary functions in order to make the S_α's transitive. Let $\varphi = \exists v\, \theta(v, \vec{w})$ be Σ_1, let $\vec{x} \in J_\gamma$ and suppose that $y \in J_\beta$ is such that $\theta(y, \vec{x})$. We show that there is a $y' \in J_\gamma$ such that $\theta(y', \vec{x})$.

Let η^* be the least $\eta < \check{\beta}$ such that there is a $y \in S_\eta$ with $\theta(y, \vec{x})$, and let y^* be some such y. So $\eta^* = \check{\alpha} + n$ for some $\alpha < \beta$ and $0 < n < \omega$. Without loss of generality, assume $\gamma \leqslant \alpha$. But then $y^* = f(\vec{w}^*, J_\alpha)$ for some $\vec{w}^* \in J_\alpha$, and some rudimentary f. Further, there is a first order $\theta^*(\vec{v}', \vec{v})$ such that for all $\vec{a}, \vec{b}$, $J_\alpha \models \theta^*(\vec{a}', \vec{b})$ iff $\theta(f(\vec{a}, J_\alpha), \vec{b})$. This follows easily from the fact that rudimentary functions are simple (viz. Lemma 1.2 of **[5]**) and the well-known reduction of properties which are Σ_0 in parameters from $J_\alpha \cup \{J_\alpha\}$ to properties which are first-order definable over J_α.

Let g be a J_α-definable Σ_k-Skolem function for J_α, where θ^* is Σ_k (g will be $\Sigma_{k+2}(J_\alpha)$ or $\Sigma_{k+3}(J_\alpha)$). Let s be the function of (1.10) of **[5]** (suitably modified to reflect the enlarged finite basis necessary to make the S_ξ's transitive). Let $\xi < \gamma$ be such that $\vec{x} \in J_\xi$ (so $\vec{x} \cup \{\vec{x}\} \in J_\xi$). Let $\tilde{s}(a) = s(a, J_\alpha)$. Then $\tilde{s}^n \circ g \in \underset{\sim}{J_\beta}$ and so $y^* \in X \in J_\beta$, where $X = \tilde{s}^n \circ g''(\omega \times (J_\xi \times \{\vec{w}^*\}))$. Let $\pi\colon (X, \in) \leftrightarrow (\overline{X}, \in)$, where $\overline{X}$ is transitive, so π is the collapsing map. Further, there is an $\overline{\eta} \leqslant \eta^*$ such that $\overline{X} = S_{\overline{\eta}}$. This is an instance of a general condensation lemma for the S-hierarchy (see **[3]**); the crucial observation was already made in **[5]**: $\pi(\tilde{s}(a)) = s(\pi(a), \pi(J_\alpha))$. Thus $\overline{X}$ is $S_{\overline{\eta}}$ for some $\overline{\eta}$, and clearly this $\overline{\eta}$ must be $\leqslant \eta^*$. Also, $\pi(\vec{x}) = \vec{x}$, by our choice of ξ, and $S_{\overline{\eta}} \models \theta(\pi(y^*), \vec{x})$, so $\theta(\pi(y^*), \vec{x})$ in fact holds.

But η^* was chosen minimal for the existence of such a y, so $\overline{\eta} = \eta^*$. But then it is easily seen that $\pi = \mathrm{id}|\, S_{\eta^*}$. Then, however, $\tilde{s}^n \circ g \in J_\beta$ and maps a subset of $\omega \times J_\xi \times \{\vec{w}^*\}$ onto S_{η^*}. But $\omega \cdot \xi < \omega \cdot \gamma < \eta^*$, and $J_\beta \models$ "$\omega \cdot \gamma$ is a cardinal", a contradiction, so in fact, $\eta^* < \omega \cdot \gamma$, as required.

(1.8) COROLLARY. *Suppose* $\nu \in S \setminus \mathrm{CARD}$, $\beta = \beta(\nu)$ *and* $n = n(\nu) = 0$. *Then*
(a) $p(\nu) \notin J_\gamma$, *whenever* $J_\beta \models$ "*γ is a cardinal*" *and* $\gamma > \alpha_\nu$,
(b) $J_\beta \models$ "*there is a largest cardinal*",
(c) *if* $f\colon J_{\overline{\beta}} \to_{\Sigma_1} J_\beta$, $p(\nu) \in \mathrm{range}\, f$, *then* "*the largest cardinal in the sense of* J_β" $\in$ *range* f.

PROOF. If $p(\nu) \in J_\gamma$, then we would have $J_\beta = h''(\omega \times (\alpha_\nu \cup p(\nu))) \subseteq J_\gamma$, since by Lemma (1.7), $J_\gamma \prec_{\Sigma_1} J_\beta$ (here h is the canonical Σ_1-Skolem function for J_β). (b) is clear from (a). For (c), let δ be the largest cardinal in the sense of J_β. Then $\delta = \mathrm{card}^{J_\beta}(\zeta)$ for some $\zeta \in p(\nu)$ (else we would have $p(\nu) \subseteq J_\delta$ so $p(\nu) \in J_\delta$). So δ is Σ_1-definable in J_β from some member of $p(\nu)$, which makes (c) clear.

2. An auxiliary square system.

(2.1) In this section, we shall construct a square system on a class containing the $\mathfrak{A}^+(\nu)$ for $\nu \in S$. This is Jensen's crucial improvement over Donder's methods. The approach is similar in spirit to that of **[6]**. It should be noted that α_ν has not been *entirely* banished from the picture: $p(\nu)$ is defined in terms of α_ν and it is possible that $\mathfrak{A}(\nu) = \mathfrak{A}(\tau)$, but $\mathfrak{A}^+(\nu) \neq \mathfrak{A}^+(\tau)$. It should, however, be observed that if $\nu < \tau$, $\mathfrak{A}(\nu) = \mathfrak{A}(\tau)$, then $p(\tau) \leqslant_* p(\nu)$ and so

$$(*) \qquad \{p(\tau): \mathfrak{A}(\tau) = \mathfrak{A}(\nu)\} \quad \text{is finite.}$$

(2.2) DEFINITION. Let $\mathscr{A} = \{\mathfrak{A}(\nu): \nu \in S\}$; $\mathscr{A}^+ = \{\mathfrak{A}^+(\nu): \nu \in S\}$. For $\mathfrak{A} \in \mathscr{A}$ (resp. $\mathfrak{A}^+ \in \mathscr{A}^+$), let $N(\mathfrak{A}) = \{\nu \in S: \mathfrak{A} = \mathfrak{A}(\nu)\}$ (resp. $N^+(\mathfrak{A}^+) = \{\nu \in S: \mathfrak{A}^+ = \mathfrak{A}^+(\nu)\}$). If $A \in \mathscr{A}^+$, let $p^{\mathfrak{A}}$, $\mathfrak{A}^- \in \mathscr{A}$ be such that $\mathfrak{A} = (\mathfrak{A}^-, p^{\mathfrak{A}})$.

Then, rephrasing (2.1)(*), we have $\{\mathfrak{B} \in \mathscr{A}^+: \mathfrak{B}^- = \mathfrak{A}^-\}$ is finite.

(2.3) DEFINITION. Suppose $\mathfrak{C} = (J_\eta, E)$ is amenable. Let

$$\text{CARD}(\mathfrak{C}) = \{\tau: \mathfrak{C} \models \text{“}\tau \text{ is an uncountable cardinal”}\},$$

$$\text{RCARD}(\mathfrak{C}) = \{\tau: \mathfrak{C} \models \text{“}\tau \text{ is a regular uncountable cardinal”}\}.$$

(2.4) Suppose $\mathfrak{A}^+ = (\mathfrak{A}, p) \in \mathscr{A}^+$, say $\mathfrak{A}^+ = \mathfrak{A}^+(\nu)$. Suppose $f: (\overline{\mathfrak{A}}, p) \to_{\Sigma_1} \mathfrak{A}^+$, where $|\overline{\mathfrak{A}}|$ is transitive (so $|\overline{\mathfrak{A}}| = J_{\bar\rho}$, $\overline{\mathfrak{A}} = (J_{\bar\rho}, \bar A)$ is amenable). Fine structure theory then gives us the canonical extension $\tilde f: J_{\bar\beta} \to \Sigma_{n(\nu)+1} J_{\beta(\nu)}$, $\tilde f \supseteq f$, where $\bar\rho = \rho_{\bar\beta}^{n(\nu)}$, $\bar A = A_{\bar\beta}^{n(\nu)}$.

If $n(\nu) = 0$, then $\text{CARD}(\overline{\mathfrak{A}}) \neq \varnothing$. Also, if $f \neq \text{id}|\ J_{\bar\rho}$, then, letting $\bar\tau$ be the critical point of f (i.e. letting $\bar\tau$ be such that $f|\bar\tau = \text{id}|\bar\tau$, $f(\bar\tau) > \bar\tau$), we have $\bar\tau \in \text{RCARD}(\overline{\mathfrak{A}})$. In any case, $f''\text{CARD}(\overline{\mathfrak{A}}) \subseteq \text{CARD}(\mathfrak{A})$, $f''\text{RCARD}(\overline{\mathfrak{A}}) \subseteq \text{RCARD}(\mathfrak{A})$.

Further, let $\bar\alpha = f^{-1}[\alpha_\nu]$. Then clearly, letting $\bar h$ = the canonical Σ_1-Skolem function for $\overline{\mathfrak{A}}$, we have

(1) $J_{\bar\rho} = \bar h''(\omega \times (\bar\alpha \cup p))$, and $\bar p$ if the $<_*$-least finite $q \in \mathscr{P}(\omega\bar\rho)$ with this property.

Now suppose $n(\nu) > 0$. Then $\tilde f: J_{\bar\beta} \to_{\Sigma_2} J_\beta$, so

(2) $\text{CARD}(\overline{\mathfrak{A}})$ has limit order type iff $\text{CARD}(\mathfrak{A})$ has limit order type and if $\bar\sigma$ is the largest element of $\text{CARD}(\overline{\mathfrak{A}})$, then $f(\bar\sigma)$ is the largest element of $\text{CARD}(\mathfrak{A})$.

So, suppose $n(\nu) = 0$. We claim

(3) $\text{CARD}(\overline{\mathfrak{A}})$, $\text{CARD}(\mathfrak{A})$ have largest elements, say $\bar\sigma$, σ and $f(\bar\sigma) = \sigma$. This is clear by (1.8).

Finally, in all cases, we have

(4) ($\bar\alpha < \check{\bar\rho}$ and $f(\bar\alpha) \geqslant \alpha_\nu$) or ($\bar\alpha = \check{\bar\rho}$, $\bar p = p = \varnothing$, and $\text{CARD}(\overline{\mathfrak{A}})$, $\text{CARD}(\mathfrak{A})$ have limit order type). This is clear since $p \subseteq \alpha \Rightarrow p = \varnothing$.

Our inability to rule out the second case of (4) is one of the reasons we must consider structures other than the $\mathfrak{A}^+(\nu)$. Another reason, leading to yet another type of structure, is discussed below.

(2.5) DEFINITION. $(\mathfrak{B}, q) \in \tilde{\mathscr{A}}^+ \Leftrightarrow \mathfrak{B} = (B, \in, A')$, $q \in B$, B is transitive and

$$(*) \quad (\exists(\mathfrak{A}, p) \in \mathscr{A}^+)(\exists U \subseteq |\mathfrak{A}|)(U \text{ is rud}(\mathfrak{A})\text{-closed} \wedge p \in U \wedge (\mathfrak{B}, q) \cong (\mathfrak{A}, p)|\ U).$$

(2.6) Thus, $(\mathfrak{B}, q) \in \tilde{\mathscr{A}}^+ \Rightarrow B = J_\eta$, $\check\eta = OR \cap B$, $(B, \in, A')$ is amenable. Further, if $(\mathfrak{A}, p) \in \mathscr{A}^+$, then either of (1), (2), below, guarantees that $(\mathfrak{B}, q) \in \tilde{\mathscr{A}}^+$:

(1) $(\mathfrak{B}, q) = (J_{\rho'}, \in, A', q) \subseteq (J_\rho, \in, A, p) = (\mathfrak{A}, p)$; $\mathfrak{B}$ is amenable.

(2) $(\mathfrak{B}, q) \cong (\mathfrak{A}', p) \prec_{\Sigma_1} \mathfrak{A}$, $|\mathfrak{B}|$ is transitive.

(2.7) We now show that (2.5) holds with $\tilde{\mathscr{A}}^+$ replacing $\mathscr{A}^+$.

PROPOSITION (rud $\mathfrak{B}$-CONDENSATION FOR $\tilde{\mathscr{A}}^+$). *Suppose* $(\mathfrak{B}, q) \in \tilde{\mathscr{A}}^+$, $\bar{U} \subseteq |\mathfrak{B}|$, $\bar{U}$ *is* rud$(\mathfrak{B})$*-closed*, $q \in \bar{U}$ *and* $(\mathfrak{C}, \bar{q}) \cong (\mathfrak{B}, q)| \bar{U}$. *Then* $(\mathfrak{C}, \bar{q}) \in \tilde{\mathscr{A}}^+$.

PROOF. Let $(\mathfrak{A}, p) \in \tilde{\mathscr{A}}^+$, $U \subseteq |\mathfrak{A}|$ witness that $(\mathfrak{B}, q) \in \tilde{\mathscr{A}}^+$, say g: $(\mathfrak{B}, q) \leftrightarrow (\mathfrak{A}, p)| U$. Let $U' = g''\bar{U}$. Then $U' \subseteq U$, U' is rud$(\mathfrak{A})$-closed $\wedge$ $p \in U'$. Also $(\mathfrak{C}, \bar{q}) \cong (\mathfrak{B}, q)| \bar{U} \cong (\mathfrak{A}, p)| U'$, so $(\mathfrak{C}, q) \in \tilde{\mathscr{A}}^+$, as required.

(2.8) REMARK. $\tilde{\mathscr{A}}^+$ is the smallest class $\mathscr{A}$ of amenable structures of the form $(J_{\rho'}, \in, A', q)$ such that $\mathscr{A} \supseteq \mathscr{A}^+$ and which satisfies (2.7).

(2.9) In what follows s, $\bar{s}$, s', etc., will be variables over $\tilde{\mathscr{A}}^+$, so we define, e.g., $\rho(s)$, $A(s)$, $p(s)$, by letting

$$s = (J_{\rho(s)}, \in, A(s), p(s)).$$

Also, if $\delta < \rho(s)$, and $p(s) \in J_\delta$, we set

$$s | \delta = (J_\delta, \in, A(s) \cap J_\delta, p(s)).$$

Thus, $s | \delta \in \tilde{\mathscr{A}}^+ \Leftrightarrow (J_\delta, A(s) \cap J_\delta)$ is amenable.

(2.10) We now begin a rather lengthy series of preliminaries prior to defining a global square system on $\tilde{\mathscr{A}}^+$. Jensen has given a more abstract treatment of this material. He axiomatizes a class of structures called *premorasses*, and constructs square systems on premorasses using an axiomatic approach. Our construction of a square system on $\mathscr{A}^+$ is a max of two themes: first the construction of a square system on a premorass, and second, the verification (which is left implicit in what follows) that $(\tilde{\mathscr{A}}^+, \mathscr{F})$ is a premorass, where $\mathscr{F}$ is as defined in (2.11). Accordingly, we sometimes prove a statement directly from the properties of our construction, rather than "factoring its proof through the notion of premorass". As a consequence, we are sometimes able to amalgamate several distinct steps of the premorass argument into one, with the result that our presentation will lag behind the premorass argument in some respects and run ahead in others. Nevertheless, the reader who is familiar with Jensen's premorass treatment should have no trouble going back and forth between such a treatment and ours. For reasons of brevity, we have decided to leave premorasses in the background, but there are real advantages to the premorass approach especially for those interested in models other than L: As is always true of axiomatic treatments, the main part of the work is done once and for all and it remains only to verify that a structure that has been constructed satisfies certain axioms (in this case, premorass axioms).

We guide the reader through the upcoming flurry of numbered items:

(2.11) defines $\mathscr{F}$; (2.12)–(2.14) establish rather elementary closure properties of $\tilde{\mathscr{A}}^+$, $\mathscr{F}$. Half of (2.14) is just (1) of (2.6), the other half is that cofinal Σ_0-elementary substructures are Σ_1-elementary substructures. This and some "Σ_0-model theory" are recurring themes.

(2.15) defines certain distinguished Σ_1-Skolem hulls whose transitive collapses (more precisely, the inverses of whose transitive collapsing maps) will play a central role in the structure of $\mathscr{F}$ and in the final construction of the square system; in a full-blown morass treatment these are referred to as basic maps.

(2.16) and (2.18) define $\beta(f)$ and $\delta(f)$ as, respectively, the critical point and the sup of the range of f. (2.17) restates that critical points are regular.

(2.19) establishes important structural features of Δ, B defined in (2.18); though the result is important, the proof is thoroughly routine and has been omitted.

(2.20)–(2.22) are also important and involve the minimality of Σ_1-Skolem hulls; once again the proofs are routine. (2.23) is somewhat more involved; it is the main tool in establishing the coherence property (2.28) of the Δ's, which in turn will yield the coherence property of the square system.

(2.25) and (2.26) involve carrying out some definitions inside the J_α's; the proofs are tedious but essentially routine.

(2.27) establishes an important technical fact which is crucial in the proof of (2.28), the coherence property.

(2.11) DEFINITION. $f = (\bar{s}, |f|, s) \in \mathscr{F} \Leftrightarrow |f|\colon \bar{s} \to_{\Sigma_1} s$, $\bar{s}, s \in \tilde{\mathscr{A}}^+$; for such f, $d(f) = \bar{s}$, $r(f) = s$. Also, if $s \in \tilde{\mathscr{A}}^+$, $\mathscr{F}_s = \{f \in \mathscr{F}\colon r(f) = s\}$. We abuse notation by writing $f(x)$ for $|f|(x)$, dom f for dom$|f|$, $f''A$ for $|f|''A$; $f^{-1} = (s, |f|^{-1}, \bar{s})$. If $f, g \in \mathscr{F}$, $d(g) = r(f)$, then $g \circ f = (d(f), |g| \circ |f|, r(g))$. Also, if $f, g \in \mathscr{F}_s$, range $f \subseteq$ range g, then $g^{-1} \circ f = (d(f), |g|^{-1} \circ |f|, d(g))$.

(2.12) PROPOSITION. *If $f, g \in \mathscr{F}$, then, when defined, $g \circ f, g^{-1} \circ f \in \mathscr{F}$.*

PROOF. Clear.

(2.13) REMARK. If $g\colon (\overline{\mathfrak{A}}, \bar{p}) \to_{\Sigma_1} s$, $s \in \tilde{\mathscr{A}}^+$, then $(\overline{\mathfrak{A}}, \bar{p})$ is the unique $\bar{s} \in \tilde{\mathscr{A}}^+$ such that $(\bar{s}, g, s) \in \mathscr{F}$.

(2.14) PROPOSITION. *If $f = (\bar{s}, |f|, s) \in \mathscr{F}$, $\delta = \sup f''\rho(\bar{s}) < \rho(s)$, then $s | \delta \in \tilde{\mathscr{A}}^+$, $(\bar{s}, |f|, s|\delta) \in \mathscr{F}$.*

PROOF. The notation makes sense since clearly $p(s) \in J_\delta$. Also, clearly $|f|\colon \bar{s} \to_{\Sigma_1} s|\delta$. But then $(J_\delta, A(s) \cap J_\delta)$ is amenable, so $s|\delta \in \tilde{\mathscr{A}}^+$ and $f \in \mathscr{F}$.

This proposition is the second reason that we had to enlarge $\mathscr{A}^+$ to $\tilde{\mathscr{A}}^+$.

(2.15) DEFINITION. If $s \in \tilde{\mathscr{A}}^+$, let h_s be the canonical Σ_1-Skolem function for s. If $\beta \leqslant OR \cap |s|$, $x \in |s|$, we set $X_{\beta,x,s} = h_s''(\omega \times \beta \times \{x\})$. We also define $|f_{\beta,x,s}|$, $\bar{s}_{\beta,x,s}$, $f_{\beta,x,s}$, by

$$|f_{\beta,x,s}|\colon \bar{s}_{\beta,x,s} \overset{\sim}{\leftrightarrow} s|\, X_{\beta,x,s},$$

is the inverse of the transitive collapse; $f_{\beta,x,s} = (\bar{s}_{\beta,x,s}, |f_{\beta,x,s}|, s)$ (so by (2.11), $f_{\beta,x,s} \in \mathscr{F}$).

(2.16) DEFINITION. If $f = (\bar{s}, |f|, s)$, $f \neq (s, \mathrm{id}|\, |s|, s)$, set $\beta(f) =$ the least β such that $f(\beta) > \beta$, $\delta(f) = \sup f''\rho(\bar{s})$.

(2.17) PROPOSITION. If f is as in (2.16), then $\beta(f) \in \mathrm{RCARD}(\bar{s})$.

PROOF. Clear, following the proof of (2.6)(1).

(2.18) DEFINITION. If $s \in \tilde{\mathscr{A}}^+$, $x \in |s|$,

$$\Delta(x, s) = \{\check{\delta}(f_{\beta,x,s}) : \delta(f_{\beta,x,s}) < \rho(s)\},$$
$$B(x, s) = \{\beta(f_{\beta,x,s}) : \delta(f_{\beta,x,s}) < \rho(s)\},$$

and for $\check{\delta} \in \Delta(x, s)$,

$$B(\check{\delta}, x, s) = \{\beta(f_{\beta,x,s}) : \check{\delta}(f_{\beta,x,s}) = \check{\delta}\}.$$

(2.19) PROPOSITION. *$\Delta(x, s)$ is a closed subset of $\check{\rho}(s)$; if $\check{\delta} \in \Delta(x,s)$, then $B(\check{\delta}, x, s)$ is the intersection of $B(x, s)$ with a closed interval.*

PROOF. Clear.

(2.20) PROPOSITION. *If $f \in \mathscr{F}_s$, $\beta \cup \{x\} \subseteq$ range f, then $X_{\beta,x,s} \subseteq$ range f.*

(2.21) PROPOSITION. *If $f = f_{\beta,x',s'}$, $g = (s', |g|, s) \in \mathscr{F}$, and $g(x') = x$, then $g \circ f = f_{\beta,x,s}$. Also, if $h = (\bar{s}, |h|, s')$, range $h \supseteq$ range f and $h(\bar{x}) = x'$, then $h^{-1} \circ f = f_{\beta,\bar{x},\bar{s}}$.*

PROOF. Clear.

(2.22) COROLLARY. *If $\bar{s} = \bar{s}_{\beta,x,s}$, and $f_{\beta,x,s}(\bar{x}) = x$, then $J_{\rho(\bar{s})} = h_{\bar{s}}(\omega \times \beta \times \{\bar{x}\})$.*

PROOF. Clear.

(2.23) COROLLARY. *If $s \in \tilde{\mathscr{A}}^+$, $\delta = \delta(f)$, and $\delta \geqslant \delta(f_{\beta,x,s})$, then $|f_{\beta,x,s}| = |f_{\beta,x,s|\delta}|$.*

PROOF. Let $\delta' = \delta(f_{\beta,x,s})$. Clearly $X_{\beta,x,s|\delta'} \subseteq X_{\beta,x,s|\delta} \subseteq X_{\beta,x,s}$ so it suffices to show $X_{\beta,x,s|\delta'} = X_{\beta,x,s}$. Let $g = f_{\beta,x,s}$, $g' = f_{\beta,x,s|\delta'}$, $\bar{s} = \bar{s}_{\beta,x,s}$ and $\bar{s}' = \bar{s}_{\beta,x,s|\delta'}$. Then $g^{-1} \circ g' = (\bar{s}', |g|^{-1} \circ |g'|, \bar{s}) \in \mathscr{F}$, and $\beta \cup \{\bar{x}\} \subseteq$ range $g^{-1} \circ g'$, where $g(\bar{x}) = x$. But $\bar{s} = \bar{s}_{\beta,x,s}$, so $X_{\beta,\bar{x},\bar{s}} = J_{\rho(\bar{s})} =$ range $g^{-1} \circ g'$; i.e. $\bar{s}' = \bar{s}$. But then $X_{\beta,x,s} = X_{\beta,x,s|\delta'}$, as required.

(2.24) COROLLARY. *If $s \in \tilde{\mathscr{A}}^+$, $f = f_{\beta,x,s} = (\bar{s}, |f|, s)$ and $\delta = \delta(f)$, then $(\bar{s}, |f|, s|\delta) = f_{\beta,x,s|\delta}$.*

PROOF. Immediate from (2.23).

(2.25) PROPOSITION. *If $s \in \tilde{\mathscr{A}}^+$, $\delta < \rho(\Delta)s|\delta \in \tilde{\mathscr{A}}^+$, then*

$$\{(\beta, x, X_{\beta,x,s|\delta}) : \beta \cup \{x\} \subseteq J_\delta\} \in J_{\rho(s)};$$

further

$$\Omega_s = \{(\check{\delta}, \beta, x, X_{\beta,x,s|\delta}) : \delta < \rho(s), s|\delta \in \tilde{\mathscr{A}}^+, \beta \cup \{x\} \subseteq J_\delta\}$$

is uniformly $\Sigma_1(s)$, and $(J_{\rho(s)}, \Omega)$ is amenable.

PROOF. Clear.

(2.26) COROLLARY. *If $f = (\bar{s}, |f|, s) \in \mathscr{F}$, $\bar{\Omega} = \Omega_{\bar{s}}$ and $\Omega = \Omega_s$, then $f''\bar{\Omega} \subseteq \Omega$; in fact $f : (\bar{s}, \bar{\Omega}) \to_{\Sigma_0} (s, \Omega)$.*

PROOF. Clear.

(2.27) PROPOSITION. *Suppose* $f = (\bar{s}, |f|, s) \in \mathscr{F}$ *and* $\delta(f) = \rho(s)$, $\beta < \rho(\bar{s})$, $\bar{x} \in J_{\rho(\bar{s})}, f(\bar{\beta}) = \beta, f(\bar{x}) = x$. *Then* $\delta(f_{\bar{\beta},\bar{x},\bar{s}}) < \rho(\bar{s}) \Leftrightarrow \delta(f_{\beta,x,s}) < \rho(s)$.

PROOF. ($\Rightarrow$) Let $\bar{\xi} < \check{\rho}(\bar{s})$ be such that $\bar{\xi} \geqslant \check{\delta}(f_{\bar{\beta},\bar{x},\bar{s}})$. If $s \vDash \varphi_n(f(\bar{\xi}), \beta, x)$, then $\bar{s} \vDash \varphi_n(\bar{\xi}, \bar{\beta}, \bar{x})$, since φ_n is Σ_1, where $\varphi_n(\eta, \theta, y)$ asserts

"for some $\eta' \geqslant \eta$, $i < \omega$, $\theta_0', \ldots, \theta_n' < \theta$,
$\eta' = h(i, \theta_0', \ldots, \theta_n', y)$, where h is the Σ_1-Skolem function."

But then $\bar{\xi} < \check{\delta}(f_{\bar{\beta},\bar{x},\bar{s}})$, contradiction. We have, in fact, shown that $\delta(f_{\beta,x,s})) \leqslant f(\delta(f_{\bar{\beta},\bar{x},\bar{s}}))$.

($\Leftarrow$) Clear, since $X_{\beta,x,s} \supseteq f''X_{\bar{\beta},\bar{x},\bar{s}}$ and $\delta(f) = \rho(s)$.

(2.28) LEMMA. *If* $\check{\delta} \in \Delta(x, s)$, *then* $\Delta(x, s|\delta) = \check{\delta} \cap \Delta(x, s)$.

PROOF. $\check{\delta} \cap \Delta(x, s) \subseteq \Delta(x, s|\delta)$, by (2.23), so suppose $\check{\delta}' \in \Delta(x, s|\delta)$, say $\delta' = \delta(f_{\beta',x,s|\delta})$. Also, let β be such that $\delta = \delta(f_{\beta,x,s})$. Again, by (2.23), $|f_{\beta,x,s}| = |f_{\beta,x,s|\delta}|$. Therefore $\beta' < \beta$. But then $X_{\beta',x,s|\delta} \subseteq X_{\beta,x,s} = X_{\beta,x,s|\delta}$; further

$$s|\delta|X_{\beta',x,s|\delta} \prec_{\Sigma_1} s|\delta, \quad s|\delta|X_{\beta,x,s|\delta} \prec_{\Sigma_1} s|\delta,$$

so $s|\delta|X_{\beta',x,s|\delta} \prec_{\Sigma_1} s|\delta|X_{\beta,x,s|\delta}$. But

$$s|\delta|X_{\beta',x,s|\delta} = s|X_{\beta',x,s|\delta}, \quad s|\delta|X_{\beta,x,s|\delta} = s|X_{\beta,x,s|\delta} = s|X_{\beta,x,s} \prec_{\Sigma_1} s,$$

so $s|X_{\beta',x,s|\delta} \prec_{\Sigma_1} s$, and so $X_{\beta',x,s|\delta} \supseteq X_{\beta',x,s}$. But clearly $X_{\beta',x,s|\delta} \subseteq X_{\beta',x,s}$, so $X_{\beta',x,s|\delta} = X_{\beta',x,s}$. Therefore $\check{\delta}' = \sup \check{\rho}(s) \cap X_{\beta',x,s}$, so $\check{\delta}' = \check{\delta}(f_{\beta',x,s}) \in \Delta(x,s)$.

(2.29) LEMMA. *Let* $f = (\bar{s}, |f|, s) \in \mathscr{F}$, $\delta(f) = \rho(s)$. *Suppose* $\bar{x} \in J_{\rho(\bar{s})}$, $x = f(\bar{x})$. *Then*

$$\Delta(\bar{x},\bar{s}) = \varnothing \Leftrightarrow \Delta(x, s) = \varnothing.$$

PROOF. $\Delta(\bar{x}, \bar{s}) = \varnothing \Leftrightarrow \delta(f_{1,\bar{x},\bar{s}}) = \rho(\bar{s})$; $\Delta(x, s) = \varnothing \Leftrightarrow \delta(f_{1,x,s}) = \rho(s)$. But clearly $X_{1,x,s} = f''X_{1,\bar{x},\bar{s}}$, and range f is cofinal in $\rho(s)$.

(2.30) LEMMA. *If* $f = (\bar{s}, |f|, s) \in \mathscr{F}$, $\bar{\delta} = \delta(f_{\bar{\beta},\bar{x},\bar{s}}) < \rho(\bar{s})$, $\delta = f(\bar{\delta})$, $\beta = f(\bar{\beta})$, *and* $x = f(\bar{x})$, *then* $\delta = \delta(f_{\beta,x,s})$.

PROOF. By the proof of the left-to-right implication of (2.27), $\delta \geqslant \delta(f_{\beta,x,s})$. But $f(\bar{s}|\bar{\delta}) = s|\delta$, so $f|\bar{s}|\bar{\delta}$: $\bar{s}|\bar{\delta} \to_{\Sigma_\omega} s|\delta$. But

$$\bar{s}|\bar{\delta} \vDash (\forall\xi)(\exists\zeta)(\exists\vec{\eta} \text{ from } \bar{\beta})(\exists i < \omega)(\xi \leqslant \zeta = h(i, \vec{\eta}, \bar{x}))$$

so

$$s|\delta \vDash (\forall\xi)(\exists\zeta)(\exists\vec{\eta} \text{ from } \beta)(\exists i < \omega)(\xi \leqslant \zeta = h(i, \vec{\eta}, x));$$

i.e. $X_{\beta,x,s|\delta}$ is cofinal in $\check{\delta}$. Finally, $X_{\beta,x,s|\delta} \subset X_{\beta,x,s}$, so $\delta = \delta(f_{\beta,x,s})$, as required.

(2.31) LEMMA. *Let* $(\bar{s}, |f|, s) \in \mathscr{F}$, $\delta(f) = \rho(s)$. *Let* $f(\bar{x}) = x$. *Then*:
(a) *if* $\check{\bar{\delta}} = \max \Delta(\bar{x}, \bar{s})$, *then* $\check{\delta} = \max \Delta(x, s)$, *where* $f(\check{\bar{\delta}}) = \check{\delta}$,
(b) $\Delta(\bar{x}, \bar{s})$ *cofinal in* $\check{\rho}(\bar{s}) \Leftrightarrow \Delta(x, s)$ *cofinal in* $\check{\rho}(s)$, *and*
(c) $f''\Delta(\bar{x}, \bar{s}) = \text{range} f \cap \Delta(x, s)$.

PROOF. (a) Let $\bar{\beta} = \max B(\check{\bar{\delta}}, \bar{x}, \bar{s})$. Then $\bar{\delta} = \delta(f_{\bar{\beta},\bar{x},\bar{s}})$, but $\delta(f_{\bar{\beta}+1,\bar{x},\bar{s}}) = \rho(\bar{s})$. But then, letting $\beta = f(\bar{\beta})$, $X_{\beta+1,x,s} \supseteq f''X_{\bar{\beta}+1,\bar{x},\bar{s}}$ is cofinal in $\check{\rho}(s)$. Thus $B(x, s) \subset \beta + 1$, so it suffices to see that $\delta = \delta(f_{\beta,x,s})$. This is by (2.30).

(b) By (a) and (2.30) it suffices to show that $\Delta(\bar{x}, \bar{s})$ cofinal in $\check{\rho}(\bar{s}) \Rightarrow \Delta(x, s)$ cofinal in $\check{\rho}(s)$. But (2.30) guarantees that $f''\Delta(\bar{x}, \bar{s}) \subseteq \Delta(x, s)$, and so our hypothesis that $\delta(f) = \rho(s)$ guarantees that $\Delta(\bar{x}, \bar{s})$ cofinal in $\check{\rho}(\bar{s}) \Rightarrow f''\Delta(\bar{x},\bar{s})$ cofinal in $\rho(s)$.

(c) Again, (2.30) yields that $f''\Delta(\bar{x}, \bar{s}) \subseteq \Delta(x, s)$, so, suppose $\check{\delta} = f(\check{\bar{\delta}})$ and $\check{\delta} \in \Delta(x, s)$. Let $\beta = \max B(\check{\delta}, x, s)$. by (a), (b), we may suppose $\check{\delta} \neq \max \Delta(x, s)$. Let $\bar{\beta} = f^{-1}[\beta]$. Then $f(\bar{\beta}) \geqslant \beta$. Clearly $X_{\bar{\beta},\bar{x},\bar{s}} \subseteq J_{\check{\bar{\delta}}}$. Now, if $\delta(f_{\bar{\beta},\bar{x},\bar{s}}) = \bar{\delta}' < \check{\bar{\delta}}$, then $f(\bar{\delta}') < \check{\delta}$, and by (2.30),

$$f(\bar{\delta}') = \delta\big(f_{f(\bar{\beta}),x,s}\big) \geqslant \delta\big(f_{\beta,x,s}\big) = \delta,$$

contradiction! So $\check{\bar{\delta}} = \bar{\delta}(f_{\bar{\beta},\bar{x},\bar{s}}) \in \Delta(x, \bar{s})$, so $\check{\delta} \in f''\Delta(\bar{x},\bar{s})$.

(2.32) PROPOSITION. *If* $f = (\bar{s}, |f|, s) \in \mathscr{F}$, $\bar{\delta} < \rho(\bar{s})$, $\bar{s}|\bar{\delta} \in \tilde{\mathscr{A}}^+$ *and* $f(\bar{\delta}) = \delta$, *then* $s|\delta \in \tilde{\mathscr{A}}^+$ *and* $(\bar{s}|\bar{\delta}, |f||J_{\bar{\delta}}, s|\delta) \in \mathscr{F}$.

PROOF. That $s|\delta \in \tilde{\mathscr{A}}^+$ is already implicit in (2.26). But as in the proof of (2.30), $|f|\,|J_{\bar{\delta}}\colon \bar{s}|\bar{\delta} \to_{\Sigma_\omega} s|\delta$, so the conclusion follows.

(2.33) LEMMA. *If* $\bar{s} \in \tilde{\mathscr{A}}^+$ *and* $\bar{x} \in J_{\rho(\bar{s})}$, *then* $(J_{\rho(\bar{s})}, \Delta(\bar{x}, \bar{s}))$ *is amenable*; *further, if* $f = (\bar{s}, |f|, s) \in \mathscr{F}$, $\bar{\eta} < \check{\rho}(\bar{s})$ *and* $\eta = f(\bar{\eta})$, *then* $f(\Delta(\bar{x}, \bar{s}) \cap \bar{\eta}) = f(\bar{\eta}) \cap \Delta(x, s)$, *where* $x = f(\bar{x})$.

PROOF. Let $\bar{\eta} < \check{\rho}(\bar{s})$, and let $\check{\bar{\delta}} = \inf(\Delta(\bar{x}, \bar{s}) \setminus \bar{\eta})$; then $\Delta(\bar{x}, \bar{s}) \cap \bar{\eta} = \Delta(\bar{x}, \bar{s}) \cap \check{\bar{\delta}}$, so it will suffice to show $\Delta(\bar{x}, \bar{s}) \cap \check{\bar{\eta}} \in J_{\rho(\bar{s})}$, where $\check{\bar{\eta}} \in \Delta(\bar{x}, \bar{s})$. But then $\Delta(\bar{x}, \bar{s}) \cap \check{\bar{\eta}} = \Delta(\bar{x}, \bar{s}|\bar{\eta})$, so it suffices to show that for $\bar{x} \in J_{\bar{\eta}}$, $\bar{s}|\bar{\eta} \in \tilde{\mathscr{A}}^+$, $\bar{\eta} < \rho(\bar{s})$, we have $\Delta(\bar{x}, \bar{s}|\bar{\eta}) \in J_{\rho(\bar{s})}$. But this is clear by (2.25) and the fact that $\Delta(\bar{x}, \bar{s}|\bar{\eta})$ is rud in $\bar{x}$ and $\{X_{\beta,\bar{y},\bar{s}|\bar{\eta}}\colon \beta \cup \{\bar{y}\} \subset J_{\bar{\eta}}\}$.

Now let f be as in the second part of the lemma. Let $\eta = f(\bar{\eta})$ and $\check{\delta} = f(\check{\bar{\delta}})$. We first show that $\check{\delta} = \inf(\Delta(x, s) \setminus \eta)$. If $\bar{\eta} = \check{\bar{\delta}}$, the conclusion is clear by (2.31)(c). If $\bar{\eta} < \check{\bar{\delta}}$, then $\check{\bar{\delta}}$ is an immediate successor, say of $\check{\bar{\gamma}}$ in $\Delta(\bar{x}, \bar{s})$.

Let $\bar{\beta}_0 = \max B(\check{\bar{\gamma}}, \bar{x}, \bar{s})$; then $\delta(f_{\bar{\beta}_0+1,\bar{x},\bar{s}}) = \bar{\delta}$. Let $\beta_0 = f(\bar{\beta}_0)$, $\check{\gamma} = f(\check{\bar{\gamma}})$. Then, by (2.30), $\gamma = \delta(f_{\beta_0,x,s})$, $\delta = \delta(f_{\beta_0+1,x,s})$ so $\check{\delta}$ is the immediate successor in $\Delta(x, s)$ of $\check{\gamma}$, and $\check{\gamma} < \eta < \check{\delta}$. Thus $\check{\delta} = \inf(\Delta(x, s) \setminus \eta)$, as required. It will now suffice to show that $f(\Delta(\bar{x}, \bar{s}|\bar{\delta})) = \Delta(x, s|\delta)$. This is true for any $\bar{\delta} < \rho(\bar{s})$, $\delta = f(\bar{\delta})$, such that $\bar{x} \in J_{\bar{\delta}}$, $\bar{s}|\bar{\delta} \in \tilde{\mathscr{A}}^+$, $x = f(\bar{x})$, as can be seen from the argument above for the amenability of $(J_{\rho(\bar{s})}, \Delta(\bar{x}, \bar{s}))$.

(2.34) We now embark on the construction of $\tilde{C}_s$. Our method is a slight modification of Jensen's construction in his "premorass notes". Our approach is as follows. We first define a_s a finite subset of $\rho(s)$, with $0 \in a_s$. Then, we let $\tilde{C}_s = \bigcup\{\Delta(\check{\delta}, s)^*: \delta \in a_s\}$. This definition of $\tilde{C}_s$ will make it clear that:

1. $\operatorname{cf} \check{\rho}(s) > \omega \Rightarrow \tilde{C}_s$ is cofinal in $\check{\rho}(s)$,
2. $\tilde{C}_s$ is closed,
3. $(\exists \alpha < \check{\rho}(s))(J_{\rho(s)} = h_s''(\omega \times \alpha)) \Rightarrow$ o.t. $\tilde{C}_s < \check{\rho}(s)$.

It will be more substantial to verify that a_s is, in fact, finite (this is done in (2.40)), and that $(\tilde{C}_s: s \in \tilde{\mathscr{A}}^+)$ has the coherence property, i.e., that $\check{\delta} \in \tilde{C}_s \cap \check{\rho}(s) \Rightarrow \tilde{C}_{s|\delta} = \tilde{C}_s \cap \check{\delta}$. By the definition of a_s, this will follow from the coherence property for the $\Delta(x, s)$, once we have proved the coherence property for $(a_s: s \in \tilde{\mathscr{A}}^+)$, i.e. that $\check{\delta} \in \tilde{C}_s \Rightarrow a_{s|\delta} = a_s \cap \check{\delta}$. This is done in (2.42).

(2.35) Definition. For $s \in \tilde{\mathscr{A}}^+$, we define $m_s \leqslant \omega$ s.t. $m_s \geqslant 1$, and for $n < m_s$, $\check{\delta}_s^n$ by recursion: $\check{\delta}_s^0 = 0$; if $\check{\delta}_s^n$ is defined and $\Delta(\check{\delta}_s^n, s) = \varnothing$, or $\Delta(\check{\delta}_s^n, s)$ is cofinal in $\check{\rho}(s)$, then $m_s = n + 1$; otherwise $\Delta(\check{\delta}_s^n, s)$ has a largest element and we set:

$$\check{\delta}_s^{n+1} = \max \Delta(\check{\delta}_s^n, s).$$

If $\check{\delta}_s^n$ is defined for all $n < \omega$, then $m_s = \omega$. We set

$$a_s = \{\check{\delta}_s^n: n < m_s\}.$$

(2.36) Remarks. 1. $n_1 < n_2 < m_s \Rightarrow \check{\delta}_s^{n_1} < \check{\delta}_s^{n_2}$.
2. $(\forall \alpha < \check{\rho}(s))(\delta(f_{\alpha,0,s}) < \rho(s) \Rightarrow m_s = 1)$.

(2.37) Definition. If $i + 1 < m_s$, let $\beta_s^i = \max B(\check{\delta}_s^{i+1}, \check{\delta}_s^i, s)$.

(2.38). Proposition. *Suppose $f = (\bar{s}, |f|, s) \in \mathscr{F}$, $\delta(f) = \rho(s)$. Then $m_{\bar{s}} = m_s$, for $i < m_{\bar{s}}$, $f(\check{\delta}_{\bar{s}}^i) = \check{\delta}_s^i$, $f(\beta_{\bar{s}}^i) = \beta_s^i$.*

Proof. We show by induction on $i < m_{\bar{s}}$ that:
(a) $\check{\delta}_s^i = f(\check{\delta}_{\bar{s}}^i)$,
(b) $\Delta(\check{\delta}_{\bar{s}}^i, \bar{s})$ has a largest element $\Leftrightarrow$ $\Delta(\check{\delta}_s^i, s)$ does, and if so $f(\check{\delta}_{\bar{s}}^{i+1}) = \check{\delta}_s^{i+1}$,
(c) $f(\beta_{\bar{s}}^i) = \beta_s^i$, if $i + 1 < m_{\bar{s}}$.

For $i = 0$, (a) is clear; now having (a) for i, (b) follows by (2.29) and (2.31)(a), (b). Note, this establishes (a) for $i + 1$, if $i + 1 < m_{\bar{s}}$. Also, by (2.30), if $i + 1 < m_{\bar{s}}$ then $f(\check{\delta}_{\bar{s}}^{i+1}) = \check{\delta}_s^{i+1} = \delta(f_{f(\beta_{\bar{s}}^i), \check{\delta}_s^i, s})$. Finally, $X_{f(\beta_{\bar{s}}^i)+1, \check{\delta}_s^i, s} \supseteq f''X_{\beta_{\bar{s}}^i+1, \check{\delta}_{\bar{s}}^i, \bar{s}}$, but $X_{\beta_{\bar{s}}^i+1, \check{\delta}_{\bar{s}}^i, \bar{s}}$ is cofinal in $\check{\rho}(\bar{s})$, and $\delta(f) = \rho(s)$, so $X_{f(\beta_{\bar{s}}^i)+1, \check{\delta}_s^i, s}$ is cofinal in $\check{\rho}(s)$; that is, $f(\beta_{\bar{s}}^i) = \beta_s^i$.

(2.39) Corollary. *If $f = (\bar{s}, |f|, s) \in \mathscr{F}$, $\delta^* = \delta(f)$, $i < m_s$ and $\check{\delta}_s^i < \check{\delta}^*$, then $\check{\delta}_s^j = f(\check{\delta}_{\bar{s}}^j) = \check{\delta}_{s|\delta^*}^j$ for all $j \leqslant i$ and if $j < i$, then $\beta_s^j = f(\beta_{\bar{s}}^j) = \beta_{s|\delta^*}^j$.*

Proof. For $j = 0$, clearly $\check{\delta}_s^0 = 0 = f(0) = f(\check{\delta}_{\bar{s}}^0)$. So, suppose that $j < i$, that $\check{\delta}_s^j = f(\check{\delta}_{\bar{s}}^j)$, and that for all $k < j$, $\beta_s^k = f(\beta_{\bar{s}}^k)$. We show $\beta_s^j = f(\beta_{\bar{s}}^j)$, and $\check{\delta}_s^{j+1} = f(\check{\delta}_{\bar{s}}^{j+1})$. Let $\bar{\beta} = \beta_{\bar{s}}^j$, $\beta = f(\bar{\beta}) = \beta_{s|\delta^*}^j$; let $\check{\bar{\delta}} = \check{\delta}_{\bar{s}}^{j+1}$, $\check{\delta} = f(\check{\bar{\delta}}) = \check{\delta}_{s|\delta^*}^{j+1}$. Then by (2.30), $\delta = \delta(f_{\beta, \check{\delta}_s^j, s})$ so $\check{\delta} \in \Delta(\check{\delta}_s^j, s)$; i.e. $\check{\delta} \leqslant \check{\delta}_s^{j+1}$. But $X_{\bar{\beta}+1, \check{\delta}_{\bar{s}}^j, \bar{s}}$ is cofinal in $\check{\rho}(\bar{s})$ and $\sup f''\check{\rho}(\bar{s})$ is cofinal in $\check{\delta}^*$, so $f''X_{\bar{\beta}+1, \check{\delta}_{\bar{s}}^j, \bar{s}}$ is cofinal in $\check{\delta}^*$, $X_{\beta+1, \check{\delta}_s^j, s}$ is cofinal in $\check{\delta}^*$, and $\check{\delta}_s^{j+1} \leqslant \check{\delta}_s^i < \check{\delta}^*$. Thus, $\check{\delta} = \check{\delta}_s^{j+1}$, $\beta = \beta_s^{j+1}$.

(2.40) Lemma. *a_s is finite.*

PROOF. We show that $i + 2 < m_s \Rightarrow \beta_s^{i+1} < \beta_s^i$. This clearly suffices. Let $\check{\delta}' = \check{\delta}_s^i$, $\check{\delta} = \check{\delta}_s^{i+1}$, $\check{\delta}* = \check{\delta}_s^{i+2}$, $\beta' = \beta_s^i$ and $\beta = \beta_s^{i+1}$. Let $f = f_{\beta,\check{\delta},s}$. Then $\delta(f) = \delta*$. Say $f = (\bar{s}, |f|, s)$. By (2.39), for $j \leqslant i + 1$, $f(\check{\delta}_{\bar{s}}^j) = \check{\delta}_s^j$; i.e. $\check{\delta}' \in \operatorname{range} f$. Therefore $\beta' \geqslant \beta$, since otherwise $\beta' + 1 \cup \{\check{\delta}'\} \subseteq \operatorname{range} f$, so $X_{\beta'+1,\check{\delta}',s} \subseteq \operatorname{range} f$, and range f would be cofinal in $\check{\rho}(s)$. Note also that $\beta' \geqslant \beta(f)$, by the same argument, and $\beta(f) \geqslant \beta$. Now either $\beta < \beta'$, in which case the claim is proved, or $\beta' = \beta(f) = \beta$, so assume $\beta' = \beta(f) = \beta$.

Let $\check{\bar{\delta}}' = \check{\delta}_{\bar{s}}^i$, $\check{\bar{\delta}} = \check{\delta}_s^{i+1}$, $\bar{g} = f_{\beta,\check{\bar{\delta}}',\bar{s}}$, $g = f_{\beta,\check{\delta}',s}$. Thus $\bar{\delta} = \delta(\bar{g})$, $g = f \circ \bar{g}$, and $\delta(f \circ \bar{g}) = \delta(g) = \delta = f(\bar{\delta})$. Let $\tilde{\beta} = f(\beta)$. Then, since $\beta = \beta(f)$, $\tilde{\beta} > \beta$. Set $\tilde{g} = f_{\tilde{\beta},\check{\delta}',s}$. Then $\delta(\tilde{g}) = \rho(s)$. But then, $\delta = f(\bar{\delta}) = f(\delta(\bar{g})) = f(\delta(f_{\beta,\check{\bar{\delta}}',\bar{s}}))$; by (2.30),

$$f\left(\delta\left(f_{\beta,\check{\bar{\delta}}',\bar{s}}\right)\right) = \delta\left(f_{f(\beta),\check{\delta}',s}\right) = \delta(\tilde{g}) = \rho(s),$$

contradiction.

(2.41) DEFINITION. $\tilde{C}_s = \bigcup\{\Delta(\check{\delta}, s)^*: \check{\delta} \in a_s\}$.

(2.42) THEOREM. *For* $s \in \tilde{\mathscr{A}}^+$:

(a) $\tilde{C}_s$ *is closed in* $\check{\rho}(s)$,

(b) $\check{\delta}* \in \tilde{C}_s \to \tilde{C}_{s|\delta*} = \check{\delta}* \cap \tilde{C}_s$,

(c) $\operatorname{cf}(\check{\rho}(s)) > \omega \Rightarrow \tilde{C}_s$ *is cofinal in* $\check{\rho}(s)$,

(d) *if* $f = (\bar{s}, |f|, s) \in \mathscr{F}$, *then for all* $\eta < \check{\rho}(\bar{s})$, $f(\bar{\eta} \cap \tilde{C}_{\bar{s}}) = f(\bar{\eta}) \cap \tilde{C}_s$,

(e) *if* $\delta(f_{\alpha,0,s}) = \rho(s)$ *and* $\check{\delta} \in \tilde{C}_s$, *then for some* $\gamma < \alpha$ *and finite* $u \subseteq \alpha$, $\delta = \delta(f_{\gamma,\langle u\rangle,s})$, *where* $\langle u\rangle \in OR$ *codes u in some reasonable manner.*

PROOF. (a) is clear. For (b), we first show

CLAIM. $\check{\delta}* \in \tilde{C}_s \Rightarrow a_{s|\delta*} = \check{\delta}* \cap a_s$.

PROOF OF CLAIM. Let $\check{\delta}' = \max a_s \cap \check{\delta}*$. Then, $\check{\delta}* \in \Delta(\check{\delta}', s)^*$. Say $\check{\delta}' = \check{\delta}_s^i$. Then by (2.40), for $j \leqslant i$, $\check{\delta}^j_{s|\delta*} = \check{\delta}_s^j$. It remains to show that $\Delta(\check{\delta}_s^j, s|\delta*)$ has no maximum element—but this is clear since $\check{\delta}* \in \Delta(\check{\delta}', s)^*$. But then

$$\tilde{C}_{s|\delta*} = \bigcup\{\Delta(\check{\delta}, s|\delta*)^*: \check{\delta} \in a_{s|\delta*}\} = \bigcup\{\Delta(\check{\delta}, s|\delta*)^*: \check{\delta} \in \check{\delta}* \cap a_s\},$$

so it suffices to show that for $\check{\delta} \in \check{\delta}* \cap a_s$, $\Delta(\check{\delta}, s|\delta*) = \Delta(\check{\delta}, s) \cap \check{\delta}*$. For $\check{\delta} = \check{\delta}'$, this is clear by (2.28).

So, suppose $\check{\delta} \in a_s \cap \check{\delta}'$. Let $\beta \in B(\check{\delta}, s)$, so $X_{\beta,\check{\delta},x} \subseteq J_{\delta'} \subseteq J_\delta$. But then $X_{\beta,\check{\delta},s|\delta*} = X_{\beta,\check{\delta},s}$; i.e. $\check{\delta}(f_{\beta,\check{\delta},s}) \in \Delta(\check{\delta}, s|\delta*)$.

For $\check{\delta} \in a_s \cap \check{\delta}'$, suppose $\check{\tilde{\delta}}$ immediately succeeds $\check{\delta}$ in a_s. Say, e.g. $\check{\tilde{\delta}} = \check{\delta}_s^{i+1}$, $\check{\delta} = \check{\delta}_s^i$. Then, by the Claim, $\check{\tilde{\delta}} = \check{\delta}^{i+1}_{s|\delta*}$; $\check{\delta} = \check{\delta}^i_{s|\delta*}$. So $\Delta(\check{\delta}, s|\delta*) = \Delta(\check{\delta}, s|\tilde{\delta}) \cup \{\check{\tilde{\delta}}\}$, by (2.28); similarly $\Delta(\check{\delta}, s) = \Delta(\check{\delta}, s|\tilde{\delta}) \cup \{\check{\tilde{\delta}}\}$, by (2.28), since $\check{\tilde{\delta}} = \max \Delta(\check{\delta}, s|\delta*) = \max \Delta(\check{\delta}, s)$. This proves (b).

(d) now follows readily by the proof of (b) and by (2.33) and (2.38).

For (e), we indicate, for the record, one reasonable way of coding finite sets in J_ρ's. Recall that for all ρ there is a $\Sigma_1(J_\rho)$-map g of $\omega\rho$ onto J_ρ. Code a finite set, $x \in J_\rho$, e.g., by the least η s.t. $g(\eta) = x$. The $\Sigma_1(J_\rho)$-definition of g may vary for different ρ's, but this creates no difficulties; further, for large classes of "similar" ρ's, the definition is uniform within the class (see the proof of (2.10) of [5] for details). The point is that $u \in X_{1,\eta,s}$ and so $u \subseteq X_{1,\eta,s}$; further if $u \in X \prec_{\Sigma_1} J_\rho$, then $\eta \in X$, $u \subseteq X$.

Now suppose $\delta(f_{\alpha,0,s}) = \rho(s)$ and $\check{\delta} \in \tilde{C}_s$. Let $\check{\delta}' = \max(a_s \cap \check{\delta})$. Then $\delta = \delta(f_{\beta,\check{\delta}',s})$ for some β, and by the proof of (2.40), $\beta < \alpha$. Thus it will suffice to show

(*) if $\check{\delta}' \in a_s$, there is a finite $u \subseteq \alpha$ s.t. $u \subseteq X_{1,\check{\delta}',s}$, $\check{\delta}' \in X_{1,\langle u\rangle,s}$.

We prove (*) by induction on $\check{\delta}' \in a_s$. If $\check{\delta}' = 0$, this is trivial. Suppose, therefore, that $\check{\delta}'$ immediately succeeds $\check{\check{\delta}}$ in a_s and $\bar{u}$ is as required for $\check{\check{\delta}}$ by (*). So $\check{\delta}' = \max \Delta(\check{\check{\delta}}, s)$; let $\beta = \max B(\check{\delta}', \check{\check{\delta}}, s)$. Then $X_{\beta,\check{\check{\delta}},s} = X_{\beta,\langle\bar{u}\rangle,s}$, $\check{\delta}' = \sup OR \cap X_{\beta,\check{\check{\delta}},s}$, and $\check{\rho}(s) = \sup OR \cap X_{\beta+1,\check{\check{\delta}},s} = \sup OR \cap X_{\beta+1,\langle\bar{u}\rangle,s}$. Say $f_{\beta+1,\langle\bar{u}\rangle,s} = (\bar{s}, |f|, s)$. By (2.31), $\check{\delta}' \in \operatorname{range} f = X_{\beta+1,\langle\bar{u}\rangle,s}$, since $f_{\beta+1,\langle\bar{u}\rangle,s} = f_{\beta+1,\check{\check{\delta}},s}$. So, for some $\gamma \leqslant \beta$ and $i < \omega$, $\check{\delta}' = h_s(i, \gamma, \langle\bar{u}\rangle)$; further $\gamma = \beta$ since $X_{\beta,\langle\bar{u}\rangle,s} \subset J_{\check{\delta}'}$. Thus, $\check{\delta}' \in X_{1,\langle\bar{u}\cup\{\beta\}\rangle,s}$; finally $\beta + 1 \leqslant \alpha$ since $\rho(s) = \delta(f_{\alpha,0,s})$. This proves (*) and (e).

For (c), suppose $\tilde{C}_s$ is not cofinal in $\check{\rho}(s)$. Let $\check{\delta}^* = \max a_s$. Then either $\Delta(\check{\delta}^*, s) = \varnothing$, or $\Delta(\check{\delta}^*, s)$ is cofinal in $\check{\rho}(s)$. If $\Delta(\check{\delta}^*, s) = \varnothing$, then $X_{1,\check{\delta}^*,s}$ is cofinal in $\check{\rho}(s)$, so $\operatorname{cf} \rho(s) = \omega$. If $\Delta(\check{\delta}^*, s)$ is cofinal in $\check{\rho}(s)$, then for some limit ordinal θ (possibly $\theta = 0$), o.t. $\Delta(\check{\delta}^*, s) = \theta + \omega$ since $\Delta(\delta^*, s)^*$ is not cofinal in $\check{\rho}(s)$. But then, again, $\operatorname{cf} \check{\rho}(s) = \omega$.

Note that when $\tilde{C}_s$ is not cofinal in $\check{\rho}(s)$, by (a), either $\tilde{C}_s = \varnothing$ or $\tilde{C}_s$ has a largest element, say $\check{\sigma}$. If $\tilde{C}_s = \varnothing$, set $\check{\sigma} = 0$. Let $\check{\delta}' = \sup\{\check{\check{\delta}} \in a_s : \Delta(\check{\check{\delta}}, s)^* \neq \varnothing\}$. Note that possibly $\check{\delta} < \check{\delta}^*$, and that possibly $\check{\delta}' = 0$ because it is the sup of the empty set (when $\tilde{C}_s = \varnothing$). Then $\check{\sigma} = \sup \Delta(\check{\delta}', s)^*$ (again, possibly $\check{\sigma} = 0$ because it is the sup of the empty set, when $\tilde{C}_s = \varnothing$).

(2.43) We have the following corollary to the proof of (2.42)(e).

COROLLARY. *Let α, $\check{\delta}$ be as in* (2.42)(e). *Let $\beta(0) = 0$, and for $\check{\delta}' \in a_s \cap \check{\delta}$, if $\check{\delta}'$ immediately succeeds $\check{\check{\delta}}$ in a_s, let $\beta(\check{\delta}') = \max B(\check{\delta}', \check{\check{\delta}}, s)$. Then, letting $\check{\delta}^* = \max(a_s \cap \check{\delta})$, $u = \{\beta(\check{\delta}') : \check{\delta}' \in a_s \cap \check{\delta}\}$ we have*

$$(\exists\gamma < \beta(\check{\delta}^*))\big(\check{\delta} = \delta\big(f_{\gamma,\langle u\rangle,s}\big)\big).$$

(2.44) LEMMA. *Suppose $(\exists\alpha < \check{\rho}(s))(\delta(f_{\alpha,0,s}) = \rho(s))$. Let $\gamma_0 =$ the least such α. Then either*:

(a) $\gamma_0 \in \operatorname{CARD}(s)$, $\tilde{C}_s = \Delta(0, s)^*$ *and* o.t. $\tilde{C}_s \leqslant \gamma_0$, *or*

(b) $\gamma_0 = \bar{\gamma}_0 + 1$, *where* $\bar{\gamma}_0 \in \operatorname{CARD}(s) \cup \{0\}$, $\bar{\gamma}_0$ *is not the largest cardinal in the sense of s and*

(i) $\bar{\gamma}_0 = 0 \Rightarrow \tilde{C}_s = \varnothing$,

(ii) $\bar{\gamma} \neq 0 \Rightarrow$ o.t. $\tilde{C}_s < \tau$, *where* $\tau = (\bar{\gamma}_0^+)^s$.

PROOF. If $\Delta(0, s)$ has limit order type, then $\Delta(0, s)$ is cofinal in $\check{\rho}(s)$ and so $a_s = \{0\}$ and $\tilde{C}_s = \Delta(0, s)^*$. We first show that we are in case (a). For $\check{\delta} \in \Delta(0, s)$, let $\beta(\check{\delta}) \in B(\check{\delta}, 0, s)$ and let $\beta = \sup\{\beta(\check{\delta}): \check{\delta} \in \Delta(0, s)\}$. On the one hand, clearly $\beta = \gamma_0$, but also, since each $\beta(\check{\delta}) \in \mathrm{RCARD}(s)$, $\beta \in \mathrm{CARD}(s)$. Finally, for all x, o.t. $\Delta(x, s)^* \leqslant$ o.t. $\Delta(x, s) \leqslant$ o.t. $B(x, s)$, so in particular this holds for $x = 0$; i.e. o.t. $\tilde{C}_s \leqslant B(0, s)$, but $B(0, s) \subseteq \beta = \gamma_0$.

If $\Delta(0, s)$ has a greatest element, then $a_s \neq \{0\}$, so letting $(\check{\delta}_k: k \leqslant n)$ increasingly enumerate a_s, $n > 0$, and o.t. $\tilde{C}_s = \alpha_0 + \cdots + \alpha_n$, where $\alpha_k =$ o.t. $\Delta(\check{\delta}_k, s)^*$. Let $\sigma_k = \max \Delta(\check{\delta}_k, s)$, and let $\beta_k = \max B(\check{\sigma}_k, \check{\delta}_k, s)$. Then $\gamma_0 = \beta_0 + 1$, so $\bar{\gamma}_0 = \beta_0$. To see that $\beta_0 \in \mathrm{CARD}(s)$, note that $\beta_0 = \beta(f_{\beta_0, s})$, so $\beta_0 \in \mathrm{RCARD}(s)$. Further, β_0 is not the largest cardinal in the sense of s, since otherwise $\beta_0 \in X_{1,0,s}$, by (1.9), so $\beta(f_{\beta_0,0,s}) = \beta(f_{\beta_0+1,0,s}) = \check{\rho}(s)$. Finally, clearly $\alpha_k \leqslant \beta_k$ and, by (2.40), $\beta_0 > \cdots > \beta_n$, so $\alpha_0, \ldots, \alpha_n \leqslant \beta_0$; $(\beta^+)^s$ is closed for ordinal addition so o.t. $\tilde{C}_s = \alpha_0 + \cdots + \alpha_n < (\beta^+)^s$.

Finally, note that $\Delta(0, s) = \varnothing \Leftrightarrow X_{1,0,s}$ is cofinal in $\check{\rho}(s)$.

3. The real thing. In this section, for $\nu \in S$, we define $C_\nu \subseteq \nu$; $(C_\nu: \nu \in S)$ will be our global square-sequence. The condensation coherence will be dealt with in §4. Our approach to defining C_ν will be to select a final segment of $\tilde{C}_s$, where $s = \mathfrak{A}^+(\nu)$, and to each $\delta \in \tilde{C}_s$ associate *a* $\lambda < \nu$. C_ν will then consist of the λ's associated with the δ's from the appropriate final segment of $\tilde{C}_s$. The definition of the latter will involve α_ν as a parameter. (3.5) is crucial.

Our first task will be to pick up the thread left dangling after (2.4) when we turned our attention from S to an analysis of the structure of $\tilde{\mathscr{A}}^+$ independent of the fact that some of its members are collapsing structures for $\nu \in S$. Towards this end, we have

(3.1) LEMMA. *Suppose* $\mathfrak{A}^+ = \mathfrak{A}^+(\nu)$, $f: (\bar{\mathfrak{A}}, \bar{p}) \to_{\Sigma_1} \mathfrak{A}^+$, $\bar{\mathfrak{A}} = (J_{\bar{\rho}}, \bar{A})$; *suppose further that* $\alpha_\nu \in$ range f, *say* $f(\bar{\alpha}) = \alpha_\nu$ *where* $\bar{\alpha} < \alpha_\nu$, *and let* $\check{\bar{\nu}} = f^{-1}[\check{\nu}]$. *Then*:

(a) *for some* $\bar{\eta} \leqslant \bar{\alpha}$, $\bar{h}(\omega \times \bar{\eta}) = J_{\bar{\rho}}$, *where* $\bar{h}$ *is the canonical* Σ_1*-Skolem function for* $(\bar{\mathfrak{A}}, \bar{p})$,

(b) $\bar{\nu} \in S \setminus \mathrm{CARD}$, $\bar{\alpha} = \alpha(\bar{\nu})$, $\bar{\rho} = \rho(\bar{\nu})$, $\bar{A} = A(\bar{\nu})$, $n(\bar{\nu}) = n(\nu)$ *and* $\bar{p} = p(\bar{\nu})$.

PROOF. For (a), let $n = n(\nu)$, let $h =$ the canonical Σ_1-Skolem function for $\mathfrak{A}^+$. Let $\bar{x} \in J_{\bar{\rho}}$. Then, for some $k < \omega$,

$$\mathfrak{A}^+ \models \text{“}(\exists i < \omega)(\exists \xi_0, \ldots, \xi_k < \alpha)(f(\bar{x}) = h(i, \xi_0, \ldots, \xi_k))\text{.”}$$

Then for some $\bar{\xi}_0, \ldots, \bar{\xi}_k < \bar{\alpha}$ and $i < \omega$, $\mathfrak{A}^+ \models$ “$f(\bar{x}) = h(i, f(\bar{\xi}_0), \ldots, f(\bar{\xi}_k))$” so $\bar{x} = \bar{h}(i, \bar{\xi}_0, \ldots, \bar{\xi}_k)$. Thus $\bar{h}''(\omega \times \bar{\alpha}) = J_{\bar{\rho}}$.

By (4.1)(b) of [**5**], let $\bar{\beta}$, $\tilde{f}$ be s.t. $\tilde{f}: J_{\bar{\beta}} \to_{\Sigma_{n+1}} J_\beta$, $\tilde{f} \supseteq f$, with the other properties of Theorem (4.1)(b) of [**5**]. By fine structure theory, $\bar{h}$ is $\Sigma_{n+1}(J_{\bar{\beta}})$, so $\bar{h} \in J_{\bar{\beta}+1}$. Also $\bar{h}^{-1}[\check{\bar{\nu}}] \in J_{\bar{\beta}+1}$, and $\bar{h}\, \bar{h}^{-1}[\check{\bar{\nu}}]: \bar{h}^{-1}[\check{\bar{\nu}}] \to_{\text{onto}} \check{\bar{\nu}}$, so $J_{\bar{\beta}+1} \models \check{\bar{\nu}}$ is not a cardinal, so $\check{\bar{\nu}}$ is not a cardinal and $\bar{\beta} \geqslant \beta(\bar{\nu})$, and if $\bar{\beta} = \beta(\bar{\nu})$, then $n \geqslant n(\bar{\nu})$.

We distinguish two cases:

A. $\nu = \beta$. Then clearly $\bar{\nu} = \bar{\beta}$. Further,

$$J_\beta \models \text{“}(\forall \tau)(\tau \text{ is a cardinal} \Rightarrow \tau \leqslant \alpha_\nu)\text{”},$$

but this is Π_2, so $J_{\bar\nu} \models$ "$(\forall\tau)(\tau$ is a cardinal $\Rightarrow \tau \leqslant \bar\alpha)$", i.e. $J_{\bar\nu} \models$ "$\bar\alpha$ is the largest cardinal". Thus $\bar\nu \in S \setminus \text{CARD}$, $\bar\alpha = \alpha_{\bar\nu}$. We have already seen that $\bar\beta \geqslant \beta(\bar\nu)$, but $\bar\beta = \bar\nu \leqslant \beta(\bar\nu)$, so $\bar\beta = \beta(\bar\nu)$. Also, $\bar\nu \leqslant \rho(\bar\nu) \leqslant \beta(\bar\nu)$, so $\bar\rho = \rho(\bar\nu)$. We have already seen that $n \geqslant n(\bar\nu)$. If $n = 0$, then $n = n(\bar\nu)$, so suppose $n > 0$ and $n(\bar\nu) = m' < n$. By (1.5)(c) for $\bar\nu$ and m, $\check\rho_{\bar\beta}^{m+1} \leqslant \bar\alpha < \bar\nu = \rho_{\bar\beta}^m = \bar\beta$, but $m + 1 \leqslant n$, and $\bar\nu = \rho_{\bar\beta}^n = \bar\beta$, contradiction. The verification that $A = A_{\bar\beta}^n = A(\bar\nu)$, $\bar p = p(\bar\nu)$ will be the same in this case as in B; see below.

B. $\nu < \beta$. The crucial observation is that $\check\nu \in \text{range} f$. If $n > 0$, this is easy, since then ν is the unique witness in J_β to

$$(*) \qquad (\exists\eta)(\check\eta \text{ is a cardinal} \wedge J_\eta \models \text{“}\alpha \text{ is the largest cardinal”})$$

so $\nu \in \text{range} f$. On the other hand if $n = 0$, we use (1.8) to conclude that if τ is the largest cardinal in the sense of J_β, then $\tau \in \text{range} f$. Thus $\check\nu \leqslant \tau$ and if $\check\nu < \tau$, then ν is the unique witness in J_τ to $(*)$. If $\check\nu = \tau$, the conclusion is clear. Otherwise, note that $J_\tau|(\text{range} f \cap J_\tau) \prec_{\Sigma_\omega} J_\tau$ (since $\tau \in \text{range} f$), so again $\check\nu \in \text{range} f$.

Thus, $\bar\nu < \bar\beta$. Also, clearly $f(\check{\bar\nu}) = \check\nu$, so $f|\, J_{\bar\nu}$: $J_{\bar\nu} \to_{\Sigma_\omega} J_\nu$. Thus $J_{\bar\nu} \models$ "$\bar\alpha$ is the largest cardinal", i.e. $\bar\nu \in S \setminus \text{CARD}$.

By elementarity $J_{\bar\beta} \models$ "$\check{\bar\nu}$ is a cardinal", and therefore, having already seen that $\bar\beta \geqslant \beta(\bar\nu)$, $\bar\beta = \beta(\bar\nu)$ and $n \geqslant n(\bar\nu)$; this completes Case B.

Now we show that $n = n(\bar\nu)$. Again if $n = 0$, this is clear. Suppose then that $n(\bar\nu) < n$, say $n(\bar\nu) = m$. Then, by (1.5)(c) for $\bar\nu$ and m, $\check\rho_{\bar\beta}^{m+1} \leqslant \bar\alpha < \bar\nu < \rho_{\bar\beta}^m$, but $m + 1 \leqslant n$ and $\bar\nu \leqslant \rho_{\bar\beta}^n = \bar\rho$, contradiction. So $n = n(\bar\nu)$. But then $\bar A = A_{\bar\beta}^n = A_{\beta(\bar\nu)}^{n(\bar\nu)} = A(\bar\nu)$.

Finally, we show $\bar p = p(\bar\nu)$. Clearly $p(\bar\nu) \leqslant_* \bar p$. If $p(\bar\nu) <_* \bar p$, let $q = f(p(\bar\nu))$. Then $\bar p = \bar h(i, \xi, p(\bar\nu))$ for some $\xi < \bar\alpha$, so $p = h(i, f(\xi), q)$ and $q <_* p$, contradicting the defining property of $p(\nu)$, so in fact $p(\bar\nu) = \bar p$.

The next three lemmas are companions to (3.1). In (3.2)–(3.5) we let $\nu \in S$, $\alpha = \alpha_\nu$, $s = \mathfrak{A}^+(\nu) = (J_\rho, A, p)$ and $\check\delta \in \tilde C_s \setminus \alpha$. We let $X = X_{\alpha,0,s|\delta}$ and let π: $(J_{\rho'}, A', p') \tilde\leftrightarrow (s|\delta)|\, X$. In (3.2)(c), we assume $\delta > \alpha$; in (3.2)(d) and (3.4) we assume $\sup OR \cap X > \alpha$.

(3.2) Lemma. (a) *X is cofinal in $\check\delta$.*

(b) *Suppose $\nu = \rho$. Then $\alpha \in X$ and α is the largest cardinal of J_δ.*

(c) *If $\nu < \rho$, then $\check\alpha < \delta \Rightarrow \nu < \delta$ and $\nu = (\alpha^+)^{J_\delta}$.*

(d) *If $\nu < \rho$ and* $\sup OR \cap X > \alpha$, *let $\eta = \inf((OR \cap X) \setminus \alpha)$. Then $\pi(\alpha) = \eta$ and so $\eta \in \text{CARD}(J_\delta)$. Further, if $\eta = \alpha$ then $\check\nu \in X$ and $X \cap \check\nu$ is transitive.*

Proof. Since $\delta(f_{\alpha,0,s}) = \rho$ and $\check\delta \in \tilde C_s$, by (2.42)(e), there is a $\gamma < \alpha$ and finite $u \subseteq \alpha$ s.t. $\delta = \delta(f_{\gamma,\langle u\rangle,s})$. By (2.24), if $f_{\gamma,\langle u\rangle,s} = (\bar s, |f|, s)$ then $f_{\gamma,\langle u\rangle,s|\delta} = (\bar s, |f|, s|\delta)$.

Then, for (a), $X \supseteq X_{\gamma,\langle u\rangle,s|\delta} = X_{\gamma,\langle u\rangle,s}$ which is cofinal in $\check\delta$.

For (b), (c), we prove:

(∗) $\mathrm{CARD}(J_\rho)$ has a largest member iff $\mathrm{CARD}(J_\delta)$ has a largest member, and if so, $\max \mathrm{CARD}(J_\rho) = \max \mathrm{CARD}(J_\delta)$, $\mathrm{CARD}(J_\rho) = \mathrm{CARD}(J_\delta)$.

(∗∗) If $\mathrm{CARD}(J_\rho)$ has limit order type, then $\mathrm{CARD}(J_\delta) = \mathrm{CARD}(J_\rho) \cap J_\delta$ and $\mathrm{CARD}(J_\delta) \cap X_{\gamma,\langle u\rangle,s|\delta}$ is cofinal in $\check{\delta}$.

We first prove (b) (c) from (∗), (∗∗). For (b), by (∗), α is the largest cardinal of J_δ and so $\alpha \in X$. Note that this means

$$(***) \qquad \delta \in S_\alpha.$$

For (c), if $\mathrm{CARD}(J_\rho)$ has a largest member, say $\check{\sigma}$, then $\check{\sigma} \geqslant \check{\nu}$, since otherwise $\check{\sigma} = \check{\alpha}$, i.e. $\rho = \nu$. But then $\check{\nu} \leqslant \check{\sigma} < \delta$, since by (∗), $\check{\sigma} = \max \mathrm{CARD}(J_\delta)$. By (∗∗), $\check{\nu} = (\alpha^+)^{J_\rho} = (\alpha^+)^{J_\delta}$.

If $\mathrm{CARD}(J_\rho)$ has limit order type, then by (∗), $\mathrm{CARD}(J_\delta)$ has limit order type and is therefore cofinal in $\check{\delta}$. Further, by (∗∗), $\mathrm{CARD}(J_\delta) = \mathrm{CARD}(J_\rho) \cap J_\delta$. Finally, since $\check{\alpha} < \delta$, there is an $\eta \in \mathrm{CARD}(J_\delta)$, $\eta > \alpha$, i.e. $(\alpha^+)^{J_\delta}$ exists. But then $(\alpha^+)^{J_\delta} = (\alpha^+)^{J_\rho} = \check{\nu} < \delta$.

Finally, for (d), the first assertion is clear. If $\eta = \alpha \in X$ then $\check{\nu} \in X$, since either $\check{\nu} = \max \mathrm{CARD}(J_\delta) \in X$, or $\mathrm{CARD}(J_\delta)$ has a largest member $\check{\sigma} > \check{\nu}$, in which case $\check{\nu} = (\alpha^+)^{J_\sigma}$, $\check{\sigma} \in X$, so $\check{\nu}$ is $\Sigma_1(J_\delta)$-definable from $\check{\alpha}, \check{\sigma}$; finally, if $\mathrm{CARD}(J_\delta)$ has limit order type,by (c), $\check{\nu} < \delta$, so, by (∗∗), let $\check{\eta} \in \mathrm{CARD}(J_\delta) \cap X_{\gamma,\langle u\rangle,s|\delta}$, $\check{\eta} > \check{\nu}$. Then $\check{\eta} \in X$, so, as above, $\check{\nu} = (\alpha^+)^{J_\eta}$ and $\check{\nu} \in X$ since it is $\Sigma_1(J_\delta)$-definable from $\check{\alpha}, \check{\eta}$. The transitivity of $X \cap \check{\nu}$ follows easily from $\check{\alpha} + 1 \subseteq X$, $\alpha = \alpha_\nu$.

We prove (∗), (∗∗). For (∗), $\mathrm{CARD}(J_\rho)$ has a largest member $\Leftrightarrow$ $\mathrm{CARD}(\bar{s})$ has a largest member $\Leftrightarrow$ $\mathrm{CARD}(J_\delta)$ has a largest member, and if so $\max \mathrm{CARD}(J_\rho) = |f|\,(\max \mathrm{CARD}(\bar{s})) = \max \mathrm{CARD}(J_\delta)$. Finally if $\check{\sigma} = \max \mathrm{CARD}(J_\rho)$, then $\mathrm{CARD}(J_\rho) = \mathrm{CARD}(J_\sigma) \cup \{\sigma\} = \mathrm{CARD}(J_\delta)$.

For (∗∗), $|f|''\mathrm{CARD}(\bar{s}) \subseteq \mathrm{CARD}(J_\rho) \cap \mathrm{CARD}(J_\delta)$, and by the proof of (∗), $\mathrm{CARD}(\bar{s})$ has limit order type and is therefore cofinal in $\check{\rho}(\bar{s})$. But then $|f|''\mathrm{CARD}(\bar{s})$ is cofinal in $\check{\delta}$, so $\mathrm{CARD}(J_\rho) \cap \mathrm{CARD}(J_\delta)$ is cofinal in $\check{\delta}$, so, in fact, $\mathrm{CARD}(J_\delta) = \mathrm{CARD}(J_\rho) \cap \check{\delta}$. Finally, $|f|''\mathrm{CARD}(\bar{s}) = \mathrm{CARD}(J_\delta) \cap X_{\gamma,\langle u\rangle,s|\delta}$.

(3.3) LEMMA. *If $\rho = \nu$, then $s|\delta = \mathfrak{A}^+(\delta)$.*

PROOF. By (3.2)(a), X is cofinal in $\check{\delta}$. We now show $X = J_\delta$. It suffices, of course, to show that $OR \cap X = \check{\delta}$, so it suffices, in fact, to show that $OR \cap X$ is transitive. This follows easily from $\check{\alpha} + 1 \subseteq X$ and $J_\delta \models$"$\check{\alpha}$ is the largest cardinal".

Now id: $s|\delta \to_{\Sigma_0} s$, so applying the downward extension of embeddings lemma to $\mathrm{id}|_{J_\delta}$, we get g, β', $g|J_\delta = \mathrm{id}$, $g\colon J_{\beta'} \to_{\Sigma_n} J_\beta$, where $\beta = \beta(\nu)$, $n = n(\nu)$, and such that $\delta = \rho^n_{\beta'}$, $A \cap J_\delta = A^n_{\beta'}$. Thus, either $\beta' = \delta$ or $J_{\beta'} \models$"$\check{\delta}$ is a cardinal", i.e., $\beta' \leqslant \beta(\delta)$. But $h''_{s|\delta}(\omega \times \alpha) = J_\delta$ and $h_{s|\delta}$ is $\Sigma_{n+1}(J_{\beta'})$, in appropriate parameters, so $\beta' = \beta(\delta)$ and $n \geqslant n(\delta)$. To see that $n = n(\delta)$, assume, without loss of generality, that $n > 0$. Then, if $n > n(\delta)$, we would have, on the one hand,

$\rho_{\beta'}^{n(\delta)+1} \leqslant \alpha < \delta \leqslant \rho_{\beta'}^{n(\delta)}$, but on the other hand, $\rho_{\beta'}^{n(\delta)+1} > \rho_{\beta'}^{n} = \delta$, contradiction! So $n = n(\delta)$, $\delta = \rho(\delta)$ and $A \cap J_\delta = A(\delta)$, i.e. $(J_\delta, A \cap J_\delta) = \mathfrak{A}(\delta)$. Since $X = J_\delta$, now-familiar arguments show that $p = p(\delta)$.

(3.4) LEMMA. *Suppose* $\rho > \nu$ *and* $\sup X > \alpha$. *Then* $(\exists! \lambda \in S_\alpha)(\exists! \pi)$

$$\pi\colon \mathfrak{A}^+(\lambda) \tilde{\leftrightarrow} (s|\delta) X.$$

PROOF. By (3.2)(d), there is a unique $\lambda \in S_\alpha$ such that either $\lambda = \rho'$ (this will be the case if η of (3.2)(d) is the largest cardinal of J_ρ) or $J_{\rho'} \models$ "$\check{\lambda}$ is a cardinal" (in this case $\pi(\check{\lambda}) = (\eta^+)^{J_\delta}$). It remains to show that $(J_{\rho'}, A', p') = \mathfrak{A}^+(\lambda)$.

Towards this end, apply the downward extension of embeddings lemma to

$$\pi\colon (J_{\rho'}, A', p') \to_{\Sigma_0} s,$$

obtaining β', $\tilde{\pi}$ such that $\tilde{\pi} \supseteq \pi$, $\rho' = \rho_{\beta'}^n$, $A' = A_{\beta'}^n$, and $\tilde{\pi}\colon J_{\beta'} \to_{\Sigma_n} J_\beta$, again, where $\beta = \beta(\nu)$, $n = n(\nu)$. The argument that $\beta' = \beta(\lambda)$, $n = n(\lambda)$, $\rho' = \rho(\lambda)$, $A' = A(\lambda)$, $p' = p(\lambda)$ is now exactly as in (3.3), but with ρ' replacing some occurrences of δ, A' replacing $A \cap J_\delta$, p' replacing p, π replacing $\mathrm{id}|J_\delta$, $\tilde{\pi}$ replacing $\tilde{g}$, and λ replacing the other occurrences of δ.

(3.5) We now describe how a final segment of $\tilde{C}_s$ is "projected" onto a subset of ν. Let $\check{\delta} \in \tilde{C}_s$, and if $\mathrm{CARD}(J_\rho)$ has limit order type assume $\delta > \check{\alpha}$.

DEFINITION. Let $Y_{\delta\nu} = X_{\alpha,0,s|\delta}$. Set $\check{\delta} \in C_\nu^\#$ iff $\sup OR \cap Y_{\delta\nu} > \alpha$ (so, by (3.2), if $\rho = \nu$ then $C_\nu^\# = \tilde{C}_s$, while if $\rho > \nu$ then $C_\nu^\#$ is a final segment of $\tilde{C}_s$).

For $\check{\delta} \in C_\nu^\#$, let $\lambda(\delta, \nu) =$ the unique $\lambda \in S_\alpha$ such that $\exists \pi\colon \mathfrak{A}^+(\lambda) \tilde{\leftrightarrow} (s|\delta)|\ Y_{\delta\nu}$ (this exists by (3.3), (3.4), and by (3.3), if $\rho = \nu$, then $\lambda(\delta, \nu) = \delta$).

Also, let $\hat{C}_\nu = \{\check{\delta} \in C_\nu^\#\colon \alpha \in Y_{\delta\nu}\}$. Finally, set $C_\nu = \{\check{\lambda}(\delta, \nu)\colon \check{\delta} \in \hat{C}_\nu\}$.

(3.6) LEMMA. (a) *$\hat{C}_\nu$ is a final segment of $\tilde{C}_s$,*

(b) o.t. $C_\nu \leqslant$ o.t. $\tilde{C}_s \leqslant \alpha_\nu$,

(c) $(\check{\delta}_1, \check{\delta}_2 \in C_\nu^\# \wedge \delta_1 < \delta_2) \Rightarrow \lambda(\delta_1, \nu) < \lambda(\delta_2, \nu)$,

(d) *if* $\check{\delta} \in \hat{C}_\nu$, $\lambda(\delta, \nu) = \beta(f_{\alpha,0,s|\delta}) = \nu \cap Y_{\delta\nu}$,

(e) *C_ν is closed.*

PROOF. (a), (b), (c) are trivial. For (d), the conclusion is clear if $\rho = \nu$, and follows from (3.2)(d) if $\rho > \nu$, since $\check{\delta} \in \hat{C}_\nu$. (e) follows immediately from (d); in fact, the only difficulty with $\{\check{\lambda}(\delta, \nu)\colon \check{\delta} \in C_\nu^\#\}$ is that it may fail to be closed. This is remedied by thinning out to C_ν.

(3.7) We now establish the coherence property for the C_ν's.

LEMMA. $\lambda \in C_\nu \Rightarrow C_\lambda = \lambda \cap C_\nu$.

PROOF. Let $\lambda = \lambda(\delta, \nu)$, where $\check{\delta} \in \hat{C}_\nu$. Let $s = \mathfrak{A}^+(\nu)$, $s' = \mathfrak{A}^+(\lambda)$, and let $\pi\colon s \tilde{\leftrightarrow} s|\delta|Y_{\delta\nu}$. The main point to be verified is that

$$C_\nu \cap \check{\delta} = \pi'' C_\lambda. \tag{1}$$

PROOF OF (1). By (2.31), (2.38) and (2.42)(b), $\tilde{C}_s \cap \check{\delta} = \pi'' \tilde{C}_{s'}$, so we need only verify that if $\check{\gamma}' \in \tilde{C}_{s'}$ and $\check{\gamma} = \pi(\check{\gamma}')$, then $\alpha_\nu \in Y_{\gamma'\lambda} \Leftrightarrow \alpha_\nu \in Y_{\gamma\nu}$. But this is clear since $Y_{\gamma\nu} = \pi'' Y_{\gamma'\lambda}$ and since $\alpha_\nu \in Y_{\delta\nu}$ (and so $\alpha_\nu = \pi(\alpha_\nu)$).

It remains only to recall that, with γ, γ' as in (1),

$$(2) \qquad \check{\gamma}' \in \hat{C}_{s'} \Rightarrow \check{\gamma}' \subseteq Y_{\gamma'\lambda} \Rightarrow \pi'\check{\gamma}' = \check{\gamma}'.$$

Thus, since $Y_{\gamma\nu} = \pi''Y_{\gamma'\lambda}$, $\lambda(\gamma, \nu) = \lambda(\gamma', \lambda)$.

(3.8) LEMMA. *C_ν cofinal in $\check{\nu} \Leftrightarrow \tilde{C}_s$ cofinal in $\check{\rho}(s)$, where $s = \mathfrak{A}^+(\nu)$.*

PROOF. Suppose $\tilde{C}_s$ is bounded in $\check{\rho}(s)$. If $\hat{C}_\nu = \varnothing$ then $C_\nu = \varnothing$, so suppose $\delta = \max \hat{C}_\nu$. Then, by (3.6), $\check{\lambda}(\delta, \nu) = \max C_\nu$, so C_ν is bounded in $\check{\nu}$. Now suppose $\tilde{C}_s$ is cofinal in $\check{\rho}(s)$. Then clearly $\hat{C}_\nu$ is cofinal in $\check{\nu}$, since then

$$\alpha_\nu \in J_{\rho(s)} = X_{\alpha_\nu, 0, s} = \bigcup \left\{ X_{\alpha_\nu, 0, s|\delta} \colon \delta \in \tilde{C}_s \right\}.$$

But again

$$\begin{aligned} \check{\nu} &= \check{\nu} \cap X_{\alpha_\nu, 0, s} = \check{\nu} \cap \bigcup \left\{ X_{\alpha_\nu, 0, s|\delta} \colon \delta \in \tilde{C}_s \right\} \\ &= \bigcup \left\{ \check{\nu} \cap X_{\alpha_\nu, 0, s|\delta} \colon \delta \in \tilde{C}_s \right\} = \bigcup \left\{ \check{\nu} \cap X_{\alpha_\nu, 0, s|\delta} \colon \check{\delta} \in \hat{C}_s \right\}, \end{aligned}$$

i.e. C_ν is cofinal in ν.

4. Squared scales. In this section, after some preliminary work, we present and prove the global combinatorial principle Squared Scales formulated by Avraham and Shelah, who suggested to Stanley that he prove it holds in L. The condensation coherence property, (4.10)(b), can be traced back to work of Donder in [**4**], which was "globalized" by Jensen. In its present form it is due to Donder and Stanley, as is the case for all of the material of this section except for the principle itself.

(4.1) We begin with a simple observation which extends material of §2.

PROPOSITION. *Let $f = (\bar{s}, |f|, s) \in \mathscr{F}$. If $\bigcup\{\Delta(\check{\delta}, s) \cup \{\check{\delta}\} \colon \check{\delta} \in a_s\} \subseteq$ range f, then* range f *is cofinal in $\check{\rho}(s)$.*

PROOF. This is trivial, unless, letting $\check{\delta} = \max\ a_s$, $\Delta)\check{\delta}, s)$ is bounded in $\check{\rho}(s)$. But since $\check{\delta} = \max a_s$, and $\Delta(\check{\delta}, s)$ is bounded in $\check{\rho}(s)$, in fact $\Delta(\check{\delta}, s) = \varnothing$. Now, since $\check{\delta} \in$ range f, range $f \supseteq X_{1, \check{\delta}, s}$ which is cofinal in $\check{\rho}(s)$, since $\Delta\ (\check{\delta}, s) = \varnothing$.

(4.2) DEFINITION. For $\nu \in S$, $s = \mathfrak{A}^+(\nu)$, let $\Gamma(\nu) = \bigcup\{\Delta(\check{\delta}, s) \cup \{\check{\delta}\} \colon \check{\delta} \in a_s\}$.

(4.3) LEMMA. *Let μ be a limit cardinal, $\nu \in S_\mu$, $s = \mathfrak{A}^+(\nu)$, and suppose* o.t. *$C_\nu < \mu$. Then, for sufficiently large cardinals $\mu' < \mu$, $\Gamma(\nu) \subseteq X_{\mu', 0, s}$.*

PROOF. Clearly o.t. $C_\nu < \mu \Rightarrow$ o.t. $\Gamma(\nu) < \mu$. Equally clearly, for sufficiently large $\mu' < \mu$, $a_s \subseteq X_{\mu', 0, s}$. Since o.t. $\Gamma(\nu) < \mu$, for each $\check{\delta} \subset a_s$, o.t. $\Delta(\check{\delta}, s) < \mu$. Further, since, in particular o.t. $\Delta(0, s) < \mu$, for some $\mu_0 < \mu$, $X_{\mu_0, 0, s}$ is cofinal in $\check{\rho}(s)$. Let $\mu' \in \mathrm{CARD} \cap \mu$ be such that $a_s \subset X_{\mu', 0, s}$, $\mu' \geqslant \mu_0$, and for all $\check{\delta} \in a_s$, $\mu' \geqslant$ o.t. $\Delta(\check{\delta}, s)$. By (2.33), letting $f_{\mu', 0, s} = (s', |f|, s)$ for each $\check{\delta} \in a_s$ and letting $f(\delta') = \delta$, we have

$$f \colon (s', \Delta(\check{\delta}', s)) \rightarrow_{\Sigma_0} (s, \Delta(\check{\delta}, s)).$$

But, in fact, f is Σ_1-elementary since range $f = X_{\mu',0,s} \supseteq X_{\mu_0,0,s}$ is cofinal in $\check{\rho}(s)$. But then for all $\xi <$ o.t. $\Delta(\check{\delta}, s) \leqslant \mu'$, $(s', \Delta(\check{\delta}, s'))$ models the Σ_1-sentence in parameter ξ which asserts the existence of a ξth element of $\Delta(\check{\delta}', s')$. If $\check{\eta}'$ is this ξth element and $f(\check{\eta}') = \check{\eta}$, then $\check{\eta}$ is the ξth element of $\Delta(\check{\delta}, s)$; i.e. $\Delta(\check{\delta}, s) \subseteq$ range f, as required.

(4.4) DEFINITION. If μ is a limit cardinal, $\nu \in S_\mu$, $s = \mathfrak{A}^+(\nu)$ and o.t. $C_\nu < \mu$, let

$$\mu_0(\nu) = \text{the least } \mu' \in \text{CARD} \cap \mu \text{ s.t. } \Gamma(s) \subseteq X_{\mu',0,s}$$

and o.t. $\Gamma(s) < \mu'$, and let

$$D(\nu) = [\mu_0(\nu), \nu) \cap \text{CARD}.$$

(4.5) LEMMA. *If μ, ν, s are as in* (4.4), *if $\mu' \in D(\nu)$ and $f_{\mu',0,s} = (s', |f|, s)$, then there is a unique $\nu' \in S_{\mu'}$ such that either $s' \models$"ν' is a cardinal", or $\rho(s') = \nu'$; further $s' = \mathfrak{A}^+(\nu')$.*

PROOF. Uniqueness is clear, and existence of such a ν' follows readily from the fact that sup $X_{\mu',0,s} = \check{\rho}(s)$, so in particular, $\mu' <$ sup $X_{\mu',0,s}$. It remains to prove that $s' = \mathfrak{A}^+(\nu')$.

Applying the downward extension of embeddings lemma to f, we obtain β', $\tilde{f}, \tilde{f}\colon J_{\beta'} \to_{\Sigma_{n(\nu)+1}} J_{\beta(\nu)}$, s.t. $s' = (J_{\rho_{\beta'}^{n(\nu)}}, A_{\beta'}^{n(\nu)}, p')$ for some p'. Since $h''_{s'}(\omega \times \mu') = |s'|$, as usual $\beta' = \beta(\nu')$ and $n(\nu) \geqslant n(\nu')$. If $n(\nu) = 0$, then clearly $n(\nu') = n(\nu)$. But exactly as in (3.1), (3.3), (3.4), $n(\nu') \geqslant n(\nu)$, since otherwise we would have $\rho(\nu') \geqslant \rho_{\beta'}^{n(\nu')+1} \geqslant \rho'$ on the one hand, but $\rho_\beta^{n(\nu')+1} \leqslant \mu' < \nu' \leqslant \rho'$ on the other. Finally, we need only verify that $p' = p(\nu')$; clearly $p' \geqslant_* p(\nu')$. But if $p' >_* p(\nu')$ then for some $n < \omega$ and $\xi_1, \ldots, \xi_n < \mu'$, p' is $\Sigma_1(s')$-definable from $\xi_1, \ldots, \xi_n$ and $p(\nu')$. This carries up to s and $_*f(p(\nu'))$, $p(\nu) = f(p') >$ contradicting the defining property of $p(\nu)$.

(4.6) DEFINITION. If μ, ν, s, μ', etc. are as in (4.5), then set

$$\Phi^\nu(\mu') = \text{the unique } \nu' \text{ as in } (4.5).$$

(4.7) PROPOSITION. *Let μ, ν, s, etc. be as in* (4.5). *Then, for sufficiently large $\mu' \in \text{CARD} \cap \mu$, $\mu' \in \bigcap\{D(\lambda)\colon \check{\lambda} \in C_\nu \cup \{\check{\nu}\}\}$.*

PROOF. The only difficulty here is that if $\check{\lambda} = \lambda(\delta, \nu) \in C_\nu$, then, in general, $X_{\mu',0,s|\delta} \subsetneqq X_{\mu',0,s} \cap J_\delta$. Let $\check{\delta}_0 = \inf C_\nu$. Let $\mu_1 \geqslant \mu_0(\nu)$ s.t. $\mu \in X_{\mu_1,0,s|\delta_0}$ (*exists, since* $\check{\delta}_0 \in \hat{C}_\nu$). Then, for all $\check{\delta} \in \hat{C}_\nu \cup \{\check{\rho}(\nu)\}$, sup $X_{\mu',0,s|\delta} > \mu'$, whenever $\mu_1 \leqslant \mu' \in \text{CARD} \cap \mu$.

Now, let $\check{\delta} \in \hat{C}_\nu$, $\mu_1 \leqslant \mu' \in \text{CARD} \cap \mu$. Let $Y = Y_{\delta\nu}$, $\lambda = \lambda(\delta, \nu)$; let π: $(J_{\rho'}, A', p') \leftrightarrow (s|\delta|)|Y$, so, by §3, $(J_{\rho'}, A', p') = s' = \mathfrak{A}^+(\lambda)$. Let $f = (\bar{s}, |f|, s) = f_{\mu',0,s}$. Then $\delta \in$ range f, say $f(\bar{\delta}) = \delta$. Also, by (4.1), $f''\check{\bar{\delta}}$ is cofinal in $\check{\delta}$, since $\mu_1 \leqslant \mu'$.

Let $\bar{\nu} = \Phi^\nu(\mu')$, so $\bar{s} = (J_{\bar{\rho}}, \bar{A}, \bar{p}) = \mathfrak{A}^+(\bar{\nu})$. By (3.3) or (3.4), $\check{\bar{\delta}} \in C_{\bar{\nu}}^\#$, since $X_{\mu',0,\bar{s}|\bar{\delta}} = f^{-1}[X_{\mu',0,s|\delta}]$ and sup $X_{\mu',0,s|\delta} > \mu'$. Therefore, $\bar{Y} = X_{\mu',0,\bar{s}|\bar{\delta}}$ is cofinal in $\check{\bar{\delta}}$. Let $\bar{\lambda} = \lambda(\bar{\delta}, \bar{\nu})$, and let $\bar{\pi}$: $(J_{\bar{\rho}'}, \bar{A}', \bar{p}') \leftrightarrow (\bar{s}|\bar{\delta})|\bar{Y}$. So, $\bar{s}' = (J_{\bar{\rho}'}, \bar{A}', \bar{p}') = \mathfrak{A}^+(\bar{\lambda})$. Then, by (2.21), $(\bar{s}', |f| \circ \bar{\pi}, s|\delta) = f_{\mu',0,s|\delta}$. Thus range $|f| \circ \bar{\pi}$ is cofinal in $\check{\delta}$. But then, arguing as in (4.3), $\Gamma(s|\delta) \subset X_{\mu',0,s|\delta}$.

This completes the proof, but we note, for future reference, that we have shown, in fact, that range $|f| \circ \bar{\pi} \subseteq Y$, that $(\bar{s}', \pi^{-1} \circ |f| \circ \bar{\pi}, s') = f_{\mu',0,s'}$, and that $\bar{\lambda} = \Phi^{\lambda}(\mu')$.

(4.8) Suppose ν, μ, $s = (J_{\rho}, A, p)$, etc., area as in (4.5) and that $\mu_1 \leqslant \mu'$ are as in the proof of (4.7). We investigate the circumstances under which $\mu' \notin X_{\mu',0,s}$.

LEMMA. *Let $f = (\bar{s}, |f|, s) = f_{\mu',0,s}$. If $\mu' \notin$ range f, then $\bar{s} \models$ "μ' is inaccessible" and $f(\mu')$ is inaccessible.*

PROOF. The first assertion is clear. For the second, note that $s \models$ "$f(\mu')$ is inaccessible" on the one hand, but on the other, since $\mu \in X_{\mu',0,s}$, $f(\mu') \leqslant \mu$, so $f(\mu')$ is, in fact, inaccessible.

(4.9) DEFINITION. Let $\mu_1(\nu) =$ the least $\mu' \in (\mathrm{CARD} \cap \mu) \setminus \mu_0(\nu)$ s.t. $\mu \in X_{\mu',0,s|\check{\delta}}$ for all $\check{\delta} \in \hat{C}_{\nu}$.

(4.10) We now prove the crucial Condensation Coherence property.

LEMMA. *If $\nu \in S_{\mu}$, etc., are as in (4.5) and if $\mu_1(\nu) \leqslant \mu' \in \mathrm{CARD} \cap \mu$, then, letting $\bar{\nu} = \Phi^{\nu}(\mu')$,*

(a) *for all $\check{\lambda} \in C_{\nu}$, $\mu_1(\lambda) \leqslant \mu'$,*

(b) *$\{\check{\Phi}^{\lambda}(\mu') : \check{\lambda} \in C_{\nu}\}$ and $C_{\bar{\nu}}$ are comparable under $\subseteq$ and the smaller is a final segment of the larger,*

(c) *if μ' is a limit cardinal, then $\mu' > \mu_1(\nu)$, $\mu_0(\bar{\nu}) = \mu_0(\nu)$ and $\Phi^{\bar{\nu}} = \Phi^{\nu}|\mu$.*

PROOF. (a) and (c) are clear. For (b), let $\check{\delta} \in \hat{C}_{\nu}$, $\lambda = \lambda(\delta, \nu)$. Adopt the notation of (4.7). Recall that we proved there that $\Phi^{\lambda}(\mu') = \lambda(\bar{\delta}, \bar{\nu})$, where $f_{\mu',0,s}(\check{\bar{\delta}}) = \check{\delta}$ (and that, a fortiori, $\check{\bar{\delta}} \in C_{\bar{\nu}}^{\#}$). Thus $\{\check{\Phi}^{\lambda}(\mu') : \check{\lambda} \in C_{\nu}\} \subseteq \{\check{\lambda}(\bar{\delta}, \bar{\nu}) : \check{\bar{\delta}} \in C_{\bar{\nu}}^{\#}\}$. Further, $f''C_{\bar{\nu}}^{\#} = C_{\nu}^{\#}$, $\hat{C}_{\bar{\nu}}$ is a final segment of $C_{\bar{\nu}}^{\#}$, $\hat{C}_{\nu}$ is a final segment of $C_{\nu}^{\#}$. Thus if $\check{\bar{\delta}}_0 = \inf \hat{C}_{\bar{\nu}}$, $\check{\delta}_0 = \inf \hat{C}_{\nu}$, either $f(\check{\bar{\delta}}_0) = \check{\delta}_0$, in which case, equality holds in (b), or $f(\check{\bar{\delta}}_0) < \check{\delta}_0$, in which case $f(\mu') \notin X_{\mu',0,s|\delta_0}$ and $\{\check{\Phi}^{\lambda}(\mu') : \check{\lambda} \in C_{\nu}\}$ is a final segment of $C_{\bar{\nu}}$ or $f(\check{\bar{\delta}}_0) > \check{\delta}_0$, in which case $f^{-1}(\mu) \notin X_{\mu',0,\bar{s}|\bar{\delta}_0}$ and $C_{\bar{\nu}}$ is a final segment of $\{\check{\Phi}^{\lambda}(\mu') : \check{\lambda} \in C_{\nu}\}$.

(4.11) We finally state the global combinatorial principle Squared Scales, asserting the existence of a global, condensation coherent square system with built-in scales. The scales and their properties embody the condensation coherence.

DEFINITION. For limit cardinals κ, let $g \in \mathfrak{G}(\kappa)$ iff dom g is a nonempty final segment of $\mathrm{CARD} \cap (\omega, \kappa)$, and $\bar{\kappa} \in \operatorname{dom} g$, $g(\bar{\kappa}) \in (\bar{\kappa}, \bar{\kappa}^{+})$. If $f, g \in \mathfrak{G}(\kappa)$, set $f <_{\kappa}^{*} g$ iff there is $\kappa_0 \in \operatorname{dom} f \cap \operatorname{dom} g$ s.t. $\kappa_0 \leqslant \bar{\kappa} \in \mathrm{CARD} \cap (\omega, \kappa) \Rightarrow f(\bar{\kappa}) < g(\bar{\kappa})$.

Let $g \in \bar{\mathfrak{G}}(\kappa)$ iff dom $g \in [\mathrm{CARD} \cap (\omega, \kappa)]^{<\kappa}$ and for $\bar{\kappa} \in \operatorname{dom} g$, $g(\bar{\kappa}) \in (\bar{\kappa}, \bar{\kappa}^{+})$. If $g \in \bar{\mathfrak{G}}(\kappa)$, canonically extend g to $\tilde{g} \in \mathfrak{G}(\kappa)$ by letting $\tilde{g}(\bar{\kappa}) = \bar{\kappa} + 1$, for $\bar{\kappa} \notin \operatorname{dom} g$, $\bar{\kappa} \in \mathrm{CARD} \cap (\inf \operatorname{dom} g, \kappa)$. We will write $\mathfrak{G}$, $\bar{\mathfrak{G}}$, $<^{*}$ when the reference to κ is clear.

Squared Scales (Shelah) is the assertion: There is a sequence $(C_{\nu} : \nu \in S)$, and for each limit cardinal κ, a sequence $\vec{\Phi}_{\kappa} = (\Phi^{\nu} : \nu \in S_{\kappa} \wedge \text{o.t. } C_{\nu} < \kappa)$ such that:

A. $(C_{\nu} : \nu \in S)$ is a global square system, i.e. for all $\nu \in S$, letting $\alpha = \alpha_{\nu}$:

(1) C_{ν} is a closed subset of $S_{\alpha} \cap \check{\nu}$, $\sup C_{\nu} < \check{\nu} \Rightarrow \operatorname{cf} \check{\nu} = \omega$,

(2) $\check{\lambda} \in C_\nu \Rightarrow C_{\check{\lambda}} = \check{\lambda} \cap C_\nu$,

(3) o.t. $(C_\nu) \leqslant \alpha$.

B. For all limit cardinals κ, all $\nu \in S_\kappa$, o.t. $C_\nu < \kappa \Rightarrow \Phi^\nu \in \mathfrak{G}(\kappa)$ and:

(1) $\bar{\kappa} \in \operatorname{dom} \Phi^\nu \Rightarrow \Phi^\nu(\bar{\kappa}) \in S_{\bar{\kappa}}$,

(2) $\bar{\kappa} \in \operatorname{dom} \Phi^\nu \Rightarrow (\forall \check{\lambda} \in C_\nu)(\bar{\kappa} \in \operatorname{dom} \Phi^\lambda)$,

(3) $(\forall \tau \in S_\kappa \cap \nu)(\text{o.t. } C_\tau < \kappa \Rightarrow \Phi^\tau <^* \Phi^\nu)$,

(4) $\sup C_\nu = \check{\nu} \Rightarrow \bar{\kappa} \in \operatorname{dom} \Phi^\nu \Rightarrow \check{\Phi}^\nu(\bar{\kappa}) = \sup\{\check{\Phi}^\lambda(\bar{\kappa}): \check{\lambda} \in C_\nu\}$,

(5) κ singular $\Rightarrow (\Phi^\nu: \nu \in S_\kappa \wedge \text{o.t. } C_\nu < \kappa)$ is a scale, i.e., whenever $g \in \overline{\mathfrak{G}}(\kappa)$, there is a $\nu_0 \in S_\kappa$ s.t. o.t. $C_\nu < \kappa$ and $\tilde{g} <^* \Phi^\nu$.

C. For limit κ and $\nu \in S_\kappa$ such that o.t. $C_\nu < \kappa$, if $\bar{\kappa} \in \operatorname{dom} \Phi^\nu$ and $\bar{\nu} = \Phi^\nu(\bar{\kappa})$:

(1) $C_{\bar{\nu}}$ and $\{\check{\Phi}^\lambda(\bar{\kappa}): \check{\lambda} \in C_\nu\}$ are comparable under $\subseteq$ and the smaller is a final segment of the larger,

(2) $\Phi^{\bar{\nu}} = \Phi^\nu | \text{CARD} \cap (\omega, \bar{\kappa})$,

(3) $\{\check{\Phi}^\lambda(\bar{\kappa}): \check{\lambda} \in C_\nu\} \in J_{\beta(\bar{\nu})+1}$.

(4.12) THEOREM ($V = L$). *Squared Scales holds.*

PROOF. The C_ν have already been defined. In verifying the principle, we shall use $\Phi^\nu | [\mu_1(\nu), \kappa)$ instead of Φ^ν. A and B(1) are immediate. B(2), B(4), C(1), (2) are by (4.10). For B(5) the main observation is

(1) if $s = \mathfrak{A}^+(\nu)$, o.t. $C_\nu < \kappa, f \in |s| \cap \overline{\mathfrak{G}}(\kappa)$ then $\tilde{f} <^* \Phi^\nu$.

PROOF OF (1). Clearly there is a $\bar{\kappa} \geqslant \mu_1(\nu)$, card dom f, $\bar{\kappa} \in \text{CARD} \cap \kappa$ such that $f \in X_{\bar{\kappa},0,s}$. Clearly dom $f \subseteq X_{\bar{\kappa},0,s}$. Now let $\mu' \in \operatorname{dom} f \setminus \bar{\kappa}$. Then $\mu' \in X_{\mu',0,s} \supseteq X_{\bar{\kappa},0,s}$, and so $f(\mu') \in X_{\mu',0,s}$; in fact $s \models$ "card $f(\mu') = \mu'$". Therefore, $g^{-1}(f(\mu')) = f(\mu')$ and $\bar{s} \models$ "card $f(\mu') = u'$", where $g = (\bar{s}, |g|, s) = f_{\mu',0,s}$, by (3.6)(b). But then $f(\mu') < \check{\Phi}^\nu(\mu) = \beta(g)$. The proof of B(5) is now completed by the obvious remark that if $f \in \overline{\mathfrak{G}}(\kappa)$, then there is a $\nu \in S_\kappa$ s.t. o.t. $C_\nu < \kappa$ and $f \in \mathfrak{A}^+(\nu)$, in fact $f \in J_\nu$.

The proof of B(3) is very similar to (4.7). First note that $\tau < \nu \Rightarrow \mathfrak{A}^+(\tau) \in |\mathfrak{A}^+(\nu)|$, in fact $\mathfrak{A}^+(\tau) \in J_\nu$. Therefore, there is a $\bar{\kappa} \geqslant \mu_1(\nu), \mu_1(\tau)$ s.t. $\mathfrak{A}^+(\tau) \in X_{\bar{\kappa},0,s}$, where $s = \mathfrak{A}^+(\nu)$. Now let $\bar{\kappa} \leqslant \mu' \in \text{CARD} \cap \kappa$. Let $f = (\bar{s}, |f|, s) = f_{\mu',0,s}$, $s' = \mathfrak{A}^+(\tau)$ and $f(\bar{s}') = s'$. Then $f_{\mu',0,s} = f \circ f_{\mu',0,\bar{s}'}$, and $X_{\mu',0,\bar{s}'} \in X_{\mu',0,\bar{s}}$. So

$$\beta(f_{\mu',0,\bar{s}'}) = \beta(f_{\mu',0,s'}) = \check{\Phi}^\tau(\mu') < ((\mu')^+)^{\bar{s}} = \check{\Phi}^\nu(\mu')$$

(where $((\mu')^+)^{\bar{s}} = \check{\rho}(\bar{s})$, if $\bar{s} \models$ "μ' is the largest cardinal"). This completes the proof of B(3) and of (4.12), since C(3) follows readily from the proof of C(1) in (4.10) and some definability computations.

(4.13) REMARKS. 1. If there are no inaccessible cardinals (or if $\kappa <$ the first inaccessible cardinal), then B(3) can be strengthened by asserting that whenever $g \in \mathfrak{G}(\kappa)$, there is a $\nu \in S_\kappa$, o.t. $C_\nu < \kappa$ such that $g <^* \Phi^\nu$. This can be seen as follows. As in the proof of B(3) we take $\nu \in S_\kappa$, s.t. o.t. $C_\nu < \kappa$ and $g \in |s| = |\mathfrak{A}^+(\nu)|$. We choose $\mu_1(\nu) \leqslant \bar{\kappa} \in \text{CARD} \cap \kappa$ s.t. $g \in X_{\bar{\kappa},0,s}$. Now, if $\bar{\kappa} \leqslant \mu' \in \text{CARD} \cap \kappa$, by (4.8) and our additional hypotheses, $\mu' \in X_{\mu',0,s}$. Then the argument of (1) of (4.12) goes through to show $f(\mu') < \Phi^\nu(\mu')$.

2. We conjecture that the existence of small large cardinals (of the type whose existence, if consistent, is consistent with $V = L$) yields counterexamples to the strengthened version, above, of B(3). While we can not yet prove this, Donder has remarked that if κ is a limit of Mahlo cardinals then the set $\{\mu' \in \text{CARD} \cap \kappa: \mu' \notin X_{\mu',0,s}\}$ is unbounded in κ for all $\nu \in S_\kappa$, where $s = \mathfrak{A}^+(\nu)$. This does give a counterexample to the method of proof of the strengthened version of B(3)—the only plausible method of proof known to us.

3. It would be possible, at the cost of greatly complicating the statement of Squared Scales to define Φ^ν for $\nu \in S_\kappa$ even when o.t. $C_\nu = \kappa$. Note that if o.t. $C_\nu = \kappa$, then $a_s = \{0\}$ and $\tilde{C}_s = \Delta(0, s)^*$. The domain of Φ^ν for such ν would be $\{\mu' \in \text{CARD} \cap (\omega, \kappa): \kappa \in X_{\mu',0,s}\}$, where $s = \mathfrak{A}^+(\nu)$; $\Phi^\nu(\mu')$ would be defined as before. Note that letting $\check{\delta} = \sup OR \cap X_{\mu',0,s}$, and $\lambda = \lambda(\delta, s)$, then $\check{\lambda} = \sup \check{\nu} \cap X_{\mu',0,s}$, and $\Phi^\nu(\mu') = \Phi^\lambda(\mu')$. Thus $\tilde{C}_s \cap X$, $\Gamma(\nu) \cap X$, $C_\nu \cap X$ are initial segments of $\tilde{C}_s$, $\Gamma(\nu)$, C_ν, respectively, so, e.g., letting $\bar{\nu} = \Phi^\nu(\mu')$, $\bar{s} = \mathfrak{A}^+(\bar{\nu})$, $s' = \mathfrak{A}^+(\lambda)$, o.t. $\tilde{C}_{\bar{s}} \leqslant \mu'$, since $X_{\mu',0,s'}$ is cofinal in $\check{\rho}(s')$ and since $f_{\mu',0,s}|\mu' = \text{id}|\mu'$, o.t. $\tilde{C}_{s'} = $ o.t. $C_{\bar{s}} \leqslant \mu'$. But possibly $f(\mu') < \kappa$, where $f = f_{\mu',0,s}$; in fact, if κ is singular, we will always have $f(\mu') < \kappa$, since either $f(\mu') = \mu'$ or $f(\mu')$ is inaccessible.

References

1. A. Beller and A. Litman, *A strengthening of Jensen's □-principles*, J. Symbolic Logic **45** (1980), 251–264.

2. K. Devlin, *Aspects of constructibility*, Lecture Notes in Math., no. 354, Springer-Verlag, Berlin and New York, 1973.

3. ______, *Hierarchies of constructible sets*, Ann. of Math. Logic **11** (1977), 195–202.

4. H. D. Donder, *Regularity and indecomposibility of uniform ultrafilters* (to appear).

5. R. B. Jensen, *The fine structure of the constructible hierarchy*, Ann. of Math. Logic **4** (1972), 229–308.

6. R. B. Jensen, A. Beller and P. Welch, *Coding the universe*, London Math. Soc. Lecture Notes Ser., Cambridge Univ. Press, London, 1982.

7. K. L. Prikry and R. M. Solovay, *On partitions into stationary sets*, J. Symbolic Logic **40** (1975), 75–80.

8. S. Shelah, *Proper forcing*, Lecture Notes in Math., no. 940, Springer-Verlag, Berlin and New York, 1982.

9. L. J. Stanley, *L-like models of set theory: forcing, combinatorial principles and morasses*, Ph. D. dissertation, University of California, 1977.

10. ______, *A short course on gap-*1 *morasses with a review of the fine structure of L*, Surveys in Set Theory (A. R. D. Mathias, ed.), London Math. Soc. Lecture Notes Ser., Cambridge University Press, London, 1983.

Institut für Mathematik II, Freie Universität Berlin, Arnimalle 3, 1000 Berlin 33, West Germany

All Soul's College, Oxford University, Oxford, United Kingdom

Department of Mathematics, Lehigh University, Bethlehem, Pennsylvania 18015

Proceedings of Symposia in Pure Mathematics
Volume **42**, 1985

Fine Structure Theory and Its Applications

SY D. FRIEDMAN

LECTURE 1. INTRODUCTION, THE DIAMOND PRINCIPLE

The fine structure of L is an exciting theory due almost entirely to the efforts of Ronald Jensen. By developing a thorough analysis of definability in L, he established combinatorial principles which can be used to solve many important problems in set theory, under the hypothesis $V = L$.

These lectures constitute an introduction to this work from a recursion-theorist's point of view. This is not wholly inappropriate, for a recursion-theoretic spirit is prevalent throughout the fine structure theory. The essential intuition is to view Σ_1-definability as a generalized form of recursive enumerability, an idea which has also been key to the development of higher recursion theory.

Thus a recursion-theoretic idea has been of importance in a set-theoretic context. In recent years there has been a significant flow in the opposite direction as well. It is our purpose in these lectures to describe the main techniques of the fine structure theory and to describe how these techniques have been applied to recursion theory.

The starting point for any discussion of the fine structure of L must be the Gödel Collapse Lemma. The uniformity of the L-hierarchy which is illustrated by this lemma is of fundamental importance.

We assume basic familiarity with Gödel's L-hierarchy $\langle L_\alpha | \alpha \in ORD \rangle$. Now fix a limit ordinal λ and suppose that $a \in L_\lambda$, $X \subseteq L_\lambda$. We say that a is *definable in* L_λ with parameters from X if for some formula in set theory $\phi(v)$ with parameters from X, $L_\lambda \models$ "a is the unique solution to $\phi(v)$". We write $a \in H_\lambda(X)$ in this case.

1980 *Mathematics Subject Classification*. Primary 03E10.

An alternative description of $H_\lambda(X)$ can be provided using Skolem functions. For each formula $\phi(x, x_1, \ldots, x_n)$ let f_ϕ be defined on

$$\overbrace{L_\lambda \times \cdots \times L_\lambda}^{n \text{ times}}$$

by $f_\phi(x_1, \ldots, x_n) = <_{L_\alpha}$-least $x \in L_\lambda$ such that $L_\lambda \models \phi(x, x_1, \ldots, x_n)$ if such an x exists; $= 0$ otherwise. Then $H_\lambda(X) =$ closure of X under the f_ϕ's. It follows that $\langle H_\lambda, \varepsilon \rangle \prec \langle L_\lambda, \varepsilon \rangle$.

GÖDEL COLLAPSE LEMMA. *For all $X \subseteq L_\lambda$, $\langle H_\lambda(X), \varepsilon \rangle$ is isomorphic to $\langle L_\beta, \varepsilon \rangle$ for some unique $\beta \leqslant \lambda$. If $t \subseteq H_\lambda(X)$ is transitive then the isomorphism is the identity on t.*

By the Mostowski Isomorphism Theorem we know that $\langle H_\lambda(X), \varepsilon \rangle$ is isomorphic to $\langle T, \varepsilon \rangle$ for some transitive T. Gödel absoluteness shows that there is a sentence ϕ of set theory such that for transitive T, $\langle T, \varepsilon \rangle \models \phi$ iff $T = L_\beta$ for some β. This proves the first statement of the Gödel Collapse Lemma. The second statement follows from the fact that any isomorphism of transitive sets is the identity.

The original use of the Gödel Collapse Lemma was to establish the Generalized Continuum Hypothesis (GCH) in L. It is worthwhile to review that argument as it is the model for much of what comes later.

THEOREM (GÖDEL). *Assume $V = L$ and suppose κ is an infinite cardinal, $\gamma < \kappa^+$. If $A \subseteq \gamma$ then $A \in L_{\kappa^+}$.*

PROOF OF THEOREM. Pick a limit λ so that $A \in L_\lambda$. Then by the Collapse Lemma, π: $H_\lambda(\gamma \cup \{A\}) \simeq L_\beta$ for some π, β and $\pi \restriction \gamma = \mathrm{id} \restriction \gamma$. (We are dropping the ε-relation from our structures to save writing.) But then $\pi(A) = A$ so $A \in L_\beta$. As $\mathrm{card}(H_\lambda(\gamma \cup \{A\})) \leqslant \kappa$ we have $\mathrm{card}(L_\beta) = \mathrm{card}(\beta) \leqslant \kappa$ so $\beta < \kappa^+$. So $A \in L_{\kappa^+}$. Q.E.D.

Thus we see that the Gödel Collapse Lemma enforces a restriction on the possible subsets of an infinite cardinal κ, the GCH being a consequence of this restriction. By taking a closer look at this technique we can uncover a deeper type of restriction, which is embodied in Jensen's Diamond Principle. This type of restriction is most easily described by making use of the notion of "cutoff" function.

DEFINITION. Suppose $V = L$, κ an infinite cardinal and $A \subseteq \kappa^+$. Define f_A: $\kappa^+ \to \kappa^+$ (the cutoff function of A) by

$$f_A(\gamma) = L\text{-rank}(A \cap \gamma) = \text{least } \delta(A \cap \gamma \in L_\delta).$$

Note that by the previous Theorem, $f_A(\gamma) < \kappa^+$ for all $\gamma < \kappa^+$, $C \subseteq \kappa^+$ is *closed unbounded* (CUB) if $\sup C = \kappa^+$, and for all $\gamma < \kappa^+$, $\sup(C \cap \gamma) \in C$.

THEOREM (JENSEN). *Assume $V = L$. There is a fixed f^*: $\kappa^+ \to \kappa^+$ such that for all $A \subseteq \kappa^+$, f^* dominates f_A on a CUB set; i.e., $\{\gamma | f^*(\gamma) \geqslant f_A(\gamma)\}$ contains a CUB set for each $A \subseteq \kappa^+$.*

Note. Of course, the CUB set depends on A!

PROOF. $f^*(\gamma) =$ least $\gamma'(L_{\gamma'} \vDash \text{card}(\gamma) \leqslant \kappa)$. Now suppose that $A \subseteq \kappa^+$ and pick a limit λ, $A \in L_\lambda$. For any γ between κ and κ^+ we may form $H_\lambda(\gamma \cup \{A\}) = H(\gamma)$.

Claim. $H(\gamma) \cap \kappa^+ =$ an ordinal γ_A.

PROOF OF CLAIM. We have to show that if $\delta \in H(\gamma) \cap \kappa^+$ then $\delta \subseteq H(\gamma)$. As $H(\gamma) \prec L_\lambda$ we have $g: \kappa \leftrightarrow \delta$, $g \in H(\gamma)$. As $\kappa \subseteq H(\gamma)$ it follows that $g[\kappa] = \delta \subseteq H(\gamma)$. Q.E.D. (of Claim)

The reason why the ordinals γ_A are useful is that $f^*(\gamma_A) > f_A(\gamma_A)$: Indeed we have $\pi: H(\gamma) \simeq L_\beta$ where $\pi \restriction \delta_A = \text{id} \restriction \gamma_A$. So $\pi(A) = \{\pi(\delta) | \delta < \gamma_A\} = A \cap \gamma_A \in L_\beta$. Thus $f_A(\gamma_A) \leqslant \beta$, by definition of f_A. But $\beta < f^*(\gamma_A)$ as $\gamma_A = \pi(\kappa^+) = (\kappa^+)^{L_\beta}$.

To complete the proof we need only check that $D = \{\gamma_A | \kappa < \gamma < \kappa^+\}$ contains a CUB set. In fact D is CUB: Clearly D is unbounded in κ^+. Suppose δ is a limit of elements of D, $\delta < \kappa^+$. Then $\delta_A = \sup_{\gamma < \delta} \gamma_A = \sup_{\gamma_A < \delta} (\gamma_A)_A$. But as $H_\lambda(\gamma \cup \{A\}) = H_\lambda(\gamma_A \cup \{A\})$ for all γ, we have $(\gamma_A)_A = \gamma_A$. So $\delta_A = \sup_{\gamma_A < \delta} \gamma_A = \delta$ and $\delta \in D$. Q.E.D.

We can now convert what we have into a tidy combinatorial principle:

$\diamondsuit^*_{\kappa^+}$: There exists $\langle D_\gamma | \gamma < \kappa^+ \rangle$ such that:

(a) For each $\lambda < \kappa^+$, $D_\gamma \subseteq \mathscr{P}(\gamma)$, $\text{card}(D_\gamma) \leqslant \kappa$.

(b) For any $A \subseteq \kappa^+$, $\{\gamma | A \cap \gamma \in D_\gamma\}$ contains a CUB set.

$\diamondsuit^*_{\kappa^+}$ is obtained by letting $D_\gamma = \mathscr{P}(\gamma) \cap L_{f^*(\gamma)}$.

The $\diamondsuit_{\kappa^+}$-principle is a variant of the above: Say that $X \subseteq \kappa^+$ is *stationary* if $X \cap C \neq \varnothing$ whenever $C \subseteq \kappa^+$ is CUB. Then $\diamondsuit_{\kappa^+}$ is obtained from $\diamondsuit^*_{\kappa^+}$ by replacing "$\text{card}(D_\gamma) \leqslant \kappa$" by "$\text{card}(D_\gamma) = 1$" and "CUB" by "stationary". A lemma of Kunen shows that $\diamondsuit^*_{\kappa^+} \to \diamondsuit_{\kappa^+}$ (in ZFC). (In fact, $\diamondsuit_{\kappa^+}$ is equivalent to $\diamondsuit'_{\kappa^+}$, where only the second of the above two changes is made.) A stronger principle, $\diamondsuit^+_{\kappa^+}$ also requires that the CUB set in (b) have the property $\gamma \in C \to C \cap \gamma \in D_\gamma$. Our proof of $\diamondsuit^*_{\kappa^+}$, in fact, demonstrates $\diamondsuit^+_{\kappa^+}$ as well. There are also versions of these principles for inaccessible cardinals. (In the case of $\diamondsuit^*_\kappa$, $\diamondsuit^+_\kappa$ one must require $\text{card}(D_\gamma) \leqslant \text{card}(\gamma)$ for $\gamma < \kappa$.) As it turns out, if $V = L$ then $\diamondsuit_\kappa$ holds for all regular κ, but $\diamondsuit^*_\kappa$, $\diamondsuit^+_\kappa$ hold exactly for those κ which are regular but not ineffable (see Kunen [**5**]).

The major application of $\diamondsuit$ $(= \diamondsuit_{\omega_1})$ in set theory is to the construction of a Souslin tree in L. For this we refer the reader to Devlin [**3**]. In recursion theory an effectivized version of $\diamondsuit^*$ has been used in α-recursion theory. We conclude this lecture by sketching this latter application.

Given any notion of RE-ness a standard question to consider is Post's Problem: Do there exist RE sets of incomparable degree? Often the priority method is used. The typical set-up is that one wishes to recursively enumerate sets A, B so as to satisfy requirements $R_0, R_1, \ldots,$ which are designed to guarantee that A, B are of incomparable degree. One hopes that each proper initial segment of requirements is permanently satisfied beyond some stage in the construction. In general, the ordering of stages may have ordertype α greater than the length ρ of the listing of requirements. If cofinality(α) $\geqslant \rho$ then the above hope is fulfilled.

If cofinality(α) $< \rho$ then what is needed is a method by which requirement R_i can "guess" at the activity of the higher proprity requirements $\langle R_j | j < i \rangle$. In case $\rho = \omega_1$ this "guess" can be provided by a $\diamondsuit$*-sequence $\langle D_\gamma | \gamma \in \omega_1 \rangle$, so that R_i uses D_i to lay out countably many possibilities. Fodor's Theorem is also relevant as the function ($i \to$ least j s.t. R_j injures R_i) is regressive.

It is worthwhile to point out that in the recursion-theoretic setting $\diamondsuit$*, and not $\diamondsuit$, is the appropriate principle to adapt. It is typical of fine structure applications to recursion theory that combinatorial principles cannot be applied directly, but must be tailor-made for the problem at hand.

Lecture 2. The Box Principle and Master Codes

The full flavor of the fine structure theory becomes first apparent through consideration of Jensen's Box Principle ($\square$). It is here as well that a recursion-theoretic intuition begins to play an important role.

If λ is a limit ordinal of cardinality $\kappa < \lambda$, then there exists a closed unbounded (CUB) set $C_\lambda \subseteq \lambda$ of ordertype at most κ. Such a subset of λ is called a *cofinalization* of λ. The Box Principle asserts that such C_λ's can be chosen so as to cohere nicely for many different λ. For any $C \subseteq$ ORD, Lim(C) denotes the set of ordinals which are limits of elements of C.

$\square(\kappa)$: There exists a sequence $\langle C_\lambda | \kappa < \lambda < \kappa^+, \lambda\ \textit{limit} \rangle$ such that for all λ

(a)C_λ is a closed unbounded subset of λ of ordertype $\leqslant \kappa$,

(b) $\lambda' \in \mathrm{Lim}(C_\lambda) \to C_{\lambda'} = C_\lambda \cap \lambda'$.

Note that we do not have *perfect coherence*: $\lambda' \in C_\lambda \to C_\lambda \cap \lambda' = C_{\lambda'}$. Perfect coherence contradicts (a) as it implies that $C_{\lambda'}$ is bounded in λ' when λ' is a successor element of C_λ. However, perfect coherence can be required if one will allow C_λ to be bounded in λ when cofinality(λ) $= \omega$.

We now proceed to describe the main points of Jensen's proof of $\square(\kappa)$ in L. It is convenient to work with a slightly modified version of $\square(\kappa)$, which we call $\square'(\kappa)$. Assume $V = L$ and let $S = \{\lambda | \kappa < \lambda < \kappa^+, \lambda$ limit and $L_\lambda \models \kappa$ is the largest cardinal$\}$.

$\square'(\kappa)$: There exists a sequence $\langle C_\lambda | \lambda \in S \rangle$ s.t. for all $\lambda \in S$

(a) C_λ is a closed unbounded subset of λ of ordertype $\leqslant \kappa$,

(b) $\lambda' \in \mathrm{Lim}(C_\lambda) \to \lambda' \in S$ and $C_{\lambda'} = C_\lambda \cap \lambda'$.

Thus essentially we have here a $\square(\kappa)$-sequence based on the ordinals in S only. Using the fact that S is a CUB subset of κ^+, it is not difficult to derive $\square(\kappa)$ from $\square'(\kappa)$ in L.

Thus we will actually describe a proof of $\square'(\kappa)$. The proof will make use of a refinement of the Skolem hull operation $H_\lambda(X)$ from the preceding lecture. For any limit ordinal λ, $X \subseteq L_\lambda$ and $n \in \omega$, let $H^n_\lambda(X)$ consist of all $y \in L_\lambda$ which are Σ_n-definable in L_λ with parameters from X. Thus $H_\lambda(X) = \bigcup_n H^n_\lambda(X)$.

Now let $\nu \in S$; we wish to locate a "canonical" cofinalization C_ν of ν. To do so first consider $\beta(\nu) =$ least β s.t. some cofinalization of ν is L_β-definable. A more useful characterization of $\beta(\nu)$ is $\beta(\nu) =$ least β s.t. there is an L_β-definable function from a subset of κ cofinally into ν. Also, let $n(\nu) =$ least n s.t. some function from a subset of κ cofinally into ν is Σ_n over $L_{\beta(\nu)}$. In all fine structure arguments the pair $(\beta(\nu), n(\nu))$ plays a fundamental role in determining what happens at ν.

It is worthwhile to first consider the case $\beta(\nu) = \nu$, $n(\nu) = 1$. In this situation we can provide a natural recursion-theoretic definition of C. The key idea is to define what it means for a Σ_1 sentence ϕ with parameters from L_ν to be *true at stage* σ, where $\sigma < \nu$. This holds if the parameters in ϕ belong to L_σ and $L_\sigma \models \phi$. The essential feature of Σ_1 statements is their persistence: $\sigma_1 < \sigma_2 < \nu$, ϕ true at $\sigma_1 \to \phi$ true at σ_2.

We are given the existence of a $\Sigma_1(L_\nu)$ function g from a subset of κ cofinally into ν. We can pick a parameter $p \in L_\nu$ so that some Σ_1 formula with parameter p defines g over L_ν. Thus for any γ, δ the statement "$g(\gamma) = \delta$" is Σ_1 with parameters γ, δ, p and, therefore, we have assigned meaning to the assertion: $g(\gamma) = \delta$ is true at stage σ. A natural cofinalization of ν can be described as follows:

$$\nu_0 = 0, \quad \gamma_0 = \text{least } \gamma \in \mathrm{Dom}(g) \text{ s.t. } g(\gamma) > 0, \quad \delta_0 = g(\gamma_0).$$

For $i > 0$

$$\nu_i = \text{least stage } \sigma\big(\text{“}g(\gamma_j) = \delta_j\text{” is true at stage } \sigma \text{ for all } j < i\big),$$
$$\gamma_i = \text{least } \gamma \in \mathrm{Dom}(g) \text{ s.t. } g(\gamma) > \nu_i,$$
$$\delta_i = g(\gamma_i).$$

Note that the sequence $\nu_0 < \nu_1 < \cdots$ is continuous and increasing. The sequence $\gamma_0 < \gamma_1 < \cdots$ is increasing. If ν_i is defined, so is γ_i and therefore so is ν_{i+1}. Thus $i_0 =$ least i so that ν_i is not defined is a limit ordinal $\leqslant \kappa$ and the sequence $\langle \nu_i | i < i_0 \rangle$ is cofinal in ν. Most importantly, if λ is a limit ordinal less than i_0, then the sequences $\langle \nu_i | i < \lambda \rangle$, $\langle \gamma_i | i < \lambda \rangle$ have the same definition over L_{ν_λ} as they do in L_ν. This is precisely the type of coherence property that we are looking for.

It is tempting to define C_ν to be $\{\nu_i | i < i_0\}$. The only difficulty is that what we have done depends on our choice of g. A final step (which will not be provided here) is required to make a canonical choice for g. The idea for doing this is to construe $H^1_\nu(\kappa \cup \{p\})$ as the range of some $\Sigma_1(L_\nu)$ function with domain contained in κ, for the least p such that $H^1_\nu(\kappa \cup \{p\})$ is unbounded in ν.

This completes the definition of C_ν in the case $\nu = \beta(\nu)$, $n(\nu) = 1$. When $\beta(\nu) > \nu$, $n(\nu) = 1$, a natural modification of the above argument can be used to define C_ν. The main point is that the phrase "ϕ is true at stage σ" when ϕ is Σ_1 makes equally good sense for $L_{\beta(\nu)}$ as it did earlier for L_ν.

But what if $n(\nu) > 1$? We have now come to a crucial point in our discussion of the fine structure theory. The natural interpretation of Σ_1 predicates as being enumerated in stages appears to break down when Σ_1 is replaced by Σ_n, $n > 1$. The essential problem is the lack of persistence: for $n > 1$ a Σ_n sentence true in L_σ need not be true in $L_{\sigma'}$ for all $\sigma' > \sigma$. The notion of Σ_{n-1} *Master Code* is designed to deal with this difficulty. Using it, Jensen showed that a Σ_n property can often be viewed as a property which is Σ_1 "relative" to a Σ_{n-1}-definable predicate. It is then possible to extend our earlier recursion-theoretic interpretation from Σ_1 to Σ_n, $n > 1$, thereby providing the missing ingredient needed to complete the proof of $\Box'(\kappa)$.

We should explain what is meant by "relativization". Fix an ordinal β and let $A \subseteq L_\beta$. A formula is Σ_1^A if it is obtained from a Σ_1 formula by replacing some free variable by A. Now $B \subseteq L_\beta$ is Σ_1 *relative to* A if B is definable over the structure $\langle L_\beta[A], \varepsilon, A\rangle$ by a Σ_1^A formula. The most manageable situation is where $L_\beta[A] = L_\beta$, in which case we say that $\langle L_\beta, A\rangle$ is *amenable*.

Fix $n = m + 1 > 1$ with a view toward obtaining a $\Sigma_m(L_\beta)$-definable $A \subseteq L_\beta$ such that (a) any $B \subseteq L_\beta$ which is $\Sigma_{m+1}(L_\beta)$ is Σ_1 relative to A, and conversely, (b) any $B \subseteq L_\beta$ which is Σ_1 relative to A is $\Sigma_{m+1}(L_\beta)$. Property (a) is easy to arrange (at least for limit β): choose A to be any universal Σ_m predicate for L_β. However, (b) is much more difficult to obtain. It would help to at least choose A so that $\langle L_\beta, A\rangle$ is amenable. Then it can be shown that (b) reduces to: If $B = \{x \in L_\beta | x \subseteq A\}$ then B is $\Sigma_{m+1}(L_\beta)$. Even this is a problem unless we know that $\langle L_\beta, A\rangle$ is amenable for *all* $\Sigma_m(L_\beta)$ sets A, a property which fails for many β.

Jensen deals with this problem by working not with β itself but with its "Σ_m projectum". This is defined to be ρ_m^β = least ρ s.t. there is a $\Sigma_m(L_\beta)$ injection from β into ρ. We now work with $\Sigma_m(L_\beta)$ subsets of $\rho_m^\beta = \rho$ instead of $\Sigma_m(L_\beta)$ subsets of β. Thus we want $A \subseteq L_\rho$ such that (a′) any $B \subseteq L_\rho$ which is $\Sigma_{m+1}(L_\beta)$ is Σ_1 relative to A and (b′) any $B \subseteq L_\rho$ which is Σ_1 relative to A is $\Sigma_{m+1}(L_\beta)$.

A subset of $L_{\rho_m^\beta}$ obeying (a′), (b′) is called a Σ_m *Master Code* for β. Jensen's fundamental Uniformization Theorem implies that $\langle L_\rho, A\rangle$ is amenable for all $\Sigma_m(L_\beta) A \subseteq L_\rho$; consequently, our earlier obstacles to demonstrating (b) have vanished and thus (b′) can be proved. To obtain (a′) let g be a $\Sigma_m(L_\beta)$ injection from L_β into ρ and define A to be the range of g on a universal Σ_m predicate for L_β. Thus a Σ_m Master Code for β exists.

Now let us return to our proof of $\Box'(\kappa)$ and reconsider the case $n(\nu) = m + 1 > 1$. Thus, by definition, there is a $\Sigma_{m+1}(L_{\beta(\nu)})$ function g from a subset of κ cofinally into ν. Now view g instead as a function which is Σ_1 relative to some Σ_m Master Code A for $\beta(\nu)$. (This is possible as the leastness of $n(\nu)$ implies that $\nu \leqslant \rho_m^{\beta(\nu)}$ and, therefore, property (a′) applies with B = Graph(g).) It is now possible to define C_ν recursion-theoretically as before, this time over the structure $\mathscr{A}(\nu) = \langle L_{\rho^{\beta(\nu)}_{n(\nu)-1}}, A\rangle$. The only new point is that A must in fact be a *canonical* Σ_m *Master Code* for $\beta(\nu)$ to guarantee the coherence property of $\Box'(\kappa)$. The structure $\mathscr{A}(\nu)$, when A is chosen to be the canonical $\Sigma_{n(\nu)-1}$ Master Code for $\beta(\nu)$, plays a major role in all fine structure arguments.

Lastly, the theory of Master Codes has had an impact on recursion theory. Assume $V = L$ and let κ be a cardinal. If $\kappa < \beta < \kappa^+$ and $\rho_n^\beta = \kappa$, then the canonical Σ_n Master Code for β is a subset of κ. The κ-degrees of these Master Codes are well ordered just as these Master Codes themelves are well ordered by $<_L$. In this way one obtains a κ-jump hierarchy $0 < 0' < 0'' < \cdots$ cofinal in the κ-degrees. When $\kappa = \omega$, Jockusch and Simpson [**6**] and Hodes [7] have shown that the Turing degree 0^λ can often be degree-theoretically characterized in terms of $\{0^\alpha | \alpha < \lambda\}$. In case $\kappa = \aleph_{\omega_1}$ we have the following *Master Code Theorem*: *Every κ-degree $\geqslant 0'$ is of the form 0^α for some α*. This result is useful in the study of uncountable admissible ordinals.

LECTURE 3. STRONG CODING

In this lecture we apply fine structure theory to the study of the Admissibility Spectrum.

An ordinal $\alpha > \omega$ is *admissible* if L_α is a model of the Σ_1-Replacement scheme, obtained from the usual Replacement scheme by restricting it to Σ_1 formulas. For any $A \subseteq \text{ORD}$, α is *A-admissible* or *admissible relative to A* if $\langle L_\alpha[A], \varepsilon, A \cap \alpha \rangle$ obeys Σ_1^A-Replacement. (Note that if A is a bounded subset of α, then this reduces to $L_\alpha(A)$ is a model of Σ_1-Replacement.) The *Admissibility Spectrum* of A is the class of all A-admissible ordinals, denoted by $\Lambda(A)$. We also introduce the notation $\Lambda(\varnothing) = \Lambda$ and $\alpha(A) = \min(\Lambda(A))$.

Observations. (1) $\alpha(\varnothing) = \omega_1^{\text{ck}}$, the least nonrecursive ordinal. The notion of admissible ordinal arose from the generalization of recursion theory on ω_1^{ck}, metarecursion theory, to recursion theory on α, α-recursion theory.

(2) If $A \in L_\alpha$, α admissible, then α is A-admissible. The converse is false even for constructible subsets of ω; indeed, Sacks showed that $\{R \subseteq \omega | \alpha(R) = \omega_1^{\text{ck}}\}$ has measure 1 in Cantor Space (provably in ZF).

(3) If α is admissible, $\mathscr{P} \in L_\alpha$ is a partial-ordering and G is $\mathscr{P}$-generic over L_α, then α is G-admissible.

The Admissibility Spectrum Problem is that of determining which Admissibility Spectra can occur. We shall focus here on the Admissibility Spectra of subsets of ω, or *reals*. This was studied by Sacks and by Jensen who showed:

(Sacks) Any countable admissible ordinal $> \omega$ is $\alpha(R)$ for some real R.

(Jensen) Suppose $\omega < \alpha_0 < \alpha_1 < \cdots$ is a countable sequence of countable admissibles and for each i, α_i is admissible relative to $\{\alpha_j | j < i\}$. Then for some real R, $\{\alpha_0, \alpha_1, \ldots\}$ is an initial segment of $\Lambda(R)$.

Note that by (2), if $R \in L$ then $\Lambda(R)$ must agree with Λ beyond some L-countable ordinal. Moreoever, (3) implies that if R belongs to a set-generic extension of L then $\Lambda(R)$ agrees with Λ beyond *some* ordinal.

Are other Admissibility Spectra possible? Yes, for $\Lambda(0\#)$ is contained in the L-cardinals. But are large cardinals necessary? This led Solovay to pose the following:

Solovay's Question. Is it consistent that for some real R, $\Lambda(R) = \mathrm{RI}$ = The Recursively Inaccessible Ordinals? (α is *recursively inaccessible* if α is admissible and the limit of admissible ordinals.)

The theorem that we wish to discuss is

THEOREM. Con ZF $\to$ Con(ZF $+ \exists R \subseteq \omega(\Lambda(R) = \mathrm{RI})$). *Specifically there is a class-generic extension N of the minimal model of* ZF *and a real $R \in N$ s.t. $N \vDash \Lambda(R) = \mathrm{RI}$.*

The proof of this theorem is based on a refinement of the technique of Jensen Coding which we call "Strong Coding". It is worthwhile to first consider the following weakened version of the Jensen Coding Theorem, in order to illustrate Jensen's technique.

THEOREM (JENSEN). *Suppose $A \subseteq \mathrm{ORD}$ and $A \cap \alpha \in L$ for all $\alpha(\langle L, A\rangle$ is* amenable). *Then there is a class-generic extension M of L s.t. $M \vDash$* ZF *and for some real $R \in M$: $M = L(R)$, A is definable over $L(R)$.*

We will now give a very rough description of Jensen's proof. First consider the simpler problem of "coding" a subset of ω_1 by a real: Thus let $B \subseteq \omega_1$. We wish to devise a forcing notion $\mathscr{R}_0^B$ s.t. if G is $\mathscr{R}_0^B$-generic then for some real R: $V[G] = V[R]$, B is definable in $V[R]$ from the parameter R. This can be done with "almost disjoint forcing", a method due to Solovay. We assume $V = L$. Using this hypothesis we can choose a "canonical" method of assigning a real r_ξ to each countable ordinal ξ so that if $\xi_1, \ldots, \xi_n$ are distinct then r_{ξ_1} is almost disjoint from $r_{\xi_2} \cup \cdots \cup r_{\xi_n}$ ($a, b \subseteq \omega$ are *almost disjoint* if $a \cap b$ is finite). Then our forcing $\mathscr{R}_0^B$ is defined so as to produce a generic real R s.t. $\xi \in B \leftrightarrow R$ is almost disjoint from r_ξ. In this way R "codes" B. More specifically: A condition $p \in \mathscr{R}_0^B$ is a pair $(s, \bar{s})$ where s is a finite subset of and $\bar{s}$ is a finite subset of $\{r_\xi | \xi \in B\}$. And, $(t, \bar{t})$ extends $(s, \bar{s})$ if $t \supseteq s$, $\bar{t} \supseteq \bar{s}$, all elements of $t - s$ are greater than $\max(s)$ and $(t - s) \cap r_\xi = \varnothing$ for $r_\xi \in \bar{s}$. It is easy to check that if G is $\mathscr{R}_0^B$-generic then $R_G = \bigcup\{s | (s, \bar{s}) \in G\}$ has the desired property. Moreover, $\mathscr{R}_0^B$ has the countable chain condition.

The idea for proving Jensen's Theorem is to reason as follows: The forcing $\mathscr{R}_0^{A \cap \omega_1}$ provides us with a "canonical procedure" for coding $A \cap \omega_1$ by a real. By generalizing almost disjoint forcing one cardinal higher, we can similarly code $A \cap \omega_2$ by a subset of ω_1, using an analogous forcing $\mathscr{R}_1^{A \cap \omega_2}$. But then by combining these two forcings we have a method for coding $A \cap \omega_2$ into a real. Jensen coding allows one to "iterate" this procedure through all the cardinals so as to code A by a real; an appropriate modification of almost disjoint forcing is needed at limit cardinals. The fine structure theory ($\diamondsuit, \square$, gap-1 morass) is needed for the proof.

Now the above ideas suggest the following approach to Solovay's problem ($\Lambda(R) = \mathrm{RI}$):

(1) Find $A \subseteq \mathrm{ORD}$ such that $\langle L, A\rangle$ is amenable and $\Lambda(A) = \mathrm{RI}$.

(2) Code A by a real R in such a way that $\Lambda(R) = \Lambda(A)$.

(1) is easy to arrange by choosing canonical cofinalizations of each successor admissible and then putting them together into a single predicate A (in fact, A is Δ_1-definable). The problem is with (2). What we want to arrange is

(2a) $A \cap \alpha$ is Δ_1 over $L_\alpha(R)$ for every admissible α,

(2b) α A-admissible $\to$ α R-admissible.

The property (2a) guarantees that $\Lambda(R) \subseteq \Lambda(A)$ and (2b) states that $\Lambda(A) \subseteq \Lambda(R)$. Intuitively, (2a) states that the "decoding" of A from R is so efficient that it can be carried out inside $L_\alpha(R)$ for every admissible α.

Unfortunately the Jensen Coding method does not provide this. For, to determine whether or not ξ belongs to $A \cap \omega_1$, we must first determine the "code" r_ξ and then ask if R is almost disjoint from r_ξ. However, not every countable admissible α will be closed under the operation $\xi \mapsto r_\xi$, as in general, r_ξ will not appear in L until a level much greater than ξ. In fact, the best that can be done is to obtain r_ξ inside L_{μ_ξ} where $\mu_\xi =$ least β s.t. $L_\beta \vDash \xi$ is countable. As a result the best that we can appear to obtain is: If α is a countable admissible then $A \cap (\omega_1)^{L_\alpha}$ is Δ_1 over $L_\alpha(R)$.

However, the following trick enables us to get around this problem and thereby establish (2a): Note that we would have what we want if $A \cap \alpha$ were Δ_1 over $L_\alpha[A \cap (\omega_1)^{L_\alpha}]$. The idea now is to apply the coding technique not to A but to a predicate A' which has the preceding property and is such that A is simply coded into A' (say $A =$ even part of A'). Then A', and hence A, can be "decoded" from the generic real R at every admissible ordinal. In actual fact, the predicate A' must be built generically and "simultaneously" with the generic real coding it. (The assumption that A is Σ_1 is needed here.) In this way we get (2a) and hence the partial result

$$(*) \qquad \mathrm{Con\,ZF} \to \mathrm{Con}(\mathrm{ZF} + \exists R(\Lambda(R) \subseteq \mathrm{RI})).$$

This result was obtained independently by René David [**10**] who also showed using a similar approach that RI can be replaced in (*) by any Σ_1 class of admissibles $X \supseteq L$-cardinals.

Notice however that by passing from A to A' we may have destroyed the admissibility of some of the A-admissible ordinals. In other words, $\Lambda(A')$ may be a proper subset of $\Lambda(A)$. The purpose of the Strong Coding method is to remove this defect. Once again we need to "improve" A to a predicate A' with the property that for any admissible α of cardinality $\kappa < \alpha$: $A' \cap \alpha$ is Δ_1 over $L_\alpha[A' \cap (\kappa^+)^{L_\alpha}]$. But now we want to also guarantee that the admissibility of any A-admissible ordinal is preserved by A'.

Note that the problem we are dealing with can itself be viewed as a "coding" problem: in this case we want to build $A' \cap (\kappa^+)^{L_\alpha}$ so as to code (at least) $A \cap \alpha$.

This suggests the solution: We should make $A' \cap (\kappa^+)^{L_\alpha}$ itself generic for a Jensen-style coding of the "universe" $\langle L_\alpha[A], \varepsilon, A \cap \alpha\rangle$ by a subset of $(\kappa^+)^{L_\alpha}$. Thus the desired forcing for coding the predicate A by a real should be constructed from conditions which are themselves generic solutions of the problem of coding initial segments $A \cap \alpha$ of A.

More explicitly, let $\overline{\text{Adm}} = \{\beta | \beta$ is admissible or the limit of admissibles$\}$ and for $\beta \in \overline{\text{Adm}}$, β-Card $= \{\kappa | \kappa = 0$ or κ is an infinite β-cardinal$\}$. Let $0^+ = \omega$. For each $\beta \in \overline{\text{Adm}}$, $\kappa \in \beta$-Card, a forcing $\mathscr{P}_\kappa^\beta$ is defined so as to produce a subset of $(\kappa^+)^{L_\beta}$ which codes $A \cap \beta$. A condition $p \in \mathscr{P}_\kappa^\beta$ is typically a function on an initial segment of (β-Card) $- \kappa$ which assigns a pair $p(\delta) = (p_\delta, \bar{p}_\delta)$ to each $\delta \in \text{Dom}(p)$. The pair $(p_\delta, \bar{p}_\delta)$ is a condition in the forcing $\mathscr{R}^{p_{\delta+}}$ for coding $A \cap \delta^+$, $p_{\delta+}$ into a subset of δ^+ (δ^+ denotes $(\delta^+)^{L_\beta}$). There are also requirements at limit β-cardinals as well which will not be discussed here. *Finally*, the set $p_\delta \cap (\delta^+)^{L_\alpha}$ must be a $\mathscr{P}_\delta^\alpha$-generic coding of $p_\delta \cap \alpha$, for every admissible α between δ and $\sup(p_\delta)$.

Thus our conditions are much like those used in Jensen Coding, but with the added restriction that the "coding elements" p_δ be themselves generic for (strong) codings at smaller ordinals. This restriction complicates the proofs of the basic lemmas from Jensen's argument; in particular, more fine structure is needed. There are two principal difficulties to handle, which we now describe.

The first is raised by the question: Do nontrivial conditions exist? To provide a positive answer one must build generic sets in L for forcings $\mathscr{P}_\kappa^\beta$ where κ is uncountable, card$(\beta) = \kappa$. If we knew that both L_β and $\mathscr{P}_\kappa^\beta$ were $< \kappa$-closed, then this would of course be easily done, but this is not generally the case. However, $\mathscr{P}_\kappa^\beta$ can be shown to be κ-distributive in L_β; thus one could hope to get a $\mathscr{P}_\kappa^\beta$-generic set by decomposing L_β into pieces $\langle B_i | i < i_0\rangle$, $i_0 \leqslant \kappa$ and then defining a sequence of conditions $p_0 \geqslant p_1 \geqslant \cdots$ where p_i meets all dense sets in B_i, each $B_i \in L_\beta$ having cardinality $\leqslant \kappa$. This approach works, provided the above type of decomposition can be obtained. It cannot in general, but instead a series of decompositions can be used, as defined by the *critical projecta* of β.

For any limit ordinal $\nu \notin$ Card, define $\beta(\nu) =$ least β s.t. there is an L_β-definable $f\colon \nu \to^{1\text{-}1} \nu'$, some $\nu' < \nu$. Also, $n(\nu) =$ least n s.t. such an f is $\Sigma_n(L_{\beta(\nu)})$, $\rho(\nu) = (\Sigma_{n(\nu)}$-projectum of $\beta(\nu)) < \nu$ and $\rho'(\nu) = \Sigma_{n(\nu)-1}$-projectum $(\beta(\nu)) \geqslant \nu$. Define the sequence $\rho_0 = \rho_0' = \nu$; $\rho_{i+1} = \rho(\rho_i)$, $\rho_{i+1}' = \rho'(\rho_i)$;…. The ρ_i's are descending so for some least $k(\nu)$, $\rho_{k(\nu)} = \text{card}(\nu)$. The *principal projecta* of ν are the ρ_i's, the *auxiliary projecta* of ν are the ρ_i''s and the *critical projecta* of ν are both. Thus we are describing the way in which ν becomes collapsed in L. For the sake of the present discussion, we deal only with the principal projecta.

Suppose now that $\kappa = \text{card}(\beta)$ and we wish to obtain a $\mathscr{P}_\kappa^\beta$-generic set G. Let $\beta = \rho_0 > \rho_1 > \cdots > \rho_k = \kappa$ be the principal projecta of β. G is constructed in k steps: First build a $\mathscr{P}_{\rho_1}^\beta$-generic set G_1, then a $\mathscr{P}_{\rho_2}^{G_1}$-generic $G_2, \ldots,$ a $\mathscr{P}_\kappa^{G_{k-1}}$-generic G_k. Here, $\mathscr{P}_{\rho_i}^{G_{i-1}}$ denotes a forcing designed to strongly code $A \cap \rho_i^+$, G_{i-1} by a subset of ρ_i^+. The desired G is obtained by "glueing together" the G_i's: $p \in G$ iff

$p \restriction (\mathrm{Dom}(p) - \rho_i) \in G_i$. Our success in obtaining G_i is based on the fact that L_β can be decomposed into an i-dimensional array of size $\leqslant \rho_i$, consisting of elements of L_β of size $\leqslant \rho_i$. This array is constructed using $\Box(\rho_1), \Box(\rho_2), \ldots, \Box(\rho_i)$.

The second difficulty is that the usual type of almost disjoint forcing does not mix well with the genericity restriction that we have placed on the p_δ's. Specifically, suppose $(p_\delta, \bar{p}_\delta)$ is a condition for coding $A \cap \delta^+$, $p_{\delta+}$ into a subset of δ^+ and we wish to extend $(p_\delta, \bar{p}_\delta)$ to a condition $(q_\delta, \bar{q}_\delta)$, $\sup(q_\delta) = \alpha$. We may assume that $\alpha > \sup(p_\delta)$ is admissible. Our restriction says that q_δ must be generic for $\mathscr{P}_\delta^\alpha$, yet the almost disjoint forcing also requires $q_\delta - p_\delta$ is disjoint from $\bigcup\{r_\xi | r_\xi \in \bar{p}_\delta\}$. We must justify the compatibility of these two requirements. Doing this requires that the restrictions of the codes $\{r_\xi \cap (\delta^+)^{L_\alpha} | r_\xi \in \bar{p}_\delta\}$ be generic for some forcing defined over L_α.

Thus the r_ξ's must be chosen so that $r_\xi \cap \alpha$ is "generic" for every admissible $\alpha < \delta^+$. Cohen genericity cannot be used as if r_ξ were Cohen generic then $r_\xi \cap [\alpha_1, \alpha_2] = \varnothing$ for many large intervals $[\alpha_1, \alpha_2]$ below δ^+, and thus $r_\xi \cap \alpha_2$ could not be generic. Instead both the r_ξ's and the forcings for which the r_ξ's are generic must be defined by induction on a gap-1 morass at δ^+. $\Box(\delta)$ is needed to get through limit stages of this induction.

This completes our outline of the Strong Coding technique. It is our hope that these lectures have helped to suggest a fruitful interplay between recursion theory and ideas in the fine structure of L.

REFERENCES

1. G. Boolos and H. Putnam, *Degrees of unsolvability of constructible sets of integers*, J. Symbolic Logic **33** (1968), 497–513.

2. R. Björn Jensen, *The fine structure of the constructible hierarchy*, Ann. Math. Logic **4** (1972), 229–308.

3. K. Devlin, *Aspects of constructibility*, Lecture Notes in Math., vol. 354, Springer-Verlag, Berlin and New York, 1973.

4. S. Friedman, *Post's problem without admissibility*, Adv. in Math. **35** (1980), 30–49.

5. K. Kunen, *Set theory*, North-Holland, Amsterdam, 1980.

6. C. Jockusch and S. Simpson, *A degree-theoretic definition of the ramified analytical hierarchy*, Ann. Math. Logic **10** (1976), 1–32.

7. H. Hodes, *Jumping through the transfinite*, J. Symbolic Logic **45** (1980), 204–220.

8. S. Friedman, *Negative solutions to Post's problem*. II, Ann. of Math. **113** (1981), 25–43.

9. A. Beller, R. B. Jensen and P. Welch, *Coding the universe*, London Math. Soc. Lecture Notes Ser., vol. 47, Cambridge Univ. Press, Cambridge, 1982.

10. R. David, *A functional* Π_2^1 *singleton*, Adv. in Math.

11. S. Friedman, *Uncountable admissibles. I: Forcing*, Trans. Amer. Math. Soc. **270** (1982), 61–73. (For critical projecta)

12. ______, *Strong coding*, in preparation.

DEPARTMENT OF MATHEMATICS, MASSACHUSETTS INSTITUTE OF TECHNOLOGY, CAMBRIDGE, MASSACHUSETTS 02139

Proceedings of Symposia in Pure Mathematics
Volume **42**, 1985

Determinacy and the Structure of $L(\mathbf{R})$

ALEXANDER S. KECHRIS[1]

Let $\omega = \{0, 1, 2, \ldots\}$ be the set of natural numbers and $\mathbf{R} = \omega^\omega$ the set of all infinite sequences from ω, or for simplicity *reals*. To each set $A \subseteq \mathbf{R}$ we associate a two-person infinite game, in which players I and II alternatively play natural numbers

$$\begin{array}{lcccccc} \text{I} & x(0) & & x(2) & & & \ldots \\ \text{II} & & x(1) & & x(3) & & \end{array}$$

$x(0)$, $x(1)$, $x(2), \ldots$ and if x is the real they eventually produce, then I wins iff $x \in A$. The notion of a winning strategy for player I or II is defined in the usual way, and we call A *determined* if either player I or player II has a winning stategy in the above game. For a collection Γ of sets of reals let Γ-DET be the statement that all sets $A \in \Gamma$ are determined. Finally AD (The *Axiom of Determinacy*) is the statement that *all* sets of reals are determined.

As usual we denote by $L(\mathbf{R})$ the smallest inner model of Zermelo-Fraenkel (ZF) set theory which contains $\mathbf{R}$, i.e. the constructible from the reals universe.

If DC is the *Axiom of Dependent Choice*, which asserts that all binary relations with no finite descending chains are well founded, then it is easy to verify (using of course that DC holds in the universe V of all sets) that $L(\mathbf{R}) \models$ DC as well. It has been proposed as a strong hypothesis of set theory that $L(\mathbf{R}) \models$ AD, i.e. the Axiom of Determinacy holds in the realm of the sets constructible from $\mathbf{R}$. In this paper we give an overview of some recent results concerning the nature of this hypothesis and its implications for the structure theory of $L(\mathbf{R})$.

1. The equivalence of determinacy and partition properties in L(R). Except for §2 we will always work in ZF + DC in the following. When stronger hypotheses are needed we will mention them explicitly.

1980 *Mathematics Subject Classification*. Primary 03E47, 90D05, 90D13.

[1]Research partially supported by NSF Grant MCS81-17804.

1.1. The study of the inner model $L(\mathbf{R})$ from AD splits into two seemingly different areas:

(a) descriptive set theory in an extended sense, i.e., the study of the structure of sets of reals in $L(\mathbf{R})$,

(b) "pure" set theory, i.e., the study of the structure of cardinals in $L(\mathbf{R})$. It turns out, however, that these two aspects of the theory of $L(\mathbf{R})$ are intimately related in a most subtle and surprising fashion. This intriguing interplay is our main subject in this paper.

A fundamental constant in the study of the continuum is the cardinal Θ defined as follows:

$$\begin{aligned}\Theta &= \sup\{\xi\colon \text{there is a surjection from } \mathbf{R} \text{ onto the ordinal } \xi\}\\ &= \sup\{\xi\colon \xi \text{ is the length of a pre-well-ordering on } \mathbf{R}\}.\end{aligned}$$

When we work in $L(\mathbf{R})$ (so that our underlying theory becomes ZF + DC + $V = L(\mathbf{R})$), the structure of cardinals $> \Theta$ is well understood without any further hypotheses. This is because for such cardinals $L(\mathbf{R})$ behaves roughly as if it is an $L[A]$ for $A \subseteq \Theta$. This will be made precise and discussed in §3. So the analysis of cardinals in $L(\mathbf{R})$ from AD concentrates on those which are $\leqslant \Theta$.

1.2. The Axiom of Determinacy, even without assuming $V = L(\mathbf{R})$, leads to a rich structure theory of the cardinals smaller than Θ. One of the important phenomena discovered is the existence of many cardinals below Θ with "large cardinal properties," for example measurability. Recall here Solovay's archetypical result that

$$\text{AD} \Rightarrow \aleph_1 \text{ is measurable.}$$

(For the basic results in descriptive set theory and the theory of determinacy we refer to Moschovakis' monograph [**19**], where references to the original papers can be found.)

Particularly important among these "large cardinal properties" are strong infinite exponent partition relations.

DEFINITION. For an ordinal λ and cardinals $\lambda \leqslant \kappa$, $\mu < \kappa$ the notation $\kappa \to (\kappa)^\lambda_\mu$ means that for every partition $F\colon [\kappa]^\lambda \to \mu$, of the set of increasing λ-sequences from κ into μ pieces, there is a set $H \subseteq \kappa$ of cardinality κ, which is homogeneous for F, i.e., $F \restriction [H]^\lambda$ is constant.

For infinite λ the Axiom of Choice (AC) disproves the existence of such κ. Martin and Kleinberg originated in the late 1960s the study of infinite exponent partition relations in choiceless situations, in particular in the context of AD. Martin proved that

$$\text{AD} \Rightarrow \aleph_1 \to (\aleph_1)^{\aleph_1}_\mu, \quad \forall \mu < \aleph_1,$$

a vast generalization of Solovay's theorem (see here [**11** and **3**]).

DEFINITION. A cardinal κ has the *strong partition property* if $\kappa \to (\kappa)^\kappa_\mu$, $\forall \mu < \kappa$.

Now in [7] it is shown that

$$\mathrm{AD} \Rightarrow \forall\lambda < \Theta\, \exists\kappa(\kappa > \lambda \wedge \kappa \text{ has the strong partition property}).$$

The main theorem in this section is the converse of this result in $L(\mathbf{R})$.

THEOREM (KECHRIS AND WOODIN [8]). *Assume $ZF + DC + V = L(\mathbf{R})$. Then the following are equivalent*:

(i) *AD*,

(ii) *there are arbitrarily large below Θ cardinals with the strong partition property.*

This is the first purely set-theoretic formulation of the Axiom of Determinacy in $L(\mathbf{R})$. It emphasizes again the strong interconnections between this hypothesis and large cardinal properties of sets. It is worth pointing out here that by a recent result of Woodin (private communication) (ii) above does not imply alone (i.e. without $V = L(\mathbf{R})$) AD, in facst it is consistent (modulo ZF + DC) with the existence of a nonprincipal ultrafilter on ω.

1.3. Some of the key ingredients in the proof of this theorem are: (1) A uniform version of the Moschovakis Coding Lemma [**19**, p. 426] (needed for the direction (i) $\Rightarrow$ (ii) proved in [7]), which is, of course, ultimately based on the Recursion Theorem; (2) Steel's analysis of scales in $L(\mathbf{R})$ via the fine structure of this inner model (see [**20**]); (3) reflection properties of admissible sets and pointclasses; and finally (4) the Martin measure on the set of Turing degrees.

We briefly sketch now the main steps of the argument.

Part I. $\mathrm{AD} \Rightarrow \forall\lambda < \Theta\, \exists\kappa(\kappa > \lambda \wedge \kappa$ has the strong partition property) (see [7]).

For each pointclass Λ let

$$o(\Lambda) = \sup\{\xi\colon \xi \text{ is the length of a pre-well-ordering in } \Lambda\}.$$

One shows that for certain Λ's, if $\kappa = o(\Lambda)$ then κ has the strong partition property. These conditions on Λ are reasonable enough so that every $A \subseteq \mathbf{R}$ belongs to at least one such Λ, so that these κ's are cofinal in Θ. For instance it is sufficient to have $\Lambda = \mathbf{\Delta}$, where $\mathbf{\Delta}$ is the ambiguous part of a pointclass Λ which contains all open sets and is closed under continuous preimages, countable intersections and unions, existential and universal quantification over $\mathbf{R}$ but not complements, and has the pre-well-ordering property. Given any $A \subseteq \mathbf{R}$, the pointclass $\Gamma = \mathbf{IND}(A)$ of all *inductive* in A sets has these properties and A belongs to its ambiguous part $\mathbf{\Delta} = \mathbf{HYP}(A)$ (= the *hyperprojective* in A sets).

To show the strong partition property for $\kappa = o(\mathbf{\Delta})$, where Γ satisfies the above conditions, one uses an appropriate uniform version of the Coding Lemma [**19**, p. 426] to code functions $f\colon \kappa \to \kappa$ by reals, and then associates to each partition $F\colon [\kappa]^\kappa \to \{0,1\}$ (for simplicity into two pieces) a game $G(F)$ on ω, in which apart from some side, but *very crucial*, conditions, players I and II collaborate to produce via this coding a function $f\colon \kappa \to \kappa$ and I wins iff $F(f) = 0$. Say I has a winning strategy σ. Then using σ one can construct a homogeneous set H of cardinality κ such that $F''[H]^\kappa = \{0\}$. The construction is by a boundedness argument in which the structural and closure properties of Γ play a basic role. Symmetrically if II has a winning strategy, one can construct a homogeneous set H of cardinality κ with $F''[H]^\kappa = \{1\}$.

Part II. Assume ZF + DC + $V = L(\mathbf{R})$. Then if there are arbitrarily large below Θ cardinals with the partition property, AD holds.

Step 1. Let us recall first a standard concept.

DEFINITION. A set of reals A is called *Souslin* if for some ordinal λ there is a tree T on $\omega \times \lambda$, with

$$A = p[T] = \{\alpha \in \mathbf{R}: \exists f \in \lambda^\omega \forall n(\alpha \upharpoonright n, f \upharpoonright n) \in T\},$$

where by a standard abuse of notation, for $s \in \omega^n$, $u \in \lambda^n$ we let $(s, u) \in T$ stand for $((s(0), u(0)), \ldots, (s(u-1), u(n-1))) \in T$.

One now has the following fact proved in [7]:

$$\forall \lambda < \Theta \exists \kappa(\kappa > \lambda \wedge \kappa \text{ has the strong partition property})$$
$$\Rightarrow \text{Every Souslin set of reals is determined.}$$

For the proof let $A \subseteq \mathbf{R}$ be Souslin, say $A = p[T]$, where T is a tree on $\omega \times \lambda$, where, of course, λ can be assumed to be $< \Theta$. Then if

$$T(\alpha) = \{u \in \lambda^{<\omega}: (\alpha \upharpoonright \text{length}(u), u) \in T\},$$

we have

$$\alpha \notin A \Leftrightarrow T(\alpha) \text{ is well founded}$$
$$\Leftrightarrow T(\alpha) \text{ with the Kleene-Brouwer ordering is well ordered.}$$

For each $s \in \omega^{<\omega}$, let

$$T(s) = \{u \in \lambda^{<\omega}: \text{length}(u) \leqslant \text{length}(s) \wedge (s \upharpoonright \text{length}(u), u) \in T\}.$$

We think of $T(s)$ as well-ordered by the Kleene-Brouwer ordering.

To show that A is determined choose first a cardinal $\kappa > \lambda$ satisfying the strong partition property. Then consider the auxiliary game in which player II also plays functions from various $T(s)$ into κ:

$$\begin{array}{llllll} \text{I} & a_0 & & a_2 & & \\ & & & & & \cdots \\ \text{II} & & a_1, f_0 & & a_3, f_2 & \end{array}$$

and II wins iff each f_n is an order-preserving map of $T((a_0, a_1, \ldots, a_{2n+1}))$ into κ and $f_0 \subseteq f_1 \subseteq f_a \subseteq \cdots$.

This is a closed game for player II, so it is determined. If II has a winning strategy he easily wins A by forgetting about the f_i's. If I has a winning strategy then use the partition property of κ to make his moves independent of the f_i's and thus obtain a winning strategy for I in A.

Step 2. This is the main result of [**8**].

THEOREM. *Assume* $ZF + DC + V = L(\mathbf{R})$. *Then if every Souslin set is determined, every set is determined.*

This result has some relevance to the question of the plausibility of the hypothesis $L(\mathbf{R}) \models \mathrm{AD}$, since it shows that determinacy of arbitrary sets in $L(\mathbf{R})$ is reduced to that of "nice" sets, i.e., ones having the Souslin property in $L(\mathbf{R})$.

For a rough sketch of the proof, let $J_\xi(\mathbf{R})$ be the ξth stage of the Jensen hierarchy of $L(\mathbf{R})$. This is defined in the usual way relativizing the analogous one for L; for details see [**20**, §1]. Abbreviate by $J_\xi(\mathbf{R})$-DET the statement that all sets of reals in $J(\mathbf{R})$ are determined. Then it is enough to show, assuming

$$\mathrm{ZF} + \mathrm{DC} + V = L(\mathbf{R}) + \text{Every Souslin set is determined},$$

that for every ordinal ξ,

$$(*) \qquad J_\xi(\mathbf{R})\text{-DET} \Rightarrow J_{\xi+1}(\mathbf{R})\text{-DET}.$$

We recall first the following concept from Steel [**20**].

DEFINITION. A Σ_1-*gap* is an interval $[\xi, \eta]$, $\xi \leqslant \eta$, of ordinals such that $J_\xi(\mathbf{R})$ is a Σ_1-substructure of $J_\eta(\mathbf{R})$ for statements involving parameters in $\mathbf{R} \cup \{\mathbf{R}\}$ only, and $[\xi, \eta]$ is maximal with this property. We also view $[\sigma_0, \infty]$ as a Σ_1-gap, where σ_0 is the least ordinal σ for which $J_\sigma(\mathbf{R})$ has this property relative to $L(\mathbf{R})$ itself.

The Σ_1-gaps partition the ordinals, so for an ordinal ξ for which we want to prove $(*)$, let $[\eta, \zeta]$ be the Σ_1-gap containing ξ. We consider now cases.

Case 1. $\eta = \xi$, i.e. ξ is the beginning of a Σ_1-gap.

Here we have two subcases.

Subcase 1.1. $J_\xi(\mathbf{R})$ is not admissible. Let then Γ be the pointclass of all sets of reals which are Σ_1 over $L_\xi(\mathbf{R})$ with parameters in $\mathbf{R} \cup \{\mathbf{R}\}$ only. Then by [**20**, §2], Γ has the scale property, using $J_\xi(\mathbf{R})$-DET. Moreover, by nonadmissibility every $A \subseteq \mathbf{R}$ in $J_{\xi+1}(\mathbf{R})$ belongs to one of the classes $\forall^{\mathbf{R}}\Gamma$, $\exists^{\mathbf{R}}\forall^{\mathbf{R}}\Gamma$, $\forall^{\mathbf{R}}\exists^{\mathbf{R}}\forall^{\mathbf{R}}\Gamma$, etc. Since every set in Γ being scaled is Souslin, we have Γ-DET, so the Second Periodicity Theorem [**19**, p. 311] applies to show that $\forall^{\mathbf{R}}\Gamma$ has the scale property, thus so does $\exists^{\mathbf{R}}\forall^{\mathbf{R}}\Gamma$ and by the same argument $\forall^{\mathbf{R}}\exists^{\mathbf{R}}\forall^{\mathbf{R}}\Gamma$ does as well, etc. So A is Souslin, therefore determined. For obvious reasons we will refer to this method of showing that every $A \in J_{\xi+1}(\mathbf{R})$ is determined as the *bootstrap argument*.

TECHNICAL REMARK. With the definition of $J_\xi(\mathbf{R})$ in [**20**] which starts with $J_1(\mathbf{R}) = V_{\omega+1}$, this argument is not literally correct for $\xi = 1$, since then $\Gamma = \Sigma_1^1$ and so Γ does not have the scale property. But clearly the same bootstrap argument works in this case by starting instead with $\Gamma = \Sigma_2^1 = \Sigma_2(J_1(\mathbf{R}))$, which does have the scale property (assuming just ZF + DC).

Subcase 1.2. $J_\xi(\mathbf{R})$ is admissible. Since ξ is the beginning of a Σ_1-gap clearly $J_\xi(\mathbf{R})$ is Σ_1-projectible into $\mathbf{R}$, so we have to deal here with the following special case of $(*)$.

KEY LEMMA. *Let $J_\xi(\mathbf{R})$ be admissible and Σ_1-projectible into* $\mathbf{R}$. *Then* $J_\xi(\mathbf{R})$-DET $\Rightarrow J_{\xi+1}(\mathbf{R})$-DET.

The proof of this lemma is in ZF + DC only and does not use our hypothesis of determinacy of Souslin sets (in $L(\mathbf{R})$). The main idea is to start with an undetermined game $A \in J_{\xi+1}(\mathbf{R})$, write down explicitly the statement that it is not determined as a sentence in the first-order language of $J_\xi(\mathbf{R})$, and then "reflect" a suitable approximation of this statement down to $J_\xi(\mathbf{R})$, which can be used to cook up an undetermined game belonging to $J_\xi(\mathbf{R})$. For the implementation of this idea we use also a technique of Martin [**16**] for handling alternating strings of existential and universal quantifiers over **R** via the Martin measure on the Turing degrees. We refer to this method of showing games in $J_{\xi+1}(\mathbf{R})$ determined as the *reflection method*.

Case 2. $\eta \leqslant \xi < \zeta$. Then by the Σ_1-elementarity of $J_\eta(\mathbf{R})$ in $J_\zeta(\mathbf{R})$, an undetermined game in $J_{\xi+1}(\mathbf{R})$ would give rise to an undetermined game in $J_\eta(\mathbf{R})$.

Case 3. $\eta < \xi = \zeta$, i.e., ξ is at the end of a gap. Let n be the least integer $\geqslant 1$ for which there is a new $\Sigma_n(J_\xi(\mathbf{R}))$ set of reals in $J_{\xi+1}(\mathbf{R})$. Then let us call following Steel [**20**, §3], $[\eta, \zeta]$ a *strong* Σ_1-*gap* if every Σ_n-type realized in $J_\zeta(\mathbf{R})$ is also realized in some $J_{\zeta'}(\mathbf{R})$, $\zeta' < \zeta$. (By a Σ_n-*type* in any $J_\rho(\mathbf{R})$ we mean the set of all Σ_n and Π_n formulas satisfied by an element of $J_\rho(\mathbf{R})$). We now have two subcases.

Subcase 3.1. $[\eta, \zeta]$ is a strong Σ_1-gap. Then one can use a version of the reflection method (as in the proof of the Key Lemma) to show that if there is an undetermined game in $J_{\xi+1}(\mathbf{R})$, then there is already one in $J_\eta(\mathbf{R})$.

Subcase 3.2. $[\eta, \zeta]$ is not a strong gap. Then by Steel [**20**, §3], using $J_\xi(\mathbf{R})$-DET, we have that if $\Gamma =$ the pointclass of all $\Sigma_n(J_\xi(\mathbf{R}))$ sets of reals, Γ has the scale property and every set of reals in $J_{\xi+1}(\mathbf{R})$ is in one of the following pointclasses: $\forall^{\mathbf{R}}\Gamma$, $\exists^{\mathbf{R}}\forall^{\mathbf{R}}\Gamma$, $\forall^{\mathbf{R}}\exists^{\mathbf{R}}\forall^{\mathbf{R}}\Gamma$, etc. Then by the bootstrap method of Subcase 1.1 we conclude that every $A \in J_{\xi+1}(\mathbf{R})$ is determined.

This finishes our sketch of the proof of the Theorem. A corollary of this proof is the following

COROLLARY. *Assume ZF + DC. If there is a cardinal* κ *with* $\kappa \to (\kappa)^\lambda_2$, $\forall \lambda < \Theta^{L(\mathbf{R})}$, *then* $L(\mathbf{R}) \models AD$.

This may be relevant to the problem of proving the consistency of the hypothesis that AD holds in $L(\mathbf{R})$ from appropriate large cardinal hypotheses.

2. The Axiom of Determinacy implies Dependent Choices in $L(\mathbf{R})$. The extensive theory of $L(\mathbf{R})$ from AD developed over the years provides a clear impression that "ZF + DC + AD + $V = L(\mathbf{R})$" is a "complete" theory for $L(\mathbf{R})$ in the similar vague sense in which "ZF + $V = L$" is thought of as a "complete" theory of L. There is one apparent difference however. Although ZF + $V = L$ implies AC, one seems to need to add to ZF + AD + $V = L(\mathbf{R})$ the choice principle DC needed in several crucial places in the development of the theory of $L(\mathbf{R})$. However, as the next result shows this is not necessary.

THEOREM [**4**]. *ZF + AD + $V = L(\mathbf{R}) \Rightarrow DC$.*

Thus one has the full analogy

$$\frac{L}{ZF + V = L} \sim \frac{L(\mathbf{R})}{ZF + V = L(\mathbf{R}) + AD}.$$

As an immediate corollary we have the following solution to an early open problem in the theory of determinacy.

COROLLARY. $\mathrm{Con}(ZF + AD) \Rightarrow \mathrm{Con}(ZF + AD + DC)$.

It should be noted again that by a result of Woodin (see [**4**]) ZF + AD does not even imply the countable Axiom of Choice (AC^{ω}). Thus $V = L(\mathbf{R})$ is necessary above.

A rough outline of the proof of the theorem is as follows. First notice that since $L = L(\mathbf{R})$, it is enough (assuming ZF + AD) to prove $\mathrm{DC}_{\mathbf{R}}$, i.e., DC for sets of reals. If now $\mathrm{DC}_{\mathbf{R}}$ fails, then it fails in $J_{\sigma_0}(\mathbf{R})$, where σ_0 is the least σ for which $J_\sigma(\mathbf{R})$ is a Σ_1-substructure of $L(\mathbf{R})$ for statements involving parameters in $\mathbf{R} \cup \{\mathbf{R}\}$ only. To get a contradiction, it is enough to show that every relation $R \subseteq \mathbf{R} \times \mathbf{R}$ which belongs in $J_{\sigma_0}(\mathbf{R})$ can be uniformized. Since $J_{\sigma_0}(\mathbf{R})$ is Σ_1-projectible into $\mathbf{R}$, every $A \subseteq \mathbf{R}$, $A \in J_{\sigma_0}(\mathbf{R})$, is Σ_1-definable in $J_{\sigma_0}(\mathbf{R})$ using only parameters in $\mathbf{R} \cup \{\mathbf{R}\}$. So it is enough to show that the class of $\Sigma_1(J_{\sigma_0}(\mathbf{R}))$ sets of reals has the scale property, since scales imply easily uniformization. This follows immediately by Steel [**20**, §2], but unfortunately his proof uses DC. The way out of this vicious circle is to weaken the notion of scale, call it *quasiscale*, then show that Steel's proof goes through with this weaker notion *without DC*, and finally show that quasiscales are enough to prove uniformization.

3. The Continuum Problem in $L(\mathbf{R})$. The Continuum Problem asks for the place of the cardinality of the continuum $2^{\aleph_0}$ in the series of alephs (= initial ordinals, called *cardinals* in the sequel). Since assuming AD the set of reals cannot be well ordered, this formulation does not make sense in our context. On the other hand it is an early theorem of AD, due to Morton Davis, that every set of reals is either countable or contains a perfect set. So a different interpretation of the Continuum Problem can be settled from AD.

Even in the absence of a well-ordering of $\mathbf{R}$, we can measure the "length" of the continuum by the cardinal

$$\Theta = \sup\{\xi : \xi \text{ is the length of a pre-well-ordering of } \mathbf{R}\}.$$

Note that with AC, $\Theta = (2^{\aleph_0})^+$. So we can reformulate the Continuum Problem as the question of the place of Θ in the series of cardinals. When understood this way the Continuum Problem leads naturally to the corresponding question about arbitrary pointclasses Λ. Let as before

$$o(\Lambda) = \sup\{\xi : \xi \text{ is the length of a pre-well-ordering in } \Lambda\}.$$

Assuming we are in $L(\mathbf{R})$ and AD holds in that universe, it can be shown (see [**19**, p. 430]) that for most interesting Λ, $o(\Lambda)$ will be a cardinal. So we can ask for the place of $o(\Lambda)$ in the series of cardinals of $L(\mathbf{R})$. We will concentrate here on some particularly important examples of Λ's for which we recall the following more or less standard notation:

(a) $o(\mathrm{power}(\mathbf{R})) = \Theta$.

(b) Let **IND** be the class of all inductive sets of reals (equivalently all sets of reals which are Σ_1 over the smallest admissible set $\mathbf{R}^+$ containing **R**), let **HYP** = $(\mathbf{IND})\check{}\cap\mathbf{IND}$ be the ambiguous part of **IND** (i.e. the *hyperprojective* sets) and let $o(\mathbf{HYP}) = \kappa^{\mathbf{R}}$. Clearly $\kappa^{\mathbf{R}}$ is the ordinal of the admissible set $\mathbf{R}^+$.

(c) Let $\mathbf{ENV}({}^3E)$ be the class of all sets of reals which are (*Kleene-*)*semirecursive in* ${}^3E$ *and a real* and let $\mathbf{SEC}({}^3E)$ be the ambiguous part of $\mathbf{ENV}({}^3E)$, i.e., the sets of reals which are recursive in ${}^3E$ and a real. Put

$$o\big(\mathbf{SEC}({}^3E)\big) = \kappa^{KL}.$$

(d) Finally let $o(\mathbf{\Delta}^m_n) = \mathbf{\delta}^m_n$. Of particular interest are the projective ordinals $\mathbf{\delta}^1_n$ and the ordinal $\mathbf{\delta}^2_1$.

The order relationship between the above ordinals is

$$\mathbf{\delta}^1_n < \kappa^{KL} < \kappa^{\mathbf{R}} < \mathbf{\delta}^2_1 < \Theta.$$

Assuming AD they are all cardinals and $\mathbf{\delta}^1_n < \mathbf{\delta}^1_{n+1}$. We now proceed to discuss in more detail their properties.

3.1. *The structure of* Θ *in* $L(\mathbf{R})$. The following simple fact is proved in ZF + DC + $V = L(\mathbf{R})$ only.

(a) Θ is a regular cardinal.

Can Θ have any large cardinal properties? The result below, again proved in just ZF + DC + $V = L(\mathbf{R})$, imposes some severe limitations.

(b) THEOREM (KECHRIS AND WOODIN [**9**]). *The cardinal* Θ *cannot be weakly compact* (*for instance there is a* (Θ, Θ)*-tree with no* Θ*-branch and thus* $\Theta \to (\Theta)^2_2$ *fails*).

The proof of this Theorem uses among other things the following result, which extends a theorem of Vopenka (see Jech [**2**]).

Assume ZF + DC + $V = L(\mathbf{R})$. Then HOD = $L[A]$, where $A \subseteq \Theta$. Moreover there is a notion of forcing **P** in HOD such that in HOD **P** has cardinality Θ (the Θ of the universe, not of HOD), and the Θ-chain condition, and $L(\mathbf{R})$ is a symmetric generic extension of HOD via **P**.

From this it follows that the cardinals and cofinalities of $L(\mathbf{R})$ above Θ are the same as those of HOD = $L[A]$ (recall here our remarks in §1). Also there are no measurable (or Ramsey cardinals) $\geqslant \Theta$ in $L(\mathbf{R})$.

On the other hand AD implies that Θ is very large.

(c) THEOREM (KECHRIS AND WOODIN [**9**]). *Assume ZF + AD +* $V = L(\mathbf{R})$. *Then* Θ *is weakly* Θ*-Mahlo.*

The proof of this Theorem uses recursion-theretic ideas. Let us show for instance that Θ is weakly Mahlo.

We first need the following simple choice Lemma.

LEMMA (ZF + DC + $V = L(\mathbf{R})$). *Let $f\colon \Theta \to V$ be such that $f(\xi) \neq \varnothing, \forall \xi < \Theta$. Then there is $A \subseteq \Theta$, A of cardinality Θ and $g\colon A \to V$ such that $\forall \xi \in A$, $g(\xi) \in f(\xi)$.*

PROOF OF THE LEMMA. It is enough to assume $f(\xi) \subseteq \mathbf{R}$, $\forall \xi < \Theta$, since there is a definable F: ORD $\times\ \mathbf{R} \overset{\text{onto}}{\to} V$. But then we claim that there is an A as above that $\bigcap_{\xi \in A} f(\xi) \neq \varnothing$, which finishes the proof. If not then for each $x \in \mathbf{R}$, $h(x) = \sup\{\xi < \Theta\colon x \in f(\xi)\} < \Theta$, so let $\lambda < \Theta$ be bigger than all the $h(x)$, $x \in \mathbf{R}$. Then $f(\lambda) = \varnothing$, a contradiction.

Now let, using this Lemma, $P\colon \Theta \to$ power($\mathbf{R} \times \mathbf{R}$) be a function such that $P(\xi)$ is a pre-well-ordering of $\mathbf{R}$ of length ξ. Fix a function $g\colon \Theta \to \Theta$. We have to find a regular cardinal $\kappa < \Theta$ such that $g(\xi) < \kappa$, $\forall \xi < \kappa$. For that consider the following type 3 object on $\mathbf{R}$:

For $X \subseteq \mathbf{R} \times \mathbf{R}$, $x, y \in \mathbf{R}$

$$^3F(X, x, y) = \begin{cases} 0 & \text{if } X \text{ is a pre-well-ordering of } \mathbf{R} \text{ of length, say, } \xi \\ & \text{and } (x, y) \in P(g(\xi)), \\ 1 & \text{otherwise.} \end{cases}$$

Let $\Gamma = \mathbf{ENV}(^3E, {}^3F)$ be the pointclass of all sets of reals which are semirecursive in ^3E, ^3F and a real. Let $\Delta = \mathbf{SEC}(^3E, {}^3F)$ be its ambiguous part and $o(\Delta) = \kappa$. Then by a result of Moschovakis (see [**19**, p. 430]) κ is regular (in fact weakly inaccessible). If $\xi < \kappa$, fix a pre-well-ordering X of length ξ which is in Δ, i.e., it is recursive in ^3E, ^3F and a real. Then

$$(x, y) \in P(g(\xi)) \Leftrightarrow {}^3F(X, x, y) = 0,$$

so $P(g(\xi))$ is also recursive in ^3E, ^3F and a real, thus $g(\xi) < \kappa$.

Actually as in §1 it turns out that for Γ as above, the corresponding $\kappa = o(\Delta)$ satisfies the strong partition property. By a result of Kleinberg [**11**] even if $\kappa \to (\kappa)_2^{\omega+\omega}$, κ is a measurable cardinal, so every closed unbounded subset of Θ contains measurable cardinals. As a simple corollary we have the following purely set-theoretic description of Θ in $L(\mathbf{R})$.

(d) Assume ZF + AD + $V = L(\mathbf{R})$. Then Θ is the supremum of the measurable cardinals.

We conclude by mentioning that it appears that nothing is known about the consistency strength of ZF + DC + $V = L(\mathbf{R})$ + Θ is large (e.g. weakly inaccessible, Mahlo, etc.).

3.2. *Measures in $L(\mathbf{R})$*. We will see that the study of the Continuum Problem in the context of AD is intimately related with the structure of measures on cardinals. Let us start with some basic facts.

(a) THEOREM (FOLKLORE). *Assume ZF + DC + AD. Then every ultrafilter is countably complete.*

Otherwise there would be a nonprincipal ultrafilter on ω, thus a nonmeasurable set of reals.

(b) THEOREM (KUNEN). *Assume ZF + DC + AD. For each $\lambda < \Theta$, the set of ultrafilters on λ is well-orderable.*

PROOF. By the Coding Lemma there is $F\colon \mathbf{R} \overset{\text{onto}}{\to} \text{power}(\lambda)$. Let $\mathscr{U}$ be an ultrafilter on λ. Let $\mathscr{D}$ be the set of Turing degrees and define $f\colon \mathscr{D} \to \lambda$ by

$$f(d) = \text{the least ordinal in } \bigcap\{F(x)\colon x \leqslant_T d \wedge F(x) \in \mathscr{U}\}.$$

Then if μ is the Martin measure on $\mathscr{D}$, i.e., the one generated by cones, $f_*\mu =^{\text{def}} \{X \subseteq \lambda\colon f^{-1}(X) \in \mu\} = \mathscr{U}$. Now for $f, g\colon \mathscr{D} \to \lambda$

$$[f]_\mu = [g]_\mu \Rightarrow f_*\mu = g_*\mu,$$

so we can map $\lambda^{\mathscr{D}}/\mu$ onto the set of all ultrafilters on λ.

Let us introduce now the following

DEFINITION. For each cardinal κ, $M(\kappa)$ = the set of all ultrafilters on κ, and $\beta(\kappa)$ = the cardinal of $M(\kappa)$.

This makes sense also in the context of AD by the preceding result.

Note that in the AC context $\beta(\kappa) = 2^{2^\kappa}$. In the AD context 2^κ does not make sense as a cardinal, and $\beta(\kappa)$ is the closest analog to a power-type operation for cardinals that we can get. However the reader should be cautioned that the usual rules of power do not necessarily apply, e.g., we can have κ with $\beta(\kappa) = \kappa$.

We will see that the Continuum Problem for pointclasses, in the way we have described it in the beginning, is intimately related with the behavior of the function $\beta(\kappa)$, whose computation can be understood as a version of the Generalized Continuum Problem in the context of AD.

The function $\beta(\kappa)$ allows us to define the concept of strong inaccessibility in a way that makes sense even with AD.

DEFINITION. We call a cardinal κ *strongly inaccessible* if

(i) κ is regular,

(ii) $\lambda < \kappa \Rightarrow \lambda^+ < \kappa$,

(iii) $\lambda < \kappa \Rightarrow \beta(\lambda) < \kappa$.

Are there any such cardinals in $L(\mathbf{R})$? The following answers that positively.

(c) THEOREM (KECHRIS [5]). *Assume ZF + AD + V = L(**R**). Then $\delta^2_1 = o(\Delta^2_1)$ is strongly inaccessible.*

On the other hand we have

(d) THEOREM (MARTIN). *Assume ZF + AD + V = L(**R**). Then for $\delta^2_1 \leqslant \kappa < \Theta$, $\beta(\kappa) = \Theta$, so Θ is not strongly inaccessible.*

As a corollary we have the following set-theoretic characterization of δ^2_1, in

ZF + AD + $V = L(\mathbf{R})$:

$$\boldsymbol{\delta}_1^2 = \text{the largest measurable stongly inaccessible cardinal.}$$

3.3. *The structure of* $\kappa^{\mathbf{R}}$. We will characterize now the ordinal $\kappa^{\mathbf{R}}$, assuming just ZF + DC + AD ($V = L(\mathbf{R})$ is not needed here). It can be seen as in §1 that $\kappa^{\mathbf{R}}$ has the strong partition property, as in 3.1(c) that it is weakly Mahlo and as in 3.2(c) that it is strongly inaccessible. To obtain a precise set-theoretic description of $\kappa^{\mathbf{R}}$ we need the following notion of indescribability.

DEFINITION. Let $B_\lambda = \{X\colon \exists \xi < \lambda (X \subseteq L_\xi)\}$ for each limit ordinal λ. Then $L_\lambda \subseteq B_\lambda$ and B_λ is transitive. Call λ *weakly* ${}^b\Pi^1_2$-*indescribable* (b for *bounded*) if for each $X \subseteq L_\lambda$ and each Π_2 formula φ with parameters in B_λ we have

$$\langle B_\lambda, \varepsilon, X\rangle \models \varphi \Rightarrow \exists \xi < \lambda, \qquad \langle B_\xi, \varepsilon, X \cap L_\xi\rangle \models \varphi.$$

Then by **[6]**, $\kappa^{\mathbf{R}}$ is weakly ${}^b\Pi^1_2$-indescribable. We now have

THEOREM. *Assume ZF + DC + AD. Then* $\kappa^{\mathbf{R}}$ *is the least cardinal* κ *such that*
(i) κ *is strongly inaccessible, and*
(ii) κ *is weakly* ${}^b\Pi^1_2$-*indescribable.*

Conjecture. In this theorem (i) can be replaced by: κ is weakly inaccessible.

3.4. *The Kleene ordinal* κ^{KL}. Again assuming ZF + DC + AD, κ^{KL} is a cardinal with the strong partition property and it is also strongly inaccessible. In fact, we have the following elegant characterization of κ^{KL}. It follows by combining results of Kechris **[5]**, Martin, and Steel **[21]**.

THEOREM (KECHRIS, MARTIN AND STEEL). *Assume ZF + DC + AD. Then* κ^{KL} *is the least strongly inaccessible cardinal.*

The following is an old conjecture in this area.

Conjecture (Moschovakis). κ^{KL} is the least weakly inaccessible cardinal.

3.5. *On the* $\boldsymbol{\delta}^1_n$*'s.* From Moschovakis **[19]** it follows that $\boldsymbol{\delta}^1_n$ is a regular cardinal. In fact, it can be shown that each $\boldsymbol{\delta}^1_n$ is measurable (Kunen **[13]**, Martin **[17]**) and, in fact, each $\boldsymbol{\delta}^1_n$ with n odd satisfies partition properties with fairly large exponents (Kunen **[14]**, Kechris and Woodin **[10]**), but it is not known whether any $\boldsymbol{\delta}^1_n$ for $n > 1$ odd has the strong partition property (the $\boldsymbol{\delta}^1_n$ with n even cannot, Kunen **[12]**). This is all assuming, of course, that ZF + DC + AD. Here are two further general rules, under the same hypothesis:

(a) Assume ZF + DC + AD. Then $\boldsymbol{\delta}^1_{2n+1} = (\lambda_{2n+1})^+$, where λ_{2n+1} is cardinal of cofinality ω.

(b) (Kunen **[15]** and Martin **[12]**) Assume ZF + DC + AD. Then $\boldsymbol{\delta}^1_{2n+2} = (\boldsymbol{\delta}^1_{2n+1})^+$.

So it is enough todetermine the δ^1_{2n+1}'s. The following are known:

$$\begin{aligned} &\delta^1_1 = \aleph_1 \quad \text{(classical)} \\ &\left.\begin{aligned} \delta^1_2 &= \aleph_2 \\ \delta^1_3 &= \aleph_{\omega+1} \\ \delta^1_4 &= \aleph_{\omega+2} \end{aligned}\right\} \text{Martin [18], in ZF + DC + AD.} \end{aligned}$$

The δ^1_n for $n \geqslant 5$ are not known in terms of the alephs. On the other hand, the δ^1_n's are almost precisely computed in terms of the β function, in view of the following result which generalizes an earlier theorem of Kunen [**14**] (for $n = 0, 1$). Its proof combines results of Becker [**1**], Kechris [**5**], Kunen [**14**], Martin and Woodin.

THEOREM (BECKER, KECHRIS, KUNEN, MARTIN AND WOODIN). *Assume ZF + DC + AD. Then for each $n \geqslant 1$,*

$$\lambda_{2n+3} \leqslant \beta(\delta^1_{2n+1}) \leqslant \delta^1_{2n+3}.$$

Conjecture. For all n, $\beta(\delta^1_{2n+1}) = \lambda_{2n+3}$. (This is true for $n = 0, 1$ by Kunen.)

From the preceding results it is now clear that the final resolution of these aspects of the Continuum Problem in $L(\mathbf{R})$ and other models of AD, reduces to the question of the computation of the function β.

REFERENCES

1. H. Becker, *A property equivalent to the existence of scales*, Trans. Amer. Math. Soc. (to appear).

2. T. Jech, *Set theory*, Academic Press, New York, 1978.

3. A. S. Kechris, *AD and projective ordinals*, Cabal Seminar '76–'77 (A. S. Kechris and Y. N. Moschovakis, eds.), Lecture Notes in Math., vol. 689, Springer-Verlag, Berlin and New York, 1978, pp. 91–132.

4. ______, *The Axiom of Determinacy implies Dependent Choices in* $L(\mathbf{R})$, J. Symbolic Logic **49** (1984), 255–267.

5. ______, *A coding theorem for measures*, mimeographed notes, June 1981.

6. ______, *Souslin cardinals, κ-Souslin sets and the scale property in the hyperprojective hierarchy*, Cabal Seminar '77–'79 (A. S. Kechris. D. A. Martin and Y. N. Moschovakis, eds.), Lecture Notes in Math., vol. 839, Springer-Verlag, Berlin and New York, 1981, pp. 127–146.

7. A. S. Kechris, E. M. Kleinberg, Y. N. Moschovakis and W. H. Woodin, *The Axiom of Determinacy, strong partition properties and nonsingular measures*, Cabal Seminar '77–'79 (A. S. Kechris, D. A. Martin and Y. N. Moschovakis, eds.), Lectures Notes in Math., vol. 839, Springer-Verlag, Berlin and New York, 1981, pp. 75–100.

8. A. S. Kechris and W. H. Woodin, *The equivalence of partition properties and determinacy*, Proc. Nat. Acad. Sci. U.S.A. **80** (1983), 1783–1786.

9. ______, *The structure of Θ in inner models of AD* (to appear).

10. ______, *Generic codes for uncountable ordinals, partition properties and elementary embeddings*, mimeographed notes, December 1980.

11. E. M. Kleinberg, *Infinitary combinatorics and the Axiom of Determinateness*, Lecture Notes in Math., vol. 612, Springer-Verlag, Berlin and New York, 1977.

12. K. Kunen, *Some more singular cardinals*, mimeographed notes, September 1971.

13. ______, *Measurability of* δ^1_n, mimeographed notes, April 1971.

14. ______, *On* δ^1_5, mimeographed notes, August 1971.

15. ______, *A remark on Moschovakis' Uniformization Theorem*, mimeographed notes, March 1971.

16. D. A. Martin, *The largest countable this, that and the other*, Cabal Seminar '79–'81 (A. S. Kechris, D. A. Martin and Y. N. Moschovakis, eds.) Lecture Notes in Math., vol. 1019, Springer-Verlag, Berlin and New York, 1983, pp. 97–106.

17. ______, *Determinateness implies many cardinals are measurable*, mimeographed notes, May 1971.

18. ______, *Projective sets and cardinal numbers: Some questions related to the continuum problem*, J. Symbolic Logic (to appear).

19. Y. N. Moschovakis, *Descriptive set theory*, North-Holland, Amsterdam, 1980.

20. J. R. Steel, *Scales in* $L(\mathbf{R})$, Cabal Seminar '79–'81 (A. S. Kechris, D. A. Martin and Y. N. Moschovakis, eds.) Lecture Notes in Math., vol. 1019, Springer-Verlag, Berlin and New York, 1983, pp. 107–156.

21. ______, *Closure properties of pointclasses*, Cabal Seminar '77–'79 (A. S. Kechris, D. A. Martin and Y. N. Moschovakis, eds.), Lecture Notes in Math., vol. 839, Springer-Verlag, Berlin and New York, 1981, pp. 147–164.

DEPARTMENT OF MATHEMATICS, CALIFORNIA INSTITUTE OF TECHNOLOGY, PASADENA, CALIFORNIA 91125

Proceedings of Symposia in Pure Mathematics
Volume **42**, 1985

Recursivity and Capacity Theory

ALAIN LOUVEAU

Introduction. The aim of this paper is to present the main effective results in the theory of capacities: outer approximation of Σ_1^1 sets by Δ_1^1 sets, and inner approximation of Δ_1^1 sets by compact Δ_1^1 sets, when it is possible. So this paper can be viewed as an extension of the work of many people, particularly G. E. Sacks [**13**] and A. S. Kechris [**6**], about effective results in the theory of Lebesgue measure.

But unlike measures, capacities are not only a mathematical object interesting in their own right. They can serve as tools for getting existence results, and their status can be compared in some sense to that of closed games: As the existence of winning strategies for ad hoc closed games can be used to derive existence results (in which games are not mentioned at all), the capacitability theorem of Choquet, for ad hoc capacities, leads to existence results. And in both cases, the result is effective, and can be used to get the existence of Δ_1^1 objects. I shall emphasize this second aspect of capacity theory in some applications of the main results.

The paper is organized as follows: The first section, *Capacities and capacitability*, deals with the noneffective aspects of capacities. I recall the definition of capacities, give some examples and prove Choquet's capacitability theorem.

In the second section, *Effective outer capacitability*, I prove a general lemma on Δ_1^1 capacities, which, in particular, implies that the outer approximation is effective: If f is a Δ_1^1 capacity and A is a Σ_1^1 set, its capacity $f(A)$ can be approximated arbitrarily close by the capacity of a Δ_1^1 set containing A. But as I said before, the lemma can also be used for various Δ_1^1 existence results. I concentrate at the end of the section on one particular application, a problem of separation of convex sets of measures, for which no other proof (i.e. without reference to effective notions) is known, and where the technique of application of the lemma is particularly clear. Other applications can be found in [**9**].

In the third section, *Effective inner capacitability*, I study the problem of effectivizing the inner approximation, thus extending the basis result for Δ_1^1 sets of

1980 *Mathematics Subject Classification*. Primary 28A12, 03D80.

positive measure. It turns out that this is not always possible, and depends on how close to a measure the capacity is. I characterize precisely, among a large class of capacities, those capacities for which the inner approximation by Δ_1^1 compact sets is true. This leads to a Δ_1^1 basis result which encompasses both the measure case and the effective perfect set theorem. And for proving the result, I establish a result on the existence of measures controlling capacities, answering a question of Dellacherie [5], and I also prove a lemma on Π_1^1 symmetric relations, much in the spirit of Harrington's result on Π_1^1 equivalence relations, which seems new.

Of course, this paper is mainly measure theory, and I have tried to be as complete as possible when using measure theoretic concepts. In particular, the basic (well-known) facts on measures that are used in this paper have been recorded in §0. On the other hand, I have assumed that the reader has some familiarity with recursion in a Polish topological space, as it is developed in Chapters 3 and 4 of Moschovakis' book, *Descriptive set theory*. I should better say "hyperarithmetic recursion", as I shall only be interested in Δ_1^1, Σ_1^1 and Π_1^1 sets and functions. Again, some facts, concerning mainly compact sets, are given in §0.

I finally want to thank C. Dellacherie and G. Mokobodzki for the illuminating discussions we had on capacity theory, and G. Debs who allowed me to publish here the joint result 2.9.

0. Preliminaries.

0.1. In the sequel, E will always stand for a compact metrizable space. Such spaces are the ones where measure theory is easily developed, and we shall restrict ourselves to this frame, although a large part of what follows could be done in more general situations.

0.2. $K(E)$ denotes the space of compact subsets of E. This space is given the Hausdorff topology, i.e. the topology generated by all sets of the form $\{K\colon K \subset U\}$ or $\{K\colon K \cap U \neq \varnothing\}$, for U open in E. As is well known, $K(E)$ is then compact and metrizable.

0.3. A numerical function $f\colon E \to \overline{\mathbf{R}}_+ = \mathbf{R}_+ \cup \{\infty\}$ is a usc function if its subgraph $G^f_\leqslant = \{(x, t)\, f(x) \geqslant t\}$ is closed (hence compact) in $E \times \overline{\mathbf{R}}_+$. It is an lsc function if $G^f_< = \{(x, t)\, f(x) > t\}$ is open. usc functions are closed under the taking of pointwise infima, and can be characterized as pointwise limits of decreasing sequences of continuous functions. The space USC(E) of all usc functions on E is given the Hausdorff topology inherited from $K(E \times \overline{\mathbf{R}}_+)$ (identifying f and its subgraph). It is then a compact metrizable space, for which the evaluation function: USC(E) $\times E \to \overline{\mathbf{R}}_+$ is usc. Note that $K(E)$ can be viewed as a compact subset of USC(E)—by identifying a compact subset of E and its characteristic function.

0.4. Letters μ, ν will denote (nonnegative) Radon measures on E. We shall not distinguish between a measure μ on the σ-field of Borel subsets of E, its outer extension to $\mathscr{P}(E)$, defined, for $A \subset E$, by $\mu(A) = \inf\{\mu(U)\colon U \text{ open}, A \subset U\}$, and its extension as an (outer) integral to numerical functions on E.

For technical simplicity, we shall only consider subprobabilities, i.e. measures on E satisfying $\mu(E) \leqslant 1$. The space of such measures is denoted by $M(E)$, and is given the topology of pointwise convergence on continuous functions on E, for which it is again compact and metrizable. This topology can be described alternatively by remarking that a measure is usc, as a function on USC(E), and that the topology of $M(E)$ is the topology inherited from the Hausdorff topology on USC(USC(E)).

0.5. Dini-Cartan lemma. *Let $(f_n)_{n\in\omega}$ be a decreasing sequence of usc functions on E, and g an lsc function such that* $\inf_n f_n < g$. *Then for some $n \in \omega$, $f_n < g$.*

0.6. Lemma. *Let φ be an increasing usc function on* USC(E), *and $(f_n)_{n\in\omega}$ a sequence of usc functions on E. Then*

$$\limsup \varphi(f_n) \leqslant \varphi(\limsup f_n)$$

where $\limsup f_n = \inf\{g \in \mathrm{USC}(E)\colon \exists p \forall n \geqslant p\ g \geqslant f_n\}$.

0.7. We shall use the following consequence of the finite-dimensional Hahn Banach theorem:

Let K be a compact convex set in a locally convex topological vector space, and A a convex set of affine functions on K. If c is a real number such that $\forall f \in A \exists x \in K\ f(x) \geqslant c$, then there is some $x \in K$ such that $\forall f \in A\ f(x) \geqslant c$.

0.8. Concerning recursivity in compact metrizable spaces, we refer the reader to Moschovakis' basic book [**12**], of which we follow the terminology. Hence a (Δ_1^1-) recursively presented space E is a space E together with a complete metric d on E and a dense sequence $(r_n)_{n\in\omega}$ of points of E such that the distance function, between the r_n's, is (Δ_1^1-) recursive. Without loss of generality, we assume in the sequel that the space E is Δ_1^1-recursively presented.

0.9 *Basis result for compact sets.* If K is a nonempty Δ_1^1 compact subset of E, K has a Δ_1^1 member. Consequently, the Δ_1^1 members of K are dense in K, and K can be Δ_1^1-recursively presented in such a way that the notions of Σ_1^1, Π_1^1 and Δ_1^1 sets associated with the presentations of K and E coincide on subsets of K (cf. [7]).

0.10. The space USC(E) can be recursively presented so that a usc function f on E is Δ_1^1-recursive iff it is a Δ_1^1 point in USC(E) (similarly for $K(E)$). This implies, using 0.9, that $M(E)$ admits a Δ_1^1-recursive presentation such that a measure μ on E is Δ_1^1-recursive from USC(E) (or $K(E)$) into $[0,1]$ iff it is a Δ_1^1 point in $M(E)$.

0.11. We shall say that a sequence $(A_n)_{n\in\omega}$ of subsets of E is Δ_1^1 ($\Sigma_1^1,\ldots$) is the relation $x \in A_n$ is Δ_1^1 ($\Sigma_1^1,\ldots$) in $\omega \times E$. Similarly for functions and points.

0.12. *Principles of Δ_1^1 choice.* (a) If $P \subset X \times Y$ is Π_1^1 and $Q(x) \leftrightarrow \exists y \in \Delta_1^1(x)\ P(x, y)$, then Q is Π_1^1, and there is a Π_1^1-recursive function $f\colon Q \to Y$ such that $\forall x \in Q\ P(x, f(x))$.

(b) If $P \subset X \times X$ is Π_1^1, x_0 is Δ_1^1 in X and $\forall x \in X \exists y \in \Delta_1^1(x)\, P(x, y)$, then there is a Δ_1^1 sequence $(x_n)_{n\in\omega}$ beginning with x_0 s.t. $\forall n\, P(x_n, x_{n+1})$.

0.13. Seq ω denotes the set of finite sequences of integers. By a privileged Suslin scheme for a $\boldsymbol{\Sigma}_1^1$ (resp. Σ_1^1) set $A \subset E$, we mean a sequence (resp. a Δ_1^1 sequence) $(K_s)_{s\in \mathrm{Seq}\,\omega}$ of compact subsets of E satisfying

(a) If $s \subset t$, or if $|s| = |t|$ and $\forall i < |s|\, t(i) \leqslant s(i)$, $K_t \subset K_s$.

(b) $x \in A \leftrightarrow \exists\alpha\forall n\, x \in K_{\alpha|n}$.

All $\boldsymbol{\Sigma}_1^1$ (resp. Σ_1^1) sets admit privileged Suslin schemes.

Associated with such a scheme is a (Σ_1^1) sequence $(A_s)_{s\in \mathrm{Seq}\,\omega}$, defined by $x \in A_s \leftrightarrow \exists\alpha(s \subset \alpha \wedge \forall n\, x \in K_{\alpha|n})$, and which satisfies

(i) $A_\varnothing = A$.

(ii) For all s, A_s is the union of the increasing sequence $(A_{s^\frown n})_{n\in\omega}$.

(iii) For all s, $A_s \subset K_s$ and, for all $\alpha \in \omega^\omega$, $\bigcap_{n\in\omega} A_{\alpha|n} = \bigcap_{n\in\omega} K_{\alpha|n}$.

1. Capacities and capacitability.

1.1. A *capacity* on E is a function $f\colon \mathscr{P}E \to \overline{\mathbf{R}}_+$ satisfying

(i) f is increasing: If $A \subset B$, $f(A) \geqslant f(B)$.

(ii) f "goes up" on increasing sequences: If $(A_n)_{n\in\omega}$ is an increasing sequence of sets, $f(\bigcup_n A_n) = \sup_n f(A_n)$.

(iii) f "goes down" on decreasing sequences of *compact* sets: If $(K_n)_{n\in\omega}$ is a decreasing sequence of compact sets, then $f(\bigcap_n K_n) = \inf_n f(K_n)$.

1.2. REMARKS. (a) Assuming f is increasing, property (iii) of the definition is equivalent to: the restriction of f to $K(E)$ is usc (this would be the correct notion for generalizations to more complicated spaces E).

(b) We shall be interested only in the restrictions of capacities to $\boldsymbol{\Sigma}_1^1$ subsets of E, and the "going up" property (ii) could be restricted to $\boldsymbol{\Sigma}_1^1$ sequences (and even Borel sequences, as we shall see in a while).

1.3. EXAMPLES. (1) It is easily seen that an (outer) measure μ on E is a capacity. In fact, the notion of capacity can be viewed in this respect as the nonadditive generalization of the notion of measure.

(2) Let Λ be some measure on $K(E)$, and define a function f_Λ on $\mathscr{P}(E)$ by $f_\Lambda(A) = \Lambda(\{K \in K(E)\colon K \cap A \neq \varnothing\})$. Then f_Λ is a capacity in E.

Particular cases of this construction are (a) Let E, F be two metric compact spaces, μ be a measure on E, and let $f(A) = \mu(\pi_E(A))$, for $A \subset E \times F$, where π_E denotes projection on E. (b) The historical case—put here only for the sake of completeness—the electrostatic capacity, which corresponds to the Wiener measure on the trajectories of the Brownian motion.

The capacities on E arising from this process have been intensely studied in the fundamental paper of Choquet [**2**], and characterized by a sequence of inequalities. They are called "alternating of order infinity".

(3) Let H be some subset of $M(E)$, and associate with H a function $\tilde{H}$ on $\mathscr{P}(E)$ by $\tilde{H}(A) = \sup\{\mu(A)\colon \mu \in H\}$. If H is a compact subset of $M(E)$, then $\tilde{H}$ is a capacity, called the *normal* capacity associated with H. Reciprocally, if f is a

capacity with values in $[0,1]$ on E, the set H_f of measures majorized by f (i.e. $\mu(K) \leqslant f(K)$ for all $K \in K(E)$) is always a (convex, hereditary) compact subset of $M(E)$, and the capacity f is said to be normal if $f = \tilde{H}_f$. Normal capacities play an important role in potential theory and probabilities. There is no known "algebraic" characterization of normality. A necessary condition is *subadditivity* $(f(A \cup B) \leqslant f(A) + f(B))$, and a sufficient condition is *strong subadditivity* $(f(A \cup B) + f(A \cap B) \leqslant f(A) + f(B))$. In particular, all capacities of Example (2) are normal.

A particular example of a normal capacity is the "Horowicz case", which can serve as a paradigm for studying normal capacities: It is the function defined by

$$f(A) = \begin{cases} 0 & \text{if } A = \varnothing, \\ 1 & \text{if } A \neq \varnothing, \end{cases}$$

which corresponds to $H_f = M(E)$.

(4) There exist interesting capacities which are in some sense very irregular—not even subadditive. One such example is the "separation" capacity of Sion: Let π_1 and π_2 be the two projections of $E \times E$ on E, and define f on $\mathscr{P}(E \times E)$ by

$$f(A) = \begin{cases} 0 & \text{if } \pi_1(A) \text{ and } \pi_2(A) \text{ can be separated by a Borel set,} \\ 1 & \text{if not.} \end{cases}$$

It can be seen that f is a capacity, and an application of the capacitability theorem to this example immediately gives the Suslin separation theorem for Σ_1^1 sets.

1.4. THE CAPACITABILITY THEOREM (CHOQUET [**2**]). *Let f be a capacity on E, and A be a Σ_1^1 subset of E. Then*

(a) (*inner approximation*) $f(A) = \sup\{f(K): K \in K(E), K \subseteq A\}$,

(b) (*outer approximation*) $f(A) = \inf\{f(B): B \text{ Borel in } E,\ A \subseteq B\}$.

PROOF. (a) Let $(K_s)_{s \in \operatorname{Seq} \omega}$ be a privileged scheme for A (see 0.13), and $(A_s)_{s \in \operatorname{Seq} \omega}$ its associated sequence of Σ_1^1 sets. If $f(A) > t$, then using the fact that $A_\varnothing = A$, A_s is the union of the increasing sequence $(A_{s^\frown n})_{n \in \omega}$, and f is going up on increasing sequences, one easily constructs a sequence $\alpha \in \omega^\omega$ such that $\forall n\, f(A_{\alpha|n}) > t$. As f is increasing, $f(K_{\alpha|n}) > t$ for all n, and as f goes down on decreasing sequences of compact sets, $K_\alpha = \bigcap_n K_{\alpha|n} = \bigcap_n A_{\alpha|n}$ is a compact subset of A such that $f(K) \geqslant t$.

(b) Define, for $X \subset E$, $f^*(X) = \inf\{f(B): B \text{ Borel in } E,\ X \subset B\}$. Then f^* is again a capacity, and f^* agrees with f on compact (in fact on Borel) sets, hence by (a) f^* agrees with f on Σ_1^1 sets. This gives (b). To see that f^* is a capacity, the only nontrivial thing is to show the "going up" of f^*. But if $(X_n)_{n \in \omega}$ is an increasing sequence of sets, and for all n, $f^*(X_n) < t$, choose $B_n \supseteq X_n$, B_n Borel, with $f(B_n) < t$, and let $B = \bigcup_m \bigcap_{n \geqslant m} B_n$. Clearly B is Borel and $\bigcup_n X_n \subseteq B$. Now for each m, $B_m^* = \bigcap_{n \geqslant m} B_n$ is contained in B_m, hence $f(B_m^*) < t$, and B is the union of the increasing sequence $(B_m^*)_{m \in \omega}$, hence $f(B) \leqslant t$. This shows $f^*(\bigcup_n X_n) \leqslant t$, and the "going up" of f^*. Remark that in this proof, we have used the "going up" of f only on Borel sequences to prove that f^* is a capacity. Replacing f by f^* in (a) still gives (a) and (b) under this a priori weaker assumption. $\square$

1.5. Instead of considering capacities with arguments in $\mathscr{P}(E)$, one can consider functional capacities, i.e. defined on functions on E. The definition is the same, with unions replaced by suprema, intersections replaced by infima, and compact sets replaced by usc functions. The capacitability theorem is then still true. Note that the restriction of any functional capacity to (characteristic functions of) sets is a capacity. On the other hand, any capacity admits an extension (not necessarily unique) to a functional capacity.

1.6. One can also extend the notion of capacity by considering functions with values in $\mathscr{P}(F)$ (or the functions on F), the ordering $\leqslant$ on $\overline{\mathbf{R}}_+$ being replaced by inclusion (resp. the ordering $\leqslant$ on functions). Then one has to add to (i), (ii) and (iii) of the definition of a capacity a fourth condition:

(iv) f transforms compact sets (or usc functions) into compact sets (or usc functions).

One obtains what Dellacherie calls a capacitary operator. These objects are important in potential theory.

1.7. One can also define capacitary operators (or capacities) of many (even countably) variables (sets or functions), by asking the "going up" (ii) only separately in each variable, but the "going down" (iii) simultaneously on all (compact or usc) variables. Note that these objects are closed under composition.

As an example of a bicapacity, defined on mixed arguments, let us quote the one associated with normal capacities: If H is a subset of $M(E)$, and f is a function on E, define $J(H, f)$ by

$$J(H, f) = \tilde{H}(f) = \sup\{\mu(f) : \mu \in H\}.$$

1.8. Pursuing one step further, one can consider operators obtained by identifying arguments, or fixing some Σ_1^1 arguments, in capacitary multi-operators. One obtains the notion of "analytic operator" of Dellacherie [**3**]. Although the "going up" and the "going down" of the operator are usually lost during this process, the inner approximation (by K_σ sets or functions) and the outer approximation (by Borel sets or functions) are still true. We shall not enter this subject anymore, but a lot of what we shall do in the subsequent sections could be generalized (in a rather simple way, although notationally awkward) to the general notions just described.

2. Effective outer capacitability.

2.1. A function $f: E \to \overline{\mathbf{R}}_+$ is called upper semi-Σ_1^1 (resp. upper semi-Σ_1^1), abbreviated in usa (**usa**), if the subgraph of f, $G_\leqslant^f = \{(x, t)\, f(x) \geqslant t\}$, is Σ_1^1 (resp. Σ_1^1) in $E \times \overline{\mathbf{R}}_+$. It is lower semi-$\Sigma_1^1$ (lsa) if $G_\leqslant^f$ is Π_1^1. usa and lsa functions play at level one a similar role to the one played at level 0 by usc and lsc functions. Note that a function is both lsa and usa iff it is Δ_1^1-recursive, and that if f is usa, g is lsa and $f \leqslant g$, there is a Δ_1^1-recursive function h with $f \leqslant h \leqslant g$.

2.2. PROPOSITION. *Let f be a capacity on E. Then f is a* **usa** *function on* (*codes for*) Σ_1^1 *sets. More precisely, if f is* **usa** *on $K(E)$, then f is* **usa** *on* (*codes for*) Σ_1^1 *sets.*

PROOF. This follows immediately from the proof of 1.4: Let G be a Σ_1^1 set in $\omega^\omega \times E$ which is universal for $\boldsymbol{\Sigma}_1^1$ subsets of E, and $(H_s)_{s \in \mathrm{Seq}\,\omega}$ be a Δ_1^1 sequence of $\boldsymbol{\Pi}_1^0$ subsets of $\omega^\omega \times E$ which form a privileged Suslin scheme for G. From 1.4, one has

$$f(G_\alpha) > t \leftrightarrow \exists m \exists \beta \forall n\, f\big((H_{\beta|n})_\alpha\big) > t + 2^{-m}. \qquad \square$$

2.3. REMARK. One can similarly define usa and **usa** operators, as those operators which uniformly (and effectively for usa) transform $\boldsymbol{\Sigma}_1^1$ sets or **usa** functions into $\boldsymbol{\Sigma}_1^1$ sets or **usa** functions. It turns out that all capacitary and analytic operators introduced in 1.6–1.8 are of this kind. In the terminology of Moschovakis, the **usa** operators are called $\boldsymbol{\Sigma}_1^1$ on $\boldsymbol{\Sigma}_1^1$.

2.4. We are ready to state the effective version of the outer capacitability result. But because of the applications we have in mind, we prefer writing it in a bit more technical way, separating the role played by the "going down" and the "going up" properties.

LEMMA. *Let E be a compact recursively presented space, and f and g two functions satisfying*

(a) *f is an increasing usc function from $K(E)$ into $\overline{\mathbf{R}}_+$.*

(b) *g is defined and increasing from $\Delta_1^1 \cap \mathscr{P}(E)$ into $\overline{R}_+$; g goes up on increasing Δ_1^1 sequences of sets, and the relation $g(C) < t$ is Π_1^1, in codes for Δ_1^1 subsets of E.*

(c) *f and g agree on Δ_1^1 compact subsets of E. Moreover, if K is compact, B is Δ_1^1 and $K \subset B$, then $f(K) \leqslant g(B)$.*

Then if A is a Σ_1^1 subset of E

$$\sup\{f(K): K \in K(E),\, K \subseteq A\} = \inf\{g(B): B \in \Delta_1^1,\, A \subseteq B\}.$$

PROOF. The inequality $\leqslant$ follows from (c). Define, for $A \subseteq E$, $g^*(A) = \inf\{g(B): B \in \Delta_1^1, A \subseteq B\}$. We want to show that if A is Σ_1^1 and $r \in \mathbf{Q}_+$ satisfy $g^*(A) > r$, there is a compact $K \subseteq A$ such that $f(K) \geqslant r$. Let $(K_s)_{s \in \mathrm{Seq}\,\omega}$ be a Δ_1^1 and privileged Suslin scheme for A, with associated $(A_s)_{s \in \mathrm{Seq}\,\omega}$, and $(r_n)_{n \in \omega}$ a sequence of rationals with $g^*(A) > r_0 > r_1 > \cdots > r$. We construct by induction a sequence $\alpha \in \omega^\omega$ s.t. $\forall n\, g^*(A_{\alpha|n}) > r_n$. By hypothesis, $g^*(A_\varnothing) = g^*(A) > r_0$. Assume $s = \alpha|n$ is constructed. Then A_s is the union of the increasing Σ_1^1 sequence $(A_{s^\frown n})_{n \in \omega}$. In order to choose $\alpha(n)$, it is enough to show there is $k \in \omega$ s.t. $g^*(A_{s^\frown k}) > r_{n+1}$. If not, then $\forall k\, g^*(A_{s^\frown k}) \leqslant r_{n+1} < r_n$. By property (b) of g and Δ_1^1 choice, there is a Δ_1^1 sequence $(B_k)_{k \in \omega}$ s.t. $A_{s^\frown k} \subseteq B_k$ and $g(B_k) < r_n$. But then $B = \bigcup_m \bigcap_{n \geqslant m} B_n$ is Δ_1^1, contains A_s and, again by property (b) of g, $g(B) = \sup_m g(\bigcap_{n \geqslant m} B_n) \leqslant \sup_n g(B_n) \leqslant r_n$, contradicting $g^*(A_s) > r_n$. This contradiction proves the existence of an α such that $\forall n\, g^*(A_{\alpha|n}) > r_n$. As $K_{\alpha|n}$ is a Δ_1^1 set containing $A_{\alpha|n}$, $f(K_{\alpha|n}) = g(K_{\alpha|n}) > r_n$. But then $K_\alpha = \bigcap_n K_{\alpha|n}$ is a compact subset of A and, from property (a) of f, $f(K_\alpha) = \inf f(K_{\alpha|n}) \geqslant r$. $\square$

2.5. COROLLARY (EFFECTIVE OUTER CAPACITABILITY). *Let f be a capacity on E which is* usa *on $K(E)$. If A is a Σ_1^1 subset of E, then $f(A) = \inf\{f(B): A \subset B, B \in \Delta_1^1\}$.*

PROOF. Apply 2.4 to $f|_{K(E)}$ and $g = f|_{\Delta_1^1}$. The only nontrivial thing is that $g(C) < t$ is Π_1^1 on Δ_1^1 codes, which is given by Proposition 2.2. □

REMARK. The boldface version of this corollary seems new—and not obtainable from Dellacherie's techniques of analytic operators. So we quote it:

Let W be some Polish space, and V be an operator from functions on E into functions on W which is **usa** *on* USC(E), *and such that that for each fixed $w \in W$, the function $V_w(f) = V(f)(w)$ is a capacity on E. If f is a* **usa** *function on E and g is an* **lsa** *function on W s.t. $V(f) \leqslant g$, there is a Borel function h on E* s.t. *$f \leqslant h$ and $V(h) \leqslant g$.*

2.6. Lemma 2.4 can be used directly to obtain separation results for capacities, e.g. for strongly subadditive capacities, and for Hausdorff measures (see [**9**]). But it can also be used in an indirect way, in much the same manner as closed games are used for the study of Baire property or the perfect set property by introducing ad hoc capacities. In the remainder of this section, we shall illustrate this indirect use of 2.4 on one example, the problem of separating convex sets of measures. The main result is a joint work with G. Debs, and extends results of Mauldin, Preiss and von Weizsäcker [**10**], using techniques of G. Mokobodzki (cf. [**11**]).

2.7. If f is a continuous function on E, let $\tilde{f}$ be its continuous extension to $M(E)$, defined by $\tilde{f}(\mu) = \mu(f)$. If Λ is a probability measure on $M(E)$, its barycenter $b(\Lambda)$ is the unique measure on E satisfying

$$\Lambda(\tilde{f}) = \tilde{f}(b(\Lambda)) = b(\Lambda)(f)$$

for all positive continuous f on E.

A set $A \subset M(E)$ is said to be strongly convex if for any probability measure Λ with $\Lambda(A) = 1$, $b(\Lambda) \in A$. Clearly, any strongly convex set is convex, and the two notions coincide on compact sets.

2.8. If μ and ν are measures on E, we denote by $\mu \wedge \nu$ the supremum of the measures less than both μ and ν. The measure $\mu \wedge \nu$ can be defined by $\mu \wedge \nu(f) = \inf\{\mu(f \cdot \varphi) + \nu(f \cdot (1 - \varphi))$: φ continuous, $0 \leqslant \varphi \leqslant 1\}$, for positive continuous functions f on E.

If $\mu \wedge \nu = 0$, the measures μ and ν are said to be mutually singular. This, of course, is equivalent to the existence of a Borel set $B \subseteq E$ which separates μ of ν, i.e. $\mu(B) = \mu(E)$ and $\nu(B) = 0$. The general separation problem is to extend this to sets of measures, consisting of pairwise mutually singular measures. A result of A. Goullet de Rugy insures that the separation is possible in case of strongly convex K_σ sets of measures. Here, we shall study the case of one strongly convex Σ_1^1 set, and one measure, and show that the separation holds uniformly. Let us first state precisely the effective version and its boldface corollaries.

2.9. THEOREM. *Let E be a compact recursively presented space, $A \subseteq M(E)$ a Σ_1^1 strongly convex set of measures, and μ a Δ_1^1 measure on E such that for all $\nu \in A$, μ and ν are mutually singular. Then there is a Δ_1^1 set $B \subset E$ which separates μ from A, i.e. $\mu(B) = \mu(E)$ and $\nu(B) = 0$ for all $\nu \in A$.*

2.10. COROLLARY. *Let W be some Polish space, A a Σ^1_1 subset of $W \times M(E)$, $w \mapsto \mu_w$ a Borel function from W into $M(E)$, and D a Σ^1_1 subset of W. Assume that for each $w \in D$, A_w is strongly convex, and consists of measures which are mutually singular with μ_w. Then there is a Borel set $B \subset W \times E$ s.t. for all w in D, B_w separates μ_w from A_w. In particular, if A_1 and A_2 are two Σ^1_1 subsets of $M(E)$, A_1 is strongly convex and for all $\mu \in A_1$, $\nu \in A_2$, μ and ν are mutually singular, then there is a Borel set B in $M(E) \times E$ s.t. for any $\nu \in A_2$, B_ν separates ν from A_1.*

The last statement in this corollary extends results of Mauldin, Preiss and von Weizsäcker, obtained using quite different ideas [**10**].

2.11. We first begin with a lemma on Borel functions. Its classical version is a well-known fact from measure theory.

Let $(f_n)_{n\in\omega}$ be a sequence of functions on E. We say that a sequence $(f'_n)_{n\in\omega}$ is subordinated to $(f_n)_{n\in\omega}$ if for any p, f'_p is in the convex hull of the set $\{f_{n+p}, n \in \omega\}$.

LEMMA. *Let μ be a Δ^1_1 measure on E, and $(f_n)_{n\in\omega}$ be a Δ^1_1 sequence of functions on E, with $0 \leqslant f_n \leqslant 1$. Then there is a Δ^1_1 sequence $(f'_n)_{n\in\omega}$ subordinated to $(f_n)_{n\in\omega}$, and such that*

$$\lim_p \mu\Big(\sup_n f'_{n+p}\Big) = \lim_p \sup_n \mu\big(f'_{n+p}\big).$$

PROOF. The first step is to extract a Δ^1_1 subsequence of $(f_n)_{n\in\omega}$ which weakly converges in $L^2(\mu)$. Let F be the closed unit ball of $L^2(\mu)$, which is weakly compact, and (Δ^1_1-) recursively presented by the usual metric and a dense Δ^1_1 sequence of continuous functions on E. The sequence (of equivalence classes of) $(f_n)_{n\in\omega}$ is still Δ^1_1 in F, hence the set of limit points of this sequence in F is a nonempty Δ^1_1 compact subset of F. By 0.9 and Δ^1_1 choice, there is an infinite Δ^1_1 subset X_0 of ω, and a Δ^1_1 point f in F s.t. $(f_n)_{n\in X_0}$ weakly converges to f.

The second step is to use convex combinations of the $(f_n)_{n\in X_0}$—in fact Cesaro means—to get a strongly convergent sequence. For u a finite subset of ω, let

$$g_u = \frac{1}{\bar{\bar{u}}}\Big(\sum_{n\in U} f_{x_n}\Big).$$

As is well known, as $f_{x_n} - f$ converges weakly to 0 in $L^2(\mu)$, then

$$\forall k\, \exists u \subseteq \omega \dot{-} k\ \|g_u - f\|_{L^2(\mu)} \leqslant 2^{-k}.$$

By Δ^1_1 choice, we can choose such a sequence in a Δ^1_1 way, say $(g_{u_k})_{k\in\omega}$. Note that $(g_{u_k})_{k\in\omega}$ is subordinated to the original sequence $(f_n)_{n\in\omega}$, and is a Cauchy sequence in $L^2(\mu)$. Finally, by refining the sequence (g_{u_k}) if necessary (in a Δ^1_1 way), we can get a subordinated Δ^1_1 sequence $(f'_n)_{n\in\omega}$ such that $\|f'_m - f'_n\|_{L^2(\mu)} \leqslant 1/(4n \cdot 2^n)$ for $n \in \omega$ and $m > n$.

But then it is easy to see that the sequence f'_n converges μ.a.e. on E. Let then $g = \lim_k \sup_n(f'_{n+k})$. The function g is Δ^1_1 on E, and is a representative of

$f \in L^2(\mu)$. [This shows in particular that any Δ^1_1 point in F admits a Δ^1_1 representative.] And by Lebesgue's convergence theorem,

$$\lim_k \mu\Big(\sup_n f'_{n+k}\Big) = \mu(g) = \lim_k \sup_n \mu(f'_{n+k}). \quad \square$$

2.12. We now introduce the ad hoc capacity, which is due to Mokobodzki: We fix a Δ^1_1 measure μ on E, and for $A \subseteq M(E)$ we set

$$c(A) = \inf\{\tilde{A}(f) + \mu(1-f) \colon 0 \leqslant f \leqslant 1, f \text{ Borel on } E\}$$

where as before $\tilde{A}(f) = \sup\{\nu(f) \colon \nu \in A\}$.

Some little changes in the arguments which follow would yield that c is indeed a capacity on $M(E)$. Instead of proving that, we shall establish that c, together with its Δ^1_1 version γ, defined by

$$\gamma(A) = \inf\{\tilde{A}(f) + \mu(1-f) \colon 0 \leqslant f \leqslant 1, f\, \Delta^1_1 \text{ on } E\}$$

satisfy the hypothesis of 2.4.

LEMMA. (a) *Let K be compact in $M(E)$. Then*

$$c(K) = \inf\{\tilde{K}(f) + \mu(1-f) \colon 0 \leqslant f \leqslant 1, f\, \Delta^1_1 \text{ and continuous on } E\}.$$

In particular, c is increasing and usc on $K(M(E))$, and $c = \gamma$ on Δ^1_1 compact sets.

(b) *If K is compact and convex, $c(K) = \sup\{\nu \wedge \mu(1) \colon \nu \in K\}$.*

PROOF. (a) Let $c^*(K) = \inf\{\tilde{K}(f) + \mu(1-f) \colon 0 \leqslant f \leqslant 1, f$ continuous and Δ^1_1 on $E\}$. Then $c(K) \leqslant c^*(K)$. Now if $\varepsilon > 0$ is given, and f is Borel on E, with $0 \leqslant f \leqslant 1$, we let h be some usc function $\leqslant f$, with $\mu(h) \geqslant \mu(f) - \varepsilon/2$. Now $\tilde{K}$ is usc and increasing on USC(E). Hence we can find a continuous and Δ^1_1 function g on E such that $h \leqslant g$ and $\tilde{K}(g) \leqslant \tilde{K}(h) - \varepsilon/2$. Finally

$$\begin{aligned}
\tilde{K}(g) + \mu(1-g) &\leqslant \tilde{K}(h) + \mu(1-g) + \varepsilon/2 \\
&\leqslant \tilde{K}(h) + \mu(1-h) + \varepsilon/2 \quad \text{as } g \geqslant h \\
&\leqslant \tilde{K}(f) + \mu(1-h) + \varepsilon/2 \quad \text{as } f \geqslant h \\
&\leqslant \tilde{K}(f) + \mu(1-f) + \varepsilon
\end{aligned}$$

which shows $c(K) \geqslant c^*(K)$. This clearly implies $c = \gamma$ on compact Δ^1_1 sets (in fact on compact sets), and also that c is usc on $K(M(E))$, since for continuous f, the function $K \mapsto \tilde{K}(f) + \mu(1-f)$ is usc.

(b) is a direct application of 0.7, with compact convex set K, and convex set of affine continuous functions $A = \{\tilde{f}|_K + \mu(1-f) \colon f$ continuous on E, $0 \leqslant f \leqslant 1\}$. From (a), one has

$$c(K) = \inf_{g \in A} \sup_{\nu \in K} g(\nu)$$

and, on the other hand,

$$\begin{aligned}
\sup\{\nu \wedge \mu(1) \colon \nu \in K\} &= \sup_{\nu \in K} (\inf\{\nu(f) + \mu(1-f) \colon \\
&\qquad\qquad f \text{ continuous on } E, 0 \leqslant f \leqslant 1\}) \\
&= \sup_{\nu \in K} \inf_{g \in A} g(\nu),
\end{aligned}$$

and 0.7 gives the desired result. $\square$

2.14. LEMMA. (a) *$\gamma(C) < t$ is Π^1_1 in codes for Δ^1_1 subsets of $M(E)$, and γ goes up on increasing Δ^1_1 sequences of sets in $M(E)$.*

(b) *If A is Σ^1_1 and $t = \inf\{\gamma(C)\colon A \subset C,\ C \in \Delta^1_1\}$ is a Δ^1_1 real, there is a Δ^1_1 function g on E such that*

$$\tilde{A}(g) + \mu(1 - g) \leqslant t.$$

PROOF. (a) $\gamma(C) < t \leftrightarrow \exists f \in \Delta^1_1(0 \leqslant f \leqslant 1 \wedge \tilde{C}(f) + \mu(1-f) < t)$, so it is enough to show that the relations $\mu(f) < t$ and $\tilde{C}(f) < t$ are Π^1_1 relations on codes for Δ^1_1 subsets of $M(E)$ and Δ^1_1 functions on E. But this follows from 2.2 applied to the capacity J of 1.7.

Consider now an increasing Δ^1_1 sequence of Δ^1_1 sets $(A_n)_{n\in\omega}$ in $M(E)$, and let $r \in \mathbf{Q}^+$ be such that $\sup_n \gamma(A_n) < r$. We want to show that $\gamma(\bigcup_n A_n) \leqslant r$. By Δ^1_1 choice, there is a Δ^1_1 sequence (f_n) of functions on E, with $0 \leqslant f_n \leqslant 1$, such that $l = \sup_n(\tilde{A}_n(f_n) + \mu(1 - f_n)) < r$.

Using Lemma 2.11, let (f'_n) be a Δ^1_1 sequence subordinated to $(f_n)_{n\in\omega}$ and such that

$$(*) \qquad \lim_n \mu\Big(\sup_k f'_{n+k}\Big) = \limsup \mu(f'_n),$$

and let $g = \liminf f'_n$. We claim that $\tilde{A}(g) + \mu(1 - g) \leqslant r$, where $A = \bigcup_n A_n$, which proves (a). It is enough to show that for any k and $\nu \in A_k$,

$$\nu(g) + \mu(1 - g) \leqslant r.$$

Now for any n, $\nu \in A_{k+n}$, hence $\nu(f_{n+k}) + \mu(1 - f_{n+k}) < r$. As (f'_n) is subordinated to (f_n), this implies for any n

$$\nu(f'_{n+k}) + \mu(1 - f'_{n+k}) < r,$$

and a fortiori, for all m

$$\nu\Big(\inf_{n\geqslant m} f'_{n+k}\Big) + \mu\Big(1 - \sup_{n\geqslant m} f'_{n+k}\Big) < r.$$

Using $(*)$, this immediately gives $\nu(g) + \mu(1 - g) \leqslant r$.

(b) is very similar. Let $t = \inf\{\gamma(C)\colon A \subset C \text{ and } C \in \Delta^1_1\}$. Assuming t is Δ^1_1, one can get by Δ^1_1 choice a Δ^1_1 sequence $(f_n)_{n\in\omega}$ of functions on E with $0 \leqslant f_n \leqslant 1$, such that

$$A(f_n) + \mu(1 - f_n) < t + 2^{-n}.$$

But if $(f'_n)_{n\in\omega}$ is a Δ^1_1 sequence subordinated to $(f_n)_{n\in\omega}$ and satisfying $(*)$, one can easily show, as in (a), that $g = \liminf f'_n$ satisfies, for all $\nu \in A$ and k in ω, $\nu(g) + \mu(1 - g) \leqslant t + 2^{-k}$, hence $\tilde{A}(g) + \mu(1 - g) \leqslant t$. □

2.15. PROOF OF THEOREM 2.9. We now fix a strongly convex Σ^1_1 subset A of $M(E)$, and a Δ^1_1 measure μ on E, mutually singular with each measure in A. The functions c and γ associated with μ satisfy the hypothesis of 2.4, by Lemmas 2.13(a) and 2.14(a). Hence, by 2.4

$$\sup\{c(K)\colon K \in K(M(E)),\ K \subset A\} = \inf\{\gamma(C)\colon C \in \Delta^1_1,\ A \subset C\}.$$

Call this number t.

Now let K be compact, $K \subset A$, and let H be its closed convex hull, i.e. $H = \{b(\Lambda)\colon \Lambda$ probability measure on $M(E)$, $\Lambda(K) = 1\}$. As A is strongly convex, H is still a subset of A and is convex and compact. By Lemma 2.13(b), $c(H) = \sup\{\nu \wedge \mu(1),\ \nu \in H\} = 0$. This shows $t = 0$. But then, we can apply Lemma 2.14(b): There is a Δ^1_1 function g on E such that $\tilde{A}(g) + \mu(1 - g) = 0$. If $B = \{x \in E\, g(x) = 1\}$, then clearly B is a Δ^1_1 set which separates μ from A. □

3. Effective inner capacitability.

3.1. The aim of this section is to try to extend to capacities the well-known Sacks-Tanaka basis result for measures.

THEOREM (SACKS **[13]**, TANAKA **[14]**). *Let μ be some Δ^1_1 measure on E.*

(a) *If A is Δ^1_1 in E, there is a Δ^1_1 sequence $(K_n)_{n \in \omega}$ of compact subsets of A s.t. $\mu(A) = \sup_n \mu(K_n)$.*

(b) *If A is Π^1_1 in E and $\mu(A) > 0$, then A has a Δ^1_1 member.*

Let us make some comments on this result. First, (b) is an easy consequence of (a), using the outer capacitability for Σ^1_1 sets, the basis result for compact sets, and the additivity of measure. One can argue as follows: Starting with A in Π^1_1, consider $E - A$. It is Σ^1_1, and its measure is less than $\mu(E)$. Using the outer capacitability 2.5 for μ, one gets a Δ^1_1 set B with $E - A \subset B$ and $\mu(B) < \mu(E)$. Then $E - B$ is Δ^1_1, contained in A, and of positive measure. From (a), one gets a Δ^1_1 compact subset K of $E - B$ with $\mu(K) > 0$, and a fortiori $K \neq \varnothing$. Then K has a Δ^1_1 member. □

Note that this proof uses additivity of measure only to replace the Π^1_1 set A by a Δ^1_1 subset of positive measure. So for capacities, we can hope to prove it for Δ^1_1 sets.

Let us sketch a proof of (a): First, one can show that if A is Δ^1_1, A admits a privileged Suslin scheme $(K_s)_{s \in \mathrm{Seq}\,\omega}$ such that the associated sequence $(A_s)_{s \in \mathrm{Seq}\,\omega}$ is still Δ^1_1. Second, using additivity of measure, one remarks that μ is Δ^1_1 uniformly on Δ^1_1 sets. Hence the function $s \mapsto \mu(A_s)$ is Δ^1_1. If $\mu(A) > r \in \mathbf{Q}_+$, it is then easy to see, following the proof of the capacitability theorem, that the sequence α such that $\forall n\, \mu(A_{\alpha|n}) > r$ can be chosen Δ^1_1. This gives a Δ^1_1 compact subset $K = \bigcap_n K_{\alpha|n}$ of A with $\mu(K) \geqslant r$. Again, we have used here the additivity of measure, in fact via a consequence of it, that μ is Δ^1_1 on Δ^1_1 sets. Note that reciprocally (a) easily implies that fact.

3.2. Capacities are not additive, so the preceding proofs cannot be generalized directly. In fact, it is easily seen that the extension fails in general.

PROPOSITION. *Let f be the Horowicz capacity on 2^ω (cf. 1.3(4)), defined by*

$$f(A) = \begin{cases} 0 & \text{if } A = \varnothing, \\ 1 & \text{if not.} \end{cases}$$

Then there is a Π^0_2 set A with $f(A) > 0$, but without any Δ^1_1 member, and s.t. $f(A) = 1$, but for all Δ^1_1 compact $K \subset A, f(K) = 0$.

PROOF. Take any nonempty Π_2^0 set A without Δ_1^1 member in 2^ω. □

So the remaining problem is: For which capacities on E is it true that

(a) Any Δ_1^1 set of positive capacity has a Δ_1^1 member.

(b) The capacity of Δ_1^1 sets can be approximated by Δ_1^1 compact subsets.

It turns out that this problem can be solved completely for normal capacities, i.e. suprema of measures.

3.3. Before giving the solution, let us make some remarks on the Horowicz case, i.e. when nonemptiness is considered a notion of "largeness". Then one has a basis result: namely, that countable Δ_1^1 (in fact Σ_1^1) nonempty sets have Δ_1^1 members. And intuitively, countability corresponds here to "sets on which the Horowicz capacity is not too far from being additive." This idea could be made precise in many ways, yielding interesting generalizations to all capacities, but we shall consider only one of them here, the notion of thin sets, which gives the solution to the inner approximation problem.

3.4. DEFINITION. Let f be some function on $\mathscr{P}(E)$. A set $A \subset E$ is *f-thin* if any family $\mathscr{F}$ of pairwise disjoint compact subsets of A, with $f(K) \neq 0$ for all $K \in \mathscr{F}$, is at most countable.

Note that for f the Horowicz capacity, thinness corresponds exactly to countability, and for f a measure, thinness is always true. We shall say that f is thin if E itself is f-thin.

3.5. THEOREM. *Let f be some Δ_1^1 normal capacity, i.e. $f = \tilde{H}$, where H is a Δ_1^1 compact subset of $M(E)$.*

(a) *If $A \subset E$ is $\Delta_1^1(\alpha)$ and f-thin,*

$$f(A) = \sup\{f(K): K \in \Delta_1^1(\alpha), K \in K(E), K \subset A\}.$$

In particular, f is Δ_1^1 uniformly on Δ_1^1 and f-thin sets, and if A is Δ_1^1 and f-thin, and $f(A) > 0$, A has a Δ_1^1 member.

(b) *The following are equivalent:*

(i) *For any $\Delta_1^1(\alpha)$ set A, $f(A) = \sup\{f(K): K \in \Delta_1^1(\alpha), K \in K(E), K \subset A\}$.*

(ii) *The function f is uniformly Δ_1^1 on Δ_1^1 subsets of E.*

(iii) *f is thin.*

3.6. REMARKS. (a) From the preceding discussion, the proof of Theorem 3.5. can be reduced to the proof of the first statement in (a), and a proof that if f is not thin, the function f is not Δ_1^1 on Δ_1^1 subsets of A.

(b) We shall not write the boldface versions of Theorem 3.5. But note that even in the simplest case of Horowicz capacity, the corollaries are not trivial: For example, the statement that f is uniformly Δ_1^1 on Δ_1^1 f-thin sets translates in that case to the fact that the projection of a Borel set with countable sections is Borel.

3.7. In order to prove Theorem 3.5, we extend the problem a bit. From now on we fix a Σ_1^1 subset H of $M(E)$; the corresponding function $\tilde{H}$ is not necessarily a normal capacity—but still analytic in Dellacherie's terminology, obtained by fixing the Σ_1^1 set H in the bicapacity J of 1.7.

DEFINITION. A measure $\mu \in M(E)$ is said to *control H* on a subset A of E if for any compact subset K of A, $\mu(K) = 0$ implies $\tilde{H}(K) = 0$.

We say for short that μ controls H if it controls H on E. In case $H = \{\nu\}$, this corresponds to the notion of absolute continuity of measures. Note that a set A is always μ-thin, for any measure μ, because of the c.c.c. property of measures. Hence it is easy to see that if A is such that some measure controls H on A, A is necessarily $\tilde{H}$-thin. The main step in the proof of 3.5 is a strong converse to this result, which answers a question of C. Dellacherie (cf. [**4**, **5**]).

3.8. THEOREM. *Let H be a Σ_1^1 subset of $M(E)$, and A a Σ_1^1 subset of E. If A is $\tilde{H}$-thin, there is a Δ_1^1 measure μ on E which controls H on A.*

3.9. COROLLARY. *Let H be a Σ_1^1 subset of $M(E)$, A a Δ_1^1 subset of E which is $\tilde{H}$-thin. Then there is a Δ_1^1 sequence $(K_n)_{n\in\omega}$ of compact subsets of A with $\tilde{H}(A) = \sup_n \tilde{H}(K_n)$. [This gives, in particular,* 3.5(a).]

PROOF. By 3.8, there is a Δ_1^1 measure μ which controls H on A. By Theorem 3.1, there is an increasing Δ_1^1 sequence $(K_n)_{n\in\omega}$ of compact subsets of A with $\mu(A) = \sup_n \mu(K_n)$. Then $\mu(A - \bigcup_n K_n) = 0$, hence, by control, for any compact subset K of $A - \bigcup_n K_n$, $\tilde{H}(K) = 0$. By capacitability, $\tilde{H}(A - \bigcup_n K_n) = 0$ and a fortiori $\tilde{H}(A) = \sup_n \tilde{H}(K_n)$. $\square$

3.10. In order to prove Theorem 3.8, we shall first establish a lemma concerning symmetric relations. We shall not use this lemma in its full generality, but it seems worth putting it down as it seems to be the best possible result along these lines. In the case of equivalence relations, this result is an easy consequence of the Silver-Harrington result on Π_1^1 equivalence relations.

LEMMA. *Let X be some recursively presented Polish space, and R be some symmetric relation on X, which is Π_1^1 with K_σ sections. Let D be some Σ_1^1 subset of X. Then either*

(i) *there is a nonempty perfect subset P of D, such that for all distinct x, y in P, $(x, y) \notin R$, or*

(ii) *there is a Δ_1^1 sequence $(x_n)_{n\in\omega}$ in X such that for any $y \in D$ there is some n s.t. $(x_n, y) \in R$.*

PROOF. We shall use Harrington's topology T on X: the topology generated by the Σ_1^1 subsets of X. As is well known, X equipped with T is a Baire space. Consider then $G = \{y \in X \exists x \in \Delta_1^1 (x, y) \in R\}$. This set is Π_1^1, and there are two cases: If $D \subset G$, then by Δ_1^1 choice, it is easy to find a Δ_1^1 sequence $(x_n)_{n\in\omega}$ with $D \subset \bigcup_n R_{x_n}$; so we are in case (ii).

Suppose now $D \not\subset G$, i.e. the Σ_1^1 set $H = D - G$ is nonempty. Then we shall construct a perfect set P as in (i) in the set H. By a standard technique, it is enough to show that $R \cap H^2$ is meager for the topology $T \times T$ on H^2. We argue by contradiction. Assume $R \cap H^2$ is not meager in H^2. By the Kuratowski-Ulam lemma, $F = \{x \in H R_x \cap H$ is not meager in H for $T\}$ is not meager in H for T. For any nonempty Σ_1^1 set A in H, define $R_A = \{x \in X \exists K$ compact and $\Delta_1^1(x)$

s.t. $A \subset K \subset R_x\}$. Each R_A is clearly a Π_1^1 set, hence is T-closed, and we claim that $F \subset \bigcup\{R_A : A\ \Sigma_1^1 \text{ in } H, A \neq \varnothing\}$. For if $x \in F$, then $R_x \cap H$ is a nonmeager subset of H, and as R_x is K_σ in X, there must be some nonempty Σ_1^1 subset A of H and some compact set K with $A \subset K \subset R_x$; and by a well-known separation result, K can be chosen $\Delta_1^1(x)$. Now F is nonmeager in H, and the $R_A \cap H$'s are closed subsets of H covering F, so one of them is nonrare. So there are two nonempty Σ_1^1 subsets A and B of H with $B \subset R_A$. By definition of R_A and Δ_1^1 choice, there is a Δ_1^1 subset C of X^2 s.t. for all $x \in B$, C_x is a compact subset of X and $A \subset C_x \subset R_x$. Let $L = \bigcap_{x \in B} C_x$. L is a Π_1^1 compact set, and $A \subset L$, hence by separation there is a Δ_1^1 compact set K with $A \subset K \subset L$. As A is nonempty, K has a Δ_1^1 member, say z. Now $z \in L$, hence $z \in C_x$ for any $x \in B$, and as $C_x \subset R_x$, $z \in R_x$ for any $x \in B$. This means that $B \subset B_z$, and a fortiori $R_z \cap H \neq 0$. But this contradicts the definition of H, and proves the lemma. □

A similar (but less effective) result could be obtained for Π_1^1 symmetric relations with $\mathbf{\Sigma}_2^0$ sections. But by a counterexample of Burgess and Mauldin [**1**], the lemma fails for Π_2^0 symmetric relations: it is enough to consider on 2^ω: $R(x, y)D \leftrightarrow x$ and y are Turing incomparable, with $D = 2^\omega$.

3.11. Proof of 3.8. We let H be a Σ_1^1 subset of $M(E)$, and A an $\tilde{H}$-thin Σ_1^1 subset of E. Consider on $M(E)$ the relation R defined by

$$(\mu, \nu) \in R \leftrightarrow \mu \wedge \nu \neq 0.$$

R is symmetric, K_σ and Δ_1^1, for

$$\begin{aligned}\mu \wedge \nu \neq 0 &\leftrightarrow \exists \mu'(\mu' \leqslant \mu \wedge \mu' \leqslant \nu \wedge \mu' \neq 0)\\ &\leftrightarrow \exists \mu' \in \Delta_1^1(\mu' \leqslant \mu \wedge \mu' \leqslant \nu \wedge \mu' \neq 0),\end{aligned}$$

remarking that $\{(\mu, \mu')\, \mu' \leqslant \mu\}$ is compact Δ_1^1, and $\mu' \neq 0$ is K_σ and Δ_1^1. (For the second equivalence, we have used the basis result for Δ_1^1 compact sets.) Let $D \subset M(E)$ be defined by

$$\nu \in D \leftrightarrow (\nu(A) = \nu(E) > 0 \wedge \exists \nu' \in H(\nu \leqslant \nu')).$$

D is a Σ_1^1 set in $M(E)$, and we can apply Lemma 3.10 to R and D. Suppose we are in case (i), i.e. there is some perfect subset P of D consisting of pariwise mutually singular measures. Let $\tilde{P}$ be the capacity associated with P. One has $\tilde{P} > 0$, $\tilde{P} \leqslant \tilde{H}$ and $\tilde{P}(B) = \tilde{P}(B \cap A)$ for all sets B by the definition of D. We now claim that $\tilde{P}$ is not thin, hence A is not $\tilde{P}$-thin (by capacitability) and a fortiori is not $\tilde{H}$-thin, a contradiction which proves that we are not in case (i).

To prove the claim, we note that if U is open in E and $\tilde{P}(U) > a$, there are open sets U_0 and U_1 such that $\overline{U}_0 \subset U$, $\overline{U}_1 \subset U$, $\overline{U}_0 \cap \overline{U}_1 = \varnothing$, and $\tilde{P}(U_0) > a$, $\tilde{P}(U_1) > a$. For the set $\{\mu \in P\, \mu(U) > a\}$ is a nonempty open subset of P, hence contains two mutually singular measures μ_0 and μ_1. So there are two disjoint compact subsets K_0 and K_1 of U with $\mu_0(K_0) > a$ and $\mu_1(K_1) > a$, and we can take as U_0 and U_1 two open subsets of U with $K_0 \subset U_0$, $K_1 \subset U_1$ and $\overline{U}_0 \cap \overline{U}_1 = \varnothing$.

Using that fact, it is easy to construct a family $(K_\alpha)_{\alpha\in 2^\omega}$ of pairwise disjoint compact subsets of E with $\tilde{P}(K_\alpha) \geqslant \frac{1}{2}\tilde{P}(E) > 0$, which proves the claim. So we are in case (ii), and there is a Δ_1^1 sequence $(\mu_n)_{n\in\omega}$ in $M(E)$ such that for any $\nu \in D$ there is an n such that $\nu \wedge \mu_n \neq 0$. Let $\mu = \Sigma_n 2^{-n}\mu_n$. We claim that the Δ_1^1 measure μ controls H on A. For let K be a compact subset of A with $\mu(K) = 0$, and let $\nu \in H$. If $\nu|_K > 0$, $\nu|_K \in D$, hence, for some n, $\nu|_K \wedge \mu_n \neq 0$, which contradicts that $\mu_n(K) = 0$ for all n. So $\nu(K) = 0$ for all $\nu \in H$ and $\tilde{H}(K) = 0$. □

3.12. Remark. The preceding proof gives a stronger result: If $\tilde{H}$ is not thin, there is a compact set $L = \{K_\alpha: \alpha \in 2^\omega\}$ of pairwise disjoint compact subsets of E such that $\forall\alpha \in 2^\omega\, \tilde{H}(K_\alpha) > 0$. (In Horowicz case, this is the perfect set theorem.) It follows that the function $\tilde{H}$ cannot be Borel uniformly on Borel subsets on E, otherwise the relation, on Borel subsets B of 2^ω,

$$B \neq \varnothing \leftrightarrow \tilde{H}\Big(\bigcup_{\alpha\in B} K_\alpha\Big) > 0,$$

would be Borel too. This ends the proof of 3.5.

3.13. Remark. It is possible, using the preceding remark, to give a quantitative version of the qualitative notion of nonthinness, by introducing a "thickness" function e by

$$e(H, A) = \sup\{a \in \mathbf{R}_+: \exists L = \{K_\alpha: \alpha \in 2^\omega\} \text{ compact in } K(E) \text{ consisting of pairwise disjoint subsets of } A \wedge \forall\alpha\, \tilde{H}(K_\alpha) \geqslant a\},$$

and quantitative versions of noncontrol by

$$e_\mu(H, A) = \sup\{\tilde{H}(K): K \subset A, K \in K(E), \mu(K) = 0\}, \quad \text{where } \mu \in M(E).$$

It is then possible to generalize 3.8 and prove that if H, A are Σ_1^1 and $e(H, A) < r$, there is a Δ_1^1 measure μ such that $e_\mu(H, A) < r$. As it is written, this looks like a separation result which could be obtained via the techniques of §2. But $e(H, A)$ is not a bicapacity (although it is again analytic in the sense of 1.8, and a supremum of measures), and I do not know how to use §2 to get this result on e.

References

1. J. P. Burgess and R. D. Mauldin, *Conditional distributions and orthogonal measures*, Ann. Probab. **9** (1981), 902–906.

2. G. Choquet, *Theory of capacities*, Ann. Inst. Fourier (Grenoble) **5** (1953), 131–295.

3. C. Dellacherie, *Capacities and analytic sets*, Cabal Seminar 77–79, Lecture Notes in Math., vol. 839, Springer-Verlag, Berlin and New York, 1981, pp. 1–32.

4. C. Dellacherie, D. Feyel and G. Mokobodzki, *Intégrales de capacités fortement sous additives*, Sém. Probab. XVI, Lecture Notes in Math., vol. 920, Springer-Verlag, Berlin and New York, 1982, pp. 8–28.

5. C. Dellacherie, *Appendice à l'exposé précédent*, Sém. Probab. XVI, Lecture Notes in Math., vol. 920, Springer-Verlag, Berlin and New York, 1982, pp. 29–40.

6. A. S. Kechris, *Measure and category in effective descriptive set theory*, Ann. Math. Logic **5** (1973), 337–384.

7. A. Louveau, *Recursivity and compactness*, Higher Set Theory, Lecture Notes in Math., vol. 669, Springer-Verlag, Berlin and New York, 1977, pp. 303–337.

8. ______, *Construction de mesures de contrôle pour des capacités minces*, C. R. Acad. Sci. Paris Sér. A **294** (1982), 353–355.

9. ______, *Capacitabilité et sélections boréliennes*, Sém. Init. Analyse 1981/82, Publ. Math. Univ. Pierre et Marie Curie, no. 54, exposé no. 19.

10. R. D. Mauldin, D. Preiss and H. von Weizsäcker, *Orthogonal transition kernels*, preprint, 1982.

11. G. Mokobodzki, *Ensembles à coupes dénombrables et capacités dominées par une mesure*, Sém. Probab. XII, Lecture Notes in Math., vol. 649, Springer-Verlag, Berlin and New York, 1978, pp. 491–508.

12. Y. N. Moschovakis, *Descriptive set theory*, North-Holland, Amsterdam, 1980.

13. G. E. Sacks, *Measure-theoretic uniformity in recursion theory and set theory*, Trans. Amer. Math. Soc. **142** (1969), 381–420.

14. H. Tanaka, *A basis theorem for* Π_1^1 *sets of positive measure*, Comment. Math. Univ. St. Paul, **16** (1968), 115–127.

EQUIPE D'ANALYSE, E. R. A. 294, UNIVERSITÉ P. ET M. CURIE, 4, PLACE JUSSIEU, 75230 PARIS, FRANCE

Proceedings of Symposia in Pure Mathematics
Volume **42**, 1985

A Purely Inductive Proof of Borel Determinacy

DONALD A. MARTIN[1]

In [**2**] we proved that all infinite Borel games of perfect information are determined. The proof had two parts: (1) a basic construction showing that Σ_k determinacy implies Σ_{k+1} determinacy; (2) for each Borel set of Borel rank α, an α-fold iteration of the basic construction, reducing the corresponding Borel game to an open game. In the basic construction there occurred an auxiliary game involving nested sequences of trees. This was fairly complex and necessitated a priority argument. Step (2) produced more serious difficulties for the reader, since the argument was not purely inductive but rather involved directly considering the α-fold iteration of the basic construction.

In this paper we present a new proof in which we have made two important changes:

(a) The basic construction handles only a single closed set (instead of infinitely many closed sets). Thus our auxiliary game has only two auxiliary moves.

(b) We state a property of sets which implies determinacy and prove, by transfinite induction on Borel rank, that all Borel sets have the property.

Our "new" proof is really only the old proof reorganized, but we expect that the reader will find it much simpler.

By a *tree* we mean a set T of finite sequences such that

(i) $(\sigma \in T \,\&\, \tau \subseteq \sigma\, [\sigma \text{ extends } \tau]) \Rightarrow \tau \in T$;

(ii) $\sigma \in T \Rightarrow \exists\tau(\sigma \subsetneqq \tau \,\&\, \tau \in T)$.

(ii) says that T has no terminal nodes. If T is a tree, $[T]$ is the set of all infinite sequences x such that $\forall n(x \restriction n \in T)$, where $x \restriction n = \langle x(0),\ldots,x(n-1)\rangle$. A *game on T* is played as follows:

$$\begin{array}{llllll} \text{I} & a_0 & & a_2 & & \cdots \\ \text{II} & & a_1 & & a_3 & \cdots \end{array}$$

1980 *Mathematics Subject Classification*. Primary 03E05; Secondary 03E60.
[1]Supported by NSF Grant MCS83-02555.

It is required that $\langle a_0,\ldots,a_n\rangle \in T$ for each n. If $A \subseteq [T]$, then $G(A, T)$ is the game on T with the following winning condition: I wins a play $\langle a_0, a_1,\ldots\rangle$ if and only if $\langle a_0, a_1,\ldots\rangle \in A$. Otherwise II wins. The notions of *strategies* and *winning strategies* for I and II are defined in the obvious way. $G(A, T)$ is *determined* if either I or II has a winning strategy. Let $S(T)$ be the set of all strategies for either player for games on T.

We give $[T]$ a topology by letting the basic open sets be those of the form $\{x: \sigma \subseteq x\}$ for $\sigma \in T$. For α a countable ordinal > 0, we define the classes Σ_α and Π_α as follows: Σ_1 = the class of all open sets. $\Pi_\alpha = \{A: A$ is the complement of a set in $\Sigma_\alpha\}$. For $\alpha > 1$, $\Sigma_\alpha = \{A: A$ is a countable union of sets in $\bigcup_{\beta<\alpha}\Pi_\beta\}$. A is *Borel* if $A \in \Sigma_\alpha$ for some $\alpha < \omega_1$.

A *covering* of a tree T is a triple $(\tilde{T}, \pi, \varphi)$ where

(1) $\tilde{T}$ is a tree;

(2) $\pi: [\tilde{T}] \to [T]$;

(3) $\varphi: S(\tilde{T}) \to S(T)$ and each $\varphi(\tilde{s})$ is a strategy for the same player as $\tilde{s}$;

(4) if x is a play consistent with $\varphi(\tilde{s})$, there is a play $\tilde{x}$ consistent with $\tilde{s}$ such that $\pi(\tilde{x}) = x$.

A covering $(\tilde{T}, \pi, \varphi)$ of T *unravels* a set $A \subseteq [T]$ if $\pi^{-1}(A)$ is clopen (closed and open).

We recall that Gale and Stewart [**1**] proved that $G(A, T)$ is determined if A is an open or closed subset of $[T]$.

LEMMA 1. *Let $(\tilde{T}, \pi, \varphi)$ be a covering which unravels $A \subseteq [T]$. $G(A, T)$ is determined.*

PROOF. $\pi^{-1}(A)$ is clopen, so $G(\pi^{-1}(A), \tilde{T})$ is determined. Let $\tilde{s}$ be a winning strategy, say for I. We show that $\varphi(\tilde{s})$ is a winning strategy for I for $G(A, T)$. Let x be a play consistent with $\varphi(\tilde{s})$. Let $\tilde{x}$ be as given by (4). Since $\tilde{x}$ is consistent with $\tilde{s}$, $\tilde{x} \in \pi^{-1}(A)$. Thus $x = \pi(\tilde{x}) \in A$.

LEMMA 2. *Let (T_1, π_1, φ_1) be a covering of T_0 and let (T_2, π_2, φ_2) be a covering of T_1. $(T_2, \pi_1 \circ \pi_2, \varphi_1 \circ \varphi_2)$ is a covering of T_0.*

Note that if $(\tilde{T}, \pi, \varphi)$ is a covering of T, $A \subseteq [T]$, $A \in \Sigma_\alpha$, and π is continuous, then $\pi^{-1}(A) \in \Sigma_\alpha$. In particular, if we compose coverings as in Lemma 2 and π_2 is continuous, then if (T_1, π_1, φ_1) unravels A it follows that $(T_2, \pi_1 \circ \pi_2, \varphi_1 \circ \varphi_2)$ unravels A.

We want to prove by induction on α (simultaneously for all T) that every set in Σ_α can be unraveled by a covering. Continuity helps, but to carry out an induction we need a stronger condition:

A covering $(\tilde{T}, \pi, \varphi)$ of T is a *k-covering* if

(a) $(\pi(\tilde{x})) \upharpoonright n$ depends only on $x \upharpoonright n$;

(b) $\varphi(\tilde{s})$ restricted to positions of length $\leqslant n$ depends only on $\tilde{s}$ restricted to positions of length $\leqslant n$.

(c) By (a) we may think of π as π: $\tilde{T} \to T$. We demand that $\pi \restriction \tilde{T}^k$ be one-one and onto T^k, where $T^k = \{\sigma \in T\colon \text{length}(\sigma) = k\}$.

LEMMA 3. *Let $A \subseteq [T]$ be closed and let $k \in \omega$. There is a k-covering of T which unravels A.*

PROOF. We describe $\tilde{T}$ implicitly by describing how games on $\tilde{T}$ are played. By increasing k if necessary, we may assume k is even.

I	a_0		a_2	$\cdots$	a_{k-2}		(a_k, T_{I})		a_{k+2}	$\cdots$
II		a_1		$\cdots$		a_{k-1}		(T_{II}, a_{k+1})		a_{k+3} $\cdots$

All $\langle a_0,\ldots,a_j\rangle$ must belong to T. T_{I} must be a I-*imposed subtree* of T: i.e., if $\langle b_0,\ldots,b_j\rangle \in T_{\text{I}}$ and $\langle b_0,\ldots,b_j, b_{j+1}\rangle \in T$ and j is even, then $\langle b_0,\ldots,b_j, b_{j+1}\rangle \in T_{\text{I}}$. (I-imposed subtrees are now often called *quasi-strategies* for I.) Furthermore we require that $\sigma \subseteq \langle a_0,\ldots,a_k\rangle$ or $\langle a_0,\ldots,a_k\rangle \subseteq \sigma$ for each $\sigma \in T_{\text{I}}$. There are two options for II:

First Option. T_{II} can be II-imposed subtree of T_{I} such that $[T_{\text{II}}] \subseteq A$. (II-*imposed* is defined as is I-*imposed*, with "odd" replacing "even".)

Second Option. T_{II} can be $\{\sigma \in T\colon \sigma \subseteq \tau \text{ or } \tau \subseteq \sigma\}$ for some $\tau \in T_{\text{I}}$ such that $\langle a_0,\ldots,a_k\rangle \subseteq \tau$ and $\{x\colon x \supseteq \tau\} \cap A = \varnothing$.

For $j > k$, $\langle a_0,\ldots,a_j\rangle$ must belong to T_{II}.

Note that each player has a legal move at every position, so we have indeed described a tree $\tilde{T}$ in our sense of "tree". The function π is the obvious one. Note that (a) and (c) in the definition of a k-covering are satisfied.

First let $\tilde{s} \in S(\tilde{T})$ be a strategy for I. Let $\varphi(\tilde{s})$ agree with $\tilde{s}$ on positions of length $< k$. Let $\langle a_0,\ldots,a_{k-1}\rangle$ be a position consistent with $\varphi(\tilde{s})$, and so with $\tilde{s}$. Let (a_k, T_{I}) be the move given by $\tilde{s}$. Let $\varphi(\tilde{s})$ play a_k at $\langle a_0,\ldots,a_{k-1}\rangle$.

Consider the game $G([T_{\text{I}}] - A, T_{\text{I}})$. $[T_{\text{I}}] - A$ is open, so this game is determined. If II has a winning strategy, let T_{II} be the II-imposed subtree of T_{I} consisting of positions in T_{I} which are not lost for II in $G([T_{\text{I}}] - \text{A}, \text{T}_{\text{I}})$. If II plays a_{k+1} at $\langle a_0,\ldots,a_k\rangle$, $\varphi(\tilde{s})$ assumes that the First Option is taken at $\langle a_0,\ldots,a_{k-1},(a_k, T_{\text{I}})\rangle$ and (T_{II}, a_{k+1}) is played and follows $\tilde{s}$ until (if ever) a position $\sigma \notin T_{\text{II}}$ is reached. When this happens, or immediately if T_{II} does not exist or $\langle a_0,\ldots,a_{k+1}\rangle \notin T_{\text{II}}$, $\varphi(\tilde{s})$ proceeds as follows: First a winning strategy for $G([T_{\text{I}}] - A, T_{\text{I}})$ is played, reaching—since A is closed—a position $\tau \in T_{\text{I}}$ such that $A \cap \{x\colon x \supseteq \tau\} = \varnothing$. Now $\varphi(\tilde{s})$ assumes that II took the Second Option at $\langle a_0,\ldots,a_{k-1},(a_k, T_{\text{I}})\rangle$ and played $T_{\text{II}} = \{\sigma \in \tau\colon \sigma \subseteq \tau \text{ or } \tau \subseteq \sigma\}$. $\varphi(\tilde{s})$ proceeds according to $\tilde{s}$.

Now let $\tilde{s} \in S(\tilde{T})$ be a strategy for II. At positions of length $\leqslant k$, $\varphi(\tilde{s})$ follows $\tilde{s}$. Let $\langle a_0,\ldots,a_k\rangle$ be consistent with $\varphi(\tilde{s})$, so that $\langle a_0,\ldots,a_{k-1}\rangle$ is consistent with $\tilde{s}$.

Consider the game $G(B, T)$ where

$$B = \{x\colon \neg\exists\tau[\tau \subseteq x \,\&\, \exists T'_{\mathrm{I}}\ (\text{if I plays } (a_k, T'_{\mathrm{I}}) \text{ at } \langle a_0,\ldots,a_{k-1}\rangle, \text{ then } \tilde{s} \text{ calls for II to take the Second Option and play } T_{\mathrm{II}} = \{\sigma \in T\colon \sigma \subseteq \tau \text{ or } \tau \subseteq \sigma\})]\}.$$

II then wins $G(B, T)$ if a position $\tau \supseteq \langle a_0,\ldots,a_k\rangle$ is reached such that $\{x\colon x \supseteq \tau\} \cap A = \varnothing$ and

$$\langle a_0,\ldots,a_{k-1},(a_k, T'_{\mathrm{I}}),(\{\sigma \in T\colon \sigma \subseteq \tau \text{ or } \tau \subseteq \sigma\}, a_{k+1})\rangle$$

is consistent with $\tilde{s}$ for some T'_{I} and some a_{k+1}. B is closed. Suppose that $\langle a_0,\ldots,a_k\rangle$ is a winning position for I in $G(B, T)$. Let $T_{\mathrm{I}} = \{\sigma \in T\colon \langle a_0,\ldots,a_k\rangle \supseteq \sigma$ or $(\langle a_0,\ldots,a_k\rangle \subseteq \sigma$ and σ is not lost for I in $G(B, T))\}$. Suppose I plays (a_k, T_{I}) at $\langle a_0,\ldots,a_{k-1}\rangle$. Clearly $\tilde{s}$ cannot call for II to take the Second Option, since the associated τ would be a loss for I in $G(B, T)$ but would belong to T_{I}. $\varphi(\tilde{s})$ thus proceeds by assuming that (a_k, T_{I}) is played and following $\tilde{s}$ (omitting of course actually to play T_{II}). If I ever departs from T_{I}, or immediately if $\langle a_0,\ldots,a_k\rangle$ is a winning position for II in $G(B, T)$, $\varphi(\tilde{s})$ proceeds to play a winning strategy for $G(B, T)$. Thus a position τ is reached such that, for some T'_{I} and a_{k+1}, if I plays (a_k, T'_I) at $\langle a_0,\ldots,a_{k-1}\rangle$ then $\tilde{s}$ calls for II to take the Second Option and play $(\{\sigma \in T\colon \sigma \subseteq \tau \text{ or } \tau \subseteq \sigma\}, a_{k+1})$. $\varphi(\tilde{s})$ then follows $\tilde{s}$, assuming that (a_k, T'_{I}) was played by I.

It is easy to check that φ satisfies (b) in the definition of a k-covering. Our construction of φ has in effect shown that π and φ satisfy (4) in the definition of a covering.

We must finally show that $\pi^{-1}(A)$ is clopen. $\tilde{x} \in \pi^{-1}(A) \Rightarrow$ II takes the First Option $\Leftrightarrow [T_{\mathrm{II}}] \subseteq A$.

LEMMA 4. *Let $(T_{i+1}, \pi_{i+1}, \varphi_{i+1})$ be $(k + i)$-coverings of T_i, for each $i \in \omega$. There is a tree $\tilde{T}$ and there are $\tilde{\pi}_i$, $\tilde{\varphi}_i$ for $i \in \omega$ such that for each $i \in \omega$: $(\tilde{T}, \tilde{\pi}_i, \tilde{\varphi}_i)$ is a $(k + i)$-covering of T_i and $\tilde{\pi}_i = \pi_{i+1} \circ \tilde{\pi}_{i+1}$ and $\tilde{\varphi}_i = \varphi_{i+1} \circ \tilde{\varphi}_{i+1}$.*

PROOF. We show in effect that the inverse limit of the system of coverings exists. Using (c) in the definition of a $(k + i)$-covering, we may assume—replacing the T_i's by isomorphic trees if necessary—that $(T_i)^{k+1} = (T_{i+1})^{k+i}$ and $\pi_{i+1} \restriction (T_{i+1})^{k+i} =$ identity for each i. Let $\sigma \in \tilde{T} \Leftrightarrow \sigma \in T_i$ for all large i ($\Leftrightarrow \sigma \in T_i$ for i minimal such that $k + i \geqslant \text{length}(\sigma)$). Let $j \in \omega$. Let $\tilde{\pi}_j(\sigma) = \pi_{j+1} \circ \cdots \circ \pi_{i-1} \circ \pi_i(\sigma)$, where $k + i \geqslant \text{length}(\sigma)$. (If $j = i$, we intend $\tilde{\pi}_j(\sigma) = \sigma$.) Condition (4) in the definition of a covering and condition (b) in the definition of a $(k + i)$-covering imply (since $\pi_{i+1} \restriction (T_{i+1})^{k+i} =$ identity), that $\varphi_{i+1}(s_{i+1})$ agrees with s_{i+1} on positions of length $\geqslant k + i$. Thus we can let $\tilde{\varphi}_j(\tilde{s}) = \varphi_{j+1} \circ \cdots \circ \varphi_i(\tilde{s})$ on positions of length $< k + i$. (Once again we intend $\tilde{\varphi}_j(\tilde{s}) = \tilde{s}$ on positions of length $< k + i$, when $j = i$. Note also that "$\varphi_i(\tilde{s})$" makes sense on positions of length $< k + i$, since $\tilde{T}^{k+i} = (T_i)^{k+i}$.)

We need only check (4). Let x_j be consistent with $\tilde{\varphi}_j(\tilde{s})$. Let $x_{j+1}, x_{j+2}, \dots$ be successively given by (4) for the coverings $(T_{j+1}, \pi_{j+1}, \varphi_{j+1}), (T_{j+2}, \pi_{j+2}, \varphi_{j+2}), \dots$. Since $\pi_{i+1} \upharpoonright (T_{i+1})^{k+i}$ is the identity, we may let $\sigma \subseteq \tilde{x} \Leftrightarrow (\sigma \subseteq x_i$ for all large $i)$. Since $\tilde{s}$ agrees with $\tilde{\varphi}_i(\tilde{s})$ on positions of fixed length, for all large i, $\tilde{x}$ is consistent with $\tilde{s}$. Also $\tilde{\pi}_i(\tilde{x} \upharpoonright n) = \tilde{x}_i \upharpoonright n$ for all large i, so $\tilde{\pi}_j(\tilde{x} \upharpoonright n) = \pi_{j+1} \circ \cdots \circ \pi_i(x_i \upharpoonright n) = x_j \upharpoonright n$.

THEOREM. *If A is a Borel subset of $[T]$ and $k \in \omega$, there is a k-covering of T which unravels A.*

PROOF. By Lemma 3, the theorem holds for all $A \in \Pi_1$, for all T. Obviously any covering which unravels A unravels the complement of A. Assume that $\alpha < \omega_1$ and that, for all T, the theorem holds for each set in Σ_{β_i} for $\beta < \alpha$. Let $A \in \Sigma_\alpha$. Then $A = \bigcup_{i \in \omega} A_i$ with each $A_i \in \Pi_\beta$, $\beta_i < \alpha$. Let (T_1, π_1, φ_1) be a k-covering of $T_0 = T$ which unravels A_0. Let (T_2, π_2, φ_2) be a $(k + 1)$-covering of T_1 which unravels $\pi_1^{-1}(A_1)$. In general, let $(T_{i+1}, \pi_{i+1}, \varphi_{i+1})$ be a $(k + i)$-covering of T_i which unravels $\pi_i^{-1} \circ \pi_{i-1}^{-1} \circ \cdots \circ \pi_1^{-1}(A_i)$. Let $\tilde{T}$ and the $\tilde{\pi}_i$, $\tilde{\varphi}_i$ be given by Lemma 4. $(\tilde{T}, \tilde{\pi}_0, \tilde{\varphi}_0)$ unravels each of $A_0, A_1, \dots$. Since $\tilde{\pi}_0^{-1}(A) = \bigcup_{i \in \omega} \tilde{\pi}_0^{-1}(A_i)$, $\tilde{\pi}_0^{-1}(A)$ is open. Let (T^*, π^*, φ^*) be a k-covering of $\tilde{T}$ which unravels $\tilde{\pi}_0^{-1}(A)$. $(T^*, \tilde{\pi}_0 \circ \pi^*, \tilde{\varphi}_0 \circ \varphi^*)$ is a k-covering of T which unravels A.

COROLLARY. *If $A \subseteq [T]$ is Borel, $G(A, T)$ is determined.*

REMARKS. (1) The priority construction of [**2**] has disappeared. It seems possible that considering the infinitely many closed sets at once, as in [**2**], might be necessary for a sharp calculation of complexity of strategies. Superficially, it might also appear that this could be necessary for getting a sharp bound on the size of the covering trees, but a little thought shows this is not the case.

(2) In [**2**] we said that our proof did not need the axiom of choice in the case T is countable. Several people have pointed out that countable choice is necessary to get a Borel code for each Borel subset of $[T]$. That is also true here, though our definition of "Borel" is more restrictive here.

(3) Moschovakis (who incidentally kept suggesting that a purely inductive proof of Borel determinacy should be possible) simplified our original proof (see [**3**]) by using trees with terminal nodes, removing the necessity for $G([T_{\mathrm{I}}] - A, T_{\mathrm{I}})$ and $G(B, T)$ used above. We do not know whether this idea can be mixed with out new proof.

(4) Does the unraveling property hold for any class beyond the Borel sets? We first note the following curiosity.

Curiosity. Assume the Axiom of Determinacy plus Uniformization for sets of pairs of elements of $\omega^{<\omega}$. There is a single 0-covering $(\tilde{T}, \pi, \varphi)$ which unravels every $A \subseteq \omega^\omega$.

PROOF. We describe games on $\tilde{T}$. I begins by playing a strategy s for I a game on T. II next plays an element x of ω^ω consistent with s. The players then amuse themselves playing, say, natural numbers to satisfy our definition of a tree. Let $\pi(s, x, \ldots) = x$. (a) is fulfilled.

If $\tilde{s}$ is a strategy for I, let $\varphi(\tilde{s})$ be the first move given by $\tilde{s}$.

Suppose $\tilde{s}$ is a strategy for II. Consider the following game $G^{\tilde{s}}$ on T. II wins a play x of $G^{\tilde{s}}$ if and only if there is an s such that, if I plays s then $\tilde{s}$ calls for II to play x. If s is a strategy for I for $G^{\tilde{s}}$, then II can defeat s by playing the x given by $\tilde{s}$. Thus, by AD, II has a winning strategy for $G^{\tilde{s}}$. By uniformization, we can pick for each $\tilde{s}$ a winning strategy $\varphi(\tilde{s})$ for II for $G^{\tilde{s}}$.

Uniformization is needed only because we required that φ is single-valued.

We know of no proof, from any large cardinal assumption consistent with choice, that every Π^1_1 set can be unraveled by a covering. If we could show that, for any countable family $\mathscr{A}$ of Π^1_1 sets, there is a covering which unravels every number of $\mathscr{A}$, then we could prove determinacy for the σ-algebra generated by the Π^1_1 sets. Results of J. Steel show that one needs at least (approximately) a measurable cardinal κ of order κ^{++}.

References

1. D. Gale and F. M. Stewart, *Infinite games with perfect information*, Contributions to the theory of games, Ann. of Math. Studies, No. 28, Princeton Univ. Press, 1953, pp. 245–266.
2. Donald A. Martin, *Borel determinacy*, Ann. of Math. (2) **102** (1975), 363–371.
3. Y. N. Moschovakis, *Descriptive set theory*, North-Holland, Amsterdam, 1980.

Department of Mathematics, University of California, Los Angeles 90024

IV. EFFECTIVE MATHEMATICS

Proceedings of Symposia in Pure Mathematics
Volume **42**, 1985

Decidable Ehrenfeucht Theories

T. MILLAR[1]

Vaught [**1**] proved that there is no complete theory with exactly two countable models up to isomorphism. Ehrenfeucht (see [**1**]) produced a family of complete theories such that, for each finite $n > 2$, there was a member of the family with exactly n countable models. The underlying structure of these examples is a dense linear order without endpoints. In order to escape the ω-categoricity of this theory, an infinite number of distinguished classes were introduced. To prevent a proliferation of nonprincipal types, these classes were given order type ω. This alone produces a theory with exactly three countable models: the prime, weakly saturated and saturated models. In the same order, these models correspond to the nonprincipal 1-type having no relations, a first realization and a realization but no first such. In order to produce a theory with $n > 3$ countable models, the universe is partitioned into $n - 2$ dense distinguished pieces. This has the effect of creating $n - 2$ weakly saturated models, differing by which partition class the least element realizing a nonprincipal type falls into.

DEFINITIONS. 1. If P is a property of theories, then a theory T is *persistently* P iff$_{\text{df}}$ for every complete n-type $\Gamma(\bar{x})$ of T, the theory $\Gamma(\bar{c})$ has property P, $n < \omega$.

2. If P is a property of structures, then a theory T is *strongly* P iff$_{\text{df}}$ every countable model of T has property P.

3. An *Ehrenfeucht theory* is a complete theory with exactly n countable models, for some $1 < n < \omega$.

4. A structure is *almost homogeneous* iff$_{\text{df}}$ some finite expansion by constants of the structure is ω-homogeneous.

5. A theory is *almost homogeneous* iff$_{\text{df}}$ each of its models is almost homogeneous.

Trivally, if a theory is strongly homogeneous, then every model of the theory is ω-homogeneous. Almost trivially,

LEMMA 1. *A theory is almost homogeneous iff it is strongly almost homogeneous.*

1980 *Mathematics Subject Classification*. Primary 03D45; Secondary 03C15.

[1] Research for this paper was paper was partially supported by NSF Grant 144-R-976.

PROOF. → is immediate. The other direction follows by an absoluteness argument; or if you prefer classical model theory, fix a model $\mathscr{A}$ that is not almost homogeneous. For each $\bar{b} \in A^{<\omega}$ fix $\bar{a}_b^1, \bar{a}_{\bar{b}}^2, \bar{c}_{\bar{b}} \in A^{<\omega}$ satisfying:

(1) $ty(\bar{b}\,\hat{}\,\bar{a}_b^1, \mathscr{A}) = ty(\bar{b}\,\hat{}\,\bar{a}_b^2, \mathscr{A})$ and

(2) $\forall \bar{d} \in A^{<\omega}$

$$ty\left(\bar{b}\,\hat{}\,\bar{a}_b^1\,\hat{}\,\bar{c}_b, \mathscr{A}\right) \neq ty\left(\bar{b}\,\hat{}\,\bar{a}_b^2\,\hat{}\,\bar{d}, \mathscr{A}\right).$$

This is possible, since $\mathscr{A}$ is not almost homogeneous. Let $\mathscr{B}_0$ be any countable elementary substructure of $\mathscr{A}$. Choose $\mathscr{B}_{n+1}$ satisfying:

(1) $\mathscr{B}_{n+1} < \mathscr{A}$,

(2) $B_n \cup \{\bar{a}_b^i, \bar{c}_b | i = 1, 2; \bar{b} \in B_n\} \subset B_{n+1}$, and

(3) B_{n+1} is countable.

In the usual way, $\mathscr{B} =_{\mathrm{df}} \bigcup_{n<\omega} \mathscr{B}_n$ is a countable model of the same theory which is not almost homogeneous. For suppose $\bar{b} \in \mathscr{B}$. Then fix n such that $\bar{b} \in B_n$. Since $\mathscr{B}_n \prec \mathscr{B} \prec \mathscr{A}$,

$$ty\left(\bar{b}\,\hat{}\,\bar{a}_b^1, \mathscr{B}\right) = ty\left(\bar{b}\,\hat{}\,\bar{a}_b^2, \mathscr{B}\right).$$

Similarly, by the above choices, for all $\bar{d} \in B^{<\omega}$,

$$ty\left(\bar{b}\,\hat{}\,\bar{a}_b^1\,\hat{}\,\bar{c}_b, \mathscr{B}\right) \neq ty\left(\bar{b}\,\hat{}\,\bar{a}_b^2\,\hat{}\,\bar{d}, \mathscr{B}\right).$$

The original Ehrenfeucht theories were:

(1) unstable,

(2) almost homogeneous,

(3) persistently Ehrenfeucht,

(4) persistently decidable, and

(5) strongly decidable.

It is easy to see that if a theory is unstable, then it is persistently unstable. The same remark applies to almost homogeneous theories. Of course, property (5) implies property (4), since any complete type realized in a decidable model must be recursive, and every type is realized in some countable model. Similarly note that any strongly decidable theory is persistently strongly decidable.

A number of results in the literature concern the necessity of these properties for Ehrenfeucht theories. Lachlan [2] showed that no Ehrenfeucht theory is superstable; but whether or not there is a stable Ehrenfeucht theory is open. Also open is whether or not every Ehrenfeucht theory is almost homogeneous. If every Ehrenfeucht theory is almost homogeneous, then they have very low Scott rank. Woodrow [3] and Millar [4] produced examples of Ehrenfeucht theories that were not persistently Ehrenfeucht.

Nerode first asked if every decidable Ehrenfeucht theory was strongly decidable. Morely produced the first counterexample by constructing a theory that was not persistently decidable. Other examples were constructed, but each failed to be persistently decidable, and all of them were strongly 0′-decidable. The nicest one was constructed by Peretyat′kin [5] using a new ω-categorical underlying theory. His example also had the least possible number of countable models for such an example, i.e. three. This theory is the model completion of the theory of a binary branching tree order with a greatest lower bound operator. The recursion-theoretic object coded into the type structure of the theory in order to produce only nonrecursive nonprincipal types is an r.e. tree with exactly one infinite branch, such that that branch is not recursive. The coding is accomplished by, say, adding infinitely many constant symbols to the language, one for each node of the tree, and then declaring in the theory that the constants are ordered just as the corresponding nodes in the tree are ordered. The nonrecursive complete 1-type is then the one which describes an element larger than all constants on the infinite branch. The three countable models obtained are similar to those of the original Ehrenfeucht theory.

These circumstances suggested two more questions. First, is every decidable Ehrenfeucht strongly $0'$-decidable? If not, is every such theory at least strongly arithmetic? (It is easy to show that they must all be strongly hyperarithmetic.) Millar [**6**] provided a negative answer with

THEOREM 1. *For every $n < \omega$ there is a decidable, persistently Ehrenfeucht theory with countable models decidable in exactly $H(n)$. Here $H(0) =_{\mathrm{df}} 0$, $H(2^x) =_{\mathrm{df}} H(x)'$ and $H(3 \cdot 5^e) =_{\mathrm{df}} \{\langle x, y\rangle | x \in H(\mu_e(y))\}$.*

What are the difficulties in producing such examples First, the Cantor Bendixson (CB) rank of any nonarithmetic complete type must be high. For example,

LEMMA 2. *If $\Gamma(x)$ is a complete type of a decidable complete theory and $\mathrm{CB}(\Gamma) < \omega$, then $\Gamma(x)$ is arithmetic.*

PROOF. Let $\{\theta_i(x) | i < \omega\}$ be an effective enumeration of the formulas in 'x' of the language of the theory T. Define

$$D_0 =_{\mathrm{df}} \{\varphi(x) | T \vdash \exists x\, \varphi(x)\}$$

and

$$D_{n+1} =_{\mathrm{df}} \Big\{\varphi(x) \big| \forall r' \exists r > r' \Big[\forall s < s' < \mathrm{lh}(r) \big(T \vdash \forall x \neg\big(\theta_{r_{(s)}}(x) \wedge \theta_{r_{(s')}}(x)\big);$$
$$T \vdash \forall x \big(\theta_{r_{(s)}}(x) \rightarrow \varphi(x)\big); \text{ and } \theta_{r_{(s)}}(x) \in D_n\big)\Big]\Big\},$$

where $r_{(s)} =_{\mathrm{df}} \alpha_s$ if $r = \prod_{i<m} P_i^{\alpha_i}$, P_i the ith prime. By induction, since T is decidable, it is easy to see that each D_n is arithmetic. Suppose $\mathrm{CB}(\Gamma) = n$. Then for the $f \in 2^\omega$ satisfying ($\theta_i \in \Gamma$ iff $f(i) = 0$), there must be an s such that for any formula $\varphi(x)$, if

$$\bigwedge_{i<s} \theta_i(x)^{f(i)} \wedge \varphi(x) \in D_n \big[\theta^0 =_{\mathrm{df}} \theta \text{ and } \theta^1 =_{\mathrm{df}} \neg\theta\big],$$

then

$$\bigwedge_{i<s} \theta_i(x)^{f(i)} \wedge \varphi(x) \in \Gamma(x).$$

From this it is easy to see that

$$\Gamma(x) = \left\{\theta_i(x)^{f(i)} \mid i < s\right\} \cup \left\{\theta_j(x) \mid \bigwedge_{i<s} \theta_i(x)^{f(i)} \wedge \theta_j(x) \in D_n\right\}.$$

Thus Γ is arithmetic.

This implies that if there is to be a nonarithmetic complete type, then the spectrum of Cantor Bendixson ranks of complete types of the theory must be infinite. Therefore, for such a theory to be Ehrenfeucht, the realizations of these various types must be intimately connected. But not too intimately! For instance,

LEMMA 3. *Let Γ, Σ be complete 1-types of a complete theory T, and suppose that there is a function f, definable in the theory, such that*

$$\forall \mathscr{A} \models T \forall a \in A[\mathscr{A} \models \Gamma(a) \rightarrow \mathscr{A} \models \Sigma(f(a))].$$

Then $\mathrm{CB}(\Gamma) \geqslant \mathrm{CB}(\Sigma)$.

PROOF. If $\mathrm{CB}(\Sigma) = 0$ then the conclusion is immediate. Suppose $\mathrm{CB}(\Sigma) = \alpha + 1$. Let $\{\sigma_n | n < \omega\}$, $\{\gamma_n | n < \omega\}$ enumerate Σ, Γ, respectively. For each $n < \omega$, fix a complete type Σ_n satisfying:

(1) $\sigma_i \in \Sigma_n$, $i < n$,
(2) $\mathrm{CB}(\Sigma_n) - \alpha$, and
(3) $\exists x(y = f(x) \wedge \bigwedge_{i \geqslant n} \gamma_i(x)) \in \Sigma_n(y)$.

Then for each $n < \omega$, fix Γ_n such that:

(1) $\gamma_i \in \Gamma_n$, $i < n$, and
(2) the image of $\Gamma_n(x)$ under f is $\Sigma_n(y)$.

By the induction hypothesis, $\mathrm{CB}(\Gamma_n) \geqslant \mathrm{CB}(\Sigma_n)$. By the choice of the Γ_n's, $\mathrm{CB}(\Gamma) \geqslant \alpha + 1$. The limit case is similar.

The final problem is that there must be a large gap in the Turing degrees of the complete types of a decidable Ehrenfeucht theory that is not persistently arithmetic. This is easy to see from the fact that, for each complete type of a theory, there is a model of the theory decidable in exactly the degree of that type.

A class of theories with the desired properties has the following description. First of all, the recursion-theoretic object that is coded in order to obtain a complex type is a recursive, ω-branching tree with exactly one infinite branch, such that that infinite branch has the same Turing degree as $H(n)$. The underlying ω-categorical theory is a dense linear order without endpoints. The tree is coded into the linear order lexicographically, using unary relation symbols. This theory alone has the desired 1-type, unfortunately the lower rank types (all of which are recursive) are completely unrelated through their realizations. Therefore a set of binary relation symbols is introduced that is used to tie realizations of lexicographically nonprincipal recursive types together. One such symbol $R_{\alpha,\beta}$ is introduced for each such pair of 1-types Γ_α, Γ_β. Their interpretations are symmetric and transitive in the subscripts α, β. There are other technical details that must be addressed, but they will be ignored here.

Each such theory has eighteen countable models. The order type of each 1-type's realizations is an interval. Of course, as always, the principal types have order type η. The recursive nonprincipal 1-type's realizations are of order 0, $1 + \eta$ or η, depending on the model. For the complex 1-type, the possible order types are 0, 1, $1 + \eta$, $\eta + 1$, $1 + \eta + 1$ and η. All of the recursive nonprincipal types have the same order to their realizations in any particular model, whereas there is no connection between the recursive types and the complex type. Each countable model is completely characterized by the order type of any one of its recursive nonprincipal type's realizations and that of its realizations of the complex type.

The second question alluded to above concerns the necessity of a nonrecursive complete type in a decidable Ehrenfeucht theory that fails to be strongly decidable. Specifically, is every persistently decidable Ehrenfeucht theory strongly decidable? This question is open. There are two avenues of possible attack. The first is to consider coding a non-r.e. set of types as the set of types realized by some finite expansion by constants of a model of the theory. This seems unlikely. The other possibility would be to code by creating a model with a nasty lack of homogeneity. To illuminate these remarks, consider

LEMMA 4. *If T is an almost homogeneous, persistently decidable, persistently Ehrenfeucht theory, then T is strongly decidable.*

PROOF. First of all, any decidable model realizes a Σ_1^0 set of recursive complete types. Next, easy arguments show that any decidable Ehrenfeucht theory has a decidable prime model, and any persistently decidable, Ehrenfeucht theory has a decidable, saturated model. The final ingredient was proved in Gonchorov [7] and Millar [**8**]:

FACT. If the set of recursive complete types of a theory is Σ_1^0, and $\mathscr{A}$ is a countable, homogeneous model of the theory realizing exactly a Σ_1^0 set of (recursive) complete types, then $\mathscr{A}$ is decidable.

Let $\mathscr{A}$ be any countable model of T. By assumption, we can fix an $\bar{a} \in A^{<\omega}$ such that $\langle \mathscr{A}, \bar{a}\rangle$ is homogeneous. Since T is persistently decidable and persistently Ehrenfeucht, the same is true of $\mathrm{Th}(\langle \mathscr{A}, \bar{a}\rangle)$. Therefore, by all of the above, it is enough to see that $\langle \mathscr{A}, \bar{a}\rangle$ realizes exactly a Σ_1^0 set of types. Let $\{\Gamma_i | i < \omega\}$ be the types realized in $\langle \mathscr{A}, \bar{a}\rangle$. For each $n < \omega$, fix $\Psi_n(\bar{x}_0, \bar{x}_1, \ldots, \bar{x}_{n-1}) \in \{\Gamma_i | i < \omega\}$ such that

$$\Gamma_0(\bar{x}_0), \ldots, \Gamma_{n-1}(\bar{x}_{n-1}) \subset \Psi_n(\bar{x}_0, \ldots, \bar{x}_{n-1}).$$

For each $n < \omega$, let $\langle \mathscr{B}_n, \bar{a}, \bar{b}_n\rangle$ be the prime model of $\Psi_n(\bar{d}_n)$. By the above, each of these is decidable. Since $\mathrm{Th}(\langle \mathscr{A}, \bar{a}\rangle)$ has only finitely many countable models, fix an $n < \omega$ such that there are infinitely many m's satisfying

$$\langle \mathscr{B}_n, \bar{a}\rangle \cong \langle \mathscr{B}_m, \bar{a}\rangle.$$

Since $\langle \mathscr{B}_n, \bar{a}, \bar{b}_n\rangle$ is prime, $\langle \mathscr{B}_n, \bar{a}\rangle$ does not realize any types not in $\{\Gamma_i | i < \omega\}$. However, by the choice of n, it is also clear that $\langle \mathscr{B}_n, \bar{a}\rangle$ realizes every type in $\{\Gamma_i | i < \omega\}$. Since $\langle \mathscr{B}_n, \bar{a}\rangle$ is decidable, it realizes a Σ_1^0 set of types and thus $\{\Gamma_i | i < \omega\}$ is Σ_1^0.

Thus, in Lemma 4 the property of being persistently Ehrenfeucht ruled out the possibility of some complete extension of the theory having a model realizing some non-r.e. set of types, and the almost homogeneity prevented any coding through nonhomogeneity.

Although the second question is open, some progress has been made. Ash and Millar **[9]** prove

THEOREM 2. *Every persistently arithmetic, persistently Ehrenfeucht theory is strongly arithmetic.*

A nice property of persistently Ehrenfeucht theories exploited in the proof of this theorem is

LEMMA 5. *If T is persistently Ehrenfeucht and $\mathscr{A}, \mathscr{B} \models T$, then*

$$\mathscr{A} \equiv_{\omega \cdot 2} \mathscr{B} \Rightarrow \{\forall a \in A\, \exists b \in B[\langle \mathscr{A}, a\rangle \equiv_{\omega \cdot 2} \langle \mathscr{B}, b\rangle]\}.$$

PROOF. The key observation is that for any model of T and for any complete type, the number of orbits of that type under automorphisms of that model is finite. This follows directly from the fact that T is persistently Ehrenfeucht. So fix $a \in A$. Let Γ be the type realized by a in $\mathscr{A}$. By the observation, there are $b_0, \dots, b_{n-1} \in B$ satisfying:

(1) b_i realizes Γ in $\mathscr{B}$, $i < n$, and

(2) $\forall b \in B\, \exists i < n$ [if b realizes Γ in $\mathscr{B}$, then $\langle \mathscr{B}, b\rangle \cong \langle \mathscr{B}, b_i\rangle$].

Since $\mathscr{A} \equiv_{\omega \cdot 2} \mathscr{B}$, we have that for every $m < \omega$ there is a $b \in B$ such that

$$\langle \mathscr{A}, a\rangle \equiv_{\omega + m} \langle \mathscr{B}, b\rangle.$$

Clearly each such b realizes Γ. Therefore, by (1) and (2) there is an $i < n$ such that, for infinitely many $m < \omega$,

$$\langle \mathscr{A}, a\rangle \equiv_{\omega + m} \langle \mathscr{B}, b_i\rangle.$$

So take $b =_{\mathrm{df}} b_i$ for such an i.

Thus, if $\mathscr{A}$ and $\mathscr{B}$ are countable models of such a theory and satisfy $\mathscr{A} \equiv_{\omega \cdot 2} \mathscr{B}$, then the usual back and forth argument shows that they are isomorphic. The next step is to expand the language and models so that different models of the original theory satisfy different first order sentences of the new language. In order to accomplish this, symbols $\{R_\Gamma(x_1, \dots, x_m) | \Gamma$ is a complete m-type of $T\}$ are introduced. Similarly,

$$\mathscr{A}^* =_{\mathrm{df}} F\langle \mathscr{A}, R_\Gamma^{\mathscr{A}}\rangle_{\Gamma \text{ complete type of } T},$$

where

$$R_\Gamma^{\mathscr{A}} =_{\mathrm{df}} \{\bar{a} | \bar{a} \text{ realizes } \Gamma \text{ in } \mathscr{A}\}.$$

The proof of the theorem then proceeds by assuming that the theory does have a nonarithmetic countable model, in order to obtain a contradiction. By the above remarks it follows that there is a first order sentence φ^* of the expanded language that is satisfied by the nonarithmetic model but not by any arithmetic model. The rest of the proof is essentially an effectivized omitting types construction used to produce an arithmetic model satisfying φ^*, which gives the desired contradiction. A few technical details must be addressed along the way, but again these will be passed over here.

The difficulty in obtaining further results centers on a dirth of techniques known for producing effective structures that are *not* isomorphic to a given collection of other structures. We end here by demonstrating one such simple technique.

THEOREM 3. *If T is a decidable Ehrenfeucht theory with a countable model that is not $0''$-decidable, then T has at least five countable models.*

PROOF. Since T is an Ehrenfeucht theory, it must have only countably many complete types. Thus every set of formulas in finitely many free variables that is consistent with T and Σ_1^0 in some degree is contained in a complete type of T that is recursive in that degree. Since T is not ω-categorical it must have a nonprincipal type; assume for notational simplicity that it has a nonprincipal 1-type. It is not difficult to see that

$$d =_{\mathrm{df}} \{\neg\theta(x) \mid \theta(x) \text{ is a complete formula of } T\}$$

is Π_1^0 and consistent with T. Let $\Gamma(x)$ be a complete 1-type of T containing D and recursive in $0'$. Since Γ contains the negations of all of the complete formulas, it must be nonprincipal. By the same argument let $\Sigma(x, y)$ be a complete type satisfying:

(1) $\Gamma(x) \subset \Sigma(x, y)$,

(2) Σ is recursive in $0''$, and

(3) $\Sigma(a, y)$ is a nonprincipal type in the theory $\Gamma(a)$.

Now the prime model of T and the reduct of the prime model of $\Gamma(a)$ are not isomorphic. If the latter is homogeneous, then it must omit the type $\Sigma(x, y)$. Then the reduct of the prime model of $\Sigma(a, b)$ gives us a third model. If this model is homogeneous, then we get yet another model that is prime over a finite expansion. But then we are done since the countable saturated model is not prime over any finite expansion and is thus a fifth model. If the reduct of the prime model of $\Sigma(a, b)$ is not homogenous, then, by the results in Millar [**10**], T also has a homogeneous model realizing $\Sigma(x, y)$ decidable in $0''$. In this case all four of the countable models mentioned are $0''$-decidable and thus the countable model of T that is not $0''$-decidable is a fifth structure.

We are thus reduced to the case where the reduct of the prime model of $\Gamma(a)$ is not homogeneous. In this case a third structure can be obtained as above, realizing $\Gamma(x)$, homogeneous, and $0'$-decidable. If this structure does not realize all the types of T, then the saturated model and the reduct of the prime model of such an omitted type again give five models. So assume that the saturated model is $0'$-decidable. The model $\mathscr{A}$ that is not $0''$-decidable accounts for four. If $\mathscr{A}$ does not realize Γ, then since it is not prime, there is a nonprincipal type Σ such that the prime model of $\Sigma(a)$ is a fifth model. Similarly if $\mathscr{A}$ does not realize all the types of T we get a fifth model. So assume $\mathscr{A}$ realizes all the types of T. By the above, $\mathscr{A}$ is not prime over any finite expansion. This suggests a strategy for producing the fifty model. We will build a model that is $0''$-decidable, not saturated, and not prime over any finite expansion. The requirement satisfied in order to avoid the saturated model will be to have one realization of a Γ not have a particular consistent extension. The other requirements will be that each finite subset of the model does have some consistent, nonprincipal extension.

We use a Henkin construction with a $0''$ oracle. Fix an infinite-to-one listing $\{\Gamma_i(\bar{x})|i < \omega\}$ of all the complete types of T. By the above, we can assume at this point that the listing is Σ_1^0 in $0'$. Let $\mathscr{A}$ be a countable model of T that is not $0''$-decidable. Since $\mathscr{A}$ is not saturated, fix a $\Gamma(x) \subset \Sigma(x, y)$ such that some realization of Γ in $\mathscr{A}$ has no Σ extension. Assume $\Gamma_0 = \Gamma$. Let $\{c_i|i < \omega\}$ be new constant symbols, and let $\{\bar{c}_i|i < \omega\}$ and $\{\theta_i|i < \omega\}$ be effective enumerations of $\{c_i|i < \omega\}^{<\omega}$ and the sentences of $L(T) \cup \{c_i|i < \omega\}$, respectively. Assume $\bar{c}_0 = \langle c_0\rangle$. At stage t of the construction a ψ_t will be defined, as well as auxiliary functions $g(x, t), f(x, t)$ for $x \leqslant t + 1$. Define

$$\chi_t =_{\mathrm{df}} \bigwedge_{i<t} \psi_t.$$

At stage t if $f(g)$ is *trivially extended*, this means that

$$f(i, t+1) =_{\mathrm{df}} \begin{cases} f(i, t) & i \leqslant t, \\ f(t, t), & i = t+1. \end{cases}$$

THE CONSTRUCTION. *Stage* $t = 0$. $\psi_0 =_{\mathrm{df}} (c_0 = c_0)$ and $g(i, j) =_{\mathrm{df}} f(i, j) =_{\mathrm{df}} 0$ for $i \leqslant j \leqslant 1$.

Stage $t = 7s + 1$. If $\psi_s = \exists x \theta$, then $\psi_t =_{\mathrm{df}} \theta(x/c_N)$ for the least N such that 'c_N' does not occur in χ_t; otherwise $\psi_t =_{\mathrm{df}} (c_0 = c_0)$. In either case f and g are trivially extended.

Stage $t = 7s + 2$. If

$$\Gamma_{f(t,t)}(\bar{c}_{g(t,t)}) \vdash (\chi_t \to \theta_s),$$

then $\psi_t =_{\mathrm{df}} \theta_s$; otherwise $\psi_t =_{\mathrm{df}} \neg\theta_s$. In either case f and g are trivially extended.

Stage $t = 7s + 3$. $\psi_t =_{\mathrm{df}} (c_0 = c_0)$. Fix the least $n < t$ such that $f(n, t) = f(n+1, t)$. Find the least m such that:

(1) $\Gamma_{f(n,t)}(\bar{c}_{g(n,t)}) \subset \Gamma_m(\bar{c}_{g(n,t)} \wedge \bar{x})$, and

(2) $f(n+1, t') < m$ for all $t' < t$ for which $f(n+1, t')$ is defined.

Then

$$f(i, t+1) =_{\mathrm{df}} \begin{cases} f(i, t), & i \leqslant n, \\ m, & n < i \leqslant t+1. \end{cases}$$

Fix the least k such that:

(1) $\mathrm{lh}(\bar{c}_k) = \mathrm{lh}(\bar{c}_{g(n,t)} \wedge \bar{x})$,

(2) $\bar{c}_k = \bar{c}_{g(n,t)} \wedge \bar{c}_r$, and

(3) nothing in $rg(\bar{c}_r)$ occurs in χ_t.

Then $g(i, t+1)$ is defined in the manner that f was, but with 'k' in for 'm'.

Stage $t = 7s + 4$. $\psi_t =_{\mathrm{df}} (c_0 = c_0)$. Determine the least $n < t$ such that

$$\forall \varphi \in \Gamma_{f(n+1,t)}(\bar{c}_{g(n+1,t)})\left[\Gamma_{f(n,t)}(\bar{c}_{g(n,t)}) \vdash (\chi_t \to \varphi)\right],$$

and $f(n+1, t') \neq f(n+1, t)$ for any $t' = 7s' + 7$, $t' < t$. Then

$$f(i, t+1) =_{\mathrm{df}} \begin{cases} f(i,t), & i \leqslant n, \\ f(n,t), & n < i \leqslant t+1, \end{cases}$$

and similarly for g.

Stage $t = 7s + 5$. $\psi_t =_{\mathrm{df}} (c_0 = c_0)$. Determine the greatest $n < t$ for which

$$\forall i < t \exists \sigma(c_0, c_i) \in \Sigma(c_0, c_i)\left[\Gamma_{f(n,t)}(\bar{c}_{g(n,t)} \nvdash \chi_t \to \sigma)\right].$$

Now proceed as in the previous case for that n.

Stage $t = 7s + 6$. Fix the least $i < t$ such that

$$\Gamma_{f(t,t)}(c_{g(t,t)}) \cup \{\chi_t\} \cup \Sigma(c_0, c_i)$$

is consistent, if such an i exists; otherwise $\psi_t =_{\mathrm{df}} (c_0 = c_0)$. By the previous stage's instructions, there is a

$$\theta_k(c_0, c_i) \in \Sigma(c_0, c_i)$$

such that

$$\Gamma_{f(t,t)}(c_{g(t,t)}) \cup \{\chi_t, \neg\theta_k(c_0, c_i)\}$$

is consistent. For the least such k, let $\psi_t =_{\mathrm{df}} \neg\theta_k(c_0, c_i)$. In either case, trivially extend f and g.

Stage $t = 7s + 7$. Fix the least n such that $f(n, t) = f(n+1, t)$. If there is no m, $2m < n$, such that 'c_m' does not occur in $\bar{c}_{g(t,t)}$, then trivially extend f and g and let $\psi_t =_{\mathrm{df}} (c_0 = c_0)$. Otherwise fix the least such m. Since the theory $\Gamma_{f(t,t)}(\bar{c}_{g(t,t)})$ has only countably many 1-types, there is a complete formula $\theta_k(\bar{c}_{g(t,t)} \wedge \langle x \rangle)$ of this theory that is consistent with χ_t. Fix the least such k (note that this can be done effectively in the jump of the degree of $\Gamma_{f(t,t)}$) and then fix the least r such that

$$\{\theta_k(\bar{c}_{g(t,t)} \wedge \langle x \rangle)\} \cup \Gamma_{f(t,t)}(\bar{c}_{g(t,t)}) \subset \Gamma_r(\bar{c}_{g(t,t)} \wedge \langle x \rangle).$$

Define

$$f(i, t+1) =_{\mathrm{df}} \begin{cases} f(i,t), & i \leqslant n, \\ r, & n < i \leqslant t+1. \end{cases}$$

Fix the least v such that $\bar{c}_v = \bar{c}_{g(t,t)} \wedge \langle c_m \rangle$ and define g in the same way as f, except for 'v' in for 'r'. This ends the description of the construction.

It is easy to check that the construction is effective in $0''$. It is also straightforward to check that an elementary diagram for a model of T is the end product, i.e.

$\{\psi_t | t < \omega\}$. Next we prove that

$$f^*(n) =_{\mathrm{df}} \lim f(n, t) \quad \text{and} \quad g^*(n) =_{\mathrm{df}} \lim g(n, t)$$

exist. Fix an element $a_0 \in A$ that realizes Γ and that has no $\Sigma(a, x)$ extension. By the construction, $f(0, t) = f(0, 0)$ for all t, and similarly for g. So inductively assume that $f^*(i)$ and $g^*(i)$ exist for all $i \leqslant n$, and that there is an $\bar{a} \in A$ such that $\langle a_0 \rangle \wedge \bar{a}$ realizes $\Gamma_{f^*(n)}$ in $\mathscr{A}$. Fix a t_0 such that $f(i, t) = f^*(i)$ for all $i \leqslant n$ and for all $t > t_0$. The first claim is that if $f(n + 1, t)$ is ever defined to satisfy the construction at a stage $t = 7s + 7 > t_0$, then its value never subsequently changes. This is by the induction hypothesis and the observation that at such a stage the extension taken over $\Gamma_{f^*(n)}$ is principal and therefore is realized in $\mathscr{A}$ by some $\langle a_0 \rangle \wedge \bar{a} \wedge \langle a' \rangle$. Thus $f(n + 1, t')$ can never be redefined to satisfy a stage $t' = 7s' + 5$ for $t' > t$. And, of course, a change at a stage $7s' + 4$ is ruled out by the choice of t, and so the claim is proved from this case.

The only other possibility is that infinitely often $f(n + 1, t)$ is redefined for the sake of stages $7s' + 3$ and redefined (injured) infinitely often at stages $7s' + 4$ or $7s' + 5$. However, $\mathscr{A}$ is not prime over $\langle a_0 \rangle \wedge \bar{a}$, and therefore there is a $\bar{a}'$ such that $\langle a_0 \rangle \wedge \bar{a} \wedge \bar{a}'$ realizes a type Γ_k and $\Gamma_k(\langle a_0 \rangle \wedge \bar{a} \wedge \bar{x})$ is not a principal type in the theory $\Gamma_{f'(n)}(\langle a_0 \rangle \wedge \bar{a})$. So in the above circumstances, eventually $f(n + 1, t)$ is defined to be such a k for some $t > t_0$. But then $f(n + 1, t')$ is never redefined for the sake of some stage $t' = 7s' + 4$, $t' > t$. Also, it is easy to see that since $\langle a_0 \rangle \wedge \bar{a} \wedge \bar{a}'$ realizes Γ_k in $\mathscr{A}$ and yet there is no a_i such that $\langle a_0, a_i \rangle$ realizes Γ, then $f(n + 1, t')$ is never redefined for a $t' = 7s' + 5$. Thus the limits exist.

By the construction, for each n there is an $m < 2n$ such that $\Gamma_{f^*(m)}(\bar{c}_{g^*(n)} \wedge \langle \bar{x} \rangle)$ is a nonprincipal type in the theory $\Gamma_{f^*(n)}(\bar{c}_{g^*(n)})$. Since $\bar{c}_{g^*(m)}$ realizes this type in the constructed model and $\bar{c}_{g^*(m)}$ extends $c_{g^*(n)}$ it follows that the constructed model is not prime over any $\bar{c}_{g^*(n)}$. However, by the stages $7s + 7$, each c_i is eventually included in some $\bar{c}_{g^*(n)}$, and therefore the model is not prime over any finite subset. Finally, by the stages $7s + 6$, there is no c_i such that $\langle c_0, c_i \rangle$ realizes Γ. Therefore we have a fifth model.

BIBLIOGRAPHY

1. R. Vaught, *Denumerable models of complete theories*, Infinistic Methods, Pergamon Press, Oxford, New York, 1959, pp. 303–321.

2. A. H. Lachlan, *On the number of countable models of a countable superstable theory*, Logic, Methodology, and Philosophy of Sciences. III, North-Holland, 1973, pp. 45–56.

3. R. E. Woodrow, *Theories with finite number of countable models*, J. Symbolic Logic **43** (1978), 442–455.

4. T. S. Millar, *Stability, complete extensions and the number of countable models*, Aspects of Effective Algebra, Upside Down A Book Co., 1981, pp. 196–205.

5. M. G. Peretyat'kin, *Complete theories with a finite number of countable models*, Algebra i Logika **12** (1973), 550–576.

6. T. S. Millar, *Persistently finite theories with hyperarithmetic models*, Trans. Amer. Math. Soc. **278** (1983), 91–99.

7. S. S. Goncharov, *Strong constructivizability of homogeneous models*, Algebra i Logika **17** (1978), 363–389.

8. T. S. Millar, *Type structure complexity and decidability*, Trans. Amer. Math. Soc. **271** (1981), 73–81.

9. C. Ash and T. Millar, *Persistently finite, persistently arithmetic theories*, Proc. Amer. Math. Soc., **89** (1983), 487–492.

10. T. S. Millar, *Foundations of recursive model theory*, Ann. Math. Logic **13** (1978), 45–72.

DEPARTMENT OF MATHEMATICS, UNIVERSITY OF WISCONSIN, MADISON, WISCONSIN 53706

Proceedings of Symposia in Pure Mathematics
Volume **42**, 1985

A Survey of Lattices of R. E. Substructures

A. NERODE[1] AND J. REMMEL[2]

1. Introduction. Some constructions in mathematics are effective, some are not. Recursion theory was invented to make precise what it means to be effective. Its principal tool, the priority method, can be used to resolve questions of effectiveness. Metakides and Nerode [**1975**, **1977**, **1979**] adapted the priority method to algebra in order to determine the effective contents of algebraic constructions and theorems. The latter work has been extended and deepened by many authors. A bibliography to 1980 is to be found in Crossley [**1981**]. The latter volume is representative of the work in the area. The introduction of noneffective methods in mathematics is discussed from an historical standpoint in Metakides and Nerode [**1982**]. Such methods also apply to the rest of mathematics, and there are results in analysis and topology as well.

A significant by-product of these investigations is the introduction as an object for study of the lattice of all recursively enumerable substructures of a fixed recursive structure. (For this purpose a recursive structure is one with a recursive domain, equality, operations, and atomic relations.) Classical recursion theory (Rogers [**1967**]) dealt only with the lattice $\mathscr{E}$ of recursively enumerable (hereafter r.e.) subsets of the integers. This can be viewed as the lattice of substructures of the system $(N, =)$ of the integers N with $=$ as the only relation. Even ordinal and higher recursion theory and admissible sets as a whole deal only with substructures of equality structures.

In algebra, the subjects of vector spaces, fields, orderings, Boolean algebras, lattices of open sets, etc., are distinct subjects. There is no deep common theory.

1980 *Mathematics Subject Classification*. Primary 03D25, 03D45; Secondary 03D50, 06A99, 06B99, 06E99.

[1] Supported in part by NSF Grant MCS-8301850.

[2] Supported in part by NSF Grant MCS-8202333.

Depth is achieved at the cost of developing separately analogous notions in each area, notions which take a somewhat different form in each. The situation is similar for the lattices of r.e. substructures of a fixed recursive vector space, field, ordering, Boolean algebra, lattice of open sets. A natural notion in $\mathscr{E}$ may have many variants in one of these lattices, only one of which gives a good theory.

One justification for the study of such lattices is that usually the question of effective content of a theorem or construction can be answered once a suitable structural property of the appropriate lattice of r.e. substructures is known. As an example, consider the question of the construction of a transcendence base for a field from the domain of the field and its operations and equality predicate. In Metakides and Nerode [**1979**] a recursive algebraically closed field F of infinite transcendence degree is constructed such that the only r.e. algebraically closed subfield of infinite transcendence degree is the whole F. If one could obtain an infinite r.e. algebraically independent set T in F, taking the algebraically closed subfield generated by the odd members in an effective enumeration of T would contradict the property above. So even though F has infinite independent sets, none are r.e. The lattice of r.e. algebraically closed subfields of F is very narrow. It consists only of F and its finite-dimensional algebraically closed subfields. How is such an F constructed? If P is a recursive polynomial domain over the prime field, P has an infinite recursive transcendence base and we can construct a suitable maximal element I of its lattice $\mathscr{L}I(P)$ of r.e. ideals by the priority method and algebraic geometry so that $P/I = F$. The required properties of I are too long to enumerate here, and the proof is too long to sketch. Instead, we consider the corresponding theorem for vector spaces. There is a recursive infinite-dimensional vector space V over any given recursive field F such that the only r.e. infinite-dimensional subspace of V is V_∞. How is this proven? Let V_∞ be an infinite-dimensional recursive vector space with recursive basis over F. Construct a maximal element W of the lattice $\mathscr{L}(V_\infty)$ of r.e. subspaces of V_∞ such that W is a recursive set, $V_\infty \bmod W$ is infinite dimensional, and the only r.e. superspace V' of W (with $V' \bmod W$ infinite dimensional) is $V' = V_\infty$. Then the required $V = V_\infty \bmod W$. This is a slight improvement on the Kalantari-Retzlaff construction of a supermaximal space [**1977**]. For a much stronger result see Theorem 45.

The mixture of algebra and recursion theory precipitates out new algebra questions, new recursion theory questions, and new algorithms in constructive algebra. A new algebra theorem is Remmel's construction of complements in lattices of subalgebras of Boolean algebras [**1980c**]. A new recursion theoretic notion is that of a universal splitting property, arising out of r.e. vector space theory (Lerman and Remmel [**1982**]). An example of a new algebra algorithm is Rosenthal [**1985**]. In Nerode and Remmel [**1983**] the question was raised of splitting elements of $\mathscr{L}(F_\infty)$ (the lattice of r.e. algebraically closed subfields of a recursive algebraically closed field F_∞ with infinite recursive transcendence basis) as direct sums of strongly nowhere simple elements. The only obstacle was the lack of a known algorithm to go from explicit finite bases of finitely generated elements of $\mathscr{L}(F_\infty)$ to an explicit basis for their intersection. John Rosenthal supplies such an algorithm (see discussion following Theorem 92), so such splittings exist. We survey lattices of r.e. substructures by examining how familiar (Post [**1944**]) notions of classical recursion theory apply. We look at recursive elements in §2; creative elements in §3; simple, r-maximal, and maximal elements in §4; splitting theorems, degrees of bases, and nowhere simple elements in §5; undecidability questions in §6. We apply them to familiar structures—vector spaces, algebraically closed fields, Boolean algebras, linear orders, open sets in Euclidean space R^n. We use the following notation for the six lattices discussed.

(1) The lattice $\mathscr{L}(V_\infty)$ of r.e. subspaces of V_∞. Here V_∞ is an infinite-dimensional recursive vector space over a recursive field which has a recursive basis. Having a recursive basis is equivalent to having an r.e. basis, or to the existence of a dependence algorithm which determines whether or not an n-tuple of vectors is dependent. The study of this lattice was introduced by Metakides and Nerode [**1977**].

(2) The lattice $\mathscr{L}(F_\infty)$ of r.e. algebraically closed subfields of F_∞. Here F_∞ is an infinite-dimensional algebraically closed field with recursive transcendence base; the same remarks about r.e. bases and dependence algorithms as above apply here, with linear dependence replaced by algebraic dependence. The study of this lattice was initiated by Metakides and Nerode [**1980**].

(3) The lattice $\mathscr{L}(B)$ of r.e. subalgebras of B, where B is a recursive Boolean algebra. These have been studied by Remmel [**1978a**, **1979**, **198 b**].

(4) The lattice of $\mathscr{L}(Q)$ of r.e. suborderings of the order of Q. Here Q is the rationals with their usual order. The study of $\mathscr{L}(Q)$ begins in Metakides and Nerode [**1975**] and continues in Metakides and Remmel [**1979**] and Remmel [**1980a**].

(5) The lattice $\mathscr{L}I(\tilde{Q})$ of r.e. ideals in an infinite recursive free Boolean algebra $\tilde{Q}$. Here the notation $\tilde{Q}$ is for the particular Boolean algebra consisting of finite unions of left closed right open intervals in the rationals Q. The study of this lattice is identical with the study of r.e. theories in countable propositional logic. There are old theorems by Martin and Pour-El [**1970**], and others.

(6) The lattice $\mathscr{L}(R^n)$ of r.e. open sets in Euclidean n-space. The study of this lattice was initiated by Kalantari and Retzlaff [**1979**]; many isolated results about it exist in the 1950–1970 recursive analysis and topology literature, especially of the Leningrad School. We take as given an enumeration $\{\mathscr{E}_i\}$ of rational boxes $X_{i=1}^n(a_{ij}, b_{ij})$, where $\varepsilon_0 = R^n$, $\varepsilon_1 = \varnothing$; and an open set X is r.e. open if there is an r.e. sequence of ε_i with union X. The enumeration $\{\varepsilon_i\}$ is assumed to have algorithms for determining whether or not $\varepsilon_{i_0} \subseteq \varepsilon_{i_1} \vee \cdots \vee \varepsilon_{i_n}$ or $\overline{\varepsilon_{i_0} \subseteq \varepsilon_{i_1} \vee \cdots \vee \varepsilon_{i_n}}$. Closely related is the lattice of r.e. full subsets of integers. These are the sets K such that K is r.e. and $\varepsilon_{i_0} \subseteq \varepsilon_{i_1} \vee \cdots \vee \varepsilon_{i_{1_n}}$ and $\varepsilon_{i_1}, \ldots, \varepsilon_{i_n} \in K$ imply $\varepsilon_{i_0} \in K$. A moment's reflection shows every r.e. open set is of the form $\bigcup\{\varepsilon_i | i \in K\}$ with K full. Each r.e. set K which is full yields an r.e. open set

$\mathscr{K} = \bigcup\{\varepsilon_i | i \in K\}$. Each r.e. set of integers W_i induces a generated full set K_i and hence an r.e. open set $\mathscr{K}_i$, also denoted as $\mathscr{W}_i$.

It would be instructive to give as a seventh example lattices of r.e. ideals in polynomial rings. This would require a book-length treatment, and will have to be dealt with separately.

The last section (§7) introduces three generalizations. The first is a generalization to recursive models where the algebraic closure of a set is that set itself. This is due to Metakides and Remmel [**1979**] and covers $\mathscr{E}$ and $\mathscr{L}(Q)$. The second is a generalization to Steinitz closure systems (transitive dependence relations, matroids). This is due to Metakides and Nerode [**1980**] and Nerode and Remmel [**1982, 1983**] and covers $\mathscr{L}(V_\infty)$ and $\mathscr{L}(F_\infty)$. This area has been further developed by Baldwin [**1982**] and Downey [**1982**]. The third is a generalization to recursive closure operations with a suitable independent sequence. This is due to Remmel [**1980b**] and covers all examples above. No global exchange principle is present in this last generalization.

Model theory is a natural way to get generalizations of results in algebra. Similarly, recursive model theory is a natural way to get results in recursive algebra, including results on effective content of mathematical constructions. This has been done by Ash and Nerode and by Smith in Crossley [**1981**], and by the Ershov School. The study of lattices of r.e. substructures for a general recursive model is as yet undeveloped.

There are other approaches to effective content. One is the Reverse Mathematics of [**1983**]. No cousins of lattices of r.e. substructures have thus far emerged from reverse mathematics. There is no reason that they should not.

We assume that in every case the domain of our structure is the integers. If S is a subset of a structure, then $(S)^*$ is the substructure generated by S. Let $U_0, U_1, \ldots$ be an effective list of all r.e. subsets of the natural numbers. It is easy to see that for any of our structures if S is r.e., then $(S)^*$ is r.e. and, moreover, we can uniformly find an index for $(S)^*$ from an index of S. Thus if $W_i = (U_i)^*$, then $W_0, W_1, \ldots$ is an effective list of all r.e. substructures. Let W_i^s denote the finite set of elements enumerated in W_i^s after s steps of the effective procedure to list W_i. Let $D_i = \{x_0, \ldots, x_k\}$, where $i = 2^{x_0} + 2^{x_1} + \cdots + 2^{x_k}$ and let $D_0 = \varnothing$. For any set S, $\deg(S)$ will always denote the Turing degree of S. $\mathbf{0}$ and $\mathbf{0}'$ will denote the Turing degrees of the recursive sets and the jump of the recursive sets, respectively. In general, given a Turing degree a, a' will denote the jump of a. We let $N^{<\omega}$ denote the set of all finite sequences from N and let $\langle, \ldots, \rangle$ denote some fixed effective pairing function from $N^{<\omega}$ onto N. Given a recursive set B we will say that $M \subseteq B$ is a simple (maximal, etc.) subset of B if there is a recursive bijection $f\colon N \to B$ and a simple (maximal, etc.) subset M' of N such that $f(M') = M$.

We note that despite the length of this paper, we have not attempted to give an exhaustive survey of the subject. Instead, we concentrate on the six structures listed above because of their familiarity to the working mathematician and because they are rich enough to demonstrate the varied kinds of results and new phenomena that have arisen in the study of the lattice of r.e. substructures. Indeed many of the results we state for our specific lattices have been proved in more general settings. Finally, we have for the most part ignored the applications of recursion theory to the study of the effective content of theorems in mathematics. Surveys of the effective content of theorems in algebra, analysis, combinatorics, and model theory are needed and are planned for the future.

2. Recursive elements. For any of our six lattices, we say an element W is *recursive* if it is recursive as a set, i.e., $\deg(W) = \mathbf{0}$. The fundamental fact about recursive sets from the point of view of $\mathscr{E}$ is that the recursive sets are the complemented elements of $\mathscr{E}$, i.e. a set R is recursive if and only if both R and $N - R$ are r.e. In our six lattices, the notions of being complemented and recursive are not always identical.

First consider the lattice $\mathscr{L}(V_\infty)$. Metakides and Nerode [**1977**] introduced the *dependence degree* of $V \in \mathscr{L}(V_\infty)$ as the degree of $D(V) = \{\langle v_0, \ldots, v_n\rangle \mid v_0, \ldots, v_n$ is dependent mod $V\}$. We say that V is *decidable* if $D(V)$ is a recursive set. It is not difficult to prove the following

PROPOSITION 1 (METAKIDES AND NERODE [**1977**]). *Let V be an r.e. subspace of V_∞. Then the following are equivalent.*

(i) *V is decidable.*

(ii) *There exists an r.e. $W \in \mathscr{L}(V_\infty)$ such that $V + W = V_\infty$ and $V \cap W = (\varnothing)^*$.*

(iii) *V is generated by a recursive subset of a recursive basis for V_∞.*

Thus the decidable elements are the complemented elements of $\mathscr{L}(V_\infty)$. There are several notions which arise naturally out of Proposition 1, which we should mention at this point. First note that using the dependence algorithm for V_∞, one can easily see that an r.e. subspace V of V_∞ always has an r.e. basis, and that any r.e. basis of V_∞ must automatically be recursive.

Call an independent subset B of V_∞ *extendible* if there exists an infinite r.e. set $I \supseteq B$ with $I - B$ infinite. We say that B is *fully extendible* if B is contained in a recursive basis for V_∞. We can also refine the dependence degree of V by defining the *kth dependence degree* of V to be the Turing degree of $D_k(V) = \{\langle v_0, \ldots, v_{k-1}\rangle \mid v_0, \ldots, v_{k-1}$ is dependent mod $V\}$. Note that the degree of V as a set is just the first dependence degree of V. Now if the underlying field F of V_∞ is finite, then to decide if $\{v_0, \ldots, v_{n-1}\}$ is dependent mod V, we need only ask if the finite set $(\{v_0, \ldots, v_{n-1}\})^*$ intersects $V - \{\vec{0}\}$. So if F is finite, V is recursive if and only if V is decidable, and for any subspace V of V_∞, $V \equiv_T D_k(V)$ for all k. However, if the underlying field F is infinite, the situation is quite different as shown by the following theorems.

THEOREM 2 (METAKIDES AND NERODE [**1977**]). *If the underlying field F of V is infinite, then there exists a recursive $V \in \mathscr{L}(V_\infty)$ such that V_∞/V is infinite dimensional and yet no basis of V is extendible.*

COROLLARY 3 (METAKIDES AND NERODE [**1977**]). *If the underlying field F of* V_∞ *is infinite, then there exists a recursive* $V \in \mathscr{L}(V_\infty)$ *such that V is not decidable.*

THEOREM 4 (SHORE [**1978a**]). *Suppose the underlying field F of* V_∞ *is infinite and* $A_0, A_1, A_2, \ldots$ *is an effective sequence of r.e. sets such that* $A_i \leqslant_T A_{i+1}$ *for all* $i \geqslant 1$ *and, for all j,* $A_j \leqslant_T A_0$ *uniformly. Then there exists a* $V \in \mathscr{L}(V_\infty)$ *such that* $D(V) \equiv_T A_0$ *and* $D_i(V) \equiv_T A_i$.

Note that Theorem 4 says that the sequence of dependence degrees of a subspace V is not much constrained. (It is easy to see that for all $i \geqslant 1$, $D_i(V) \leqslant_T D(V)$ uniformly.)

How is $D_i(V)$ related to $D_{i+1}(V)$? For any $i \geqslant 1$, let $\{a_0,\ldots,a_i\}$ be any set which is independent mod V. Then we claim that $\langle v_0,\ldots,v_{i-1}\rangle \in D_i(V)$ iff for all $j = 0,\ldots,1$, $\langle v_0,\ldots,v_{i-1}, a_j\rangle \in D_{i+1}(V)$. Clearly if $v_0,\ldots,v_{i-1}$ are dependent mod V, then each sequence $v_0,\ldots,v_{i-1}, a_j$ is dependent mod V. If for all j, $v_0,\ldots,v_{i-1}, a_j$ is dependent mod V, then $v_0,\ldots,v_{i-1}$ being independent mod V would imply that, modulo V the $(i+1)$-dimensional space $(\{a_0,\ldots,a_i\})^*$ is contained in the i-dimensional space $(\{v_0,\ldots,v_{i-1}\})^*$ which is impossible. This shows that $D_i(V)$ is bounded truth-table reducible to $D_{i+1}(V)$, $D_i(V) \leqslant_{\text{btt}} D_{i+1}(V)$ for all $i \geqslant 1$. This is the best possible as the following theorem shows.

THEOREM 5 (NERODE AND REMMEL [**1983**]). *There exists a* $V \in \mathscr{L}(V_\infty)$ *such that for all* $i, j \geqslant 1$ *with* $i \neq j$, $D_i(V)$ *is many-one incomparable to* $D_j(V)$. *(Here the underlying field may be finite or infinite.)*

There are no obvious analogues of Theorems 2, 4 and 5 in $\mathscr{E}$ or $\mathscr{E}^*$ ($\mathscr{E}$ mod finite sets). The phenomena of having an r.e. basis which is not extendible or fully extendible are new in $\mathscr{L}(V_\infty)$. We shall see later on (when we discuss nowhere simple and supermaximal spaces) that there are spaces $V_1 \in \mathscr{L}(V_\infty)$ such that every r.e. basis of V is extendible but no basis of V_1 is fully extendible; and that there are spaces $V_2 \in \mathscr{L}(V_\infty)$ so that no basis of V_2 is extendible (over any recursive field). Thus the notions of fully extendible and extendible are different.

In the lattice $\mathscr{L}(F_\infty)$, we have concepts and results analogous to Theorems 2, 4 and 5.

In $\mathscr{L}(F_\infty)$, we replace the notion of linear dependence by algebraic dependence, the notion of vector space basis by transcendence basis, and the notion of V being a complement of W by the conditions that $W + V = F_\infty$ and W is independent over V (W is independent over V if whenever B and C are independent sets contained in V and W, respectively, then $B \cup C$ is an independent set). The analogues of Proposition 1, Theorem 2, and Theorem 4i n $\mathscr{L}(F_\infty)$ are due to Metakides and Nerode [**1980**]. The analogue of Theorem 5 is due to Nerode and Remmel [**1983**].

Now consider recursive Boolean algebras. Here we shall see that the question of whether the complemented elements of $\mathscr{L}(\mathscr{B})$ are identical with the recursive subalgebras leads to some new results in pure algebra. The problem is that there is more than one concept of complement.

Fix a recursive Boolean algebra $\mathscr{B}$ and a subalgebra B of $\mathscr{B}$. We say a subalgebra C of $\mathscr{B}$ is a *pseudo-complement* of B if $B \cap C = \{0_{\mathscr{B}}, 1_{\mathscr{B}}\}$ and for any $x \notin C$, $\mathscr{B} \cap (C \cup \{x\})^* = \{0_{\mathscr{B}}, 1_{\mathscr{B}}\}$. This says C is a maximal element of the class of subalgebras of $\mathscr{B}$ which intersect B trivially. Note that if B is r.e. and B has an r.e. pseudo-complement C, then C is recursive. That is, $\mathscr{B} - C = \{x \in \mathscr{B} | (C \cup \{x\})^* \cap B \neq \{0_{\mathscr{B}}, 1_{\mathscr{B}}\}\}$ is easily seen to be r.e. Moreover, it is easy to see that every recursive subalgebra B of $\mathscr{B}$ has an r.e. pseudo-complement. Thus we would have a complete analogue of the fact that an r.e. set is recursive iff it is complemented in $\mathscr{E}$ if we only knew that whenever C is a pseudo-complement of B, then B is a pseudo-complement of C. Let us say that C is a *bicomplement* of B if C is a pseudo-complement of B and B is a pseudo-complement of C. The question that we are naturally let to is "When is a pseudo-complement a bicomplement?" Finally, let us say that C is a *full complement* of B if C is a bicomplement of B and $B + C = \mathscr{B}$. The following purely algebraic results are due to Remmel and were motivated precisely by the considerations above.

THEOREM 6 (REMMEL [**1980c**]). (i) *If $\mathscr{B}$ is a finite Boolean algebra, every pseudo-complement is a full complement.*

(ii) *A Boolean algebra $\mathscr{B}$ (of any cardinality) has the property that every pseudo-complement is a bicomplement if and only if $\mathscr{B}$ is the Boolean algebra of finite and cofinite subsets of some set S.*

(iii) *If $\mathscr{B}$ is a countable Boolean algebra, then every subalgebra B of $\mathscr{B}$ has a full complement.*

Now the proof of Theorem 6(iii) is effective so that for recursive Boolean algebras, we have the following

THEOREM 7 (REMMEL [**1978a**]). *If $\mathscr{B}$ is a recursive Boolean algebra, then $B \in \mathscr{L}(B)$ is a recursive subset of $\mathscr{B}$ if and only if B is r.e. and has an r.e. full complement.*

Obviously, in $\mathscr{L}(Q)$, a subset of Q is recursive if and only if it is complemented. In the lattice $\mathscr{L}I(\tilde{Q})$ the notion of recursive and complemented are different.

Given an ideal I of $\tilde{Q}$, we say that J is a *full complement* of I if $J \cap I = \{\mathbf{0}_{\tilde{Q}}\}$ and $J + I = \tilde{Q}$. Only principal ideals in $\tilde{Q}$ have full complements. That is, if $J + I = \tilde{Q}$ and $J, I \neq \tilde{Q}$, then there exist $j \in J$ and $i \in I$ such that $i \vee j = 1_{\tilde{Q}}$. But then $\neg(j) \subseteq i$ and hence $\neg(j) \in I$. Since $I \cap J = \{0_{\tilde{Q}}\}$, then clearly $I = [\neg(j)]$ and $J = [j]$, where $[x]$ denotes the ideal generated by x in $\tilde{Q}$. We say that I is a *pseudo-complement* of J if $I \cap J = \{0_{\tilde{Q}}\}$ and, for any $x \notin I$, $([x] + I) \cap J \neq \{0_{\tilde{Q}}\}$. It is not difficult to show that if I is a pseudo-complement of J then $I = \{x \in \tilde{Q} \,|\, [x] \cap J = \{Q_{\tilde{Q}}\}\}$. The following result shows that not every recursive $J \in \mathscr{L}I(\tilde{Q})$ has an r.e. pseudo-complement.

THEOREM 8 (REMMEL, UNPUBLISHED). *There exists a recursive ideal I of $\tilde{Q}$ such that the pseudo-complement of I is not r.e.*

For the lattice $\mathscr{L}(R^n)$, there are several possible notions of recursiveness. If K, L, $M, \ldots$ are full r.e. subsets of N, we let $\mathscr{K}, \mathscr{L}, \mathscr{M}, \ldots$ denote the corresponding r.e. open sets in R^n, i.e., $\mathscr{K} = \bigcup\{\varepsilon_i | i \in K\}$, etc. Thus we say an open $\mathscr{K} \subseteq R^n$ is r.e. (recursive), etc., if there is a full r.e. (recursive) K with $\mathscr{K} = \bigcup\{\varepsilon_i | i \in K\}$.

DEFINITION. Given $\mathscr{K} \subseteq R^n$, we define

$$\text{int } \mathscr{K} = \{i | \varepsilon_i \subseteq K\}, \quad \text{ext } \mathscr{K} = \{i | \varepsilon_i \cap \mathscr{K} = \varnothing\},$$
$$\text{bnd } \mathscr{K} = \{i | \varepsilon_i \cap \mathscr{K} \neq \varnothing \text{ and } \varepsilon_i \cap (R^n - \mathscr{K}) \neq \varnothing\},$$
$$\text{bor}(\mathscr{K}) = \{i | \varepsilon_i \cap \mathscr{K} \neq \varnothing \text{ and } \varepsilon_i \cap \text{ext } \mathscr{K} \neq \varnothing\}.$$

Note that for any $\mathscr{K}$, int $\mathscr{K}$, ext $\mathscr{K}$, and bnd $\mathscr{K}$ partition N and bnd $\mathscr{K} \supseteq$ bor $\mathscr{K}$. We should note, however, that, other than obvious relations among the degrees of int $\mathscr{K}$, ext $\mathscr{K}$ and bnd $\mathscr{K}$ which are forced because these three sets partition N, the degrees of int $\mathscr{K}$, ext $\mathscr{K}$, bor $\mathscr{K}$ and bnd($\mathscr{K}$) are essentially independent. For example, Kalantari and Remmel have proved the following

THEOREM 9 (KALANTARI AND REMMEL [**1983**]). *There exists an r.e. open set $\mathscr{K}$ in R^n such that the degrees of* int $\mathscr{K}$, ext $\mathscr{K}$, bor $\mathscr{K}$ *and bnd $\mathscr{K}$ are pairwise distinct.*

DEFINITION. Given $\mathscr{K} \subseteq \mathscr{L}(R^n)$, we say

(i) $\mathscr{K}$ is *complemented* if there exists an $\mathscr{M} \in \mathscr{L}(R^n)$ such that $\mathscr{M} \cap \mathscr{K} = \varnothing$ and $\mathscr{M} \cup \mathscr{K}$ is dense in R^n,

(ii) $\mathscr{K}$ is *bi-r.e.* if ext $\mathscr{K}$ is r.e.,

(iii) $\mathscr{K}$ is *completely recursive* if $\mathscr{K}$, ext $\mathscr{K}$ and bnd $\mathscr{K}$ are all recursive,

(iv) $\mathscr{K}$ is a *recursive region* if $\mathscr{K}$ is completely recursive and bor $\mathscr{K}$ = bnd $\mathscr{K}$. (Note that in this case, $\mathscr{K}$ is a regular open set.)

Kalantari and Retzlaff then proved the following

THEOREM 10 (KALANTARI AND RETZLAFF [**1979**]). *There exists a recursive $\mathscr{K} \in \mathscr{L}(R^n)$ which is not complemented.*

THEOREM 11 (KALANTARI AND RETZLAFF [**1979**]). *$\mathscr{K}$ is a recursive region $\Rightarrow$ $\mathscr{K}$ is completely recursive $\Rightarrow$ $\mathscr{K}$ is bi-r.e. $\Rightarrow$ $\mathscr{K}$ is complemented* (*and none of the reverse implications hold*).

3. Creative elements. A set $C \subseteq N$ is *creative* if C is r.e. and there exists a partial recursive function f such that if $C \cap U_e = \varnothing$, then $f(e)\downarrow$ and $f(e) \notin C \cup U_e$. From the point of view of $\mathscr{E}$, the creative sets are the effectively noncomplemented elements of $\mathscr{E}$. The main results in $\mathscr{E}$ on creative sets are

THEOREM 12 (MYHILL [**1955**]). (i) *A set C is creative iff C is one-one complete iff C is many-one complete.*

(ii) *If C_1 and C_2 are creative, then there exists a recursive permutation p of N such that $p(C_1) = C_2$.*

Taking the view that the creative sets are the uniformly noncomplemented elements, the natural definition of creative in $\mathscr{L}(V_\infty)$ is that a subspace $V \in \mathscr{L}(V_\infty)$ is *creative* if there is a partial recursive function f such that if $V \cap W_e = \{\vec{0}\}$, then $f(e)\downarrow$ and $f(e) \notin V + W_e$. It is not difficult to construct creative spaces. In fact the following is true.

THEOREM 13 (TE KOLSTE, IN METAKIDES AND NERODE [**1977**]). *Let $V \in \mathscr{L}(V_\infty)$. Then V is a creative subspace iff V is a creative subset of $V_\infty (= N)$.*

COROLLARY 14. *If B is a recursive basis for V_∞ and C is a creative subset of B, then $(C)^*$ is a creative subspace of V_∞.*

PROOF. If C is a creative subset of B, then $(C)^*$ is many-one complete since $(C)^* \cap B = C$ and C is many-one complete. Thus $(C)^*$ is a creative subset of V_∞ and hence by Theorem 12, $(C)^*$ is a creative subspace of V_∞. □

However, the second half of Myhill's Theorem fails for creative spaces. That is, there are creative spaces which do not differ by a recursive automorphism of V_∞. Nevertheless, there is a natural subclass of creative spaces which are unique up to recursive automorphism. To define such a subclass, we need to consider the notion of an r.e. presented vector space.

DEFINITION. An *r.e. presented* vector space V over a recursive field F consists of (i) an r.e. set $|V|$ of N, (ii) operations of vector addition and scalar multiplication which are partial recursive, and (iii) an r.e. congruence relation $\equiv$ on V such that $V \bmod \equiv$ is a vector space. By an index k of an r.e. presented space $W = \langle |W|, +, \{\lambda_i\}_{i \in F}, \equiv \rangle$, we mean an r.e. index of the following: (i) an r.e. index for $|W|$, (ii) partial recursive indices for $+$ and λ_i, and (iii) an r.e. index for the r.e. equivalence relation $\equiv$. Another way to think of r.e. presented spaces is given by the following

PROPOSITION 15 (METAKIDES AND NERODE [**1977**]). *Every r.e. presented space is recursively isomorphic to V_∞ mod W with $W \in \mathscr{L}(V_\infty)$ and every V_∞ mod W with $W \in \mathscr{L}(V_\infty)$ is r.e. presented.*

Now given two $\aleph_0$-dimensional vector spaces V and V' over the same field F, say $\langle a_1, \dots, a_t \rangle$ from V has the same type as $\langle a'_1, \dots, a'_t \rangle$ from V', written $\langle a_1, \dots, a_t \rangle \sim \langle a'_1, \dots, a'_t \rangle$, if the map $a_i \to a'_i$ for $i = 1, \dots, t$ can be extended to an isomorphism of the subspaces $(\{a_1, \dots, a_t\})^*$ and $(\{a'_1, \dots, a'_t\})^*$.

DEFINITION. (i) An r.e. presented space E is *effectively universal homogeneous* (EUH) if whenever k is an index of an r.e. presented W, $a_1, \dots, a_t \in W$, $a'_1, \dots, a'_t \in E$, and $\langle a_1, \dots, a_t \rangle \sim \langle a'_1, \dots, a'_t \rangle$, we can effectively compute for any $a_{t+1} \in W$, an $a'_{t+1} \in E$ such that $\langle a_1, \dots, a_t, a_{t+1} \rangle \sim \langle a'_1, \dots, a'_t, a'_{t+1} \rangle$.

(ii) If E is EUH and E is recursively isomorphic to V_∞ / V for $V \in \mathscr{L}(V_\infty)$, then V is called an EUH kernel.

The definition of an EUH space is due to Metakides and Nerode [**1977**] and may be viewed as an analogue of the essential property that creative sets have to prove the second half of Myhill's Theorem.

THEOREM 16 (METAKIDES AND NERODE [**1977**]).

(i) *Any two EUH spaces (over the same field) are recursively isomorphic.*

(ii) *Given any r.e. presented* W_k, *any* $\langle a_1,\ldots,a_t\rangle$ *from* $|W_k|$, *and* $\langle a'_1,\ldots,a'_t\rangle$ *from E with* $\langle a_1,\ldots,a_t\rangle \sim \langle a'_1,\ldots,a'_t\rangle$, *there exists a partial recursive isomorphism* ϕ *from* W_k *onto a subspace of* V_∞ *such that* $\phi(a_i) = a'_i$ *for* $i = 1,\ldots,t$.

THEOREM 17 (METAKIDES AND NERODE [**1977**]).

(ii) *If* S_1 *and* S_2 *are EUH kernels, then there exists a recursive automorphism* ϕ *of* V_∞ *such that* $\varphi(S_1) = S_2(S_1) = S_2$.

(iii) *If S is an EUH kernel, then no basis of S is fully extendible.*

Thus the EUH kernels satisfy the exact analogue of Theorem 11. Moreover, we get two interesting corollaries from Theorems 16 and 17.

COROLLARY 18 (METAKIDES AND NERODE [**1977**]). *There exist creative subspaces* C_1 *and* C_2 *in* $\mathscr{L}(V_\infty)$ *such that there is no recursive automorphism* ϕ *of* V_∞ *with* $\phi(C_1) = C_2$.

PROOF. Let C_1 be generated by a creative subset of a recursive basis and C_2 be an EUH kernel. Clearly the property of being generated by a subset of a recursive basis is preserved under any recursive automorphism ϕ so that $\phi(C_1) \neq C_2$. □

COROLLARY 19 (METAKIDES AND NERODE [**1977**]). *There exists a* $V \in \mathscr{L}(V_\infty)$ *such that no basis of V is fully extendible but that any r.e. basis B of V is extendible.*

PROOF. Let V be an EUH kernel. That fact that V is creative allows us to generate an infinite-dimensional W such that $W \cap V = \{\vec{0}\}$. Thus any r.e. basis B of V can be extended by the addition of an r.e. basis C of W. □

In fact, $V \in \mathscr{L}(V_\infty)$ with extendible bases but not fully extendible bases occur in every r.e. degree possible; see Theorem 79.

We see from V_∞ that there are two possible ways to try to define the notion of creative in other lattices. That is, we can take the point of view that creative elements are the effectively noncomplemented elements in the lattice or we can take the point of view that creative objects produce effectively universal homogeneous objects. We shall see that effectively noncomplemented elements always exist in our lattices, while EUH objects do not always exist or make sense. For example, Metakides and Nerode have shown that there are no EUH algebraically closed fields. However, we can define an algebraically closed subfield $C \in \mathscr{L}(F_\infty)$ to be *creative* if there exists a partial recursive function f such that if W_e is independent over C, then $f(e)\downarrow$ and $f(e) \notin C + W_e$. Creative algebraically closed subfields do exist; in fact, the analogues of Te Kolste's result and its corollary continue to hold. Again, creative spaces are not unique up to isomorphism as our next result will show. Indeed, our next result holds both for $\mathscr{L}(V_\infty)$ and $\mathscr{L}(F_\infty)$ with the same proof and shows that the obvious converse of Theorem 13 fails in a rather spectacular way.

THEOREM 20 (REMMEL). *There exists a simple subset S of a recursive besis B for V_∞ (a recursive transcendence basis B of F_∞) such that $(S)^*$ is a creative subspace of V_∞ (is a creative algebraically closed subfield of F_∞).*

PROOF. Let $B_0, B_1, \ldots$ be an effective list of all r.e. subsets of B. To ensure that S is a simple subset of B, we shall meet the following requirements.

$$\mathrm{R}_e\text{:}\quad \text{If } B_e \text{ is infinite, then } B_e \cap S \neq \varnothing.$$

Effectively partition B into pairwise disjoint sets $D_0, D_1, \ldots,$ where D_i has $i+2$ elements and let f be the recursive function such that $f(i) = \sum_{b \in D_i} b$. We shall build S in stages so that $(S)^*$ is creative with productive function f. We say that requirement R_e is *satisfied* at stage s if $B_e^s \cap S^s \neq \varnothing$, where S^s denotes the finite set of elements enumerated into S at the end of stage s. In the construction to follow, the even stages of our construction will ensure that $(S)^*$ is creative and the odd stages will ensure that S is simple in B.

CONSTRUCTION. *Stage* $2s$. Let

$$S^{2s} = S^{2s-1} \cup \{ D_i | i \leqslant 2s \text{ and } f(i) \in (S^{2s-1} \cup W_e^{2s})^* \}.$$

(Here $S^{-1} = \varnothing$.)

Stage $2s+1$. Look for the least $e \leqslant 2s+1$ such that requirement R_e is not satisfied at stage $2s$, and there is an $x \in B_e^s$ such that $x \notin \bigcup_{j \leqslant e} D_e$. If there is no such e, let $S^{2s+1} = S^{2s}$. Otherwise pick the least such e and the least x corresponding to e and let $S^{2s+1} = S^{2s} \cup \{x\}$.

We note that $f(i)$ can never enter $(S)^*$ at an odd stage since at most i elements of D_i can enter S at odd stages, i.e., only requirements $R_0, \ldots, R_{i-1}$ can force elements in D_i into S and at most one element is forced into S by each requirement. Thus $f(i) \in (S)^*$ if and only if $f(i) \in (S)^* + W_e$. Moreover, it is easy to see that if $f(i) \in (S)^*$, then at the stage s such that $f(i) \in S^s - S^{s-1}$, we have ensured that $(W_e^s)^*$ is not independent over $(S^s)^*$. Thus $(S)^*$ is creative. Since only finitely many elements are ever restrained from entering S for the sake of any fixed requirements R_e, the usual priority argument will show that all the requirements R_e are met. Finally since for infinitely many i, $W_i = (\varnothing)^*$, it is easy to see that for infinitely many i, $f(i) \notin (S)^*$ and hence $B - S$ is infinite. Thus S is simple in B. □

COROLLARY 21. *There exist creative subspaces C_1 and C_2 in $\mathscr{L}(F_\infty)$ such that there is no recursive automorphism ϕ of F_∞ such that $\phi(C_1) = \phi(C_2)$.*

PROOF. Let C_1 be generated by a creative subset of a recursive transcendence basis B and C_2 be generated by a simple subset of a recursive transcendence basis. □

At first glance, it may seem that the lattice $\mathscr{L}(Q)$ exhibits no new phenomena that do not already occur in $\mathscr{E}$. However, we shall see that there are natural analogues of Myhill's theorem. We say $W \in \mathscr{L}(Q)$ is *bidense* if both W and $Q - W$ are dense in the rationals.

DEFINITION. We say $C \in \mathscr{L}(Q)$ is *bidense creative* if (i) C is bidense, and (ii) there is a partial recursive function $f(x, y, z)$ such that for any rationals $q_1 < q_2 \in Q$ and any r.e. W_e, if $W_e \cap C \cap [q_1, q_2] = \varnothing$, then $f(e, q_1, q_2)\downarrow$ and

$$f(e, q_1, q_2) \in [q_1, q_2] - (W_e \cup C),$$

where

$$[q_1, q_2] = \{q \in Q | q_1 \leqslant q \leqslant q_2\}.$$

Thus C is bidense creative if C is effectively noncomplemented in every interval. The definition of bidense creative was given by Metakides and Nerode [**1975**]. The following theorem was first proved in the literature as a corollary to a later more general result of Metakides and Remmel [**1979**].

THEOREM 22 (METAKIDES AND NERODE [**1975**], PROOF IN METAKIDES AND REMMEL [**1979**]). (i) *There exist bidense creative subsets of Q.*

(ii) *Given any two bidense creative subsets C_1 and C_2 of Q, there exists a recursive order automorphism ϕ of Q such that $\phi(C_1) = C_2$.*

The notions of an EUH linear ordering and EUH Boolean algebras can be defined by analogy with the definition above.

An *r.e. presented linear ordering* consists of a recursive set $|L|$ together with an r.e. preorder $\leqslant$ on L ($\leqslant$ is transitive, reflexive, but it may be that $x \neq y$ and yet both $x \leqslant y$ and $y \leqslant x$). Another way to think of r.e. presented linear orderings is as quotients of the rationals $(Q, \leqslant)$ mod r.e. congruence relations $\sim$. Given two r.e. presented linear orderings $L_1 = (|L_1|, \leqslant_1)$ and $L_2 = (|L_2|, \leqslant_2)$, we say $\vec{x} = \langle x_1, \ldots, x_n \rangle$ from $|L_1|$ has the *same order type* as $\vec{y} = \langle y_1, \ldots, y_n \rangle$ from $|L_2|$, denoted $\vec{x} \sim \vec{y}$, if the map $x_i \to y_i$ for $i = 1, \ldots, n$ satisfies $x_i \leqslant_1 x_j$ iff $y_i \leqslant_2 y_j$ for all $i, j \leqslant n$. We would say that an r.e. presented linear ordering $L = (|L|, \leqslant)$ is *effectively universal homogeneous* (EUH) if there is a uniform effective procedure which given any r.e. index k of an r.e. presented linear ordering L_k, $\langle x_1, \ldots, x_n \rangle$ from L_k and $\langle y_1, \ldots, y_n \rangle \in L$ such that $\langle x_1, \ldots, x_n \rangle \sim \langle y_1, \ldots, y_n \rangle$, and $x \in L_k$, effectively produces a $y \in L$ such that $\langle x_1, \ldots, x_n, x \rangle \sim \langle y_1, \ldots, y_n, y \rangle$. Similarly, an *r.e. presented Boolean algebra* $\mathscr{B}$ consists of a recursive set B, an r.e. equivalence relation $\equiv$, and partial recursive operations $\wedge$ (meet), $\vee$ (join), and $\neg$ (complementation) under which $B/\equiv$ becomes a Boolean algebra. Another way to think of r.e. Boolean algebras $\mathscr{B}$ is as $\tilde{Q}$ modulo an r.e. ideal $\mathscr{I}$ or propositional logic mod an r.e. theory. Given two r.e. presented Boolean algebras $\mathscr{B}_1$ and $\mathscr{B}_2$, we say that $\vec{x} = \langle x_1, \ldots, x_n \rangle$ from B_1 has the same type as $\vec{y} = \langle y_1, \ldots, y_n \rangle$ from $\mathscr{B}_2$, written $\vec{x} \sim \vec{y}$, if the map $x_i \to y_i$ for $i = 1, \ldots, n$ can be extended to an isomorphism between the subalgebras generated by $\langle x_1, \ldots, x_n \rangle$ and $\langle y_1, \ldots, y_n \rangle$ in $\mathscr{B}_1$ and $\mathscr{B}_2$, respectively. By an index k of an r.e. presented Boolean algebra $\mathscr{B} = \langle B, \equiv, \wedge, \vee, \neg \rangle$ we mean that k codes r.e. indices for B and $\equiv$ and partial recursive indices for $\vee$, $\wedge$ and $\neg$.

DEFINITION. (i) An r.e. presented Boolean algebra $\mathscr{B}$ is *effectively universal homogeneous* (EUH) if there is a uniform effective procedure which given the index k of an r.e. presented Boolean algebra $\mathscr{B}_k$, $\langle x_1,\ldots,x_n\rangle$ from $\mathscr{B}_k$ and $\langle y_1,\ldots,y_n\rangle$ from $\mathscr{B}$ such that $\langle x_1,\ldots,x_n\rangle \sim \langle y_1,\ldots,y_n\rangle$, and x from $\mathscr{B}_k$, produces $y \in \mathscr{B}$ such that $\langle x_1,\ldots,x_n, x\rangle \sim \langle y_1,\ldots,y_n, y\rangle$.

(ii) If $\mathscr{B}$ is an EUH Boolean algebra and $\mathscr{B}$ is recursively isomorphic to $\tilde{Q}/I$, where I is an r.e. ideal of $\tilde{Q}$, then I is called an *EUH kernel*.

The next two theorems show that there is a rather sharp contrast between linear orderings and Boolean algebras with regards to the EUH property.

THEOREM 23 (ROY [**1978**, **1983**], HINGSTON [**1982**]). *There does not exist an EUH r.e. presented linear ordering.*

THEOREM 24 (METAKIDES AND NERODE, UNPUBLISHED). (i) *There exist EUH r.e. presented Boolean algebras.*

(ii) *Any two EUH r.e. presented Boolean algebras are recursively isomorphic.*

(iii) *Any two EUH kernels in $\mathscr{L}I(\tilde{Q})$ differ by a recursive automorphism of $\tilde{Q}$.*

We can also look at creative subalgebras and ideals from the effectively noncomplemented point of view. That is, given a recursive Boolean algebra $\mathscr{B}$, let $W_0, W_1,\ldots$ be an effective list of all r.e. subalgebras of $\mathscr{B}$ and $I_0, I_1, I_2,\ldots$ an effective list of all r.e. ideals of $\mathscr{B}$. We say that a subalgebra $C \in \mathscr{L}(\mathscr{B})$ is *creative* if there is a partial recursive function such that if $W_i \cap C = \{0_{\mathscr{B}}, 1_{\mathscr{B}}\}$, then $f(i)\downarrow$, $f(i) \notin W_i \cup C$, and $(W_i \cup \{f(i)\})^* \cap C = \{0_{\mathscr{B}}, 1_{\mathscr{B}}\}$. Similarly, an ideal $I \in \mathscr{L}I(\mathscr{B})$ is *creative* if there is a partial recursive function g such that if $I_e \cap I = \{0_{\mathscr{B}}\}$, then $g(e)\downarrow$, $g(e) \notin I_e \cup I$, and $(I_e \cup \{g(e)\})^* \cap I = \{0_{\mathscr{B}}\}$. (Here $0_{\mathscr{B}}$ and $1_{\mathscr{B}}$ denote the zero and one of $\mathscr{B}$, respectively.)

The study of the lattice of subalgebras and the lattice of ideals of recursive Boolean algebras is in some sense more difficult and richer than the study of $\mathscr{L}(V_\infty)$ due to the fact that there are lots of isomorphism types of recursive Boolean algebras with widely different properties and, even within a classical isomorphism type, recursive Boolean algebras have widely different effective properties.

Given a linear ordering L with a least element, let $\mathscr{B}_L$ denote the Boolean algebra which is generated by the left-closed right-open intervals of $\mathscr{B}$. Then if η is the order type of the rationals and ω is the order type of the nonnegative integers, then $\mathscr{B}_{1+\eta}$ is the atomless Boolean algebra and $\mathscr{B}_\omega$ is the Boolean algebra of finite and cofinite subsets of N. For Boolean algebra $\mathscr{B}$, let $\mathscr{A}(\mathscr{B})$ denote the set of atoms of $\mathscr{B}$ and $\mathscr{I}(\mathscr{A}(\mathscr{B}))$ denote the ideal generated by the atoms of $\mathscr{B}$. We say $\mathscr{B}$ is *atomic* if for all $y \in \mathscr{B}$ with $y \neq 0_{\mathscr{B}}$, there exists an $x \in \mathscr{A}(\mathscr{B})$ with $x \leqslant_{\mathscr{B}} y$.

The simplest infinite atomic Boolean algebra is $\mathscr{B}_\omega$. If $\tilde{N}$ is a recursive Boolean algebra isomorphic to $\mathscr{B}_\omega$ with $\mathscr{A}(\tilde{N})$ recursive, then $\mathscr{I}(\mathscr{A}(\tilde{N}))$ is also recursive. In contrast, Remmel has proved the following

THEOREM 25 (REMMEL [**198 a**]). *There exists a recursive Boolean algebra $\tilde{N}_0$ isomorphic to $\mathscr{B}_\omega$ such that $\mathscr{I}(\mathscr{A}(\tilde{N}_0))$ is immune.*

The only ideals of $\mathscr{B}_\omega$ are either principal or generated by some infinite subset of $\mathscr{A}(\mathscr{B}_\omega)$. It follows that the only r.e. ideals of $\tilde{N}_0$ are principal. Also if we consider $\tilde{N}_0$ as a Boolean ring, then $\tilde{N}_0$ is a ring which is recursively Noetherian but not Noetherian. In $\tilde{N}$, it is not difficult to show that every creative ideal is generated by a creative subset of the atoms and any creative subsets of the atoms generates a creative ideal. Using these facts it is not difficult to prove the following

THEOREM 26 (REMMEL [**198 a**]). (i) *There exist no creative ideals in* $\tilde{N}_0$.

(ii) *There exist creative ideals in* $\tilde{N}$ *and any two creative ideals in* $\tilde{N}$ *differ by a recursive automorphism of* $\tilde{N}$.

Despite such varied examples, we can make some general statements about the lattice of r.e. subalgebras and the lattice of r.e. ideals of recursive Boolean algebras. To this end, introduce three specific recursive Boolean algebras:

$\tilde{N}$ = the recursive Boolean algebra isomorphic to $\mathscr{B}_\omega$, the Boolean algebra of finite cofinite subsets of N, where $\mathscr{A}(\tilde{N})$ is recursive.

$\tilde{Q}$ = the recursive Boolean algebra isomorphic to $\mathscr{B}_{1+\eta}$, the atomless Boolean algebra.

$\tilde{C}$ = the recursive Boolean algebra isomorphic to $\mathscr{B}_{1+\omega\cdot\eta}$, where both $\mathscr{A}(\tilde{C})$ and $I(\mathscr{A}(\tilde{C}))$ are recursive. (One can think of $\tilde{C}$ as the subalgebra generated by the left-closed right-open intervals of the rationals together with $\{\{q\}\,|\,q \in Q\}$.)

One can also show that the properties of $\tilde{N}$, $\tilde{Q}$ and $\tilde{C}$ above determine these Boolean algebras up to recursive isomorphism. Given any recursive Boolean algebra $\mathscr{B}$ and $x \in \mathscr{B}$, we let $[x]$ denote the principal ideal generated by x. Note that $[x]$ may be considered a recursive Boolean algebra where for the operations $\vee_{[x]}$(join) and $\wedge_{[x]}$(meet), we use the restriction of the join and meet operations from $\mathscr{B}$ and for $\neg_{[x]}$(complement), we say that $\neg_{[x]}(z) = x \wedge \neg_{\mathscr{B}}(z)$. For recursive Boolean algebras $\mathscr{B}$ and $\mathscr{C}$ we write $\mathscr{B} \approx \mathscr{C}$ if $\mathscr{B}$ is isomorphic to $\mathscr{C}$ and $\mathscr{B} \approx_r \mathscr{C}$ if $\mathscr{B}$ is recursively isomorphic to $\mathscr{C}$. It is clear that $\mathscr{B} \approx_r [x] \times [\neg(x)]$ for any $x \in \mathscr{B}$, so we say that *$\mathscr{C}$ is a factor of $\mathscr{B}$*, if $\mathscr{C} \approx_r [x]$ for some $x \in \mathscr{B}$. Now it is often the case that constructions which produce elements with various properties such as maximal, creative, etc. in some factor $\mathscr{C}$ of $\mathscr{B}$ can easily be modified to produce elements with the same property in $\mathscr{B}$ itself. For example, if C is a creative ideal in the Boolean algebra $[x]$ for $x \in \mathscr{B}$, then $C + [\neg(x)]$ is a creative ideal in $\mathscr{B} = [x] \times [\neg(x)]$. Thus, the following theorem allows us to restrict our attention to constructions in $\tilde{N}$, $\tilde{Q}$ and $\tilde{\mathscr{C}}$ if we are just trying to show the existence of various types of elements in $\mathscr{L}(\mathscr{B})$ or $\mathscr{L}I(\mathscr{B})$ for at least one recursive Boolean algebra $\mathscr{B}$ in each classical isomorphism type which contains a recursive Boolean algebra.

THEOREM 27 (REMMEL [**1978a**]). *For any recursive Boolean algebra $\mathscr{D}$, there exists a recursive Boolean algebra $\mathscr{B}$ isomorphic to $\mathscr{D}$ such that either $\tilde{N}$ and $\tilde{Q}$ is a factor of $\mathscr{B}$ or $\mathscr{B} = \tilde{\mathscr{C}}$.*

Now Theorem 26 shows that we cannot construct creative ideals in every recursive Boolean algebra, but using Theorem 27 we can show that there is at least one recursive Boolean algebra in every classical isomorphism type which contains a recursive Boolean algebra in which creative ideals exist.

THEOREM 28 (REMMEL [**198 a**]). *For any recursive B.A. $\mathscr{D}$, there exists a recursive B.A. $\mathscr{B}$ isomorphic to $\mathscr{D}$ such that there are creative ideals I_1 and I_2 in $\mathscr{B}$ such that no recursive automorphism of $\mathscr{B}$ carries I_1 to I_2.*

We note that in light of Theorem 25, Theorem 27 asserts that there is yet a third recursive Boolean algebra $\tilde{N}_1$ isomorphic to $\mathscr{B}_\omega$ which is not recursively isomorphic to either $\tilde{N}$ or $\tilde{N}_0$.

THEOREM 29 (REMMEL [**1978a**]). *For any recursive B.A. $\mathscr{D}$, there exists a recursive B.A. $\mathscr{B}$ isomorphic to $\mathscr{D}$ such that there exist creative subalgebras B_1, B_2 of $\mathscr{B}$ such that no recursive automorphism of $\mathscr{B}$ carries B_1 to B_2.*

We can say some things in general about creative subalgebras.

THEOREM 30 (REMMEL [**1978a**]). *If C is a creative subalgebra of a recursive Boolean algebra $\mathscr{D}$, then C is a creative subset of N.*

Finally, we consider the lattice $\mathscr{L}^*(R^n)$. We say an r.e. open set $\mathscr{K}$ is *creative* if there exists a partial recursive function f such that if $W_i \cap \mathscr{K} = \varnothing$ (where W_i is the r.e. open set corresponding to the ith full subset), then $f(i)\downarrow$ and $\varepsilon_{f(i)} \cap (\mathscr{K} \cup W_i) = \varnothing$. Thus a creative open set is effectively noncomplemented. A new phenomenon occurs in $\mathscr{L}(R^n)$ due to the fact that if $\mathscr{K}$ is creative and $\mathscr{M}$ is an r.e. open subset of $\mathscr{K}$ which is dense in $\mathscr{K}$, then clearly $\mathscr{M}$ is also creative with the same productive function as $\mathscr{K}$. That is, in $\mathscr{L}(R^n)$ there exist creative open sets which are recursive or of any r.e. degree whatsoever. To prove the existence of such r.e. open sets, we need the following

THEOREM 31 (KALANTARI AND REMMEL [**1983**]). *If $\mathscr{K}$ is a nonempty r.e. open set in $\mathscr{L}(R^n)$, then for any r.e. degree δ, there exists an r.e. open subset $\mathscr{M}_\delta$ of $\mathscr{K}$ which is dense in $\mathscr{K}$ such that* $\deg(\mathscr{M}_\delta) = \delta$.

We note that the special case of Theorem 31, where $\delta = \mathbf{0}$ or $\delta = \mathbf{0}'$, was first proved by Kalantari and Retzlaff in [**1979**]. Now combining Theorem 31 with the remark above, we get the following

COROLLARY 32. *For any r.e. degree δ, there exists a creative r.e. open set $\mathscr{M}_\delta$ such that* $\deg(\mathscr{M}_\sigma) = \delta$.

PROOF. Let f be any one-one recursive function such that $\{\varepsilon_{f(i)}\}_{i=1}^{\infty}$ is a sequence of pairwise disjoint sets. Then $C = \bigcup\{\varepsilon_{f(i)} | W_i \cap \varepsilon_{f(i)} \neq \varnothing\}$ is a creative open set with productive function f. Let $\mathscr{M}_\delta$ be an r.e. open set of degree δ which is dense in $\mathscr{C}$. $\mathscr{M}_\delta$ exists by Theorem 31 and $\mathscr{M}_\delta$ is creative with productive function f. □

3. Simple, R-maximal and maximal elements. We recall the following definitions for sets of positive integers. A set $A \subseteq N$ is

(i) *immune* if A is infinite and A contains no infinite r.e. set,

(ii) *h-immune* if A is infinite and there is no recursive function f such that $\forall i, j$ $(i \neq j \to D_{f(i)} \cap D_{f(j)} = \varnothing)$ and $\forall j\,(D_{f(j)} \cap A \neq \varnothing)$,

(iii) *h-h-immune* if A is infinite and there is no recursive function f such that $\forall i$, $j\,(i \neq j \to U_{f(i)} \cap U_{f(j)} = \varnothing)$ and $\forall i\,(U_{f(i)}$ is finite & $U_{f(i)} \cap A \neq \varnothing)$,

(iv) *r-cohesive* if A is infinite and for any recursive set R either $R \cap A =^* \varnothing$ or $R \cap A =^* A$, and

(v) *cohesive* if A is infinite and for any r.e. set W either $W \cap A =^* \varnothing$ or $W \cap A =^* A$.

We say an infinite r.e. set S is (1) *simple* if $\bar{S}$ is immune, (2) *h-simple* if $\bar{S}$ is h-immune, (3) *h-h-simple* if $\bar{S}$ is h-h-immune, (4) *r-maximal* if $\bar{S}$ is r-cohesive, and (5) *maximal* if $\bar{S}$ is cohesive.

The following, and only those, implications hold:

$$\begin{array}{ccccccc} 1) & \Rightarrow & 2) & \Rightarrow & 3) & \Rightarrow & 5) \\ & & & \nwarrow & & \nearrow & \\ & & & & 4) & & \end{array}$$

When we approach such concepts from the point of view of our six lattices, we run into a number of problems. First of all, the notion of complement is not unique. This causes no great problems from the point of view of the definitions (1)–(5) above since they can be reformulated to avoid explicit mention of the complement. For example, the following definitions are clearly natural analogues of the definition of simple, maximal and r-maximal.

DEFINITION. We say that $V \in \mathscr{L}(V_\infty)$ is

(i) *simple* if $\dim(V_\infty/V) = \infty$ and for every infinite-dimensional $W \in \mathscr{L}(V)$, $W \cap V \neq (\varnothing)^*$,

(ii) *maximal* if $\dim(V_\infty/V) = \infty$ and for every $W \in \mathscr{L}(V_\infty)$ such that $W \supseteq V$, either $\dim(W/V) < \infty$ or $\dim(V_\infty/W) < \infty$,

(iii) *r-maximal* if $\dim(V_\infty/V) = \infty$ and for every $R_1, R_2 \in \mathscr{L}(V_\infty)$ such that $R_1 + R_2 = V_\infty$, either $\dim(V_\infty/R_1 + V)$ or $\dim(V_\infty/R_2 + V)$ is finite.

(*Note*. To define the notions above for $\mathscr{L}(F_\infty)$ simply replace V_∞ by F_∞ and replace the notions of dimension by transcendence degree.)

One factor that makes such definitions easy in $\mathscr{L}(V_\infty)$ or $\mathscr{L}(F_\infty)$ is that the analogue of being a coinfinite set is clearly having infinite codimension or cotranscendence degree. However, in such lattices as $\mathscr{L}(R^n)$ or even $\mathscr{L}I(\tilde{Q})$, the analogue of being coinfinite is not so clear. If we concentrate on the complements, we get another set of definitions.

DEFINITION. We say that a subspace U of $\mathscr{L}(V_\infty)$ is

(i) immune of U contains no infinite-dimensional r.e. subspace,

(ii) *cohesive* if for any r.e. subspace W either $\dim(W \cap U) < \infty$ or $\dim(U/(W \cap U)) < \infty$,

(iii) *co-r.e.* if $V_\infty - U$ is an r.e. set.

Now co-r.e. immune or co-r.e. cohesive subspaces do not have the same relationships to simple and maximal subspaces as co-r.e. immune and co-r.e. cohesive subsets have to simple and maximal sets as the following results show.

THEOREM 33 (DOWNEY [**1984a**]). Let B be any recursive basis for V_∞.
(i) *A is a co-r.e. immune subset of B iff $(A)^*$ is a co-r.e. immune subspace of V_∞.*
(ii) *If S is any co-r.e. subset of B, then there exists a decidable space R such that $R \oplus (S)^* = V_\infty$.*

THEOREM 34 (SHORE). *Let B be any recursive basis for V_∞.*
(i) *If M is a maximal subset of B, then $(M)^*$ is a maximal subspace.*
(ii) *If C is a co-r.e. cohesive subset of B, then $(C)^*$**is not** a co-r.e. cohesive subspace.*

THEOREM 35 (REMMEL [**1977a**]). (i) *If V is a co-r.e. cohesive subspace and W is an r.e. subspace such that $V \oplus W = V_\infty$, then W is a maximal subspace.*
(ii) *There exist co-r.e. cohesive subspaces in every high r.e. degree.*

Note that Downey's results show that an r.e. complement of an immune space is not necessarily simple, while by Theorem 34 an r.e. complement of a co-r.e. cohesive space is always a maximal subspace. We note that part (i) of Theorem 34 is published in Metakides and Nerode [**1977**], while part (ii) was not published by Shore. However, Theorem 34 (ii) easily follows from Theorem 33(ii) and Theorem 35(i). That is, by Theorem 33(ii) any subspace $(C)^*$ generated by a co-r.e. cohesive subset of a recursive basis has a decidable and hence nonmaximal complement and so $(C)^*$ is not a co-r.e. cohesive subspace by Theorem 35(i). Finally, we should note that while Theorem 20 shows that a simple subset S of a recursive basis B does not generate a simple space, hypersimple subsets H of B do generate simple subspaces.

THEOREM 36 (GUHL [**1973**]). (i) *If S is a simple but not h-simple subset of a recursive basis B for V_∞, then $(S)^*$ is not a simple space.*
(ii) *If H is an h-simple subset of a recursive basis B for V_∞, then $(H)^*$ is a simple space.*

The fact that such properties as above do not always transfer the desired properties to a complement shows that we must be careful in transferring constructions that are designed for $\mathscr{E}$ to our other lattices. In most priority constructions in $\mathscr{E}$ to build r.e. sets with properties such as (1)–(5) above, one usually concentrates on ensuring that the complement $\bar{S}$ of the r.e. set S has the desired properties and regards S as those elements which we have thrown out of $\bar{S}$. Such a point of view does not always work in our lattices and at the very least must be modified.

We shall now look at each of our lattices in turn and survey some of the types of results which are in contrast to results in $\mathscr{E}$. Here are several basic results from $\mathscr{E}$.

THEOREM 37 (MARTIN [**1966**]). *An r.e. degree δ contains a maximal set iff δ is high (i.e., $\delta' = \mathbf{0}''$).*

THEOREM 38 (SOARE [**1974**]). *Given any two maximal sets, there is an automorphism ϕ of $\mathcal{E}$ such that $\phi(M_1) = M_2$.*

THEOREM 39 (LACHLAN [**1968**]). *There is an r-maximal set R which is not maximal but which is contained in a maximal set.*

THEOREM 40 (ROBINSON [**1967**] AND LACHLAN [**1968**]). *There is an r-maximal set R which is not contained in any maximal set.*

In $\mathcal{L}(V_\infty)$, the analogues of both Theorems 37 and 38 fail due to the existence of a new type of maximal space introduced by Kalantari and Retzlaff [**1977**].

DEFINITION. (i) $v \in \mathcal{L}(V_\infty)$ is *supermaximal* if $\dim(V_\infty/V) = \infty$ and for any r.e. subspace $W \supset V$ either $\dim(W/V) < \infty$ or $W = V_\infty$.

(ii) $V \in \mathcal{L}(V_\infty)$ is *k-thin* if $\dim(V_\infty/V) = \infty$, and for any r.e. subspace $W \supset V$ either $\dim(W/V) < \infty$ or $\dim(V_\infty/W) \leqslant k$, and there exists a $U \in \mathcal{L}(V_\infty)$ such that $U \subseteq V$ and $\dim(V_\infty/U) = k$.

Now it is clear that if M is a maximal subset of a recursive basis B, $(M)^*$ is a maximal space which is not supermaximal or k-thin for any k so that $(M)^*$ cannot be the image of a supermaximal subspace under any automorphism of $\mathcal{L}(V_\infty)$. Thus a supermaximal space has no fully extendible basis and hence has no extendible basis (since for maximal spaces V, it is easy to see that a basis R of V is extendible iff it is fully extendible). The other interesting fact about supermaximal subspaces is that they are easier to construct than maximal sets. That is, maximal sets are constructed by the usual e-state construction. Now the e-state construction can be modified to produce maximal subspaces in V_∞. Indeed, the existence of a maximal subspace of V_∞ was first proved by Metakides and Nerode [**1977**] using an e-state construction and the existence of maximal subspaces with no extendible basis was first proved by Remmel [**1977a**] using an e-state construction. But supermaximal subspaces can be constructed via a rather simple finite injury priority argument.

THEOREM 41 (KALANTARI AND RETZLAFF [**1977**]). *There exists a supermaximal subspace M.*

PROOF. We shall build M in stages. At each stage s, we shall specify a finite independent set I^s and an r.e. sequence $b_0^s, b_1^s, b_2^s, \dots$ such that $I^s \cup \{b_0^s, b_1^s, \dots\}$ is a basis for V_∞. Our construction will ensure that $I^s \subseteq I^{s+1}$ for all x, $\lim_s b_i^s = b_i$ exists for all i, and that $I \cup \{b_0, b_1, \dots\}$ is a basis for V_∞. We shall let $M = (I)^*$ so that $\dim(V_\infty/M)$ will be infinite. To ensure that M is supermaximal, we meet the following set of requirements.

$$P_{\langle e,n\rangle}\colon \quad \text{If } W_e \supseteq M \text{ and } \dim(W_e/M) = \infty, \text{ then } \varepsilon_n \in W_e + M$$

(where $\varepsilon_0, \varepsilon_1, \ldots$ is some fixed recursive basis for V_∞). Note that if we meet all the requirements $P_{\langle e,n\rangle}$ we will have ensured that if $W_e \supset M$ and $\dim(W_e/M) = \infty$, then $W_e + M = W_e = V_\infty$ so that M will be supermaximal.

DEFINITION. $P_{\langle e,n\rangle}$ requires attention at stage s if

(i) $\varepsilon_n \notin (W_e^s)^* + M^s$, where $M^s = (I^s)^*$, and

(ii) $(\exists x)[x \in W_e^s$ and $I^s \cup \{b_1^s, \ldots, b_{\langle e,n\rangle}^s\} \cup \{x + \varepsilon_n\}$ is independent].

It is easy to see that using the dependence algorithm of V_∞, we can effectively decide if $P_{\langle e,n\rangle}$ requires attention.

CONSTRUCTION. *Stage* 0. Let $I_0 = \varnothing$ and $b_i^0 = \varepsilon_i$ for all i.

Stage $s+1$. If no requirement $P_{\langle e,n\rangle}$ with $\langle e,n\rangle \leqslant s+1$ requires attention, then let $I^{s+1} = I^s$ and $b_i^{s+1} = b_i^s$ for all i. Otherwise assume $\langle e, n\rangle$ is the least number $\leqslant s+1$ such that $P_{\langle e,n\rangle}$ requires attention. Let x be the least element such that $x \in W_e^s$ and $I^s \cup \{b_0^s, \ldots, b_{\langle e,n\rangle}^s\} \cup \{x + \varepsilon_n\}$ is independent and let $j = \mu i(b_j^s \in \operatorname{supp}_s(x + \varepsilon_n)$ and $i > \langle e, n\rangle)$, where $\operatorname{supp}_s(x)$ denotes the support of x relative to the basis $I^s \cup \{b_0^s, b_1^s, \ldots\}$. Note that j exists by our assumptions and, by the exchange property, $I^s \cup \{b_0^s, \ldots, b_{j-1}^s, x + \varepsilon_n, b_{j+1}^s, \ldots\}$ is a basis for V_∞. Then set $I^{s+1} = I^s \cup \{x + \varepsilon_n\}$ and define

$$b_i^{s+1} = \begin{cases} b_i^s & \text{if } i < j, \\ b_{i+1}^s & \text{if } i \geqslant j. \end{cases}$$

(Note in the latter case we have ensured that $\varepsilon_n \in M^{s+1} + (W_e^s)^*$.)

This completes the construction. It is easy to see by induction that the $\lim_s b_i^s = b_i$ exists for all i since once we have reached a stage t such that $b_j^s = b_j$ for all $s \geqslant t$ and $j < i$, then $b_i^{s+1} \neq b_i^s$ for $s \geqslant t$ at most once for each of the requirements $P_0, \ldots, P_{i-1}$.

Now suppose that $W_e \supset M$ and $\dim(W_e/M) = \infty$. Then clearly for any ε_n, there is an $x \in W_e$ such that $\{\varepsilon_n + x\} \cup I \cup \{b_0, \ldots, b_{\langle e,n\rangle}\}$ is independent. It then follows by the usual priority argument that each requirement $P_{\langle e,n\rangle}$ will eventually require attention and that requirement $P_{\langle e,n\rangle}$ will be satisfied. Thus M will be supermaximal. □

COROLLARY 42 (KALANTARI AND RETZLAFF [**1977**]). *For each $k \geqslant 0$, there exists a $V_k \in \mathscr{L}(V_\infty)$ such that V is k-thin.*

PROOF. Fix k and let V be a decidable subspace of V_∞ of codimension k. Build a supermaximal subspace V_k relative to k as in Theorem 40. It is not difficult to show that if $W \in \mathscr{L}(V_\infty)$, $W \supseteq V_k$ and $\dim(W/V_k) = \infty$, then $\dim(W \cap V/V_k) = \infty$ so that $W \cap V = V$ by the supermaximality of V_k in V. Thus $\dim(V_\infty/W) \leqslant k$ and hence V_k is k-thin. □

COROLLARY 43 (KALANTARI AND RETZLAFF [**1977**]). *There exists an infinite sequence of maximal spaces $V_0, V_1, \ldots$ such that if $i \neq j$, then there is no automorphism ϕ of $\mathscr{L}(V_\infty)$ with $\phi(V_i) = V_j$.*

PROOF. Let V_k be k-thin. □

Thus Corollary 42 contrasts with Soare's automorphism result for maximal sets (Theorem 37). Because the construction of Theorem 40 is a finite injury priority argument, it can be modified with a Yates permitting argument to prove the following

THEOREM 44 (REMMEL [**1980d**]). *For any nonzero r.e. degree* δ, *there exists a supermaximal* $V \in \mathcal{L}(V_\infty)$ *such that* $\deg(V) = \deg(D(V)) = \delta$.

Thus Theorem 44 shows that maximal spaces exist in every degree in contrast to maximal sets which exist in only the high degrees by Martin's result (Theorem 37). Remmel and Metakides and Nerode also showed that if the underlying field F of V_∞ is infinite then there are supermaximal spaces which are recursive as sets. In fact, supermaximal spaces can have an arbitrary sequence of dependence degrees if the underlying field of F is infinite.

THEOREM 45 (NERODE AND REMMEL [**1983**]). *Let* $A_0, A_1, A_2, \ldots$ *be any sequence of r.e. sets such that* $A_i \leqslant_T A_0$ *for all i uniformly*, $A_i \leqslant_T A_{i+1}$ *for all* $i \geqslant 1$, *and* A_0 *is not recursive. Then if the underlying field F of* V_∞ *is infinite, there exists a supermaximal space* $M \in \mathcal{L}(V_\infty)$ *such that* $D_i(M) \equiv_T A_i$ *for* $i \geqslant 1$ *and* $D(M) \equiv_T A_0$.

COROLLARY 46. *There exists a recursively presented vector space V such that for each* $k \geqslant 1$, *there is a uniform effective procedure which will determine for any* $v_0, \ldots, v_{k-1}$ *in V if* $\{v_0, \ldots, v_{k-1}\}$ *is an independent set and yet the only r.e. independent sets of V are finite.*

PROOF. Let $V = V_\infty/M$, where M is a supermaximal space of Theorem 45 and $A_1, A_2, \ldots$ are all recursive sets. □

Since it is not the case that all maximal sets differ by an automorphism of $\mathcal{L}(V_\infty)$, one might ask whether all supermaximal sets differ by an automorphism of $\mathcal{L}(V_\infty)$. However, Guichard showed that Soare's Theorem also fails for supermaximal sets. A *semilinear transformation* of V_∞ is a bijective map ϕ from V_∞ to V_∞ such that for all $v, w \in V$, $\phi(v + w) = \lambda\phi(v) + \beta\phi(w)$ for some λ and β in F. Kalantari [**1979**] showed that every automorphism ϕ of $\mathcal{L}(V_\infty)$ is induced by some semilinear transformation of V_∞. Guichard strengthened Kalantari's result to prove the following

THEOREM 47 (GUICHARD [**1983**]). *Every automorphism of* $\mathcal{L}(V_\infty)$ *is induced by a recursive semilinear transformation.*

Note that Theorem 47 also contrasts with both $\mathcal{E}$ and $\mathcal{E}^*$ where there are a $2^{\aleph_0}$-automorphisms as shown by Kent (see Rogers [**1967**]) and Lachlan, respectively; (see Soare [**1974**]). Since recursive semilinear transformations preserve the Turing degrees of r.e. vector spaces, automorphisms of $\mathcal{L}(V_\infty)$ preserve Turing degrees and hence it follows immediately from Theorem 44 that there are supermaximal spaces which do not differ by an automorphism of $\mathcal{L}(V_\infty)$. By diagonalizing over the recursive semilinear transformations, Guichard was also able to prove the following

THEOREM 48 (GUICHARD [**1983**]). *For every nonzero r.e. degree δ, there exist $V_0, V_1, \ldots$ in $\mathcal{L}(V_\infty)$ such that each V_i is supermaximal and has degree δ and, if $i \neq j$, then there is no automorphism of $\mathcal{L}(V_\infty)$ which carries V_i to V_j.*

If the underlying field F of V_∞ is infinite, Nerode and Remmel have shown that for any sequences of r.e. sets $A_0, A_1, \ldots$ (as in Theorem 45) there exists a sequence $V_0, V_1, \ldots$ of supermaximal subspaces such that for each k, $D_i(V_k) \equiv_T A_i$ if $i \geqslant 1$ and $D(V_k) \equiv_T A_0$; and if $j \neq k$, then there is no automorphism of $\mathcal{L}(V_\infty)$ which carries V_j to V_k.

The first r-maximal space which is not a maximal space was constructed by Kalantari [**1978**]. Given subspaces $A \subseteq B$ in $\mathcal{L}(V_\infty)$, we say that A is a *major subspace* of B if $\dim(B/A) = \infty$ and for any $W \in \mathcal{L}(V_\infty)$, $W + A =^* V_\infty$ (i.e., $\dim(V_\infty/W + A) < \infty$) whenever $W + B = V_\infty$. Of course, the notion of a major subspace is the analogue of the notion major subset as defined by Lachlan [**1968**].

THEOREM 49 (KALANTARI [**1978**]). *If B is an r.e. nondecidable subspace of V_∞, then there exists an r.e. subspace $A \subseteq B$ such that A is a major subspace of V_∞.*

COROLLARY 50. *There exists an r-maximal subspace A which is not maximal.*

PROOF. Let B be a maximal space and A a major subspace of B. Then clearly A is not maximal. But A is r-maximal since if $W + U = V_\infty$, where $W, U \in \mathcal{L}(V_\infty)$, then either $W + B =^* V_\infty$ or $U + B =^* V_\infty$. If $W + B =^* V_\infty$, there exists a finite-dimensional space E such that $(W + E) + B = V_\infty$. Then $(W + E) + A =^* V_\infty$ and hence $W + A =^* V_\infty$. Similarly if $U + B =^* V_\infty$, $U + A =^* V_\infty$. □

Remmel [**1980b**] showed that if B is a recursive basis for V_∞ and C is any nonrecursive r.e. subset of B, then there exists a major subset A of C such that $(A)^*$ is a major subspace of $(C)^*$. Thus if C is a maximal subset of B, then $(A)^*$ is an r-maximal but not maximal space with an extendible basis. In fact, Downey showed the following

THEOREM 51 (DOWNEY [**1983**]). *If B is a recursive basis for V_∞ and A is a major subset of C, where C is r.e. and $C \subseteq B$, then $(A)^*$ is major subspace of $(C)^*$.*

COROLLARY 52. *If B is a recursive basis for V_∞ and C is a maximal subset of B, then any major subset A of C generates an r-maximal subspace of V_∞ which is not maximal.*

We note that the question of whether or not every r-maximal subset of a recursive basis generates an r-maximal subspace is still open. However, by modifying Robinson's construction of an r-maximal set which is not contained in any maximal set, Remmel has proved the following

THEOREM 53 (REMMEL [**1978b**].) (i) *There exists an r-maximal subset R of a recursive basis B of V_∞ such that $(R)^*$ is an r-maximal subspace which is not contained in any maximal subspace of V_∞.*

(ii) *There exists an r-maximal subspace V of V_∞ such that no basis of V is extendible and V is not contained in any maximal subspace of V_∞.*

Inspired by the definition of supermaximal, Guichard defined $R \in \mathcal{L}(V_\infty)$ to be *super-r-maximal* if $\dim(V_\infty/R) = \infty$ and, for any $W, V \in \mathcal{L}(V_\infty)$ with $W + V = V_\infty$, either $W + R$ or $V + R$ equals V_∞. If $A, B \in \mathcal{L}(V_\infty)$ we say that A is a *supermajor* subspace of B if $A \subseteq B$, $\dim(B/A) = \infty$; and whenever $W + B = V_\infty$, $W \in \mathcal{L}(V_\infty)$, we have $W + A = V_\infty$.

THEOREM 54 (GUICHARD [**1984**]). *Let δ be any nonzero r.e. degree.*

(i) *If B is any r.e. nondecidable space, there exists a supermajor subspace A of B such that* $\deg(A) = \delta$.

(ii) *Every supermajor subspace of a supermaximal subspace is a super-r-maximal subspace.*

(iii) *There exists a super-r-maximal subspace R which is not contained in any maximal subspace such that* $\deg(R) = \delta$.

Note that Corollary 52 and Theorem 54(i), (ii) imply that r-maximal but not maximal spaces which are contained in maximal spaces exist with either fully extendible bases or with no extendible bases. Also the degree theoretic results of Theorem 54 contrast with the fact that in $\mathcal{E}$, r-maximal sets exist only in the high degrees.

The situation for an algebraically closed field with regard to the concepts of simple, maximal, and r-maximal differs from the vector space case in several respects. First of all, there are no simple elements in $\mathcal{L}(F_\infty)$. That is, suppose that $A \in \mathcal{L}(F_\infty)$ and $a_0, a_1, \ldots$ is an r.e. transcendence basis for A. Then if $x \notin A$, Ash observed that if $B = (\{x, a_0 + a_1x, a_1 + a_2x, a_2 + xa_3, \ldots\})^*$, then $B \cap S = (\varnothing)^*$. Thus if A has an infinite transcendence basis and $A \neq F_\infty$, then A is not simple. Certainly if A has a finite transcendence basis, A is not simple so that there are no simple elements in $\mathcal{L}(F_\infty)$. However, maximal elements exist. In fact, Metakides and Nerode [**1980**] and Remmel [**1980b**] showed that every maximal subset of a recursive transcendence basis for F_∞ generates a maximal space. Supermaximal spaces also exist; indeed, the exact analogue of Theorem 45 holds for fields as Remmel and Nerode proved Theorem 45 for a general class of matroids which include both F_∞ and V_∞, where the underlying field is infinite. Also Guichard's Theorem 52 holds for fields. However, it is not known whether k-thin elements exist for any $k \geqslant 1$. Also, it is not known how many automorphisms of $\mathcal{L}(F_\infty)$ there are. But once again a maximal element generated by a maximal subset of a recursive transcendence basis of F_∞ and a supermaximal element cannot differ by an automorphism of $\mathcal{L}(F_\infty)$. Thus, as with vector spaces, the analogues of Martin's and Soare's theorems fail for algebraically closed fields.

Remmel [**1980b**] has shown that there are r-maximal elements R_1 and R_2 of $\mathcal{L}(F_\infty)$ with fully extendible basis such that R_1 is not maximal but is contained in a maximal element and R_2 is not contained in any maximal element. Guichard's result (Theorem 54) on super r-maximal elements shows that r-maximal spaces exist with the same properties as R_1 and R_2 except that their bases are not extendible.

Next, let us consider the lattice $\mathscr{L}(Q)$. If we concentrate on the bidense suborderings of Q, we have the following definitions.

DEFINITION. Let A be an r.e. bidense subset of Q. We say A is

(i) *bidense simple* if $\bar{A}$ contains no infinite r.e. dense subset of Q.

(ii) *bidense maximal* if for any r.e. $W \supseteq A$, either $W - A$ or $Q - W$ is not dense.

(iii) *bidense r-maximal* if for any recursive bidense subset R and its complement $\bar{R}$, either $Q - (R \cup A)$ or $Q - (\bar{R} \cup A)$ is not dense.

The existence of bidense maximal sets was proved by Metakides and Nerode, and by Remmel. In fact, the following is proved in Metakides and Remmel [**1979**].

THEOREM 55 (METAKIDES AND REMMEL [**1979**]). (i) *There exist bidense maximal elements of $\mathscr{L}(Q)$ in every high r.e. degree.*

(ii) *There exist bidense simple elements of $\mathscr{L}(Q)$ in every nonzero r.e. degree.*

In fact, the constructions used to prove Theorem 55 in Metakides and Remmel [**1979**] actually produced elements with an even stronger property. Remmel made the following definitions in Remmel [**1980a**].

DEFINITION. Let A be any r.e. bidense subset of Q. Assume $q_1 < q_2$ and $[q_1, q_2] = \{q \in Q | q_1 \leqslant q \leqslant q_2\}$. We say that:

(i) A is *uniformly of degree* δ if $A \cap [q_1, q_2] \equiv_T \delta$ for all q_1, q_2.

(ii) A is *uniformly bidense maximal* if for all $W \supset A$ and all q_1, q_2, either $W - A$ or $Q - A$ is not dense in $[q_1, q_2]$.

(iii) A is *uniformly bidense simple* if for all q_1, q_2, there is no infinite r.e. $W \subseteq [q_1, q_2] - A$ such that W is dense in $[q_1, q_2]$.

(iv) A is *uniformly bidense r-maximal* if for all recursive bidense sets R and for all $[q_1, q_2]$ either $[q_1, q_2] - (R \cup A)$ or $[q_1, q_2] - (\bar{R} \cup A)$ is not dense in $[q_1, q_2]$.

(v) A is a *uniformly major subset* of an r.e. bidense subset B of Q if $B - A$ is dense in Q and for all q_1, q_2, and r.e. sets W, if $W \cup B \supseteq [q_1, q_2]$, then there exist

$$q_1 < r_1 < r_2 < q_2 \quad \text{such that } W \cup A \supseteq [r_1, r_2].$$

Now the construction of Theorem 55 produced elements which were of uniform degree and which were uniformly simple and uniformly maximal. However, suppose we carry out the same construction within some recursive bidense subset B of Q. If M is a uniformly bidense maximal subset of Q and N is a uniformly bidense maximal subset relative to B, then it is easy to see that

$$P = [(-\infty, 0) \cap N] \cup [(0, \infty) \cap M]$$

is "maximal" but not "uniformly maximal" and is of uniform degree. Similarly $P' = [(-\infty, 0) \cap B] \cup [(0, \infty) \cap M]$ is also a bidense maximal element but is not of uniform degree. Using the arguments above and Theorem 55, we have the following

COROLLARY 56 (REMMEL [**1980a**]). *Let δ be any high r.e. degree; then there exist bidense maximal elements M_1, M_2 and M_3 such that M_1 is uniformly bidense maximal and of uniform degree, M_2 is not uniformly bidense maximal and of uniform degree, and M_3 is neither uniformly bidense maximal nor of uniform degree.*

A similar corollary can be derived from Theorem 53 for simple sets. Moreover, Remmel has proved the following

THEOREM 57 (REMMEL [**1980a**]). *Let δ be any high r.e. degree; then there exists a bidense maximal element M of $\mathscr{L}(Q)$ such that M is not contained in any uniformly bidense maximal elements.*

Corollary 56 and Theorem 57 show that the notion of "uniformly" having a property with respect to all intervals is something new which arises in $\mathscr{L}(Q)$. However, we shall see that from the point of view of the general setting introduced in Metakides and Remmel [**1979**] and Remmel [**1980a**] which has both $\mathscr{E}$ and $\mathscr{L}(Q)$ as special cases, the notions of uniformly bidense maximal, simple, etc. are the natural analogues of maximal, simple, etc. sets in $\mathscr{E}$.

With regard to r-maximal elements, Remmel has proved the following

THEOREM 58 (REMMEL [**1980a**]). (i) *Every nonrecursive r.e. bidense subset B of uniform degree of Q has a uniformly major subset.*

(ii) *Every uniformly major subset of a maximal bidense subset is bidense r-maximal and every uniformly major subset of a uniformly maximal subset is uniformly bidense r-maximal.*

(iii) *There exists a uniformly bidense r-maximal subset of Q which is not contained in any uniformly bidense maximal element of Q.*

Next we turn to the lattice of r.e. subalgebras of a recursive Boolean algebra $\mathscr{B}$. Before we can define simple, maximal and r-maximal subalgebras of $\mathscr{B}$, we need to define the appropriate analogue of being coinfinite. To this end, we say that two subalgebras B and D of $\mathscr{B}$ are *equivalent mod finite sets*, $B =^* D$, if there exist finite sets S_1 and S_2 in $\mathscr{B}$ such that $(B \cup S_1)^* = (D \cup S_2)^*$. It is not difficult to show that $=^*$ is an equivalence relation and even that it is a congruence relation with respect to $+$ in $\mathscr{L}(\mathscr{B})$. That is, if $B_1 =^* B_2$ and $D_1 =^* D_2$, then $B_1 + D_1 =^* B_2 + D_2$. However, it is not a congruence relation with respect to intersection. Remmel has shown in [**1978a**] that if $\mathscr{B}$ is not isomorphic to $\mathscr{B}_\omega$, then there exist r.e. subalgebras B_1, B_2, D_1, D_2 of $\mathscr{B}$, such that $B_1 =^* B_2$ and $D_1 =^* D_2$ but $B_1 \cap D_1 \neq^* B_2 \cap D_2$. Nevertheless, we can use the notion finitely equivalent to $\mathscr{B}$ as the analogue of being coinfinite in $\mathscr{E}$.

DEFINITION. Let B be an r.e. subalgebra of a recursive Boolean algebra $\mathscr{B}$. We say that B is

(i) *simple* if $B \neq^* \mathscr{B}$ and, for any infinite r.e. $W \in \mathscr{L}(\mathscr{B})$, $W \cap B \neq \{0_{\mathscr{B}}, 1_{\mathscr{B}}\}$,

(ii) *maximal* if $B \neq^* \mathscr{B}$ and, for any r.e. $W \in \mathscr{L}(\mathscr{B})$ with $W \supseteq B$, either $W =^* B$ or $W =^* \mathscr{B}$,

(iii) *r-maximal* if $B \neq^* \mathscr{B}$ and, for any $W, U \in \mathscr{L}(\mathscr{B})$ with $U + W = \mathscr{B}$, either $W + B =^* \mathscr{B}$ or $U + B =^* \mathscr{B}$.

It is an open question whether or not there exist maximal or r-maximal subalgebras in every infinite recursive Boolean algebra $\mathscr{D}$. However, we can prove that for any recursive Boolean algebra $\mathscr{D}$, there exists a recursive Boolean algebra $\mathscr{B}$ isomorphic to $\mathscr{D}$ such that $\mathscr{L}(\mathscr{B})$ contains simple, maximal and r-maximal elements. Recall that Theorem 27 tells us that any recursive Boolean algebra $\mathscr{D}$ is isomorphic to a recursive Boolean algebra of the form $\tilde{N} \times \mathscr{B}$, $\tilde{Q} \times \mathscr{B}$, or $\tilde{C}$ for some recursive Boolean $\mathscr{B}$. Now within recursive Boolean algebras of the form $\tilde{N} \times \mathscr{B}$, $\tilde{Q} \times \mathscr{B}$, or $\tilde{C}$, we can construct simple, maximal and r-maximal subalgebras as the next results will show.

Recall that $\mathscr{A}(\mathscr{D})$ denotes the atoms of $\mathscr{D}$, $\mathscr{A}(\tilde{N})$ is a recursive set, and $(S)^*$ denotes the subalgebra generated by S.

THEOREM 59 (REMMEL [**1978a**]). *Let $\mathscr{B}$ be any recursive Boolean algebra.*

(i) *If S is a maximal subset of $\mathscr{A}(\tilde{N})$, then $(S)^*$ is a maximal subalgebra of $\tilde{N}$ and $(S)^* \times \mathscr{B}$ is a maximal subalgebra of $\tilde{N} \times \mathscr{B}$.*

(ii) *If S is a hypersimple subset of $\mathscr{A}(\tilde{N})$, then $()^*$ is a simple subalgebra of $\tilde{N}$ and $(S)^* \times \mathscr{B}$ is a simple subalgebra of $\tilde{N} \times \mathscr{B}$.*

(iii) *There exists $S_1, S_2 \subseteq \tilde{\mathscr{A}}(N)$ such that for $i = 1, 2$, $(S_i)^*$ and $(S_i)^* \times \mathscr{B}$ are r-maximal but not maximal subalgebras of $\tilde{N}$ and $\tilde{N} \times \mathscr{B}$, respectively, $(S_1)^*$ and $(S_1)^* \times \mathscr{B}$ are not contained in any maximal subalgebras of $\tilde{N}$ and $\tilde{N} \times \mathscr{B}$, respectively, and $(S_2)^*$ and $(S_2)^* \times \mathscr{B}$ are contained in maximal subalgebras of $\tilde{N}$ and $\tilde{N} \times \mathscr{B}$, respectively.*

Recall that $\tilde{Q}$ is the Boolean algebra generated by the left-closed right-open intervals of the rationals Q. Let $\tilde{R} = \{[i, i+1) | i = 0, 1, 2, \ldots\}$. Then $\tilde{R}$ is a recursive subset of Q. Let $\mathscr{H}$ denote the subalgebra of $\tilde{Q}$ generated by all intervals $[a, b)$ such that either (i) $[a, b) \subseteq (-\infty, 0)$ or (ii) for some n, $[a, b) \subseteq [n, n+1)$ and $b < n + 1$. Now $\mathscr{H}$ is a recursive subalgebra of $\tilde{Q}$. For any set $S \subseteq \tilde{R}$, let $S_{\mathscr{H}}$ denote $(S \cup \mathscr{H})^*$.

THEOREM 60 (REMMEL [**1978a**]). *Let $\mathscr{B}$ be any recursive Boolean algebra.*

(i) *If S is a maximal subset of $\tilde{R}$, then $S_{\mathscr{H}}$ is a maximal subalgebra of $\tilde{Q}$ and $S_{\mathscr{H}} \times \mathscr{B}$ is a maximal subalgebra of $\tilde{Q} \times \mathscr{B}$.*

(ii) *If S is a hypersimple subset of $\tilde{R}$, then $S_{\mathscr{H}}$ is a simple subalgebra of $\tilde{Q}$ and $S_{\mathscr{H}} \times \mathscr{B}$ is a simple subalgebra of $\tilde{Q} \times \mathscr{B}$.*

(iii) *There exist $S^1, S^2 \subseteq \tilde{R}$ such that for $i = 1, 2$, $S^i_{\mathscr{H}}$ and $S^i_{\mathscr{H}} \times \mathscr{B}$ are r-maximal but not maximal subalgebras of $\tilde{Q}$ and $\tilde{Q} \times \mathscr{B}$, respectively, $S^1_{\mathscr{H}}$ and $S^1_{\mathscr{H}} \times \mathscr{B}$ are not contained in any maximal subalgebras of $\tilde{Q}$ and $\tilde{Q} \times \mathscr{B}$, respectively, and $S^2_{\mathscr{H}}$ and $S^2_{\mathscr{H}} \times \mathscr{B}$ are contained in maximal subalgebras of $\tilde{Q}$ and $\tilde{Q} \times \mathscr{B}$, respectively.*

Recall that $\tilde{\mathscr{C}}$ is the Boolean algebra generated by $\tilde{Q}$ and all the singletons $\{q\}$ for $q \in Q$. Thus $\tilde{Q}$ can be thought of as a recursive subalgebra of $\tilde{\mathscr{C}}$.

THEOREM 61 (REMMEL [**1978a**, **1979**]). *Let S be an r.e. subalgebra of $\tilde{Q}$.*

(i) *If S is a maximal subalgebra of $\tilde{Q}$, $(S \cup \mathscr{A}(\tilde{C}))^*$ is a maximal subalgebra of $\tilde{\mathscr{C}}$.*

(ii) *If S is a simple subalgebra of $\tilde{Q}$, $(S \cup \mathscr{A}(\tilde{\mathscr{C}}))^*$ is a simple subalgebra of $\tilde{\mathscr{C}}$.*

(iii) *If S is an r-maximal subalgebra of $\tilde{Q}$, $(S \cup \mathscr{A}(\tilde{C}))^*$ is an r-maximal element of $\tilde{\mathscr{C}}$.*

(iv) *If S is an r.e. subalgebra of $\tilde{Q}$ not contained in any maximal element of $\tilde{Q}$, $(S \cup \mathscr{A}(\tilde{\mathscr{C}}))^*$ is not contained in any maximal element of $\tilde{\mathscr{C}}$.*

Now between Theorems 59, 60 and 61, it follows that there exist simple, maximal and r-maximal elements in every recursive Boolean algebra of the form $\tilde{N} \times \mathscr{B}$, $\tilde{Q} \times \mathscr{B}$, and $\tilde{\mathscr{C}}$. Moreover, if $S \subseteq \mathscr{A}(\tilde{N})$, $S \equiv_T (S)^* \equiv_T (S)^* \times \mathscr{B}$ and if $S \subseteq \tilde{R}$, $S \equiv_T S_{\mathscr{H}} \equiv_T S_{\mathscr{H}} \times \mathscr{B} \equiv_T (S_{\mathscr{H}} \cup \mathscr{A}(\tilde{\mathscr{C}}))^*$ so that the maximal and r-maximal elements exist in all high r.e. degrees and the simple elements exist in all nonzero r.e. degrees in the lattices $\mathscr{L}(\tilde{N} \times \mathscr{B})$, $\mathscr{L}(\tilde{Q} \times \mathscr{B})$, and $\mathscr{L}(\tilde{\mathscr{C}})$. However, things are not as straightforward as Theorems 59–61 might suggest. For example, a maximal subalgebra need not be simple. Now all the maximal subalgebras of Theorems 59–61 are simple subalgebras. However, the following holds.

THEOREM 62 (REMMEL [**1978a**]). (i) *Every maximal subalgebra M of $\tilde{N}$ is a simple subalgebra of $\tilde{N}$.*

(ii) *For any recursive Boolean algebra $\mathscr{D}$ which is not isomorphic to $\tilde{N}$, there exists a recursive Boolean algebra $\mathscr{B} \approx \mathscr{D}$ such that there exists a maximal subalgebra $M \in \mathscr{L}(\mathscr{B})$ which is not a simple subalgebra of $\mathscr{B}$.*

Finally, we should remark that one can also define the notions of a supermaximal subalgebra in analogy with the definition following Theorem 40. It is not difficult to show that there are no supermaximal subalgebras in $\tilde{N}$ but the existence of supermaximal subalgebras remains an open question for other recursive Boolean algebras. In particular, it is an open question whether supermaximal subalgebras exist even in $\tilde{Q}$. Nevertheless, the analogue of Soare's result still fails for the lattice of r.e. subalgebras, at least in most cases. It is a theorem of D. Sachs [**1962**] that every automorphism of the full lattice of all subalgebras of a Boolean algebra $\mathscr{B}$ is induced by an automorphism of $\mathscr{B}$. Guichard noticed that the Sachs result also holds for $\mathscr{L}(\mathscr{B})$ so that the following is true.

THEOREM 63 (GUICHARD [**1983**]). *Let $\mathscr{B}$ be any recursive Boolean algebra; then every automorphism of $\mathscr{L}(\mathscr{B})$ is induced by an automorphism of $\mathscr{B}$.*

It is easy to modify the constructions of Theorems 59–61 to produce maximal subalgebras of $\mathscr{B}$ which do not differ by an automorphism of $\mathscr{B}$ and hence do not differ by automorphisms of $\mathscr{L}(\mathscr{B})$. Thus the analogue of Soare's Theorem 38 fails for all Boolean algebras of the form $N \times \mathscr{B}$, $\tilde{Q} \times \mathscr{B}$, and $\tilde{\mathscr{C}}$ for any recursive Boolean algebra $\mathscr{B}$. Moreover, for recursive Boolean algebras with strong enough properties there are only countably many automorphisms.

THEOREM 64 (KALANTARI AND REMMEL, UNPUBLISHED). *Every automorphism of $\mathcal{L}(\tilde{N})$ is induced by a recursive automorphism of $\tilde{N}$.*

THEOREM 65 (GUICHARD [**1983**]). *Every automorphism of $\mathcal{L}(\tilde{Q})$ is induced by a recursive automorphism of $\tilde{Q}$.*

Remmel was able to combine the ideas of proofs of Theorems 64 and 65 to prove the following

THEOREM 66 (REMMEL [**198 b**]). (i) *If $\mathcal{B}$ is an atomic recursive Boolean algebra such that $\mathcal{A}(\mathcal{B})$ is recursive, then every automorphism of $\mathcal{L}(\mathcal{B})$ is induced by a recursive permutation of $\mathcal{A}(\mathcal{B})$.*

(ii) *If $\mathcal{B}$ is a recursive Boolean algebra such that $\mathcal{A}(\mathcal{B})$ is recursive and the set of atomic elements $\mathcal{A}t(\mathcal{B})$ is recursive, then there are only countably many automorphisms of $\mathcal{L}(\mathcal{B})$.*

We note that $x \in \mathcal{B}$ is *atomic* if $[x]$, considered as a Boolean algebra, is an atomic Boolean algebra. In case (ii) of Theorem 66, it is actually shown that every automorphism ϕ of $\mathcal{L}(\mathcal{B})$ is induced by a pair $\langle \phi_1, \phi_2 \rangle$, where ϕ_1 is a recursive permutation of $\mathcal{A}(\mathcal{B})$ and ϕ_2 is a recursive automorphism of $\mathcal{B} \bmod \mathcal{A}t(\mathcal{B})$. Again, it is an open question whether for every recursive Boolean algebra $\mathcal{B}$, there are only countably many automorphisms of $\mathcal{L}(\mathcal{B})$.

The situation for the lattice of r.e. ideals of $\tilde{Q}$ with regard to the notions of simple, maximal and r-maximal is more difficult. Take, for example, the notion of maximal. First we must decide what it means to be "coinfinite". It seems natural to say that an ideal I is *coinfinite* if $\tilde{Q}/I$ is an infinite Boolean algebra. But then we have the problem of defining the analogue of what it means for $W \supseteq I$ to be either "finitely equivalent" to I or "finitely equivalent" to $\tilde{Q}$. If we try to say that *I is "finitely equivalent" to W, $I =^* W$,* if there exists finite sets F_1 and F_2 such that $(I \cup F_1)^* = (I \cup F_2)^*$ as with subalgebras, we run into the problem that all ideals are finitely equivalent since $\tilde{Q} = (\{1_Q\})^*$. One way out of the problem is to fix some recursive nonprincipal maximal ideal M and consider only r.e. I ideals in M. Then not all ideals $I \subseteq M$ are finitely equivalent, in particular $M \neq^* \{0_{\mathcal{B}}\}$. In such a situation, we can define $I \subset M$ to be *maximal* if $I \neq^* M$ and for any r.e. ideal J with $I \subseteq J$, either $I =^* J$ or $J =^* M$. Remmel [**1980b**] took this point of view and produced a maximal ideal I in any recursive nonprincipal maximal ideal M. The construction of I produced the following

THEOREM 67 (REMMEL [**1980b**]). *There exists an r.e. ideal I such that $\tilde{Q}/I \approx \mathcal{B}_\omega$ and, for any r.e. ideal $J \supseteq I$, the ideal J/I in $\tilde{Q}/I$ is principal, i.e., $J/I = [x]$ where either x is a finite union of atoms in $\tilde{Q}/I$ or x is the complement of a finite union of atoms.*

It seems reasonable to call the r.e. ideal I maximal, and indeed Remmel constructed these elements as a special case of a construction which works in a much broader setting and which yields also as special cases maximal sets, vector spaces, subalgebras, etc. in every high r.e. degree. This setting will be described at the end of this paper. However, it should be noted that every recursive Boolean algebra can be realized as $\tilde{Q}/I$ for some recursive ideal I of $\tilde{Q}$. Thus, in particular, if $\mathscr{B} = \tilde{N}_0$, the recursive Boolean algebra of Theorem 25, where $\mathscr{B}_\omega \approx \tilde{N}_0$ and $\mathscr{I}(\mathscr{A}(\tilde{N}_0))$ is immune, then I, where $\tilde{N}_0 \approx_r \tilde{Q}/I$, also satisfies the properties of Theorem 66. Thus the I of Theorem 67 also can be recursive. We should also point out an old result of Martin and Pour-El, first expressed in terms of r.e. theories of the propositional calculus, yields the following

THEOREM 68 (MARTIN AND POUR - EL [**1970**]). *There exists an r.e. ideal I_0 such that $\tilde{Q}/I_0 \approx \tilde{Q}$, and, for any r.e. ideal $J \supseteq I_0$, the ideal J/I_0 in $\tilde{Q}/I$ is principal.*

The ideal I_0 of Theorem 68 might also be called "maximal" and since $\tilde{Q}$ has nonprincipal r.e. ideals, it follows that such ideals cannot be recursive.

If we return to ideals I contained in a fixed recursive nonprincipal maximal ideal M, Downey has shown that if we define $I \subseteq M$ to be *supermaximal* if for any r.e. ideal $J \supset I$ either $J =^* I$ or $J = M$, then supermaximal ideals do not exist. Downey [**1982**] has also explored other notions of maximal for r.e. filters of $\tilde{Q}$ which is, of course, equivalent to exploring r.e. ideals.

The notions of simple and r-maximal have similar problems in $\mathscr{L}I(\tilde{Q})$. As with the case of maximal, there are special cases of general results due to Remmel [**1980a**] which apply to r.e. ideals I contained in some fixed recursive nonprincipal maximal ideal M for the notions of simple and r-maximal. However, other than such general results such notions have not really been explored in $\mathscr{L}I(\tilde{Q})$. Finally, we should note that with regard to the number of automorphisms, there is a sharp contrast between the lattice of r.e. ideals and the lattice of r.e. subalgebras.

THEOREM 69 (REMMEL [**198 b**]). *Let $\mathscr{D}$ be a recursive Boolean algebra of the form $\tilde{N} \times \mathscr{B}$, $\tilde{Q} \times \mathscr{B}$, or $\tilde{\mathscr{C}}$, where $\mathscr{B}$ is a recursive Boolean algebra; then there are $2^{\aleph_0}$ automorphisms of the lattice of r.e. ideals of $\mathscr{D}$.*

In the lattice $\mathscr{L}(R^n)$, the first problem in defining the notions of simple, maximal and r-maximal is to decide what is the analogue of coinfinite. In this case, the notion of fragment number due to Kalantari [**1982**] (Kalantari and Leggett [**1982**]) provides a uniform theory. The *fragments* of a set $\mathscr{A} \subseteq R_n$ are defined to be the components of its interior. Then define

$$\operatorname{Frag}(\mathscr{A}) = \begin{cases} n & \text{if } \mathscr{A} \text{ has exactly } n \text{ fragments for some } n < \infty, \\ \infty & \text{otherwise;} \end{cases}$$

$$\operatorname{Frag}(\mathscr{A}, \mathscr{B}) = \begin{cases} n & \text{if exactly } n \text{ fragments of } \mathscr{A} \text{ have nonempty} \\ & \text{intersection with the interior of } \mathscr{B},\ n < \infty, \\ \infty & \text{otherwise.} \end{cases}$$

DEFINITION. Let $\mathscr{K}$ be an r.e. open set of R^n. We say that $\mathscr{K}$ is

(i) *simple* if $\operatorname{Frag}(R^n - \mathscr{K}) = \infty$ and for any r.e. open $U \subseteq R^n$, $\operatorname{Frag}(R^n - \mathscr{K}, U) = \infty$ implies $U \cap \mathscr{K} \neq \varnothing$,

(ii) *maximal* if $\mathrm{Frag}(R^n - \mathscr{K}) = \infty$ and for any r.e. open $U \supseteq \mathscr{K}$ either $\mathrm{Frag}(R^n - U) < \infty$ or $\mathrm{Frag}(R^n - \mathscr{K}, R^n - U) < \infty$,

(iii) *r-maximal* if $\mathrm{Frag}(R^n - \mathscr{K}) = \infty$ and for any r.e. open U_1 and U_2 such that U_1 is a complement of U_2, either $\mathrm{Frag}(R^n - \mathscr{K}, U_1) < \infty$ or $\mathrm{Frag}(R^n - \mathscr{K}, U_2) < \infty$.

There are also analogues of extendible and nonextendible bases of vector spaces in this setting. If $\mathscr{K}$ is an open set we say that $\Pi = \{\varepsilon_i | i \in S\}$ is a *partition* of $\mathscr{K}$ if the ε_i for $i \in S$ are pairwise disjoint and $\bigcup_{i \in S} \varepsilon_i$ is dense in $\mathscr{K}$. Π is said to be r.e. if S is r.e. Π is said to be *extendible* if there is an r.e. partition Π_1 of R^n such that $\Pi \subseteq \Pi_1$. We say an r.e. open set $\mathscr{K}$ is *extendible* if $\mathscr{K}$ has an r.e. partition which is extendible and $\mathscr{K}$ is *nonextendible* otherwise. Kalantari and Retzlaff produced the first example of an r.e. open set which is not extendible. This result was strengthened by Kalantari and Leggett.

THEOREM 70 (KALANTARI AND LEGGETT [**1982**, **1983**]). *There exist nonextendible r.e. open sets $\mathscr{K}_1$, $\mathscr{K}_2$ and $\mathscr{K}_3$ such that*

(i) *$\mathscr{K}_1$ is simple but not r-maximal or maximal,*

(ii) *$\mathscr{K}_2$ is r-maximal but not maximal,*

(iii) *$\mathscr{K}_3$ is maximal.*

In contrast to Theorem 70, the following holds for extendible r.e. open sets.

THEOREM 71 (KALANTARI AND LEGGETT [**1982**, **1983**]). (i) *There exists an extendible simple open set in R^n.*

(ii) *There are no extendible r-maximal or maximal open sets in R^n.*

Also in R^n, we cannot produce r-maximal sets not contained in any maximal sets since the following holds.

THEOREM 72 (KALANTARI AND LEGGETT [**1983**]). *Every nondense subset of R^n has a maximal superset.*

We should also note that if $\mathscr{K}$ is simple (maximal, r-maximal) and U is a dense subset of $\mathscr{K}$, then U is also simple (maximal, r-maximal). Since every r.e. open subset $\mathscr{K}$ has dense r.e. open subsets in every r.e. degree by Theorem 31, it follows that there are simple (maximal, r-maximal) subsets in every r.e. degree.

4. Splitting theorems, degrees of bases and nowhere simple elements. In this section, we shall explore the analogues of the splitting theorems of $\mathscr{E}$ in our six lattices. If A, B and C are r.e. sets such that $A = B \cup C$ and $B \cap C = \varnothing$, we say that B and C are an r.e. splitting of A. Given any set A, we let $\mathscr{E}_A$ denote the lattice $\{W \cap A | W \text{ is r.e.}\}$ under intersection and union. We say that an r.e. set W is *complemented* in $\mathscr{E}_A$ if there exists an r.e. set V such that $(W \cup V) \cap A = A$ and $W \cap V \cap A = \varnothing$. There are five splitting theorems in $\mathscr{E}^*$ which we shall consider. First, Friedberg [**1958**] showed that if A is an r.e. nonrecursive set, then A has an r.e. splitting $A = B \cup C$, where B and C are nonrecursive sets. Sacks [**1963**] strengthened this to show that if A is a nonrecursive r.e. set, there exists an r.e. splitting $A = B \cup C$ such that $B \mid_T C$. Owings [**1967**] also generalized Friedberg's splitting theorem to show that if $D = W - U$, where W and U are r.e. sets and A is noncomplemented in $\mathscr{E}_D$, then there exists an r.e. splitting $A = B \cup C$ such that both B and C are noncomplemented in $\mathscr{E}_D$. Morley and Soare generalized all these theorems.

THEOREM 73 (MORLEY AND SOARE [**1975**]). *Let D be any Δ_2^0-set and A an r.e. set such that A is noncomplemented in $\mathscr{E}_D$. Then there exists an r.e. splitting $A = B \cup C$ such that $B \mid_T C$ and both B and C are noncomplemented in $\mathscr{E}_D$.*

Another type of splitting theorem was proved by Shore [**1978b**]. Shore's splitting theorem involves the notion of nowhere simple sets. An r.e. set A is said to be *nowhere simple* if for any r.e. set W such that $W - A$ is infinite, there exists an infinite r.e. set $U \subseteq W - A$. A is said to be *effectively nowhere simple* if there is a recursive function f such that for all e, $W_{f(e)} \subseteq W_e - A$ and $W_{f(e)}$ is infinite if and only if $W_e - A$ is infinite. It turns out that both the notions of nowhere simple and effectively nowhere simple are elementary definable in $\mathscr{E}$ and $\mathscr{E}^*$. A class $\mathscr{C}$ of r.e. sets is elementary definable in $\mathscr{E}$ ($\mathscr{E}^*$) if there is a formula in one free variable $\phi(x)$ in the language with $=$, $\cap$, $\cup$, 0 and 1, where 0 and 1 represent the zero and one of the lattices $\mathscr{E}$ ($\mathscr{E}^*$) such that $W \in \mathscr{C}$ if and only if $\mathscr{E} \models \phi(w)$ ($\mathscr{E}^* \models \phi(w)$).

Nowhere simplicity is obviously definable in $\mathscr{E}$ and $\mathscr{E}^*$. The fact that effective nowhere simplicity is also definable in $\mathscr{E}$ and $\mathscr{E}^*$ follows from the fact that Miller and Remmel [**1984**] have proved that an r.e. set A is effectively nowhere simple if and only if there exists an r.e. set W called a witness set for A such that $W \cap A = \varnothing$ and for any r.e. set W_i, $W_i \cap W$ is infinite iff $W_i - A$ is infinite. The following results are now known about nowhere simple sets.

THEOREM 74. (i) (*Shore* [**1978a**]). *Every r.e. set A can be split into two nowhere simple sets B and C. If, also, A is nonrecursive, then there exist such B and C with $B \mid_T C$.*

(ii) (*Miller and Remmel* [**1984**]). *In every nonzero r.e. degree, there exist r.e. sets A and B such that A cannot be split into two effectively nowhere simple sets and B cannot be split into two noneffectively nowhere simple sets.*

THEOREM 75 (i) (*Shore* [**1978b**]). *There exist noneffectively nowhere simple sets in every nonzero r.e. degree.*

(ii) (*Miller and Remmel* [**1984**]). *There exist effectively nowhere simple sets in every nonzero r.e. degree.*

In $\mathscr{L}(V_\infty)$, the first splitting theorem was due to Retzlaff, who proved the analogue for vector spaces of the Friedberg Splitting Theorem.

THEOREM 76 (RETZLAFF [**1979**]). *Let V be a nondecidable space in $\mathscr{L}(V_\infty)$. Then there exist r.e. spaces U and W such that $U \oplus W = V$, and U and W are nondecidable.*

Shore proved the analogue for vector spaces of the Sacks Splitting Theorem in terms of dependence degrees.

THEOREM 77 (SHORE [**1978a**]). *If V is any nondecidable space in $\mathcal{L}(V_\infty)$, then in $\mathcal{L}(V_\infty)$ there exist U and W such that $U \oplus W = V$ and $D(U) \mid_T D(W)$.*

Nerode and Remmel proved the analogue for vector spaces of Shore's Splitting Theorem 75 for sets. Define a $V \in \mathcal{L}(V_\infty)$ to be *nowhere simple* if for any r.e. space W such that $\dim(W/V) = \infty$, there exists an r.e. space $U \subseteq W$ such that $\dim(U) = \infty$ and $U \cap V = \{\vec{0}\}$. Then $V \in \mathcal{L}(V_\infty)$ is said to be *effectively nowhere simple* if there is a recursive function f such that for all i, $W_{f(i)} \subseteq W_i$, $W_{f(i)} \cap U = \{\vec{0}\}$, and $\dim(W_{f(i)}) = \infty$ iff $\dim(W_i/V) = \infty$.

THEOREM 78 (NERODE AND REMMEL [**1982**]). *Let $V \in \mathcal{L}(V_\infty)$ be infinite dimensional. Then there exist nowhere simple W and U such that $W \oplus U = V$. If, in addition, V is nondecidable, then one can ensure that W and U also satisfy $D(W) \mid_T D(U)$.*

In contrast to $\mathcal{E}$, Retzlaff [**1979**] proved that if the underlying field F of V_∞ is infinite, then any $V \in \mathcal{L}(V_\infty)$ is the direct sum of two recursive spaces. Ash and Downey strengthened this result to prove the following (no matter whether F is finite or infinite)

THEOREM 79 (ASH AND DOWNEY [**1984**]). *If $V \in \mathcal{L}(V_\infty)$, then V is the direct sum of two decidable spaces.*

Note that any decidable space is effectively nowhere simple so that every $V \in \mathcal{L}(V_\infty)$ is a direct sum of two effectively nowhere simple spaces in contrast to Theorem 74(ii). There are several results on nowhere simple and effectively nowhere simple spaces in $\mathcal{L}(V_\infty)$.

THEOREM 80 (NERODE AND REMMEL [**1982**]). (i) *If B is a recursive basis for V_∞ and E is a nowhere simple subset of B, then $(E)^*$ is a nowhere simple space. If, in addition, E is an effectively nowhere simple subset of B, $(E)^*$ is an effectively nowhere simple space*

(ii) *In every nonzero r.e. degree δ, there exists an effectively nowhere simple space V such that no basis of V is fully extendible.*

THEOREM 81 (DOWNEY AND REMMEL [**198 a**]). *Let B a recursive basis for V_∞.*

(i) *There exists a simple subset S of B such that $(S)^*$ is an effectively nowhere simple space.*

(ii) *There exists a noneffectively nowhere simple subset N of B such that $(N)^*$ is an effectively nowhere simple space.*

(iii) *In every nonzero r.e. degree δ, there exists a noneffectively nowhere simple subset N_δ of B such that $(N_\delta)^*$ is a noneffectively nowhere space.*

(iv) *In every nonzero r.e. degree δ, there exists a noneffectively nowhere simple space V such that no basis of V is fully extendible.*

We note that every r.e. basis of a nowhere simple space V is extendible if $\dim(V_\infty/V) = \infty$ since there must exist an infinite-dimensional $U \in \mathscr{L}(V_\infty)$ with $U \cap V = \{\vec{0}\}$. Thus nowhere simple spaces with no fully extendible bases provide examples of spaces with extendible bases but with no fully extendible bases in every nonzero r.e. degree. We should note that if the underlying field F of V_∞ is infinite, then Nerode and Remmel have shown that there exist recursive nowhere simple spaces V such that no basis of V_∞ is fully extendible. Also effectively nowhere simple subspace are elementary definable in $\mathscr{L}(V_\infty)$ since Downey and Remmel [**198 a**] have shown that $V \in \mathscr{L}(V_\infty)$ is effectively nowhere simple if and only if there exists an r.e. subspace W such that $W \cap V = \{\vec{0}\}$ and, for any r.e. subspace U, $\dim(U \cap (W \oplus V)/V) = \infty$ iff $\dim(U/V) = \infty$.

Shore used nowhere simple sets to give a simple proof of his result that any class $\mathscr{C}$ of r.e. sets which is closed under finite equivalence and recursive permutations and contains an infinite and coinfinite set is an automorphism base for $\mathscr{E}$ and $\mathscr{E}^*$. That is, if ϕ is an automorphism of $\mathscr{E}$ $(\mathscr{E})^*$ and $\phi \restriction \mathscr{C}$ is the identity, then ϕ is the identity. A similar result holds for $\mathscr{L}(V_\infty)$ and $\mathscr{L}^*(V_\infty)$ (the lattice of r.e. subspaces mod finite-dimensional subspaces). Define a class $\mathscr{C}$ to be an *automorphism basis* for $\mathscr{L}(V_\infty)$ $(\mathscr{L}^*(V_\infty))$ if for any automorphism ϕ of $\mathscr{L}(V_\infty)$ $(\mathscr{L}^*(V_\infty))$, $\phi \restriction \mathscr{C}$ equals the identity if and only if ϕ equals the identity. One fact that makes life easy in $\mathscr{E}$ and $\mathscr{E}^*$ is that every automorphism of $\mathscr{E}^*$ is induced by an automorphism of $\mathscr{E}$. However, this fact fails for $\mathscr{L}(V_\infty)$ and $\mathscr{L}^*(V_\infty)$.

THEOREM 82 (GUICHARD [**1982**]). *Not every automorphism of $\mathscr{L}^*(V_\infty)$ is induced by an automorphism of $\mathscr{L}(V_\infty)$.*

In fact, the question remains open how many automorphisms of $\mathscr{L}^*(V_\infty)$ there are. Nevertheless, analogues of Shore's automorphism base theorem for $\mathscr{E}$ and $\mathscr{E}^*$ do hold. Define $V =^* W$ if there exist finite sets F_1 and F_2 such that $(V \cup F_1)^* = (W \cup F_2)^*$. We say $\mathscr{C}$ is **-closed* if $V \in \mathscr{C}$ and $W =^* V$ implies $W \in \mathscr{C}$.

THEOREM 83. *Let $\mathscr{C}$ be a class of r.e. subspaces of V_∞ such that $\mathscr{C}$ is *-closed and $\mathscr{C}$ is closed under invertible recursive linear transformations. Then*

(i) (*Nerode and Remmel* [**1982**]) *$\mathscr{C}$ is an automorphism base for $\mathscr{L}(V_\infty)$ and*

(ii) (*Downey and Remmel* [**198 c**]) *$\mathscr{C}$ is an automorphism base for $\mathscr{L}^*(F_\infty)$.*

We should note that the proof of Therorem 83 is more than a slavish imitation of Shore's proof. The first proof of Theorem 83(i) used not only the splitting theorem for nowhere simple subspaces (Theorem 78), but also the fact that every automorphism of $\mathscr{L}(V_\infty)$ is induced by a recursive semilinear transformation (Theorem 47) while the proof of Theorem 83(ii) required the splitting theorem for decidable spaces (Theorem 79).

There is also a splitting theorem for $\mathscr{L}(V_\infty)$ which has no analogue back in $\mathscr{E}$.

THEOREM 84 (DOWNEY [**1984b**]). *If V is an r.e. nondecidable subspace of $\mathscr{L}(V_\infty)$, then there exist W and U in $\mathscr{L}(V_\infty)$ such that $W \oplus U = V$ and no basis of W or U is fully extendible.*

There is an intimate relation between the degrees of splittings of r.e. sets and the degrees of r.e. bases for an r.e. space $V \in \mathcal{L}(V_\infty)$. We note that if $V \in \mathcal{L}(V_\infty)$ and R is an r.e. basis for V, then $R \leqslant_T V$. That is, to decide if $x \in R$, we first decide if $x \in V$. Of course, if $x \notin V$, then $x \notin R$. If $x \in V$, then we simply enumerate R until we find at stage s that $x \in (R^s)^*$, in which case we know $x \in R$ iff $x \in R^s$. The following result proves that the extremes for the possible degrees of R are realized.

THEOREM 85. *Let $V \in \mathcal{L}(V_\infty)$.*
(i) (*Dekker* [**1971**]) *There exists a recursive basis R of V.*
(ii) (*Remmel* [**1980d**]) *There exists an r.e. basis R of V such that $R \equiv_T V$.*

Now suppose V is a nonrecursive space, R is an r.e. basis of V of degree V, and $R = D \cup E$ is an r.e. splitting of R. Then we claim that we can produce r.e. bases B_1 and B_2 of V such that $B_1 \equiv_T D$ and $B_2 \equiv_T E$. That is, by Theorem 85, there exist recursive bases R_D of $(D)^*$ and R_E of $(E)^*$. Thus if $B_1 = D \cup R_E$ and $B_2 = E \cup R_D$, B_1 and B_2 are r.e. bases of V which are Turing equivalent to D and E, respectively. If we now apply the Sacks Splitting Theorem repeatedly to R (split $R = R_1 \cup R_2$ with $R_1 |_T R_2$, then split $R_1 = R_{11} \cup R_{12}$ with $R_{11} |_T R_{12}$ and split $R_2 = R_{21} \cup R_{22}$ with $R_{21} |_T R_{22}$, etc.), then we can easily prove the following

THEOREM 86 (REMMEL [**1980d**]). *Let V be a nonrecursive space in $\mathcal{L}(V_\infty)$; then there exist r.e. bases of V in infinitely many distinct r.e. degrees.*

It is natural question to ask whether for any r.e. set $A \leqslant_T V$, V has an r.e. basis B such that $B \equiv_T A$. The answer to this question is positive if R has the property that for any r.e. set $A \leqslant_T R$, there exists an r.e. splitting $R = B \cup C$ such that $B \equiv_T A$. It is precisely this reasoning that lead Lerman and Remmel to define the following new concept back in $\mathcal{E}$.

DEFINITION. An r.e. set A has the *universal splitting property* if for any r.e. set $E \leqslant_T A$, there exist r.e. sets B and C such that $A = B \cup C$, $B \cap C = \varnothing$ and $B \equiv_T E$.

A similar concept in $\mathcal{L}(V_\infty)$ is the following

DEFINITION. An r.e. subspace V of V_∞ has the *universal basis property* if for every r.e. set $E \leqslant_T V$, there exists an r.e. basis B of V such that $B \equiv_T E$.

By our argument above, if every r.e. set has the universal splitting property, then every r.e. subspace would have the universal basis property. However, Lerman and Remmel [**1982, 1984**] have proved that there exist sets without the universal splitting property; in fact, the degrees of such sets are dense. Moreover, there is an r.e. degree δ such that every r.e. set A in δ fails to have the universal splitting property. Also every creative set has the universal splitting property and there are nonrecursive r.e. sets with the universal splitting property below a nonzero r.e. degree δ. Similar results hold for the universal basis property.

THEOREM 87 (LERMAN AND REMMEL [**1982**, **1984**]). (i) *The degrees of r.e. subspaces V of $\mathscr{L}(V_\infty)$ which fail to have the universal basis property include* $\mathbf{0}'$ *and are dense in the r.e. degrees.*

(ii) *There exists a nonzero r.e. degree δ such that no $V \in \mathscr{L}(V_\infty)$ of degree δ has the universal basis property.*

(iii) *There exist $V \in \mathscr{L}(V_\infty)$ with the universal basis property of degree* $\mathbf{0}'$ *and of degree strictly between* $\mathbf{0}$ *and any nonzero r.e. degree δ.*

One could also ask about a "universal splitting property" for an r.e. subspace of V_∞. However, our next result shows that such a property would be identical with the universal basis property.

THEOREM 88 (DOWNEY, REMMEL AND WELCH [**198**]). *Let $V \in \mathscr{L}(V_\infty)$; then δ is the degree of an r.e. basis B of V if and only if there exist $W, U \in \mathscr{L}(V_\infty)$ with $W \oplus U = V$ and $W \equiv_T \delta$. In fact, γ and δ are r.e. degrees such that there exist disjoint r.e. sets C of degree γ and D of degree δ with $C \cup D$ an r.e. basis of V if and only if there exist r.e. subspaces W of degree γ, and U of degree δ such that $W \oplus U = V$.*

Note that as a corollary of Theorem 88, we can prove the following analogue of the Sacks Splitting Theorem.

COROLLARY 89. *Let V be any nonrecursive r.e. subspace of V_∞. Then there exist U and W in $\mathscr{L}(V_\infty)$ such that $U \oplus W = V$ and $U|_T W$.*

PROOF. Just let A be an r.e. basis of V such that $B \equiv_T V$ is guaranteed by Theorem 85. By the Sacks Splitting Theorem we can split A into r.e. sets B and C such that $B|_T C$. The result now follows by Theorem 88. □

Another concept with regards to the restrictions of r.e. splittings is that of mitotic sets. An r.e. set A is *mitotic* if there exist r.e. sets B and C which split A such that $B \equiv_T C \equiv_T A$ and A is said to be nonmitotic otherwise. Lachlan [**1967**] proved that nonmitotic sets exist and Ladner [**1973a**, **1973b**] studied the possible degrees of mitotic sets. Following some work of Downey and Welch for sets [**198**], we say an r.e. subspace V is *strongly atomic* if whenever A_1 and A_2 in $\mathscr{L}(V_\infty)$ split A, then $\inf(\deg(A_1), \deg(A_2)) = \mathbf{0}$. Clearly a strongly atomic space is "nonmitotic" in the sense that there are no r.e. subspaces B and C which split A such that $B \equiv_T C \equiv_T A$. Moreover by applying Theorem 88, it is easy to see that every r.e. basis R of A is nonmitotic. The existence of a variety of superatomic spaces follows from our next result and Theorem 92.

Define a subspace Q to be *fully co-r.e.*. if Q is generated by a co-r.e. subset of a recursive basis for V_∞

THEOREM 90 (DOWNEY, REMMEL AND WELCH [**198**]). *There exists a fully co-r.e. space C such that C is of high degree and if $W \in \mathscr{L}(V_\infty)$ and $W \oplus C = V_\infty$, then W is strongly atomic. Moreover, there exists an r.e. complement V of C such that $V \equiv_T C$ and there is a nonzero r.e. degree α less than* $\deg(V)$ *such that for any r.e. set B with $\mathbf{0} <_T \deg(B) \leqslant_T \alpha$ either V has no r.e. basis R with $R \equiv_T B$ or V has an* *r.e. subspace U with $U \equiv_T B$ such that there is no $W \in \mathscr{L}(V_\infty)$ satisfying $U \oplus W = V$ and* $\deg(U) \vee \deg(W) = \deg(V)$.

Remmel [**1977a**] noted that an infinite injury argument can be applied to $\mathscr{L}(V_\infty)$. Theorem 90 represents the first results in $\mathscr{L}(V_\infty)$ that at present require the machinery of infinite injury constructions.

Another property that is closely related to the universal splitting and basis properties arises from the fact that complementation is not unique in $\mathscr{L}(V_\infty)$. For example, Remmel has proved the following

THEOREM 91 (REMMEL [**1980d**]). *In $\mathscr{L}(V_\infty)$, there exist r.e. subspaces V_1 and V_2 and co-r.e. subspaces Q_1 and Q_2 such that for all $i, j \in \{1,2\}$, $V_1 \oplus Q_j = V_\infty$ and V_1 is supermaximal, V_2 is not maximal, Q_1 has a fully extendible basis, and Q_2 has no basis which is extendible.*

THEOREM 92. (DOWNEY [**1984a**]). (i) *Every r.e. subspace $V \in \mathscr{L}(V_\infty)$ has a fully co-r.e. complement.*

(ii) *Every fully co-r.e. subspace Q of V_∞ has a decidable complement and a nondecidable r.e. complement.*

(iii) *If Q is a fully co-r.e. subspace generated by an immune subset of a recursive basis, then Q has supermaximal, k-thin complements for every $k \geqslant 1$, and super-r-maximal but not supermaximal complements.*

Thus from Theorem 92, there seem to be relatively few lattice-theoretic restrictions on the r.e. complements of fully co-r.e. subspaces. One could ask for the possible degrees of r.e. complements of a fully co-r.e. space Q. Say $Q = (C)^*$, where C is a co-r.e. subset of a recursive basis B of V_∞. It is not difficult to show that if $W \in \mathscr{L}(V_\infty)$ and $W \oplus Q = V_\infty$, then $W \leqslant_T Q$. Thus the extremes of the possible degrees of r.e. complements of Q are possible since Q has a decidable complement by Theorem 92(ii) and $W = (B - C)^*$ is an an r.e. complement such that $W \equiv_T Q$. We say that Q has the *universal complementation property* if for every r.e. degree $\delta \leqslant_T \deg(Q)$, there exists an r.e. complement W of Q such that $\deg(W) = \delta$. Downey and Remmel have proved the following

THEOREM 93 (DOWNEY AND REMMEL [**198 b**]). (i) *Every fully co-r.e. subspace Q which is nonrecursive has r.e. complements in infinitely many r.e. degrees.*

(ii) *The degrees of fully co-r.e. subspaces of V_∞ which fail to have the universal complementation property include* $\mathbf{0}'$ *and are dense in the r.e. degrees.*

(iii) *There are fully co-r.e. subspaces of V_∞ with the universal complementation property in every r.e. degree which contain an r.e. set with the universal splitting property.*

Once again there are several contrasts between $\mathscr{L}(V_\infty)$ and $\mathscr{L}(F_\infty)$ with regards to the type of results in §4. All the splitting theorems, i.e., the analogues of Theorems 76–79 and 84, continue to hold for $\mathscr{L}(F_\infty)$. However, we should note that the notion of nowhere simplicity has no real content in $\mathscr{L}(F_\infty)$. That is, the obvious notion of nowhere simplicity is to define a coinfinite-dimensional $V \in \mathscr{L}(F_\infty)$ to be *nowhere simple* if whenever $W \in \mathscr{L}(F_\infty)$ and $\dim(W/V) = \infty$, there exists an $R \in \mathscr{L}(F_\infty)$ such that $\dim(R) = \infty$, $W \supseteq R$ and $R \cap V = (\varnothing)^*$. Recall Ash's example that if $x \notin U$ and $a_0, a_1, \ldots$ is an r.e. infinite transcendence basis for U, then $U_x = (\{x, a_0 + a_1 x, a_1 + a_2 x, \ldots\})^*$ is an r.e. algebraically closed subfield of F_∞ such that $U_x \cap U = (\varnothing)^*$. Now suppose $V \in \mathscr{L}(F_\infty)$. If $\dim(V) < \infty$, then clearly V is nowhere simple. If $\dim(V) = \infty$ and $W \in \mathscr{L}(F_\infty)$ is such that $\dim(W/V) = \infty$, then either $\dim(W \cap V) < \infty$ in which case it is easy to construct $R \subseteq W$ such that R is independent over $W \cap V$ and $\dim(R) = \infty$; or $\dim(W \cap V) = \infty$ in which case if $U = W \cap V$ and $x \in W - U$, then U_x as defined above satisfies $\dim(U_x) = \infty$ and $U_x \cap V = (\varnothing)^*$. Thus with nowhere simplicity defined as above, every $V \in \mathscr{L}(F_\infty)$ is nowhere simple so that the analogue of Theorem 78 has no real content in $\mathscr{L}(F_\infty)$. To remedy this, we define $V \in \mathscr{L}(F_\infty)$ to be *strongly nowhere simple* if whenever $W \in \mathscr{L}(F_\infty)$ and $\dim(W/V) = \infty$, there exists an $R \in \mathscr{L}(V_\infty)$ such that $R \subseteq W$, $\dim(R) = \infty$, and R is independent over V. We can then define the notion of effectively strongly nowhere simple in the obvious way.

In Nerode and Remmel [**1983**] it was shown that every infinite dimensional V in $\mathscr{L}(F_\infty)$ is the direct sum of two strongly nowhere simple elements *provided* that there is an algorithm which determines, from bases of finitely generated A, $B \in \mathscr{L}(F_\infty)$, a basis for $A \cap B$. No such algorithm was known at the time. John Rosenthal has since given one, so this splitting theorem does hold for $\mathscr{L}(F_\infty)$.

Moreover, the analogue of Theorem 80 as to the existence of strongly nowhere simple algebraically closed subfields with either fully extendible or no fully extendible bases continues to hold. Also, the analogue of parts (iii) and (iv) of Theorem 81 concerning the existence of noneffectively strongly nowhere simple algebraically closed subfields with either a fully extendible or no fully extendible basis also holds in F_∞. Parts (i) and (ii) of Theorem 81 remain open in F_∞.

We should also remark that $=^*$ is not a congruence relation with respect to intersection in F_∞. Now, $W =^* V$ means there exists finite sets F_1 and F_2 in F_∞ such that $(W \cup F_1) =^* (V \cup F_2)$. We see that with U and U_x as defined in the previous paragraph, $U =^* U_x$ and $U \cap U_x = (\varnothing)^*$. Thus there is no lattice $\mathscr{L}(F_\infty)$ of $\mathscr{L}(F_\infty)$ modulo finite-dimensional fields as there is for V_∞. As we mentioned previously, the question of how many automorphisms of $\mathscr{L}(F_\infty)$ exist remains open. We can prove an automorphism base result for $\mathscr{L}(F_\infty)$. That is, the Ash-Downey splitting theorem, Theorem 79, holds for $\mathscr{L}(F_\infty)$ so that every $V \in \mathscr{L}(F_\infty)$ is the direct sum of two decidable algebraically closed subfields. Using this result, Downey and Remmel have proved the following

THEOREM 94 (DOWNEY AND REMMEL [**198 c**]). *Let $\mathscr{C}$ be any class of elements of $\mathscr{L}(F_\infty)$ such that there exists a $V \in \mathscr{C}$ with* $\dim(V) = \dim(F_\infty/V) = \infty$ *and $\mathscr{C}$ is closed under any recursive automorphism of $\mathscr{L}(F_\infty)$. Then $\mathscr{C}$ is an automorphism base for $\mathscr{L}(F_\infty)$, i.e., if ϕ is an automorphism of $\mathscr{L}(F_\infty)$ and $\phi \restriction \mathscr{C} =$ identity, then $\phi =$ identity.*

Finally, we should note that all analogues of the theorems about the degrees of r.e. bases of r.e. subspaces of V_∞ carry over for the degrees of r.e. transcendence bases of r.e. algebraically closed subfields of $\mathscr{L}(F_\infty)$. Similarly, all the analogues of Theorems 89–91 concerning the uniqueness of complements and the degrees of r.e. complements of fully co-r.e. subspaces hold for $\mathscr{L}(F_\infty)$.

Splitting theorems in recursive Boolean algebras have not been explored a great deal for either the lattice of r.e. ideals or the lattice of r.e. subalgebras. Once again, the problem is dependent on the isomorphism type of the Boolean algebra. For example, let us consider the lattice of r.e. ideals. Recall that $\tilde{N}$ is the Boolean algebra of finite and cofinite subsets of N, where the set of atoms $\mathscr{A}(\tilde{N})$ is recursive. Now in $\tilde{N}$, there are only two types of r.e. ideals, namely principal ideals $[x]$ or ideals generated by subsets of atoms $I = (B)^*$ for $B \subseteq \mathscr{A}(\tilde{N})$. Of course, principal ideals can only be split into principal ideals. If $I = (B)^*$ with $B \subseteq \mathscr{A}(\tilde{N})$ and $I = I_1 + I_2$, where $I_1 \cap I_2 = \{0_N\}$, then $I_1 = (B_1)^*$ and $I_2 = (B_2)^*$, where B_1 and B_2 split B. Since $\deg(I) = \deg(B)$ whenever $I = (B)^*$ and $B \subseteq \mathscr{A}(\tilde{N})$, it follows that all results about r.e. splitting of I into r.e. ideals are equivalent to results about r.e. splittings of B into r.e. sets. Thus, there are no really new phenomena or results about r.e. splittings of r.e. ideals in $\tilde{N}$. However, in constrast, we have the following result about r.e. ideals in the atomless Boolean algebra $\tilde{Q}$.

THEOREM 95 (REMMEL [**198 a**]). *For any nonrecursive nonprincipal r.e. ideal I in $\tilde{Q}$, there exist recursive ideals I_1 and I_2 such that $I_1 + I_2 = I$ and $I_1 \cap I_2 = \{0_{\tilde{Q}}\}$.*

We should note that Theorem 95 points to problems in trying to prove an analogue of the Sacks Splitting Theorem for ideals. That is, given a nonrecursive r.e. set A, the strategy for the Sacks Splitting Theorem is to build r.e. sets B and C which split A so that the following requirements are met for $e = 0, 1, \ldots$:

$$\mathrm{R}_{e,B}\colon \phi_e^B \neq A \quad \text{and} \quad \mathrm{R}_{e,C}\colon \phi_e^C \neq A,$$

where $\phi_e^B \neq A$ means that the function computed by the eth oracle machine with oracle B is not the characteristic function of A. For sets, it is easy to see that since $\deg(B) \vee \deg(C) = \deg(A)$, meeting all the requirements $\mathrm{R}_{e,B}$ and $\mathrm{R}_{e,C}$ ensures that $B \not\leqslant_T C$ and $C \not\leqslant_T B$. However, by Theorem 95, we see that if $I = I_1 + I_2$, where I, I_1 and I_2 are r.e. ideals of $\tilde{Q}$ and $I_1 \cap I_2 = \{0_{\mathscr{B}}\}$, we can ensure $\phi_e^{I_1} \neq I$ and $\phi_e^{I_2} \neq I$ for all e and yet we cannot conclude that $I_1 |_T I_2$. Actually the same problem arises in vector spaces or algebraically closed fields, but in those cases we were able to reduce the splittings of r.e. elements of $\mathscr{L}(V_\infty)$ or $\mathscr{L}(F_\infty)$ to splitting of r.e. bases via Theorem 88. In the case of r.e. ideals and in r.e. subalgebras, the analogue of the Sacks splitting remains open.

One nice property that r.e. ideals have is that if I_0 and I_1 is an r.e. splitting of an r.e. ideal I, then $I_0 \leqslant_T I$ and $I_1 \leqslant_T I$. That is, $I = I_0 + I_1$ and $I_0 \cap I_1 = \{0_{\mathscr{B}}\}$, then $I - I_j = \{x \in I \mid [x] \cap I_{1-j} \neq \{0_{\mathscr{B}}\}\}$ is r.e. for $j = 0$ or 1 so that I_j and $I - I_j$ is a splitting of I into r.e. sets. This fact allows one to prove the analogue of

Lerman and Remmel's results on the universal splitting property. However, this property is not shared by r.e. subalgebras. That is, given an r.e. subalgebra B of a recursive Boolean algebra $\mathscr{B}$, we say that r.e. subalgebras B_1 and B_2 of $\mathscr{B}$ *split* B if $B_1 + B_2 = B$ and $B_1 \cap B_2 = \{0_{\mathscr{B}}, 1_{\mathscr{B}}\}$. It does not follow that if B_1 and B_2 split B that $B_1 \leqslant_T B$ and $B_2 \leqslant_T B$. In fact, Remmel has proved the following

THEOREM 96 (REMMEL, UNPUBLISHED). *Let $\mathscr{B}$ be a recursive Boolean algebra, and α, β and γ any r.e. degrees. Then there exist a recursive Boolean algebra $\mathscr{D}$ isomorphic to $\mathscr{B}$, and r.e. subalgebras A, B and C of $\mathscr{D}$ such that B and C split A and* $\deg(A) = \alpha$, $\deg(B) = \beta$ *and* $\deg(C) = \gamma$.

About the only general theorem for recursive Boolean algebras with regard to the lattice of r.e. subalgebras is the following for which we have seen only an abstract.

THEOREM 97 (SHI NIANDONG, PERSONAL COMMUNICATION). *Let B be any nonrecursive r.e. subalgebra of a recursive Boolean algebra $\mathscr{B}$; then B can be split into two nonrecursive r.e. subalgebras B_1 and B_2.*

Next we turn to the lattice $\mathscr{L}(Q)$. In $\mathscr{L}(Q)$, we have a full analogue of the Morley-Soare Splitting Theorem.

DEFINITION. Let S be a dense subset of Q; then we say that an r.e. set A is *uniformly noncomplemented* in $\mathscr{E}_S$ if for any $a < b$ in Q, there is no r.e. set $B \subseteq Q$ such that $(A \cup B) \cap S \supseteq [a, b] \cap S$ and $A \cap B \cap S \cap [a, b] = \varnothing$. Note that if A is uniformly noncomplemented in a dense subset S of Q, then automatically $A \cap S$ is bidense. Remmel has proved the following analogue of Theorem 73.

THEOREM 98 (REMMEL [**1980a**]). *If S is a Δ_2^0 dense subset of Q and A is an r.e. set which is uniformly noncomplemented in $\mathscr{E}_S$, then there exist r.e. sets B and C such that*

(i) $B \cup C = A$ *and* $B \cap C = \varnothing$,

(ii) *B and C are uniformly noncomplemented in $\mathscr{E}_S$, and*

(iii) $B \mid_T C$.

We note that the hypotheses of A being uniformly noncomplemented in $\mathscr{E}_S$ is essential in Theorem 98. That is, if we try to weaken Theorem 98 by replacing the condition that A is uniformly noncomplemented with the condition $A \cap S$ is dense and A is noncomplemented in $\mathscr{E}_S$, we may not even be able to split A into two r.e. sets B and C so that $B \cap S$ and $C \cap S$ are dense.

THEOREM 99 (REMEL [**1980a**]). *There exist a co-r.e. dense set $S \subseteq Q$ and an r.e. $A \subseteq Q$ such that $A \cap S$ is dense in Q, A is noncomplemented in $\mathscr{E}_S$, but there are no r.e. sets B and C such that $B \cup C = A$, $B \cap C = \varnothing$, and both $B \cap S$ and $C \cap S$ are dense in Q.*

PROOF. Let R be the set of Gödel numbers of the integers $\mathbf{Z}$ as they sit inside Q and let M be the uniformly bidense maximal subset as in Corollary 56. Then it is easy to see that $U = M - R$ is still a bidense maximal element of $\mathcal{L}(Q)$. Let $S = Q - U$ and T be a nonrecursive r.e. subset of R. Note that $R \subseteq S$ and that if $A = T \cup (Q - R)$, then $A \cap S$ is dense in Q and is noncomplemented in $\mathcal{E}_S$ since T is nonrecursive. Then if B and C split A, where $B \cap S$ and $C \cap S$ are dense, then $M \cup B$ would be an r.e. set such that both $(M \cup B) - M$ and $Q - (M \cup B)$ are dense, which would violate the bidense maximality of M. □

Using Theorem 98, one can also generalize Lachlan's remarkable characterization of *hh*-simple sets S [**1968**] as being those r.e. sets S such that $\mathcal{E}_{\bar{S}}$ is a Boolean algebra. That is, we define an r.e. set $A \subseteq Q$ to be *bidense hh-simple* if A is bidense and there is a recursive function $h(x, y, z)$ such that for all e and all $a < b$ in Q, $W_{h(e,a,b)}$ is finite, $W_{h(e,a,b)} \cap \bar{A} \cap [a, b] \neq \varnothing$, and

$$W_{h(e,a,b)} \cap W_{h(x,y,z)} = \varnothing \quad \text{if } (e, a, b) \neq (x, y, z).$$

We then have the following

THEOREM 100 (REMMEL [**1980a**]). *For an r.e. bidense subset $A \subseteq Q$, A is bidense hh-simple if $\mathcal{E}_{\bar{A}}$ contains no uniformly noncomplemented elements.*

Finally, we consider the lattices $\mathcal{L}(R^n)$. Given an r.e. open set $\mathcal{K}$ in $\mathcal{L}(R^n)$, we say that r.e. open sets $\mathcal{L}_1$ and $\mathcal{L}_2$ split $\mathcal{K}$ if $\mathcal{L}_1 \cup \mathcal{L}_2$ is dense in $\mathcal{K}$ and $\mathcal{L}_1 \cap \mathcal{L}_2 = \varnothing$. Due to the fact that every r.e. open set $\mathcal{K}$ has r.e. open dense subsets in every r.e. degree by Theorem 31, we can easily prove the following result which is in sharp contrast to the Lerman-Remmel result that there exist r.e. sets which fail to have the universal splitting property.

THEOREM 101 (KALANTARI AND REMMEL [**1983**]). *Let $\mathcal{K}$ be a nonempty r.e. open set in R^n, and δ and η any two r.e. degrees; then $\mathcal{K}$ can be split into two r.e. open sets $\mathcal{L}_1$ and $\mathcal{L}_2$ such that* $\deg(\mathcal{L}_1) = \delta$ *and* $\deg(\mathcal{L}_2) = \eta$.

5. Undecidability of the lattice of r.e. substructures. A major recent open question in classical recursion theory was whether the theory of the lattices $\mathcal{E}$ and $\mathcal{E}^*$ are undecidable. Harrington, and independently Hermann, have announced that the theories of the lattices of $\mathcal{E}$ and $\mathcal{E}^*$ are undecidable.

All but two of the theories of the lattices of recursively enumerable substructures we have discussed have been proved undecidable. Nerode and Smith [**1982**], proved the lattices $\mathcal{L}(V_\infty)$ and $\mathcal{L}^*(V_\infty)$ to be undecidable. Carroll [**1983**] announced the undecidability of the lattice $\mathcal{L}(\mathcal{B})$ for any infinite recursive Boolean algebra $\mathcal{B}$. We shall see that the undecidability of $\mathcal{L}I(Q)$ follows from the undecidability of $\mathcal{E}$. At this time the undecidability of the theories of $\mathcal{L}(F_\infty)$ and $\mathcal{L}(R^n)$ remain open. As for $\mathcal{L}(Q)$, if we restrict ourselves to the collection $\mathcal{D}$ of bidense open r.e. sets, then $\mathcal{D}$ is clearly not a lattice while if we consider all r.e. open sets of Q, then the theory of $\mathcal{L}(Q)$ is identical with the theory of $\mathcal{E}$.

Nerode and Smith [**1982**] showed that the theories of $\mathcal{L}(V_\infty)$ and $\mathcal{L}^*(V_\infty)$ are undecidable. The lattice $\mathcal{L}^*(V_\infty)$ is definable in $\mathcal{L}(V_\infty)$ since a subspace $V \in \mathcal{L}(V_\infty)$ is finite dimensional iff every subspace $U \subseteq V$ is complemented in $\mathcal{L}(V_\infty)$. Thus if the theory of $\mathcal{L}^*(V_\infty)$ is undecidable, then the theory of $\mathcal{L}(V_\infty)$ is undecidable. The undecidability of $\mathcal{L}^*(V_\infty)$ follows from the result below which was proved with a modification of the supermaximal subspace construction of Theorem 41.

THEOREM 102 (NERODE AND SMITH [**1982**]). *Every finite distributive lattice is a filter in* $\mathcal{L}^*(V_\infty)$.

Note that the undecidability of $\mathcal{L}^*(V_\infty)$ follows from Theorem 102 since Ershov and Taitslin [**1963**] have shown that the Gödel numbers of the sentences in the theory of distributive lattices are recursively inseparable from the Gödel numbers of those sentences which are refutable in some finite distributive lattices.

COROLLARY 103 (NERODE AND SMITH [**1982**]). *The theories of* $\mathcal{L}^*(V_\infty)$ *and* $\mathcal{L}(V_\infty)$ *are undecidable.*

No analogues of Theorem 102 are known for any of the five other lattices we have discussed. Indeed, the construction of Theorem 102 fails even in $\mathcal{L}(F_\infty)$. Downey [**1982**] has observed that Theorem 102 works in a special class of matroids which includes V_∞ but not F_∞. The analogue of Theorem 101 fails in $\mathcal{E}$ or $\mathcal{E}^*$ since Lachlan has shown that the only finite distributive lattices which are filters in $\mathcal{E}$ or $\mathcal{E}^*$ are finite Boolean algebras. (This does not help in $\mathcal{E}$ or $\mathcal{E}^*$ since the theory of Boolean algebras is decidable.)

There are some other results about what types of filters there are in our lattices. For example, Remmel has proved the following analogues of Lachlan's result that the lattice of supersets of *hh*-simple sets consists of all ∃-∀-∃-Boolean algebras.

THEOREM 104 (REMMEL [**1977b**]). *Given any* ∃∀∃*-Boolean algebra* $\mathcal{B}$, *there is a* $V \in \mathcal{L}(V_\infty)$ *such that the lattice of r.e. superspaces of* V mod *finite-dimensional space*, $\mathcal{L}\uparrow(V)$, *is isomorphic to* $\mathcal{B}$.

The proof of Theorem 104 fails in $\mathcal{L}(F_\infty)$. However, the analogue of Theorem 104 holds for the lattice of r.e. subalgebras of a recursive Boolean algebra $\mathcal{B}$. Recall that if B and C are subalgebras of $\mathcal{B}$, we defined $B =^* C$ if there exist finite sets F_1 and F_2 such that $(B \cup F_1)^* = (C \cup F_2)^*$. Now $\mathcal{L}(\mathcal{B})/=^*$ is not a lattice since $=^*$ is not a congruence relation with respect to intersection. Nevertheless, given $B \in \mathcal{L}(\mathcal{B})$, define $\mathcal{L}^*\uparrow(\mathcal{B}) = \{C \in \mathcal{L}(V_\infty) | C \supseteq B\}/=^*$. Then the following holds.

THEOREM 105. (REMMEL [**198 c**]). *Let* $\mathcal{B}$ *be any* ∃∀∃*-Boolean algebra and let* $\mathcal{C}$ *be any recursive infinite Boolean algebra. Then there exists a recursive Boolean algebra* $\mathcal{D} \approx \mathcal{C}$ *and a* $B \in \mathcal{L}(\mathcal{D})$ *such that* $\mathcal{L}^*\uparrow(B)$ *is isomorphic to* $\mathcal{B}$.

We should note that Feiner [**1970**] proved that there exists an r.e. Boolean algebra $\mathcal{B}$ (i.e., $\mathcal{B} = \tilde{Q}/\mathcal{I}$, where $\mathcal{I}$ is an r.e. ideal) such that $\mathcal{B}$ is not isomorphic to any recursive Boolean algebra. Using that fact, Feiner was able to conclude that $\mathcal{E}$ and $\mathcal{E}^*$ are not recursively presentable because by Lachlan any ∃∀∃-Boolean algebra is a filter in $\mathcal{E}^*$. Similarly, by Theorems 104 and 105, we have the following

COROLLARY 106 (REMMEL [**198 c**]). (i) *$\mathscr{L}(V_\infty)$ and $\mathscr{L}^*(V_\infty)$ cannot be recursively presented.*

(ii) *For every infinite recursive Boolean algebra $\mathscr{E}$, there is a recursive Boolean algebra $\mathscr{D} \approx \mathscr{E}$ such that $\mathscr{L}(\mathscr{D})$ cannot be recursively presented.*

Consider the lattice of r.e. ideals $\mathscr{L}I(\mathscr{B})$ of a Boolean algebra $\mathscr{B}$. Unlike Carroll's result that the lattice of an r.e. subalgebra of $\mathscr{B}$ always has an undecidable theory, the question of whether $\mathscr{L}I(\mathscr{B})$ is undecidable depends on the recursive presentation of $\mathscr{B}$. Recall $\tilde{N}_0$, the recursive presentation of $\mathscr{B}$ such that $\mathscr{A}(\tilde{N}_0)$ is immune (see Theorem 25). In $\tilde{N}_0$, all r.e. ideals are principal and it is easy to see that $\mathscr{L}I(\tilde{N}_0)$ is isomorphic to $\mathscr{B}_\omega$ which has a decidable theory. However, if we consider $\tilde{N}$, the recursive presentation of $\mathscr{B}_\omega$ such that $\mathscr{A}(\tilde{N})$ is recursive, then $\mathscr{L}I(\tilde{N})$ is undecidable. To see this let I be the ideal generated by the atoms of $\tilde{N}$. It is not difficult to see that the lattice $\mathscr{L} = \{J \in \mathscr{L}I(\tilde{N}) | J \subseteq I\}$ is isomorphic to $\mathscr{E}$. But I is elementary definable in $\mathscr{L}I(\tilde{N})$ since I is the largest ideal which is not fully complemented in $\mathscr{L}I(\tilde{N})$. Thus $\mathscr{L}I(\tilde{N})$ is undecidable by the undecidability of $\mathscr{E}$.

THEOREM 107 (REMMEL [**198 c**]). *Let $\mathscr{B}_\omega$ denote the Boolean algebra of finite and cofinite subsets of N; then there exist recursive presentations of $\mathscr{B}_\omega$, $\tilde{N}_0$ and $\tilde{N}$, such that the theory of $\mathscr{L}I(\tilde{N}_0)$ is decidable and the theory of $\mathscr{L}I(\overline{N})$ is undecidable.*

To prove that the theory of the lattice of r.e. ideals of the atomless Boolean algebra is undecidable, all we have to observe is that we can define when an r.e. ideal J is such that the lattice of superideals of

$$J, \mathscr{L}I(J)\uparrow = \{J' \in \mathscr{L}I(\tilde{Q}) | J' \supseteq J\},$$

is isomorphic to $\mathscr{L}I(\tilde{N})$. But $\mathscr{L}I(J)\uparrow$ is isomorphic to $\mathscr{L}I(\tilde{N})$ if:

(1) $\tilde{Q}/J$ is atomic, i.e., for each ideal $I' \supset J$, there is an ideal I'' such that $I' \supseteq I''$ and I''/J is an atom in $\tilde{Q}/J$, i.e. I'' is such that $I' \supseteq I'' \supset J$ and for all ideals I''', $I'' \supset I''' \supseteq J \Rightarrow J = I'''$.

(2) There is a unique ideal $I_0 \in \mathscr{L}I(J)\uparrow$ such that

(a) for all $I' \supset I_0$, I' is fully complemented over J,

(b) I_0 is not fully complemented over J,

(c) I_0/J is the ideal generated by the atoms in $\tilde{Q}/J$, i.e. I_0 is the smallest ideal in $\mathscr{L}(J)\uparrow$ which contains all ideal I'' in $\mathscr{L}(J)\uparrow$ such that I''/J is an atom in $\tilde{Q}/J$.

It is now not difficult to prove that I_0 plays the role in $\tilde{Q}/J$ of the ideal generated by the atoms in $\tilde{N}$ and that $\mathscr{L}(\tilde{Q}/J) \approx \mathscr{L}(\tilde{N})$. (We should also note that Downey [**1982**] showed that $\mathscr{E}$ could be imbedded into the lattice of r.e. filters of $\tilde{Q}$ which is entirely equivalent to the analogous result for $\mathscr{L}I(\tilde{Q})$.) Thus as a corollary of the Harrington-Hermann result we have

THEOREM 108. *The theory of $\mathscr{L}I(\tilde{Q})$ is undecidable.*

6. General settings. In this section we shall discuss three general settings in which theorems about the lattice of r.e. substructures can be proved so that the results can be specialized to give theorems for many recursively presented structures.

We shall begin by giving a general setting due to Metakides and Remmel [**1979**] which covers all the results we mentioned for the lattice $\mathcal{L}(Q)$ when the results are specialized to the model $\langle Q, \leqslant \rangle$, and which also gives results for $\mathcal{E}$ when specialized to the model $\langle N, = \rangle$. One of the basic insights which gave rise to the model theoretic setting which we are about to describe is that the infinite-coinfinite subsets of the natural numbers and the bidense subsets of Q have a common property with respect to their respective models. That is, the only sets which are definable by first order formulas in $\langle N, = \rangle$ are the finite and the cofinite sets so that a set $A \subseteq N$, where both A and $N - A$ are infinite, has the property that both A and $N - A$ intersect every infinite definable set in $\langle N, = \rangle$. Similarly in $\langle Q, \leqslant \rangle$, the infinite definable sets are just intervals plus or minus a finite set of elements so that a bidense subset A of Q has the property that both A and $N - A$ intersect every infinite definable subset of $\langle Q, \leqslant \rangle$. Basically, we will say that a set A in a model $\mathcal{M}$ is "in general position" if A and its complement in $\mathcal{M}$ intersect every infinite definable subset of $\mathcal{M}$. Thus if we prove theorems about sets "in general position" in $\mathcal{M}$, then when we specialize to $\mathcal{M} = \langle N, = \rangle$ we get results about infinite-coinfinite subsets and when we specialize to $\mathcal{M} = \langle Q, \leqslant \rangle$, we get results about bidense subsets of Q.

More formally, Metakides and Remmel define a model $\mathcal{M}$ of a complete decidable theory T over a countable language L to be decidable if the universe M of $\mathcal{M}$ is a recursive subset of N and the satisfaction relation "$\vec{a}$ satisfies Φ in $\mathcal{M}$" is recursive (where $\vec{a}$ is a finite sequence of elements from M and Φ is a formula in L). Assume that whenever $\mathcal{M}$ is decidable, then $M = N$ and L contains equality. A theory T is said to have decidable atoms if for each n, we can effectively decide if a formula $\phi(v_0, \ldots, v_n)$ is an atom in the Lindenbaum algebra $B^n(T)$ of formulas of $L(T)$ with n free variables. A model $\mathcal{M}$ of a theory T is atomic if each sequence $a_1, \ldots, a_n$ from M satisfies some atom in $B^n(T)$.

A subset X of a model $\mathcal{M}$ is said to be definable from $A \subseteq M$ if there is a formula $\Phi(v_0, \ldots, v_n)$ and a finite sequence of elements $a_1, \ldots, a_n$ in A such that $X = \{x \in M \mid \mathcal{M} \models \Phi(x, a_1, \ldots, a_n)\}$. If $A \subseteq M$, then the algebraic closure of A, written cl(A), is the union of all finite sets definable from A. A is said to be algebraically closed if cl(A) $= A$.

Metakides and Remmel then study $\mathcal{C}$, the class of decidable atomic models of a complete decidable theory T with decidable atoms such that the algebraic closure of every subset of M is itself. The simplest examples are

the natural numbers with equality, $\langle N, = \rangle$, and

the rationals Q under the usual ordering, $\langle Q, \leqslant \rangle$.

Many other examples may be found in Metakides and Remmel [**1979**].

Let $\mathcal{M}$ denote a model in $\mathcal{C}$. Because the algebraic closure of every subset is itself, every set $B \subset M$ determines a substructure $\mathcal{B}$ of $\mathcal{M}$ such that the inverse of

$\mathscr{B}$ is B. Thus we will specify substructure by simply giving a subset of M. If $B \subseteq M$, $\bar{B}$ will denote $M - B$. Since $\mathscr{M}$ is decidable, we can effectively list all pairs $\langle \phi, a \rangle$, where $\phi(v_0, \ldots, v_n)$ is in $L(T)$, $\vec{a} = \langle a_1, \ldots, a_n \rangle$ is an n-tuple of elements of M, and $\mathscr{M} \models \exists x(\phi(x, a_1, \ldots, a_n) \& x \neq a_1 \& \cdots \& x \neq a_n)$. Let $\langle \phi_0, \vec{a}_0 \rangle, \langle \phi_1, \vec{a}_1 \rangle, \ldots$ be such an effective list. Since for all $A \subseteq M$, $\text{cl}(A) = A$, it follows that if $\mathscr{M} \models \exists x(\phi(x, a_1, \ldots, a_n) \& x \neq a_1 \& \cdots \& x \neq a_n)$, then there are infinitely many $a \in M$ such that $\mathscr{M} \models \phi(a, a_1, \ldots, a_n)$. For each i, let $X_i = \{a \in M \mid \mathscr{M} \models \phi_i(a, a_1, \ldots, a_n) \& x \neq a_1 \& \cdots \& x \neq a_n\}$, where $\vec{a}_i = \langle a_1, \ldots, a_n \rangle$. Thus each X_i is an infinite recursive subset of M. It is clear that for any infinite set $X \subseteq M$ definable with finitely many parameters from M, there is an i and a finite set $S \subseteq \{a_1, \ldots, a_n\}$ where $\vec{a}_i = \{a_1, \ldots, a_n\}$ such that $X_i \cup S = X$. The X_i's have several properties which make them convenient for constructions. Suppose that $X_i \cap X_j \neq \varnothing$, $\vec{a}_i = \langle a_1, \ldots, a_n \rangle$ and $\vec{a}_j = \langle b_1, \ldots, b_k \rangle$, then $\mathscr{M} \models \exists y(\phi_i(y, a_1, \ldots, a_n) \& \phi_j(y, b_1, \ldots, b_k) \& (\&_{i=1}^{n} y \neq a_1) \& (\&_{i=1}^{k} y \neq b_i))$ so that $|X_i \cap X_j| = \infty$ and there is a k such that $X_i \cap X_j = X_k$. Similarly, we can show that if $a \in X_i$, there is a j such that $X_j = X_i - \{a\}$.

DEFINITION. A substructure $M' \subseteq \mathscr{M}$ is said to be *in general position* (i.g.p.) if for every infinite set $X \subseteq M$ definable from finitely many parameters from M both $|X \cap M'| = \infty$ and $|X - M'| = \infty$ or, equivalently, if for all i, $|X_i \cap M'| = |X_i - M'| = \infty$.

EXAMPLES. (1) In $\langle N, = \rangle$, the only finite definable sets are cofinite so that $S \subseteq N$ is i.g.p. iff S and $N - S$ are infinite.

(2) In $\langle Q, < \rangle$, the sets X_i are just unions of open intervals so that $S \subseteq Q$ is i.g.p. iff S and $Q - S$ are dense.

DEFINITION. A substructure $M' \subseteq \mathscr{M}$ is said to have a *creative presentation* if M' is r.e., M' is i.g.p, and there is a partial recursive function $f\colon N \times N \to M$ such that if $W_i \cap X_j \cap M' = \varnothing$, then $f(i, j)$ is defined and $f(i, j) \in X_j - (W_i \cap M')$. f is then said to be a productive function for M'.

It is easy to see that for $\mathscr{M} = \langle N, = \rangle$, a set S has a creative presentation iff S is creative in the usual sense and for $\mathscr{M} = \langle Q, \leqslant \rangle$, a set $S \subseteq Q$ has a creative presentation if and only if S is dense creative. Thus, the following theorem (due to Metakides and Remmel) generalizes both Myhill's Theorem on creative sets and Theorem 22.

THEOREM 107 (METAKIDES AND REMMEL [**1979**]). *Let $\mathscr{M}$ be a decidable model of an $\aleph_0$-categorical decidable theory with decidable atoms and suppose that the algebraic closure of any subset of $\mathscr{M}$ is itself. Then*:

(i) *There is a substructure M_1 of $\mathscr{M}$ with a creative presentation.*

(ii) *If M_1 and M_2 are substructures of $\mathscr{M}$ with creative presentations, then there is a recursive automorphism $f\colon \mathscr{M} \to \mathscr{M}$ such that $f(M_1) = M_2$.*

We can generalize other results about $\mathscr{E}$ and $\mathscr{L}(Q)$ in a similar manner. For example, we make the following definitions.

DEFINITION. (i) A substructure $M' \subseteq \mathcal{M}$ is said to be a *cohesive substructure* in general position if M' is i.g.p. and for any r.e. set $W \subseteq M$ either $W \cap M'$ or $M' - W$ is not i.g.p.

(ii) A substructure $M' \subseteq \mathcal{M}$ is said to be a *maximal substructure* in general position if M' is r.e. and $M - M'$ is a cohesive substructure i.g.p.

EXAMPLES. (1) In $\langle N, = \rangle$, if $M' \subseteq N$ is i.g.p. and for any r.e. set either $W \cap M'$ or $M' - W$ is not i.g.p., then M' is infinite and for any r.e. set W either $W \cap M'$ or $M' - W$ is finite which is the usual definition of cohesive sets. Thus the cohesive and maximal substructures i.g.p. are just the cohesive and maximal sets, respectively.

(2) In $\langle Q, \leqslant \rangle$, the maximal substructures i.g.p. are just the bidense maximal suborderings of $\langle Q, \leqslant \rangle$.

Then the following theorem generalizes Martin's Theorem on maximal sets, Theorem 37 and Theorem 55 (i).

THEOREM 108 (METAKIDES AND REMMEL [**1979**]). *Let $\mathcal{M}$ be as in Theorem* 107 *and δ be any high r.e. degree; then there exists a maximal substructure in general position of $\mathcal{M}$ with degree δ.*

In a similar manner, all the results about $\mathcal{L}(Q)$ mentioned in this paper as well as many other results are formulated and proved for models in $\mathcal{C}$. Such results can be found in Metakides and Remmel [**1979**] and Remmel [**1980a**].

Our second general setting is that of recursively presented Steinitz closure systems or matroids which were first studied by Metakides and Nerode [**1980**] so that they could generalize results about V_∞ and F_∞.

DEFINITION. A Steinitz closure system (U, cl) is a set U together with an operation cl mapping the power set of U, $\mathcal{P}(U)$, into $\mathcal{P}(U)$ such that for all subsets A and B of U

(i) $A \subseteq \mathrm{cl}(A)$,

(ii) $A \subseteq B$ implies $\mathrm{cl}(A) \subseteq \mathrm{cl}(B)$,

(iii) $\mathrm{cl}(\mathrm{cl}(A)) = \mathrm{cl}(A)$,

(iv) $x \in \mathrm{cl}(A)$ implies that for some infinite set $A' \subseteq A$, $x \in \mathrm{cl}(A')$,

(v) (exchange) $x \in \mathrm{cl}(A \cup \{y\}) - \mathrm{cl}(A)$ implies $y \in \mathrm{cl}(A \cup \{x\})$.

The most natural examples of Steinitz closure systems are V_∞ and F_∞. We can even consider the natural numbers N a Steinitz closure system where $\mathrm{cl}(A) = A$. Now Steinitz closure systems were invented by van der Waerden [**1949**] exactly to capture the common features of linear independence in vector spaces and algebraic independence in fields. Whitney [**1935**] gave the completely equivalent notion of matroid, couched in the language of independent sets; a matroid consists of a set U and a collection I of subsets of U (called the independent sets of U) such that:

(1) The null set is independent.

(2) A set is independent if and only if all its finite subsets are independent.

(3) If A, B are independent and the cardinality of A is less than that of B, then there is an x in $B - A$ with $A \cup \{x\}$ independent. (Say x is in $\mathrm{cl}(A)$ iff $A \cup \{x\}$

is not independent.) A particular construction, well motivated from Whitney's point of view is that of submatroid. Namely, if (U, cl) is given, and $X \subseteq U$, define the independent sets of the submatroid X to be those independent sets in U which are subsets of X. Equivalently, if $A \subseteq X$, define the closure of A in the submatroid X to be the intersection of X with the closure of A in U. A *linear matroid* is one which is isomorphic to a submatroid of a vector space matroid. Similarly, another example of a Steinitz closure system is V_∞ under affine dependence where the closure operator kl is defined by $x \in \mathrm{kl}(\{y_1,\ldots,y_n\})$ iff there exist λ_i in the underlying field F such that $x = \sum_{i=1}^n \lambda_i y_i$ and $\sum_{i=1}^n \lambda_i = 1$.

DEFINITION. A Steinitz closure system (U, cl) is *recursively presented* if:

(i) U is a recursive set.

(ii) For any a, $b_1,\ldots,b_n$ in U, it can be effectively determined whether or not $a \in \mathrm{cl}(\{b_1,\ldots,b_n\})$.

Given a recursively presented Steinitz closure system, one can then study $\mathcal{L}(U)$, the lattice of r.e. closed subsets of U, where a set $A \subseteq U$ is *closed* if $\mathrm{cl}(A) = A$. Keeping V_∞ and F_∞ in mind as examples, it is then routine to define the analogues of the various notions we introduced by $\mathcal{L}(V_\infty)$ and $\mathcal{L}(F_\infty)$ such as simple, maximal, supermaximal, r-maximal, extendible and fully extendible bases, etc. One can then give constructions of maximal elements for a general recursively presented Steinitz closure system so that one theorem can cover our lattices $\mathcal{L}(V_\infty)$ and $\mathcal{L}(F_\infty)$. However, for the various constructions to go through, one requires some additional properties of (U, cl). For example, there are supermaximal elements in $\mathcal{L}(V_\infty)$ and $\mathcal{L}(F_\infty)$ but there are no supermaximal elements in $\mathcal{E}$ even though $\mathcal{E} = \mathcal{L}(U, \mathrm{cl})$, where $U = N$ and $\mathrm{cl}(A) = A$ for all $A \subseteq N$. Similarly, there exist recursive supermaximal elements in $\mathcal{L}(F_\infty)$ and $\mathcal{L}(V_\infty)$ if the underlying field is infinite but there are no recursive supermaximal elements in $\mathcal{L}(V_\infty)$ if the underlying field is finite. Thus various extra axioms have been introduced which are required for various constructions to go through. To date, seven general axioms have been introduced and studied.

AXIOM I. *Let V be closed and let I be an infinite basis for (U, cl_V). Then there exists a z such that in (U, cl_V), $\mathrm{supp}_I z$ has at least two members. (Here $\mathrm{cl}_V(X) = \mathrm{cl}(V \cup X)$.)*

AXIOM II. *Let V be closed and let J be infinite independent in (U, cl_V). Then in (U, cl_V) the dimension of $\mathrm{cl}(J \cup \{x\}) - \mathrm{cl}(J)$ is infinite.*

AXIOM III. *For some k: for all infinite independent I, all independent J in (U, cl_I), all y in J, all finite $F \subseteq J$ with $\{y\} \subseteq F \subseteq J$ and $|F| = k$, all $v_1,\ldots,v_n \notin \mathrm{cl}_I(\varnothing)$, there is an $x \in \mathrm{cl}_I(F)$, with $v_1,\ldots,v_n \notin \mathrm{cl}_I(\{x\})$, and in (U, cl_I), we have $|\mathrm{supp}_F x| \geqslant 2$ and $y \in \mathrm{supp}_F x$.*

AXIOM IV (DOWNEY'S SEMIREGULARITY **[1982]**). *No finite-dimensional closed set is the union of two proper closed subsets.*

Axiom V (Federation over A of Baldwin **[1982]**). *There is a finite set A such that for any finite independent set B in (U, cl_A), there exist x in $\mathrm{cl}_A(B)$ not in any $\mathrm{cl}_A(B')$ for any proper subset B' of B.*

Axiom VI (Weak regularity of Baldwin **[1982]**). *No k-dimensional closed set is a union of k $(k-1)$-dimensional closed subsets.*

Axiom VII (Regularity of Metakides and Nerode **[1982]**). *No finite-dimensional closed set is a finite union of proper closed subsets.*

Axiom VII was the first of the axioms introduced and it occurred in Metakides and Nerode's paper **[1980]**. It is the strongest of the axioms given above and is sufficient for all of the theorems on $\mathscr{L}(V_\infty)$ and $\mathscr{L}(F_\infty)$ we have mentioned. The drawback of Axiom VII is that it is satisfied by F_∞ and V_∞ where the underlying field is infinite but not by N or by V_∞ where the underlying field is finite. Later Downey **[1982]** and Baldwin **[1982]** introduced the axioms attributed to them and finally Nerode and Remmel **[1982]** introduced what turns out to be the three weakest Axioms I, II and III. Nerode and Remmel **[1982]** completely classified the relative strength of these axioms. All the following implications hold, and by example no others listed hold:

$$\begin{array}{ccccc} \mathrm{VII} \rightarrow & \mathrm{VI} \rightarrow \mathrm{V} \leftrightarrow \mathrm{IV} \rightarrow & & \mathrm{II} \rightarrow \mathrm{I} \\ & \searrow \qquad \nearrow & & \\ & \mathrm{III} & & \end{array}$$

In order to give the reader a feel for the types of results which need the various axioms, we list a series of results and the axioms required below. We note that we are not claiming that the axioms are necessary and sufficient for the corresponding results. We are only saying that the axioms are sufficient and that at present no one knows how to prove the results with any weaker axioms. An interesting project for further research would be to find necessary and sufficient conditions for such results. Also we should note that all of our axioms are purely algebraic. The theorems that follow all can be somewhat generalized, but at the cost of more detail in proofs, by restricting attention to recursively presented matroids, and then asserting that there is a decidable infinite-coinfinite-dimensional closed set A such that the given axiom holds in (U, cl_A), where $\mathrm{cl}_A(B) = \mathrm{cl}(A \cup B)$. Some of these theorems, however, simply follow in this more general case from the observation that if A is a decidable closed subset of recursively presented (U, cl), then (U, cl_A) is recursively presented too.

(a) *No extra axiom required.*

Theorem 109 (Metakides and Nerode **[1980]**, Remmel **[1980b]**). *There exists a maximal element of $\mathscr{L}(U)$ with a fully extendible basis in every high r.e. degree.*

Theorem 110 (Nerode and Remmel **[1982]**). *There exist effectively nowhere simple elements of $\mathscr{L}(U)$ in every r.e. degree.*

(b) *Axiom* I *required.*

THEOREM 111 (NERODE AND REMMEL [**1983**]). *There exists a maximal element of $\mathscr{L}(U)$ with no extendible basis in every high r.e. degree.*

(c) *Axiom* II *required.*

THEOREM 112. (NERODE AND REMMEL [**1982**]). *There exist supermaximal elements in $\mathscr{L}(U)$ in every nonzero r.e. degree.*

THEOREM 113. (NERODE AND REMMEL [**1983**]). *There exist effectively nowhere simple elements in $\mathscr{L}(U)$ with no fully extendible basis in every r.e. degree.*

(d) *Axiom* III *required.*

THEOREM 114 (NERODE AND REMMEL [**1982**]). *There exist recursive supermaximal elements in $\mathscr{L}(U)$.*

THEOREM 115 (NERODE AND REMMEL [**1982**]). *There exist recursive effectively nowhere simple elements in $\mathscr{L}(U)$ with no fully extendible basis.*

(e) *Axiom* IV *required.*

THEOREM 116 (NERODE AND REMMEL [**1983**]). *There exists a $V \in \mathscr{L}(U)$ such that for all $i \neq k$, $D_i(V)$ is many-one incomparable with $D_k(V)$.*

(f) *Axiom* VI *required.*

THEOREM 117 (NERODE AND REMMEL [**1983**]). *Let $A_0, A_1, A_2, \ldots$ be any sequence of r.e. sets such that $A_i <_T A_0$ for all i uniformly, $A_i \leqslant_T A_{i+1}$ for all $i \geqslant 1$, and A_0 is not recursive. Then there exists a supermaximal $V \in \mathscr{L}(U)$ such that $D_i(V) \equiv_T A_i$ for $i \geqslant 1$ and $D(V) \equiv_T A$.*

Finally, we come to our third and most general setting which was introduced by Remmel [**1980b**]. One considers closure systems with a considerably weakened exchange property. This will cover such examples as Boolean algebras and distributive lattices. Due to the lack of exchange, one is not able to discuss a general notion of independence as in Steinitz closure systems. Instead, one assumes that there is something analogous to a recursive basis of V_∞ or a recursive transcendence basis of F_∞. These are special sequences. It turns out that Boolean algebras also have special sequences; for example, the set of atoms of $\tilde{N}$ is a special sequence for $\tilde{N}$. By using this notion of special sequence, Remmel is able to give general constructions which specialize to results in $\mathscr{E}$, results in $\mathscr{L}(V_\infty)$ and $\mathscr{L}(F_\infty)$ concerning elements with fully extendible bases, results in $\mathscr{L}(\mathscr{B})$ for certain recursive Boolean algebras $\mathscr{B}$, results in $\mathscr{L}I(Q)$, as well as results in many other models. However, because exchange is so weakened so as to cover Boolean algebras and the like, one is unable to talk about results in $\mathscr{L}(V_\infty)$ and $\mathscr{L}(F_\infty)$ concerning elements with no extendible bases or no fully extendible bases. For such results, the appropriate setting is really the Steinitz closure systems of Metakides and Nerode.

The basic objects which Remmel studies are models that possess a closure operation. We define a closure system $\mathscr{M}$ as an ordered pair (M, cl), where M is a set (called the universe of $\mathscr{M}$) and cl is a map from the set of all subsets of M, $\mathscr{P}(M)$, into $\mathscr{P}(M)$ such that for all $A, B \subseteq M$

(i) $A \subseteq \mathrm{cl}(A)$,

(ii) $A \subseteq B$ implies $\mathrm{cl}(A) \subseteq \mathrm{cl}(B)$,

(iii) $\mathrm{cl}(\mathrm{cl}(A)) = \mathrm{cl}(A)$,

(iv) $x \in \mathrm{cl}(A)$ if and only if there is some finite set $A' \subseteq A$ such that $x \in \mathrm{cl}(A')$.

An effective closure system consists of a closure system (M, cl) such that M is a recursive subset of N and the operation cl is effective on finite sets; that is, given a finite $A \subseteq M$, $\mathrm{cl}(A)$ is always a recursive set and moreover there is a recursive function f such that if x is the canonical index of A, then

$$\phi_{f(x)}(y) = \chi_{\mathrm{cl}(A)}(y) = \begin{cases} 1 & \text{if } y \in \mathrm{cl}(A), \\ 0 & \text{if } y \notin \mathrm{cl}(A). \end{cases}$$

We say a set $A \subseteq M$ is closed if $\mathrm{cl}(A) = A$.

EXAMPLES. (1) $\langle N, = \rangle$, where $\mathrm{cl}(A) = A$.

(2) V_∞, where $\mathrm{cl}(A)$ equals the subspace generated by A.

(3) F_∞, where $\mathrm{cl}(A)$ equals the algebraically closed subfield generated by A.

(4) Any recursive Boolean algebra $\mathscr{B}$, where $\mathrm{cl}(A)$ equals the subalgebra generated by A.

Many other examples may be found in Remmel [**1980b**].

From now on we will always assume that $\mathscr{M} = \langle M, \mathrm{cl} \rangle$ is an effective closure system such that $M = N$. Also, instead of saying that $B \subseteq M$ is a closed set, we will say that B is a substructure of $\mathscr{M}$.

Let $\mathscr{L}(\mathscr{M})$ be the lattice of recursively enumerable substructures of $\mathscr{M}$, under the operations of intersection and sum where given B, $C \in \mathscr{L}(\mathscr{M})$, $B + C = \mathrm{cl}(B \cup C)$. In the set case, one usually considers a lattice closely related to $\mathscr{E}$, namely $\mathscr{E}^*$, the lattice of r.e. subsets of N modulo finite sets. In our case we generalize the notion of being equivalent modulo finite sets as follows. Given substructures B and C in $\mathscr{M}$, we say B is equivalent to C modulo finite sets, $B =_F C$, if there exists finite sets T_1 and T_2 in M such that $\mathrm{cl}(B \cup T_1) = \mathrm{cl}(C \cup T_2)$. Clearly $=_F$ is an equivalence relation. While $=_F$ is a congruence relation with respect to sum, in general, $=_F$ is not a congruence relation with respect to intersection: Thus we cannot talk about $\mathscr{L}^*(\mathscr{M})$, the lattice of r.e. substructures of $\mathscr{M}$ modulo finite sets, but, the relation $=_F$ does provide a natural way to generalize the notions of infinite and coinfinite sets. The analogue of a substructure B being infinite is $B \neq_F \mathrm{cl}(\varnothing)$. The analogue of a substructure B being coinfinite is $B \neq_F M$.

Next we turn our attention towards defining the notion of an "independent sequence" in our closure structures which will allow us to generalize many of the constructions that produce examples in the lattice of r.e. sets $\mathscr{E}$. For example, one of the main techniques for lifting constructions on the natural numbers to constructions for vector spaces has been to manipulate the elements of a recursive basis for V_∞ in much the same way as one manipulated the natural numbers. That is, one could not always lift constructions on N directly since, for example, if a construction on N involved some sort of diagonalization over all r.e. sets, then a direct lifting of the construction, i.e., by substituting basis elements for natural numbers and then taking the space generated by the subset of the recursive basis so constructed, would diagonalize only over all r.e. subspaces which are generated by r.e. subsets of the recursive basis and not over all r.e. subspaces. But usually, with a few modifications, such a construction produced the desired examples. What is somewhat more surprising is that similar techniques could be applied to Boolean algebras, even in the case of the atomless Boolean algebra, where at first glance there does not seem to be any natural analogue of basis. In [**198** a], Remmel was able to find certain "independent" sequences modulo some recursive subalgebra which acted enough like a basis to allow one to lift many constructions for $\mathscr{E}$ to the lattices $\mathscr{L}(\mathscr{B})$. More generally, Remmel showed that whenever a model has such an "independent" sequence modulo a closed recursive set, one can lift many of the constructions on N to the model.

DEFINITION. Let $\mathscr{M} = (M, \mathrm{cl})$ be an effective closure system. A recursive set $S \subseteq M$ is said to be *special* over A, where A is a recrusive substructure of $\mathscr{M}$, if:

(a) There is an effective algorithm which, given any finite set $D \subseteq M$ and any $x \in M$, will decide if $x \in \mathrm{cl}(A \cup D)$.

(b) For all $B \subseteq S$, $\mathrm{cl}(A \cup B) \cap S = B$ (in particular $A \cap S = \varnothing$).

(c) $\mathrm{cl}(A \cup S) = M$.

(d) For all $B_1, B_2 \subseteq S$, $\mathrm{cl}(A \cup B_1) \cap \mathrm{cl}(A \cup B_2) = \mathrm{cl}(A \cup [B_1 \cap B_2])$.

We note that by property (iv) of our definition of closure system and the fact that $\mathrm{cl}(A \cup S) = M$, there is for any given $x \in M$, a finite set B' such that $x \in \mathrm{cl}(A \cup B')$. It follows from condition (d) above that there is a unique smallest finite set B such that $x \in \mathrm{cl}(A \cup B)$ which we call the support of x relative to S over A and denote by $\mathrm{supp}(x)$. It also easily follows from our definitions that given x, we can effectively calculate the canonical index of $\mathrm{supp}(x)$.

EXAMPLES. (i) In the set case where $\mathscr{M} = N$ and $\mathrm{cl}(B) = B$ for all $B \subseteq N$, we simply let $S = N$ and $A = \varnothing$.

(ii) In the case where $\mathscr{M} = V_\infty$, we let S be any recursive basis for V_∞ and let $A = \{\vec{0}\}$, where $0 \to$ is the zero vector of V_∞.

(iii) If $\mathscr{M} = F_\infty$, we let S be any recursive transcendence basis for F_∞ and $A = \mathrm{cl}(\varnothing)$ be the base field.

(iv) The case of Boolean algebras is more complicated. Recall that Theorem 27 tells us that within the classical isomorphism type of any weakly recursively presented Boolean algebra, there must be a Boolean algebra of the form $\tilde{N} \times \mathscr{D}$, $\tilde{Q} \times \mathscr{D}$, or $\tilde{\mathscr{C}}$, where $\mathscr{D}$ is a weakly recursively presented Boolean algebra. Thus we consider only Boolean algebras of the form $\tilde{N} \times \mathscr{D}$, $\tilde{Q} \times \mathscr{D}$, or $\tilde{C}$. Given a Boolean algebra $\mathscr{B}$, we let $0_{\mathscr{B}}$ and $1_{\mathscr{B}}$ denote the zero and one of $\mathscr{B}$, respectively.

Recalling Theorems 59–61, it is not difficult to see the motivations for the following examples of special sequences.

(iv.a) If $\mathscr{B} = \tilde{N} \times \mathscr{D}$, then $S = \{\langle a, 0_{\mathscr{D}}\rangle | a \in \mathscr{A}(\tilde{N})\}$ and let $A = \{\langle 0_{\tilde{N}}, d\rangle, \langle 1_{\tilde{N}}, d\rangle | d \in \mathscr{D}\}$.

(iv.b) If $\mathscr{B} = \tilde{Q} \times \mathscr{D}$, then we let $S = \{\langle [i, i+1), 0_{\mathscr{D}}\rangle | i \in N\}$. We let I denote the ideal in Q which is generated by x of the form (i) $x \subseteq (0, -\infty)$ or (ii) $x \subseteq [i, i+1)$ for some i and $x = [j_0, j_1) \cup \cdots \cup [j_{2n}, j_{2n+1})$, where $i \leqslant j_0 < j_1 < \cdots < j_{2n} < j_{2n+1} < i+1$. Then we let $A = \mathrm{cl}(\{\langle x, d\rangle | x \in I \text{ and } d \in \mathscr{D}\})$.

(iv.c) If $\mathscr{B} = \tilde{\mathscr{C}}$, then we let $S = \{[i, i+1) | i \in N\}$ and let $A = \mathrm{cl}(I \cup \mathscr{A}(\tilde{C}))$, where I is the ideal of $\tilde{Q}$ as it sits inside $\tilde{C}$ mentioned above.

In each of the examples above, it is not difficult but somewhat lengthy to check that S and A have the desired properties. We refer the interested reader to Remmel [**1978a**] for more details.

It is not difficult to show that if S is a special sequence over A for a recursive closure system $\mathscr{M} = (M, \mathrm{cl})$, then, for all $W_1, W_2 \subseteq S$, (i) $\mathrm{cl}(A \cup W_i) \equiv_T W_i$, (ii) if $S - W_i$ is infinite, then $\mathrm{cl}(A \cup W_i) \neq_F M$, and (iii) $W_1 \subseteq W_2$ and $|W_2 - W_1| = \infty$ implies $\mathrm{cl}(A \cup W_1) \neq_F \mathrm{cl}(A \cup W_2)$. Moreover, we can generalize the definition of simple, maximal and r-maximal in $\mathscr{M}$ as follows.

DEFINITION. An r.e. closed set $V \subseteq \mathscr{M}$ such that $V \neq_F M$ is

(i) *simple* if there is no r.e. closed set $B \subseteq \mathscr{M}$ such that $B \neq_F \mathrm{cl}(\varnothing)$ and $B \cap V = \mathrm{cl}(\varnothing)$,

(ii) *maximal* if for any r.e. closed set $W \supseteq V$ either $W =_F M$ or $W =_F V$,

(iii) *r-maximal* if for any pair of r.e. closed sets W_1 and W_2 with $W_1 + W_2 = M$ either

$$\mathrm{cl}(V \cup W_1) =_F M \quad \text{or} \quad \mathrm{cl}(V \cup W_2) =_F M.$$

It is easy to see that the definitions above generalize the notions of simple, maximal and r-maximal in the lattices $\mathscr{E}, \mathscr{L}(V_\infty), \mathscr{L}(F_\infty), \mathscr{L}(\mathscr{B})$ and $\mathscr{L}I(\tilde{Q})$.

One requires some additional axioms to prove the existence of simple, maximal and r-maximal objects. To give the reader a feel for the type of results of Remmel [**1980b**] we end this section with three examples. In each of the three theorems to follow, assume that $\mathscr{M}$ is a recursive closure system, $A \subseteq \mathscr{M}$ is recursive, and S is a special sequence over A.

THEOREM 118 (REMMEL [**1980b**]). *Suppose $\mathscr{M}$, A and S satisfy the following:*
(∗) *Let $D \subseteq S$ be such that $S - D$ is infinite and let V be a substructure of $\mathscr{M}$ such that $V \neq_F \mathrm{cl}(\varnothing)$ and $V \cap \mathrm{cl}(A \cup D) = \mathrm{cl}(\varnothing)$; then for any finite set $E \subseteq S$, there exists $v \in V$ such that $\mathrm{supp}(v) \neq \varnothing$ and $\mathrm{supp}(v) \cap E = \varnothing$. Then if H is a hypersimple subset of S, then $\mathrm{cl}(A \cup H)$ is a simple substructure of $\mathscr{M}$.*

If $S, A \subseteq \mathscr{M}$ and S is special over A, we say S has the local exchange property (L.E.P.) over A if for all $x \in \mathscr{M}$ with $\mathrm{supp}(x) = \{b_1, \ldots, b_n\}$, we have for all $j = 1, \ldots, n$, $b_j \in \mathrm{cl}\{x, b_1, \ldots, b_{j-1}, b_{j+1}, \ldots, b_n\} \cup A$.

THEOREM 119 (REMMEL [**1980b**]). *Suppose that S has the local exchange property over A and E is a maximal subset of S; then* $\mathrm{cl}(A \cup E)$ *is a maximal substructure of* $\mathcal{M}$.

THEOREM 120 (REMMEL [**1980b**]). *Suppose that S has the local exchange property over A. Then there exist r-maximal substructures R_1 and R_2 such that R_1 is not contained in any maximal substructure of $\mathcal{M}$, and R_2 is not maximal but is contained in a maximal substructure of $\mathcal{M}$.*

We note that there is no claim that the extra hypothesis of Theorems 118–120 are necessary for the existence of simple, maximal and *r*-maximal substructures, respectively. Nevertheless, for each of Theorems 118–120, there are examples of $\mathcal{M}$, S and A which fail to satisfy the above hypotheses for which either the theorem or the construction of the theorem fails. For example, F_∞ fails to satisfy condition $(*)$ of Theorem 117 and indeed F_∞ does not have any simple substructure. Similar counterexamples for Theorems 118 and 119 can be found in Remmel [**1980b**].

Note added in proof. We note that the undecidability of the theory of $\mathcal{L}(R^n)$ for $n \geqslant 2$ has recently been established by S. Brady in her Ph.D. Thesis [1984], Cornell University.

BIBLIOGRAPHY

C. J. Ash and R. Downey [**1983**], *Decidable subspaces and recursively enumerable subspaces*, J. Symbolic Logic (to appear).

J. T. Baldwin [**1982**], *Recursion theory and abstract dependence*, Patras Logic Symposiun (G. Metakides, ed.), North-Holland Studies in Logic, vol. 109, North-Holland, New York, pp. 67–76.

J. S. Carroll [**1983**], *The undecidability of the lattice of r.e. subalgebras of a recursive Boolean algebra*, Notices Amer. Math. Soc. *83T-03-281.

J. N. Crossley [**1981**], *Aspects of effective algebra* (J. N. Crossley, ed.), Upside Down A Book Co., Yarra Glen, Victoria, Australia.

J. C. E. Dekker [**1971**], *Two notes on vector spaces with recursive operations*, Notre Dame J. Formal Logic **12**, 329–334.

R. Downey [**1982**], *Abstract dependence, recursion theory, and the lattice of r.e. filters*, Ph.D. Thesis, Monash University.

______ [**1983**], *On a question of A. Retzlaff*, Z. Math. Logik Grundlag. Math. **29**, 379–384.

______ [**1984a**], *Co-immune subspaces and complementation in* V_∞, J. Symbolic Logic **49**, 528–538.

______ [**1984b**], *A note on decompositions of recursively enumerable subspaces* Z. Math. Logik Grundlag. Math. (to appear).

R. Downey and J. Hird [**1985**], *Certain classes of supermaximal spaces and automorphisms of the lattice of r.e. subspaces*, J. Symbolic Logic (to appear).

R. Downey and J. B. Remmel [**198a**], *Effectively and noneffectively nowhere simple vector spaces*, in preparation.

______ [**198 b**], *The universal complementation property*, preprint.

______ [**198 c**], *Automorphism bases for recursive matroids*, in preparation.

R. Downey and L. Welch [**198**], *Decomposition properties and r.e. degrees*, in preparation.

R. Downey, J. B. Remmel and L. Welch [**198**], *Degrees of r.e. vector spaces* in preparation.

J. Ershov and M. A. Taitslin [**1963**], *The undecidability of certain theories*, Algebra i Logika **2**, 37–41.

H. M. Friedman, S. G. Simpson and R. L. Smith [**1983**], *Countable lgebra and set existence axioms*, Ann. Pure Appl. Logic **25**, 103–127.

L. Feiner [**1970**], *Heierarchies of Boolean algebras*, J. Symbolic Logic **35**, 365–373.

R. M. Friedberg [**1958**], *Three theorems on recursive enumeration*, J. Symbolic Logic **23**, 309–316.

D. Guichard [**1982**], *Automorphisms and large submodels in effective algebra*, Ph.D. Dissertation, University of Wisconsin, Madison.

______ [**1983**], *Automorphisms of substructure lattices in recursive algebra*, Ann. Pure Appl. Logic **25**, 47–58.

______ [**1984**], *A note on R-maximal subspaces of* V_∞, Ann. Pure Appl. Logic **26**, 1–10.

R. A. Guhl, Jr. [**1973**], *Two types of recursively enumerable vector spaces*, Ph.D. Thesis, Rutgers University, Newark, N. J.

P. Hingston [**1982**], *A note on a theorem of D. Roy*, Monash University Logic paper.

I. Kalantari [**1978**], *Major subspaces of recursively enumerable vector spaces*, J. Symbolic Logic **43**, 293–303.

______ [**1979**], *Automorphisms of the lattice of recursively enumerable vector spaces*, Z. Math. Logik Grundlag. Math. **25**, 385–401.

______ [**1982**], *Major subsets in effective topology*, Patras Logic Symposium, North-Holland Studies in Logic, North-Holland, New York, pp. 77–94.

I. Kalantari and A. Legett [**1982**], *Simplicity in effective topology*, J. Symbolic Logic **47**, 169–183.

______ [**1983**], *Maximality in effective topology*, J. Symbolic Logic **48**, 100–112.

I. Kalantari and J. B. Remmel [**1983**], *Degrees of recursively enumerable topological spaces*, J. Symbolic Logic **48**, 610–622.

I. Kalantari and A. Retzlaff [**1977**], *Maximal vector spaces under automorphisms of the lattice of recursively enumerable vector spaces*, J. Symbolic Logic **42**, 481–491.

______ [**1979**], *Recursive constructions in topological spaces*, J. Symbolic Logic **44**, 609–625.

A. H. Lachlan [**1967**], *The priority method*. I, Z. Math. Logik Grundlag. Math. **13**, 1–10.

______ [**1968**], *On the lattice of recursively enumerable sets*, Trans. Amer. Math. Soc. **130**, 1–27.

R. E. Ladner [**1973a**], *Mitotic recursively enumerable sets*, J. Symbolic Logic **38**, 199–211.

______ [**1973b**], *A completely mitotic nonrecursive r.e. degree*, Trans. Amer. Math. Soc. **183**, 479–507.

M. Lerman and J. B. Remmel [**1982**], *The universal splitting property*. I, Logic Colloquium '80, North-Holland Studies in Logic, vol. 108, North-Holland, New York, pp. 181–209.

______ [**1984**], *The universal splitting property*. II, J. Symbolic Logic **49**, 137–150.

A. Manaster and J. B. Remmel [**1981**], *Partial orderings of fixed finite dimension: model companions and density*, J. Symbolic Logic **46**, 789–802.

D. A. Martin [**1966**], *Classes of recursively enumerable sets and degrees of unsolvability*, Z. Math. Logik Grundlag. Math. **12**, 295–310.

D. A. Martin and M. B. Pour-El [**1970**], *Axiomatizable theories with few axiomatizable extensions*, J. Symbolic Logic **35**, 205–209.

G. Metakides and A. Nerode [**1975**], *Recursion theory and algebra*, Lecture Notes in Math., vol. 450, Springer-Verlag, Berlin and New York, pp. 209–219.

______ [**1977**], *Recursively enumerable vector spaces*, Ann. Math. Logic **11**, 141–171.

______ [**1979**], *Effective content of field theory*, Ann. Math. Logic **17**, 289–320.

______ [**1980**], *Recursion theory on fields and abstract dependence*, J. Algebra **65**, 36–59.

______ [**1982**], *The introduction of non-recursive methods into mathematics*, Brouwer Symposium Volume, Springer-Verlag, Berlin and New York.

G. Metakides and J. Remmel [**1979**], *Recursion theory on orderings, a model theoretic setting*, J. Symbolic Logic **44**, 383–402.

D. Miller and J. B. Remmel [**1984**], *Effectively nowhere simple sets*, J. Symbolic Logic **49**, 129–136.

M. D. Morley and R. I. Soare [**1975**], *Boolean algebras, splitting theorems, and* Δ_2^0 *sets*, Fund. Math. **90**, 45–82.

J. Myhill [**1955**], *Creative sets*, Z. Math. Logik Grundlag. Math. **1**, 97–108.

A. Nerode and J. B. Remmel [**1982**], *Recursion theory on matroids*, Patras Logic Symposium (G. Metakides, ed.), North-Holland Studies in Logic, vol. 109, North-Holland, New York, pp. 67–76.

______ [**1983**], *Recursion theory on matroids*. II, Proc. Singapore Logic Sympos. Southeast Asian Conference on Logic (C. T. Chong and M. S. Wicks, ed.), North Holland, New York, pp. 133–184.

A. Nerode and R. Smith [**1982**], *The undecidability of the lattice of r.e. subspaces*, Proc. Third Brazilian Conf. on Math. Logic (A. I. Arruda, N. C. A. Di Costa, A, M. Sette, eds.), pp. 245–252.

J. C. Owings, Jr. [**1967**], *Recursion, metarecursion, and inclusion*, J. Symbolic Logic **32**, 173–178.

E. L. Post [**1944**], *Recursively enumerable sets and their decision problems*, Bull. Amer. Math. Soc. **50**, 286–316.

J. B. Remmel [**1977a**], *Maximal and cohesive vector spaces*, J. Symbolic Logic **41**, 611–625.

______ [**1977b**], *On the lattice of r.e. superspaces of an r.e. vector space*, Notices Amer. Math. Soc., #77T-E26.

______ [**1978a**], *Recursively enumerable Boolean algebras*. Ann. Math. Logic **14**, 75–107.

______ [**1978b**], *An r-maximal vector space that is not contained in any maximal vector space*, J. Symbolic Logic **43**, 430–441.

______ [**1979**], *R-maximal Boolean algebras*, J. Symbolic Logic **44**, 533–548.

______ [**1980a**], *Recursion theory on orderings*. II, J. Symbolic Logic **45**, 317–333.

______ [**1980b**], *Recursion theory on algebraic structures with an independent set*, Ann. Math. Logic **18**, 153–191.

______ [**1980c**], *Complementation in the lattice of subalgebras of a Boolean algebra*, Algebra Universalis **10**, 48–64.

______ [**1980d**], *On r.e. and co r.e. vector spaces with nonextendible bases*, J. Symbolic Logic **45**, 20–24.

______ [**198 b**], *On the lattice of recursively enumerable ideals of a recursive Boolean algebra*, in preparation.

______ [**198 a**], *On the number of automorphisms of the lattice of r.e. ideals and subalgebras of a Boolean algebra*, in preparation.

______ [**198 c**], *On the lattice of r.e. superstructures of an r.e. structure*, in preparation.

A. Retzlaff [**1979**], *Direct summands of r.e. vector spaces*, Z. Math. Logik Grundlag. Math. **25**, 363–372.

R. W. Robinson [**1967**], *Simplicity of recursively enumrable sets*, J. Symbolic Logic **32**, 162–172.

H. Rogers, Jr. [**1967**], *Theory of recursive functions and effective computability*, McGraw-Hill, New York.

J. Rosenthal [**1985**], *Intersections of algebraically closed fields*, Ann. Pure. Appl. Logic (to appear).

D. K. Roy [**1978**], Ph.D. Dissertation, University of Rochester, New York.

______ [**1983**], *R. e. presented linear orderings*, J. Symbolic Logic **48**, 369–378.

D. Sachs [**1962**], *The lattice of lattice of subalgebras of a Boolean algebra*, Canad. J. Math. **14**, 451–460.

G. E. Sacks [**1963**], *On the degrees less that* 0′, Ann. of Math. **77**, 211–231.

R. Shore [**1978a**], *Controlling the dependence degree of a recursively enumerable vector space*, J. Symbolic Logic **43**, 13–22.

______ [**1978b**], *Nowhere simple sets, simple sets and the lattice of recursively enumerable sets*, J. Symbolic Logic **43**, 322–330.

R. I. Soare [**1974**], *Automorphisms of the lattice of recursively enumerable sets. Part* I: *maximal sets*, Ann. of Math. (2) **100**, 80–120.

______ [**1978**], *Recursively enumerable sets and degrees*, Bull. Amer. Math. Soc. **84**, 1149–1182.

B. L. van der Waerden [**1949**], *Modern algebra*, Ungar, New York.

H. Whitney [**1935**], *On abstract properties of linear dependence*, Amer. J. Math. **57**, 509–533.

Department of Mathematics, Cornell University, Ithaca, New York 14853

Department of Mathematics, University of California at San Diego, La Jolla, California 92093

Proceedings of Symposia in Pure Mathematics
Volume **42**, 1985

Survey of Constructions in Noetherian Rings

A. SEIDENBERG

By request, instead of speaking of my latest results in construction, I will survey my contributions.

My interest in constructions started in 1948 when I was working on a problem on normal algebraic varieties (cf. [**5**]). An irreducible algebraic variety in affine n-space is given by a prime ideal P in a ring of polynomials $k[X_1,\ldots,X_n]$ over a field k. The ring $k[X_1,\ldots,X_n]/P = k[x_1,\ldots,x_n]$ is called the ring of the variety; and the variety is called normal if its ring is integrally closed. A similar definition holds for projective varieties.

The problem was to show that almost all hyperplane sections $a_0 + a_1X_1 + \cdots + a_nX_n = 0$ of a normal variety V of dimension $r > 1$ are themselves (irreducible and) normal. The "point" $(a_0, a_1,\ldots,a_n)$ varies over a projective space S and by *almost always* one means all except perhaps those lying on a proper subvariety of S. The condition $r > 1$ is necessary, since in the case V is a curve, the sections will obviously, in general, consist of several points. Conversely, it was known in some important cases that if $\dim V > 1$, then the hyperplane sections were almost always irreducible; and an additional argument gave the assertion in complete generality. Thus it was the normality of the sections that was at issue.

A minor technique showed that the *generic hyperplane* $u_0 + u_1X_1 + \cdots + u_nX_n = 0$ cuts out a normal section. Here $u_0, u_1,\ldots,u_n$ are indeterminates over k.

Now the main idea of the proof was to show that by field computations over k one could decide whether V is normal. Without qualifications this idea is not correct, though over an algebraically closed field k it is. Over a nonalgebraically closed field one cannot, in general, even test the irreducibility of a variety by means of rational computations alone; think, for example, of the variety $X^2 + a = 0$ over the real field. But with appropriate assumptions on V/k the idea is correct. Applying the computations to the generic section yields, of course, a positive answer. In the computations, which are over $k(u_0,\ldots,u_n)$, certain

1980 *Mathematics Subject Classification*. Primary 13E05, 14M05.

elements in $k[u_0,\ldots,u_n]$ occur in the denominators. If $(a_0,\ldots,a_n)$ avoids the places where any of these denominators vanish, then one gets that the section by $a_0 + a_1X_1 + \cdots + a_nX_n = 0$ is normal.

It was well known that a normal variety V of dimension r is free of $(r-1)$-dimensional singularities; and, by the so-called Principal Ideal Theorem (in integrally closed rings), that any principal ideal $(a) \neq (0)$ in the ring of V is unmixed $(r-1)$-dimensional. It was also known, though not so well, that the converse holds. As for the singularities, it was known how to compute these from the ideal $P = (f_1,\ldots,f_s)$ of V; namely, at least in the case of characteristic 0, from the Jacobian $\|\partial f_i/\partial X_j\|$ of P (and similarly from the so-called mixed-Jacobian more generally). Thus the sections are almost always free of $(r-2)$-dimensional singularities.

Let $\mathcal{O} = k[x_1,\ldots,x_n]$ be the ring of V and $\bar{\mathcal{O}}$ be its integral closure. As for the unmixedness condition, if V is free of $(r-1)$-dimensional singularities and is not normal, then there must be some $(a) \neq (0)$ that is mixed; in fact, any element $c \neq 0$ in the conductor $\mathcal{O}\colon \bar{\mathcal{O}}$ will yield a mixed (c). One known, at least in the case that the field $k(x_1,\ldots,x_n)$ of V is separably generated, how to write down canonically an element of the conductor, at any rate over a certain pure transcendental extension $k(u) = k(\ldots, u_{ij},\ldots)$ of k; and this will suffice. If $z_i = \sum u_{ij}x_j$ $(i = 1,\ldots,r+1)$ and $E(Z_1,\ldots,Z_{r+1}, u)$ is the irreducible polynomial satisfied by the z_i, then $\partial E/\partial Z_{r+1}$ mod P is such an element. Thus V will be normal if and only if $(P, \partial E/\partial Z_{r+1})$ is unmixed.

At about this time I had come across Krull's papers [**2**, **3**] on "Parameterspezialisierung". Krull considers a ring $k(t)[X_1,\ldots,X_n]$, where $k(t)$ is a simple transcendental extension of k. Any element f in $k(t)[X]$ can be written as $f(t, X)/d(t)$ with $f \in k[t, X]$ and $d \in k[t]$. For an $a \in k$, if $d(a) \neq 0$, one gets an element $\bar{f} = f(a, X)/d(a)$; and similarly from an ideal A in $k(t)[X]$ one gets, by specialization, an ideal $\bar{A}$. Krull studied the relations between ideals $A, B,\ldots$ and the ideals $\bar{A}, \bar{B},\ldots$. In particular, he showed that an unmixed ideal specializes almost always to an unmixed ideal; and it was easy to extend this to several $t_1,\ldots,t_s$. This was just the result I needed. And with it, applied to the generic hyperplane section, I completed the proof.

Krull's paper in turn relied on a paper by G. Hermann [**1**] on constructions in a polynomial ring $k[X_1,\ldots,X_n]$. Her object was to show how to construct a primary ideal decomposition for a given ideal $A = (f_1,\ldots,f_s)$; by a *given ideal* we shall always mean one given via a finite basis, and *to construct* an ideal will mean to construct a finite basis for it. I.e., she wanted to show how to construct finite bases for the primary ideals; and she wanted also to construct the associated prime ideals. One assumes, of course, that one can compute, i.e., carry out the rational operations, in k; in the terminology of van der Waerden, that k is *explicitly given*, or, as I prefer to say nowadays, that it is *discrete*, the word discrete emphasizing that one can tell whether an element a is zero or not. But even to solve the problem for a principal ideal $(f) \subset k[X]$, $X = X_1$, one will need to know how to factor f completely over k. Kronecker had shown how to do this over the rational field Q; and Hermann showed how to do it over any constructed finite field. Unfortunately she persuaded herself that this factorization condition, which I call *condition* (F), holds for any field k. This error becomes serious at her Theorem 10, but with what preceded one can show (cf. [**17**]) how to write any ideal as a finite intersection of unmixed ideals. This can be done over any discrete field. In particular, one can decide, using only rational operations, whether a given ideal is unmixed. So Krull's result was untouched.

A theme or thesis that emerges from the foregoing is that one can often see things about mathematical objects simply from their having been constructed. Using this thesis, I was able (cf. [**20**]) to answer several questions left open by Krull in [**3**]. Thus he considered a field k, an extension field k' of k, the polynomial rings $k[X_1,\dots,X_n]$ and $k'[X_1,\dots,X_n]$, an ideal A in the first ring and its extension A' in the second; and asked [**3**, p. 134]: If A is unmixed, then is A' unmixed? This is easily answered in the context of Krull's considerations! The algorithm for deciding whether $A = (f_1 2,\dots,f_s)$ is unmixed involves only rational operations and obviously depends only on the coefficients of the f_i, not on the encompassing field. Hence a decision that A is unmixed is at the same time a decision that A' is unmixed. Krull's other questions, though not quite so easily answered, yield to the same leading idea. Krull appears to have missed the simple observation that what he calls canonical algorithms (cf. [**2**, pp. 60ff]) give rise to properties independent of the base field.

Hilbert's Nullstellensatz yields to the same idea. The theorem concerns a polynomial ring $k[X_1,\dots,X_n]$ over a field k and says that if $F_1,\dots,F_s$, $G \in k[X_1,\dots,X_n]$ and if the system

$$(1) \qquad F_1 = 0,\dots,F_s = 0, \qquad G \neq 0$$

has no algebraic solutions, i.e., no solution $(a_1,\dots,a_n)$ with the a_i in the algebraic closure $\bar{k}$ of k, then some power of G is in the ideal $(F_1,\dots,F_s)$. A simple reformulation, which uses the elements of ideal theory but none of the structure theory, brings the theorem to the following form: *If* (1) *has a solution in some extension field of* k, *then it has a solution in* $\bar{k}$.

Let L be an extension field of k and $\bar{L}$ its algebraic closure. By an utterly simple argument (cf. [**8**]), one can decide whether (1) has a solution in $\bar{L}$ using only rational operations. Thus a negative decision for some L, in particular $L = k$, yields a negative decision for all L, and the theorem follows.

In 1968 I noticed G. Stolzenberg's paper [**24**]. The ground had been covered more or less in my paper [**5**], though not quite, as having decided that $R = k[x_1,\dots,x_n]$ is not integrally closed, one still has to construct an element $y \in k(x_1,\dots,x_n)$ integral over R but not in it.

Stolzenberg confines himself to the case that $k(x_1,\dots,x_n)/k$ is separable (i.e., separably generated) and that k satisfies (F). The separability assumption was included in order to simplify life. However, it was fairly clear to me that

Stolzenberg's result could not be quite right. A point of a curve V is normal if and only if its local ring is integrally closed; or, equivalently, if and only if the local ring is regular. According to Zariski [**27**], it is the regularity of the local ring that defines simplicity; and the criterion for this is given by the mixed-Jacobian matrix, not, as classically, simply by the Jacobian matrix. The mixed-Jacobian involves the pth roots of the coefficients of the f_i defining V. So in testing whether a point of a curve is simple, the pth roots of the coefficients of the defining equations of the curve come in. Thus inseparability considerations enter, even if the field $k(x_1,\ldots,x_n)$ of the curve is separable over k (p = characteristic of k).

In [**6**] I gave an example of a plane algebraic curve over a field k satisfying (F) that is absolutely irreducible and has a separable function field, but such that one cannot tell whether it is normal. The curve is free of singularities except perhaps for one point P, and one cannot tell whether P is simple. A technique for showing that some constructions are not possible will be considered below.

In dealing with construction problems involving inseparability, I introduced a *condition* (P) on discrete fields k of characteristic $p \neq 0$. The condition in effect says that one can find the degree $[k^p(a_1,\ldots,a_s): k^p]$ for any given elements $a_1,\ldots,a_s \in k$.

In an earlier paper [**12**], though one subsequent to Stolzenberg's [**24**], I took up the question of the construction of the integral closure of a finite integral domain $k[x_1,\ldots,x_n]$ in complete generality. In the paper I was partly concerned with finding the exact conditions under which the integral closure can be constructed. I found that it was condition (P) on k, and that (F) is not needed; and I showed there that condition (P) does not imply condition (F). Thus it is condition (P) that is crucial in constructive normalization, not (F).

In [**12**], in the construction of the integral closure of $k[x_1,\ldots,x_n]$, I first reduced to the case that $k(x)/k$ is separable and the degree of transcendency is 1, or equivalently, that the variety V whose ring is $k[x_1,\ldots,x_n]$ is a curve, and one may even assume that it is plane ($n = 2$). Now it is classically known how to resolve a singularity of a plane curve by means of locally quadratic transformations (transformations of the form $x' = x$, $y' = y/x$). To anyone who knows this, the normalization of a plane curve is more or less of a foregone conclusion. However, in the problem of constructing an element $y \in k(x_1,\ldots,x_n)$ integral over $k[x_1,\ldots,x_n]$ but not in it, I dealt only with the case of a plane curve (i.e., $n = 2$), whereas I should have considered a curve in any affine space. This slip was pointed out (in effect) by H. Kurke and was corrected in [**21**].

I had long realized that Hermann's paper, though very valuable, was somewhat treacherous. I had often thought that her paper should be redone, and I did this in [**17**].

In her Theorem 11, Hermann asserted that one could construct (finite bases for) the associated primes of a given ideal. The proof relies on Theorem 10, where in an induction on the number n of variables $x_1,\ldots,x_n$ Hermann adjoins an (algebraic) element x_1 to the base field k. Now k was assumed to satisfy (F), but condition (F) may no longer hold for $k(x_1)$! This error could not impinge on Hermann's attention, since she thought (F) was automatic.

To overcome this difficulty I adjoined condition (P), already mentioned. Then all of Hermann's constructions go forward for any discrete field satisfying (F) and (P). However, in redoing her work, I gave the necessary and sufficient conditions for each construction. For given ideals A and B one can construct $A \cap B$ and $A : B$ over any discrete field. One can also (effectively) write any given ideal A as a finite intersection of unmixed ideals over any discrete field. And one can construct $k[X_1,\ldots,X_{n-1}] \cap A$ and find l such that $A : B^l = A : B^{l+1}$ over any discrete field. On the other hand, sufficient for writing a given ideal A as a finite intersection of primary ideals is that the (discrete) field k should satisfy (F); a slightly weaker condition than (F) is necessary and sufficient. Necessary and sufficient for getting the associated prime of a primary ideal is condition (P). A necessary and sufficient condition for finding the associated primes of any given ideal is that k should satisfy (F) and (P).

The reason that (F) and (P) come in is that the irreducibility of a polynomial $f \in k[X]$ cannot be decided by rational operations over k. This can be done over the algebraic closure $\bar{k}$ of k, but in bringing the results down to k, nonrational and inseparability phenomena must be taken into account.

If $\operatorname{ch} k = p > 0$, then for any $a_1,\ldots,a_n \in k$, the ideal $(X_1^p - a_1,\ldots,X_n^p - a_n)$ in $k[X_1,\ldots,X_n]$ is primary; and it is prime if and only if $a_1,\ldots,a_n$ are p-independent, i.e. $[k^p(a_1,\ldots,a_n) : k^p] = p^n$. In [**17**, §51] I constructed a field K satisfying (F) and containing elements a, b whose p-independence cannot be decided. Hence, too, one has a primary ideal, namely, $(X^p - a, Y^p - b)$, whose associated prime cannot be constructed. In constructivist terminology, $(X^p - a, Y^p - b)$ is a primary ideal for which no associated prime exists.

A finitely generated field extension $k(z_1,\ldots,z_n)$ of k can be given as the quotient field of a polynomial ring $k[Z_1,\ldots,Z_n]$ mod a given prime ideal. A constructive theory for such fields is given in [**17**]. For example, if $x_1,\ldots,x_r$ are given elements of $k(z_1,\ldots,z_n)$, one can construct the ideal of the relations that $x_1,\ldots,x_r$ satisfy over k. Recently Professor Nerode asked me how to construct the intersection $k(x_1,\ldots,x_r) \cap k(y_1,\ldots,y_s)$ of two such subfields, especially in the case that k is algebraically closed. I leave this as an interesting open question. (Professor John W. Rosenthal, of Ithaca College, Ithaca, has now shown how to construct the intersection of the algebraic closures of those fields.)

In reading [**17**] it may be helpful to have [**19**] at hand. There I give a synopsis of [**17**] along with commentary. There are also some improvements on [**17**] and a few new results. For example, in [**17**] I showed that $k(X_1)[X_1,\ldots,X_n]A \cap k[X_1,\ldots,X_n]$ could be constructed for any discrete field k, at least for so-called "transformed" variables $X_1,\ldots,X_n$, i.e., variables arising from originally given variables $Y_1,\ldots,Y_n$ when these have been subjected to a generic homogeneous, nonsingular, linear transformation; and I asked whether this could be done also for the untransformed, or original, variables. In [**19**] I show that the answer is yes.

In [**14**], written prior to my Transactions paper [**17**], I gave a constructive version of Hilbert's theorem on ascending chains $A_0 < A_1 < A_2 < \cdots$ of ideals in a polynomial ring $k[X_1, \ldots, X_n]$. Although by Hilbert's theorem any such chain is of bounded length, it is easy to construct for any positive integer N a chain of length $> N$ (e.g., $(X^{N+1}) < (X^N) < \cdots < (1)$). However, it seemed to me that if one put a bound $f(i)$ on the degrees of some basis elements of the ideal A_i, then one could place a bound s on the length; a special case of this, with the basis elements monomials, had already occurred in one of my papers, [**7**]. In [**14**] I establish such a bound, and even write one down in terms of f and n. The referee in his report gave a very snappy proof, a free treatment of which I included, showing the existence of a bound s. Although the referee did not ask me to with-draw my paper, still I felt uneasy. It was clear to me that the referee was violating the spirit in which I approached the problem. He not only used Hilbert's theorem itself, but even Zorn's Lemma. And he merely showed that one could write down a bound, whereas I had actually done so. More technically said, his bound was general recursive, whereas mine, though not primitive recursive, like the ones found in Hermann's paper, still was better than general recursive. But still I could not deny that the referee had established the existence of a bound s and had done it in less than a sixth of the space I had.

I had to clarify the situation, at least to myself. At the time I wrote [**14**], my approach was a simple, classical one. The essential difference between the classical and the constructivist points of view is that the classical mathematician makes a distinction between existence and construction. The classicist can first obtain the existence of some object by means of convictions very widely, though not universally, held, and then it is a question of constructing this object; but for the constructivist, the object must first be constructed, and only then is it declared to exist. So I saw nothing to do but renounce the classical point of view and go over to the constructivist (or insist on nuances, as for example that my bound was better than general recursive). This I did in my paper [**15**]. At the end, I again enter into the proof of the referee of [**14**], and explain why, from a constructivist point of view, it is begging the question.

The theorem of [**14**] probably holds with a ring of finite length at the base instead of a field, but I have not written out a proof.

In accordance with the new point of view, in [**17**] I also give a strictly finitist treatment of the problems considered by Hermann.

I may remark that there has been nothing of a fervor of conversion in my shift of point of view. Although I have been for a long time philosophically inclined to what might be called constructivism, this inclination was never strong enough for me not to want to do classical mathematics; nor is it now. It is simply that when it comes to construction, the constructivist point of view is better—more adapted to what one is trying to do.

In 1973 I came upon F. Richman's paper [**4**], a strictly constructivist work. Richman's paper is as clear, brief, and elegant as one could wish, but at the very beginning there stands a basic and vital result that is hidden away in an unpublished thesis of J. B. Tannenbaum [**25**]. However inconvenient such a procedure may be for the reader, I suppose it can be scientifically justified; and surely I was glad to have Richman's considerations. Anyway, I really wanted to understand Richman's work, and I found myself trying to fill the gap. In the course of this I made some improvements; these appear in [**18**]. Aside from the main result, I found Richman's definition of a Noetherian ring intriguing. Consider the following conditions on a ring R:

(a) Given a chain $A_1 \subset A_2 \subset \cdots$ of finitely generated ideals in R, one can find an i such that $A_i = A_{i+1}$.

(b) The finitely generated ideals are finitely presented (or, what is equivalent to this by [**18**, Theorem 2], the set of finitely generated ideals is closed under intersection and quotient).

(c) The finitely generated ideals are *detachable*, i.e., given a finitely generated ideal A and an a in R, one can tell whether a is in A. Richman calls a ring *Noetherian* if it satisfies conditions (a), (b), (c); and his main result is that if R is Noetherian, then so is $R[X]$, where X is an indeterminate. I was able to show (in [**18**]) that also conditions (a) and (b) alone transfer to $R[X]$ and so propose that these two conditions are the right conditions for defining Noetherian.

The very first proposition in [**17**] reads in part:

1. Given a system of homogeneous equations $f_{i1}g_1 + \cdots + f_{is}g_s = 0$, $i = 1,\dots,r$, $f_{ij} \in k[X_1,\dots,X_n] = k[X]$, one can construct a $k[X]$-module basis for the solutions $(g_1,\dots,g_s)$, $g_j \in k[X]$.... In the proof I first observe that one may adjoin an indeterminate u to k and later remove it, so that one may suppose k infinite. We may suppose the r equations linearly independent over $k(X)$ and by notation that

$$\Delta = \begin{vmatrix} f_{11} & \cdots & f_{1r} \\ \vdots & & \vdots \\ f_{r1} & \cdots & f_{rr} \end{vmatrix} \neq 0.$$

Since k is infinite, by a homogeneous nonsingular linear transformation we may assume that the coefficient of the highest power of X_n in Δ is free of $X_1,\dots,X_{n-1}$. Then we can bring the equations to the equivalent form

$$\begin{aligned} \Delta g_1 &= (\cdots)g_{r+1} + \cdots + (\cdots)g_s \\ &\vdots \\ \Delta g_r &= (\cdots)g_{r+1} + \cdots + (\cdots)g_s \end{aligned}$$

with the $(\cdots)$ in $k[X]$. Now we note the existence of solutions

$$(g_{11},\dots,g_{1r},\Delta,0,\dots,0),\dots,(g_{s-r1},\dots,g_{s-rr},0,\dots,0,\Delta).$$

By the property of Δ mentioned, we may bound the degree in X_n of the $g_{r+1},\dots,g_s$ sought, and hence also the degrees of $g_1,\dots,g_r$. Now we can rewrite the equations in terms of $X_1,\dots,X_{n-1}$.

Note that the argument fails for the ring of integers Z instead of k at the base. For we need not only that the coefficient of the highest power of X_n in Δ should be free of $X_1, \ldots, X_{n-1}$, but that it should even be a *unit*. Thus I could not take even one step in solving the problems at issue for $Z[X_1, \ldots, X_n]$. From Richman's paper **[4]** I got an idea for overcoming this difficulty and was able to get started; and then, using the methods of **[17]**, I was able to solve the main problems for Z at the base also; see **[22]**.

If k is a discrete field satisfying (F), then a simple separably algebraic extension of k continues to satisfy (F). This has long been well known, classically, but the classical proof should have been better scrutinized for the strictly finitist part of **[17]**. This was done in **[22**, p. 701, Lemma 2].

Let A be a given ideal in $k[X_1, \ldots, X_n]$. In **[17]** I had been concerned with constructing the ideal $k(X_1)[X_2, \ldots, X_n]A \cap k[X_1, \ldots, X_n]$—this had been done by Hermann, at least in the case that the X_i are "transformed" variables. To solve this, in **[17]** I considered, following Hermann, the analogous problem in a free module over $k[X_1, \ldots, X_n]$; this was done for the sake of an induction. Since at the time ring theory kept itself very much to rings and rarely went to module theory to solve a problem for rings, the proof of [**17**, §8] is noteworthy. The same method, suitably modified, also plays a central role when Z is at the base; see **[22]**.

In the course of trying to make a construction, if the work does not go forward, one may begin to wonder whether the construction is possible. There is a technique for attacking such problems that goes back to a paper **[26]** by van der Waerden. As already remarked, Hermann convinced herself that every discrete field satisfies (F). One can, of course, locate the error, but van der Waerden gave an argument to show that it is highly unlikely, to say the least, that one should be able to prove (F). The main idea is to have at hand a proposition $E(n)$, $n \geqslant 1$, such that we can check whether $E(n)$ is true for any given $n \geqslant 1$ but do not know whether $E(n)$ is true for some n. For example, let $E(n)$ say that in the decimal expansion of π there are 99 successive nines beginning with the nth digit. Let $Q(x_1, \ldots, x_i)$, $i = 1, 2, \ldots,$ be a sequence of fields defined as follows: if $E(i)$ is false, then place $x_i = 0$, while if $E(i)$ is true, then place $x_i = \sqrt{-1}$. Let $K = Q(x_1, x_2, \ldots)$. Then K is discrete and $X^2 + 1$ factors properly over K if and only if $E(n)$ is true for some n. Hence if we could factor $X^2 + 1$ completely over K, we could decide whether $E(n)$ is true for some n. Instead of the stated $E(n)$, one could use any other $E(n)$ corresponding to an outstanding conjecture (for a certain kind).

One might criticize the above line of argument in that although at present we do not know whether $E(n)$ is true for some n, still it is easily conceivable that tomorrow we shall. The point is, however, that if we could give a method for factoring $X^2 + 1$ over any discrete field, one would also have a method of resolving all outstanding conjectures $E(n)$ of the described type. Presumably no one believes that that can be done.

In [**16**] I explained how to construct, from a classical point of view, a function $E^*(n)$ from nonnegative integers to nonnegative integers having the following properties: we can compute $E^*(n)$ for every n but there is not, and cannot be, a way of deciding for every m whether there exists an n such that $E^*(n) = m$; and I then modified van der Waerden's argument to construct a discrete field K not satisfying (F). This would probably be considered an improvement by the classicist. The construction of E^* depends, however, on a hypothesis as to what is *computable*, and although the hypothesis is very plausible, still in the nature of things it cannot be proved and so remains a hypothesis. In retrospect I think van der Waerden's argument is just as good, or better rather, since it is at least as convincing; and it does not commit us to a classical point of view.

My papers [**6**, **9**, **10**, **11**, **13**] have a constructive aspect to them (and one of them [**6**] very nicely illustrates a theme mentioned above), but since it is difficult to get these under the heading of Noetherian, I will say no more about them for the present.

Finally a word on my most recent work. In Richman's work [**4**] and in my corresponding paper [**18**], a definition of a Noetherian ring R is given and it is shown that if R is Noetherian, then so is $R[X]$. But in these considerations the Lasker-Noether theorem that every ideal has a normal decomposition into primary ideals never occurred. Since this theorem is a cornerstone of Noetherian ring theory, in [**23**] I ask: Suppose that for any given ideal of the Noetherian ring R one is given, or can construct, a normal decomposition of it into primary ideals together with its asssociated primes. Then can one do the same for any given ideal in $R[X]$? Some necessary conditions on R for this problem to have a positive resolution are formulated; and then these are shown to be sufficient. Since Noether's proof is highly nonconstructive, it is far from a foregone conclusion that the problem would have a positive outcome.

Noether's work is usually acclaimed for its transparency and rigor. Perhaps the ultimate judgement will be that she had an astounding intuition.

REFERENCES

1. G. Hermann, *Die Frage der endlich vielen Schritte in der Theorie der Polynomideale*, Math. Ann. **95** (1926), 736–788.

2. W. Krull, *Parameterspezialisierung in Polynomringen*, Arch. Math. **1** (1948), 56–64.

3. ______, *Parameterspezialisierung in Polynomringen*. II. *Das Grundpolynom*, Arch. Math. **1** (1948), 129–137.

4. F. Richman, *Constructive aspects of Noetherian rings*, Proc. Amer. Math. Soc. **44** (1974), 436–441.

5. A. Seidenberg, *The hyperplane sections of normal varieties*, Trans. Amer. Math. Soc. **69** (1950), 357–386.

6. ______, *A new decision method for elementary algebra*, Ann. of Math. (2) **60** (1954), 365–374.

7. ______, *An elimination theory for differential algebra*, Univ. of Calif. Publ. in Math. (N.S.) **3** (1956), 31–66.

8. ______, *Some remarks on Hilbert's Nullstellensatz*, Arch. Math. (Basel) **7** (1956), 235–240.

9. ______, *Abstract differential algebra and the analytic case*. Proc. Amer. Math. Soc. **9** (1958), 159–164.

10. ______, *Comments on Lefschetz's Principle*, Amer. Math. Monthly **65** (1958), 685–690.

11. ______, *Abstract differential algebra and the analytic case*. II, Proc. Amer. Math. Soc. **23** (1969), 689–691.

12. ______, *Construction of the integral closure of a finite integral domain*, Rend. Sem. Mat. Fis. Milano **40** (1970), 101–120.

13. ______, *On k-constructable sets, elementary formulae, and elimination theory*, J. Reine Angew. Math. **239 / 240** (1970), 256–267.

14. ______, *On the length of a Hilbert ascending chain*, Proc. Amer. Math. Soc. **29** (1971), 443–450.

15. ______, *Constructive proof of Hilbert's theorem on ascending chains*, Trans. Amer. Math. Soc. **174** (1972), 305–312.

16. ______, *On the impossibility of some constructions in polynomial rings*, Atti del Convegno Internationale di Geometria, Accademia Naxionale dei Lincei, 1973, pp. 77–85.

17. ______, *Constructions in algebra*, Trans. Amer. Math. Soc. **197** (1974), 273–313.

18. ______, *What is Noetherian*?, Rend. Sem. Mat. Fis. Milano **44** (1974), 55–61.

19. ______, *Constructions in algebra*, Accad. Nazionale Lincei (Contributi del Centro Interdisciplinare di Scienze Mat. e Loro Appl.) **9** (1975), 3–17.

20. ______, *The prime ideals of a polynomial ideal under extension of the base field*, Ann. Mat. Pura Appl. (4) **102** (1975), 57–59.

21. ______, *Construction of the integral closure of a finite integral domain*. II, Proc. Amer. Math. Soc. **52** (1975), 368–372.

22. ______, *Constructions in a polynomial ring over the ring of integers*. Amer. J. Math. **100** (1978), 685–703.

23. ______, *On the Lasker-Noether decomposition theorem*, Amer. J. Math. (to appear).

24. G. Stolzenberg, *Constructive normalization of an algebraic variety*, Bull. Amer. Math. Soc. **74** (1968), 595–599.

25. J. B. Tannenbaum, *A constructive version of Hilbert's basis theorem*, Ph.D. Dissertation, Univ. of Calif., San Diego, Calif., 1973.

26. B. L. van der Waerden, *Eine Bemerkung über die Unzerlegbarkeit von Polynomen*, Math. Ann. **102** (1930), 738–739.

27. O. Zariski, *The concept of a simple point of an abstract algebraic variety*, Trans. Amer. Math. Soc. **62** (1947), 1–52.

Department of Mathematics, University of California, Berkeley, California 94720

V. FOUNDATIONS AND COMPLEXITY THEORY

Proceedings of Symposia in Pure Mathematics
Volume **42**, 1985

Elements de Logique Π^1_n

JEAN-YVES GIRARD AND JEAN PIERRE RESSAYRE

Commençons avec un slogan:

la logique Π^1_2 *c'est l'indiscernabilité plus la bonne fondation.*

(i) *l'indiscernabilité* correspond grosso modo à l'aspect *algébrique* de la théorie: on sait par exemple que les dilatateurs peuvent être remplacés par des *systèmes de dénotation*, et la propriété essentielle de ces systèmes est que la comparaison entre

$$(z_0;\, x_0,\ldots,x_{n-1}x) \quad \text{et} \quad (z_1;\, x'_0,\ldots,x'_{m-1};\, x),$$

se fait (z_0, n, z_1, m étant fixés) au moyen de la comparaison entre les x_i et les y_j: c'est bien là une condition d'indiscernabilité.

Tout l'aspect algébrique (i.e. finitaire) de la logique Π^1_2 pourra donc être traduit en termes d'indiscernables; en particulier, les techniques de théorie des modèles pourront être appliquées pour donner une approche alternative à la logique Π^1_2, avec des résultats variables, mais toujours avec l'avantage de diversifier les points de vue possibles. Il semble que cette approche soit particulièrement intéressante dans le cas des *ptykes* de type > 2, où la plus grande partie de nos connaissances [**G2**, Chapter 12], est en fait de nature algébrique, et donc du ressort de l'indiscernabilité; cette approche se justifie peut-être moins dans le cas des dilatateurs (ptykes de type 1), où l'ensemble des résultats [**G1**; **G2**, Chapters 8, 9] dépend pour beaucoup d'une analyse trés détaillée où le choix de la formulation (dilatateurs, systèmes de dénotation, et les modèles d'Ehrenfeucht-Mostowski considérés ici) ne joue aucun rôle appréciable. Rappelons que l'étude des ptykes de type n forme le cœur de ce que nous appellerons la logique Π^1_{n+1}.

Il est important de remarquer que réciproquement, les questions d'indiscernabilité peuvent souvent être traduites en termes de foncteurs; par exemple, $0^\#$, qui correspond à un critère d'indiscernabilité, induit un foncteur de ON dans ON qui préserve les limites directes. (A ce propos, un mot sur la préservation des

Universite Paris VII, U.E.R. de Mathematique et Informatique, Tour 45-55-5eme Etage, 2, Place Jussieu, 75251 Paris Cedex 05, France.

1980 *Mathematics Subject Classification*. Primary 03D60; Secondary 03F15.

produits fibrés: cette propriété, qui en termes de modèles E.M. correspond à l'existence de supports minima, joue un rôle essentiel, en simplifiant à l'extrême l'aspect algébrique des théories des dilatateurs et autres ptykes . Or c'est une propriété qui coûte peu: par exemple si D est un foncteur de ON dans ON qui préserve les limites directes, alors $D \circ (\omega \cdot (\mathbf{1} + \mathrm{Id}))$ est un dilatateur, voir [**G2**, Example 8.G.15].

(ii) *La bonne fondation*. C'est l'aspect qui contient la "force" des méthodes, au sens brutal de "complexité logique" et à opposer à "subtilité", qui serait plutôt liée à l'aspect algébrique. Une comparaison: on peut étudier les séries formelles en tant qu'anneau, c'est l'aspect algébrique; mais évidemment, savoir qu'une série formelle converge pour toute ou certaine valeur de la variable, nous donne tout de suite un certain nombre de résultats qui ne sont pas du ressort de l'étude algébrique. La bonne fondation d'un ordre linéaire obtenu comme limite directe d'entiers est quelque chose qui rappelle une condition de convergence.

Les conditions de bonne fondation concentrent en elles la complexité logique: Π^1_2 pour les dilatateurs, et plus généralement Π^1_{n+1} pour les ptykes de type n.

Cette estimation grossière de la "force" des ptykes de type n, conduit à penser qu'il est possible de les utiliser de manière à caractériser des bons ordres de complexité logique voisine, l'intérêt résidant dans le caractère algébrique simple des ptykes, qui fait que ces caractérisations nous donnent l'approche la plus "effective" possible de ces bons ordres très généraux. Les résultats s'organisent autour de l'égalité

$$\pi^1_n = \Xi_n(\Xi_{n-1})$$

où π^1_n désigne le premier ordinal non Π^1_n, et Ξ_n dénote la somme des ptykes effectifs de type n. Les Ξ_n sont en quelque sorte les paramètres non effectifs en lesquels une large portion de l'univers dénombrable set effective: on pourrait dire que c'est des "ordinaux admissible de type fini", ou encore des "ptykes admissibles"!

Ces résultats nous donnent un certain nombre de points de repère; mais ils ne font qu'effleurer le sujet. Il va de soi que si π^1_n, pour $n > 2$, est assez loin de nos préoccupations, par contre $\pi^1_2 = \sigma_0$ est un objet raisonnable, et on est en droit d'attendre une description rationnelle de tout ou partie de $\sigma_0 = \Xi_2(\Xi_1)$. Pour celà, nous savons que tout ordinal $x < \sigma_0$ s'écrit $\Phi(\Xi_1)$ pour un certain Φ effectif de type 2 (non unique), et on peut chercher, à partir de la définition infinitaire de x en termes de propriétés de réflexion,..., à trouver un Φ effectif qui le caractérise au moyen de $x = \Phi(\Xi_1)$; les résultats de cet article assurent l'existence d'une solution, mais le problème est de trouver des solutions mémorables dans des cas concrets, avec si possible des applications.[1] Ce qu'on peut espérer à terme, c'est une théorie "effective" de la récursion généralisée *totale*; une telle théorie serait

---

[1] Pour x = le premier α non set-récursif en une fonction g de ON dans ON donnée, g Δ_0, la question a été résolue dans [**GN**]; pour $x = I_0$ (le premier récursivement inaccessible, cas particulier du précédent), pour $x = M_0$ (le premier récursivement Mahlo) la question a été résolue dans [**GV3**].

nécessairement assez différente de la récursion généralisée telle que nous la connaissons.

1. Modèles D'Ehrenfeucht-Mostowski colorés et foncteurs continus. Ce paragraphe introduit une version généralisée des *modèles E.M. et foncteurs E.M.* (E.M. = d'Ehrenfeucht-Mostowski); cette généralisation permet de représenter par des modèles E.M. beaucoup de foncteurs continus, notamment les "ptykes". Nous donnons successivement deux définitions équivalentes des modèles E. M. généralisés: nous n'utiliserons que la deuxième, plus adaptée à nos besoins; mais son équivalence avec la première montre le lien étroit avec la notion classique de modèle E. M.

Abus de notation. Nous écrivons ā au lieu de $a_0,\ldots,a_{n-1}$, $\bar{a} \in M$, au lieu de $\{a_0,\ldots,a_{n-1}\} \subset M$; et ceci s'étend à toutes les lettres à la place de a, M. Souvent, ce sera le lecteur qui, en fonction du contexte, devra juger si la longueur n de la suite notée ā est quelconque ou est d'une valeur fixée; et si cette longueur est égale ou non à la longueur d'autres suites $\bar{b}$, $\bar{x}$....

Soient N_1, N_2 deux stuctures dont les langages respectifs contiennent le langage $\mathscr{L}$, et soient $\bar{a} \in N_1$ et $\bar{b} \in N_2$; on écrira $\bar{a} \cong_{\mathscr{L}} \bar{b}$ ssi les suites $\bar{a}$, $\bar{b}$ ont même longueur et satisfont les mêmes formules atomiques de $\mathscr{L}$-autrement dit, si l'application: $a_i \mapsto b_i$ est un $\mathscr{L}$-isomorphisme partiel de N_1 vers N_2.

1.1. *Modèles E.M., première définition.* Soient $\mathscr{L}^0$, $\mathscr{L}$ deux langages, X une $\mathscr{L}^0$-structure et M une $\mathscr{L}$-structure.

(a) On dira que *M est engendré par X* si $|X| \subset |M|$ et tout élément de M est de la forme $t^M(\bar{x})$, où t est un terme de $\mathscr{L}$ et $\bar{x} \in X$;

(b) on dira que *X est $\mathscr{L}^0$-indiscernable dans M* si $|X| \subset |M|$ et pour toutes suites $\bar{x}$, $\bar{y} \in X$, $\bar{x} \simeq_{\mathscr{L}^0} \bar{y}$ entraîne $\bar{x} \simeq_{\mathscr{L}} \bar{y}$;

(c) on dira que *M est un modèle E.M. engendré par X* s'il existe un langage $\mathscr{L}^\# \supset \mathscr{L}$ et une $\mathscr{L}^\#$-structure $M^\#$ telle que $M = M^\# \upharpoonright \mathscr{L}$, *$M^\#$ est engendré par X* et X est $\mathscr{L}^0$-indiscernable dans $M^\#$.

REMARQUES. (a) La seule différence, de la notion classique de modèle E.M. à la notion ci-dessus, est le remplacement de l'ensemble indiscernable *ordonné* X par un ensemble "$\mathscr{L}^0$-indiscernable". Or, cette notion d'ensemble $\mathscr{L}^0$-indiscernable" X est la généralisation des indiscernables ordonnés où *X est une structure quelconque* au lieu d'être nécessairement un ordre total. Cette généralisation est si naturelle que son introduction va de soi. D'ailleurs, le rôle qu'elle a déjà joué est important: le travail de Abramson et Harrington sur les modèles et l'arithmétique[2] est centré sur les "*indiscernables colorés*" et les modèles de l'arithmétique qui sont engendrés par ces indiscernables, autrement dit, qui sont des "*modèles E.M. colorés*". *Or, ces notions sont en gros équivalentes à celles de l.l.* (Nota Bene: cependant, à part ces notions et le "théorème d'étirement", cf. 1..4, le travail d'Abramson-Harrington n'a rien de commun avec le présent travail.)

(b) *Lien entre indiscernabilité et foncteurs.* Pour toute structure N, soit $\mathscr{S}_\omega(N)$ la catégorie des sous-structures finîment engendrées de N et des isomorphismes

[2] Cf. [**A-H**], et pour un bref résumé cf. [**R2**].

simples entre elles; alors X est $\mathscr{L}^0$-indiscernable dans $M \Leftrightarrow$ l'application identité est un foncteur de $\mathscr{S}_\omega(X)$ dans $\mathscr{S}_\omega(M)$.

(c) La définition 1.1(b) de $\mathscr{L}^0$-indiscernabilité s'exprime également par la condition suivante:

(1.1.bis) $$M \models \theta(t^0(\bar{x}) \cdots t^k(\bar{x})) \Leftrightarrow \theta(t^0(\bar{y}) \cdots t^k(\bar{y})),$$

pour toute formule atomique $\theta(t^0[\dot{v}] \cdots t^k[\dot{v}])$ de $\mathscr{L}$, et toutes suites $\bar{x}, \bar{y} \in X$ telles que $\bar{x} \simeq_{\mathscr{L}^0} \bar{y}$.

Notations (*déjà utilisées dans* 1.1.bis). Nous écrivons $t[\cdots]$ pour désigner des termes *formels*, alors que $t(\cdots)$ désignera la valeur du terme formel $t[\cdots]$ dans une interprétation donnée. Par ailleurs, lorsque t est un terme à k variables et $\bar{x}$ une suite de longueur $n \geqslant k$, nous écrivons souvent $t[\bar{x}]$, en *sous-entendant* que si $n > k$, alors $t[\bar{x}]$ *représente le terme* $t[x_{i_1} \cdots x_{i_k}]$ où $i_1 \leqslant \cdots \leqslant i_k < n$ sont fixés: alors pour toute autre suite $\bar{y}$, $t[\bar{y}]$ représente $t[y_{i_1} \cdots y_{i_k}]$ (et de même avec $t(\cdots)$ à la place de $t[\cdots]$).

1.2. *Modèles E.M., deuxième définition.* Soient X une $\mathscr{L}^0$-structure, M une $\mathscr{L}$-structure, $|F|$ un ensemble de termes d'un langage (généralement distinct de $\mathscr{L}^0, \mathscr{L}$):

(a) On dira que M est engendré par X et $|F|$ si $|M| = \{t[\bar{x}]: t \in |F|, \bar{x} \in X\}$.

(b) Alors on dira que *X est $\mathscr{L}^0$-indiscernable dans M* si $M \models \theta(t^0[\bar{x}] \cdots t^k[\bar{x}]) \Leftrightarrow \theta(t^0[\bar{y}] \cdots t^k[\bar{y}])$, pour toute formule atomique $\theta(u^0 \cdots u^k)$ de $\mathscr{L}$, tous termes $t^0[\bar{v}] \cdots t^k[\bar{v}]$ de $|F|$ et toutes suites $\bar{x}, \bar{y} \in X$ telles que $\bar{x} \simeq_{\mathscr{L}^0} \bar{y}$ (c'est l'analogue de 1.1.bis).

(c) Enfin on dira que *M est un modèle E.M. engendré par X et* $|F|$ si (a) et (b) sont réalisés.

Notation. Un tel modèle E.M. sera généralement noté $F[X]$: quand on dira "soit $F[X]$ un modèle E.M.", il sera *sous-entendu* que ce modèle est engendré par X et par un ensemble de termes *qui sera noté* $|F|$.

Il y a *presque* équivalence entre la première et la seconde définition de modèle E.M.; plus précisément:

(a) Soit M un modèle E.M. au sens 1.1: $M = M^\sharp \upharpoonright \mathscr{L}$, $M^\sharp$ engendré par X, qui est $\mathscr{L}^0$-indiscernable dans $M^\sharp$. Posons $|F|$ = ensemble des termes de $\mathscr{L}^\sharp$; de 1.1.bis, résulte que M est un modèle E.M. engendré par X et $|F|$ au sens 1.2, *à un seul point près*: dans 1.2, les éléments de M *sont* des termes formels, alors que dans 1.1 on leur demande seulement d'être les interprétations dans $M^\sharp$ de termes formels.

(b) Inversement, soit $F[X]$ un $\mathscr{L}$-modèle E.M. au sens 1.2; nous posons $\mathscr{L}^\sharp = \mathscr{L}^0 \cup \mathscr{L} \cup |F|$, avec les conventions suivantes: on suppose que les variables de $\mathscr{L}^0$ et celles de $\mathscr{L}$ sont deux sortes disjointes de variables de $\mathscr{L}^\sharp$; et chaque terme à k variables de $|F|$ est traité comme un symbole de fonction à k arguments de $\mathscr{L}^\sharp$, qui a pour domaine la sorte $\mathscr{L}^0$ de variables, et pour image la sorte $\mathscr{L}$.

Alors, soit $M^\sharp$ la $\mathscr{L}^\sharp$-structure $(X, F[X], (t^*)_{t \in |F|})$, où t^* est "l'interprétation canonique": $\bar{x} \in X \mapsto t[\bar{x}]$. Alors $M^\sharp$ est un modèle E.M. engendré par X, au sens 1.1; et $M^\sharp \upharpoonright \mathscr{L} = F[X]$, *donc* $F[X]$ *est un modèle E.M. au sens* 1.1.

1.3. *Nota Bene.* Nos modèles $F[X]$ ne sont pas nécessairement égalitaires: il peut exister deux termes distincts $t[\bar{x}]$, $s[\bar{y}]$ dans $|F[X]|$ tels que $F[X] \models t[\bar{x}] = s[\bar{y}]$. Mais, bien entendu, *on suppose toujours que $F[X]$ satisfait les axiomes de l'égalité.* Et on sait que si $\sim$ désigne $=_{F[X]}$, alors $\sim$ est une relation d'équivalence, et le quotient $F[X]/\sim$ est un modèle égalitaire que la plupart du temps *on peut identifier avec* $F[X]$, vu le peu de différence entre $F[X]$ et $F[X]/\sim$: pour tous $a_1 \cdots a_n$ dans $F[X]$ et toute formule $\psi(\bar{v})$ du langage de $F[X]$, on a $F[X] \models \psi(a_1,\ldots,a_n) \Leftrightarrow F[X]/\sim \;=\; \psi(a_1/\sim \cdots a_n/\sim)$.

Notations. Pour tout langage $\mathscr{L}$, mod $\mathscr{L}$ désigne la catégorie des $\mathscr{L}$-structures, munie des $\mathscr{L}$-isomorphismes simples d'une structure *dans* une autre. Et si $M \in$ mod $\mathscr{L}$, $\overline{M}$ désigne la restriction de la catégorie mod $\mathscr{L}$ à toutes les structures N qui sont *finîment plongeables dans* M ($\Leftrightarrow \forall \bar{a} \in N$, $\exists \bar{b} \in M$, tels que $\bar{a} \simeq_{\mathscr{L}} \bar{b}$; ce qui équivaut à dire que N est la $\varinjlim$ d'une famille inductive de sous-structures de M).

Exemple. Soit OL la catégorie des ordres totaux et morphismes d'ordre; alors $OL = \overline{X}$ pour tout ordre total *infini* X—en particulier $OL = \bar{\omega} = \overline{Q}$.

Nous fixons une $\mathscr{L}^0$-structure Q^0, et un modèle E.M. $F[Q^0] \in$ mod $\mathscr{L}$. Soit alors X quelconque $\in \overline{Q}^0$: il se peut que $X \subset Q^0$, mais le cas intéressant sera celui où X est plus grand que Q^0 (par exemple $Q^0 = \omega$, donc $\overline{Q}^0 = OL$, et X est un ordinal $> \omega$). Alors comme X est finîment isomorphe à Q^0, l'usage est de penser X comme obtenu en "étirant" Q^0. On va faire correspondre à X un modèle E.M. noté $F[X]$, qui intuitivement est obtenu en "étirant" pareillement $F[Q^0]$ (dans le cas $\overline{Q}^0 = OL$, c'est l'"étirement" bien connu des modèle E.M. *classiques*):

1.4. Théorème d'étirement. *Soit $F[Q^0]$ un $\mathscr{L}$-modèle E.M.; pour tout $X \in \overline{Q}^0$, on note $F[X]$ la $\mathscr{L}$-structure définie par $|F[X]| = \{t[\bar{x}]\colon\ t \in |F|,\ \bar{x} \in X\}$ et $F[X] \models \theta(t_1[\bar{x}] \cdots t_k[\bar{x}]) \Leftrightarrow$ il existe $\bar{a} \in Q^0$ tel que $\bar{a} \simeq_{\mathscr{L}^0} \bar{x}$ et $F[Q^0] \models \theta(t_1[\bar{a}] \cdots t_k[\bar{a}])$; ceci pour toute formule atomique $\theta(v_1 \cdots v_k) \in \mathscr{L}$, et tous termes $t_1 \cdots t_k \in |F|$.*

Ceci définit la $\mathscr{L}$-structure $F[X]$ sans ambiguïté, ni contradiction; d'autre part, pour toute application $f\colon X \to Y$, soit $F[f]$ l'application: $t[\bar{x}] \mapsto t[f\bar{x}]$ ($t \in |F|$, $\bar{x} \in X$) de $F[X]$ dans $F[Y]$. Si f est un $\mathscr{L}^0$-morphisme de X dans Y, alors $F[f]$ est un $\mathscr{L}$-morphisme de $F[X]$ dans $F[Y]$. Et l'application $F[\cdot]$ ainsi définie est un foncteur de la catégorie $\overline{Q}^0$ dans mod $\mathscr{L}$.

Preuve. Tout comme dans le cas d'un modèle E.M. classique.

1.5. Définitions. (a) On appellera *foncteur E.M.* tout foncteur $F[\cdot]$ obtenu par étirement comme ci-dessus d'un modèle E.M. $F[Q^0]$. $F[\cdot]$ sera également noté F pour abréger.

(b) Si $F'[Q^0]$ est un autre modèle E.M., on dit que le *foncteur $F'[\cdot]$ est une restriction du foncteur* $F[\cdot]$, ce qu'on note $F' \subset F$, si $|F'| \subset |F|$ et $F'[Q^0] \subset F[Q^0]$ *—ce qui entraîne $F'[X] \subset F[X]$, pour tout $X \in Q^0$.*

(c) Soit $G[Q^0]$ un modèle E.M. et ϕ une application de $|F|$ dans $|G|$; pour tout $X \in \overline{Q}^0$, on note ϕ^X l'application: $t[\bar{x}] \mapsto \phi(t)[\bar{x}]$ de $F[X]$ dans $G[X]$. Et l'on dit que ϕ *est une interprétation de* $F[Q^0]$ *dans* $G[Q^0]$ si ϕ^{Q^0} est un $\mathscr{L}$-isomorphisme de $F[Q^0]$ dands $G[Q^0]$.

1.4.bis. THÉORÈME D'ÉTIREMENT (SUITE). *Si* ϕ^{Q^0} *est un* $\mathscr{L}$*-isomorphisme de* $F[Q^0]$ *dans* $G[Q^0]$, *alors* ϕ^X *en est un de* $F[X]$ *dans* $G[X]$ *pour tout* $X \in \overline{Q}^0$, *et la famille* $(\phi^X)_{X \in \overline{Q}^0}$ *est une transformation naturelle* (*abrégé t.n.*) *du foncteur* $F[\cdot]$ *dans le foncteur* $G[\cdot]$. *Ainsi, toute interprétation* ϕ *de* $F[Q^0]$ *dans* $G[Q^0]$ *détermine une t.n. de* $F[\cdot]$ *dans* $G[\cdot]$. (La preuve est évidente.)

1.6. REMARQUES ET NOTATIONS. (a) La notion de restriction F' d'un foncteur E.M. F est un cas particulier de t.n. entre ces foncteurs: si $F' \subset F$, alors l'application d'inclusion j de $|F'|$ dans $|F|$ est une interprétation de $F'[Q^0]$ dans $F[Q^0]$, et la famille $(j^X)_{X \in \overline{Q}^0}$ une t.n. de F' dans F.; et réciproquement.

(b) Notez par ailleurs que *le foncteur F possède autant de restrictions que* $|F|$ *possède de sous-ensembles*: à chaque ensemble $X \subset |F|$ est associé *le foncteur noté* $F \upharpoonright X$, qui vérifie $F \upharpoonright X \subset F$ et $|F \upharpoonright X| = X$; ce foncteur est appelé *restriction du foncteur F à X*.

Dans la suite, nous considérons des foncteurs de $\overline{Q}^0$ dans mod $\mathscr{L}$, qui ne sont pas forcément des foncteurs E.M.; ces foncteurs seront notés F, $F(\cdot)$, $F[\cdot]$, et de même avec les symboles, G, F', etc. à la place de F. Mais *plus précisément nous employons*: la notation $F[\cdot]$ *uniquement* pour des foncteurs E.M.; la notation $F(\cdot)$ pour marquer que le foncteur n'est pas forcément E.M., et la notation F si peu importe la nature de F ou si celle-ci a déjà été indiquée auparavant.

DÉFINITIONS. Soit $F(\cdot)$ un foncteur de $\overline{Q}^0$ dans mod $\mathscr{L}$; on dit que F *est continu*, ou F commute aux $\varinjlim$, si pour toute famille inductive $(X_i, f_{ij})_{i<j\in I}$ de structures dans $\overline{Q}^0$, on a

$$\varinjlim \left(F(X_i), F(f_{ij}) \right)_{i<j\in I} = F\left(\varinjlim \left(X_i, f_{ij} \right)_{i<j\in I} \right).$$

Si $X \in \overline{Q}^0$, $X' \subset X$, $z \in F(X)$, on dit que X' *est un support de* z, si z est dans l'image de l'application $F(\mathrm{id}_{X'})$: $F(X') \to F(X)$; et on note $F^X(X')$ l'ensemble des $z \in F(X)$ qui ont X' pour support, cf. Figure 1.

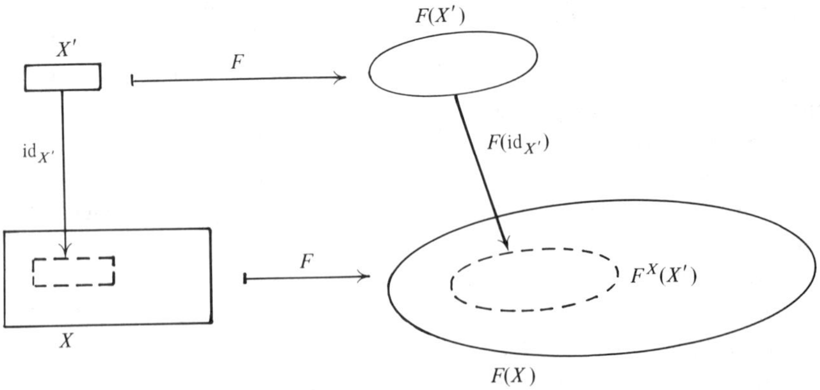

FIGURE 1

1.7. PROPOSITION. (a) *Un foncteur F de $\overline{Q}^0$ dans* mod $\mathscr{L}$ *est continue ssi F possède la propriété de support fini*: $\forall X \in \overline{Q}^0$, $\forall z \in F(X)$, *z possède un support fini.*

(b) *Et F commute aux produits fibrés ssi $\forall X \in \overline{Q}^0$, $\forall z \in F(X)$, l'intersection de deux supports de z est encore un support de z.*

PREUVE. Simple et bien connue; nous n'avons même pas rappelé la définition de commutation aux produits fibrés, mais on peut s'en passer et considérer que cette propriété est définie par la caractérisation donnée en (b).

1.8. PROPOSITION. (a) *Soit $F[\cdot]$ un foncteur E.M. de $\overline{Q}^0$ dans* mod $\mathscr{L}$; *si $X' \subset X \in \overline{Q}^0$, et $z \in F[X]$, alors X' est support de z ssi il existe $t \in |F|$, $\bar{x} \in X'$ tel que $z = t[\bar{x}]$ (donc $F^X[X']$ égale simplement $|F[X']|$).*

(b) *Tout foncteur E.M. est continu.*

(c) *$F[\cdot]$ commute aux produits fibrés ssi $\forall X \in \overline{Q}^0$, $\forall z \in F[X]$, z a un support minimum (ssi $\forall X \in \overline{Q}^0$, $F[X] \models s[\bar{y}] \Rightarrow 2$ il existe un terme $r \in |F|$ et une suite $\bar{z}$ dont chaque coordonnée z_i est à la fois dans $\bar{x}$ et $\bar{y}$, tels que $F[X] \models t[\bar{x}] = r[\bar{z}]$).*

PREUVE. (a) est évident; si $X \in \overline{Q}^0$, $z \in F[X]$, z est un terme $t[x_1 \cdots x_k]$, alors $\{x_1, \ldots, x_k\}$ est un support *fini* de z. Donc par 1.7(a), $F[\cdot]$ est un foncteur continu, et (b) est montré. Enfin, (c) est conséquence de (a) et de 1.7(b).

Résumons ce qui précède: les modèles E.M. et leurs étirements sont un moyen de définir des foncteurs continus de $\overline{Q}^0$ dans mod $\mathscr{L}$, et des t.n. entre ces foncteurs. Le théorème de représentation qui suit va montrer qu'à isomorphismes près, on obtient ainsi *tous* les foncteurs continus de $\overline{Q}^0$ dans mod $\mathscr{L}$, et *toutes* les t.n. entre eux.

1.9. THÉORÈME DE REPRÉSENTATION. (a) *A tout foncteur continu $F(\cdot)$ de $\overline{Q}^0$ dans* mod $\mathscr{L}$, *on peut associer canoniquement un foncteur E.M. $F[\cdot]$, qui est isomorphe à$F(\cdot)$.*

(b) *Si $F[\cdot]$, $G[\cdot]$ sont deux foncteurs E.M. de $\overline{Q}^0$ dans* mod $\mathscr{L}$, *toute t.n. de $F[\cdot]$ dans $G[\cdot]$ est de la forme $(\phi^X)_{X \in \overline{Q}^0}$, où l'application ϕ: $|F| \to |G|$ est une interprétation de $F[Q^0]$ dans $G[Q^0]$. Donc si $F(\cdot)$, $G(\cdot)$ sont des foncteurs continus quelconques de $\overline{Q}^0$ dans* mod $\mathscr{L}$, *et si l'on identifie $F(\cdot)$ avec $F[\cdot]$, $G(\cdot)$ avec $G[\cdot]$ (où $F[\cdot]$, $G[\cdot]$ sont donnés par* (a)), *alors les t.n. de $F(\cdot)$ dans $G(\cdot)$ s'identifient avec les familles $(\phi^X)_{X \in \overline{Q}^0}$, ϕ interprétation de $F[Q^0]$ dans $G[Q^0]$.*

PREUVE DE (a). Nous supposons fixé l'ensemble $|F|$ de symboles de fonction; et pour tous $t \in |F|$, $X \in \overline{Q}^0$ et $\bar{x} \in X$, nous supposons choisi un élément de $F(X)$ noté $t^X(\bar{x})$, de façon que

$$(*) \qquad F(f)(t^X(\bar{x})) = f^Y(f\bar{x}),$$

pour tout $\mathscr{L}^0$-morphisme f: $X \to Y$ et tout élément $t^X(\bar{x})$ de $F(X)$; de plus tout élément $z \in F(X)$ est de la forme $z = t^X(\bar{x})$ pour un $t \in |F|$ et un $\bar{x} \in X$.

Nous définissons alors un modèle $F[X]$ par

$$(**)\qquad \begin{array}{l} F[X] \text{ est la } \mathscr{L}\text{-structure de domaine } \{t[\bar{x}];\ t \in |F|,\ \bar{x} \in X\} \\ \text{telle que } F[X] \models \theta(t_1[\bar{x}] \cdots t_k[\bar{x}]) \Leftrightarrow_{df} F(X) \models \theta(t_1^X(\bar{x}) \cdots \\ t_k^X(\bar{x})), \end{array}$$

pour toute formule atomique $\theta(u_1, \ldots, u_k) \in \mathscr{L}$, tous $\bar{x} \in X$ et $t_1 \cdots t_k \in |F|$.

Soit alors I_X l'application: $t[\bar{x}] \mapsto t^X(\bar{x})$, et soit $\sim$ l'interprétation de $=$ dans $F[X]$. La condition $(**)$ revient à dire que $I_X/\sim$ est un $\mathscr{L}$-isomorphisme de $F[X]/\sim$ sur $F(X)$. Vu le Nota Bene 1.3; nous identifions $F[X]/\sim$ et $F[X]$:[3] nous ferons comme si I_X était un isomorphisme de $F[X]$ sur $F(X)$.

Alors pour tout $\mathscr{L}^0$-isomorphisme f: $X \to Y$, soit $F[f]$ l'application de $F[X]$ dans $F[Y]$ définie par $F[f](t\bar{x}) = t[f\bar{x}]$. La condition $(*)$ entraîne la commutativité du carré

$$\begin{array}{ccc} F[X] & \xrightarrow{\;I_X\;} & F(X) \\ {\scriptstyle F[f]}\downarrow & & \downarrow{\scriptstyle F(f)} \\ F[Y] & \xrightarrow[\;I_Y\;]{} & F(Y) \end{array}$$

laquelle entraîne que $F[f]$ est un isomorphisme. En faisant varier f, on en déduit facilement:

pour tout $X \in \bar{Q}^0$, $F[X]$ est un modèle E.M., et est l'étirement de $F[Q^0]$; donc $F[\cdot]$ est un foncteur E.M.;

la famille $(I_X)_{X \in \bar{Q}^0}$ est un isomorphisme de ce foncteur $F[\cdot]$ sur le foncteur $F(\cdot)$.

Donc (a) est prouvé, *sous réserve* de choisir $|F|$ et réaliser la condition $(*)$. L'idée pour cela est que cette condition elle-même impose dans une certaine mesure le choix de $|F|$ et la définition de $t^X(\bar{x})$, par l'intermédiaire des remarques qui suivent.

Soient $\bar{a} \in Q^0$ et $z \in F(Q^0 \upharpoonright \bar{a})$ tels que $\bar{a}$ est un support minimal de z; on voit facilement que:

(1) pour réaliser $(*)$, il *faut* introduire un symbole t dans $|F|$ avec k arguments, où $k = l(\bar{a})$ et poser $z = t^X(\bar{a})$;

(2) cela étant, soient $\bar{x} \in X \in \bar{Q}^0$ et $z_1 \in F(X)$; on pose $(z, \bar{a}) \sim (z_1, \bar{x})$ si l'application f: $a_i \mapsto x_i$ est un $\mathscr{L}^0$-isomorphisme de $Q^0 \upharpoonright \bar{a}$ dans X, et si $F(f)(z) = z_1$. Pour réaliser $(*)$, il *faut* également poser $t^X(\bar{x}) = z_1$, pour tout couple $(z_1, \bar{x}) \sim (z, \bar{a})$. Donc l'introduction du symbole t dans $|F|$ faite pour $(z, \bar{a})$ sert également pour tous ces couples $(z_1, \bar{x})$.

Moyennant quoi, pour realiser $(*)$ nous dévons introduire un symbole t comme dans (1) autant de fois qu'il y a de classes d'équivalences $(z, \bar{a})/\sim$; ce que font les notations ci-dessous.

[3] Nous le faisons d'autant plus volontiers que pour tous les foncteurs $F(\cdot)$ qui nous intéressent par la suite, les Théorèmes 1.13 et 2.13 entraînent que $\sim$ est l'identité, donc $F[X]$ est égalitaire, et $F[X]\sim$, $I_X/\sim$ sont vraiment identiques à $F[X]$, I_X, etc.

1.10. *Notations.* (a) Pour toute $\mathscr{L}$-structure N et tout $\bar{a} \in N$, on appelle $\mathscr{L}$-type de $\bar{a}$ dans N l'ensemble des formules atomiques et de leurs négations qui sont satisfaites par $\bar{a}$ dans N—ce qui entraîne que $\bar{b} \simeq_{\mathscr{L}} \bar{a}$ si et seulement si le $\mathscr{L}$-type de $\bar{b}$ est égal à celui de $\bar{a}$.

(b) Nous supposons fixée une famille $(p_i)_{i \in I}$ de $\mathscr{L}^0$-types satisfaits dans Q^0, et vérifiant: pour toute suite de points distincts $\bar{a} \in Q^0$, il existe un *unique* indice $i \in I$ tel que p_i est le $\mathscr{L}^0$-type de l'une au moins des permutations de la suite $\bar{a}$ (autrement dit, $(p_i)_{i \in I}$ énumère *tous* les $\mathscr{L}^0$-types satisfaits dans Q^0, à ceci près que si $p_i(v_1 \cdots v_n)$ est énuméré, alors sont omises toutes les variantes telles que $p_i(v_{\sigma_1} \cdots v_{\sigma_n})$, $p_i(v_1 \cdots v_n) \cup \{v_n = v_{n+1}, \dots\}$, etc., qui font double emploi avec $p_i(v_1 \cdots v_n)$).[4]

(c) Nous posons $|F| = E/\sim$, où $\sim$ est définie en (2) ci-dessus, et $E = \{(z, \bar{a})$: $\bar{a} \in Q^0$, $z \in F(Q^0 \upharpoonright \bar{a})$, $\bar{a}$ est support minimal de z, enfin $\exists i \in I$ tel que p_i est le $\mathscr{L}^0$-type de $\bar{a}\}$. Et si $t = (z, \bar{a})/\sim$ appartient à $|F|$, nous considérons t comme un symbole de fonction à k arguments, $k = l(\bar{a})$.

(d) Si $t = (z, \bar{a})/\sim \in |F|$, $\bar{x} \in X \in \overline{Q}^0$, $z_1 \in F(X)$ et $(z, \bar{a}) \sim (z_1, \bar{x})$, alors nous posons $t^X(\bar{x}) = z_1$ $(= F(f)(z)$, où f est l'application: $a_i \mapsto x_i)$.

Nous allons voir que ces définitions réalisent en grande partie la condition $(*)$.

En premier lieu, la définition (d) n'est pas ambigüe: elle ne dépend pas du représentant $(z, \bar{a})$ de t choisi. En effet, soit $(z', \bar{b}) \in E$ tel que $(z', \bar{b}) \sim (z, \bar{a})$, donc $\bar{b} \simeq_{\mathscr{L}^0} \bar{a}$ et si ϕ est le $\mathscr{L}^0$-morphisme: $b_i \mapsto q_i$, alors $F(\phi)(z') = z$. Puisque $t = (z, \bar{a})/\sim\, = (z', \bar{b})/\sim$, en vertu de (d), nous avons posé à la fois:

$$t^X(\bar{x}) = F(g)(z') \quad \text{et} \quad t^X(\bar{x}) = F(f)(z),$$

où f est le $\mathscr{L}^0$-morphisme: $a_i \mapsto x_i$, et g le $\mathscr{L}^0$-morphisme: $b_i \mapsto x_i$. Mais $g = f \circ \phi$ donne $F(g) = F(f \circ \phi) = F(F) \circ F(\phi)$, donc $F(g)(z') = F(f) \circ F(\phi)(z') = F(f)(z)$: ainsi $t^X(\bar{x})$ est bien unique.

En second, lieu, un raisonnement tout semblable montre la "formule" de $(*)$: $F(f)(t^X(\bar{x})) = t^Y(f\bar{x})$, pour tout $\mathscr{L}^0$-morphisme f: $X \to Y$.

En troisième lieu, tout point $z_1 \in F(X)$, où $X \in \overline{Q}^0$, est de la forme $t^X(\bar{x})$, où $t \in |F|$, $\bar{x} \in X$: ceci utilise la continuité du foncteur $F(\cdot)$, qui par 1.7, entraîne l'existence d'un support fini de z_1. Ce support $\bar{x} \in X$ de z_1 étant fini peut être supposé minimal: le $\mathscr{L}^0$-type de $\bar{x}$ dans X est satisfait par une suite $\bar{a} \in Q^0$ puisque $X \in \overline{Q}^0$, et par 1.10(b), quitte à réordonner $\bar{x}$ et $\bar{a}$, on peut supposer que leur $\mathscr{L}^0$-type est de la forme p_j pour un $j \in I$. Soit f le $\mathscr{L}^0$-morphisme: $a_i \mapsto x_i$ de $Q^0 \upharpoonright \bar{a}$ dans X. $F(f)$ est un $\mathscr{L}$-isomorphisme de $F(Q^0 \upharpoonright \bar{a})$ dans $F(X)$, dont l'image contient z_1 puisque $\bar{x} = f\bar{a}$ est support de z_1. Il existe donc $z \in F(Q^0 \upharpoonright \bar{a})$ tel que $F(f)(z) = z_1$. Comme $\bar{x}$ est support minimal de z_1, $\bar{a}$ est support minimal

[4] Lorsque $\overline{Q}^0 = OL$, il est naturel de prendre pour famille $(p_i)_{i \in I}$ celle où $I = \omega$, et $p_i =$ "$v_0 < \cdots < v_{i-1}$". Et si $\overline{Q}^0$ est quelconque, les types p_i, $i \in I$, joueront le même rôle que ces types "$v_0 < \cdots < v_{i-1}$" vis-à-vis des types d'ordres quelconques.

de z; donc $(z, \bar{a}) \in F$, $(z, \bar{a}) \sim (z_1, \bar{x})$, d'où $z_1 = t^X(\bar{x})$, où $t = (z, \bar{a})/\sim$ appartient à F.

Il ne manque plus qu'une chose à la réalisation de $(*)$: si $t = (z, \bar{a})/\sim\ \in |F|$ et p_i est le $\mathscr{L}^0$-type de $\bar{a}$, la clause 1.10(d) définit $t^X(\bar{x})$ seulement sur $\{\bar{x} \in X\colon \bar{x}$ satisfait $p_i\}$; il faut donc une clause définissant $t^X(\bar{x})$ lorsque $\bar{x} \in X$, $l(\bar{x}) = l(\bar{a})$ mais $\bar{x}$ ne satisfait pas p_i, autrement dit, $\bar{x} \not\simeq_{\mathscr{L}^0} \bar{a}$. Voici une telle clause, mais qui est valable avec une restriction sur le foncteur $F(\cdot)$: elle suppose que $F(\varnothing)$ *est défini et* $F(\varnothing) \neq \varnothing$.

(e) On fixe un élément ε de $F(\phi)$; l'ensemble vide étant support minimum de ε, $|F|$ contient $(\varepsilon, \phi)/\sim$, "symbole de fonction à 0 arguments", autrement dit, constante, que nous notons encore ε. Nous définissons alors $t^X(\bar{x})$, où $t = (z, \bar{a})/\sim\ \in |F|$ et $\bar{x} \in X \in \overline{Q}^0$, lorsque $\bar{a} \not\simeq_{\mathscr{L}^0} \bar{x}$: *dans ce cas, nous convenons que* $t^X(\bar{x}) = \varepsilon^X$ (où ε^X est déjà défini par 1.10(d)).

On laisse au lecteur le soin de vérifier que la "formule" $F(f)(t^X(\bar{x})) = t^Y(f\bar{x})$ reste satisfaite par cette extension de t^X; la condition $(*)$ est donc réalisée, ce qui achève la preuve du théorème de représentation (a), mais en utilisant l'hypothèse supplémentaire $F(\varnothing) \neq \varnothing$.

C'est une hypothèse peu restrictive, mais de toutes façons la clause 1.10(e) est artificielle, et nous introduisons une légère modification de la notion 1.2 de modèle E.M., *qui permet de supprimer la clause* 1.10(e), *donc l'hypothèse* $F(\varnothing) \neq \varnothing$.

DÉFINITIONS. (a) *La notion de modèle E.M.* $F[X]$ est définie comme en 1.2, excepté le point suivant: au lieu de poser $|F[X]| = \{t[\bar{x}];\ t \in |F|,\ \bar{x} \in X\}$ on demande seulement:

$$(1.2.\text{bis})\qquad \begin{aligned} &|F[X]| \subset \{t[\bar{x}]\colon t \in |F|, \bar{x} \in X\} \\ &\text{si } t[\bar{a}] \in |F[X]|, \text{ alors } t[\bar{x}] \in |F[X]| \\ &\text{pour toute suite } \bar{x} \in X \text{ telle que } \bar{x} \simeq_{\mathscr{L}^0} \bar{a}. \end{aligned}$$

(b) On modifie en conséquence la Définition 1.5 de l'étirement $F[X]$ d'un modèle E.M. $F[Q^0]$: on pose

$$(1.5.\text{bis})\qquad \begin{aligned} |F[X]| = \{t[\bar{x}];\ t \in |F|, \bar{x} \in X &\text{ et il existe } \bar{a} \in Q^0 \\ &\text{tel que } \bar{a} \simeq_{\mathscr{L}^0} \bar{x} \text{ et } t[\bar{a}] \in F[Q^0]\}; \end{aligned}$$

le reste de la définition de $F[X]$ est conservé et on appelle foncteur E.M. tout foncteur qui s'obtient par l'étirement ainsi modifié d'un modèle E.M. $F[Q^0]$.

Tous les résultats démontrés jusqu'ici restent valables, avec cette notion modifiée de modèle et de foncteur E.M. C'est facile à vérifier, nous le faisons en ce qui concerne le théorème de représentation (a): étant donné $F(\cdot)$, foncteur continu de $\overline{Q}^0$ dans mod $\mathscr{L}$, on a vu que les conditions 1.10(b)–(d) réalisent la condition $(*)$, mais définissent $t^X(\bar{x})$ uniquement pour les suites $\bar{x} \in X$ telles que t est de la forme $t = (z, \bar{a})/\sim$ avec $\bar{a} \simeq_{\mathscr{L}^0} \bar{x}$. Pour $X \in \overline{Q}^0$, on définit alors la structure

$F[X] \in \operatorname{mod} \mathscr{L}$ par la condition (**), modifiée en posant

$$|F[X]| = \{t[\bar{x}];\ t = (z, \bar{a})/\sim\ \in |F|, \bar{x} \in X, \bar{x} \simeq_{\mathscr{L}^0} \bar{a}\};$$

et l'on définit I_X, $F[f]$ comme avant. On voit facilement que l'application $F[\cdot]$ ainsi redéfinie est un foncteur E.M. au sens modifié, isomorphe à $F(\cdot)$ par la t.n. $(I_X)_{X \in \bar{Q}^0}$.

La clause 1.10(e) est donc supprimée, et son hypothèse $F(\varnothing) \neq \varnothing$ est inutile.

PREUVE DU THÉORÈME DE REPRÉSENTATION (b). Laissée au lecteur.

1.11. *Notations.* Désormais, pour tout foncteur continu $F(\cdot)$ de $\bar{Q}^0$ dans $\operatorname{mod} \mathscr{L}$, on note $F[\cdot]$ le foncteur E.M. isomorphe à $F(\cdot)$ défini ci-dessus, et l'on note $|F|$ son ensemble de symboles de fonction. Et si $G(\cdot)\colon \bar{Q}^0 \to \operatorname{mod} \mathscr{L}$ est un autre foncteur continu, on identifiera les t.n. de $F(\cdot)$ dans G avec celles de $F[\cdot]$ dans $G[\cdot]$—donc avec les familles $(\phi^X)_{X \in \bar{Q}^0}$, où $\phi\colon |F| \to |G|$ est une interprétation de $F[Q^0]$ dans $G[Q^0]$—identification justifiée par le théorème de représentation (b).

1.12. REMARQUES.[5] (a) Pour $X \in \bar{Q}^0$, $t \in |F|$, appelons *domaine de t dans $F[X]$* l'ensemble des suites $\bar{x} \in X$, telles que $t[\bar{x}] \in F[X]$. Alors *pour chaque $t \in |F|$, il existe un $\mathscr{L}^0$-type que nous noterons p_t, tel que le domaine de t dans $F[X]$ est constitué par toutes les suites de X ayant p_t comme $\mathscr{L}^0$-type* (en effet, si $t = (z, \bar{a})/\sim$, p_t est le $\mathscr{L}^0$-type satisfait par $\bar{a}$ dans Q^0).

(b) Nous avons déjà signalé que $F[X]$ n'est pas toujours égalitaire: il peut arriver que $t^X(\bar{x}) = s^X(\bar{y})$, donc $F[X] \models t[\bar{x}] = s[\bar{y}]$, même si $t[\bar{x}]$ et $s[\bar{y}]$ sont des termes distincts. Mais les clauses 1.10(b) et 1.10(c) réduisent à un minimum cette duplication des termes, comme va le montrer la suite des remarques.

(c) Ainsi, supposons $t^X(\bar{x}) = s^X(\sigma\bar{x}) = z_1 \in F(X)$, où σ est une permutation de $\bar{x}$; donc $F[X] \models t[\bar{x}] = s[\sigma\bar{x}]$. Nous allons voir qu'alors $t = s$ et la permutation σ est un $\mathscr{L}^0$-morphisme, $\bar{x} \simeq_{\mathscr{L}^0} \sigma\bar{x}$.

Soit $(z, \bar{a})$ un élément de E tel que $t = (z, \bar{a})/\sim$, donc $(z, \bar{a}) \sim (z_1, \bar{x})$ et $\bar{a} \simeq_{\mathscr{L}^0} \bar{x}$; soit f le $\mathscr{L}^0$-morphisme: $a_i \mapsto x_i$. Alors $\sigma\bar{a}$ et $\sigma\bar{x}$ se correspondent par ce même $\mathscr{L}^0$-morphisme f, et de $s^X(\sigma\bar{x}) = z_1$ on déduit par 1.10(c) que s est de la forme $(z', \sigma\bar{a})/\sim$, avec $F(f)(z') = s^X(\sigma\bar{x})$. Mais $s^X(\sigma\bar{x}) = z_1$, et $F(f)(z) = z_1$, donc $z' = z$ et $s = (z, \sigma\bar{a})/\sim$. Comme $t, s \in |F|$, $(z, \bar{a})$ et $(z, \sigma\bar{a})$ sont dans E, donc $\exists i, j \in J$ tels que p_i est le $\mathscr{L}^0$-type de $\bar{a}$ et p_j celui de $\sigma\bar{a}$. Mais $\sigma\bar{a}$ étant une permutation de $\bar{a}$, $p_i = p_j$ à une permutation près, donc $i = j$ vu le choix 1.10(b) de la famille (p_i); *bref*: $F[X] \models t[\bar{x}] = s[\sigma\bar{x}]$ *entraîne $t = s$ et σ est un $\mathscr{L}^0$-morphisme*, pour toute permutation σ de $\bar{x}$.

(d) Si $X \in \bar{Q}^0$ et $z \in F(X)$, z a toujours un support minimal puisque $F(\cdot)$ est continu; ici nous le supposons toujours *minimum*. Alors chaque fois que $F[X] \models t[\bar{x}] = s[\bar{y}]$, $\bar{x}$ et $\bar{y}$ sont par 1.10 des supports minimaux du même élément $z = t^X(\bar{x}) = s^X(\bar{y})$ de $F(X)$, et comme z a un support minimum, $\bar{x}$ et $\bar{y}$ sont deux

[5] Nos conventions initiales concernant la notation $t[\bar{x}]$ permettaient que $l(\bar{x}) >$ nombre d'arguments de t; dans 1.12 et 1.13 au contraire, nous convenons que cette notation implique l'égalité.

énumérations de ce support: $\bar{y} = \sigma\bar{x}$, où σ est une permutation. Alors de plus (c) s'applique, d'où résulte que $t = s$ et σ est un $\mathscr{L}^0$-morphisme. D'où

1.13. THÉORÈME. *Soit* $F(\cdot)$: $\overline{Q}^0 \to \text{mod}\ \mathscr{L}$ *un foncteur continu.*

(a) *Si* $F(\cdot)$ *commute aux produits fibrés, alors pour tout* $X \in \overline{Q}^0$ *et tous éléments* $t[\bar{x}]$, $s[\bar{y}]$ *de* $F[X]$ *on a*

$$F[X] \vDash t[\bar{x}] = s[\bar{y}] \Rightarrow t = s,$$

$\bar{y}$ *est une permutation de* $\bar{x}$ *et cette permutation est un* $\mathscr{L}^0$*-morphisme.*

Réciproquement, la propriété $[F[X] \vDash t[\bar{x}] = s[\bar{y}] \Rightarrow \bar{y}$ *permutation de* $\bar{x}]$ *entraîne déjà que F commute aux produits fibrés.*

(b) *On suppose que* $\mathscr{L}^0$ *contient un symbole* $<$ *qui est un ordre total sur* Q^0; *alors F commute aux produits fibrés ssi* $F[X]$ *est un modèle égalitaire pour tout* $X \in \overline{Q}^0$ *—donc tout élément* $z \in F(X)$ *est de la forme* $z = t^X(\bar{x})$ *pour un unique symbole* $t \in |F|$ *et une unique suite* $\bar{x} \in X$.

PREUVE. (a) La partie réciproque résulte facilement de 1.8(c), et la partie directe est conséquence immédiate de 1.12(d) et 1.8(c).

(b) Conséquence de 1.13(a) en notant que si $\bar{x} \in X \in Q^0$, la seule permutation de $\bar{x}$ qui soit un $\mathscr{L}^0$-morphisme est l'identité, vu qu'elle respecte l'ordre $<^X$ sur X, qui est total puisque $X \in \overline{Q}^0$ et $<^{Q^0}$ est total.

1.14. REMARQUES. Le théorème de représentation fait apparaître que les t.n. entre foncteurs continus de $\overline{Q}^0$ dans mod $\mathscr{L}$ ont une structure très simple:

(a) Soit $(\phi^X)_{X \in \overline{Q}^0}$ une t.n. du foncteur $F[\cdot]$ dans le foncteur $G[\cdot]$, et posons G' = restriction du foncteur $G[\cdot]$ à $|G'| = \{\phi(t)\colon t \in |F|\}$; alors $(\phi^X)_{X \in \overline{Q}^0}$ est un *isomorphisme* de F sur G', et $G' \subset G$. Bref, *toutes les t.n. entre foncteurs continus de* $\overline{Q}^0$ *dans* mod $\mathscr{L}$ *se ramènent, à isomorphismes près au cas particulièrement simple où* $G' \subset G$.

(b) De leur côté les isomorphismes $(\phi^X)_{X \in \overline{Q}^0}$ d'un foncteur $F[\cdot]$ *sur* un foncteur $G[\cdot]$ sont également de nature particulièrement simple: essentiellement ce sont des changements de notations—pour chaque $t \in |F|$, on écrit partout $\phi(t)$ à la place de t; une fois ce changement de notation effectué, $F[\cdot]$ devient *identique* à $G[\cdot]$.

Le Théorème 1.9 et les Notations 1.11 nous ont permis de représenter tout foncteur continu $F(\cdot)$: $\overline{Q}^0 \to \text{mod}\ \mathscr{L}$ par un foncteur E.M. $F[\cdot]$, qui a l'avantage de vérifier 1.12(a) et surtout 1.13. Mais *même* si $F(\cdot)$ avait *déjà* toutes ces propriétés, $F[\cdot]$ ainsi choisi est toujours distinct de $F(\cdot)$, ce qui complique inutilement; aussi nous convenons

1.11.bis. Pour tout foncteur continu $F(\cdot)$: $\overline{Q}^0 \to \text{mod}\ \mathscr{L}$, on pose

$F[\cdot] =_{\text{df}} F(\cdot)$ si $F(\cdot)$ est un foncteur E.M. vérifiant 1.12 et 1.13,

$F[\cdot] =_{\text{df}}$ le foncteur associé à $F(\cdot)$ par 1.9 sinon.

Et on note $|F|$ l'ensemble de symboles de fonction du foncteur $F[\cdot]$.

2. Foncteurs continus de type > 1.

2.1. *Notations.* Si **C**, **D** sont des catégories, nous notons $\mathscr{C}^1(\mathbf{C},\mathbf{D})$ la catégorie dont les objets sont les foncteurs de **C** dans **D** qui sont continus et de plus commutent aux produits fibrés; et dont les morphismes sont les transformations naturelles entre ces foncteurs. Et l'on définit par récurrence sur n la catégorie $\mathscr{C}^{n+1}(\mathbf{C},\mathbf{D}) =_{\mathrm{df}} \mathscr{C}^1(\mathscr{C}^n(\mathbf{C},\mathbf{D}),\mathbf{D})$.

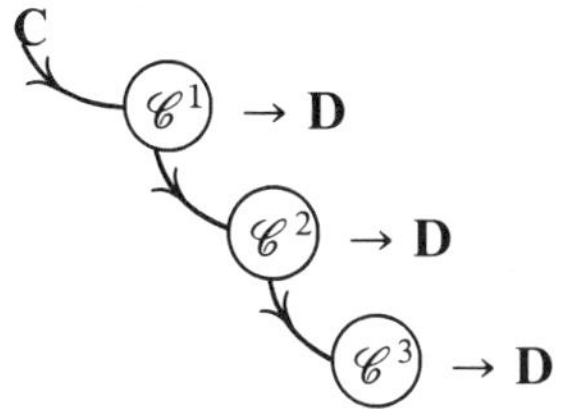

FIGURE 2. $\mathscr{C}^n$ *désigne* $\mathscr{C}^n(\mathbf{C},\mathbf{D})$.

Nous voulons étudier, pour nous en servir, les catégories $\mathscr{C}^n(OL, OL)$, notées $\mathscr{C}^n(OL)$ pour abréger; les foncteurs de $\mathscr{C}^n(OL)$ seront aussi appelés *foncteurs de type n*. Les modèles et foncteurs E.M. du paragraphe 1 fournissent déjà une représentation concrète, et de ce fait utile, des foncteurs de type 1; le théorème du codage qui est le but de ce paragraphe permettra d'appliquer cette même représentation aux foncteurs de type > 1—comme l'indiqueront la Remarque 2.4 et les Notation 2.11(b).

2.A. *Le théorème du codage.*

2.2. DÉFINITION. Une structure Q de langage L est dite (*simplement*) ω*-homogène* si $\forall \bar{a}, \bar{b} \in Q$ tels que $\bar{a} \simeq_L \bar{b}$, on a

$$\forall \bar{c} \in Q\, \exists \bar{d} \in Q\colon \bar{a}^\frown \bar{c} \simeq_L \bar{b}^\frown \bar{d}.$$

EXEMPLE TYPIQUE. La structure $(\mathbf{Q}, <)$ est ω-homogène.

2.3. THÉORÈME DU CODAGE. (a) *Soit* Q^0 *une* $\mathscr{L}^0$*-structure* ω*-homogène; alors il existe un langage* $\mathscr{L}^1$ *et une* $\mathscr{L}^1$*-structure* ω*-homogène* Q^**, telle que la catégorie* $\mathscr{C}^1(\overline{Q}^0, OL)$ *est équivalente à la catégorie* $\overline{Q}^*$.

(b) *Pour tout* $n \in \omega$, *il existe un langage* $\mathscr{L}_n$ *et une* $\mathscr{L}_n$*-structure* ω*-homogène* Q_n^*, *telle que* $\mathscr{C}^n(OL)$ *est équivalente à* $\overline{Q}_n^*$.

PREUVE DE (b) (À PARTIR DE (a)). Par récurrence sur n: Pour $n = 0$, nous convenons de poser $\mathscr{C}^0(OL) = OL$; comme $OL = \overline{\mathbf{Q}}$, l'hypothèse de récurrence est vérifiée en posant $Q_0^* = \mathbf{Q}$. Nous supposons donc l'hypothèse vérifiée pour n donné $\mathscr{C}^n(OL) \sim \overline{Q}_n^*$, où Q_n^* est ω-homogène; alors

$$\mathscr{C}^{n+1}(OL) =_{\mathrm{df}} \mathscr{C}^1(\mathscr{C}^n(OL), OL) \sim \mathscr{C}^1\big(\overline{Q}_n^*, OL\big),$$

et par (a) appliqué en posant $\mathscr{L}^0 = \mathscr{L}_n$, $Q^0 = Q_n^*$ et $Q^* = Q_{n+1}^*$, $\mathscr{C}^1(\overline{Q}_n^*, OL) \sim \overline{Q}_{n+1}^*$. D'où $\mathscr{C}^{n+1}(OL) \sim \overline{Q}_{n+1}^*$.

2.4. REMARQUE. Le théorème de codage permet d'identifier $\mathscr{C}^{n+1}(OL)$ avec

$\mathscr{C}^1(\overline{Q}_n^*, OL)$ donc, en appliquant le théorème de représentation 1.9 à $\mathscr{C}^1(\overline{Q}_n^*, OL)$, de *représenter les foncteurs de type* $n + 1$ *par des foncteurs E.M.*

Reste à prouver le (a) du théorème de codage; deux choses vont nous servir:

(I) les propriétés d'amalgame dans les catégories de modèles, et leurs liens avec les modèle ω-homogènes;

(II) le *codage* des modèle E.M., qui fournit une seconde représentation des foncteurs $\in \mathscr{C}^1(\overline{Q} \bmod \mathscr{L})$. Cette représentation s'ajoute à celle du paragraphe 1 (par modèle E.M.). La Remarque 2.11(b) indiquera l'utilité *respective* de ces deux représentations.

Nous examinons successivement (I) et (II).

(I) *Amalgame et modèles ω-homogènes.*

2.5. Définitions. (a) Une classe de structures $\mathscr{A} \subset \bmod \mathscr{L}$ a l'*amalgame* si chaque fois que B_1, B_2 sont deux structures de $\mathscr{A}$ ayant une partie commune $A \in \mathscr{A}$, on peut les plonger dans une structure $C \in \mathscr{A}$ par deux isomorphismes $\phi_1 \colon B_1 \to C$ et $\phi_2 \colon B_2 \to C$ *tels que* $\phi_1 \restriction A = \phi_2 \restriction A$.

(b) Une catégorie **C** a l'amalgame si chaque fois que le diagramme

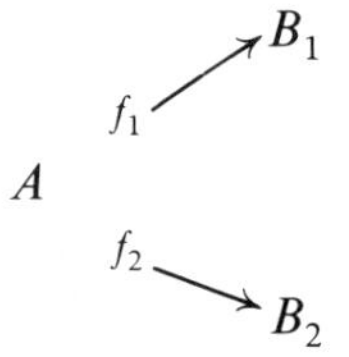

est dans **C**, onpeut le compléter dans **C** par un diagramme

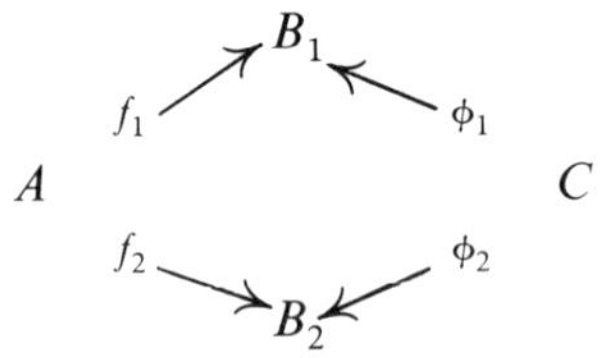

qui commute.

Nota Bene. Une classe de modèle $\mathscr{A} \subset \bmod \mathscr{L}$ close par isomorphismes a l'amalgame (au sens du (a)) ssi la catégorie $\bmod \mathscr{L} \restriction \mathscr{A}$ a l'amalgame (au sens du (b)); donc (a) est un cas particulier de (b) (nous l'avons mis à part pour donner un exemple concret, et qui va nous servir, d'amalgame).

2.6. Proposition. (a) *Soit M une $\mathscr{L}$-structure; la catégorie $\overline{M}$ a l'amalgame si et seulement si il existe $Q \in \overline{M}$, Q ω-homogène et tel que $\overline{M} = \overline{Q}$.*

Example. $\omega = OL$ a l'amalgame donc égale $\overline{Q}$ pour une structure ω-homogène Q—en l'occurence $Q = \mathbf{Q}$.

2.6. Proposition. (b) *Soit $\mathscr{A}$ une classe de $\mathscr{L}$-structures qui est close par restrictions et par limites directes de structures; alors $\mathscr{A}$ possède l'amalgame ssi $\mathscr{A}$ contient une structure ω-homogène Q, telle que $\mathscr{A}$ est l'ensemble des structures finîment plongeables dans Q (autrement dit, ssi* $\bmod \mathscr{L} \restriction \mathscr{A} = \overline{Q}$).

Preuve admise. C'est de la théorie des modèles très classique.

2.7. Théorème d'amalgame. (a) *Pour toute* $\mathscr{L}^0$*-structure* Q^0, *la catégorie* $\mathscr{C}^1(\overline{Q}^0, \mathrm{mod}\ \mathscr{L})$ *a la propriété d'amalgame.*

(b) *Si la structure* Q^0 *est* ω*-homogène (ou si* $\overline{Q}^0$ *a l'amalgame, ce qui revient au même), alors la catégorie* $\mathscr{C}^1(\overline{Q}^0, OL)$ *a l'amalgame.*

(c) *Les catégories* $\mathscr{C}^n(OL)$, $n \in \omega$, *ont l'amalgame.*

La preuve de (a) est omise, elle est semblable à celle de (b) mais plus simple.

Preuve de (b). Nous rappelons d'abord l'amalgame dans OL: soient $<_i$ $(i \leqslant 2)$ ordres totaux tels que $<_0 = <_1 \cap <_2$. La Figure 3 représente ces trois ordres et *une* façon de les "amalgamer" dans un ordre $<_3$.

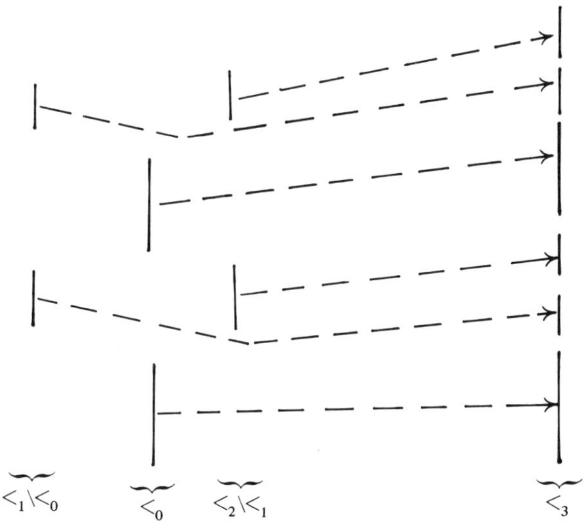

Figure 3

Cette manière consiste à poser $|<_3| = |<_1| \cup |<_2|$ et définir $<_3$ par les conditions: $<_1 \subset <_3$, $<_2 \subset <_3$, enfin si $e_1 \in <_1 \setminus <_2$ et $e_2 \in <_2 \setminus <_1$, $e_2 >_3 e_1$ $\Leftrightarrow_{\mathrm{df}}$ il existe $e_0 \in <_0$ tel que $e_2 >_2 e_0 >_1 = e_1$. ($e_2 <_3 e_1$ sinon.)

Supposons que l'ordre $<_0$ est la relation d'ordre d'un modèle E.M. $F_0[Q^0]$—donc $|<_0| \subset \{t^0[\bar{a}];\ t^0 \in |F_0|,\ \bar{a} \in Q^0\}$; supposons de même que $<_i = F_i[Q^0]$ $(i = 1, 2)$ avec $|F_1| \cap |F_2| = |F_0|$. Et considérons dans ce cas l'ordre $<_3$ ci-dessus; alors si $e_1 \in |<_1| \setminus |<_2|$, $e_2 \in |<_2| \setminus |<_1|$, il existe $t^1 \in |F_1| \setminus |F_2|$ et $t^2 \in |F_2| \setminus |F_1|$ tels que $e_1 = t^1[\bar{a}]$, $e_2 = t^2[\bar{b}]$ $(\bar{a}, \bar{b} \in Q^0)$. Par la définition de $<_3$, $t^2[\bar{b}] >_3 t^1[\bar{a}] \Rightarrow$ il existe $t^0[\bar{c}] \in F_0[Q^0]$ tel que $t^2[\bar{b}] >_2 t^0[\bar{c}] >_1 t^1[\bar{a}]$. Soient alors $\bar{x}, \bar{y} \in Q^0$ tels que $\bar{x} \cap \bar{y} \simeq_{\mathscr{L}^0} \bar{a}^\frown \bar{b}$, où $\mathscr{L}^0$ est la langage de Q^0; puisque Q^0 est ω-homogène, il existe $\bar{z} \in Q^0$ tel que $\bar{x}^\frown \bar{y}^\frown \bar{z} \simeq_{\mathscr{L}^0} \bar{a}^\frown \bar{b}^\frown \bar{c}$. Et par $\mathscr{L}^0$-indiscernabilité de Q^0 dans $F_2[Q^0]$ et $F_1[Q^0]$ nous en concluons $t^2[\bar{y}] >_2 t^0[\bar{z}] >_1 t^1[\bar{x}]$, soit $t^2[\bar{y}] >_3 t^1[\bar{x}]$; d'où $[t^2[\bar{b}] >_3 t^1[\bar{a}]$ et $\bar{a}^\frown \bar{b} \simeq_{\mathscr{L}^0} \bar{x}^\frown \bar{y}] \Rightarrow t^2[\bar{y}] >_3 t^1[\bar{x}]$. Nous venons de vérifier la $\mathscr{L}^0$-indiscernabilité de Q^0 dans $F_3[Q^0]$, dans un cas particulier. Les autres cas ($t^1, t^2 \in F_1$ ou $t^1, t^2 \in F_2$) sont triviaux, par indiscernabilité de Q^0 dans $F_1[Q^0]$ et $F_2[Q^0]$. Donc $F_3[Q^0]$ est un modèle E.M.; et comme visiblement l'identité induit une t.n. de $F_i[\cdot]$ dans $F_3[\cdot]$ ($i = 1, 2$), $F_3[\cdot]$ amalgame $F_1[\cdot]$ et $F_2[\cdot]$. Nous avons montré l'amalgame dans $\mathscr{C}^1(\overline{Q}^0, OL)$.

PREUVE DE (c). C'est une conséquence immédiate de (b) et du théorème de codage (b).

(II) *Le codage des modèles E.M.*

Le lemma qui suit précise certains aspects du théorème du codage (a):

2.8. LEMME DU CODAGE. *Soit Q^0 une $\mathscr{L}^0$-structure; pour un langage approprié $\mathscr{L}^1$, on peut associer à tout foncteur $F \in \mathscr{C}^1(\overline{Q}^0, \mathrm{mod}\,\mathscr{L})$ une $\mathscr{L}^1$-structure F^* de domaine $|F|$, de façon que pour tout autre foncteur G de $\mathscr{C}^1(\overline{Q}^0, \mathrm{mod}\,\mathscr{L})$ et toute application ϕ de $|F|$ dans $|G|$, ϕ est une interprétation de $F[Q^0]$ dans $G[Q^0]$ ssi c'est un $\mathscr{L}^1$-isomorphisme de F^* dans G^*. Autrement dit, si $\mathscr{A}$ désigne la classe $\{F^*: F \in \mathscr{C}^1(Q^0, \mathrm{mod}\,\mathscr{L})\}$, alors l'application qui envoie F sur F^*, et envoie la t.n. $(\phi^X)_{X \in \overline{Q}^0}$ sur ϕ, est un foncteur de $\mathscr{C}^1(\overline{Q}^0, \mathrm{mod}\,\mathscr{L})$ dans* $\mathrm{mod}\,\mathscr{L}^1 \restriction \mathscr{A}$, *et ce foncteur est une équivalence des catégories $\mathscr{C}^1(\overline{Q}^0, \mathrm{mod}\,\mathscr{L})$ et* $\mathrm{mod}\,\mathscr{L}^1 \restriction \mathscr{A}$.

PREUVE DU THÉORÈME DE CODAGE 2.3(a) (À PARTIR DU LEMME). Comme dans le lemme de codage, excepté que mod $\mathscr{L}$ est remplacé par OL, nous posons

$$\mathscr{A} = \left\{F^*: F \in \mathscr{C}^1\left(\overline{Q}^0, OL\right)\right\}.$$

On voit facilement que $\mathscr{C}^1(\overline{Q}^0, OL)$ est close par $\varinjlim$, donc la catégorie $\mathrm{mod}\,\mathscr{L}^1 \restriction \mathscr{A}$ qui lui est équivalente par le lemme de codage est également close par $\varinjlim$. D'autre part, le cas particulier du lemme où ϕ est une application d'inclusion montre que pour $F, G \in \mathscr{C}^1(\overline{Q}^0, OL)$, $F \subset G \Leftrightarrow F^* \subset G^*$; d'où résulte que $\mathscr{A}$ est close par restrictions. Enfin le théorème d'amalgame 2.7(b) entraîne que $\mathscr{C}^1(\overline{Q}^0, OL)$, donc $\mathscr{A}$, possèdent l'amalgame (car par hypothèse la structure Q^0 est ω-homogène). Alors par la Proposition 2.6(b) il existe $Q^* \in \mathscr{A}$, structure ω-homogène telle que la catégorie $\mathrm{mod}\,\mathscr{L}^1 \restriction \mathscr{A}$ est équivalente à $\overline{Q}^*$; moyennant le lemme de codage, $\overline{Q}^*$ est équivalente à $\mathscr{C}^1(\overline{Q}^0, OL)$ par le foncteur: $F \mapsto F^*$, et la preuve est faite.

Reste à prouver le lemme du codage, ce à quoi la fin de ce paragraphe 2.A est consacrée: il nous faut "coder" chaque foncteur $F \in \mathscr{C}^1(\overline{Q}^0, \mathrm{mod}\,\mathscr{L})$ par une structure F^*, de façon que les t.n. à valeur dans F soient également "codées" par les isomorphismes à valeur dans F^*. Or, le Théorème de Représentation 1.9 associe déjà à F une structure $F[Q^0]$, qui est un candidat naturel pour être F^*. Mais si G est un autre foncteur de $\mathscr{C}^1(\overline{Q}^0 \mathrm{mod}\,\mathscr{L})$, et $G[Q^0]$ est la structure associée, il y a généralement des isomorphismes de $F[Q^0]$ dans $G[Q^0]$ qui ne sont *pas* de la forme ϕ^{Q^0}, pour une application $\phi: |F| \to |G|$, et donc qui ne correspondent *pas* à une t.n. de F dans G. C'est la raison pour laquelle F^* ne peut être choisi égal à $F[Q^0]$, et le "codage" de F par F^* ne fait pas double emploi avec sa "représentation" par $F[Q^0]$. Avant de coder F par une *structure* F^* nous allons rappeler la manière classique, en Théorie des modèles et Théorie des ensembles, de coder un modèle E.M. et ses étirements—donc un foncteur E.M.— par une *théorie*:

Codage d'un modèle E.M. par une théorie. Soit pour commencer M un modèle E.M. au sens *classique*: M contient un ensemble ordre-indiscernable X_0, et est engendré par X_0, que nous supposons infini. $\mathscr{L}$ désignant le langage de M, on fixe une suite sans répétition $(v_i)_{i\in\omega}$ de variables de $\mathscr{L}$, et on pose $\Sigma_M = \{\theta(t^1[\bar{v}] \cdots t^k[\bar{v}]);\ \theta(t^1[\bar{v}] \cdots t^k[\bar{v}])$ formule atomique de $\mathscr{L}$ telle que $M \vDash \theta(t^1(\bar{x}) \cdots t^k(\bar{x}))$ pour une (donc pour toute) suite strictement croissante $\bar{x} \in X_0$ de longueur $l(\bar{v})\}$.

Il est bien connu que $\forall X \in OL$, l'étirement $M[X]$ de M est complètement déterminé par X et cette théorie Σ_M. Nous voulons étendre ceci aux modèle E.M. géréralisés, de la forme $F[Q^0]$, où Q^0 est une $\mathscr{L}^0$-structure donnée. La différence est seulement la suivante: dans un ordre total X, à une permutation près il n'y a que les types d'ordres "$v_0 < \cdots < v_{n-1}$" qui soient réalisés. Dans Q^0, ce qui remplace ces types d'ordres, ce sont tous les $\mathscr{L}^0$-types satisfaits dans Q^0.

2.9. Définition. Pour tout $F \in \mathscr{C}^1(\overline{Q}^0, \operatorname{mod} \mathscr{L})$, on pose $\Sigma_F = \{p(\bar{v}) \to \theta(t^1[\bar{v}] \cdots t^k[\bar{v}]);\ p(\bar{v})$ est un $\mathscr{L}_0$-type, $\theta(u^1 \cdots u^k)$ est formule atomique de $\mathscr{L}$, $t^1 \cdots t^k \in |F|$, et $F[Q^0] \vDash \theta(t^1[\bar{a}] \cdots t^k[\bar{a}])$ pour une (donc pour toute) suite $\bar{a} \in Q^0$ telle que $Q^0 \vDash p(\bar{a})\}$.

Exactement comme dans le cas classique de Σ_M, pour tout $X \in \overline{Q}^0$, l'étirement $F[X]$ de $F[Q^0]$ est complètement déterminé par X et par Σ_F; une version précise et une preuve de ceci sont données par le Théorème 3.11(a) et sa preuve.

Codage d'un modèle E.M. par une structure. Nous revenons au problème de coder par une structure F^* tout foncteur $F \in \mathscr{C}^1(\overline{Q}^0, \operatorname{mod} \mathscr{L})$, donc essentiellement tout modèle $F[Q^0]$. La définition de F^* comporte nécessairement une part de conventions arbitraires; nous allons fixer celles-ci de manière que la diagramme simple de F^* soit *quasi-identique à la théorie* Σ_F *ci-dessus*:

2.10. Définitions. (a) Soit Q^0 une $\mathscr{L}^0$-structure; on définit alors un langage $\mathscr{L}^1$, et pour tout $F \in \mathscr{C}^1(\overline{Q}^0, \operatorname{mod} \mathscr{L})$ une $\mathscr{L}^0$-structure F^*, par les conditions suivantes:

Pour chaque $n \in \omega$, $\mathscr{L}^1$ comporte des "variables de sorte n"; le domaine de ces variables dans F^* est $\{t \in |F|$: t est à n arguments$\}$. Donc $|F^*|$ (réunion des domaines de F^*) égale $|F|$.

Pour tout $\mathscr{L}^0$-type $p(\bar{v})$ réalisé dans Q^0, et pour toute formule atomique $\theta(u^1 \cdots u^k) \in \mathscr{L}$, $\mathscr{L}^1$ comporte un symbole de relation θ_p à k arguments et si $t^1 \cdots t^k \in |F|$ on pose

$$F^* \vDash \theta_p(t^1 \cdots t^k) \Leftrightarrow_{\mathrm{df}} \text{la formule } \left[p(\bar{v}) \to \theta\left(t^1[\bar{v}] \cdots t^k[\bar{v}]\right)\right]$$

est dans Σ_F.

($\Leftrightarrow F[Q^0] \vDash \theta(t^1[\bar{a}] \cdots t^k[\bar{a}])$ pour une—donc pour toute—suite $\bar{a} \in Q^0$ telle que $Q^0 \vDash p(\bar{a})$.)

(b) Souvent on identifiera la formule atomique $\theta_p(t^1 \cdots t^k)$ avec la formule "$p(\bar{v}) \to \theta(t^1[\bar{v}] \cdots t^k[\bar{v}])$" écrivant celle-ci entre guillemets à la place de celle-là.

Alors le diagramme de F^* *s'identifie* avec Σ_F; F^* n'est donc qu'un déguisement de Σ_F. Mais ce qui nous intéresse ici est que la *conclusion du lemme de codage des* *vérifiée*. En effet, soient $G \in \mathscr{C}^n(\overline{Q}^0 \bmod \mathscr{L})$, et $\phi\colon |F| \to |G|$ un $\mathscr{L}^1$-isomorphisme de F^* dans G^*; alors si $\theta(u^1 \cdots u^k)$ est formule atomique de $\mathscr{L}$, $t^1[\bar{a}] \cdots t^k[\bar{a}]$ sont éléments de $F[Q^0]$, on a

$$F[Q^0] \vDash \theta(t^1[\bar{a}] \cdots t^k[\bar{a}])$$

$$\Leftrightarrow F^* \vDash \theta_p(t^1 \cdots t^k),\ p \text{ désignant le } \mathscr{L}^0\text{-type de } \bar{a} \text{ dans } Q^0$$

(par définition de F^*);

$$\Leftrightarrow G^* \vDash \theta_p(\phi(t^1) \cdots \phi(t^k)) \quad (\text{par hypothèse sur } \phi);$$

$$\Leftrightarrow G[Q^0] \vDash \theta(\phi(t^1)[\bar{b}] \cdots \phi(t^k)[\bar{b}]),$$

pour une suite $\bar{b} \in Q^0$ de $\mathscr{L}^0$-type p (par définition de G^*);

$$\Leftrightarrow G[Q^0] \vDash \theta(\phi(t^1)[\bar{a}] \cdots \phi(t^k)[\bar{a}])$$

(vu $\bar{a} \simeq_{\mathscr{L}^0} \bar{b}$ et la $\mathscr{L}^0$-indiscernabilité de Q^0 dans $G[Q^0]$).

Les termes extrêmes de ces équivalences montrent que ϕ^{Q^0} est un $\mathscr{L}$-isomorphisme de $F[Q^0]$ dans $G[Q^0]$, donc $(\phi^X)_{X \in \overline{Q}^0}$ une t.n. de F dans G. La réciproque se montre de même, donc l'application: $(\phi^X)_{X \in \overline{Q}^0} \mapsto \phi$ envoie bijectivement les t.n. de F dans G sur les $\mathscr{L}^1$-isomorphismes de F^* dans G^*. Le lemme de codage est ainsi démontré.

Le théorème de codage est ce qui permet d'appliquer aux foncteurs de type > 1 le théorème de représentation (par foncteurs E.M.) des foncteurs de type 1; les notations qui suivent précisent ces rôles respectifs des Théorèmes de Codage 2.3 et de Représentation 1.9:

2.11. *Notations*. (a) Par le Théorème de Codage, $\mathscr{C}^n(OL)$ est équivalent à $\overline{Q}_n^*$, donc $\mathscr{C}^{n+1}(OL) = \mathscr{C}^1(\mathscr{C}^n(OL), OL)$ est équivalente à $\mathscr{C}^1(\overline{Q}_n^*, OL)$; nous notons i cette dernière équivalence. Alors pour tout foncteur $\Phi = \Phi(\cdot)$ de $\mathscr{C}^{n+1}(OL)$, nous posons

$$\Phi^* =_{\mathrm{df}} (i\Phi)^*, \qquad |\Phi| =_{\mathrm{df}} |i\Phi|, \qquad \Phi[\cdot] =_{\mathrm{df}} i\Phi[\cdot]$$

(vu que $i\Phi \in \mathscr{C}^1(\overline{Q}_n^*, OL)$, le codage $(i\Phi)^*$ de $i\Phi$ est défini par 2.10 appliqué avec $Q_n^* = Q^0$, de même, le foncteur E.M. $i\Phi[\cdot]$, et son ensemble $|i\Phi|$ de symboles de fonction, sont définis par 1.11.bis. Notez que Φ^* a pour domaine $|\Phi|$ et $\Phi^* \in \overline{Q}_{n+1}^*$).

(b) $\Phi[\cdot]$ fournit une représentation de Φ par un foncteur E.M.; quand on voudra étudier $\mathscr{C}^{n+1}(OL)$, c'est cette représentation qui sera le plus utile, et on identifiera $\Phi(\cdot)$ et $\Phi[\cdot]$. En revanche, quand c'est $\mathscr{C}^{n+2}(OL)$ que l'on veut étudier (donc Φ est non pas un des foncteurs étudiés, mais un élément de leur domaine), alors c'est l'identification de Φ avec Φ^* qui servira; c'est par elle que $\mathscr{C}^{n+1}(OL)$ est identifié avec $\overline{Q}_{n+1}^*$, grâce à quoi $\mathscr{C}^{n+2}(OL)$ s'identifiera avec $\mathscr{C}^1(\overline{Q}_{n+1}^*, OL)$.

Par le Théorème 1.13(b), si $F \in \mathscr{C}^1(OL)$, alors pour tout $x \in OL$, le modèle $F[X]$ est égalitaire: si $F[X] \vDash t[x_0 \cdots x_k] = s[y_0 \cdots y]$, c'est que $t = s$, $k = l$ et $\bar{x} = \bar{y}$. La fin de ce §2 va étendre ce résultat à $\mathscr{C}^n(OL)$.

2.12. THÉORÈME. *Pour tout $n \geqslant 0$, il existe un foncteur p.r. $\tau \in \mathscr{C}^{n+1}(OL)$, tel que $\forall F \in \mathscr{C}^n(OL)$, l'ordre $\tau(G)$ a pour domain $|F|$.*

2.13. THÉORÈME. *Pour tout* $n \geqslant 0$ *et tout* $\Phi \in \mathscr{C}^{n+1}(OL)$, *le modèle* $\Phi[Q_n^*]$ *est égalitaire (de même que* $\Phi[X^*]$ *pour tout* $X \in \overline{Q}_n^*$); *en d'autres termes: pour* $F \in \mathscr{C}^n(OL)$ *et* $z \in \Phi(F)$, *disons que* $t[\bar{x}]$ *est une notation pour* z *si* $t[\bar{x}] \in \Phi[F^*]$ *et si l'isomorphisme de* $\Phi[\cdot]$ *sur* $\Phi(\cdot)$ (*cf.* 2.11) *envoie* $t[\bar{x}]$ *sur* z. *Alors si* $t[x_0 \cdots x_{k-1}]$ *et* $s[y_0 \cdots y_{-1}]$ *sont des notations pour* z, *c'est que* $t = s$, $k = l$ *et* $\bar{x} = \bar{y}$.

La preuve de ces deux théorèmes se fait par récurrence *simultanée* sur n.

Case $n = 0$. $\mathscr{C}^0(OL) = OL$, donc 2.12 est trivialement vrai dans ce cas, en prenant $\tau =$ identité sur OL. D'autre part, nous avons noté ci-dessus que le Théorème 1.13(b) entraîne 2.13 pour $n = 0$.

Cas quelconque. Comme hypothèse de récurrence nous supposons $\tau \in \mathscr{C}^n(OL)$ déjà obtenu, et 2.13 vrai pour. Nous définissons τ: $C^n(OL)\mathscr{C}^n(OL) \to OL$; bien que ça ne soit pas strictement nécessaire, pour plus de clarté nous faisons une définition particulière quand $n = 1$.

Sous-cas $n = 1$. Soit $F \in \mathscr{C}^1(OL)$; pour $t, s \in |F|$, on pose

$$t <_{\tau(F)} s \Leftrightarrow_{\mathrm{df}} F[\omega] \models t[0 \cdots k-1] < s[0 \cdots l-1];$$

où k est le nombre d'arguments de t, l celui de s. L'application: $t \mapsto t[0 \cdots k-1]$ de $|F|$ dans $F[\omega]$ est un isomorphisme de $\tau(F) =_{\mathrm{df}} (|F|, <_{\tau(F)})$ dans $F[\omega]$, donc $\tau(F) \in OL$. De plus, si $(\phi^X)_{X \in OL}$ est un morphisme de F dans G, donc ϕ une application de $|F|$ dans $|G|$ telle que ϕ^ω est un isomorphisme de $F[\omega]$ dans $G[\omega]$, il est clair que ϕ est un isomorphisme de $\tau(F)$ dans $\tau(G)$. Et on voit facilement qu'en posant $\tau((\phi^X)_{X \in OL}) = \phi$, on définit un foncteur τ de $\mathscr{C}^1(OL)$ dans OL.

Sous-cas $n > 1$. Soit $\Phi \in \mathscr{C}^n(OL)$, nous voulons définir $\tau(\Phi) \in OL$; par 2.11, nous avons défini $\Phi[Q_{n-1}^*] \in OL$. Cet ordre $\Phi[Q_{n-1}^*]$ va jouer, pour définir $\tau(\Phi)$, le rôle de $F[\omega]$ pour définir $\tau(F)$ dans le sous-cas $n = 1$. Rappelons que $\Phi[Q_{n-1}^*]$ a son domaine inclus dans $\{T[\bar{t}]: T \in |\Phi|, \bar{t} \in |Q_{n-1}^*|\}$; et fixons une énumération p.r. $(\bar{t}^i)_{i \in \omega}$ de $|Q_{n-1}|^{<\omega}$. Nous posons, pour $T, S \in |\Phi|$,

$$T <_{\tau(\Phi)} S \Leftrightarrow_{\mathrm{df}} \Phi[Q_{n-1}] \models T[\bar{t}^{i_0}] < S[\bar{s}^{j_0}];$$

où $i_0 = \mu i$: $T[\bar{t}^i] \in |\Phi[Q_{n-1}^*]|$, $j_0 = \mu j$: $S[\bar{t}^j] \in |\Phi[Q_{n-1}^*]|$. L'application $T \mapsto T[\bar{t}^{i_0}]$ est un isomorphisme de $\tau(\Phi) = (|\Phi|, <_{\tau(\Phi)})$ dans $\Phi[Q_{n-1}^*]$, donc $\tau(\Phi) \in OL$; posons de plus $\tau((\phi^X)_{X \in \overline{Q}_{n-1}}) = \phi$, pout tout morphisme (ϕ^X) de $\mathscr{C}^n(OL)$; ainsi τ est défini sur $\mathscr{C}^n(OL)$, et on voit facilement que $\tau \in \mathscr{C}^{n+1}(OL)$.

Pour achever la récurrence, il n'y a plus qu'à montrer 2.13 pour $\phi \in \mathscr{C}^{n+1}(OL)$: la preuve est quasi-identique à celle du Théorème 1.13(b), le seul changement consistant à utiliser, au lieu de l'ordre total $<_X$ postulé par 1.13(b) sur tout $X \in \overline{Q}^0$, l'ordre $\tau(X)$ que nous venons de définir sur tout $X \in \overline{Q}_n$.

3. Foncteurs de type fini: aspects finitaires et effectifs. Jusqu'ici nous avons cherché à mettre en évidence, au moyen de modèles E.M., uniquement les aspects structurels, voire géométriques, des foncteurs de type $\geqslant 1$. Nous examinons dans ce paragraphe comment une partie de ces foncteurs peuvent être considérés comme des objets finis ou effectifs; et comment l'on peut considérer tous les foncteurs de type $\geqslant 1$ comme opérant sur une catégorie engendrée par une structure p.r.

3.A. *Foncteurs de type fini: aspects finitaires.* Soit F un foncteur continu de $\overline{Q}^0$ dans OL; la donnée de F se ramène essentiellement à celle du modèle $F[Q^0]$. Dans les cas qui nous intéressent, Q^0 sera infini; et alors $F[Q^0]$ également sauf si F est constant: hormis ce cas trivial, il semblerait donc que la donnée de F comporte toujours une infinité d'informations. Mais nous allons voir que lorsque $|F|$ est fini, $F[Q^0]$ sera entièrement déterminé par une restriction finie de $F[Q^0]$. Donc les foncteurs F tels que $|F|$ est fini, appelés *foncteurs de dimension finie*, seront essentiellement des objets finis.

3.1. DÉFINITIONS. (a) On dit que $F \in \mathscr{C}^1(\overline{Q}^0, OL)$ est de *dimension finie* si $|F|$ est fini; cette définition n'est pas intrinsèque (puisqu'elle utilise $|F|$, donc la représentation de F par $F[\cdot]$), mais voici une définition "catégorique" qui lui est équivalente: *F est de dimension finie ssi à isomorphisme près, il n'y a qu'un nombre fini de foncteurs qui s'envoient dans F par une t.n.* (En effet, par les Remarques 1.6 et 1.12 il y a autant de tels foncteurs qu'il y a de sous-ensembles de $|F|$; et ces sous-ensembles sont en nombre fini ssi $|F|$ est fini.)

(b) On dira que $\Phi \in \mathscr{C}^{n+1}(OL)$ est de *dimension finie* ssi $|\Phi|$ est fini (où $|\Phi|$ est défini par 2.11). Plus généralement, on appellera *dimension de* Φ le cardinal de $|\Phi|$.

Le caractère essentiellement fini des foncteurs de dimension finie résultera de l'étude des *Modèles E.M. finîment engendrés*. Soit X une sous-structure *propre* de Q^0, et soit $F[X] \in OL$ un modèle E.M. Alors l'étirement $F[Y]$ de $F[X]$ a été défini pour tout $Y \in \overline{X}$; mais $F[Q^0]$ ne l'est pas, si $\overline{X} \subsetneqq \overline{Q}^0$. Si l'on essaye malgré tout de définir $F[Q^0]$ à partir de $F[X]$, on rencontre diverses situations, suivant le choix de X et $F[X]$:

(i) Il arrive qu'il existe un unique modèle E.M. $F[Q^0]$ étendant $F[X]$.

(ii) Mais il arrive qu'il n'en existe aucun, ou au contraire qu'il en existe plus d'un.

Lorsque le nombre d'arguments de tous les symboles $t \in F$ est borné par un entier k, il y a une condition naturelle sur X qui entraîne (i):

3.2. THÉORÈME. *Soit F un ensemble de symboles de fonctions à au plus k arguments, et soit $X \subset Q^0$ tel que tout $\mathscr{L}^0$-type à $3k$ variables qui est réalisé dans Q^0 l'est déjà dans X; alors pour tout modèle E.M. $F[X] \in OL$ on a*

(a) *$\forall Y \in \overline{Q}^0$ l'étirement $F[Y]$ de $F[X]$ est défini et unique, et $F[Y] \in OL$; donc à $F[X]$ est associé un unique foncteur $F[\cdot]$ de $\overline{Q}^0$ (pas seulement $\overline{X}$) dans OL.*

(b) *Pour tout modèle E.M. $G[Q^0] \in OL$ et toute application ϕ: $F \to G$, si ϕ^X est un morphisme d'ordre de $F[X]$ dans $G[X]$, alors ϕ^{Q^0} en est un de $F[Q^0]$ dans $G[Q^0]$, donc $(\phi^Y)_{Y \in \overline{Q}^0}$ est une t.n. de $F[\cdot]$ dans $G[\cdot]$, et ϕ un $\mathscr{L}^1$-isomorphisme de F^* dans G^*.*

PREUVE. Rappelons que l'étirement $F[Y]$ de $F[X]$ est défini par

(i) $|F[Y]| = \{t[\bar{y}]$: *$t \in F$, $\bar{y} \in Y$ il existe $\bar{x} \in X$ tel que $\bar{x} \simeq_{\mathscr{L}^0} \bar{y}$ et $t[\bar{x}] \in$ $|F[X]|\}$; et*

(ii) *si* $\bar{y} \in Y$, $t[\bar{y}]$ *et* $s[\bar{y}] \in |F[Y]|$, *alors* $F[Y] \models t[\bar{y}] \leqslant s[\bar{y}] \Leftrightarrow$ *il existe* $\bar{x} \in X$ *tel que* $\bar{x} \simeq_{\mathscr{L}^0} \bar{y}$ *et* $F[X] \models t[\bar{x}] \leqslant s[\bar{x}]$.

Avec toutes nos hypothèses sur k, F, dans (i) on peut se limiter à des suites $\bar{y}$ de longueur k, et dans (ii) à $\bar{y}$ de longueur $2k$; donc l'étirement $F[Y]$ de $F[X]$ est défini sans ambiguïté pour tout Y tel que tout sous-ensemble à $2k$ éléments de Y se plonge isomorphiquement dans X. Or vu notre hypothèse sur X, cela est le cas de tout $Y \in \overline{Q}^0$.

Reste à voir que $F[Y] \in OL$; soient $\bar{y} \in Y$ de longueur $\leqslant 3k$, et $t^1, t^2, t^3 \in F$ tels que

$$F[Y] \models \left(t^1[\bar{y}] \leqslant t^2[\bar{y}]\right) \text{ et } \left(t^2[\bar{y}] \leqslant t^3[\bar{y}]\right).$$

Par hypothèse sur X, si $Y \in \overline{Q}^0$, alors il existe $\bar{x} \in X$ tel que $\bar{x} \simeq_{\mathscr{L}^0} \bar{y}$; alors

$$F[X] \models \left(t^1[\bar{x}] \leqslant t^2[\bar{x}]\right) \text{ et } \left(t^2[\bar{x}] \leqslant t^3[\bar{x}]\right),$$

et comme $F[X] \in OL$, par transitivité

$$F[X] \models t^1[\bar{x}] \leqslant t^3[\bar{x}], \quad \text{d'où } F[Y] \models t^1[\bar{y}] \leqslant t^3[\bar{y}];$$

ce qui montre la transitivité de $F[Y]$.

L'antisymétrie et la linéarité de $F[Y]$ se vérifient de la même manière, ainsi que le (b) du théorème.

Dans les cas qui nous intéresseront, la structure Q^0 considérée aura la propriété:

(3.3) *Pour toute suite* $v^1 \cdots v^n$ *de variables de* $\mathscr{L}^0$, *l'ensemble des* $\mathscr{L}^0$*-types de la forme* $p(v^1 \cdots v^n)$ *réalisés dans* Q^0 *est fini.*

Dans ce cas-là, les foncteurs de dimension finie de $\mathscr{C}^1(\overline{Q}^0, OL)$ seront essentiellement des objets *finis*; c'est ce que va développer la série de remarques qui suit.

3.4. Remarques. (a) *La condition* (3.3) *entraîne évidemment pour chaque entier* k *l'existence d'une structure finie* $X_k \subset Q^0$ *qui satisfait l'hypothèse du Théorème* 3.2. Alors, soit $\mathrm{F} \in \mathscr{C}^1(Q^0, OL)$ de dimension finie; il existe donc un entier k qui majore le nombre d'arguments de t pour tout symbole $t \in |F|$. Par le théorème de représentation, F est entièrement déterminé (à isomorphisme près) par le modèle $F[Q^0]$ lequel par 3.2 est entièrement déterminé par $F[X_k]$. Mais X_k et $|F|$ étant finis, $F[X_k]$ est une structure finie. Ainsi *tout foncteur de dimension finie de* $\mathscr{C}^1(\overline{Q}^0, OL)$ *est essentiellement un modèle E.M. fini.*

(b) Nous savons (théorème de codage 2.3) l'existence d'un langage $\mathscr{L}^1$ et d'une $\mathscr{L}^1$-structure Q^* telle que $\mathscr{C}^1(\overline{Q}^0, OL) \simeq Q^*$. Comme conséquence de (a), nous verrons en (c) que *la condition* (3) *vérifiée par* Q^0 *l'est également par* Q^* *pour son langage* $\mathscr{L}^1$. Par récurrence sur n, cela entraîne que les structures Q_n^* telles que $\mathscr{C}^n(OL) \simeq \overline{Q}_n^*$ vérifient également—dans le langage $\mathscr{L}_n$—cette condition (3.3). Alors, la remarque (a) s'applique aux foncteurs de $\mathscr{C}^n(OL)$ pour tout n: *tout foncteur de type n de dimension finie est essentiellement un modèle E.M. fini.*

(c) Rappelons que la $\mathscr{L}^1$-structure Q^* telle que $\mathscr{C}^1(\overline{Q}^0, OL) \simeq \overline{Q}^*$ est définie par 2.10, à partir d'un foncteur approprié $Q \in \mathscr{C}^1(\overline{Q}^0, OL)$; et que le langage $\mathscr{L}^1$ comporte pour chaque $k \in \omega$ des variables de sortes k dont le domaine dans Q^* est $\{t \in |Q|$: t symbole de fonction à k arguments$\}$.

Soient $t^1 \cdots t^n$ des symboles de fonctions fixés; quels que soient $G \in \mathscr{C}^1(\overline{Q}^0, OL)$ et $s^1 \cdots s^n \in |G|$ tels que s^i et t^i ont même nombre d'arguments, il existe évidement un foncteur $F[\cdot]$ tel que $|F| = \{t^1,\ldots,t^n\}$ et $F \simeq G \restriction \{s^1,\ldots,s^n\}$—cf. Remarques 1.14: il suffit de renoter t^i ce qui dans $G \restriction \{s^1,\ldots,s^n\}$ est noté s^i, pour obtenir $F[\cdot]$. Et $F \simeq G \restriction \{s^1,\ldots,s^n\}$ revient à dire que le $\mathscr{L}^1$-type de $(t^1 \cdots t^n)$ dans F^* est égal au $\mathscr{L}^1$-type de $(s^1 \cdots s^n)$ dans G^*.

Bref, soient $v^1 \cdots v^n$ des variables de $\mathscr{L}^1$ telles que v^i a pour sorte le nombre d'arguments de t^i; ce qui précède entraîne facilement qu'il y a autant de $\mathscr{L}^1$-types de la forme $p(v^1,\ldots,v^n)$ réalisés dans Q^*, que de foncteurs E.M., $F[\cdot] \in \mathscr{C}^1(\overline{Q}^0, OL)$ tels que $|F| = \{t^1,\ldots,t_n\}$. Mais par la remarque (a), il y a autant de tels foncteurs F qu'il y a de modèle E.M. de la forme $F[X_k] \in OL$; or $|F| = \{t^1,\ldots,t^n\}$ étant fixé et fini, de même que X_k, il n'y a qu'un nombre fini de tels modèles $F[X_k]$. Donc il n'y a qu'un nombre fini de $\mathscr{L}^1$-types de la forme $p(v^1 \cdots v^n)$ réalisés dans Q^*; *supposant* (3.3) *réalisé par* Q^0, *nous venons d'en déduire que* Q^* *vérifie la même condition*; *ce qui justifie le* (b).

Dans le théorème de codage (a), l'isomorphisme de $\mathscr{C}^1(\overline{Q}^0 \bmod \mathscr{L})$ sur Q^* est essentiellement canonique; mais on ne s'est pas préoccupé d'un choix canonique de la structure Q^* parmi toutes celles qui engendrent $\overline{Q}^*$: on s'est contenté d'invoquer un résultat général d'existence de Q^*, la Proposition 2.6. De même dans le théorème de codage (b), l'isomorphisme de $\mathscr{C}^n(OL)$ sur $\overline{Q}^*_n$ est au fond canonique, mais pas le choix de la structure ω-homogène Q^*_n. On va remédier à cela de la manière suivante: rappelons qu'une structure est $\aleph_0$-*catégorique* si sa théorie complète l'est, c'est-à-dire ne possède qu'un modèle dénombrable (à isomorphisme près). Ainsi **Q** est $\aleph_0$-catégorique, et si l'on veut fixer une structure Q_0 telle que $\overline{Q}_0 = OL = \mathscr{C}^0(OL)$, cela rend naturel de choisir Q_0 égal à **Q**. La suite va généraliser ceci à $\mathscr{C}^n(OL)$, $n > 0$: elle permettra de choisir $\aleph_0$-catégorique (et de cardinal $\aleph_0$) la structure Q^*_n telle que $\mathscr{C}^n(OL) \simeq \overline{Q}^*_n$. *Alors l'unicité de* Q^*_n *à isomoprhisme près rend son choix canonique.*

Il nous faut quelques rappels sur l' $\aleph_0$-catégoricité.

3.5. Proposition. *Soit* Q *une structure* ω-*homogène dans un langage* L; *alors* Q *est* $\aleph_0$-*catègorique si et seulement si* Q *vérifie la condition* (3.3) *déjà considérée*: *pour toute suite finie* $\bar{u}$ *de variables de* L, *il n'y a qu'un nombre fini de* L-*types de la forme* $p(\bar{u})$ *réalisés dans* Q.

Preuve de $\Leftarrow$. Nous supposons d'abord que L contient seulement un nombre fini de symboles de relations (et pas de symboles de fonction). Grâce à quoi chaque L-type est un ensemble fini de formules, dont la conjonction est dans L. Pour chaque $n \in \omega$, soit $\{p_i(\bar{x});\ i \in I_n\}$ l'ensemble fini des L-types $p(x_0 \cdots x_{n-1})$ réalisés dans Q. Q satisfait donc $\forall \bar{x} \bigvee_{i\in I_n} p_n(\bar{x})$. D'autre part, Q satisfait la formule $\bigwedge_{i\in I_1} \exists x_0\, p_i(x_0)$; et l'$\omega$-homogénéité de Q s'exprime par les formules suivantes:

$\forall \bar{x}\big[(p_i(\bar{x})) \to \exists x_n\, q_j(\bar{x}, x_n)\big]$ où p_i, q_j varient sur les types tels que $i \in I_n, j \in I_{n+1}$ et $p_i(\bar{x}) \subset q_j(\bar{x}, x_n)$, pour un $n \in \omega$.

Soit T la théorie qui contient toutes les formules précédentes et qui est donc vraie dans Q. T est $\aleph_0$-catègorique: si Q_1, Q_2 sont deux modèles de T de cardinal $\aleph_0$, ils sont ω-homogènes vu les formules de T; alors un "va-et-vient" tout semblable à celui de Cantor entre deux ordres denses construit un isomorphisme entre Q_1 et Q_2 (notez que l' $\aleph_0$-catègoricité des ordres denses est un cas particulier *typique* de ce qui précède).

Extension à un langage L quelconque. Dans ce cas les L-types $p_i(\bar{x})$, $i \in I_n$, peuvent être infinis; mais comme ils sont en nombre fini, chacun d'eux est "engendré" par un sous-ensemble fini $p_i'(\bar{x})$. On refait alors essentiellement la preuve ci-dessus en utilisant $p_i'(\bar{x})$ au lieu de $p_i(\bar{x})$.

PREUVE DE $\Rightarrow$. Si la théorie complète de Q est $\aleph_0$-catègorique, le critère classique de Ryll-Nardzewski entraîne qu'elle n'a qu'un nombre fini de types *complets* avec n variables fixées; a fortiori, de L-types, c'est-à-dire de types *simples* avec ces variables.

2.3.bis. THÉORÈME DE CODAGE AMÉLIORÉ. (a) *Pour toute structure ω-homogène, $\aleph_0$-catègorique Q^0, il existe une structure ω-homogène $\aleph_0$-catègorique Q^* telle que*

$$\mathscr{C}^1(\overline{Q}^0, OL) \subset \overline{Q}^*.$$

(b) *Pour tout n, la structure Q_n^* telle que $\mathscr{C}^n(OL) \simeq \overline{Q}_n^*$ est $\aleph_0$-catègorique et l'isomorphisme de $\mathscr{C}^n(OL)$ vers $\overline{Q}_n^*$ envoie les foncteurs de dimension finie de $\mathscr{C}^n(OL)$ sur les structures de domaine fini de $\overline{Q}_n^*$.*

PREUVE. (a) Cette résulte est immédiatement du théorème de codage 2.3, au moyen de la Remarque 2.15(b) et de la Proposition 3.5.

Le premier point de (b) résulte de (a) par récurrence sur n. Et le second point est trivial: l'isomorphisme de $\mathscr{C}^n(OL)$ vers $\overline{Q}_n^*$ est l'isomorphisme: $\Phi \mapsto \Phi^*$ (cf. Definitions 2.10 et Notations 2.11); par définition, Φ est de dimension finie $\Leftrightarrow$ $|\Phi|$ est fini; or $|\Phi| = |\Phi^*|$.

Voici une deuxième notion liée aux aspects finitaires:

3.6 DÉFINITION.[6] On dira qu'un foncteur $F \in \mathscr{C}^1(\overline{Q}^0, OL)$ est *faiblement fini* s'il envoie les structures finies dans les ordres finis: $\forall A \in \overline{Q}^0\, A$ fini $\Rightarrow F(A)$ fini. Et si $\Phi \in \mathscr{C}^{n+1}(OL)$, on dira que Φ *est faiblement fini* si le foncteur $\Phi[\cdot]$: $\overline{Q}_n^* \to OL$ l'est; ce qui équivaut à dire que $\phi(F)$ est fini pour tout foncteur $F \in \mathscr{C}^n(OL)$ *de dimension finie*.

Il est immédiat que *F est faiblement fini $\Leftrightarrow$ pour tout $k \in \omega$, l'ensemble* $\{t \in |F|$: t est à k arguments$\}$ *est fini*. D'où résulte que tout foncteur de dimension finie est faiblement fini, mais la réciproque est fausse. Ajoutons qu'en général les foncteurs faiblement finis ne sont pas des objets essentiellement finis (comme le sont par

[6] Cette définition correspond à la première version de [**G2**, Chapitre 12], en fait, il semble plus naturel de la modifier ainsi: $\Phi \in \mathscr{C}^{n+1}(OL)$ est faiblement fini ssi $\Phi(F)$ est fini pour tout $F \in \mathscr{C}^n(OL)$ faiblement fini.

Cette redéfinition semble plus maniable que la précédente et correspond exactement à celle introduite dans la version définitive de [**G2**, Chapitre 12]. Il semble que tous les résultats obtenus avec la Définition 3.6 de faiblement fini sont aisément adaptés à la nouvelle définition, mais nous n'avons pas vérifié ceci en détail.

3.4(a) les foncteurs de dimension finie); ils ne font qu'envoyer les objets finis sur des objets finis.

3.B. *Aspects effectifs*. Les démonstrations de ce paragraphe sont simples dans leur principe, mais parfois fastidieuses, aussi nous omettons certains détails en comptant sur le lecteur pour y suppléer.

Lorsque nous dirons qu'un ensemble quelconque A est p.r., il est sous-entendu que A est contenu dans un ensemble tel que ω ou V_ω (= collection des ensembles héréditairement finis), qui possède une notion de sous-ensemble p.r. Et de même avec, à la place de p.r., les autres notions de Récursivité et Théorie descriptive: Δ^0_1, Σ^0_1, Δ^1_1, etc.

Nous dirons qu'une L-structure M est p.r. si L et $|M|$ sont p.r., de même que la fonction

$$(\bar{a}, \theta) \mapsto \begin{cases} 1 & \text{si } \bar{a} \in M \text{ et } \theta \text{ est une formule} \\ & \text{atomique de } L \text{ satisfaite par } \bar{a}, \\ 0 & \text{sinon} \end{cases}$$

3.7. THÉORÈME. *Soit Q^*_n la structure dénombrable, ω-homogène et $\aleph_0$-catègorique telle que $\mathscr{C}^n(OL) \sim \overline{Q}^*_n$; à isomorphisme près, Q^*_n est une structure p.r.*

PREUVE. La Remarque 3.4(c) montrait essentiellement ceci: suppons que quelles que soient variables $v_0 \cdots v_k$ de $\mathscr{L}_n$, il n'y a qu'un nombre fini de $\mathscr{L}_n$-types avec ces variables réalisés dans Q^*_n; alors il en est de même pour Q^*_{n+1} et $\mathscr{L}_{n+1}$. En examinant la preuve de cette remarque sous l'angle effectif, on en tire facilement la version suivante, où le nombre de types est non seulement fini, mais encore borné de façon p.r.: $(v_k)_{k\in\omega}$ énumérant les variables de $\mathscr{L}_n$, supposons que

(∗) *il existe $(p_i)_{i\in\omega}$, énumération des $\mathscr{L}_n$-types réalisés dans Q^*_n qui est p.r.. et s'accompagne d'une fonction p.r. $f\colon \omega \to \omega$, telle que pour tout k, $(p_i)_{i\leqslant f(k)}$ comprend tous les types ayant $v_0 \cdots v_k$ comme variables*;

alors $Q^*_{n+1}, \mathscr{L}_{n+1}$ possèdent la même propriété.

(*Nota Bene*. On rappelle que vu l' $\aleph_0$-catègoricité, les types considérés ici se ramènent à des ensembles *finis* de formules.)

Comme $\mathscr{C}^0(OL) = OL = \mathbf{Q}$ et $Q^*_0 = \mathbf{Q}$, $\mathscr{L}_0 = \{\leqslant\}$, il est clair que (∗) est vérifiée pour $n = 0$, et il en résulte par récurrence sur n:

(3.8) *pour tout n, Q^*_n vérifie* (∗).

Pour qu'une structure dénombrable Q'_n soit isomorphe à Q^*_n, il suffit qu'elle soit ω-homogène et réalise les mêmes $\mathscr{L}_n$-types. Or moyennant (∗), il se trouve que la construction standard d'une telle structure Q'_n va être p.r., d'où résultera le Théorème 3.7:

$(p_i)_{i\in\omega}$ désigne l'énumération du (∗), $\langle i, j\rangle$ désigne l'entier qui code le couple d'entiers (i, j) (donc l'application: $(i, j) \mapsto \langle i, j\rangle$ est une bijection p.r. de ω^2 sur ω). Par récurrence sur $N \in \omega$, on définit une chaîne $(q_N(v_0 \cdots v_N))_{N\in\omega}$ de $\mathscr{L}_n$-types en posant $q_{N+1} = p_j$, où j est le premier entier vérifiant

(i) $q_N(v_0 \cdots v_N) \subset p_j = (v_0 \cdots v_{N+1})$;

(ii) *si* $N = \langle i, k\rangle$ *et* $q_k(v_0 \cdots v_k) \subset p_i = p_i(v_0 \cdots v_k x)$, *alors* $p_i(v_0 \cdots v_k v_{N+1}) \subset p_j(v_0 \cdots v_{N+1})$; (l'existence de j résulte de l'ω-homogénéite de Q_n^*).

La condition (i) entraîne l'existence d'une structure Q_n' ayant pour diagramme $\bigcup_N q_N(c_0 \cdots c_N)$. Et (ii) entraîne pour tout $k \in \omega$ que toute extension de $q_k(v_0 \cdots v_k)$ réalisée dans Q_n^* l'est aussi dans Q_n', de manière que Q_n' est ω-homogène et réalise les mêmes $\mathscr{L}_n$-types que Q_n^*. Donc $Q_n^* \simeq Q_n'$, de plus vu que $(p_i)_{i\in\omega}$ est p.r. $(q_N)_{N\in\omega}$ l'est aussi, donc Q_n' est p.r. et le théorème est montré.

3.9. Remarques. (a) En utilisant la borne p.r. f fournie par $(*)$, il est facile d'obtenir une fonction p.r. g avec la propriété suivant: si $p = p(\bar{u}v)$ est un $\mathscr{L}_n$-type, $\bar{a} \in Q_n'$ et $p(\bar{a}v)$ est réalisé dans Q_n', alors $g(\bar{a}, p)$ est défini et est un élément b tel que $Q_n' \models p(\bar{a}, b)$.

(b) Désormais, identifiant Q_n' et Q_n^*, on supposera que Q_n^* est p.r., et comporte la fonction g du (a).

3.10. Notation et Remarque. Pour tout foncteur Φ de $\overline{Q}^0$ dans OL, on notera $\Phi \restriction sQ^0$ la restriction de Φ aux sous-structures finies de Q^0 et à leurs isomorphismes. Notez que *si le foncteur* Φ *est continu, alors à isomorphisme près il est complètement déterminé par* $\Phi \restriction sQ^0$—tout comme une fonction continue est entièrement déterminée par sa restriction à un ensemble *dense* de valeurs; ici cet ensemble "dense", ce sont les sous-structures finies de Q^0, dont tout élément de $\overline{Q}^0$ est la $\varinjlim$: si $X \in \overline{Q}^0$, alors $X = \varinjlim(X_i, f_{ij})_{i<j\in I}$, où les X_i sont parties finies de Q^0. Alors connaissant seulement $\Phi \restriction sQ^0$, on connaît la famille inductive $(\Phi(X_i), \Phi(f_{ij}))$ et par continuité de Φ, on a

$$\Phi(X) = \varinjlim\big(\Phi(X_i), \Phi(f_{ij})\big),$$

ce qui détermine $\Phi(X)$ à isomorphisme près. Le même raisonnement s'étend à $\Phi(f)$, où f est un morphisme de $\overline{Q}^0$.

Ce fait suggère, quand on veut définir la "complexité" d'un foncteur continu Φ: $\overline{Q}^0 \to OL$, de poser qu'elle est égale à celle de $\Phi \restriction sQ^0$—par exemple de dire que Φ est p.r. si la fonction $\Phi \restriction sQ^0$ peut-être codée de façon p.r. (moyennant codage dans V_ω de son domaine et de son image). Une deuxième idée pour mesurer la complexité de Φ est la suivante: un foncteur E.M. $\Phi[\cdot]$ sur $\overline{Q}^0$ est entièrement déterminé, au moyen de l'étirement, par le modèle $\Phi[Q^0]$; et celui-ci l'est par la théorie Σ_Φ, donc par le codage Φ^* qui est un déguisement de Σ_Φ (cf. §2). Alors on va mesurer la complexité de $\Phi[\cdot]$ par celle de Φ^*. Notez que l'étirement n'est que la "version E.M." des $\varinjlim$ utilisées dans la Notation et Remarque 3.10 pour reconstituer $\Phi(\cdot)$ à partir de $\Phi \restriction sQ^0$; donc cette deuxième idée est la version E.M. de la première. Ci-dessous on va utiliser les deux idées.

En vue du Théorème 3.11 ci-dessous, nous remarquons que l'application de codage: $\Phi \in \mathscr{C}^{n+1}(OL) \mapsto \overline{Q}_{n+1}^*$ est surjective: cela résulte aisément de nos conventions 1.11.bis; en particulier, *il existe un foncteur E.M.* Q_{n+1} *dont le codage* est *la structure* Q_{n+1}^*.

3.11. Théorème. (a) *Pour tout* $n \geqslant 0$, *si* $\Phi \in \mathscr{C}^{n+1}(OL)$ *et* $F \in \mathscr{C}^n(OL)$, *alors* $\Phi(F)$ *est un ordre p.r. relativement à* Φ^* *et* F^*, *de façon uniforme.*

(b) *Pour tout* $n \geqslant 0$, *si* $\Phi \subset Q_{n+1}$ *et* $F \subset Q_j$, *alors* $\Phi(F)$ *est un ordre p.r. relativement à* $|\Phi|$ *et* $|F|$, *de façon positive et uniforme.*

PREUVE DE (a). $|\Phi[F^*]| = \{t[\bar{x}]: t \in |\Phi|, \bar{x} \in |F|$ tel que $F \vDash p_t(\bar{x})\}$. En revenant à la Définition 2.10 de Φ^*, il est clair que p_t s'obtient de façon p.r. en t et Φ^*: p_t est le type $p(\bar{v})$ tel que la formule "$p(\bar{v}) \to t[\bar{v}] = t[\bar{v}]$" est dans Σ_Φ; autrement dit, si $\theta(u)$ désigne la formule $u = u$, c'est le type p tel que $\Phi^* \vDash \theta_p(t)$. Donc $|\Phi[F]|$ est p.r. relativement à Φ^* et F^*. De façon semblable, $\Phi[F^*] \vDash t[\bar{x}] < s[\bar{x}] \Leftrightarrow \Phi^* \vDash \theta_p(t, s)$, où $\theta(u_1, u_2)$ désigne la formule $u_1 < u_2$ et $p(\bar{v})$ le $\mathscr{L}_n$-type tel que $F^* \vDash p(\bar{x})$. Donc $\Phi[F^*]$ est p.r. en Φ^* et F^*; d'où (a) puisque $\Phi[F^*] \simeq \Phi(F)$.

PREUVE DE (b). Le (a) entraîne en particulier que $Q_{n+1}[Q_n^*]$ est un ordre p.r. Et si $\Phi \subset Q_{n+1}$, $F \subset Q_n$, alors

$$\Phi(f) \simeq \Phi[F^*] = Q_{n+1}[Q_n^*] \upharpoonright \{t[\bar{x}]: t \in |\Phi|, \bar{x} \in |F|\}.$$

Et ceci est clairement p.r. en $|\Phi|$ et $|F|$, de façon positive!

Le (b) n'est valable que pour des foncteurs Φ, F qui sont des restrictions de Q_{n+1} et Q_n; mais à isomorphismes près ce théorème est beaucoup plus général:

3.12. THÉORÈME. (a) *Pour tout foncteur* $F \in \mathscr{C}^n(OL)$, F *est de dimension dénombrable si et seulement si il existe* $X \subset |Q_n^*|$ *tel que* $F \simeq Q_n \upharpoonright X$.

(b) *De plus l'ensemble* X *tel que* $F \simeq Q_n \upharpoonright X$ *peut être choisi de façon p.r. en* F^*; *et donc*: F *est isomorphe à un foncteur* G *tel que* G^* *est p.r. si et seulement si il existe* X *p.r.* $\subset |Q_n^*|$, *tel que* $F \simeq Q_n \upharpoonright X$.

PREUVE. On sait que $F \in \mathscr{C}^n(OL) \Leftrightarrow F^* \in \overline{Q}_n^*$ et que $F \simeq Q_n \upharpoonright X \Leftrightarrow F^* \simeq Q_n^* \upharpoonright X$; alors pour montrer (a), il suffit de montrer que toute structure dénombrable $F^* \in \overline{Q}_n^*$ se plonge isomorphiquement dans Q_n^*. Cela résulte de ce que Q_n^* est ω-homogène; en voici la preuve, très classique: on choisit une énumération $(a_N)_{N \in \omega}$ de $|F^*|$, et par récurrence sur N, on définit une suite $(b_N)_{N \in \omega}$ de points de Q_n^* tels que l'application: $a_i \mapsto b_i$, $i < N$, est un $\mathscr{L}_n$-isomorphisme de F^* dans Q_n^*. Pour obtenir b_N, désignons par $\bar{a}$ et $\bar{b}$ les suites $a_0 \cdots a_{N-1}$ et $b_0 \cdots b_{N-1}$; il existe une suite $\bar{c}c_N \in Q_n^*$ telle que $\bar{a}a_N \simeq_{\mathscr{L}_n} \bar{c}c_N$, d'autre part, par hypothèse de récurrence $\bar{a} \simeq_{\mathscr{L}_n} \bar{b}$. Alors $\bar{b} \simeq_{\mathscr{L}_n} \bar{c}$ ce qui par ω-homogénéité de Q_n^* nous donne $b_N \in Q_n^*$ tel que $\bar{b}b_N \simeq_{\mathscr{L}_n} \bar{c}c_N$, ce qui entraîne $\bar{b}b_N \simeq_{\mathscr{L}_n} \bar{a}a_N$, montrant l'hypothèse de récurrence pour $N + 1$. Ceci prouve (a).

Supposons que la structure F^* était p.r.; alors on pouvait choisir p.r. l'énumération $(a_N)_{N \in \omega}$ de $|F^*|$; de plus, si dans la construction ci-dessus de (b_N) on utilisait la function g de la Remarque 3.9, pour choisir b_N, alors la suite (b_N) elle aussi était p.r.; et en utilisant (3.8), il est facile de faire que l'ensemble $X = \{b_N; N \in \omega\}$ soit p.r. lui aussi. Alors $F^* \simeq Q_n^* \upharpoonright X$, donc $F \simeq Q_n \upharpoonright X$, où X est p.r., ceci montre le (b).

Le Théorème 3.11(b) étendu par 3.12 constitue la propriété fondamentale, du point de vue effectif, des foncteurs de type fini: le caractère effectif *et* positif de $\Phi(F)$, en fonction de Φ et de F. Nous revenons maintenant à l'idée de résumer un foncteur continu Φ sur $\overline{Q}^0$ non par Φ^*, mais par $\Phi \restriction sQ^0$. Soit Q^0 une structure p.r.; lorsqu'un foncteur Φ: $\overline{Q}^0 \to OL$ est faiblement fini, alors l'image de $\Phi \restriction sQ^0$ est constituée d'ordres finis; donc, à isomorphisme canonique près, cette image est contenue dans V_ω, et ainsi (vu $|Q^0| \subset V_\omega$) $\Phi \restriction sQ^0$ tout entier est contenu dans V_ω. Moyennant quoi, par exemple, la notion "$\Phi \restriction sQ^0$ est p.r." a un sens.

3.13. Théorème. *Soit Φ un foncteur faiblement fini*: $\overline{Q}_n^* \to OL$;

(a) *pour tout $X \in \overline{Q}_n^*$, l'ordre $\Phi(X)$ est p.r. relativement à $\Phi \restriction sQ_n^*$ et à X.*

(b) *En modifiant légèrement la définition du domaine $|\Phi|$ de Φ^*, on a que Φ^* est une structure p.r. relativement à $\Phi \restriction sQ_n^*$.*

Preuve de (a). Ceci résulte immédiatement de (b) et 3.11(a).

Preuve de (b). Par définition, $t \in |\Phi| \Leftrightarrow t = (z, a)/\sim$, où $\bar{a} \in Q_n^*$ est support minimal de $z \in \Phi(Q_n^* \restriction \bar{a})$. Ceci fait de t une classe d'équivalence généralement infinie; pour l'éviter, on remarque que (3.8) et la Remarque 3.9(b) permettent de construire *de façon p.r.* une chaîne $(X_k)_{k\in\omega}$ de sous-structures *finies* de Q_n^*, telles que tout type à $3k$ variables réalisé dans Q_n^* l'est déjà dans X_k. On pose alors $t \in |\Phi| \Leftrightarrow \exists k \in \omega\, \exists \bar{a} \in X_k$ de longueur k, $\bar{a}$ support minimal de $z \in \Phi(Q_n^* \restriction \bar{a})$, tels que $t = (z, \bar{a})/\sim$; où cette fois $\bar{a}$ parcourt X_k et non Q_n^*, donc t est une classe d'équivalence finie.

Comme dit plus haut, $\Phi \restriction sQ_n^*$ est une partie de V_ω; on vérifie aisément que la nouvelle version ci-dessus de $|\Phi|$ est Δ_0-définissable dans V_ω à partir de $\Phi \restriction sQ_n^*$, de même que la $\mathscr{L}_{n+1}$-structure Φ^*. Or les fonctions Δ_0 dans V_ω sont p.r., d'où le (b).

Notez que *si Φ n'est pas faiblement fini*, alors $\Phi(Q_n^* \restriction \bar{a})$ peut être infini, et la question, pour $z \in \Phi(Q_n^* \restriction \bar{a})$, de savoir si $\bar{a}$ est support minimum de z, n'est peut-être pas effectivement décidable; il est douteux que 3.13 reste vrai dans ce cas.

3.14. Définitions. (a) On dira que *$F \in \mathscr{C}^n(OL)$ est un foncteur p.r.* si $F \restriction sQ_{n-1}^*$ est p.r. et si de plus F est faiblement fini.

(b) Pour chaque $k \in \omega$, on note $|Q_n^*|_k$ le domaine de sorte k de Q_n^*; et on fixe une fonction p.r. $\phi(\cdot,\cdot)$ telle que $\phi(k, \cdot)$ énumère l'ensemble des parties finies de $|Q_n^*|_k$. On suppose fixé un codage récursif des fonctions p.r. de ω dans ω par des entiers: f_e désigne la fonction p.r. de code $e \in \omega$. Alors pour tout $e \in \omega$, F_e désigne le foncteur $Q_n \restriction \bigcup_{k\in\omega} \phi(k, f_e(k))$.

Notez que F_e est un foncteur faiblement fini: en effet, le domaine de sorte k de F_e est l'ensemble $|F_e|_k = \phi(k, f_e(k))$, que est fini par choix de ϕ. D'autre part, $F_e \restriction sQ_{n-1}^*$ est p.r., car $|F_e|$ est p.r. par construction et l'on applique alors le Théorème 3.11. Donc F_e est un foncteur p.r., ce qui montre une moitié du

3.15. Théorème. *Pour tout $F \in \mathscr{C}^n(OL)$, F est un foncteur p.r. si et seulement si il existe $e \in \omega$ tel que $F \simeq F_e$.*

Preuve. La moitié qui reste se montre comme 3.12(b).

Notez qu'ainsi $(F_e)_{e\in\omega}$ est une famille de foncteurs p.r., qui est codable de façon Δ_1^0 et qui à isomorphismes près comprend tous les foncteurs p.r. de type n.

3.16. Remarque. Le foncteur τ du Théorème 2.12 est faiblement fini: en effet $\tau(F)$ a pour domaine $|F|$ par construction, donc $|F|$ fini $\Rightarrow \tau(F)$ fini. De plus, comme les structures Q_n^* et Q_{n-1}^* sont p.r., on voit immédiatement que $\tau[Q_n^*]$ est une structure p.r. (car définie au moyen d'un plongement p.r. de Q_n^* dans $Q_n[Q_{n-1}^*]$). D'où résulte que τ^* est p.r. Par conséquent, *τ est un foncteur p.r.*

4. Ptykes. Notons ON la catégorie dont les objets sont les ordinaux, et dont les morphismes sont les applications strictement croissantes. Et nous définissons par récurrence sur n des catégories notées PT^n: $PT^0 = ON$; PT^{n+1} est la catégorie ayant pour objets les foncteurs de PT^n dans ON qui préservent les $\varinjlim$ et produits fibrés, et pour morphismes les t.n. entre ces foncteurs. Notez que ces catégories PT^n sont celles qu'au §2, nous aurions notées $\mathscr{C}^n(ON)$ ou $\mathscr{C}^n(ON, ON)$; mais nous les notons PT^n pour suivre la terminologie de l'ancienne version de [**G2**, Chapitre 12]. Vu que $PT^n = \mathscr{C}^n(ON, ON)$ le dessin du §2 figurant $\mathscr{C}^n(\mathbf{C}, \mathbf{D})$ s'applique.

Les foncteurs PT^n seront appelés les *ptykes de type n* (au singulier: ptyx). Les ptykes de type n sont la généralisation des ordinaux qui est appropriée à la "Π_{n+1}^1-Logique", et leur étude est le but principal de ce travail. Les foncteurs de type n et leurs catégories $\mathscr{C}^n(OL)$ étudiées précédemment n'ont donc *ici* qu'un but auxiliaire: grâce à un plongement naturel de PT^n dans $\mathscr{C}^n(OL)$, PT^n va hériter des résultats des §§1–3. Mais le concept général de foncteur de type fini—au sens de foncteur d'une catégorie de la forme $\mathscr{C}^n$ (mod $\mathscr{L}$, mod $\mathscr{L}^1$)—est naturel, et d'autres cas particuliers utiles de ces foncteurs devraient tôt ou tard donner d'autres occasions d'employer les §§1–3.

Rappelons que pour tout catégorie $\mathscr{C}$, la fermeture de $\mathscr{C}$ par limites directes est la catégorie $\overline{\mathscr{C}}$, unique à équivalence près, qui vérifie

(i) $\mathscr{C} \subset \overline{\mathscr{C}}$, et tout élément (objet ou morphisme) de $\overline{\mathscr{C}}$ est limite directe d'une famille inductive dans $\mathscr{C}$;

(ii) $\overline{\mathscr{C}}$ est close par limites directes. Par exemple, $\overline{ON} = OL$, puisque OL est clos par limites directes, et tout ordre total est limite directe d'ordres finis, donc d'ordinaux. A l'encontre de $\mathscr{C}^n(OL)$, la catégorie PT^n n'est pas close par limites directes—déjà pour $n = 0$, vu $ON \subsetneqq OL = \overline{ON}$. Aussi il sera utile d'introduire les catégories $\overline{PT^n}$: c'est déjà fait pour $n = 0$ parce que $\overline{PT}^0 = \overline{ON} = OL = \mathscr{C}^0(OL)$. Mais dès que $n > 0$, $\overline{PT^n}$ sera une sous-catégorie *propre* de $\mathscr{C}^n(OL)$: prenant le cas $n = 1$ en exemple, considérons le foncteur Opp: $OL \to OL$ qui est l'identité sur les morphismes et envoie tout ordre sur son opposé; alors Opp $\restriction ON$ n'est pas limite directe de foncteurs de PT^1, donc Opp $\restriction ON \notin \overline{PT^1}$ bien que Opp $\in \mathscr{C}^1(OL)$. Donc $\overline{PT^n} \neq \mathscr{C}^n(OL)$; mais on trouvera une sous-structure p.r. $\mathbf{Q}_n^*$ de Q_n^* telle que $\overline{PT^n}$ est équivalent à $\mathbf{Q}_n^*$ par l'application de codage. Donc on retrouvera, avec $\overline{PT^n}$ au lieu de $\mathscr{C}^n(OL)$, la situation des §§1–3.

4.A. *Les catégories PT^1 et $\overline{PT}^1$.* Nous voulons plonger PT^1 dans $\mathscr{C}^1(OL)$; et utiliser ce plongement pour définir $\overline{PT}^1$ comme sous-catégorie de $\mathscr{C}^1(OL)$, et pour étudier PT^1 et $\overline{PT}^1$. Ce plongement découle du prolongement par continuité des foncteurs:

4.1. PROPOSITION. *Pour tout foncteur continu F d'une catégorie* $\mathbf{C}$ *vers une catégorie* $\mathbf{D}$, *il existe un foncteur continu $\bar{F}$ qui étend F et va de* $\bar{\mathbf{C}}$ *vers* $\bar{\mathbf{D}}$; *ce foncteur $\bar{F}$ est unique à isomorphismes près. Si F préserve les produits fibrés, alors $\bar{F}$ en fait autant. Si T est une t.n. de F dans un autre foncteur continu G:* $\mathbf{C} \to \mathbf{D}$, *$T$ s'étend de manière unique en une t.n. $\bar{T}$ de $\bar{F}$ dans $\bar{G}$. Et l'application: $F \mapsto \bar{F}$, $T \mapsto \bar{T}$ est un isomorphisme de* $\mathscr{C}^1(\mathbf{C}, \mathbf{D})$ *dans* $\mathscr{C}^1(\bar{\mathbf{C}}, \bar{\mathbf{D}})$.

PREUVE. Standard et laissée en lecteur; notez qu'on a déjà appliqué ce résultat au sous-section 3.B: la Notation et Remarque 3.10 est le cas particulier de 4.1 où $\mathbf{C} = sQ^0$, $\mathbf{D} = \bar{\mathbf{D}} = OL$, donc F: $sQ^0 \to OL$ est prolongé en $\bar{F}$: $\bar{Q}^0 \to OL$.

4.2. *Notations.* Dorénavant, pour tout foncteur $F \in PT^1$, nous identifions F avec son prolongement par continuité $\bar{F}$; en vertu de 4.1, $\bar{F} \in \mathscr{C}^1(\overline{ON}) = \mathscr{C}^1(OL)$, et PT^1 devient par ce plongement une sous-catégorie de $\mathscr{C}^1(OL)$. De plus, ayant $PT^1 \subset \mathscr{C}^1(OL) = \overline{\mathscr{C}^1(OL)}$, nous pouvons définit $\overline{PT^1}$ comme étant également une sous-catégorie de $\mathscr{C}^1(OL)$: nous posons

$\overline{PT^1}$ = *restriction de* $\mathscr{C}^1(OL)$ *aux foncteurs qui sont limite directe d'une famille inductive de foncteurs de* PT^1.

4.3. REMARQUES.

(a) Soit WO la restriction de la catégorie OL aux bons ordres; la catégorie WO est équivalente à ON, et donc $PT^1 = \mathscr{C}^1(ON)$ est équivalente à $\mathscr{C}^1(WO)$. Nous supposerons $\mathscr{C}^1(WO)$ plongée dans $\mathscr{C}^1(OL)$ de la même manière que PT^1 l'est pour 4.2; alors on voit facilement qu'un foncteur $F \in \mathscr{C}^1(OL)$ est isomorphe à un foncteur de PT^1 si et seulement si $F \in \mathscr{C}^1(WO)$.

(b) Pour $F \in \mathscr{C}^1(OL)$, il est facile de voir (en utilisant le §1) que F est limite directe de foncteurs de PT^1 si et seulement si toute restriction *finie* $F \restriction \{t^1, \dots, t^k\}$ de F est isomorphe à un foncteur de PT^1; joint à (a), ceci nous donne

$\overline{PT^1}$ = restriction de $\mathscr{C}^1(OL)$ à $\{F\colon F \restriction \{t^1, \dots, t^k\} \in \mathscr{C}^1(WO)$ pour tout ensemble $\{t^1, \dots, t^k\} \subset |F|\}$.

4.4. PROPOSITION. (a) *Soit G un foncteur de dimension finie de $\mathscr{C}^1(OL)$; alors $G \in \mathscr{C}^1(WO)$ si et seulement si toutes ses restrictions de dimension* un, *soit $G \restriction \{t\}$ où $t \in |G|$, sont dans $\mathscr{C}^1(WO)$.*

(b) *Donc $F \in \overline{PT^1} \Leftrightarrow F \in \mathscr{C}^1(OL)$ et $F \restriction \{t\} \in \mathscr{C}^1(WO)\ \forall t \in |F|$.*

PREUVE DE (b). C'est une conséquence immédiate de (a) et de 4.3(b).

PREUVE DE (a). Supposons $G \notin \mathscr{C}^1(WO)$: il existe $R \in WO$ et une suite strictement décroissante dans $G[R]$, soit $(t^n[\bar{x}^n])_{n\in\omega}$, *où* $t^n \in |G|$ et $\bar{x}^n \in R$. Comme $|G|$ est fini, il existe $t \in |G|$ tel que $t = t^n$ pour une infinité de n; d'où une suite extraite de la forme $(t[\bar{x}^{n_p}])_{p\in\omega}$ dans $G[R]$. Cette suite est également dans l'image de R par $G \restriction \{t\}$, donc $G \restriction \{t\} \notin \mathscr{C}^1(WO)$.

Ce plongement de PT^1, $\overline{PT^1}$ dans $\mathscr{C}^1(OL)$ nous permet de transférer à PT^1, $\overline{PT^1}$ tout le travail fait sur $\mathscr{C}^1(OL)$ aux §§1, 2 et sous-section 3.A; ainsi par exemple:

4.5. COROLLAIRES. (a) *Les catégories* $\overline{PT}^1$ *et* PT^1 *ont l'amalgame.*

(b) Q_1^* *étant la structure* ω*-homogène telle que* $\mathscr{C}^1(OL)$ *est isomorphe à* $\overline{Q}_1^*$ *par l'application codage, soit* $\mathbf{Q}_1^*$ *la restriction de* $\overline{Q}_1^*$ *à* $\{t \in |Q_1|: Q_1 \upharpoonright \{t\} \in \mathscr{C}^1(WO)\}$; *alors* $\overline{\mathbf{Q}}_1^*$ *est isomorphe à* $\overline{PT}^1$ *par l'application codage; et* $\mathbf{Q}_1^*$ *est une structure* ω*-homogène et* $\aleph_0$*-catégorique.*

PREUVE DE (a). Supposons $F_i \in \overline{PT}^1$ $(i < 3)$ et $F_0 \subset F_i$ $(i = 1, 2)$; puisque $\mathscr{C}^1(OL)$ a l'amalgame, il existe $G \in \mathscr{C}^1(OL)$ qui amalgame F_1 et F_2. En restreignant au besoin G, on peut supposer $|G| = |F_1| \cup |F_2|$. Alors $F_i \in \overline{PT}^1 \Leftrightarrow F_i \upharpoonright \{t\} \in \mathscr{C}^1(WO)$ $\forall t \in |F_i|$ (par 4.4(b)) $\Rightarrow G \upharpoonright \{t\} \in \mathscr{C}^1(WO)$ $\forall t \in |F_1| \cup |F_2|$ $(= |G|)$ $\Leftrightarrow$ (par 4.4(b)) $G \in \overline{PT}^1$. Ainsi l'amalgame est montré dans $\overline{PT}^1$; pour le montrer dans PT^1, supposons que de plus $F_1, F_2 \in PT^1$. Alors $G \in PT^1$; en effet, sinon soient $R \in WO$ et $(t^n[\bar{x}^n])_{n\in\omega}$ une suite strictement décroissante dans $G[R]$; donc $\bar{x}^n \in R$, et $t^n \in |G|$ pour tout n. Comme $|G| = |F_1| \cup |F_2|$, on peut trouver $i \in \{1, 2\}$ et une suite extraite $(t^{p_n}[\bar{x}^{n_p}])_{p\in\omega}$ telle que $t^{n_p} \in |F_i|$ $\forall p \in \omega$. Mais alors cette suite est dans $F_i[R]$ donc $F_i[R] \notin WO$ contredisant l'hypothèse $F_i \in PT^1$. Donc $G \in PT^1$, et le (a) est prouvé.

PREUVE DE (b). Si $F \in \overline{PT}^1$, alors $F \in \mathscr{C}^1(OL)$ donc $F^* \in \overline{Q}_1^*$, c'est-à-dire que toute restriction finie $F^* \upharpoonright \{t^1, \ldots, t^k\}$ de F^* s'envoie par un $\mathscr{L}_1$-isomorphisme f dans Q_1^*; mais $F \in \overline{PT}^1 \Rightarrow F \upharpoonright \{t^1, \ldots, t^k\} \in \mathscr{C}^1(WO)$, donc l'image de $F^* \upharpoonright \{t^1, \ldots, t^k\}$ par f est incluse dans $\mathbf{Q}_1^*$. Ceci montre que $F^* \in \mathbf{Q}_1^*$. Inversement, si $F^* \in \overline{\mathbf{Q}}_1^*$, alors $F \upharpoonright \{t\} \in \mathscr{C}^1(WO)$ $\forall t \in |F|$ par définition de $\mathbf{Q}_1^*$, donc $F \in \overline{PT}^1$ par 4.4(b). Enfin $\mathbf{Q}_1^*$ est une partie de Q_1^* qui est close par isomorphismes finis: si $\bar{a}, \bar{b} \in Q_1^*$ et $\bar{a} \simeq_{\mathscr{L}_1} \bar{b}$, alors $\bar{a} \in \mathbf{Q}_1^* \Rightarrow \bar{b} \in \mathbf{Q}_1^*$; alors l' ω-homogénéité de Q_1^* entraîne celle de $\mathbf{Q}_1^*$. Et $\mathbf{Q}_1^*$ est $\aleph_0$-catégorique pour les mêmes raisons que Q_1^*.

Moyennant ces corollaires, les seuls résultats des §§1–3 qui ne sont pas immédiatement transférés de $\mathscr{C}^1(OL)$ à $\overline{PT}^1$ sont les résultats "effectifs" de la sous-section 3.B. Mais il est clair que ceux-ci se transfèrent également si nous montrons que $\mathbf{Q}_1^*$ est une sous-structure p.r. de Q_1^*; comme Q_1^* est p.r., il suffit de montrer que le domaine de $\mathbf{Q}_1^*$ est p.r. Ce qui revient à trouver, pour les foncteurs de dimension *un* de $\mathscr{C}^1(OL)$, une caractérisation de ceux qui appartiennent à $\mathscr{C}^1(WO)$. Le résultat qui suit donne une telle caractérisation.

4.6. PROPOSITION.[7] *Soit* F *un foncteur de dimension* 1 *de* $\mathscr{C}^1(OL)$: $|F| = \{t\}$; *alors* $F \in \mathscr{C}^1(WO) \Leftrightarrow$

(∗) *si* $\bar{x}, \bar{y} \in X \in OL$ *et* $\bar{x}, \bar{y}$ *sont dans le domaine de* t, *alors*
$$X \models \bigwedge_{i<l(\bar{x})} x_i \leqslant y_i \Rightarrow F[X] \models t[\bar{x}] \leqslant t[\bar{y}]$$

autrement dit, t *est "croissant par rapport à chacun de ses arguments".*

Nota Bene. Soit k le nombre d'arguments de t; alors dans (∗), nécessairement $l(\bar{x}) = l(\bar{y}) = k$. Et il est clair que (∗) est vrai pour tout ordre X si et seulement si

[7] Déjà démontrée dans [**G1**].

($*$) est vrai dans *un* ordre X_0 a $2k$ éléments—vu que X_0 réalise tous les types d'ordres possibles à $2k$ variables. D'où le caractère p.r. de ($*$), qui entraîne

4.7. COROLLAIRE. *La structure* $\mathbf{Q}_1^*$ *est p.r.*

PREUVE DE 4.6. Supposons $X \in WO$, mais $F[X] \notin WO$; donc il y a une suite strictement décroissante dans $F[X]$, de la forme $(t[\bar{x}^n])_{n\in\omega}$ puisque $|F| = \{t\}$. Posons $(\bar{x}^i, \bar{x}^{i'}) \sim (\bar{x}^j, \bar{x}^{j'})$ si $i < i', j < j'$ et si $\bar{x}^i \frown \bar{x}^{i'}$ a même type d'ordre que $\bar{x}^j \frown \bar{x}^{j'}$; cette relation d'équivalence n'a qu'un nombre fini de classes, vu qu'il y a seulement un nombre fini de types d'ordres possibles. Le théorème de Ramsey entraîne alors l'existence d'une suite extraite infinie $(\bar{x}^{n_p})_{p\in\omega}$ qui est *homogène* pour $\sim$: c'est-à-dire que *toutes* les suites $\bar{x}^{n_p} \frown \bar{x}^{n_q}$ avec $p < q$ ont le même type d'ordre. Renotons $(\bar{y}^p)_{p<\omega}$ cette suite extraite; s'il existe $i < l(\bar{y}^0)$ tel que $y_i^0 > y_i^1$, alors par homogénéité $(y_i^p)_{p\in\omega}$ est une suite strictement décroissante, ce qui contredit $X \in WO$. Donc $X \models \bigwedge\!\!\!\bigwedge_{i<l(\bar{y}^0)} y_i^0 \leqslant y_i^1$; cependant par hypothèse $F[X] \models t[\bar{y}^0] > t[\bar{y}^1]$, montrant que la condition ($*$) n'est pas satisfaite. Ceci prouve une moitié de 4.6 sous forme contraposée, et la réciproque se démontre pareillement.

Les ptykes de type 1 ont déjà été étudiés en détail dans [**G1**] sous le nom de *dilatateurs*—en conséquence, la catégorie notée ici PT^1 est notée DIL dans [**G1**], les éléments de $\overline{PT}^1$ y sont appelés *pré*-dilatateurs et leur catégorie notée PIL. Le lecteur connaissant [**G1**] doit donc penser DIL et PIL chaque fois que, dans ce qui précède et la suite, il voit écrit PT^1 et $\overline{PT}^1$. Pour faire le lien entre certains résultats de base de [**G1**] et le présent exposé, nous rappelons le *Théorème des dénotations* de [**G1**], après quoi nous indiquerons la forme prise ici par ce théorème—à savoir l'application à PT^1 du Théorème de Représentation 1.9 et de 2.13.

Le théorème des dé notations associe à tout ptyx F: $ON \to ON$ et à tout ordinal α un système de notations des ordinaux $< F(\alpha)$; lorsque α est p.r. ainsi que F, ce système fournit des notations "constructives" de segments initiaux des ordinaux. On sait le rôle joué en Théorie des démonstrations par de nombreux systèmes de notations ordinales, dont les premiers sont basés sur la forme normale de Cantor. L'utilité de théorème des dénotations (joint aux travaux [**GV1**, **GV2**, **V**]) et des ptykes ou dilatateurs est de montrer que ces systèmes de la Théorie des démonstrations résultent toujours, à des détails près, de l'application du théorème des dénotations à certains ptykes particuliers. Ce qui représente une unification et un progrès conceptuel considérables dans les questions de notations ordinales.

Rappelons que si F est un ptyx: $ON \to ON$, $\alpha \in ON$ et $z \in F(\alpha)$, alors le théorème des dénotations de [**G1**] associe à z une unique suite de la forme $(z_0; x_0, \ldots, x_{n-1}; \alpha)$, où $z_0 \in F(\omega)$ et $x_0 < \cdots < x_{n-1} < \alpha$:

(i) $\{x_0, \ldots, x_{n-1}\}$ est le support minimum de z dans $F(\alpha)$,

(ii) z_0 est l'élément de support $\{0, \ldots, n-1\}$ dans $F(\omega)$, tel que $F(f)(z_0) = z$, où f est l'application: $i < n \mapsto x_i$. Pour voir le lien avec la représentation de $F(\cdot)$

par un foncteur E.M. $F[\cdot]$, rappelons que le Théorème 1.9 (appliqué au prolongement par continuité de F) a défini $F[\cdot] \in \mathscr{C}^1(OL)$ et pour chaque $X \in OL$ un isomorphisme I_X: $t[\bar{x}] \mapsto t^X(\bar{x})$ de $F[X]$ sur $F(X)$, de telle sorte que la famille $(I_X)_{X \in OL}$ est un isomorphisme des foncteurs $F[\cdot]$ et $F(\cdot)$. Dans le cas particulier où X est α, à chaque élément z de $F(\alpha)$ est ainsi associé le terme formel $t[\bar{x}] \in F(\alpha)$—unique par 2.13—tel que $z = t^\alpha(x)$. Or la définition de $t[\bar{x}]$ en fonction de z fait qu'à des questions de présentation graphique près, il y a *identité* entre la notation $t^\alpha(\bar{x})$ et la notation $(z_0; \bar{x}; \alpha)$: c'est la même suite $\bar{x}$ qui est choisie dans les deux cas (par la condition (i)), et le terme t correspond d'une manière bijective avec l'élément z_0 défini par (ii)—$t[v_0 \cdots v_{n-1}]$ et $(z_0; v_0, \ldots, v_{n-1})$ sont le même objet formel.

Bref, le foncteur E.M. $F[\cdot]$, c'est le foncteur qui pour tout α remplace $F(\alpha)$ par l'ensemble des notations des éléments $z \in F(\alpha)$. Et le Théorème de Représentation 1.9 par foncteur E.M. joint à 2.13, est une forme généralisée du théorème des dénotations: la sous-section 4.B va plonger les catégories PT^n dans $\mathscr{C}^n(OL)$ pour tout n, comme on vient de le faire pour $n = 1$; donc 1.9 et 2.13 s'appliquent aux ptyxes de type n quelconque, et leur fourniront une forme du théorème des dénotations.

4.B. *Les catégories* PT^{n+1}, $\overline{PT}^{n+1}$. En guise d'hypothèse de récurrence sur n, nous supposons *déjà* défini un plongement $PT^n \subset \overline{PT}^n \subset \mathscr{C}^n(OL)$; ce qui est le cas si $n = 1$, mais ici n est quelconque $\geqslant 1$. Nous voulons alors plonger PT^{n+1} dans $\mathscr{C}^{n+1}(OL)$: si $\Phi \in PT^{n+1}$, donc Φ: $PT^n \to ON$, nous voulons considérer Φ comme un foncteur de $\mathscr{C}^n(OL)$ dans OL. PT^n étant déjà inclus dans $\mathscr{C}^n(OL)$, il reste à prolonger Φ à $\mathscr{C}^n(OL) \setminus PT^n$, par un foncteur que nous noterons $\hat{\Phi}$. Ce prolongement se fait en deux étapes: tout d'abord on réalise le prolongement par continuité $\overline{\Phi}$ de Φ (cf. 4.1); $\overline{\Phi}$ va de $\overline{PT}^n$ dans OL, mais $\overline{PT}^n \subsetneqq \mathscr{C}^n(OL)$ (cf. le foncteur Opp, quand $n = 1$). Donc il reste à étendre $\overline{\Phi}$ à $\mathscr{C}^n(OL) \setminus \overline{PT}^n$ pour obtenir $\hat{\Phi}$. Pour cela, nous faisons une hypothèse de récurrence supplémentaire:

$(*)$ pour tout $F \in \mathscr{C}^n(OL)$, on a $F \in \overline{PT}^n \Leftrightarrow F \restriction \{t\} \in \overline{PT}^n \, \forall t \in |F|$.

Cette hypothèse permet de séparer F en une partie F^+ appartenant à $\overline{PT}^n$, et une partie F^- complètement étrangère à $\overline{PT}^n$:

4.8. *Notations*. Pour $F, G \in \mathscr{C}^n(OL)$ et T, t.n. de F dans G, on pose $F^+ = F \restriction \{t$: $F \restriction \{t\} \in \overline{PT}^n\}$, $T^+ = T \restriction F^+$. Pour $\Phi \in PT^{n+1}$, on définit $\hat{\Phi}$ par

$$\hat{\Phi}(F) = \overline{\Phi}(F^+) \quad \text{et} \quad \hat{\Phi}(T) = \hat{\Phi}(T) = \overline{\Phi}(T^+).$$

Il résulte de $(*)$ que F^+ est le plus grand foncteur $F_0 \subset F$ tel que $F_0 \in \overline{PT}^n$. Et $\hat{\Phi}$ est le prolongement de $\overline{\Phi}$ qui, bien que défini sur $\mathscr{C}^n(OL)$, ignore la partie $\mathscr{C}^n(OL) \setminus \overline{PT}^n$ aussi totalement que Φ. Et l'on voit trivialement

4.9. PROPOSITION. *Pour tout* $\Phi \in PT^{n+1}$, $\hat{\Phi} \in \mathscr{C}^{n+1}(OL)$; *et si* $\mathscr{C}$ *est une t.n. de* Φ *dans* Ψ, *alors le prolongement par continuité de* $\mathscr{C}$, *qui par* 4.1 *est une t.n.* $\overline{\mathscr{C}}$ *de* $\overline{\Phi}$ *dans* $\overline{\Psi}$, *reste une t.n. de* $\hat{\Phi}$ *dans* $\hat{\Psi}$. *Donc l'application*: $\Phi \mapsto \hat{\Phi}$, $\mathscr{C} \mapsto \overline{\mathscr{C}}$ *est un isomorphisme de* PT^{n+1} *sur une sous-catégorie de* $\mathscr{C}^{n+1}(OL)$.

4.10. DÉFINITIONS. (a) On identifiera Φ avec $\hat{\Phi}$ et $\mathscr{C}$ avec $\overline{\mathscr{C}}$, pour tous Φ, $\mathscr{C} \in PT^{n+1}$; moyennant quoi $PT^{n+1} \subset \mathscr{C}^{n+1}(OL)$.

(b) On pose $\overline{PT}^{n+1}$ = restriction de $\mathscr{C}^{n+1}(OL)$ à $\{\Phi$: Φ est limite directe d'une famille inductive dans $PT^{n+1}\}$.

4.11. THÉORÈME. *Pour tout $n \geqslant 0$ on a les résultats suivants:*

(a) *Si $\Phi \in \mathscr{C}^{n+1}(OL)$, alors $\Phi \in \overline{PT}^{n+1} \Leftrightarrow$ toute restriction de dimension* un *de Φ appartient à $\overline{PT}^{n+1} \Leftrightarrow \forall T \in |\Phi|$, $\Phi \restriction \{T\}$ envoie PT^n dans WO, et est nul en dehors de $\overline{PT}^n$.*

(b) *Les catégories $\overline{PT}^{n+1}$ et PT^{n+1} ont l'amalgame.*

(c) *Q_{1+1}^* étant la structure ω-homogène telle que $\mathscr{C}^{n+1}(OL) \simeq \overline{Q}_{1+1}^*$, soit $\mathbf{Q}_{1+1}^*$ la restriction de Q_{1+1}^* à $\{T \in |Q_{n+1}|$: $Q_{n+1} \restriction \{T\} \in \overline{PT}^{n+1}\}$; alors l'application codage est un isomorphisme de $\overline{PT}^{n+1}$ sur $\overline{\mathbf{Q}}_{1+1}^*$, et $\mathbf{Q}_{n+1}^*$ est une structure ω-homogène et $\aleph_0$-catégorique.*

PREUVE. Par récurrence sur n; le cas $n = 0$ a été traité en sous-section 4.A, donc on supposes le résultat vrai pour $n \geqslant 1$. La preuve de l'étape $n \Rightarrow n + 1$ est toute semblable aux preuves du cas $n = 1$ données en sous-section 4.A, et laissée au lecteur.

Pour généraliser le sous-section 4.A, à tout n, il reste à étendre à PT^{n+1} la Proposition 4.6 qui caractérisait les ptykes de dimension un de PT^1:

4.12. THÉORÈME. *Pour tout $n \geqslant 0$ et tout $T \in |Q_{n+1}|$, on peut déterminer de façon p.r. si $Q_{n+1} \restriction \{T\}$ est un ptyx—autrement dit, si $T \in |\mathbf{Q}_{n+1}|$; donc $\mathbf{Q}_{n+1}$ est une structure p.r.*

PREUVE. On dira qu'un $\mathscr{L}_n$-type $q(\bar{u}, \bar{v})$ est *itérable* s'il existe $C \subset |Q_n|$ et une suite $(\bar{t}^i)_{i \in \omega}$ dans C telle que $Q_n^* \models \bigwedge\!\!\!\bigwedge_{i<j<\omega} q(\bar{t}^i, \bar{t}^j)$; et que q est *fortement* itérable si de plus $Q_n \restriction C$ est un ptyx.

LEMME 1. *Si $T \in |Q_{n+1}|$, alors $T \in |\mathbf{Q}_{n+1}| \Leftrightarrow$ pour tout $\mathscr{L}_n$-type fortement itérable $q(\bar{u}, \bar{v})$ contenant $p_T(\bar{u}) \cup p_T(\bar{v})$, on a $Q_{n+1}^* \models q(\bar{u}, \bar{v}) \rightarrow T[\bar{u}] \leqslant T[\bar{v}]$.*

PREUVE DE $\Leftarrow$. Sinon, $Q_{n+1}^* \models q(\bar{u}, \bar{v}) \rightarrow T[\bar{u}] > T[\bar{v}]$, et il y a un ptyx $Q_n \restriction C$ dont l'image par le foncteur $Q_{n+1} \restriction \{T\}$ contient la suite infinie décroissante $(T[\bar{t}^i])_{i<\omega}$.

PREUVE DE $\Rightarrow$ Supposons que $T \notin |Q_{n+1}|$, donc il existe un ptyx F de type n dont l'image par $Q_{n+1} \restriction \{T\}$ n'est pas un bon ordre: elle contient une suite infinie décroissante $(T(\bar{s}^i))_{i \in \omega}$, où $\bar{s}^i \in |F|$. Par le même argument de Ramsey que dans la preuve de la Proposition 4.6, on en extrait une suite $(\bar{t}^i)_{i<\omega}$ telle que $i < j \Rightarrow \bar{t}^i \simeq_{\mathscr{L}_n} \bar{t}^j$. Quitte à restreindre F, on peut supposer que $|F|$ est dénombrable; alors par 3.12(a), on peut se ramener au cas où $F \subset Q_n^*$. Alors si $q(\bar{u}, \bar{v})$ est le $\mathscr{L}_n$-type de $\bar{t}^0$, $\bar{t}^1$ dans F^*, q est fortement itérable. Ce qui achève la preuve du Lemma 1.

Vu ce lemme, pour démontrer le théorème, il suffit de montrer que l'ensemble des $\mathscr{L}_n$-type fortement itérables est p.r.

LEMME 2. (a) *Pour toute $\mathscr{L}_n$-structure M, $M \in \overline{Q}_n^*$—c'est-à-dire toute partie finie de M se plonge dans Q_n^**—déjà *si toute partie à* trois *éléments de M se plonge dans Q_n^*.*

(b) *Pour tout $\mathscr{L}_n$-type $q(\bar{u}, \bar{v})$, q est itérable—une infinité de fois*—déjà *si q est itérable* trois *fois—c'est-à-dire $Q_n^* \models \exists \bar{u}^0 \exists \bar{u}^1 \exists \bar{u}^2 \bigwedge\!\!\!\bigwedge_{i<j<3} q(\bar{u}^i, \bar{u}^j)$.*

PREUVE DE (a). Nous définissons une structure $M[Q_{n-1}^*]$ en posant:

$$|M[Q_{n-1}^*]| = \{t[\bar{x}] : \bar{x} \in |Q_{n-1}|, t \in |M| \text{ tels que } M \models \text{“}p(\bar{u}) \to t[\bar{u}] = t[\bar{u}]\text{”}$$
$$(\text{cf. } 2.10(\text{b})), \text{ et } \quad Q_{n-1}^* \models p(x)\} \text{ et}$$
$$M[Q_{n-1}^*] \models t[\bar{x}] < s[\bar{x}] \Leftrightarrow M \models \text{“}p(\bar{u}) \to t[\bar{u}] < s[\bar{u}]\text{”},$$

où p est le $\mathscr{L}_{n-1}$-type de $\bar{x}$ dans Q_{n-1}^*. Il suffit de montrer que $M[Q_{n-1}^*] \in OL$ et est un modèle E.M.: par étirement il en résulte un foncteur E.M. $M[\cdot]$ de $\overline{Q}_{n-1}^*$ dans OL, dont le codage M^* appartient à $\overline{Q}_n^*$, par choix de Q_n^*; donc $M \in \overline{Q}_n^*$, car $M = M^*$, par définition de $M[Q_{n-1}]$.

Pour montrer la transitivité de $<_{M[Q_{n-1}^*]}$ nous supposons $M[Q_{n-1}^*] \models t[\bar{x}] < s[\bar{x}]$ et $s[\bar{x}] < r[\bar{x}]$; donc $M \models$ “$p(\bar{u}) \to t[\bar{u}] < s[\bar{u}]$” et “$p(\bar{u}) \to s[\bar{u}] < r[\bar{u}]$”, où p est le $\mathscr{L}_{n-1}$-type de $\bar{x}$. Par hypothèse sur M, il existe un $\mathscr{L}_{n-1}$-isomorphisme f: $M \restriction \{t, s, r\} \to Q_n^*$; donc $Q_n^* \models$ “$p(\bar{u}) \to f(t)[\bar{u}] < f(s)[\bar{u}]$” et “$p(\bar{u}) \to f(s)[\bar{u}] < f(r)[\bar{u}]$”.

Par définition du codage Q_n^* de Q_n, cela entraîne

$$Q_n[Q_{n-1}^*] \models f(t)[\bar{x}] < f(s)[\bar{x}] \text{ et } f(s)[\bar{x}] < f(r)[\bar{x}];$$

d'où $f(t)[\bar{x}] < f(r)[\bar{x}]$, par transitivité de $<_{Q_n[Q_{n-1}^*]}$. D'où $Q_n^* \models$ “$p(u) \to f(t)[\bar{u}] < f(r)[\bar{u}]$”, et puisque f est un isomorphisme $M \models$ “$p(\bar{u}) \to t[\bar{u}] < t[\bar{u}]$”, d'où $M[q_{n-1}^*] \models t[\bar{x}] < r[\bar{x}]$. La transitivité de $<_{M[Q_{n-1}^*]}$ est ainsi montrée; on voit de façon semblable que cet ordre est total, donc $M[Q_{n-1}^*] \in OL$. Enfin, la définition de $M[q_{n-1}^*]$ entraîne facilement que c'est un modèle E.M.—et le (a) est vu.

PREUVE DE (b). Si $q(\bar{u}, \bar{v})$ est itérable trois fois, la $\mathscr{L}_n$-structure ayant pour diagramme $\bigwedge\!\!\!\bigwedge_{i<j<\omega} q(\bar{c}^i, \bar{c}^j)$ a toutes ses parties à trois éléments plongeables dans Q_n^*, donc appartient à $\overline{Q}_n^*$ par (a); donc est plongeable dans Q_n^* moyennant 3.12(a). Ainsi q est itérable.

LEMME 3. (a) *Pour $n > 0$, un $\mathscr{L}_n$-type itérable $q(\bar{u}, \bar{v})$ est fortement itérable si et seulement si pour tout $\mathscr{L}_{n-1}$-type fortement itèrable $r(\bar{x}, \bar{y})$, on a*

$$Q_{n-1}^* \models \text{“}r(\bar{x}, \bar{y}) \to t_k[\bar{x}] \leqslant s_k[\bar{y}]\text{”},$$

chaque fois que $k < l(\bar{u})$, et t_k, s_k sont kièmes coordonnées de suites $\bar{t}$, $\bar{s}$ telles que $Q_n^* \models q(\bar{t}, \bar{s})$.

(b) *Si $n = 0$, donc $Q_n = Q_0 = \mathbf{Q}$, alors un type d'ordre $q(\bar{u}, \bar{v})$ est fortement itérable si et seulement si*

$$l(\bar{u}) = l(\bar{v}) \text{ et } q(\bar{u}, \bar{v}) \vdash \bigwedge\!\!\!\bigwedge_{i<l(\bar{u})} u_i \leqslant v_i.$$

La preuve de (a) est toute semblable à celle du Lemme 1 (le seul changement est que les suites $\bar{t}$ et $\bar{s}$ remplacent T). La preuve du (b) est un argument de Ramsey vu dans la preuve de 4.6.

Si l'on dispose déjà d'une procédure p.r. pour déterminer les $\mathscr{L}_{n-1}$-types fortement itérables, les Lemmes 2 et 3(a) en fournissent une autre pour déterminer les $\mathscr{L}_n$-types fortement itérables—compte tenu du fait que Q_{n-1}^*, Q_n^* sont p.r. et qu'il n'y a qu'un nombre fini de $\mathscr{L}_{n-1}$-types de la forme $r(\bar{x}, \bar{y})$. Alors du Lemme 3(b) résulte par récurrence sur n que les $\mathscr{L}_n$-types fortement itérables forment une famille p.r. Et le théorème 4.12 est alors démontré, vu le Lemme 1.

4.13. REMARQUES. (a) On trouvera dans [**G2**] une extension aux ptykes de type n quelconque de la Proposition 4.6; cette extension fournit une caractérisation des ptykes de dimension 1 plus précise que 4.12 ci-dessus. Et dans le cas des ptykes de type 1 (dilatateurs), on trouvera dans [**G1**] une caractérisation encore plus précise —qui actuellement manque dans le cas des ptykes de type > 1.

(b) Puisque les structures $\mathbf{Q}_n^*$ sont p.r. au même titre que Q_n^*, les résultats effectifs du sous-section 3.B restent évidemment valables quand on remplace les catégories $\mathscr{C}^n(OL)$ par $\overline{PT^n}$, et Q_n, Q_n^* par $\mathbf{Q}_n$, $\mathbf{Q}_n^*$. En particulier,

> *pour tout n il existe une famille $(F_e)_{e\in\omega}$, codable de façon récursive, de foncteurs $F_e \subset \mathbf{Q}_n$, qui à isomorphisme près énumère tous les foncteurs p.r. de $\overline{PT^n}$.*

Il est désormais clair que toute la théorie des foncteurs de type fini développée aux paragraphes 1–3 reste intégralement valide quand on se restreint aux catégories $\overline{PT^n}$.

Nous abordons maintenant les résultats spécifiques aux foncteurs de $\overline{PT^n}$ et PT^n. Tout d'abord notons que le foncteur $\mathbf{Q}_n$ qui engendre $\overline{PT^n}$ n'est pas un ptyx (par exemple $\mathbf{Q}_0 = \mathbf{Q}$ est tout le contraire d'un bon ordre...); mais $\mathbf{Q}_n$ peut être remplacé par un ptyx (qui toutefois n'a pas le caractère canonique de $\mathbf{Q}_n$—excepté dans le cas $n = 0$, où l'on peut remplacer $\mathbf{Q}$ par ω, pour engendrer OL...):

4.14. LEMME. *Pour tout n il existe un ptyx Ω_n tel que $\overline{\Omega}_n^* = \overline{\mathbf{Q}}_n^*$, de plus Ω_n^* est une structure p.r.*

PREUVE. Soit $(C_i)_{i\in\omega}$ une énumération p.r. des parties finies de $|\mathbf{Q}_n|$; la somme $\sum_{i\in I} F_i$ d'une famille ordonnée de foncteurs $(F_i)_{i\in I}$ étant définie *ci-dessous*, nous posons: $\Omega_n = \sum_{i\in\omega} \mathbf{Q}_n \upharpoonright C_i$.

Il est trivial que la somme d'une famille de ptykes *bien* ordonnée et codable de façon p.r. est elle-même ptyx codable de façon p.r. D'où les propriétés voulues de Ω_n.

4.15. *Notations*. Soit $(R_i)_{i\in I}$ une famille telle que $I \in OL$ et $R_i \in OL\ \forall i \in I$; alors $\sum_I R_i$ désigne la somme ordonnée des ordres R_i; si h_i est un morphisme de R_i dans S_i ($i \in I$), alors $\sum_I h_i$ désigne le morphisme de $\sum_I R_i$ dans $\sum_I S_i$ défini par $(\sum_I h_i)(x) = h_{i_0}(x)$, où $i_0 \in I$ est l'unique indice tel que $x \in R_{i_0}$. Alors pour toute famille $(\Phi_i)_{i\in I}$ de foncteurs de $\overline{Q}$ dans OL, $\Sigma(\Phi_i)_{i\in I}$ désigne le foncteur de $\overline{Q}$ dans OL défini par

$$\sum(\Phi_i)(X) = \sum_I \Phi_i(X), \qquad \sum(\Phi_i)(f) = \sum_I \Phi_i(f)$$

pour tout objet X et tout morphisme f de $\overline{Q}$.

4.16. THÉORÈME. *Pour tout $n \geqslant 0$ il existe un* ptyx *p.r.* τ: $\overline{PT^n} \to OL$, *tel que* $|\tau(F)| = |F|$ *pour tout* F.

PREUVE. Par Théorème 2.13 et Remarque 3.16, nous avons déjà un foncteur p.r. τ: $\mathscr{C}^n(OL) \to OL$ tel que $|\tau(F)| = |F|$ pour tout F. Par ailleurs, la construction de $\tau(F)$, cf. 2.13, fournit un isomorphisme de $\tau(F)$ dans $F[Q^*_{n-1}]$—à savoir celui qui sert à définir $<_{\tau(F)}$; et il est immédiat qu'à la place de Q^*_{n-1}, on peut utiliser n'importe quelle structure p.r. Q^* telle que $\overline{Q}^* = \overline{Q}^*_{n-1}$: dans ce cas l'isomorphisme envoie $\tau(F)$ dans $F[Q^*]$. De plus, si on ne s'intéresse qu'aux foncteurs $F \in \overline{PT^n}$, on peut se contenter d'avoir $\overline{Q}^* = \mathbf{Q}^*_{n-1}$ (au lieu de $\overline{Q}^*_{n-1}$); appliquant cela lorsque Q^* est la structure Ω^*_{n-1} fournie par le Lemme 4.14, nous obtenons un foncteur τ: $\overline{PT^n} \to OL$ avec pour tout $F \in \overline{PT^n}$ un isomorphisme de $\tau(F)$ dans $F[\Omega^*_{n-1}]$. Mais Ω_{n-1} est un ptyx, donc si F est un ptyx, $F[\Omega^*_{n-1}] \in WO$, d'où $\tau(F) \in WO$: τ est un ptyx et le théorème est prouvé.

Puisque les structures $\mathbf{Q}^*_n$ sont p.r., on peut considérer que les ensembles $X \subset |\mathbf{Q}^*_n|$ sont sous-ensembles de ω, ce que nous faisons ci-dessous:

4.17. THÉORÈME. *Pour tout $n \geqslant 0$, il existe une formule $\psi(X) \in \Pi^1_{n+1}$ telle que pour tout $X \subset |\mathbf{Q}_n|$, $\mathbf{Q}_n \restriction X$ est un ptyx $\Leftrightarrow$ $\psi(X)$ est vrai (au sens standard). En conséquence, si $(F_e)_{e\in\omega}$ est l'énumération récursive des foncteurs p.r. $F_e \subset \mathbf{Q}_n$, l'ensemble $\{e \in \omega$: F_e est un ptyx$\}$ est Π^1_{n+1}.*

(Il résultera du §6 que $\{e \in \omega$: F_e est un ptyx$\}$ est de plus Π^1_{n+1}-complet; ceci généralise à la "Π^1_{n+1}-logique" le théorème de Π^1_2-complétude des dilatateurs p.r. de [**G2**]. Lequel pour sa part généralise le théorème de Π^1_1-complétude des codes de bons ordres p.r., de Kleene et Brouwer. Ceci suggère que pour la théorie descriptive des ensembles Π^1_{n+1}, les ptykes p.r. de type n puissent jouer un rôle similaire au rôle des bons ordres p.r. dans l'étude des ensembles Π^1_1.)

PREUVE. Par récurrence sur n; pour $n = 0$ c'est le fait bien connu que la notion de bon ordre est Π^1_1. Donc nous supposons le résultat vrai pour $n - 1$, $n \geqslant 1$ fixé. Alors si $X \subset |\mathbf{Q}_n|$, $\mathbf{Q}_n \restriction X$ est un ptyx $\Leftrightarrow$ $\forall H \in PT^{n-1}$ l'image de H par $\mathbf{Q}_n \restriction X$ est un bon ordre $\Leftrightarrow$ $\forall H' \in PT^{n-1}$ de *dimension dénombrable* l'image de H' par $\mathbf{Q}_n \restriction X$ est un bon ordre (en effet, si une suite décroissante est dans l'image de H par $\mathbf{Q}_n \restriction X$, il existe $H' \subset H$, de dimension dénombrable et dont l'image par $\mathbf{Q}_n \restriction X$ contient cette même suite) $\Leftrightarrow$ (par 3.12(a)) $\forall Y \subset |\mathbf{Q}_{n-1}|$ si $\mathbf{Q}_n \restriction Y$ est un ptyx alors son image par $\mathbf{Q}_n \restriction X$ est un bon ordre $\Leftrightarrow$ $\forall Y \subset |\mathbf{Q}_{n-1}|(\psi_{n-1}(Y) \to$ l'image de $\mathbf{Q}_{n-1} \restriction Y$ par $\mathbf{Q}_n \restriction Y$ est un bon ordre); où $\psi_{n-1}(Y)$ est la formule Π^1_n qui, par hyppothèse de récurrence exprime que $\mathbf{Q}_{n-1} \restriction Y$ est un ptyx. Par 3.11, l'image de $\mathbf{Q}_{n-1} \restriction Y$ pour $\mathbf{Q}_n \restriction X$ est un ordre p.r. relativement à X et Y, donc il suffit d'une formule $\theta(X, Y) \in \Pi_1^1$ pour exprimer que cet ordre est un bon ordre; et "$\mathbf{Q}_n \upharpoonright X$ est un ptyx" prend donc la forme $\psi(X) = \forall Y[\psi_{n-1}(Y) \to \theta(X, Y)]$ avec $\theta \in \Pi_1^1$, $\psi_{n-1} \in \Pi_n^1$ et "$\forall Y$" variant sur $\mathscr{P}(\omega)$. Sous forme prénexe, ceci est Π_{n+1}^1, c.q.f.d.

5. Majorations par des foncteurs p.r. La première moitié de ce travail a développé en détail la notion de foncteur p.r. de type fini; nous entamons la deuxième moitié, qui est consacrée à *une* de ses utilisations[8]: les majorations de fonctions des ordinaux dans les ordinaux au moyen de ptykes p.r. Ces majorations sont des résultats de caractère surprenant au premier abord, dont l'exemple-type a été fourni par [**G3**]. Nous rappelons d'abord ce premier résultat et sa signification en sous-section 5.A. La notion de ptyx p.r. de type > 1 du §4 permet de vastes généralisations de ces résultats. Le présent travail va fournir quelques exemples de ces généralisations, énoncés en sous-section 5.B et démontrés à partir du paragraphe I. C. Ces exemples suggèrent l'existence de tout un *réseau de relations fonctorielles* entre les objets étudiés par la Théorie descriptive des ensembles. L'étude de cette toile d'araignée ptygienne ne fait que commencer; sa signification et son utilité en Théorie descriptive, sans doute importantes, sont à découvrir.

5.A. *Majorations de fonctions α^+-récursives.* Nous avons besoin des notions de base de la Récursivité sur les ordinaux, rappelées très brièvement ci-dessous; pour plus de détails, consulter [**B**] ou [**H**]. Pour tout ordinal γ, L_γ désigne l'ensemble des ensembles constructibles avant γ. On dit que γ est *admissible* si $L_\gamma (= (L_\gamma, \varepsilon \upharpoonright L_\gamma))$ satisfait le schéma de remplacement restreint aux fonctions qui sont Σ-définissables; si γ est admissible, on appelle *γ-récursives* les fonctions de γ dans γ qui sont Σ-définissables dans L_γ. Une propriété fondamentale des admissibles est:

5.0. (a) Si γ est admissible, pour tout bon ordre R, $R \in L_\gamma$ entraîne que le type d'ordre $\|R\|$ de R est $< \gamma$; et la fonction: $R \mapsto \|R\|$ est Σ-définissable dans L_γ.

Joint à 3.11, cela fournit une classe de fonctions γ-récursives dont on verra plus loin le rôle:

(b) Soit Φ: $PT^1 \times ON \to ON$ un ptyx p.r., et γ un ordinal admissible; pour tout ptyx $F \in PT^1$ tel que $F \upharpoonright \omega \in L_\gamma$, la fonction: $\alpha \mapsto \Phi(F, \alpha)$ est γ-récursive (où $F \upharpoonright \omega$ désigne la restriction de F à la catégorie $ON \upharpoonright \omega$ des ordinaux finis).

Preuve. Ceci résulte de 5.0(a) et 3.11. Au lieu que Φ soit p.r., il suffirait que $\Phi^* \in L_\gamma$.

Pour tout ordinal α, α^+ désigne le plus petit ordinal admissible $> \alpha$; en particulier, on sait que ω^+, noté ω_1^{ck}, est le sup des ordinaux récursifs et que

(5.1) ω^{++}, noté ω_2^{ck}, est le sup de $\{\alpha: \exists X \subset \omega$, tel que X est Π_1^1 et α récursif en $X\}$.

[8] Pour d'autres utilisations, des articles sont en préparation par le premier auteur: en Théorie descriptive des ensembles (sur les théorèmes d'uniformisation) et en Théorie des démonstrations (analyse de $\Pi_2^1 - CA$).

Soit s_0 le premier ordinal Σ_1^1-réflexif—c'est-à-dire tel que toute formule Σ_1^1 avec paramètres, vraie dans la structure L_{s_0} est déjà vraie dans L_α pour un $\alpha < s_0$; cf. [**AR**] pour plus de détails. Cet ordinal s_0 ne joue pas un rôle essentiel pour la suite et le lecteur non spécialiste peut se contenter de retenir que s_0 est déjà un "grand" ordinal dénombrable: en termes approximatifs, s_0 dépasse ω_1^{ck}, ω_2^{ck} autant que le premier cardinal inaccessible ou le premier cardinal Mahlo dépasse $\aleph_1$, $\aleph_2$.

Cette grandeur de s_0 a pour nous l'effet que le Théorème 0 ci-dessous est valable pour un "grand" nombre d'ordinaux. Ce Théorème 0 obtenu dans [**G3**][9] est le prototype des majorations par foncteurs p.r. étudiées ici:

THÉORÈME 0. *Soit α un ordinal tel que $\omega \leqslant \alpha < s_0$; alors*

(1) $\alpha^+ = \sup\{F(\alpha)$: *$F$ foncteur p.r. de ON dans ON*$\}$,

(2) *pour toute fonction α^+-récursive f il existe un foncteur p.r. F de ON dans ON qui majore f à partir de l'ordinal α: $f(\gamma) \leqslant F(\gamma)\ \forall\gamma \in [\alpha, \alpha^+[$.*

Ce type de résultat n'étant pas familier en logique, le reste du paragraphe 5.A est consacré à expliquer ce que le Théorème 0 a de remarquable; pour commencer, nous en tirons des corollaires concernant des questions plus familières: la longueur des bons ordres sur ω définissables de façon Π_1^1 ou Σ_1^1. Nous rappelons quelques faits bien connus à ce sujet.

5.2. (a) $\omega_1^{ck} = \sup\{\|R\|$: R bon ordre Σ_1^1 de domaine $\subseteq \omega\}$,

(a′) $\omega_1^{ck} = \sup\{\|R\|$: R bon ordre Π_1^1 de domaine exactement $\omega\}$:

En revanche.

(b) Il existe R, bon ordre Π_1^1 de domaine $\subset \omega$, tel que $\|R\| = \omega_1^{ck}$.

PREUVE. (a) est le Théorème de la borne Σ_1^1 de Spector.

(a′) est une conséquence de (a): en effet, si R est un bon ordre Π_1^1 de domaine exactement ω, alors $(x, y) \notin R \Leftrightarrow (y, x) \in R$, donc R est également Σ_1^1, et par (a), $\|R\| < \omega_1^{ck}$. Pour prouver (b), rappelons qu'il existe un ordre total *récursif* de domain ω, qui possède un segment initial bien ordonné de type ω_1^{ck}. La notion de bon ordre étant Π_1^1, ce segment initial constitue le bon ordre Π_1^1 cherché.

DÉFINITION. (a) Nous dirons qu'un ordinal α est Π_1^1 si $\alpha = \|R\|$, avec R bon ordre Π_1^1 de domaine inclus dans ω. (*Vu* 5.2, *il est essentiel de permettre l'inclusion stricte*.)

(b) Nous définissons semblablement la notion d'ordinal Π_n^1, et d'ordinal Σ_n^1 $(n \geqslant 1)$.

Par 5.2(b), ω_1^{ck} est un ordinal Π_1^1, et il en résulte facilement que $\omega_1^{ck} + \omega_1^{ck}$, $\omega_1^{ck} \cdot \omega_1^{ck}$, $(\omega_1^{ck})^{(\omega_1^{ck})}$ sont des ordinaux Π_1^1. On peut continuer ainsi à construire des ordinaux Π_1^1 de plus en plus grands, mais cela devient de plus en plus compliqué et il n'est pas clair où cela s'arrête. Les cas particulier $\alpha = \omega_1^{ck}$ du Théorème 0 va éclaircir ce point, et va aussi donner un résultat concernant la longueur des *pré* bons ordres Σ_1^1, qui contraste avec 5.2(a).

5.3. COROLLAIRE. (a) *ω_2^{ck} est le* sup *des ordinaux* Π_1^1.

[9] Cf. aussi [**M**, **R1**, **VW**], pour des compléments au Théorème 0.

(b) $\omega_2^{\mathrm{ck}} = \sup\{\|R\|$: *R pré bon ordre* Σ^1_1 *de domain* $\omega\}$.

PREUVE DE (a). Vu que $(\omega_1^{\mathrm{ck}})^+ = \omega_2^{\mathrm{ck}}$ et vu le Théorème 0 appliqué avec $\alpha = \omega_1^{\mathrm{ck}}$, on a

$$\omega_2^{\mathrm{ck}} = \sup\left\{F\left(\omega_1^{\mathrm{ck}}\right): F \text{ foncteur } p.r. \text{ de } ON \text{ dans } ON\right\}.$$

Par le Théorème 3.11, l'ordre $F(\omega_1^{\mathrm{ck}})$ est p.r. *positivement* en F et en l'ordre ω_1^{ck}; comme ω_1^{ck} est Π^1_1, si F est p.r., $F(\omega_1^{\mathrm{ck}})$ est Π^1_1, ce qui entraîne $\omega_2^{\mathrm{ck}} \leqslant \sup\{\alpha$: α est $\Pi^1_1\}$. L'inégalité inverse résulte de (5.1)—dont (a) est une "version fonctorielle".

PREUVE DE (b). En 5.12(b) nous rappelons comment définir un pré bon ordre Σ^1_1 de domaine ω et de type ω_1^{ck}, que nous notons R_0. Pour tout foncteur continu F de ON dans ON, définissons $F(R_0)$ exactement comme si R_0 était un ordre; alors $F(R_0)$ est un pré bon ordre, de type $F(\|R_0\|) = F(\omega_1^{\mathrm{ck}})$ et de plus, par une extension facile de 3.11, ce pré bon ordre est p.r. positivement en R_0 et F. Donc si F est p.r., comme R_0 est Σ^1_1, $F(R_0)$ est un pré bon ordre Σ^1_1 de type $F(\omega_1^{\mathrm{ck}})$. Donc

$$\begin{aligned}\omega_2^{\mathrm{ck}} &= \sup\left\{F\left(\omega_1^{\mathrm{ck}}\right): F \text{ p.r. de } ON \text{ dans } ON\right\}\\ &= \sup\left\{\|F(R_0)\|: F \text{ p.r. de } ON \text{ dans } ON\right\}\\ &\leqslant \sup\left\{\|R\|: R \text{ pré bon ordre } \Sigma^1_1\right\}.\end{aligned}$$

La preuve de la relation inverse $\geqslant$ est standard.

Nous revenons au Théorème 0, considéré en lui-même et non au travers d'applications. Pour $\alpha < s_0$, il entraîne que toute fonction f qui est p.r. sur α^+ est majorée (à partir de α), par un foncteur p.r. F; c'est déjà inattendu parce que la récursion primitive qui détermine *F est sur ω*, et non sur α^+ comme pour f! Mais les fonctions p.r. sur α^+ ne sont qu'un cas particulier très modeste de fonction α^+-récursive; pour commencer, par diagonalisation on construit facilement une fonction f_1 qui est α^+-récursive mais qui majore (sous-entendu: *finalement*) *toute* fonction p.r. sur α^+. Cependant, f_1 est majorée par un foncteur p.r., nous dit le Théorème 0. Mais il y a pire: on peut définir une chaîne de fonctions α^+-récursives de plus en plus croissantes $(f_\gamma)_{\gamma<\alpha^{++}}$ (pas seulement $\gamma < \alpha^+$!), telle que

$$\gamma > \beta \Rightarrow f_\gamma \textit{ majore } f_\beta \textit{ et toute fonction } p.r. \textit{ en } f_\beta \textit{ sur } \alpha^+.$$

Et par le Théorème 0, toutes ces fonctions f_γ seront majorées par des foncteurs définis par récursion primitive... sur ω.

Comme le Théorème 0 est surprenant, on peut se demander si, après tout, des résultats encore plus forts ne sont pas vérifiés; c'est ce que nous examinons ci-dessous.

Dans **[G3]** est montré que la conclusion du Théorème 0 est fausse pour $\alpha = s_0$; cependant, il est facile d'étendre le Théorème 0 à certains ordinaux $> s_0$. Rappelons d'abord—cf. [**B**, **H**]:

5.4. (a) Un ordinal σ est dit *stable* si L_σ est un sous-modèle Σ-élémentaire de L_σ;

(b) σ_0, le *premier* ordinal stable, est aussi l'ensemble de tous les ordinaux Σ-définissables dans L, et celui de tous les ordinaux Δ'_2.

Dans [**R**] est montré:

THÉORÈME 0′. *La conclusion du Théorème* 0 *reste vraie pour une classe d'ordinaux qui est cofinale dans* σ_0.

Mais la conclusion de Théorème 0 cesse définitivement d'étre vraie à partir de σ_0: $\alpha \geqslant \sigma_0 \Rightarrow \alpha^+ > \sup\{F(\alpha): F \text{ p.r. de } ON \text{ dans } ON\}$. Donc:

5.5. (a) Le Théorème 0′ est optimum du point de vue du domaine de validité de la conclusion.

(b) Nous avons vu en 5.0(b) que pour tout foncteur p.r. F et *tout* α, la fonction: $\gamma \mapsto F(\gamma)$ (restreinte à α^+) est α^+-récursive. Donc le Théorème 0′ est optimum en ce sens que les foncteurs p.r. ne peuvent pas majorer mieux que les fonctions α^+-récursives.

(c) Pour la même raison, on ne peut pas majorer toutes les fonctions γ-récursives, lorsque l'ordinal admissible γ n'est pas de la forme α^+, donc lorsque $\alpha < \gamma \Rightarrow \alpha^+ < \gamma$: en effet soit alors f la fonction: $\alpha \mapsto \alpha^+$ (restreinte à γ); f est γ-récursive, mais n'est pas majorée par un foncteur p.r. F—cela contredirait (b), et aussi le point (1) du Théorème 0.

(d) Encore pour la même raison, dans le point (2) du Théorème 0, on ne peut pas majorer $f(\gamma)$ par $F(\gamma)$ *avant* $\gamma = \alpha$ (prendre $\gamma = \omega < \alpha = \omega^+$ et f α^+-récursive telle que $f(\omega) = \omega^+ \ldots$).

5.B. *Majorations par des foncteurs p.r. de type* > 1. La conclusion (5.5) est qu'on ne peut aller au-delà du Théorème 0′; en fait, elle est toute provisoire: au moyen de foncteurs p.r. *de type* > 1 on va dépasser toutes les limitations indiquées par 5.5.

5.6. *Notations*. La somme ordonnée $\Sigma(\Phi_i)_{i \in I}$ d'une famille de foncteurs de type n a été définie en 4.15; il est immédiat que si I est un bon ordre et si chaque Φ_i, $i \in I$, est un ptyx, alors la somme $\Sigma(\Phi_i)_{i \in I}$, est encore un ptyx. En particulier, si $(F_e)_{e \in \omega}$ désigne l'énumération récursive de tous les foncteurs p.r. de type n (cf. 3.14(b)), la somme $\Sigma(F_e;\ F_e$ est un ptyx) est un ptyx de type n, qui sera noté Ξ_n.

5.7. REMARQUES. (a) Dans cette définition et dans les théorèmes qui suivent, ces ptykes particuliers Ξ_n "sortent d'un chapeau"; mais leur rôle sera expliqué petit à petit. Pour commencer, notons que le Théorème 0 introduit Ξ_1 de façon implicite; en effet, le point (1), $\alpha^+ = \sup\{F(\alpha): F \text{ foncteur p.r. de } ON \text{ dans } ON\}$ revient à $\alpha^+ = \Xi_1(\alpha)$ pour tout $\alpha < s_0$. Et par le Théorème 0′, cette égalité $\alpha^+ = \Xi_1(\alpha)$ s'étend à un sous-ensemble cofinal de σ_0; par 5.5(a), elle cesse d'être vraie à partir de σ_0:

$$\alpha^+ > \Xi_1(\alpha)\ \forall \alpha \geqslant \sigma_0.$$

(b) Le foncteur Ξ_n dépend de l'énumération récursive $(F_e)_{e \in \omega}$ choisie, *mais il n'en dépend que peu*: comme nécessairement chaque foncteur p.r. figure une infinité de fois dans cette énumération, la fonction: $\alpha \mapsto \Xi_n(\alpha)$ n'en dépend pas, non plus que la valeur de $\Phi(\Xi_n)$, pour tout ptyx Φ de type $n+1$, on le vérifie aisément.

THÉORÈME 1. (a) $\forall\alpha < \sigma_0^+$, *il existe un foncteur p.r.* Φ: $PT^1 \times ON \to ON$ *tel que* $\alpha \leqslant \Phi(\Xi_1, \sigma_0)$.

(b) *Pour toute fonction* σ_0^+*-récursive f il existe un foncteur p.r.* Φ: $PT^1 \times ON \to ON$ *tel que* $f(\gamma) \leqslant \Phi(\Xi_1, \gamma)\, \forall\gamma \in [\sigma_0, \sigma_0^+]$.

THÉORÈME 2. (a) *Pour tout ordinal* α *qui est* Σ_2^1, *il existe un foncteur p.r. F*: $ON \to ON$ *tel que* $\alpha \leqslant F(\sigma_0)$.

(b) *Pour tout ordinal* α *qui est* $< \sigma_0$, *il existe un foncteur p.r.* Φ: $PT^1 \to ON$ *tel que* $\alpha < \Phi(\Xi_1)$.

(c) *Il existe un ensemble D cofinal dans* σ_0 *tel que pour toute fonction* σ_0*-récursive f, il existe un foncteur p.r.* Φ: $PT^1 \times ON \to ON$ *pour lequel* $f(\gamma) \leqslant \Phi(\Xi_1, \gamma)$ $\forall\gamma \in D$.

THÉORÈME 3. *Pour tout ordinal* α *qui est* Π_{n+1}^1, *il existe un foncteur p.r.* Φ: $PT^n \to ON$ *tel que* $\alpha \leqslant \Phi(\Xi_n)$.

COROLLAIRE 1. $\sup\{\Phi(\Xi_1, \sigma_0);\ \Phi$ *p.r.*: $PT^1 \times ON \to ON\} = \sigma_0^+$.

COROLLAIRE 2. $\sup\{\alpha$: *α est le type d'un pré bon ordre* Π_2^1 *sur* $\omega\} = \sigma_0^+$.

COROLLAIRE 3. $\sup\{\alpha$: *α est* $\Sigma_2^1\} = \Xi_1(\sigma_0)$.

COROLLAIRE 4. $\sup\{\alpha$: *α est* $\Pi_2^1\} = \sigma_0 < \Xi_1(\sigma_0) < \sigma_0^+$.

COROLLAIRE 5. $\sup\{\alpha$: *α est* $\Pi_{n+1}^1\} = \Xi_{n+1}(\Xi_n)$ *pour tout* $n \geqslant 1$.

5.8. *Nota Bene*. (a) En contraste avec les Corollaires 2 et 3, si R est un bon ordre Σ_2^1 ou Π_2^1 de domaine *exactement* ω, alors R est Δ_2^1, donc $|R| < \sigma_0$; c'est l'analogue de 5.2(a′).

(b) Les Corollarires 2.3 et 4 sont l'analogue parfait de 5.2 et du Corollaire 5.3, quand on fait correspondre ω_1^{ck} avec σ_0, Σ_1^1 avec Π_2^1 et Π_1^1 avec Σ_2^1 (en particulier le Corollaire 4 est l'analogue Π_2^1 du théorème de Σ_1^1-borne). La seule différence est que $\Xi_1(\omega_1^{\mathrm{ck}}) = (\omega_1^{\mathrm{ck}})^+$, alors que $\Xi_1(\sigma_0) < \sigma_0^+$.

(c) Vu $PT^0 = ON$, il est naturel de définir Ξ_0 comme suit, par analogie avec la définition de Ξ_n, $n \geqslant 1$: $(R_e)_{e\in\omega}$ étant une énumération récursive des ordres p.r. sur ω, on pose $\Xi_0 = \Sigma\ (R_e$: R_e est un *bon* ordre). Pratiquement, cela revient à poser $\Xi_0 = \omega_1^{\mathrm{ck}}$, et alors le Corollaire 5.3 nous donne: $\sup\{\alpha$: α est $\Pi_1^1\} = \Xi_1(\Xi_0)$, ce qui devient le cas particulier $n = 0$ des Théorème 3 et Corollaire 5.

Quelques éclaircissements sur ces théorèmes avant de les prouver:

(a) ce qu'il y a de frappant dans la conclusion (1) du Théorème 0, c'est que "l'abîme" entre α et α^+ (comparé à ce qu'est un objet p.r. *sur* ω) peut être comblé en utilisant *seulement* des foncteurs p.r. Dans ces conditions, il est naturel de considérer comme une extension satisfaisante du Théorème 0 tout résultat qui prend la forme

(1*) $\qquad \alpha^+ = \sup\{\Phi(\Xi, \alpha)$: Φ foncteur p.r. de $\mathscr{C} \times ON$ dans $ON\}$,

où $\mathscr{C}$ est une catégorie de structures et Ξ une structure appartenant à $\mathscr{C}$ *qui est définissable dans* L_α.

En effet, il y a *essentiellement* le même "abîme" entre $L_{\alpha+1}$ et α^+ qu'entre α et α^+; donc (puisque $\Xi \in L_{\alpha+1}$) entre (Ξ, α) et α^+, qu'entre α et α^+; et donc le trait frappant du Théorème 0 est conservé dans (1*). De ce point de vue, le Théorème 1(a) est une *bonne* extension du Théorème 0 au cas $\alpha = \sigma_0$—cas à partir duquel justement la conclusion du Théorème 0 devient définitivement en défaut. En effet, le foncteur $\Xi_1 \upharpoonright \omega$ (avec lequel on peut identifier Ξ_1, cf. Remarque 3.10) *est définissable dans* L_{σ_0}: ce résultat facile montré ci-dessous fait que le Théorème 1(a) est un cas particulier de (1*).

Il y a bien d'autres ordinaux que $\alpha = \sigma_0$ admettant un résultat de la forme (1*); en particulier, cf. [**N**], même si l'on *renforce* (1*) en y restreignant Φ aux ptykes de type $\leqslant 2$, il faut montrer jusqu'au premier ordinal Π_2^1-réflexif pour rencontrer un ordinal stable α qui ne vérifie pas (1*).

(b) Nous pensons de même que le Théorème 2 contient une bonne extension du Théorème 0 à la majoration de fonctions γ-récursives, lorsque γ n'est pas de la forme α^+. Et le Théorème 3 étend à tout n certains aspects des Théorèmes 0 et Corollaire 5.3; mais nous n'insistons pas ici sur ces affirmations.

5.9. PROPOSITION. (a) *Le foncteur* $\Xi_1 \upharpoonright \sigma_0$ *est* Π_1-*définissable dans* L_{σ_0}.

(b) *Pour chaque n le foncteur* Ξ_n *est codable de façon* Π_{n+1}^1.

PREUVE DE (a). Notons d'abord que pour tout foncteur p.r. F de type 1, F est un ptyx ssi $F(x) \in WO\ \forall x < \sigma_0$. En effet, si F n'est pas ptyx, $L \models \exists x F(x) \notin WO$; vu que $\|F(x)\|$ est Σ-définissable à partir de x par 5.0(b), cet énoncé vrai dans L est Σ, donc est vrai dans L_{σ_0} en vertu de 5.4(b). Cela étant, la définition de Ξ_1 devient

$$\Xi_1 = \Sigma\Big(F_e\colon L_{\sigma_0} \models \forall\alpha \|F_e(\alpha)\| \in O_n\Big).$$

La famille ci-dessus a une définition Π_1 dans L_{σ_0}; d'où résulte que $\Xi_1 \upharpoonright \sigma_0$, donc $\Xi_1 \upharpoonright \omega$ et Ξ_1^*, sont Π_1-définissable dans L_{σ_0}.

PREUVE DE (b). $(F_e)_{e\in\omega}$ désigne ici l'énumération récursive de tous les foncteurs p.r. de type n — $n > 0$ fixé. Par 4.17, la famille (F_e; F_e est un ptyx) est codable de façon Π_{n+1}^1, et il en résulte aisément que Ξ_n, la somme de cette famille l'est aussi.

5.10. *Nota Bene*. Le cas particulier $n = 1$ de (b), et (a) se déduisent chacun de l'autre en appliquant le "*théorème du compagnon* Π_2^1" de Kripke-Platek, suivant lequel une relation sur ω est Π_2^1 si et seulement si elle est définissable dans L_{σ_0} par une formule Π (cf. [**B**]).

PREUVE DU COROLLAIRE 1. Le Théorème 1 peut s'écrire

$$\sup\big\{\Phi(\Xi_1, \sigma_0);\ \Phi \text{ p.r.: } PT^1 \times ON \to ON\big\} \geqslant \sigma_0^+,$$

donc il suffit de montrer l'inégalité inverse, et pour cela que $\Phi(\Xi_1, \sigma_0) < \sigma_0^+$ pour tout Φ p.r.: $PT^1 \times ON \to ON$. Or $\Xi_1 \upharpoonright \omega$ est définissable dans L_{σ_0} par 5.9, donc appartient à L_{σ_0+1} et a fortiori à $L_{\sigma_0^+}$ par 5.0(b).

PREUVE DU COROLLAIRE 5. Le Théorème 3 peut s'écrire $\Xi_{n+1}(\Xi_n) \geqslant \sup\{\alpha\colon \alpha$ est $\Pi_{n+1}^1\}$, donc il suffit de montrer l'inégalité inverse, et pour cela que chaque ordinal de la forme $\Phi(\Xi_n)$ pour un ptyx p.r. Φ de type $n+1$ est Π_{n+1}^1. Or par le

Théorème 3.11, l'ordre $\Phi(\Xi_n)$ est p.r. positivement en Φ et Ξ_n, Ξ_n est Π^1_{n+1} par 5.9, et Φ est p.r. par hypothèse; d'où résulte aisément que l'ordre $\Phi(\Xi_n)$ est codable de façon Π^1_{n+1}.

Pour prouver les Corollaires 2 et 3, il nous faut un pré bon ordre Π^1_2 et un bon ordre Σ^1_2, de type σ_0 tous les deux; en voici une construction.

5.11. NOTATION ET REMARQUE. On dira qu'un ordinal α est *L-descriptible* s'il existe $\theta \in \Sigma$ tel que $\alpha = \mu\gamma\colon L_\gamma \models \theta$. De 5.4(b) se déduit aisément que *l'ensemble des ordinaux L-descriptibles est un sous-ensemble cofinal de* σ_0.

Sur l'ensemble (des énoncés) Σ nous définissons une relation $\theta \leqslant \theta' \Leftrightarrow_{\text{df}} L_{\sigma_0} \models \forall\alpha\ \forall\beta[\alpha = \mu\alpha\colon L_\gamma \models \theta \text{ et } \beta = \mu\gamma\colon L_\gamma \models \theta' \to \alpha \leqslant \beta]$. Il est facile de voir que cette relation $\leqslant$ comporte d'abord, comme éléments minorant tous les autres, les énoncés $\theta \in \Sigma$ tels que $\forall\alpha L_\alpha \models \neg\, \theta$, et que l'application $f\colon \theta \mapsto \mu\gamma\colon L_\gamma \models \theta$ envoie tous les autres énoncés $\theta \in \Sigma$ sur les ordinaux L-descriptibles, de manière que $\theta \leqslant \theta' \Rightarrow f(\theta) \leqslant f(\theta')$. D'où résulte que $\leqslant$ est un pré bon ordre de type σ_0 sur Σ, qui est Π_1-définissable dans L_{σ_0}, donc Π^1_2 par le théorème du compagnon Π^1_2 (cf. 5.10). Au moyen d'une bijection p.r. de Σ sur ω, on peut transformer ceci en pré bon ordre sur ω. Ceci démontre le (a) de

5.12. PROPOSITION. (a) *Il existe un pré bon ordre* Π^1_2 *de type* σ_0 *sur* ω.

(b) *Il existe un pré bon ordre* Σ^1_1 *de type* ω_1^{ck} *sur* ω.

(c) *Il existe un bon ordre* Σ^1_2 *de type* σ_0 *sur une partie de* ω.

PREUVE. La preuve de (b) est toute semblable à celle de (a) déjà faite: sur Σ on considère la relation $\theta \leqslant \theta' \Leftrightarrow_{\text{df}} L_{\omega_1^{\text{ck}}} \models \forall\alpha\ \forall\beta[\alpha = \mu\gamma\colon L_\gamma \models \theta \text{ et } \beta = \mu\gamma\colon L_\gamma \models \theta' \to \alpha \leqslant \beta]$. C'est la même relation que dans (a) mais interprétée dans $L_{\omega_1^{\text{ck}}}$ au lieu de L_{σ_0}; on voit facilement que c'est un pré bon ordre de type ω_1^{ck}. Comme ce pré bon ordre est Π_1-définissable dans $L_{\omega_1^{\text{ck}}}$, il est Σ^1_1 par le "théorème de compagnon Π^1_1"—cf. [**B**]. Reste à voir (c); pour $\theta, \theta' \in \Sigma$ nous posons

$$\theta \leqslant {}^*\theta' \Leftrightarrow_{\text{df}} L_{\sigma_0} \models \exists\alpha\, \exists\beta \geqslant \alpha\big[\alpha = \mu\gamma\colon L_\gamma \models \theta \text{ et}$$
$$\cdots \forall\theta_1 \in \Sigma(\ulcorner\theta_1\urcorner < \ulcorner\theta\urcorner \to L_\alpha \models \neg\theta_1) \text{ et } \beta = \mu\gamma\colon L_\gamma \models \theta' \text{ et}$$
$$\cdots \forall\theta_1 \in \Sigma(\ulcorner\theta_1\urcorner < \ulcorner\theta'\urcorner \to L_\beta \models \neg\theta_1)\big].$$

Comme dans (a) on vérifie que $\leqslant^*$ est un bon ordre de type σ_0, qui est Σ-définissable dans L_{σ_0}, donc Σ^1_2. Et qui se ramène à un bon ordre sur une partie de ω au moyen de l'application: $\theta \in \Sigma \mapsto \ulcorner\theta\urcorner \in \omega$.

PREUVE DU COROLLAIRE 2. Si R est un pré bon ordre Π^1_2 sur ω, alors $R \in L_{\sigma_0+1}$, donc $\|R\| < \sigma_0^+$. Inversement, si $\alpha < \sigma_0^+$, par le Théorème 1(a), il existe un ptyx p.r. Φ tel que $\alpha \leqslant \Phi(\Xi_1, \sigma_0)$; soit R un pré bon ordre Π^1_2 sur ω, de type σ_0; par l'extension aux pré bons ordres du Théorème 3.11, $\Phi(\Xi_1, R)$ est un pré bon ordre de type $\Phi(\Xi_1, \|R\|) = \Phi(\Xi_1, \sigma_0) \geqslant \alpha$, qui est récursif positivement en Φ, Ξ_1 et R. Comme Φ est p.r. et Ξ_1, R sont Π^1_2, le pré bon ordre est Π^1_2.

PREUVE DU COROLLAIRE 3. Comme celle du Corollaire 2, mais en utilisant le Théorème 2(a) au lieu du Théorème 1(a).

PREUVE DU COROLLAIRE 4. Vu 5.4 évidemment $\sup\{\alpha\colon \alpha \text{ est } \Pi^1_2\} \geqslant \sigma_0$; reste donc à voir que tout ordinal Π^1_2 est majoré par σ_0. Cela va résulter du Théorème

2(b): si α est Π^1_2, il existe Φ ptyx p.r. tel que $\alpha \leqslant \Phi(\Xi_1)$. Pour tout ordinal z, notons Ξ_1^z le foncteur $\Sigma(F_e; \|F_e(z)\| \in On)$, où $(F_e)_{e\in\omega}$ est l'énumération récursive de tous les foncteurs p.r. de $\overline{PT^1}$. Il y a une transformation naturelle évidente de Ξ_1 dans Ξ_1^z (pour chaque z fixé), donc il y a morphisme d'ordre de $\Phi(\Xi_1)$ dans $\Phi(\Xi_1^z)$. En conséquence, pour majorer $\Phi(\Xi_1)$, donc α, par σ_0, il suffit de montrer: $L_{\sigma_0} \vDash \exists z \Phi(\Xi_1^z) \in ON$. Or cet énoncé Σ est vrai dans L (prendre $z = \sigma_0$), donc il est vrai dans L_{σ_0} par stabilité de σ_0.

5.C. PREUVE DU THÉORÈME 2.(a) Nous commençons avec cette preuve, parce qu'elle utilise uniquement les méthodes déjà exposées dans [**R1**], pour le Théorème 0, et elle nous permettra le rappel de ces méthodes. Les preuves des autres utiliseront ces mêmes méthodes mais en les étendant à la machinerie des ptykes de type > 1 vue aux paragraphes 2–4.

5.13. *Rappels de "β-logique"*. Soit $\mathscr{L}^{ON}$ le langage comportant le symbole $<$, et des "variables de sorte ON". Soit $\mathscr{L}$ un langage contenant $\mathscr{L}^{ON}$ (et éventuellement d'autres sortes de variables); une $\mathscr{L}$-structure $\mathscr{M}$ est un α-modèle (où $\alpha \in ON$) si α est son domaine de sorte ON, et $<^{\mathscr{M}} = \varepsilon \restriction \alpha$—c'est-à-dire $\mathscr{M} \restriction \mathscr{L}^{ON} = (\alpha, \varepsilon \restriction \alpha)$. On écrit $T \mapsto_\alpha \psi$ si tout α-modèle de la théorie T satisfait la formule ψ et $T \vdash_{\leqslant\alpha} \psi$ si $T \vdash_{\leqslant\gamma} \psi$, $\forall\gamma \leqslant \alpha$. Enfin, $T \vdash_{ON} \psi$ si $T \vdash_\alpha \psi$ pour tout $\alpha \in ON$; dans ce cas, on dit également que ψ est *β-conséquence* de T.

Le théorème suivant, qui est une forme abstraite du "théorème de β-complétude", cf. [**G2**] n'est peut-être pas très parlant, mais c'est un outil puissant pour construire des foncteurs de ON dans ON.

5.14. THÉORÈME. *Soit $T \subset \mathscr{L}$ une théorie fixée; à toute formule close ψ de $\mathscr{O}$ on peut associer un foncteur p.r. F de OL dans OL tel que*:

(1) $F_\psi(\alpha) \in WO \Leftrightarrow T \vdash_{\leqslant\alpha} \psi$ pour tout α.

(2) *L'application*: $(\psi, n, f) \mapsto (F_\psi(n), F_\psi(f))$ *(où $\psi \in \mathscr{L}$, $n \in \omega$ et f est un morphisme de $ON \restriction \omega$) qui détermine complètement les foncteurs F_ψ, est p.r. en T.*

(3) $T \vdash_{ON} \psi \Leftrightarrow F_\psi$ *est un foncteur p.r. de ON dans ON.*

PREUVE. Cf. [**R1**, §2].

5.13.bis *Notation*. Par la suite nous utilisons une variante commode des notions. 5.13: le langage $\mathscr{L}^{ON}$ est remplacé par le langage de Théorie des ensembles $\mathscr{L}_\varepsilon$ auquel on adjoint une constante d'individu C, on le note alors $\mathscr{L}$; et on dit qu'une $\mathscr{L}$-structure $\mathscr{M}$ est un α-modèle si

$$(C^{\mathscr{M}}, \varepsilon^{\mathscr{M}}) = (\alpha, \varepsilon \restriction \alpha).$$

5.15. PROPOSITION. *La notion* 5.13.bis *de α-modèle se ramène essentiellement à un cas particulier de celle de* 5.13; *en conséquence, le Théorème* 5.14 *reste vrai pour la nouvelle notion d'α-modèle.*

PREUVE. Soit $\mathscr{M}$ un α-modèle au sens modifié; on étend $\mathscr{M}$ en un α-modèle de $\mathscr{L}^{ON}$ noté $\mathscr{M}^*$, en posant: domaine de sorte ON de $\mathscr{M} = \alpha$, et $<^{\mathscr{M}} = \varepsilon \restriction \alpha$. Notez qu'alors $\mathcal{M}^*$ satisfait la théorie suivante, notée T^*, où $u \in \mathcal{L}_\varepsilon$ et $v \in \mathcal{L}^{ON}$:

$$\forall u[u \in C \leftrightarrow \exists v\, u = v],$$

$$\forall u \in C \forall u' \in C(u \in u' \leftrightarrow u < u').$$

Inversement, si $\mathcal{N}^*$ est un α-modèle au sens de 5.1.3 et $\mathcal{N}^* \models T^*$, il est clair que $\mathcal{N}^* \restriction \mathcal{L}$ est un α-modèle au sens modifié. Donc en appliquant le Théorème 5.14 à la théorie $T \cup T^*$, on déduit que ce théorème s'applique aussi à la notion modifiée d'α-modèle.

Preuve du Théorème 2(a). Soit R un bon ordre Σ_2^1 de domaine $\subset \omega$; nous allons définir une théorie T et une formule ϕ de $\mathcal{L}$ telles que $T \vdash_{ON} \phi$—donc le foncteur F_ϕ associé à ϕ par le Théorème 5.14 est un foncteur p.r. de ON dans ON; et telles que de plus $\|R\| \leqslant F_\phi(\sigma_0)$. Cela montrera une moitié du Théorème 2(a): à savoir que $\sup\{\gamma\colon \gamma \text{ est } \Sigma_2^1\} \leqslant \Xi_1(\sigma_0)$.

Choix de T. T comprend: la théorie KP (cf. [**B**]), ou une partie finie appropriée de ZF; et de plus l'axiome "$C \geqslant \omega$".

Avant de choisir ϕ, nous nous occupons d'avoir dans tout ON-modèle $\mathcal{M}$ de T une représentation, ou plutôt une approximation du bon ordre R que nous voulons majorer. Pour cela nous choisissons $\psi(uv)$, formule Σ telle que $R = \{(m, n) \in \omega^2\colon L_{\sigma_0} \models \psi(m, n)\}$ (l'existence d'un tel ψ caractérise les relations Σ_2^1 sur ω, cf. 5.4); et pour tout ordinal γ, nous notons R_γ la relation $\{(m, n) \in \omega^2\colon L_\gamma \models \psi(m, n)\}$.

Fait. (a) *R_γ est bien fondé pour tout γ.*

(b) *Si $\mathcal{M}$ est un γ-modèle de T, alors $R_C^{(\mathcal{M})} = R_\gamma$.*

Preuve de (a). Si $\gamma < \sigma_0$, L_γ est partie transitive de σ_0, alors comme ψ est préservée par extensions transitives, $R_\gamma \subset R_{\sigma_0} = R$, donc R_γ est bien fondée puisque R l'est. Et si $\gamma \geqslant \sigma_0$, $R_\gamma = R_{\sigma_0} = R$ puisque L_{σ_0} est un sous-modèle Σ-élémentaire de L_γ, par stabilité de l'ordinal σ_0.

Preuve de (b). Est évident puisque $C^{\mathcal{M}} = \gamma$, donc $L_C^{(\mathcal{M})} = L_\gamma$.

Choix de ϕ. Comme l'arithmétique s'interprète dans notre théorie T, on peut appliquer le lemme d'auto-référence de Gödel[10]: on peut supposer associé à chaque formule ψ de $\mathcal{L}$ un terme de $\mathcal{L}$ noté $\ulcorner\psi\urcorner$, de telle sorte que l'application $\psi \mapsto \ulcorner\psi\urcorner$, est injective et récursive, et que pour toute formule $A(u)$ de $\mathcal{L}$ il existe une formule close ϕ vérifiant: $T \vdash \phi \leftrightarrow A(\ulcorner\phi\urcorner)$. Nous appliquons ceci quand $A(u)$ est la formule de $\mathcal{L}$ exprimant:

$$\exists\psi \in \mathcal{L}\colon u = \ulcorner\psi\urcorner \wedge \neg\big[\|F_\psi(C)\| < \|R_C\|\big].$$

La formule ϕ ainsi obtenue vérifie donc

$$T \vdash \phi \leftrightarrow \exists\psi \in \mathcal{L}\colon \ulcorner\phi\urcorner = \ulcorner\psi\urcorner \wedge \neg\big[\|F_\psi(C)\| < \|R_C\|\big],$$

ce qui (vu $\ulcorner\phi\urcorner = \ulcorner\psi\urcorner \leftrightarrow \phi = \psi$) se ramène à

$$(*) \qquad T \vdash \phi \leftrightarrow \neg\big[\|F_\phi(C)\| < \|R_C\|\big].$$

---

[10]. En suivant une idée de Harrington, déjà exposée dans [**R1**].

5.16. LEMME. (a) *Pour tout On-modèle* $\mathscr{M}$ *de T, on a*

$$\mathscr{M} \models \|F_\phi(C)\| < \|R_C\| \Leftrightarrow V \models \|F_\phi(\gamma)\| < R_\gamma,$$

où $\gamma = C$.

(b) $T \vdash_{ON} \phi$.

PREUVE DE (a). Si $\mathscr{M}$ est un γ-modèle de T, $\gamma \in On$, alors $\gamma \geqslant \omega$ puisque T comprend "$C \geqslant \omega$". Ainsi $\mathscr{M}$ est ω-standard, donc les notions p.r. sont absolues dans $\mathscr{M}$; en particulier, $F_\phi(C)^{\mathscr{M}} = F_\phi(\gamma)$. D'autre part, $R_C^{\mathscr{M}} = R_\gamma$ par le Fait (b). Enfin, on sait que pour tout modèle $\mathscr{N}$ de KP et tout ordre $S \in \mathscr{N}$, si $V \models S$ est un bon ordre, alors $\|S\|^{\mathscr{N}} = \|S\|^V$ d'où $\|F_\phi(C)\|^{\mathscr{M}} = \|F_\phi(\gamma)\|$ et $\|R_C\|^{\mathscr{M}} = \|R_\gamma\|$, ce qui entraîne (a).

PREUVE DE (b). Soit $\mathscr{M}$ un γ-modèle de T; supposons que $\mathscr{M} \models \neg\, \phi$. Par $(*)$, $\mathscr{M} \models \|F_\phi(C)\| < \|R_C\|$; d'où par (a) $\|F_\phi(\gamma)\| < \|R_\gamma\|$. Comme R_γ est bien fondé (Fait (a)), $F_\phi(\gamma) \in WO$, et le Théorème 5.14 entraîne $T \vdash_{\leqslant\gamma} \phi$. Alors $\mathscr{M} \models \phi$ puisque $\mathscr{M}$ est γ-modèle de T: supposant que $\mathscr{M} \models \neg\, \phi$ nous en avons déduit que $\mathscr{M} \models \phi$, ce qui par l'absurde prouve (b).

Le (b) entraîne que F_ϕ est un foncteur p.r. de WO dans WO, par le Théorème 5.14(3); considérons alors l'univers standard V comme un σ_0-modèle, en interprétant C par σ_0. Puisque $V \models T$ et $T \vdash_{\sigma_0} \phi$ par le Lemma 5.16, V est un σ_0-modèle de ϕ d'où par $(*)$: $\neg\|F_\phi(\sigma_0)\| < \|R_{\sigma_0}\|$. Comme $\|F_\phi(\sigma_0)\| \in On$, c'est donc que $\|F_\phi(\sigma_0)\| \leqslant \|R_{\sigma_0}\| = \|R\|$; nous avons donc achevé la preuve du Théorème 2.(a).

REMARQUE. Si R est un bon ordre Σ_2^1, alors l'énoncé "R est un bon ordre" est un énoncé Π_2^1; c'est sans doute la raison profonde du Théorème 2(a) et de son corollaire "$\sigma_2^1 = \Xi_1(\pi_2^1)$".

§6. Π_n^1-artillerie. Ce paragraphe expose une extension à la Π_n^1-Logique, $n \geqslant 3$, de certains résultats de base de Π_2^1-Logique; tout d'abord deux extensions du Théorème 5.14, où la catégorie de structures ON est remplacée par d'autres.

6.1. DÉFINITIONS. Soient $\mathscr{L}^1 \subset \mathscr{L}$ deux langages, et M une $\mathscr{L}^1$-structure; on appelle M-modèle toute $\mathscr{L}$ structure $\mathscr{M}$ telle que $\mathscr{M} \upharpoonright \mathscr{L}^1 = M$. On écrit $T \vdash_m \psi$ si tout M-modèle de la théorie T est un modèle de ψ; $T \vdash_{\subset M} \psi$ si $T \vdash_{M'} \psi$ pour tout $M' \subseteq M$. Et lorsque $\mathscr{U}$ est une classes de $\mathscr{L}^1$-structures, $T \vdash_{\mathscr{U}} \psi$ si $T \vdash_M \psi$ pour tout $M \in \mathscr{U}$.

6.2. LEMME. *Soit* $\mathscr{U}$ *une classe de* $\mathscr{L}^1$*-structures close par restrictions, et soit T une théorie dans* $\mathscr{L}$. *A tout modèle* $M \in \mathscr{U}$ *et toute formule close* $\psi \in \mathscr{L}$ *on peut associer un arbre* $\mathscr{C}_\psi(M) \subset |M|^{<\omega}$ *tel que*:

(1) $\mathscr{C}_\psi(M)$ *est bien fondé ssi* $T \vdash_{\subset M} \psi$; *donc* $\mathscr{C}_\psi$ *est une application de* $\mathscr{U}$ *dans les arbres bien fondés ssi* $T \vdash_{\mathscr{U}} \psi$.

(2) *Si* $\bar{a} \in M$, $\bar{b} \in N$ *et* $\bar{a} \simeq_{\mathscr{L}^1} \bar{b}$, *alors* $\bar{a} \subset \mathscr{C}_\psi(M) \Leftrightarrow \bar{b} \in \mathscr{C}_\psi(N)$.

(3) *La relation* $\bar{a} \in \mathscr{C}_\psi(M)$ (*où* $\bar{a}$ *varie dans M et* ψ *dans* $\mathscr{L}$) *est p.r. relativement à T et M, d'une manière uniforme en T et M.*

REMARQUES. (a) La notion 5.1.3.bis de α-modèle est évidement le cas particulier de M-modèle où $M = (\alpha, \varepsilon \restriction \alpha)$. Donc $T \vdash_{\leqslant\alpha} \psi$ au sens du paragraphe 5.C $\Leftrightarrow T \vdash_{\subset\alpha} \psi$ au sens ci-dessus; et $T \vdash_{ON} \psi$ est le cas particulier $\mathscr{U} = On$ de la notion $T \vdash_{\mathscr{U}} \psi$.

(b) Le cas particulier $\mathscr{U} = On$ du Lemme 6.2 est très proche du Théorème 5.14: la propriété (2) de $\mathscr{C}_\psi$ est proche de la propriété qu'a F_ψ d'être un foncteur continu commutant aux p.f.; et la seule propriété de F_ψ qui manque à $\mathscr{C}_\psi$, est que $\mathscr{C}_\psi(\alpha)$ est un arbre, c'est-à-dire un ordre *partiel* et non total. Donc ce cas $\mathscr{U} = On$ de 6.2. est une étape de la preuve du Théorème 5.14, et à ce titre est démontré par exemple dans [**R1**]. Comme la preuve de 6.2 pour $\mathscr{U}$ quelconque est semblable, nous ne la détaillons pas beaucoup et renvoyons à [**R1**] pour les détails:

PREUVE DE 6.2. Soit $(v_n)_{n\in\omega}$ une suite de variables de $\mathscr{L}^1$, et soit $(\phi_n(v_0 \cdots v_n))$ une énumération des formules de $\mathscr{L}$ avec ces variables libres; on pose

$$\lambda_n(v_0 \cdots v_n) = (\exists v_n \phi_n(v_0 \cdots v_n)) \to \phi_n(v_0 \cdots v_n).$$

(λ_n) énumère donc tous les axiomes de Henkin du langage $\mathscr{L}$ pour les variables de $\mathscr{L}^1$. On fixe une énumération $(p_n)_{n\in\omega}$ de toutes les preuves du langage $\mathscr{L}$. $\mathscr{L}^1$ et toutes ces énumérations sont supposées p.r. On pose:

(*) $\mathscr{C}_\psi(M) =$ ensemble des suites $\bar{a}_n = (a_0 \cdots a_{n-1}) \in M$ telles que $(p_i)_{i\leqslant n}$ ne contient pas de preuve que la théorie $T + \psi + \bigwedge\!\!\!\bigwedge_{i\leqslant n} \lambda_i(\bar{a}_n) + d(\bar{a}_n)$ est contradictoire;

où $d(\bar{a})$ désigne la diagramme simple de $\bar{a}$ dans M.

Visiblement cette définition satisfait 6.2(2) et (3); de plus, si $\mathscr{C}_\psi(M)$ possède une branche infinie $(a_n) \in M^\omega$, alors vu (*), la théorie

$$T + \neg\psi + \bigwedge\!\!\!\bigwedge_n \lambda_n(a_0 \cdots a_n) + \text{diagramme simple de } M \restriction \{a_n;\ n \in \omega\}$$

est finîment consistante. Comme elle contient les axiomes de Henkin appropriés, elle a un modèle $\mathscr{N}$ dans lequel l'interprétation des variables de $\mathscr{L}^1$ est l'ensemble des témoins $\{a_n;\ n \in \omega\}$. Donc $\mathscr{N} \restriction \mathscr{L}^1 = M \restriction \{a_n;\ n \in \omega\}$, puisque $\mathscr{N}$ satisfait le diagramme simple de cette structure. De plus, $\mathscr{N} \models T + \neg\,\psi$, donc $T \vdash_{\subset M} \psi$ est faux: on a ainsi montré une moitié de (1).

Réciproquement, si $T \vdash_{\subset M} \psi$ est faux, soit $M' \subset M$ tel que $T + \neg\,\psi$ possède un M'-modèle η; alors il existe une énumération (a_n) de $|M'|$ telle que η satisfait les axiomes de Henkin $\bigwedge\!\!\!\bigwedge_n \lambda_n(a_0 \cdots a_n)$. D'où résulte aisément que $(a_n)_{n\in\omega}$ est une branche infinie de $\mathscr{C}_\psi(M)$, donc $\mathscr{C}_\psi(M)$ n'est pas bien fondé; ce qui montre la moitié restante de (1).

Le Théorème 5.14 était un outil puissant pour construire des foncteurs p.r. de ON; en étendant ce théorème au cas où ON $(= PT^0)$ est remplacé par PT^n, $n > 0$, nous obtiendrons un moyen de construire des foncteurs p.r. de PT^n dans ON. Pour faire ce remplacement de ON par PT^n, il nous faut définir ce qu'est un PT^n-modèle, donc pour $F \in PT^n$ ce qu'est un "F-modèle". Or nous avons représenté chaque foncteur $F \in PT^n$ par une *structure* F^* de langage $\mathscr{L}_n$; et puisque F^* est une structure, la notion de F^*-modèle est déjà définie par 6.1.

Moyennant quoi les notions relatives aux PT^n-modéles vont être des cas particuliers de 6.1:

6.4. DÉFINITIONS. Soit $F \in \overline{PT^n}$, et soit $\mathscr{L}$ un langage contenant le langage $\mathscr{L}_n$ du codage F^* de F (cf. §2). On appellera F-modèle tout F^*-modèle; et on combine cette façon de parler avec chaque définition de 6.1:

$$T \vdash_F \phi \text{ signifie } T \vdash_{F^*} \phi;$$

$$T \vdash_{\subset F} \phi \text{ signifie } T \vdash_{\subset F^*} \phi$$

($\Leftrightarrow T \vdash_G \phi$, pour tout foncteur G tel que $G \subset F$ au sens de la Définition 1.5(a));

$$T \vdash_{PT^n} \phi \text{ signifie } T \vdash_F \phi \text{ pour tout } F \in PT^n.$$

6.5. THÉORÈME DE PT^n-COMPLÉTUDE. *Soit T une théorie dans un langage $\mathscr{L} \supset \mathscr{L}_n$; à tout énoncé ψ de $\mathscr{L}$ on peut associer un foncteur continu ϕ_ψ de $\overline{PT^n}$ dans OL de façon que*:

(1) $\forall F \in PT^n \; T \vdash_{\subset F} \phi \Leftrightarrow \phi_\psi(F) \in WO$.

(2) *L'application*: $\psi \mapsto \phi_\psi$ *est codable de façon p.r.*

(3) $T \vdash_{PT^n} \psi \Leftrightarrow \phi_\psi$ *est un ptyx p.r.*

PREUVE. Nous appliquons le Lemme 6.2 quand $\mathscr{L}^1 = \mathscr{L}_n$, $\mathscr{U} = \{F^*: F \in PT^n\}$; son hypothèse que $\mathscr{U}$ est clos par restrictions est vérifiée du fait que

6.6. REMARQUE. Soit $F \in PT^n$; si $G \subset F$ (ou plus généralement si G s'envoie dans F par une t.n.), alors $G \in PT^n$.

Par 6.2. nous avons donc pour chaque énoncé ψ de $\mathscr{L}$ et chaque $F \in \overline{PT^n}$ un arbre $\mathscr{C}_\psi(F^*)$—noté par la suite $\mathscr{C}_\psi(F)$—contenu dans $|F|^{<\omega}$. Vu la Remarque 6.3(b), pour tirer de $\mathscr{C}_\psi$ le foncteur Φ_ψ cherché, il reste à *linéariser* $\mathscr{C}_\psi(F)$ de manière appropriée. Pour cela, nous utilisons l'ordre $\tau(F)$ de domaine $|F|$ (cf. le Théorème 4.16), et nous appelons *ordre de Brouwer-Kleene sur* $|F|^{<\omega}$ l'ordre bien connu *induit par l'ordre* $\tau(F)$ *sur* $|F|$: on sait que c'est un ordre total et que si $\tau(F) \in WO$, alors la restriction de cet ordre à un arbre $\mathscr{C} \subset |F|^{<\omega}$ est un bon ordre si et seulement si $\mathscr{C}$ est bien fondé. Posons alors pour $\psi \in \mathscr{L}$, $F \in \overline{PT^n}$, $\Phi_\psi(F) = \mathscr{C}_\psi(F)$ *muni de l'ordre de B.K.* (= Brouver-Kleene).

(i) Vu les propriétés de cet ordre $\Phi_\psi(F) \in OL$.

(ii) Vu 6.2(1), l'arbre $\mathscr{C}_\psi(F)$ est bien fondé ssi $T \vdash_{\subset F} \psi$; dans le cas où $F \in PT^n$, $\tau(F) \in WO$, et vu les propriétés de l'ordre de B.K., on a $\mathscr{C}_\psi(F)$ bien fondè $\Leftrightarrow \Phi_\psi(F) \in WO$. D'où: $\Phi_\psi(F) \in WO \Leftrightarrow \mathscr{C}_\psi(F)$ bien fondé $\Leftrightarrow$ (par 6.5) $T \vdash_{\subset F} \psi$, pour $F \in PT^n$.

(iii) Vu le caractère p.r. en T, ψ, F de $\mathscr{C}_\psi(F)$ (cf. 6.2(3)) et le caractère p.r. de τ (Théorème 4.16), et de l'ordre de B.K., l'ordre $\Phi_\psi(F)$ est p.r. en T, ψ, F d'une manière uniforme.

(iv) Si $(\phi^X)_{X \in \overline{PT^{n-1}}}$ est une t.n. de F dans G (donc ϕ est un $\mathscr{L}_n$-isomorphisme de F^* dans G^*, *et* un morphisme d'ordre de $\tau(F)$ dans $\tau(G)$), soit $\Phi_\psi(\phi)$ l'application: $(a_0 \cdots a_n) \in \mathscr{C}_\psi(F) \mapsto (\phi(a_0) \cdots \phi(a_n)) \in \mathscr{C}_\psi(G)$; vu que ϕ est un $\mathscr{L}_n$-isomorphisme et vu 6.2., $\Phi_\psi(\phi)$ est un isomorphisme d'arbre de $\mathscr{C}_\psi(F)$ dans $\mathscr{C}_\psi(G)$, et une fois ces arbres munis de l'ordre de B.K., $\Phi_\psi(\phi)$ devient un isomorphisme d'ordre de $\Phi_\psi(F)$ dans $\Phi_\psi(G)$. Et à l'aide de 6.2(2), on voit facilement que Φ_ψ est alors un foncteur continu commutant aux produits fibrés, de $\overline{PT^{n-1}}$ dans OL.

(v) Supposons $T \vdash_{PT^n} \psi$, et $F \in PT^n$; alors par 6.2(1), $\mathscr{C}_\psi(F)$ est bien fondé. Supposons de plus F de dimension finie, soit $|F|$ fini; comme $\mathscr{C}_\psi(F) \subset |F|^{<\omega}$, par le lemme de Koenig, $\mathscr{C}_\psi(F)$ est fini—donc $\Phi_\psi(F)$ est fini. Nous avons montré que $T \vdash_{PT^n} \psi$ entraîne: Φ_ψ foncteur p.r. de PT^n dans WO.

Ceci achève la preuve du théorème.

6.7. REMARQUES. (a) L'ordre $\tau(F)$ défini sur $|F|$ dans 4.16 ne serait qu'un *pré* ordre si le foncteur F ne commutait pas aux produits fibrés: par exemple dans le cas $n = 1$, c'est-à-dire F: $OL \to OL$, on pourrait avoir

$$F[\omega] = t[0 \cdots k-1] = s[0 \cdots l-1], \quad \text{donc } t =_{\tau(F)} s$$

même si $t \neq s$. Alors le foncteur Φ_ψ qui en résulte dans 6.5 prendrait ainsi pour valeurs des pré ordres.

(b) Pour définir $\tau(F)$ dans le cas $n > 1$ nous nous sommes servi d'une énumération $(F_i)_{i \in \omega}$ des ptykes de dimension finie; le foncteur τ est p.r. parce que cette énumération l'est: c'est à ce point précis que sert le Théorème 4.12, c'est-à-dire la caractérisation p.r. parmi tous les foncteurs continus de dimension finie, de ceux qui sont des ptykes.

(c) Le Théorème 6.5 a une extension évidente à des situations un peu plus compliquées: par exemple, si $F \in \overline{PT^n}$ et $X \in OL$, appelons (F, X)-*modèle* d'une théorie T (dans un langage $\mathscr{L}$ contenant $\mathscr{L}_n$ et $\mathscr{L}^{ON}$) tout modèle $\mathscr{M}$ de T tel que $\mathscr{M} \restriction \mathscr{L}_n = F^*$ et $\mathscr{M} \restriction \mathscr{L}^{ON} = X$. Pour ce cas le Théorème 6.5. devient:

6.5.bis. THÉORÈME. *A tout énoncé $\psi \in \mathscr{L}$, on peut associer un foncteur continu Φ_ψ de $PT^n \times OL$ dans OL de façon que*

(1) $$\forall F \in PT^n \, \forall \gamma \in ON \, \|\Phi_\psi(F, \psi)\| \in ON \Leftrightarrow T \vdash_{\subset F, \leqslant \gamma} \psi$$

ce qui signifie: $\mathscr{M} \models \psi$, *pour tout* (G, α)-*modèle* $\mathscr{M}$ *de* T *avec* $G \subset F$, $\alpha \leqslant \gamma$.

(2) *L'application*: $\psi \mapsto \Phi_\psi$ *est codable de façon p.r. en T.*

(3) $$T \vdash_{PT^n, ON} \psi \Leftrightarrow \Phi_\psi \text{ est un ptyx p.r.}$$

(d) Une autre extension facile du Théorème 6.5 est celle où l'on restreint la notion de modèle aux modèles ω-standards:

6.5. ter. THÉORÈME. *Les points* (1) *et* (2) *du Théorème* 6.5 *restent vrais lorsque* $\mathscr{L}$, *T vérifient les hypothèses suivantes*: *le langage* $\mathscr{L}$ *de T contient* $\mathscr{L}_\varepsilon$; *et T contient, outre des énoncés finis de* $\mathscr{L}$, *la formule infinie*,

$$\forall n \left[n \text{ entier} \leftrightarrow \bigvee\!\!\!\bigvee \{ x = \underline{n};\, n \in \omega \} \right]$$

qui caractérise les modèles ω-standards.

Preuve. Soit T' la théorie obtenue en remplaçant la formule infinie de T par les formules finies suivantes de $\mathscr{L} \cup \mathscr{L}^{ON}$:

$$\forall x[x \text{ est un entier} \leftrightarrow \exists u^{On} x = u^{ON}],$$

$$\forall u^{ON}\big[(\exists v^{ON} u^{ON} < v^{ON}) \wedge \big(u^{ON} \text{ possède un prédécesseur } \textit{immédiat}$$

$$\text{pour } <, \text{ excepté si } u^{ON} \text{ est le plus petit élément}\big)\big].$$

Il est clair que les γ-modèles de ces formules, $\gamma \in ON$, vérifient tous $\gamma = \omega$, donc

$$(*) \qquad T \vdash_{\subset F} \psi \Leftrightarrow T' \vdash_{\subset F,\omega} \psi \Leftrightarrow T' \vdash_{\subset F,ON} \psi.$$

En appliquant 6.2.bis à T', on associe à tout énoncé ψ de $\mathscr{L}$ un foncteur Φ_ψ de $\overline{PT^n} \times OL$ dans OL. De $(*)$ et de 6.2.bis, il est facile de déduire que le foncteur $\Phi_\psi(\cdot, \omega)$ de $\overline{PT^n}$ dans OL qui en résulte vérifie les propriétés (1), (2) de 6.2, ce qui achève la preuve.

Notez que, même si $\Phi_\psi(\cdot,\cdot)$ est *faiblement fini*, le foncteur $\Phi_\psi(\cdot, \omega)$ ne l'est généralement pas, ce pourquoi il ne vérifie pas le point (3) de 6.2.

Nous introduisons maintenant une variante commode de la notion de $\overline{PT^n}$-modèle, qui étend au cas $n > 0$ la variante 5.13.bis de ON ($= PT^0$)-modèle: soit $\mathscr{L}$ le langage constitué de $\mathscr{L}_\varepsilon$ auquel on adjoint une constante d'individu $\mathscr{F}$; si $F \in \overline{PT^n}$, on appelle *F-modèle* toute $\mathscr{L}$-structure $\mathscr{M}$ telle que $\mathscr{F}^{\mathscr{M}}$ est une $\mathscr{L}_n$-structure isomorphe à F^*.

6.8. **Proposition.** Si *on se restreint aux modèles* ω-standards, *alors la variante ci-dessus de F-modèle se ramène essentiellement à un cas particulier de la notion* 6.3; *ainsi le Théorème* 6.2.ter *reste vrai pour cette variante.*

Preuve. Sur le même principe que celui de la Proposition 5.15 (la seule différence est que le langage $\mathscr{L}^{ON}$ est remplacé par $\mathscr{L}_n$ qui est langage avec une infinité de sortes; mais la restriction ici aux ω-modèles permet de faire que $\mathscr{L}_n$ soit toujours un *élément* des modèles $\mathscr{M}$ considérés, moyennant quoi la notion de $\mathscr{L}_n$-structure y est exprimable).

6.9. **Théorème de Π^1_{n+1}-complétude (des ptykes de type n).** *La notion de ptyx de type n est* Π^1_{n+1}*-complète: d'une part vu* 4.17 *la propriété, pour un foncteur de type n, d'être un ptyx s'exprime de façon* Π^1_{n+1}; *d'autre part, à tout énoncé* $\psi \in \Pi^1_{n+1}$ *on peut associer un foncteur p.r.* F_ψ *de type n, de façon que:*

(1) $\quad \psi$ *est vrai* $\big($*dans le modèle standard* $(\omega, +, \cdot, \omega^\omega)\big) \Leftrightarrow F_\psi$ *est un ptyx.*

(2) $\quad$ *L'application:* $\psi \mapsto F_\psi$ *est codable de façon p.r.*

Preuve. Nous prouvons par récurrence sur n une version relativisée à un réel $\rho \in \omega^\omega$ de ce résultat:

à toute formule $\psi(R) \in \Pi^1_{n+1}$ (où R est variable libre du second ordre) et à tout réel $\rho \in \omega^\omega$ on peut associer un foncteur $F_{\psi(\rho)}$ de type n, de façon que:

(1)$'$ $\quad \psi(\rho)$ est vrai $\Leftrightarrow F_{\psi(\rho)}$ est un ptyx.

(2)$'$ $\quad$ L'application: $(\psi(R), \rho) \mapsto F_{\psi(\rho)}$ est codable de façon p.r.

Dans le cas particulier $n = 0$, $F_{\psi(\rho)} \in \overline{PT^0} = OL$ est un ordre et non un foncteur, et $PT^0 = WO$, et le résultat ci-dessus est le théorème bien connu de Kleene de forme normale des énoncés Π_1^1. Nous supposons donc le résultat vrai pour $n - 1$ quelconque: si $\psi(R) \in \Pi_n^1$, à tout réel ρ est donc associé un foncteur $F_{\psi(\rho)} \in \overline{PT^1}$, tel que $\psi(\rho)$ est vrai $\Leftrightarrow F_{\psi(\rho)}$ est un ptyx, et l'application: $\rho \mapsto F_{\psi(\rho)}$ est p.r. Soit $\forall R \neg\psi(R)$, où $\psi(R) \in \Pi_n^1$, un énoncé Π_{n+1}^1; alors

(*) $\forall R \neg\psi(R)$ est vrai $\Leftrightarrow \forall\rho \in \omega^\omega[\neg\psi(\rho)$ *est vrai*$] \Leftrightarrow \forall\rho \in \omega^\omega F_{\psi(\rho)}$ n'est pas un ptyx $\Leftrightarrow \forall F \in PT^{n-1}\, \forall\rho \in \omega^\omega F_{\psi(\rho)} \neq F \Leftrightarrow T \vdash_{PT^{n-1}} \phi$, lorsque T est la théorie $KP + \forall x\, [x$ est un entier $\leftrightarrow \bigvee\{x = \underline{n};\, n \in \omega\}]$ et $\phi = \phi(\mathscr{F})$ la formule "$\forall\rho \in \omega^\omega F^*_{\psi(\rho)} \neq \mathscr{F}$".

Notons $\Phi_{\forall R \neg\psi}$ le foncteur Φ_ϕ obtenu en appliquant le Théorème 6.5.ter à la théorie T et la formule ϕ de (*); par ce théorème nous avons:

(1)″ $\forall R \neg\psi$ est vrai $\Leftrightarrow \Phi_{\forall R \neg\psi}$ est un ptyx.

(2)″ L'application: $\forall R \neg\psi \in \Pi_{n+1}^1 \mapsto \Phi_{\forall R \neg\psi}$ est p.r.

Ce sont les propriétés (1)′ et (2)′ qu'il nous faut démontrer pour achever la récurrence sur n, *excepté* que (1)′ et (2)′ sont des versions avec un paramètre réel (formules de la forme $\forall R \neg\psi(R, \rho_1)$ au lieu de $\forall R \neg\psi(R)$); ces versions paramétrisées se démontrent de la même manière, aussi nous considérons (1)′ et (2)′ comme démontrés pour tout n.

Les point (1) et (2) du théorème en résultent *excepté* que les foncteurs F_ψ associés aux énoncés ψ ne sont présentement pas faiblement finis (vu que cela n'est pas le cas dans le Théorème 6.5.ter). Mais cette propriété supplémentaire se déduit de

6.10. Théorème. *Pour toute famille* $(\Phi_i)_{i\in\omega}$ *de foncteurs de* $\overline{PT^n}$ *tels que* $|\Phi_i|$ *est dénombrable* $\forall i \in \omega$, *il existe une famille* $(\Phi'_i)_{i\in\omega}$ *de foncteurs* faiblement finis *tels que*:

(1) $\Phi'_i(F) = \Phi_i(F)\ \forall i \in \omega$ *et* $\forall F \in \overline{PT^{n-1}}$ de dimension infinie.

(2) Φ'_i *est un ptyx ssi* Φ_i *en est un*, $\forall i \in \omega$.

(3) *Enfin, si* Φ_i *est p.r. relativement à un ensemble* X, Φ'_i *l'est aussi, et si* $(\Phi_i)_{i\in\omega}$ *est p.r. relativement à* X, $(\Phi'_i)_{i\in\omega}$ *l'est aussi.*

Preuve. Soit $\Phi \in PT^n$, $(t^n)_{n\in\omega}$ une énuḿration de $|\Phi|$; nous posons pour tout $F \in \overline{PT^{n-1}}$.

$$\Phi'(F) =_{\text{df}} \Phi[F] \restriction \{t^i[\bar{a}];\, \bar{a} \in |F|,\, i < \operatorname{card}|F|\}$$

et si $T = (\phi_X)_{X\in\overline{PT^{n-1}}}$ est une t.n. de F dans G, si $(\psi_I)_{X\in\overline{PT^{n-1}}}$ désigne la t.n. $\Phi[T]$ de $\Phi[F]$ dans $\Phi[G]$, alors

$$\Phi'(T) =_{\text{df}} \big((\psi_X \restriction \Phi'(X))_{X\in\overline{PT^{n-1}}}\big).$$

On laisse au lecteur le soin de vérifier que ceci définit un foncteur faiblement fini $\Phi' \in \overline{PT^n}$; que pour cette définition de Φ', (1) est vérifié ce qui entraîne (2); enfin que Φ' est p.r. en une énumération de $|\Phi|$, de façon uniforme, ce qui entraîne (3).

Le Théorème de Π^1_{n+1}-complétude nous permet d'aborder le rôle que joue dans les généralisations du Théorème 0 abordées au §5.B, la notion de ptyx de type > 1 et les ptykes particuliers Ξ_n—rôle peu expliqué au §5.

Auparavant, considérons une définition Σ^0_1 *positivement* en X, où X est un sous-ensemble de ω; il existe donc une machine de Turing M avec oracle X telle que:

M n'utilise que les réponses positives ("*oui*, $p \in X$") de l'oracle, et ne tient pas compte des autres;

un objet x satisfait la définition ssi M appliquée à x fait un calcul qui converge.

Supposons $X' \subset X$, et que la machine M appliquée à un argument x fixé a fait un calcul *qui converge*, en utilisant comme oracle X' à la place de X. Alors, la réponse de ce calcul est juste—c'est à-dire est la même que celle qu'avait donnée l'utilisation du "vrai" oracle X. Cela *même* si éventuellement X' est beaucoup plus simple que X: la complexité de X fait qu'éventuellement $M(x)$ converge pour beaucoup plus d'arguments x avec l'oracle X qu'avec l'oracle X'; mais si l'on s'intéresse à un x fixé, alors au lieu de X il nous suffit de X'—du moment que

(1) $X' \subset X$,

(2) $M(x)^{(X')}$ converge.

Tout à l'heure nous utiliserons ces idées lorsque le foncteur Ξ_n joue le rôle de l'oracle X: Ξ_n est essentiellement un ensemble Π^1_{n+1}-complet, et servira d'oracle pour toute question Π^1_{n+1}; et tout autre foncteur F de type n *qui est un ptyx* joue alors le rôle de $X' \subset X$: c'est-à-dire le fait pour F d'être un ptyx de type n fera de F un "oracle" qui, face au véritable oracle Π^1_{n+1} constitué par Ξ_n, a la propriété que toutes ses réponses *positives* seront justes, et donc utilisables à la place des réponses données par Ξ_n. Mais ceci reste dans le vague par ce que je n'ai pas indiqué le "*mode de consultation* de l'oracle" Ξ_n, et de ses approximations $F \in PT^n$:

6.11. Proposition. (a) *Pour toute formule close* $\psi \in \Pi^1_{n+1}$, ψ *est vrai* $\Leftrightarrow$ *il existe une t.n. de* F_ψ *dans* Ξ_n (*où* F_ψ *est le foncteur défini en* 6.9).

(b) *Pour tout ptyx* F *de type* n, *une moitié de* (a) *reste vraie*: *s'il existe une t.n. de* F_ψ *dans* F, *alors* ψ *est vrai*.

Preuve laissée au lecteur, résultant de la Remarque 6.6.

Ceci va servir dans la preuve des Théorème 3 et 2(c): voir le Fait 1 dans chacune de ces preuves.

7. Preuve des résultats du §5.

Preuve du Théorème 3. Soit α un ordinal Π^1_{n+1}: $\alpha = \|R\|$, où $R \subset \omega^2$ est un bon ordre Π^1_{n+1}; nous voulons définir un ptyx p.r. Φ de type $n+1$, tel que $\alpha \leqslant \Phi(\Xi_n)$—où Ξ_n est le ptyx de type n obtenu en sommant tous les ptyxes p.r. de type n, cf. 5.6. Pour obtenir Φ, nous appliquons le théorème de PT^n-complétude 6.5.ter: soit $\mathscr{L}$ le langage des PT^n-modèles (cf. Proposition 6.8: $\mathscr{L} = \mathscr{L}_\varepsilon$ auquel on ajoute la constante $\mathscr{F}$); nous allons choisir une théorie p.r. T et un énoncé ϕ dans

$\mathscr{L}$, tels que $T \vdash_{PT^n} \phi$, donc le foncteur Φ_ϕ associé à T, ϕ par 6.5 est un ptyx p.r.; et nous montrerons que $\alpha \leqslant \Phi_\phi(\Xi_n)$, ce qui résoudra notre problème avec $\Phi = \Phi_\phi$.

Choix de T.

$$T = KP + \forall x\big[x \text{ est un entier} \leftrightarrow \bigvee\!\!\!\!\bigvee \{n = \underline{n};\, n \in \omega\}\big].$$

Avant de choisir ϕ, nous nous occupons de représenter, ou plutôt approximer, la relation R de type α, dans les PT^n-modèles de T (comme déjà fait dans la preuve du Théorème 2(a) cf. sous-section 5.C): soit $\psi(uv) \in \Pi^1_{n+1}$ telle que $R = \{(i, j)\colon \psi(i, j) \text{ est vrai}\}$. Si $F \in PT^n$, dans tout F-modèle de T, nous approximerons R par la relation R_F qui se sert du ptyx F comme "oracle Π^1_{n+1}" afin de savoir si $\psi(\underline{i}\, \underline{j})$ est vrai (cf. la Proposition 6.11):

$$R_F = \big\{(i, j) \in \omega^2\colon \text{il existe un } \mathscr{L}_n\text{-isomorphisme de } F^*_{\psi(\underline{i}\,\underline{j})} \text{ dans } F^*\big\},$$

où $F_{\psi(\underline{i}\,\underline{j})}$ est le foncteur p.r. de type n associé par 6.9 à $\psi(\underline{i}\, \underline{j})$.

FAIT 1. *Pour tout $F \in PT^n$, et tout F-modèle $\mathscr{M}$ de T,*

$$R^{\mathscr{M}}_{\mathscr{F}} \subset R_F \subset R = R_{\Xi_n}.$$

PREUVE. $(i, j) \in R_F \Leftrightarrow$ il existe une t.n. de $F_{\psi(\underline{i}\,\underline{j})}$ dans $F \Rightarrow$ (vu 6.6 et $F \in PT^n$) $F_{\psi(\underline{i}\,\underline{j})} \in PT^n \Rightarrow$ (par 6.9) ψh(i, j) est vrai $\Leftrightarrow (i, j) \in R$; d'où $R_F \subset R$. $R_{\Xi_n} \subset R$ résulte de ce que si $\psi(i\, j)$ est vrai, $F_{\psi(\underline{i}\,\underline{j})}$ est un ptyx p.r., d'où une t.n. de $F_{\psi(\underline{i}\,\underline{j})}$ dans Ξ_n puisque Ξ_n est la somme de ces ptykes—d'où finalement un $\mathscr{L}_n$-isomorphisme de $F^*_{\psi(\underline{i}\,\underline{j})}$ dans Ξ^*_n. Enfin, $R^{\mathscr{M}}_{\mathscr{F}} = \{(i, j) \in \omega^2\colon \mathscr{M} \models \text{il existe un } \mathscr{L}_n\text{-isomorphisme de } F_{\psi(\underline{i}\,\underline{j})} \text{ dans } \mathscr{F}\}$ est inclus dans R_F parce que $\mathscr{F}^{\mathscr{M}} = F^*$, $(F^*_{\psi(\underline{i}\,\underline{j})})^{\mathscr{M}} = F^*_{\psi(\underline{i}\,\underline{j})}$ (vu que $\mathscr{M} \models T$ donc est ω-standard, et que l'existence du $\mathscr{L}_n$-isomorphisme dans $\mathscr{M}$ s'exprime de façon Ξ, donc est absolue.

Choix de ϕ. Nous prenons $\phi = \phi(\mathscr{F})$ tel que

$$T \vdash \phi \leftrightarrow \neg\big\|\Phi_\phi(\mathscr{F})\big\| < \|R_{\mathscr{F}}\|$$

où Φ_ϕ désigne le foncteur associé à ϕ par le théorème de PT^n-complétude 6.5 ter; l'existence de ϕ se montre comme dans la preuve du Théorème 2(a) en sous-section 5.C.

FAIT 2.

$$T \vdash_{PT^n} \phi.$$

PREUVE. Soit $\mathscr{M}$ un F-modèle de T, $F \in PT^n$, et supposons que $\mathscr{M} \models \neg\, \phi$; par choix de ϕ, $\mathscr{M} \models \|\Phi_\phi(\mathscr{F})\| <$ s$\|R_{\mathscr{F}}\|$; par un argument d'absoluité tout semblable à celui du Lemme 5.16(a) (dans la preuve du Théorème 2(a)) on en déduit: $\mathscr{M} \models \|\Phi_\phi(F)\| < \|R_F\|$. Alors comme $R_F \subset R$ est bien fondé, $\Phi_\phi(F) \in WO$; par le choix de Φ_ϕ, cela équivaut à $T \vdash_{\subset F} \phi$, d'où $\mathscr{M} \models \phi$, contradiction qui prouve le Fait 2.

Puisque, par ce fait ϕ est PT^n-conséquence de T, par 6.5.ter, Φ_ϕ est un ptyx, donc $\Phi_\phi(\Xi_n) \in WO$. Et comme V est un Ξ_n-modèle de T, V satisfait $\phi(\Xi_n)$ soit

$\neg\|\Phi_\phi(\Xi_n)\| < \|R_{\Xi_n}\|$. Ainsi $\|\Phi_\phi(\Xi_n)\| \geqslant \|R_{\Xi_n}\| = \|R\| = \alpha$; le ptyx Φ'_ϕ obtenu en appliquant le Théorème 6.10 à Φ_ϕ vérifie encore $\|\Phi'_\phi(\Xi_n)\| = \|\Phi_\phi(\Xi_n)\| \geqslant \alpha$, et il est de plus p.r., ce qui montre le Théorème 3.

PREUVE DE THÉORÈME 2(b). C'est le cas particulier $n = 1$ du Théorème 3.

PREUVE DE THÉORÈME 2(c). Soit f une fonction σ_0-récursive; dans la preuve du Théorème 2(b) (= cas $n = 1$ du Théorème 3), pour majorer un ordinal α par $\Phi(\Xi_1)$, on a utilisé une représentation de α au moyen d'un bon ordre $R \in \Pi^1_2$, de type α et de domaine $\subset \omega$; l'analogue ici, pour majorer $f(\alpha)$ par $\Phi(\Xi_1, \alpha)$, est de représenter $f(\alpha)$, et non pas α, par un bon ordre $R(\alpha) \in \Pi^1_2$:

LEMME. *A tout ordinal L-descriptible* (*cf.* 5.11) α, *on peut associer un bon ordre* $R(\alpha) \in \Pi^1_2$, *de type* $\geqslant f(\alpha)$ *et de domain* $\subset \omega$, *de façon que la formule* Π^1_2 *qui définit* $R(\alpha)$ *soit définissable à partir de* α *dans* $L_{\alpha+\omega}$, *uniformément en* α.

PREUVE. A tout ordinal L-descriptible α, nous associons d'abord le plus petit[11] énoncé $\theta \in \Sigma$ tel que $\alpha = \mu\gamma$: $L_\gamma \models \theta$. A θ, nous associons

$$\theta^* = \exists x\big(L_x \models \theta \wedge f(x)\ \textit{est}\ \text{défini}\big).$$

Evidemment, il existe γ tel que $L_\gamma \models \theta^*$, et le plus petit tel γ est $> f(\alpha)$. Nous posons $R(\alpha) = \{(m, n)\colon L_{\sigma_0} \models \psi^*(m, n)\}$, où $\psi^*(m, n) = \forall\beta\ \forall\lambda[\beta = \mu\gamma\colon L_\gamma \models \theta^*$ et $\lambda =$ plus petite bijection de ω sur β pour l'ordre $<_L \to f(\lambda(m)) < f(\lambda(n))$. ψ^* est une formule Π, donc par le théorème du compagnon Π^1_2, il existe $\psi \in \Pi^1_2$, qui s'obtient de façon p.r. à partir de ψ^*, telle que $R(\alpha) = \{(m, n)\colon \psi(m, n)$ est vrai$\}$. Comme θ est évidemment définissable à partir de α dans $L_{\alpha+\omega}$, il en est de même de θ^*, ψ^* et ψ.

Ayant représenté $f(\alpha)$ par $R(\alpha)$, nous allons approximer $R(\alpha)$: soient α, L-définissable, et $F \in PT^1$, et soit $\psi(u\, v)$ la définition Π^1_2 de $R(\alpha)$—cf. lemme ci-dessus; on pose $R_F(\alpha) = \{(m, n) \in \omega^2$: il existe un $\mathscr{L}_1$-isomorphisme de $F^*_{\psi(\underline{m}\,\underline{n})}$ dans $F^*\}$.

FAIT 1. (a) $R_F(\alpha) \subset R(\alpha) = R_{\Xi_1}(\alpha)$.

(b) *Soit* $\mathscr{L} = \mathscr{L}_\varepsilon$ *auquel on ajoute les constantes* $\mathscr{F}$ *et* C; *soit* T *la théorie* KP + "C *est un ordinal L-descriptible infini*"; *pour tout* (F, α)-*modèle de* T, *où* $F \in PT^1$, $R_{\mathscr{F}}(C)^{\mathscr{M}}$ *est inclus dans* $R_F(\alpha)$ *donc est bien fondé*.

PREUVE. Comme le Fait 1 analogue dans la preuve du Théorème 3.

Soit alors $\phi = \phi(\mathscr{F}, C)$ une formule de $\mathscr{L}$ telle que

$$T \vdash \phi \leftrightarrow \neg\big\|\Phi_\phi(\mathscr{F}, C)\big\| < \big\|R_{\mathscr{F}}(C)\big\|$$

où Φ_ϕ désigne le foncteur de $PT^1 \times OL$ dans OL associé à ϕ par 6.5.bis.

FAIT 2. $T \vdash_{PT^1, ON} \phi$.

PREUVE. Comme celle du fait analogue de la preuve du Théorème 3. Comme dans cette preuve, on en déduit que Φ_ϕ est un ptyx et que $\|\Phi_\phi(\Xi_1, \alpha)\| \geqslant \|R_{\Xi_1}(\alpha)\|$

[11] Pour un ordre quelconque fixé de type ω sur Σ.

$\geqslant f(\alpha)$, pour tout α qui est L-descriptible, achevant la preuve du Théorème 2(b).

PREUVE DU THÉORÈME 1(b). Soient $\mathscr{L}$ le langage ci-dessus, T la théorie $KP + \forall e \in \omega[\|F_e(C)\| \notin On$ ou il existe un $\mathscr{L}_1$-isomorphisme de F_e^* dans $\mathscr{F}]$

$$+\exists\alpha\big[\alpha \leqslant C < \alpha^+ \text{ et } \forall\gamma < \alpha\, \exists e\big[\|F_e(\gamma)\| \in On \text{ et } \|F_e(\alpha)\| \notin On\big]\big].$$

On vérifie aisément:

FAIT 1. *Si $\mathfrak{M}$ est un (F, γ)-modèle de T, $F \in PT^1$ et $\gamma \in On$, alors $\gamma \in [\sigma_0, \sigma_0^+[$.*

Soit f une fonction σ_0^+-récursive; soit $\phi = \phi(\mathscr{F}, C)$ tel que

$$T \vdash \phi \leftrightarrow \neg\|\Phi_\phi(\mathscr{F}, C)\| < f(C),$$

où Φ_ϕ est le foncteur p.r. associé à ϕ par 6.5.bis.

FAIT 2. $T \vdash_{PT^1, ON} \phi$.

PREUVE. Comme la preuve du Théorème 0′ dans [**R1**], en utilisant le Fait 1.

Par le Fait 2, Φ_ϕ est un ptyx; de plus si $\gamma \in [\sigma_0, \sigma_0^+[$, considérons V comme un (Ξ_1, γ)-modèle, alors il satisfait T, donc ϕ, donc

$$V \vDash \neg\|\Phi_\phi(\Xi_1, \gamma)\| < f(\gamma).$$

Comme $\Phi_\phi(\Xi_1, \gamma) \in WO$ par le Fait 2, cela entraîne

$$\|\Phi_\phi(\Xi_1, \gamma)\| \geqslant f(\gamma), \quad \text{c.q.f.d.}$$

PREUVE DU THÉORÈME 1(a). C'est conséquence immédiate du Théorème 1(b).

Ayant démontré les résultats principaux, il nous reste à noter qu'il est facile d'en déduire des résultats plus fins, où les ordinaux sont *atteints* par des ptykes p.r. au lieu d'être seulement majorés. Comme premier exemple, nous considérons $\alpha < \omega_1^{\text{ck}}$; le Théorème 0 nous donne un ptyx p.r. F: $ON \to ON$ tel que $\alpha < F(\omega)$. Nous allons en tirer un ptyx p.r. G tel que $\alpha = G(\omega)$[12]: l'ordre $F[\omega]$ étant de type $> \alpha$, il contient un terme $t[\bar{a}](t \in |F|$ et $\bar{a} \in \omega)$ tel que $F[\omega] \upharpoonright t[\bar{a}]$ est de type α—où $F[\omega] \upharpoonright t[\bar{a}]$ est la restriction de l'ordre $F[\omega]$ aux termes qui précèdent strictement $t[\bar{a}]$. On fixe alors $n_0 \in \omega$ tel que $\bar{a} \in n_0$; l'application qui à tout γ fait correspondre $n_0 + \gamma$, et à tout morphisme f: $\gamma \to \gamma'$ fait correspondre le morphisme $\text{id}_{n_0} + f$: $n_0 + \gamma \to n_0 + \gamma'$, est un foncteur de ON dans ON. En le composant avec F, on obtient un foncteur de ON dans ON qu'il est naturel de noter $F \circ (n_0 + \text{Id})$. Comme $\bar{a} \in n_0$, le terme $t[\bar{a}]$ appartient à $F \circ (n_0 + \text{Id})[\gamma]$ pour tout γ—c'est à cela que sert l'addition de n_0, et cela nous permet de définir un foncteur G par

$$G[\gamma] = F[n_0 + \gamma] \upharpoonright t[\bar{a}],$$

[12] Ce renforcement du Théorème 0 a déjà été montré dans [**G**].

et si f est un morphisme de γ dans γ',

$$G[f] = \text{restriction } \grave{a}\ G[\gamma] \text{ de } F\left[\mathrm{id}_{n_0} + \gamma\right].$$

Autrement dit, G est obtenu en tronquant à $t[\bar{a}]$ le foncteur $F \circ (n_0 + \mathrm{Id})$. De ce que F est un ptyx p.r. résulte aisément que G en est un; de plus, soit f_0 l'isomorphisme de $n_0 + \omega$ sur ω. Il laisse ā fixe, donc $F[f_0]$ est un isomorphisme de $F[n_0 + \omega]$ sur $F[\omega]$ qui laisse $t[\bar{a}]$ fixe (car il envoie $t[\bar{a}]$ sur $t[f_0\bar{a}] = t[\bar{a}]$). Donc la restriction de $F[f_0]$ est un isomorphisme de $F[n_0 + \omega] \restriction t[\bar{a}]$ sur $F[\omega] \restriction t[\bar{a}]$; ainsi

$$\alpha \simeq F[\omega] \restriction t[\bar{a}] \simeq F[n_0 + \omega] \restriction t[\bar{a}] = G(\omega).$$

Et tout ordinal $\alpha < \omega_1^{\mathrm{ck}}$ est atteint de cette façon, ce qui renforce le Théorème 0.

On peut renforcer de la même façon le Théorème 2(b): soit $\alpha < \sigma_0$; par 2(b), il existe un ptyx p.r. Φ: $PT^1 \to ON$ tel que $\alpha < \Phi(\Xi_1)$. L'ordre $\Phi[\Xi_1]$ étant de type $> \alpha$, il contient un terme $t[\bar{a}]$ ($t \in |\Phi|$, $\bar{a} \in |\Xi_1|$) tel que $\Phi[\Xi_1] \restriction t[\bar{a}]$ est de type α. Ci-dessus, nous avons considéré $F_0(n_0 + \mathrm{Id}) \restriction t[\bar{a}]$, où n_0 était tel que $\bar{a} \in n_0$; ici nous allons de même considérer $\Phi \circ (A_0 + \mathrm{Id})$, où A_0 est un "segment initial" de Ξ_1 tel que $\bar{a} \in A_0$: puisque $\Xi_1 = \sum_{e \in \omega}\{F_e\colon F_e \in PT^1\}$ et $\bar{a} \in |\Xi_1|$, il existe $e_0 \in \omega$ tel que si A_0 désigne le ptyx $\sum_{e \leqslant e_0}\{F_e;\ F_e \in PT^1\}$, alors $\bar{a} \in |A_0|$ (moyennant l'identification canonique de A_0 avec un segment initial de Ξ_1). Cela étant, le foncteur $\Phi \circ (A_0 + \mathrm{Id}) \restriction t[\bar{a}]$ est défini, nous le notons $\psi[\cdot]$.

FAIT. *Il existe des applications* f_0: $|\Xi_1| \to |A_0 + \Xi_1|$ *et* g_0: $|A_0 + \Xi_1| \to |\Xi_1|$ *telles que*:

f_0 *et* g_0 *sont l'identité sur* A_0, *donc sur* $\bar{a}$;

dans la catégorie PT^1, f_0 *est un morphisme de* Ξ_1 *dans* $A_0 + \Xi_1$, *et* g_0 *en est un de* $A_0 + \Xi_1$ *dans* Ξ_1.

(L'application f_0 est triviale, et g_0 est facile à construire, en utilisant $\Xi_1 = \sum_{e \in \omega}\{F_e\colon F_e \in PT^1\}$ et le fait que $\forall$ e il existe une infinité d'entiers e′ tels que $F_e \simeq F_{e'}$.)

De ce fait, $\Phi(f_0)$ est un isomorphisme de $\Phi[\Xi_1]$ dans $\Phi[A_0 + \Xi_1]$, qui envoie $t[\bar{a}]$ sur $t[f_0\bar{a}] = t[\bar{a}]$; donc la restriction de $\Phi(f_0)$ envoie $\Phi[\Xi_1] \restriction t[\bar{a}]$ dans $\Phi[A_0 + \Xi_1] \restriction t[\bar{a}]$. Ainsi $\alpha = \|\Phi[\Xi_1] \restriction t[\bar{a}]\| \leqslant \|\Phi[A_0 + \Xi_1] \restriction t[\bar{a}]\| = \|\psi[\Xi_1]\|$. En utilisant pareillement g_0 au lieu de f_0, on a $\alpha \geqslant \|\psi[\Xi_1]\|$, donc $\alpha = \|\psi[\Xi_1]\|$. De plus, comme Φ et A_0 sont des ptykes p.r., on voit aisément que ψ est un ptyx p.r.-définissable. Il n'est pas faiblement fini, mais, *soit* en appliquant 6.10 à ψ, *soit* en refaisant la construction ci-dessus avec $A_0 \restriction \bar{a}$ au lieu de A_0 on parvient à un ptyx p.r. ψ' tel que $\alpha = \psi'(\Xi_1)$.

7.1. COROLLAIRE. (a) *Pour tout* $\alpha < \sigma_0$, *il existe un ptyx p.r.* ψ: $PT^1 \to ON$ *tel que* $\alpha = \psi(\Xi_1)$.

(b) *Pour tout ordinal* α *qui est* Π^1_{n+1}, *il existe un ptyx* ψ: $PT^n \to ON$ *tel que* $\alpha = \psi(\Xi_n)$.

La preuve de (a) est faite, celle de (b) est toute semblable.

Bibliographie

[AH] F. Abramson et L. Harrington, *Models without indiscernibles*, J. Symbolic Logic **43** (1978), 572–600.

[AR] P. Aczel et W. Richter, *Inductive definitions and reflecting properties of ordinals*, Generalized Recursion Theory (J. Fenstadt et P. Hinman, eds.), North-Holland, Amsterdam, 1974.

[B] J. Barwise, *Admissibles sets and structures*, Springer-Verlag, Berlin and New York, 1974.

[G1] J. Y. Girard, Π_2^1 *logic*. part 1; Ann. Math. Logic **21** (1981), 75–271.

[G2] ______, *Proof theory and logical complexity*, Chapitres 8 à 12; à paraître aux éditions Bibliopolis à Napoli.

[G3] ______, *A survey of* Π_2^1 *logic*, Proc. 6th Internat. Congr. of Logic, Methodology... (Hannover, 1979), North-Holland, Amsterdam, 1982.

[GN] J. Y. Girard et D. Normann, *Set recursion and* Π_2^1 *logic*, Ann Pure Appl. Logic (to appear).

[GV1] J. Y. Girard et J. Vauzeilles, *A functorial construction of the Veblen hierarchy*, J. Symbolic Logic **49** (1984)

[GV2] ______, *A functorial construction of the Bachmann hierarchy*, J. Symbolic Logic (submitted).

[GV3] ______, *Les premiers récursivement inaccessibles et Mahlo et la théorie des dilatateurs*, Ardi. Math. Logik Grundlag. **24**.

[H] P. Hinman, *Recusion theoretic hierarchies*, Springer-Verlag, Berlin and New York, 1975.

[M] M. Masseron, *Majoration des fonctions* ω_1^{CK}*-récursives par des* ω*-échelles primitives récursives*, Thèse de 3° cycle, Université Paris-Nord, 1980.

[N] S. N'diaye, *Thèse de* 3° *cycle*, Université Paris VII, 1983.

[R1] J. P. Ressayre, *Bounding generalized recursive functions by effective functors*, Proc. Herbrand Sympos., Marseille, 1981, (J. Stern, ed.), North-Holland, Amsterdam, 1982.

[R2] ______, *Review of papers of Abramson-Harrington, F. Knight et H. Gaifman*, J. Symbolic Logic **47** (1982), 484.

[VW] ______, *Recursive dilators and generalized recursions*, Proc. Herbrand Sympos., Marseille, 1981 (J. Stern, ed.) North-Holland, Amsterdam, 1982.

[V] ______, *Dilators and gardens*, Proc. Herbrand Sympos., Marseille, 1981 (J. Stern, ed.) North-Holland, Amsterdam, 1982.

Proceedings of Symposia in Pure Mathematics
Volume **42**, 1985

Paris-Harrington Incompleteness and Progressions of Theories

KENNETH MCALOON[1]

Abstract. We introduce finite Ramsey theorems based on partition properties involving ordinals. Paris-Harrington incompleteness theorems are then given for corresponding theories in a progression obtained from Peano arithmetic by iterating complete reflection or equivalent notions. The recursive functions that the Ramsey theorems give rise to are shown to be fast-growing and are located in the Grzegorcyck-Wainer subrecursive hierarchy. Also, we give a first order "combinatorial" axiomatization of the first order consequences of ATR_0 and also of $ACA_0 + RT$, when RT is the infinite Ramsey theorem. The proofs use techniques of models of arithmetic and of proof theory.

The main result of this paper can be paraphrased as saying that there is Paris-Harrington incompleteness for an axiomatic theory of first- or second-order arithmetic once a proof-theoretic ordinal analysis of it exists. That is, we show that there are finite Ramsey Theorems, which imply over Peano arithmetic (PA) the consistency of PA plus the scheme of transfinite induction on given ordinals, cf. Theorem A and Corollary 3 below. These results extend and simplify the work in our [**Mc 3**].

We write $P\colon [X]^n \to k$ to signify that P maps the n-element subsets of X into a set of cardinality k; P is *homogeneous* on $H \subseteq X$ if P is constant on $[H]^n$. We denote the cardinality of a finite set X by $|X|$; if X is a set of positive integers, $\min X$ denotes the least element of X and $\min^2 X$ denotes the (min X)th element of X.

By recursion on α we define the partition relation $X \to^{\alpha} (m)^n_k$ for finite sets of positive integers X and numbers m, n, k ($n, k \geqslant 1$; $m \geqslant n + 1$) and countable α as follows: For $\alpha = 0$, the relation is the same as $\to_*$ of [P, H]; *viz.* $X \to^0 (m)^n_k$ $\Leftrightarrow$ for every partition $P\colon [X]^n \to k$, there is $H \subseteq X$ such that H is homogeneous

1980 *Mathematics Subject Classification*. Primary 03F30; Secondary 03C62, 03D20, 03E35.
[1] Research partially supported by NSF grant MCS 81-02859.

for P, $|H| \geqslant m$, and $|H| \geqslant \min H$. For successor ordinals $\alpha + 1$,

$$X \overset{\alpha+1}{\to} (m)^n_k \Leftrightarrow \text{ for every partition } P\colon [X]^n \to k \text{ there is } H \subseteq X$$
$$\text{such that } H \text{ is homogeneous for } P,\ |H| \geqslant m,$$
$$\text{and } H \overset{\alpha}{\to} (\min^2 H + 1)^{\min^2 H}_{\min^2 H}.$$

For limit ordinals λ with chosen fundamental sequence $\lambda(n) \nearrow \lambda$,

$$X \overset{\lambda}{\to} (m)^n_k \Leftrightarrow \text{ for every partition } P\colon [X]^n \to k \text{ there is } H \subseteq X$$
$$\text{such that } H \text{ is homogeneous for } P,\ |H| \geqslant m$$
$$\text{and } H \overset{\lambda(n)}{\to} (\min^2 H + 1)^{\min^2 H}_{\min^2 H}.$$

Using Koenig's Lemma and the infinite form of Ramsey's Theorem, as in [**Ac**], one proves

For every infinite set of natural numbers $\overline{X}$, for all m, n, k and for every countable α, there is a finite subset $X \subseteq \overline{X}$ such that $X \to^{\alpha} (m)^n_k$.

Define $\sigma_\alpha(n)$ to be the least integer p such that $[1, p] \to^{\alpha} (n + 1)^n_n$, where $[1, p] = \{1, \ldots, p\}$. We shall show, roughly speaking, that "ε_α induction is necessary and sufficient to prove σ_α total."

Throughout this paper Δ denotes an ordinal described by a primitive recursive ordering $<$ on the natural numbers. For this ordering we assume that the ordinal arithmetic operations of successor, addition and multiplication are primitive recursive as well as exponentiation to base ω, $\alpha \mapsto \omega^\alpha$, and the ε-function $\alpha \mapsto \varepsilon_\alpha$. Also, let $(\alpha, n) \mapsto \alpha(n)$ be a primitive recursive function of two variables which associates with each α a fundamental sequence $\alpha(n)$, $n \in \mathbf{N}$; by this we mean that if α is a successor ordinal, we have $\alpha(n) = \alpha - 1$ for all n and if α is a limit ordinal, then $\alpha(n)$ is an increasing sequence which converges to α. The usual arithmetic properties of these functions and operations are assumed provable in PA.

To lighten notation we write $\alpha(n_1)(n_2)$ for $(\alpha(n_1))(n_2)$, $\alpha(n_1)(n_2)(n_3)$ for $(\alpha(n_1))(n_2)(n_3)$, etc. A further property of the ordering Δ which we assume, but which we do *not* require to be provable in PA, is the following: if $m < n$ and if $\alpha(n)(n_1) \cdots (n_k) \leqslant \alpha(m)$ for some sequence $n_1, \ldots, n_k$, then for some $j \leqslant k$, we have $\alpha(n)(n_1) \cdots (n_j) = \alpha(m)$. This property is a standard one for systems of notations and forms part of the definition of a Bachman collection *cf.* [**B**] for example.

By $[TI]_\alpha$ we mean the theory obtained by adjoining the scheme of arithmetic transfinite induction on α to PA:

$$\forall x_1 \cdots \forall x_n \big[\forall \gamma < \alpha \big[\forall \delta < \gamma \Phi(\delta, x_1, \ldots, x_n) \to \Phi(\gamma, x_1, \ldots, x_n)\big] \to \forall \gamma < \alpha \Phi(\gamma, x_1, \ldots, x_n)\big], \qquad \Phi \text{ a formula.}$$

We can now give a precise statement of our main results.

THEOREM A. *For* $\alpha < \Delta$, *the function* σ_α *is not a provably recursive function of* $[TI]_\beta$ *for* $\beta < \varepsilon_\alpha$; *in fact* σ_α *eventually dominates every provably recursive function of* $[TI]_\beta$ *for* $\beta < \varepsilon_\alpha$. *However,* σ_α *is provably total in* $PA + [TI]\varepsilon_\alpha$.

The next result applies to Δ for which the Grzegorcyck-Wainer hierarchy has been developed [**W**].

COROLLARY A. *For* $\alpha < \Delta$, *the function* σ_α *eventually dominates all functions which appear in Grzegorcyck-Wainer hierarchy below level* ε_α, *but* σ_α *is primitive recursive in a function appearing below level* $\varepsilon_{\alpha+1}$.

REMARK. It would be interesting to develop proofs of these results with the "direct" method of α-large sets of Ketonen-Solovay [**K, S**].

The following theorem connects up with work of Jager [**Ja**] and of Schlipf [**S**]. It summarizes a sequence of results that appear toward the end of this paper. The system ACA_0 is basically PA formulated as a second order theory, cf. [**F, M, S**] or [**F**].

THEOREM B. *Let* RT *be the theory of second order arithmetic obtained by adjoining the infinite Ramsey Theorem to* ACA_0. *Then* RT *is also axiomatized by adjoining the axiom* $\forall X \forall n[X^{(n)}$, *the nth Turing jump of X, exists*]. *Moreover, the first order consequences of* RT *are the same as those of* $\bigcup_n Q^n$ *where* Q^n *is the theory obtained by iterating uniform reflection n-times over PA. Finally, the proof theoretic ordinal of* RT *is* ε_ω.

REMARK. In Theorem 5 below, we also give a first order axiomatization of the first-order consequences of the system ATR_0.

The proofs of Theorem A and of Corollary A use proof-theoretic and model-theoretic techniques. An alternative definition of $\to^\alpha$ is given which incorporates certain properties of the partition relation whose proofs require induction on α. The alternative notion, written $\to_\alpha$, yields a relation which is Π_1^0 in PA. Ultimately, we will show that the relation $[a, b] \to_\alpha (2c)_c^c$ gives an indicator in the sense of Kirby-Paris for models of Q^α, the theory obtained by iterating uniform reflection α-times over PA. This in turn involves giving an alternative axiomatization of Q^α in terms of $\to_\alpha$; *viz.*, Q^α will be shown equivalent to the theory P^α obtained by adjoining to PA the scheme to the effect that every infinite arithmetic set $\underline{X}$ has, for all m, n, k, finite subsets X such that $X \to_\alpha (m)_k^n$. The proof theory and the model theory are brought together in a uniform way using the elegant methods of U. Schmerl [**Sch**], which allow us to reduce things to simple "local" considerations in order to obtain results involving progressions of theories. Moreover, to come full circle, we use Schmerl's result that Q^α is equivalent to $[TI]_\alpha$, which extends a classical result of Kreisel-Levy [**K, L**].

For finite sets of positive natural numbers K, H we write $K \geqslant H$ to mean $|K| \geqslant |H|$ and the ith element of $K \leqslant$ the ith element of H for all i, $1 \leqslant i \leqslant |H|$. Let Tr_1 be a truth-predicate for Π_1^0-sentences in PA.

For successor $\alpha < \Delta$, we write $\alpha(n) = \alpha - 1$; we also consider -1 an ordinal and set $0(n) = -1$ for all n. We use the following convention: if $\ulcorner\phi\urcorner$ is the Gödel number of a formula of the language of arithmetic and if $a_1,\dots,a_n$ are natural numbers, then $\ulcorner\phi\urcorner(a_1,\dots,a_n)$ is the Gödel number of the formula $\phi(\bar{a}_1,\dots,\bar{a}_n)$ obtained by substituting terms $\bar{a}_1,\dots,\bar{a}_n$ denoting $a_1,\dots,a_n$ for the first n free variables of ϕ. In the definition that follows the quantifier $\exists R \subseteq K$ is an abbreviation for equivalent universal and bounded quantifications.

We use the Kleene Primitive Recursion Theorem [**K**] to find primitive recursive F such that for $\beta < \Delta$, $F(\beta)$ is the Gödel number of a Π^0_1-formula with free variables X, m, n, k such that

$$\begin{aligned} F(\alpha+1) = {}&\ulcorner \forall K \geqslant X\, \forall P\colon [K]^n \to k\, \exists R\; R \subseteq K,\, |R| \geqslant m \\ &\& \, Tr_1(F(\alpha)(R, \min^2 R + 1, \min^2 R, \min^2 R))\urcorner, \\ F(\lambda) = {}&\ulcorner \forall K \geqslant X\, \forall P\colon [K]^n \to k\, \exists R\; R \subseteq K,\, |R| \geqslant m \\ &\& \forall\, \forall s \leqslant n\, Tr_1(F(\lambda(s))(R, \min^2 R + 1, \min^2 R, \min^2 R))\urcorner, \\ F(-1) = {}&\ulcorner |X| \geqslant \min X \geqslant \bar{1}\urcorner. \end{aligned}$$

Now set

$$X \underset{\alpha}{\to} (m)^n_k \Leftrightarrow Tr_1(F(\alpha)(X, m, n, k)) \wedge k, n \geqslant 1 \wedge m \geqslant n + 1.$$

We have by the material adequacy of the truth-definition, $X \to_{\alpha+1} (m)^n_k \Leftrightarrow$ for every $K \geqslant X$, for every $P\colon [K]^n \to k$, there is $H \subseteq K$ such that

$$H \to_\alpha (\min^2 H + 1)^{\min^2 H}_{\min^2 H}$$

and we have $X \to_\lambda (m)^n_k \Leftrightarrow$ for every $K \geqslant X$, for every $P\colon [K]^n \to k$ there is $H \subseteq K$ such that

$$H \underset{\substack{\lambda(n') \\ n' \leqslant n}}{\to} (\min^2 H + 1)^{\min^2 H}_{\min^2 H}.$$

REMARK. It is straightforward to show, by induction on α, that $\to_\alpha$ and $\to^\alpha$ coincide. However, we hold this point in abeyance for the time being and concentrate on $\to_\alpha$ and properties that can be proved in PA.

The following four lemmas can be proved in PA.

LEMMA 1. *For $\alpha < \Delta$, we have*

$$X \underset{\alpha}{\to} (m)^n_k \Leftrightarrow \forall K \geqslant X \forall P\colon [K]^n \to k\, \exists H \subseteq K$$
$$\left(H \text{ homogeneous for } P,\ |H| \geqslant m \;\&\; H \underset{\substack{\alpha(n') \\ n' \leqslant n}}{\to} (\min^2 H + 1)^{\min^2 H}_{\min^2 H} \right).$$

PROOF. The equivalence follows from the material adequacy of the truth-definition Tr_1 and the remark that for successor α, we have $\alpha(n') = \alpha(n)$ for all $n' \leqslant n$.

LEMMA 2. *The relation $X \to_\alpha (m)^n_k$ is Π^0_1 in the variables X, α, m, n, k.*

PROOF. Immediate.

LEMMA 3. *For* $\alpha < \Delta$, $K \geqslant X \,\&\, X \to_{\alpha} (m)^n_k \Rightarrow K \to_{\alpha} (m)^n_k$; *thus in particular*, $K \supseteq X$ *and* $X \to_{\alpha} (m)^n_k \Rightarrow K \to_{\alpha} (m)^n_k$.

PROOF. Immediate.

LEMMA 4. *For* $\alpha < \Delta$, $X \to_{\alpha} (m)^n_k$, $n' + 1 \leqslant m' \leqslant m$, $1 \leqslant k' \leqslant k \,\&\, 1 \leqslant n' \leqslant n$ $\Rightarrow X \to_{\alpha} (m')^{n'}_{k'}$.

PROOF. For m' and k', the result is immediate. For n', we observe that the definition of $X \to_{\alpha} (m)^n_k$ requires the homogeneous sets H to satisfy

$$H \underset{\substack{\alpha(n') \\ n' \leqslant n}}{\to} (\min^2 H + 1)^{\min^2 H}_{\min^2 H}.$$

We define $P^{(-1)} = \mathrm{PA}$ and for $0 \leqslant \gamma < \Delta$, we set P^{γ} to be PA augmented by the following scheme:

$$\forall x_1 \cdots \forall x_n \Big[\forall x \exists y (y > x \,\&\, \Phi(y, x_1, \ldots, x_n)) \\ \to \forall m, k, l \exists X (\forall y (y \in X \to \Phi(y, x_1, \ldots, x_n))) \,\&\, X \underset{y}{\to} (m)^k_l \Big].$$

Thus P^{γ} is PA together with an axiom scheme expressing that for every infinite arithmetic $\overline{\mathrm{X}}$ and every m, n, l, there is a finite subset $X \subseteq \overline{\mathrm{X}}$ satisfying $X \to_{\gamma} (m)^k_l$.

We shall need the following definition: let $G\colon \mathbf{N} \to \mathbf{N}$ be a total strictly increasing function; then we write

$$X \overset{G}{\underset{\gamma}{\to}} (m)^n_k \Leftrightarrow \text{for every } P\colon [X]^n \to k \text{ there is } H \subseteq X, \\ H \text{ homogeneous for } P,\ |H| \geqslant m,\ |H| \geqslant G(\min H), \\ \text{and } H \underset{\substack{\gamma(n') \\ n' \leqslant n}}{\to} (2 \min^2 H)^{\min^2 H}_{\min^2 H}.$$

Now define $\overline{P}^{(-1)} = \mathrm{PA}$ and for $0 \leqslant \gamma < \Delta$, set $\overline{P}^{\gamma}$ to be PA augmented by the scheme

$$\forall x_1 \cdots \forall x_n \Big[\forall x \exists ! y \Phi(x, y, x_1, \ldots, x_n) \\ \to \forall m, k, l \exists p \big(\forall P\colon [1, p]^k \to l, \\ \exists H \subseteq [1, p], H \text{ homogeneous for } P, |H| \geqslant m, \\ \forall y \Phi(\min H, y, x_1, \ldots, x_n) \to y \leqslant |H| \big) \\ \,\&\, H \underset{\substack{\gamma(k') \\ k' \leqslant k}}{\to} (2 \min^2 H)^{\min^2 H}_{\min^2 H} \Big].$$

Thus $\overline{P}^{\gamma}$ is PA augmented by the scheme which expresses that for every arithmetic $G\colon \mathbf{N} \to \mathbf{N}$, for all m, k, l, there is p such that

$$[1, p] \underset{\gamma}{\overset{G}{\to}} (m)_l^k .$$

LEMMA 5. *For* $\gamma < \Delta$, $P^{\gamma} \vdash \overline{P}^{\gamma}$.

PROOF. This is a straightforward argument. Given $G\colon \mathbf{N} \to \mathbf{N}$ arithmetic, we can suppose G is strictly increasing. Let $\overline{\mathrm{X}} = \{G(1), G(2), G(3), \ldots\}$. Then $\overline{\mathrm{X}}$ is an infinite arithmetic set. So for integers m, k, l there is a finite subset X of $\overline{\mathrm{X}}$ which satisfies $X \to_{\gamma} (m)_l^k$. Let $X = \{x_1, \ldots, x_p\}_<$. We claim $[1, p] \to^G (m)_l^k$; in fact, if $P\colon [1, p]^k \to l$, then P induces a partition $\tilde{P}$ on X and there is a homogeneous set $\tilde{Y}$ for $\tilde{P}$ satisfying

$$\tilde{Y} \underset{\substack{\gamma(m') \\ m' \leqslant m}}{\to} (2 \min^2 \tilde{Y})^{\min^2 \tilde{Y}}_{\min^2 \tilde{Y}}.$$

Now $\tilde{Y}$ corresponds to a set $Y \subseteq [1, p]$ and we have $\min \tilde{Y} \geqslant G(\min Y)$; thus $|Y| \geqslant G(\min Y)$ and since $Y \geqslant \tilde{Y}$, we have

$$Y \underset{\substack{\gamma(m') \\ m' \leqslant m}}{\to} (2 \min^2 \tilde{Y})^{\min^2 \tilde{Y}}_{\min^2 \tilde{Y}}.$$

The result now follows from Lemma 4 since $\min^2 \tilde{Y} \geqslant \min^2 Y$.

For $\gamma < \Delta$, by $\bigcup_n P^{\gamma(n)}$ we mean the theory axiomatized by the union of the axiom sets $P^{\gamma(n)}$ for n an integer. Note that for $\gamma = \alpha + 1$, we have $\bigcup_n P^{\gamma(n)} = P^{\alpha}$.

By $RFN(\bigcup_n P^{\gamma(n)})$ we mean the axiom scheme of complete reflection over $\bigcup_n P^{\gamma(n)}$; cf. [**Sch**] for example. More precisely, let Tr_k be a truth definition for Π_k^0-sentences, $k \geqslant 1$. Then $RFN(\bigcup_n P^{\gamma(n)})$ is the scheme

$$(*) \qquad \forall \phi \Big(Tr_k(\phi) \to \phi \text{ is consistent with } \bigcup_n P^{\gamma(n)} \Big), \quad \text{all } k \in \mathbf{N}.$$

An equivalent scheme over PA is

$$\forall \phi \Big(\phi \in \pi_k^0 \;\&\; \bigcup_n P^{\gamma(n)} \vdash \phi \to Tr_k(\phi) \Big), \quad \text{all } k \in N,$$

as is the following one which is used in [**Fef**]:

$$\forall x_1 \cdots \forall x_s \Big(\bigcup_n P^{\gamma(n)} \vdash \overline{\Phi}(\tilde{x}_1, \ldots, \tilde{x}_s) \to \Phi(x_1, \ldots, x_s) \Big), \quad \text{all formulas } \Phi.$$

We identify the set of axioms for a theory with the theory itself and we write + to indicate the union of two sets of axioms or the adjunction of a formula to a set of axioms.

LEMMA 6. For $\gamma < \Delta$, $\overline{P}^{\gamma} \vdash \mathrm{PA} + RFN(\bigcup_n P^{\gamma(n)})$.

PROOF. We use model theoretic methods following [**Mc 3**]. We fix $k \in \mathbf{N}$ and work informally in a model M of $\overline{P}^{\gamma}$. We will construct a Π_{k+2}^0-elementary end-extension M' of M which has truth-definition which is Δ^0_{k+4} in M and which is a model of $P^{\gamma(n)}$ for all $n \in M$. Thus M will satisfy scheme $(*)$ above.

Let $F(x)$ be a Σ^0_k-Skolem function, i.e. F is a Δ^0_{k+1}-definable function with the property that every initial segment, of a model of PA, which is closed under F is a Σ^0_k-elementary subsystem. As in [**Mc 3**], let us use the arithmetized form of the Henkin Completeness Theorem, and let M' be a Π^0_{k+2}-elementary end extension of M which has a Δ^0_{k+4} truth definition in M (such M' exists because PA is reflexive in the sense that $\mathrm{PA} \vdash \forall\phi(\phi \in \Pi^0_{k+2} \,\&\, Tr_{k+2}(\phi) \to \phi \text{ is consistent})$). We now write $M <_{k+2} M'$ and, for $l \in M' - M$ we write $l > M$. Set $F'(x) = \sup_{y \leqslant x} F(x) + x + 1$. Since $M' >_{k+1} M$, there are l, $c > M$ such that $M' \vDash [1, l] \underset{\gamma}{\overset{F'}{\to}} (2c)^c_{3c}$. Now we claim there is $\overline{H} \subseteq [1, l]$ such that

$$M' \vDash \overline{H} = \{h_1, h_2, \ldots, h_{r+c}\}_<,$$

$$M' \vDash \overline{H} \underset{\substack{\gamma(d)\\ d \leqslant c}}{\to} (2h_{h_1})^{h_{h_1}}_{h_{h_1}},$$

$$M' \vDash F(h_1) < h_r,$$

and such that the elements of $H = \{h_1, \ldots, h_r\}$ form a collection of indiscernables for Σ^0_{k+1}-formulas. To find $\overline{H}$ and H, we construct a partition P: $[1, l] \to (2c)^c_{3c}$ in M'. Let $\phi_0, \phi_1, \ldots$ be an enumeration of the Σ^0_{k+1}-formulas of the language of arithmetic and let $e > M$ satisfy $M' \vDash 2^e < c$. Define a map f from $[1, e] \times [1, l]^c$ to $\{0, 1\}$ by $f(i, \langle x_1, \ldots, x_c\rangle) = 1$ if $Tr_{k+1}(\phi_i(\bar{x}_1, \ldots, \bar{x}_c))$, 0 otherwise. Note that for $i \in M$, if ϕ_i has s free variables, we have for all $x_1 < \cdots < x_s \leqslant l$, and any sequence $\langle x_{s+1}, \ldots, x_c\rangle \in M'$,

$$\begin{aligned} M' &\vDash f(i, \langle x_1, \ldots, x_s, x_{s+1}, \ldots, x_c\rangle) \\ &= 1 \Leftrightarrow M' \vDash \phi_i(x_1, \ldots, x_s). \end{aligned}$$

Now let $g \in M'$ be an injection of 2^e into $[1, c]$ and define P by

$$P(x_1, \ldots, x_c) = \begin{cases} x_1 & \text{if } x_1 \leqslant 2c, \\ 2c + g(f(0, \langle x_1, \ldots, x_c\rangle), f(1, \langle x_1, \ldots, x_c\rangle), \ldots, \\ \qquad\qquad f(e, \langle x_1, \ldots, x_c\rangle)). \end{cases}$$

Next let $\overline{H} \in M'$ be homogeneous for P and such that $\overline{H} = \{h_1, \ldots, h_{r+c}\}_<$ and

$$M' \vDash \overline{H} \underset{\substack{\gamma(d)\\ d \leqslant c}}{\to} (2\min^2 H)^{\min^2 H}_{\min^2 H},$$

$$M' \vDash |\overline{H}| \geqslant F'(h_1).$$

Since $2c < h_1$, we have $\gamma + c > 2c$; thus every sequence $h_{i_1}, \ldots, h_{i_s}$ with $s \in M$ and $i_s \leqslant r$ can be extended in M' to a sequence of elements of $\overline{H}$ of length c. Hence the elements of H are indiscernible in M' for Σ^0_{k+1}-formulas (of M). Finally since $\sup_{x \leqslant h_1} F(x) + h_1 < r + c$, we have $F(h_1) < h_r$ and so by indiscernability $F(h_i) < h_j$ for all $1 \leqslant i < j \leqslant r$.

By $\Delta_0^0(\Pi_1^0)$-formulas, we mean those in the set obtained by taking the closure of the Π_1^0-formulas under bounded quantification and boolean connectives. In M', the $\Delta_0^0(\Pi_1^0)$-formulas of M are equivalent to Σ_2^0-formulas and so H forms a set of indiscernibles for these formulas. Now, by using the fact that $|\overline{H}| \geq \min \overline{H}$, one can repeat the argument of **[P]**, **[Las]** or **[Mc 1]** to show the elements of H are in fact strong or diagonal indiscernibles for $\Delta_0^0(\Pi_1^0)$-formulas in the following sense: for any $\Delta_0^0(\Pi_1^0)$-formula $\Phi(u_1,\ldots,u_t, w, w_1,\ldots,w_s)$ (of M), we have, for all $h_{i_0} < h_{i_1} < \cdots < h_{i_t}, h_{i_0} < h_{j_1} < \cdots < h_{j_t}$ and for all $a_1,\ldots,a_t \leq h_{i_0}$,

$$M' \models \Phi(a_1,\ldots,a_t, h_{i_0}, h_{i_1},\ldots,h_{i_t}) \Leftrightarrow \Phi(a_1,\ldots,a_t, h_{i_0}, h_{j_1},\ldots,h_{j_t}).$$

We set $I = \{x \in M' \colon \exists n \in M,\ x < h_n\}$. Now $M < I < M'$ and since I is closed under the Skolem function F, $M <_k I <_k M'$. We claim that I has an arithmetic truth definition; in fact, I satisfies the following

TRUTH LEMMA. *For any prenex formula $Q^1x_1 \cdots Q^sx_s\Phi(x_1,\ldots,x_s, u_1,\ldots,u_t)$, we have for all $a_1,\ldots,a_t \leq h_k \in I$,*

$$I \models Q^1x_1 \cdots Q^sx_s\Phi(x_1,\ldots,x_s, a_1,\ldots,a_t)$$
$$\Leftrightarrow M' \models Q^1x_1 < h_{r-(s-1)} \cdots Q^sx_s < h_r\Phi(x_1,\ldots,x_s, a_1,\ldots,a_t).$$

The Truth Lemma follows from the fact that the elements of H are diagonal indiscernibles for Δ_0^0-formulas, *cf.* **[P, Mc 1** or **Las]**. From the Truth Lemma and the fact that M' satisfies Δ_0^0-induction, it follows that truth in I is Δ_{k+4}^0-definable, *cf.* **[Mc 3** or **Mc 4]** and that I is a model of PA, *cf.* **[P, P, H, Las** or **Mc 1]**. A bit further on we will use the fact that the elements of H are strong indiscernables for $\Delta_0^0(\Pi_1^0)$-formulas.

We now must verify the claim that $I \models \bigcup_n P^{\gamma(n)}$ for all $n \in M$. So suppose $a_1,\ldots,a_p \leq h_t$ where $p, t \in M$ and that $\Phi(x, a_1,\ldots,a_p)$ is a formula which defines an unbounded subset $\overline{\mathrm{X}}$ of the model I. We can assume that Φ is in prenex form, say

$$\Phi(x, a_1,\ldots,a_p) \equiv Q^1x_1 \cdots Q^mx_m\Psi(x, a_1,\ldots,a_p, x_1,\ldots,x_m),$$

where $\Psi(x)$ is quantifier-free. Since $\overline{\mathrm{X}}$ is unbounded in I, for some t' in M, $t < t'$, we have

$$(*) \qquad I \models \exists x\big(h_t < x \leq h_{t'+1} \wedge Q^1_{x_1} < h_{r-(m-1)} \cdots Q^m_{x_m} < h_r\Psi(x)\big).$$

And by the indiscernibility, $(*)$ holds for all $t' \in M'$ satisfying $t < t' < r - (m - 1)$. For each $\bar{h} = \{h_{i_1},\ldots,h_{i_{m+1}}\}_<$ with $i_1 > t + 1$, set

$$\overline{\mathrm{X}}(\bar{h}) = \big\{x \colon M' \models h_{t+1} < x \leq h_{i_1} \,\&\, Q^1x_1 < h_{i_2} \cdots Q^mx_m < h_{i_{m+1}}\Psi(x)\big\}.$$

Note that for $i_{m+1} < r$, the code for the set $\overline{\mathrm{X}}(\bar{h})$ is Δ_0^0-definable from $a_1,\ldots,a_p$, $h_{t+1}, h_{i_1},\ldots,h_{i_{m+1}}, h_r$. Set $\overline{\overline{H}} = H - \{h_1,\ldots,h_{t+1}, h_{r-m},\ldots,h_r\}$. By $(*)$ we see that $\overline{\mathrm{X}}(\langle h_{t+2}, h_{r-m},\ldots,h_{r-1}\rangle) \geq \overline{\overline{H}}$ and so by Lemma 3,

$$M' \models \overline{\overline{\mathrm{X}}}(\langle h_{t+2}, h_{r-m},\ldots,h_{r-1}\rangle) \underset{\substack{\gamma(d)\\ d \leq c}}{\rightarrow} (2h_{h_1})^{h_{h_1}}_{h_{h_1}-(m+t+2)}.$$

Note that $h_1 > M$ and $r - m > h_1$; so for all $s \in M$, we have $h_s < h_{h_1} < h_{r-m}$. Now we use the fact that $\to_{\gamma(d)}$ is Π_1^0 and that the elements of H are diagonal indiscernibles for $\Delta_0^0(\Pi_1^0)$-formulas, to find for $t+2 < q < s_1 < \cdots < s_m$ with $s_m \in M$, and for $d = n$ with $n \in M$,

$$M' \models \overline{X}(\langle h_{t+2}, h_{s_1}, \ldots, h_{s_m}\rangle) \underset{\gamma(n)}{\to} (2h_q)^{h_q}_{h_q - (m+t+2)},$$

and

$$I \models \overline{X}(\langle h_{t+2}, h_{s_1}, \ldots, h_{s_m}\rangle) \underset{\gamma(n)}{\to} (2h_q)^{h_q}_{h_q - (m+t+2)}.$$

Also by the Truth Lemma,

$$I \models \overline{X}(\langle h_{t+2}, h_{s_1}, \ldots, h_{s_n}\rangle) \subseteq \overline{X}.$$

Finally, since $q \in M$ can be taken arbitrarily large we have for all $n \in M$

$$I \models \forall x, y, z\, \exists x' \geqslant x\, \exists y' \geqslant y\, \exists z' \geqslant z\, \exists X \subseteq \overline{X}\left(X \underset{\gamma(n)}{\to} (x')^{y'}_{z'} \right).$$

This completes the proof of Lemma 6.

LEMMA 7. *For* $\gamma < \Delta$, $PA + RFN(\bigcup_n P^{\gamma(n)}) \vdash P^{\gamma}$.

PROOF. Here we again work informally in a model M which we suppose satisfies PA + $RFN(\bigcup_n P^{\gamma(n)})$ and we argue by contradiction thus: suppose that $\overline{X}$ is an unbounded arithmetic set definable in M by a Σ_p^0 formula, $p \in \mathbf{N}$, and that $m, k, l \in M$ are such that

$$(*) \qquad M \models \forall Y\left(Y \subseteq \overline{X} \to \neg\left(Y \underset{\gamma}{\to} (m)^k_l \right)\right).$$

So for each $s \in M$,

$$M \models \exists P\Bigg(P\colon [\overline{X} \cap s]^k \to l \,\&\, \forall H \subseteq (\overline{X} \cap s)\Bigg(|H| \geqslant m$$

$$\& H \text{ homogeneous for } P \to \neg\Bigg(H \underset{\substack{\gamma(k') \\ k' \leqslant k}}{\to} (\min{}^2 H + 1)^{\min^2 H}_{\min^2 H} \Bigg)\Bigg)\Bigg),$$

where $\overline{X} \cap s = [1, s] \cap \overline{X}$. By the arithmetic form of Koenig's Lemma, *cf.* [**J**], there is a Σ^0_{p+3} map F such that

$$M \models F\colon [\overline{X}]^k \to l,$$

and

$$(**)\qquad M \models \forall H\Big(H \subseteq \overline{X} \,\&\, H \text{ homogeneous for } F \,\&\, |H| \geqslant m \rightarrow \neg\Big(H \underset{\substack{\gamma(k') \\ k' \leqslant k}}{\rightarrow} (\min^2 H + 1)^{\min^2 H}_{\min^2 H}\Big)\Big)$$

So let M' be a Δ^0_{p+6}-definable Π^0_{p+4}-elementary extension of M which is a model of $P^{\gamma(n)}$, all $n \in M$. Note that $(*)$ and $(**)$ are both satisfied in M' and that, from M, M' is a model of $P^{\gamma(n)}$ for all n. By the arithmetic form of the infinite Ramsey theorem **[J]**, in the model M' there is a Π^0_{p+k+2}-definable infinite subset $Z \subseteq \overline{X}$ which is homogeneous for F. Since $M' \models P^{\gamma(k)}$, there is $Y \in M'$, $Y \subseteq Z$ such that

$$M' \models Y \underset{\substack{\gamma(k') \\ k' \leqslant k}}{\rightarrow} (2z_1)^{z_{z_1}}_{z_{z_1}},$$

where $z_1 = \min(Z - \{0\})$. But then $H = Y \cup \{z_1, \ldots, z_{z_1}\} \cup \{z_1, \ldots, z_m\}$ satisfies

$$M' \models H \subseteq \overline{X},\ |H| \geqslant m \,\&\, H \underset{\substack{\gamma(k') \\ k' \leqslant k}}{\rightarrow} (2\min^2 H)^{\min^2 H}_{\min^2 H}, \&\, H \text{ is homogeneous for } F,$$

which contradicts $(*)$ and $(**)$. We remark that the last part of this proof uses the fact that M' satisfies induction for all formulas in M since the integer k might well be nonstandard.

THEOREM 1. *For $\gamma < \Delta$, P^γ and Q^γ axiomatize the same theory, where $Q^{-1} = PA$ and $Q^\gamma = PA + RFN(\bigcup_n Q^{\gamma(n)})$ for $0 \leqslant \gamma < \Delta$.*

PROOF. If transfinite induction on γ is provable in PA, then this equivalence can be proved by induction on γ in the obvious way using Lemma 5, 6 and 7. To extend this to all $\gamma < \Delta$, one needs the observation that the inductive hypothesis can be weakened from $(P^{\gamma(n)} \equiv Q^{\gamma(n)})$, all n to $\mathrm{Prov}_{\mathrm{PA}}(P^{\gamma(n)} \equiv Q^{\gamma(n)})$, all n. This is what U. Schmerl calls a *reflexive progression* and he shows, using Löb's Theorem, that in this case the induction can be carried out in PA itself. More precisely, in **[Sch]** Schmerl shows that PA implies the scheme

$$\forall \alpha < \Delta\big[\big(\forall \gamma < \alpha\, \mathrm{Prov}_{\mathrm{PA}}(\overline{\Phi}(\dot{\gamma}))\big) \rightarrow \Phi(\alpha)\big] \rightarrow \forall \alpha < \Delta \Phi(\alpha).$$

So the theorem follows from the fact that Lemmas 5, 6 and 7 are provable in PA.

We then have

COROLLARY 1. *For $\alpha < \Delta$, $P^\alpha \equiv TI[\varepsilon_\alpha]$.*

PROOF. In **[Sch]**, it is shown that $Q^\alpha \equiv [TI]_{\varepsilon_\alpha}$.

We set $\tau_\gamma(n) = \mu l([1, l] \rightarrow_\gamma (n+1)^n_n)$. It follows that $P^\gamma \vdash \forall x \exists y\, \tau_\gamma(x) = y$. However

THEOREM 2. *For $\gamma < \Delta$, the map τ_γ eventually dominates all provably recursive functions of $\bigcup_n P^{\gamma(n)}$; hence $\forall x \exists y\, \tau_\gamma(x) = y$ is not a theorem of $\bigcup_n P^{\gamma(n)}$.*

PROOF. Let f be a function which is provably recursive in $\bigcup_n P^{\gamma(n)}$; thus $\bigcup_n P^{\gamma(n)} \vdash \forall x \exists y\ (y = \text{U}(\mu z T(e_0, x, z)))$ where T is the Kleene T-predicate and e_0 is a Gödel number for f. We will write $y = f(x)$ rather than $y = \text{U}(\mu z T(e_0, x, z))$. Supposing that $f(n) > \tau_\gamma(n)$ for infinitely many n, let $M' > \mathbf{N}$ be a countable elementary extension with nonstandard a such that $M \models \tau_\gamma(a) < f(a)$. Now $M' \models [1, \tau_\gamma(a)] \to_\gamma (2 \cdot a/3)_a^{a/3}$, by Lemma 3. So by the argument of the proof of Lemma 6, there is I such that $a/3 \in I$, $f(a) \notin I$ and $I \models \text{PA}$, and where I is constructed using diagonal indiscernibles for Δ_0^0-formulas. Using the notation of the proof of Lemma 6, for $\overline{\text{X}}$ an unbounded arithmetic subset of I, we have

$$M' \models \overline{\text{X}}(\langle h_{t+2}, h_{s_1}, \ldots, h_{s_m} \rangle) \underset{\substack{\gamma(d)\\ d \leqslant c}}{\to} (2h_q)_{h_q - (m+t+1)}^{h_q},$$

and, since d is nonstandard, for all $n \in \mathbf{N}$,

$$M' \models \overline{\text{X}}(\langle h_{t+2}, h_{s_1}, \ldots, h_{s_m} \rangle) \underset{\gamma(n)}{\to} (2h_q)_{h_q - (m+t+1)}^{h_q}.$$

Now, since the relation $\to_{\gamma(n)}$ is Π_1^0, we conclude that

$$I \models \overline{\text{X}}(\langle h_{t+2}, h_{s_1}, \ldots, h_{s_m} \rangle) \underset{\gamma(n)}{\to} (2h_q)_{h_q - (m+t+1)}^{h_q},$$

and finally that $I \models \bigcup_n P^{\gamma(n)}$. However, the function f is not total in I since $I \models \forall z \neg T(e_0, a, z)$.

COROLLARY 2. *The function τ_γ dominates all functions that appear before level ε_γ in the Grzegorcyck-Wainer hierarchy but is primitive recursive in a function which appears before level $\varepsilon_{\gamma+1}$.*

PROOF. Functions that appear before level ε_γ are provably total in some theory $[TI]_{\varepsilon_{\gamma(n)}}$ for $n \in \mathbf{N}$. So ε_γ dominates all such functions by Theorem 2. However, ε_γ can be proved total in the theory P^γ which is equivalent to $[TI]_{\varepsilon_\gamma}$ and so is primitive recursive in a function which appears before level $\varepsilon_{\gamma+1}$.

A variant of the argument of Theorem 2 above shows that the function $Y(x, y) =$ the maximum z such that $[x, y] \to_{\gamma(z)} (2z)_z^z$ is an indicator in the sense of Kirby-Paris [**K, P**] for models of $\bigcup_n P^{\gamma(n)}$. Hence, by [**P**] or [**Mc 2**] we have

COROLLARY 3. *The statement $\forall x \exists y (\tau_\gamma(x) = y)$ is equivalent in PA to* 1-*Consistency* ($\bigcup_n P^{\gamma(n)}$).

The next lemma is needed to show that $\to_\alpha$ and $\to^\alpha$ do in fact coincide.

LEMMA 8. *For $\gamma < \Delta$, $X \to_\gamma (m)_k^n \Leftrightarrow X \to^\gamma (m)_k^n$.*

PROOF. One first verifies that $\to^\gamma$ satisfies

(i) $X \to^\gamma (m)_k^n \Rightarrow \forall Y \geqslant X(Y \to^\gamma (m)_k^n)$

and

(ii) $X \to^{\gamma(s)} (m)_k^n \Rightarrow \forall t \leqslant s(X \to^{\gamma(t)} (m)_k^n)$.

Fact (i) is easily checked by induction on γ. For (ii), one uses the fact that for $t < s$ and any sequence $n_1, \ldots, n_q$, if $\gamma(s)(n_1) \cdots (n_q) \leqslant \gamma(t)$, then for some $p \leqslant q$, we will have $\gamma(s)(n_1) \cdots (n_p) = \gamma(t)$. Now if $X \rightarrow^{\gamma(s)} (m)^n_k$ we have $m \geqslant n+1$, $k \geqslant 1$ and $n \geqslant 1$. If $n = k = 1$, if $1 \in X$ and if $|X| \geqslant m$, then $X \rightarrow_\alpha (m)^n_k$ for all α. If $n \neq 1$ or $k \neq 1$, let Q be a partition of $[X]^n$ for which any homogeneous set Y with $\geqslant m$ elements must satisfy $\min^2 Y \geqslant n, k$. So then we have

$$Y \overset{\gamma(s)(n)}{\rightarrow} (2 \min^2 Y)^{\min^2 Y}_{\min^2 Y} \quad \text{with } \min^2 Y \geqslant n, k.$$

Then there exist sequences $n_1, n_2, \ldots, n_p$ and partitions $P_1, \ldots, P_p$ and sets $Y_1 \supseteq Y_2 \supseteq \cdots \supseteq Y_p$ such that $n_1 = n$, $Y_1 = Y$, $P_1 = Q$, Y_{i+1} is homogeneous for P_{i+1} and $n_{i+1} = \min^2 Y_{i+1}$ and such that $\gamma(n_1)(n_2) \cdots (n_p) = \gamma(t)$ and

$$Y_i \underset{\gamma(s)(n_1) \cdots (n_i)}{\rightarrow} \left(2 \min^2 Y_i\right)^{\min^2 Y_i}_{\min^2 Y_i}.$$

But then $\min^2 Y_p \geqslant m, n, k$ and so $Y_p \rightarrow_{\gamma(t)} (m)^n_k$ by Lemma 3.

THEOREM 3. *For all $\alpha < \Delta$, $\tau_\alpha = \sigma_\alpha$.*

PROOF. By induction on α and Lemma 8.

This completes the proof of Theorem A and its corollary. For our next results, we refer the reader to [**F**] or [**F, M, S**] for a description of the systems ACA_0 and ATR_0.

Let RT denote the theory (in the language of second order arithmetic) obtained by adjoining the infinite form of Ramsey's Theorem as an axiom to ACA_0:

$$\text{RT} = ACA_0 + \forall F \forall n \forall k \left[F\colon [\mathbf{N}]^n \rightarrow k \right.$$
$$\left. \Rightarrow \exists H(H \text{ infinite and homogeneous for } F) \right]$$

By the proof of Theorem 3.2 of [**F, M, S**], RT can also be axiomatized as

$$ACA_0 + \forall X \forall n \exists Y(Y \text{ is the } n\text{th Turing jump of } X).$$

The following is essentially announced in [**S**].

THEOREM 4. *The first order consequences of RT are the same as those of $\bigcup_n P^n$.*

PROOF. What we first do is to show that any first order consequence of RT is a consequence of $\bigcup_n P^n$. To argue by contradiction, suppose Φ is a Σ^0_k-sentence such that $\neg\Phi + \bigcup_n P^n$ is consistent and has countable model M. Let F be a Σ^0_k-Skolem function and let $\overline{\mathrm{X}}$ be an infinite arithmetic set such that

$$M \vDash \forall x, y\left(x, y \in \overline{\mathrm{X}} \;\&\; x < y \rightarrow \sup_{u \leqslant x} F(u) < y\right).$$

Let M' be a countable elementary end extension of M and fix $a \in M' - M$. We have for each standard n,

$$M' \vDash \exists y > a\left([a, y] \cap \overline{\mathrm{X}} \underset{n}{\rightarrow} (2a)^a_a\right)$$

by virtue of the axioms of P^n. Thus, by overspill, for some nonstandard d and b

$$M' \models [a,b] \cap \overline{X} \underset{d}{\rightarrow} (2a)^a_a.$$

Now by Kirby-Paris methods, *cf.* [**K, P**] and [**P**], there is a strong cut I, $a < I < b$, such that $X \cap I$ is unbounded in I and such that for every $e \in I$ and every partition P: $[[a,b] \cap \overline{X}]^e \rightarrow e$ which is coded in M', there is a coded subset $Y \subseteq [a,b] \cap \overline{X}$ such that Y is homogeneous for P and $Y \cap I$ is unbounded. Since I is a strong cut, $(I, \mathbf{R}_M, I) \models ACA_0$, *cf.* [**Ki**], and by the remarks just made $(I, \mathbf{R}_M, I) \models$ RT. Finally since $\overline{X}$ is unbounded in I, $I <_k M'$ and so $(I, \mathbf{R}_M, I) \models -\Phi$. Thus Φ is not a theorem of RT.

In the other direction, by induction on $n = -1, 0, 1, 2, \ldots$ one shows that RT $\vdash \forall \overline{X}[\overline{X}$ infinite $\rightarrow \forall x, y, z \exists X \subseteq \overline{X}(X \rightarrow_n (x)^y_z)]$. For $n = -1$, the result is clear. For the passage from n to $n+1$, the argument is similar to that of (i) of Theorem 2.1 of [**F, M, S**]: arguing by contradiction and using König's Lemma, one constructs F: $[\overline{X}]^y \rightarrow z$ with no finite homogeneous set X satisfying

$$X \rightarrow_n (2\min^2 X)^{\min^2 X}_{\min^2 X};$$

but then by Ramsey's Theorem, there is infinite $H \subseteq \overline{X}$ homogeneous for F. So by the induction hypothesis there is a finite initial segment X of H satisfying

$$X \underset{n}{\rightarrow} (2\min^2 H)^{\min^2 H}_{\min^2 H}$$

which is a contradiction as X is homogeneous for F and $\min^2 X = \min^2 H$.

Since RT is equivalent to $ACA_0 + \forall X \forall n (X^{(n)}$ exists), the following corollary follows from (and also reproves) a result of Jager [**Ja**].

COROLLARY 4. (i) *The function σ_ω eventually dominates all the provably recursive functions of RT.*

(ii) *ε_ω is the proof-theoretic ordinal of RT.*

PROOF. Part (i) follows at once from the theorem, as does Part (ii) since ε_ω is the ordinal of $\bigcup_n Q^n$.

This completes the proof of Theorem 3. Our final result deals with the system ATR_0, *cf.* [**F**] or [**F, M, S**].

THEOREM 5. *The following first-order scheme adjoined to PA axiomatizes the first order consequences of ATR_0: for every infinite arithmetic $\overline{X}$, there exists $X \subseteq \overline{X}$ such that X is n-dense in the sense of* [**F, M, S**], *all $n \in \mathbf{N}$.*

PROOF. In [**F, M, S**] it is shown in the proof of (i) of Theorem 2.1 that the scheme is provable in ATR_0. For the other direction, one follows the line of argument of the first part of the proof of Theorem 4, this time using the indicator method to produce a strong cut I such that $(I, \mathbf{R}_M, I) \models$ RT$(< \omega, 2)$ where RT$(< \omega, 2)$ is the version of the Galvin-Prikry Theorem for closed sets of [**F, M, S**]. By Theorem 3.2 of [**F, M, S**], $ATR_0 = ACA_0 +$ RT$(< \omega, 2)$.

BIBLIOGRAPHY

[Ac] P. Aczel, *Two notes on the Paris independence result*, Model Theory and Arithmetic (C. Berline, K. McAloon and J.-P. Ressayre, eds.), Lecture Notes in Math., vol. 890, Springer-Verlag, Berlin and New York, 1981, pp. 21–31.

[B] J. Bridge, *Some problems in mathematical logic: systems of ordinal functions and ordinal notations*, D. Phil. Thesis, Oxford, 1972.

[Fef] S. Feferman, *Transfinite recursive progressions of axiomatic theories*, J. Symbolic Logic **27** (1962), 259–316.

[F] H. Friedman, *Systems of second order arithmetic with restricted induction* (abstracts), J. Symbolic Logic **41** (1976), 557–559.

[F,M,S] H. Friedman, K. McAloon and S. Simpson, *A finite combinatorial principle which is equivalent to the* 1-*consistency of predicative analysis*, Logikon Symposion (G. Metiakiades, ed.), North-Holland, Amsterdam, 1982.

[Ja] G. Jager, *Theories for iterated jumps*, handwritten notes, Oxford, 1980.

[J] C. Jockush, Jr., *Ramsey's theorem and recursion theory*, J. Symbolic Logic **37** (1972), 268–280.

[K] S. C. Kleene, *Extension of an effectively generated class of functions by enumeration*, Colloq. Math. **6** (1958), 67–78.

[K,S] J. Ketonen, R. Solovay, *Rapidly growing Ramsey functions*, Ann. of Math. (2) **113** (1981), 314–367.

[Ki] L. A. S. Kirby, *Initial segments of models of arithmetic*, Ph.D. Thesis, Manchester, 1977.

[K,L] G. Kreisel and A. Levy, *Reflection principles and their use for establishing the complexity of axiomatic systems*, Z. für Math. Logik Grundlag. Math. **14** (1968), 97–142.

[K,P] L. A. S. Kirby and J. Paris, *Initial segments of models of Peano's axioms*, Set Theory and Hierarchy Theory V, (A. H. Lachlan, M. Srebrny and A. Zarach, eds.), Lecture Notes in Math., vol. 619, Springer-Verlag, Berlin and New York, 1977, pp. 211–226.

[Las] D. Lascar, *Une indicatrice de type Ramsey pour l'arithmétique de Peano et la formule de Paris-Harrington*, Modèles de l'Arithmétique, Astérisque **73** (1980), pp. 19°0.

[Mc 1] K. McAloon, *Formes combinatories du théorème d'incomplétude, exposé no.* 521, Séminaire Bourbaki, 1977/78, Lecture Notes in Math., vol. 719, Springer-Verlag, Berlin and New York, 1979.

[Mc 2] ______, *Le rapport entre la méthode de Gödel et la méthode des indicatrices pour obtenir des résultats d'indépendance*, Modèles de l'Arithmétique, Astérisque **73** (1980), pp. 31–40.

[Mc 3] ______, *Progressions transfinies de théories axiomatiques, formes combinatoires du théorème d'incomplétude et fonctions récursives à croissance rapide*, Modèles de l'Arithmétique, Astérisque **73** (1980), pp. 41–58.

[Mc 4] ______, *On the complexity of models of arithmetic*, J. Symbolic Logic **47** (1982), 403–415.

[P] J. Paris, *Some independence results for Peano arithmetic*, J. Symbolic Logic **43** (1978), 725–731.

[P,H] J. Paris and L. Harrington, *A mathematical incompleteness in Peano arithmetic*, Handbook of Mathematical Logic (J. Barwise, ed.), North-Holland, Amsterdam, 1977, pp. 1133–1142.

[S] J. Schlipf, *Scribblings on papers of Kirby and Paris and Paris and Harrington*, Abstract 755-E16, Notices Amer. Math. Soc. **25** (1978), A-387.

[Sch] U. Schmerl, *A fine structure generated by reflection formulas over primitive recursive arithmetic*, Logic Colloquium 79 (M. Boffa, K. McAloon and D. Van Dalen, eds.), North-Holland, Amsterdam, 1980, pp. 335–350.

[W] S. Wainer, *A classification of the ordinal recursive functions*, Arch. Math. Logik Grundlag. **13** (1970), 136–153.

DEPARTMENT OF COMPUTER AND INFORMATION SCIENCE, BROOKLYN COLLEGE, (CUNY), BROOKLYN NEW YORK, 11210.

DEPARTMENT OF MATHEMATICS, GRADUATE SCHOOL AND UNIVERSITY CENTER (CUNY), NEW YORK, NEW YORK 10036

Proceedings of Symposia in Pure Mathematics
Volume **42**, 1985

Reverse Mathematics

STEPHEN G. SIMPSON[1]

This paper is a slightly revised version of the author's talk at the AMS Summer Institute in Recursion Theory, Cornell University, June–July 1982.

The purpose of the research reported here is to obtain precise answers to the following question: *Which set existence axioms are needed to prove the theorems of ordinary mathematics*?

Our leading question is of central importance for the philosophy of mathematics, especially with respect to the foundations of mathematics and the existence of mathematical objects. Thus we are not primarily concerned here with the development of new techniques in model theory, set theory, recursion theory and proof theory, either for their own sake or for the sake of answering questions which arise from within these disciplines. Rather, we seek to apply the known techniques (developing new ones only when needed) to study the foundations of mathematics. Our motivation is philosophical rather than purely mathematical.

Our leading question referred to "ordinary mathematics". By *ordinary mathematics* we mean, roughly speaking, mainstream or non-set-theoretic mathematics, i.e. mathematics as it was before the abstract set theorists got hold of it (or perhaps: as it would have been if the abstract set theorists had never gotten hold of it). Thus ordinary mathematics includes number theory, geometry, calculus, differential equations, real and complex analysis, combinatorics, countable algebra, separable Banach spaces, computability theory, and the topology of complete separble metric spaces. It does not include abstract functional analysis, abstract set theory, universal algebra, or general topology.

A basic discovery (attributable to Hilbert and Bernays [**6**]) is that most or all of ordinary mathematics can be developed in the formal system of second order

---

1980 *Mathematics Subject Classification*. Primary 03B30, 03D80, 03E30.
Key words and phrases. Second order arithmetic.

[1]Alfred P. Sloan Research Fellow. Partially supported by NSF Grant MCS-8107867.

arithmetic. Before resuming our discussion of means and ends, we shall describe this formal system in detail.

The *language of second order arithmetic* is a two-sorted first order language with *number variables* $i, j, m, n, \ldots$ and *set variables* $X, Y, Z, \ldots$. The number variables are intended to range over $\mathbf{N} = \omega$, while the set variables are intended to range over subsets of ω. Numerical terms are built up as usual from number variables, the constant symbols 0 and 1, and binary operations $+$ and $\cdot$. Atomic formulas are $t_1 = t_2$, $t_1 < t_2$, and $t_1 \in X$, where t_1 and t_2 are numerical terms. *Formulas* are built up from atomic formulas by means of propositional connectives, number quantifiers $\forall n$ and $\exists n$, and set quantifiers $\forall X$ and $\exists X$. A formula is said to be *arithmetical* if it contains no set quantifiers.

All of the formal systems which we shall consider include the familiar ordered semiring axioms for $\mathbf{N}, +, \cdot, 0, 1, <$ as well as the *induction axiom*

$$(0 \in X \wedge \forall n(n \in X \rightarrow n + 1 \in X)) \rightarrow \forall n(n \in X).$$

At all times we assume the law of the excluded middle. The formal system $\Pi^1_\infty - \mathrm{CA}_0$ (also known as Z_2 or *second order arithmetic*) consists of the above axioms plus the *comprehension scheme*

$$\exists X \forall n(n \in X \leftrightarrow \varphi(n)),$$

where $\varphi(n)$ is any formula in which X does not occur freely.

By a *subsystem of* $\Pi^1_\infty - \mathrm{CA}_0$ we mean, of course, any formal system in the language of second order arithmetic whose theorems are a subset of those of $\Pi^1_\infty - \mathrm{CA}_0$. For instance, one especially important subsystem is obtained by restricting the comprehension scheme to arithmetical formulas $\varphi(n)$. This subsystem is known as ACA_0. (The acronym ACA stands for *arithmetical comprehension axiom*. The subscript zero denotes restricted induction, i.e. we assume only the induction axiom and not the full induction scheme

$$(\varphi(0) \wedge \forall n(\varphi(n) \rightarrow \varphi(n + 1))) \rightarrow \forall n \varphi(n)$$

for arbitrary formulas φ.)

As stated above, most or all theorems of ordinary mathematics can be stated and proved in $\Pi^1_\infty - \mathrm{CA}_0$. This can be accomplished by means of codes. For instance, rational numbers can be encoded as pairs of integers using a definable pairing function, e.g.

$$(m, n) = \tfrac{1}{2}(m + n)(m + n + 1) + m.$$

Real numbers can be encoded as Cauchy sequences or Dedekind cuts of rationals. Countable algebraic structures can be encoded as subsets of ω, again by means of the pairing function. Complete separable metric spaces, continuous functions, and countable well-orderings can also be encoded as subsets of ω, and on this basis a good theory of Borel sets and Borel functions can be developed. Alternatively, if one does not want to work with codes, one can pass to conservative extensions of $\Pi^1_\infty - \mathrm{CA}_0$ in which the codes are unnecessary (e.g. the conservative extension might have special sorts of variables ranging over real numbers, countable ordinals, etc.).

But why bother to do this? After all, ZFC is the "official language of mathematics," accepted (at least passively) by almost all pure mathematicians. Moreover, if one believes in the natural ω-model $\mathscr{P}(\omega)$ for $\Pi^1_\infty - \mathrm{CA}_0$, one will almost surely eventually come to believe in the cumulative hierarchy $V_\alpha = \bigcup\{\mathscr{P}(V_\beta)\colon \beta < \alpha\}$ which under some natural assumptions gives a natural model of ZFC. Thus nobody really believes that $\Pi^1_\infty - \mathrm{CA}_0$ is more likely to be consistent than ZFC is.

Nevertheless, we claim that there are several good reasons for studying the formalization of ordinary mathematics within $\Pi^1_\infty - \mathrm{CA}_0$, and not only in ZFC. An obvious reason is that $\Pi^1_\infty - \mathrm{CA}_0$ is a better guide or approximation to ordinary mathematical practice than ZFC is. Namely, the language and axioms of ZFC are specifically tailored to the study of high-order abstractions and uncountable cardinals such as $\aleph_\omega$. These objects do not arise in ordinary mathematics, and they do not exist in $\Pi^1_\infty - \mathrm{CA}_0$.

Another, more subtle, reason for studying $\Pi^1_\infty - \mathrm{CA}_0$ is provided by Gödel's incompleteness theorem. More precisely, we are spurred by the sharp contrast between the promise of Gödel's theorem and the actual performance up to the present time. Gödel tells us that there exist arithmetical sentences which are provable in stronger systems (such as ZFC) but not in weaker systems (such as $\Pi^1_\infty - \mathrm{CA}_0$). From this, a naive person might expect ZFC and even measurable cardinals to yield many applications in ordinary mathematics. The empirical facts are otherwise: Only in a very few cases (most of them discovered by Harvey Friedman) do we know of ordinary mathematical statements which are true but not provable in $\Pi^1_\infty - \mathrm{CA}_0$. In other words, the resources of ZFC have hardly ever been exploited in ordinary mathematics, although we know by Gödel's theorem that the resources exist.

This is where the study of $\Pi^1_\infty - \mathrm{CA}_0$ comes in. In order to learn how to exploit the resources of ZFC, we must isolate some properties of theorems of $\Pi^1_\infty - \mathrm{CA}_0$ which are not shared by theorems of ZFC which are statable in the language of $\Pi^1_\infty - \mathrm{CA}_0$. We can then reasonably hope to use these properties to discover theorems of ZFC which are significant from the viewpoint of ordinary mathematics, yet not provable in $\Pi^1_\infty - \mathrm{CA}_0$.

But, if the above arguments are cogent when applied to $\Pi^1_\infty - \mathrm{CA}_0$, they are all the more cogent when applied to *subsystems* of $\Pi^1_\infty - \mathrm{CA}_0$. This is because a carefully chosen subsystem of $\Pi^1_\infty - \mathrm{CA}_0$ may well be a better guide to ordinary mathematical practice (or a substantial part of that practice) than the full system taken uncritically.

We therefore turn to a systematic study *subsystems of second order arithmetic*. Many questions can be asked, but in this paper we want to focus on the question of what parts of ordinary mathematics can be developed in which subsystems of $\Pi^1_\infty - \mathrm{CA}_0$. We have already explained our reasons for this choice of emphasis.

It is easy to see that $\Pi^1_\infty - \mathrm{CA}_0$ has infinitely many subsystems. For example, a formula is said to be Π^1_k if it is of the form $\forall X_1 \exists X_2 \cdots X_k \theta$, $k = 0, 1, 2, \ldots,$

where θ is arithmetical. We can then consider subsystems $\Pi^1_k - CA_0$, $k = 0, 1, 2, \ldots$, in which the comprehension scheme is restricted to formulas $\varphi(n)$ which are Π^1_k. Thus $\Pi^1_0 - CA_0$ is the same as ACA_0, while $\Pi^1_{k+1} - CA_0$ is stronger than $\Pi^1_k - CA_0$ for each k. There are also many other ways to define subsystems of $\Pi^1_\infty - CA_0$, and such systems have been studied by many people, especially the proof theorists.

However, it turns out that most of the known subsystems of $\Pi^1_\infty - CA_0$ are of little or no interest from the narrow point of view of the foundations of mathematics. With only a little exaggeration, we may say that there are only five systems which are relavant to ordinary mathematical practice. These five systems, in increasing order, are as follows:

1. RCA_0. Here RCA stands for *recursive comprehension axiom*. The axioms of RCA_0 will be stated precisely later; roughly speaking, these axioms are only strong enough to prove the existence of recursive sets. In our work, RCA_0 plays the role of a formal system which is strong enough to develop some basic theory of continuous functions of a real variable and of countable algebraic structures. Thus RCA_0 serves as a basic system on which the remaining systems are built.

2. WKL_0. This system consists of RCA_0 plus a further axiom known as *weak König's lemma*: Every infinite subtree of $2^{<\omega}$ has an infinite path. An equivalent statement is the compactness of the Cantor space 2^ω or of the unit interval $[0, 1]$. This system WKL_0 gives an improved theory of continuous functions, and is also just strong enough to permit the development of ideal theory in countable commutative rings.

3. ACA_0. Here ACA stands for *arithmetical comprehension axiom*; the axioms of $ACA_0 = \Pi^1_0 - CA_0$ were described above. ACA_0 includes WKL_0 but, in addition, it is just strong enough for a smooth theory of sequential convergence. ACA_0 isolates the same portion of ordinary mathematical practice which was identified as "predicative analysis" by Herman Weyl in [**16**].

4. ATR_0. Here ATR stands for *arithmetical transfinite recursion*. The principal axiom of ATR_0 says that arithmetical comprehension can be iterated along any countable well-ordering. ATR_0 is just strong enough to permit the development of a good theory of countable well-orderings, Borel sets, analytic sets, etc.

5. $\Pi^1_1 - CA_0$. This is the system of Π^1_1 *comprehension* which has already been introduced. $\Pi^1_1 - CA_0$ is properly stronger than ATR_0 and yields an improved theory of countable well-orderings, etc. Both of these systems have numerous mathematical consequences in the realms of algebra, analysis, and combinatorics.

The rest of this paper is devoted to a more detailed discussion of each of these five systems.

In order to state the axioms of RCA_0 precisely we need a couple of definitions. An arithmetical formula is said to be Δ^0_0 if all of its number quantifiers are *bounded*, i.e. of the form $\forall m(m < t \rightarrow \cdots)$ or $\exists m(m < t \wedge \cdots)$. A Σ^0_1 (respectively Π^0_1) formula is one of the form $\exists m\theta$ (respectively $\forall m\theta$), where θ is Δ^0_0. The Δ^0_1 comprehension scheme consists of all axioms of the form

$$\forall n(\varphi(n) \leftrightarrow \psi(n)) \to \exists X \forall n(n \in X \leftrightarrow \varphi(n)),$$

where φ is Σ_1^0, ψ is Π_1^0, and X does not occur freely in φ or ψ. The Σ_1^0 induction scheme consists of all axioms of the form

$$(\varphi(0) \wedge \forall n(\varphi(n) \to \varphi(n+1))) \to \forall n \varphi(n),$$

where φ is Σ_1^0. The axioms of RCA_0 consist of the ordered semiring axioms for $\mathbf{N}$, $+, \cdot, 0, 1, <$ together with the schemes of Δ_1^0 comprehension and Σ_1^0 induction.

Within RCA_0 one can develop primitive recursive functions and prove basic number-theoretic facts such as the fundamental theorem of arithmetic. One can then define (codes for) rational numbers to be certain ordered pairs of integers and prove that $\mathbf{Q}, +, -, \cdot, 0, 1, <$ is an ordered field; here $\mathbf{Q}$ is the set of codes for rational numbers. Within RCA_0 we define a (code for a) *real number* to be a set $X \subseteq \mathbf{Q}$ such that $\varnothing \neq X \neq \mathbf{Q}$ and

(i) $\forall q \forall q'(q < q' \in X \to q \in X)$,

(ii) $\forall q \exists q'(q \in X \to q < q' \in X)$.

Here q and q' range over $\mathbf{Q}$. Thus a real number is encoded as the set of smaller rational numbers. Within RCA_0 we can then prove that $\mathbf{R}, +, -, \cdot, 0, 1, <$ is an Archimedean ordered field. Here we use $\mathbf{R}$ informally to denote the set of (codes for) real numbers, a set which, of course, does not formally exist in RCA_0.

Within RCA_0 we can talk about sequences of real numbers: We define a (code for a) *sequence of real numbers* to be a set $X \subseteq \mathbf{Q} \times \mathbf{N}$ such that, for each $n \in \mathbf{N}$, $(X)_n = \{q \in \mathbf{Q}\colon (q, n) \in X\}$ is a (code for a) real number. Notions such as *convergence*, etc. can be defined as usual. Informally we denote elements of $\mathbf{R}$ by $x, y, \ldots$ and sequences of reals by $\langle x_n \rangle_{n \in \mathbf{N}}$, etc.

Within RCA_0 we cannot prove that bounded increasing sequences of real numbers are convergent. (See Theorem 4 and Remarks 5 and 6.) However, in RCA_0 we can prove a very useful completeness property of $\mathbf{R}$ known as *nested inverval completeness*:

THEOREM 1 (RCA_0). *Let $\langle a_n \rangle_{n \in \mathbf{N}}$ and $\langle b_n \rangle_{n \in \mathbf{N}}$ be sequences of reals such that for all n, $a_n \leqslant a_{n+1} \leqslant b_{n+1} \leqslant b_n$; and $\lim_n |a_n - b_n| = 0$. Then there exists a real x such that $x = \lim_n a_n = \lim_n b_n$ (Simpson* [**14**]).

PROOF (SKETCH). The set of rationals less than x can be defined in a Σ_1^0 or Π_1^0 way as

$$\{q\colon \exists n\, q < a_n\} = \{q\colon \forall n\, q < b_n\}.$$

Hence this set exists by Δ_1^0 comprehension. This gives the existence of x, and it is then easy to prove that $x = \lim_n a_n = \lim_n b_n$.

Within RCA_0 we can develop the rudiments of the theory of continuous functions on $\mathbf{R}$. We define a (code for a partial) *continuous function* $f\colon \mathbf{R} \overset{p}{\to} \mathbf{R}$ to be a set of ordered quadruples $(a, b, u, v) \in \mathbf{Q}^4$ such that $a < b$, $u \leqslant v$, and

(i) $(a, b, u, v), (a, b, u', v') \in f \Rightarrow (a, b, u', v) \in f$,

(ii) $(a, b, u, v) \in f, u' \leqslant u \leqslant v \leqslant v' \Rightarrow (a, b, u', v') \in f$,

(iii) $(a, b, u, v) \in f, a \leqslant a' < b' \leqslant b \Rightarrow (a', b', u, v) \in f$.

Intuitively, $(a, b, u, v) \in f$ means that $u \leqslant f(x) \leqslant v$ whenever x is a real number such that $a < x < b$.

A real number x is defined to be in the *domain* of f if for all $\varepsilon > 0$ there exists $(a, b, u, v) \in f$ such that $a < x < b$ and $|u - v| < \varepsilon$. In this case $f(x)$ is defined to be the unique real number y such that $u \leqslant y \leqslant v$ for all $(a, b, u, v) \in f$ such that $a < x < b$. Here y exists by nested interval completeness.

Within RCA_0 we can prove that polynomials, e^x, $\sin x$, etc. are continuous. We can also prove the *intermediate value theorem* for continuous functions:

THEOREM 2 (RCA_0). *Let f be a continuous function whose domain includes the closed unit interval* $[0,1]$. *If* $f(0) < 0 < f(1)$, *then there exists* $x \in [0,1]$ *such that* $f(x) = 0$. (*Simpson* [**14**]).

PROOF (SKETCH). Suppose that the conclusion fails. Use primitive recursion relative to the code of f to define a nested sequence of rational intervals

$$[a_0, b_0] = [0,1],$$

$$[a_{n+1}, b_{n+1}] = \begin{cases} [(a_n + b_n)/2, b_n] & \text{if } f((a_n + b_n)/2) < 0, \\ [a_n, (a_n + b_n)/2] & \text{if } f((a_n + b_n)/2) > 0. \end{cases}$$

By nested interval convergence we have the existence of $x = \lim_n a_n = \lim_n b_n$. It is easy to see that $f(x) = 0$.

The development of real variable theory sketched above can be carried considerably farther in RCA_0. In addition, the rudiments of countable algebra and first order logic (via Gödel numbers) can be developed in RCA_0. For details see Simpson [**14**].

We now turn to systems stronger than RCA_0. One of our main themes is that often a mathematical result is equivalent (over RCA_0) to the subsystem of $\Pi^1_\infty - CA_0$ which is needed to prove that result. This is the theme of "reverse mathematics" (see Remark 5). We illustrate this theme by means of an example involving the Bolzano-Weierstrass theorem and the system ACA_0.

Within RCA_0 we can encode functions f: $\mathbf{N} \to \mathbf{N}$ as sets of ordered pairs. We can then prove

LEMMA 3 (RCA_0). *The following are equivalent*:

(1) ACA_0;

(2) $\Sigma^0_1 - CA_0$ (*i.e. the comprehension scheme restricted to Σ^0_1 formulas* $\varphi(n)$);

(3) *for all one-to-one functions* f: $\mathbf{N} \to \mathbf{N}$ *the range of f exists, i.e.*

$$\exists X \forall n (n \in X \leftrightarrow \exists m\, f(m) = n).$$

PROOF (SKETCH). In RCA_0 we can straightforwardly prove that, given a Σ^0_1 formula $\varphi(n)$, if $\{n: \varphi(n)\}$ is not finite, then there exists a one-to-one function f: $\mathbf{N} \to \mathbf{N}$ such that $\forall n(\varphi(n) \leftrightarrow \exists m\, f(m) = n)$. (This amounts to just the usual proof that every Σ^0_1 set is the range of a one-to-one recursive function.) From this the implication from (3) to (2) is immediate. The implication from (2) to (3) is trivial, and (1) and (2) are equivalent by relativization (e.g. $\Sigma_2^0 = \Sigma_1^0$ relative to Σ_1^0, etc.). For details see Simpson [**13**, **14**].

THEOREM 4 (RCA_0). *The following are equivalent*:

(1) ACA_0;

(2) *the Arzela-Ascoli lemma*: *Any bounded equicontinuous sequence of functions on* $[0,1]$ *has a uniformly convergent subsequence* (*Simpson* [**14**]);

(3) *the Bolzano-Weierstrass theorem*: *Every bounded sequence of real numbers has a convergent subsequence* (*Friedman* [**2**]);

(4) *every Cauchy sequence of real numbers is convergent*;

(5) *every bounded sequence of real numbers has a supremum* (*Friedman* [**2**]);

(6) *every bounded increasing sequence of real numbers is convergent* (*Friedman* [**2**]).

PROOF (SKETCH). The direct implications (1) $\Rightarrow$ (2) $\Rightarrow$ (3) $\Rightarrow$ (4) $\Rightarrow$ (5) $\Rightarrow$ (6) are straightforward imitations of the usual proofs of these mathematical results in ACA_0. For details see Simpson [**13**, **14**].

To prove that (6) implies ACA_0, assume (6) and note that by the lemma it suffices to prove that every one-to-one function $f\colon \mathbf{N} \to \mathbf{N}$ has a range. Define a bounded increasing sequence of real numbers by

$$a_n = \sum_{i=0}^{n} \frac{1}{2^{f(i)}}.$$

By (6) we have the existence of $a = \lim_n a_n$. Then range(f) exists by recursive comprehension since

$$n \notin \operatorname{range}(f) \leftrightarrow \exists X \subseteq \{0,1,\ldots,n-1\} \left[\sum_{i \in X} \frac{1}{2^i} < a \leqslant \sum_{i \in X} \frac{1}{2^i} + \frac{1}{2^n} \right].$$

REMARK 5. The above theorem says, in particular, that the Bolzano-Weierstrass theorem is *equivalent* to ACA_0 (provably in the weak system RCA_0). This is the strongest possible sense in which one can say that ACA_0 is "just strong enough" to prove the Bolzano-Weierstrass theorem.

The above theorem exemplifies a phenomenon which pervades our subject. *Very often, if a theorem of ordinary mathematics is proved from the weakest possible set existence axioms, it will be possible to "reverse" the theorem by proving that it is equivalent to those axioms over a weak base theory.* This phenomenon is known as *REVERSE MATHEMATICS.*

The pervasiveness of the reverse mathematics phenomenon was first emphasized by Harvey Friedman [**1**]. The equivalence of (1), (3) and (6) above, as well as several other equivalences of a similar nature, are due to Friedman [**1**, **2**, **5**]. Other examples of reverse mathematics are due to Simpson [**12**, **13**, **14**, **4**, **5**], Smith [**5**] and Steel [**15**]. Below we shall describe some of these examples.

REMARK 6. The minimum ω-model of RCA_0 consists of the recursive subsets of ω. Therefore, the development of real analysis in RCA_0 is somewhat parallel to what is called "recursive real analysis", i.e. a theory of real analysis which makes use of recursive real numbers, recursively coded continuous functions, etc.

However, the proof of Theorem 4 illustrates a key difference between our work and recursive real analysis. In recursive analysis one takes a recursive function f: $\mathbf{N} \to \mathbf{N}$ with nonrecursive range, and one then constructs a "Specker sequence" $a_n = \sum_{i=0}^{n} 2^{-f(i)}$ as above. Such a sequence is regarded by the recursive analysts as a *counterexample*, showing that a bounded increasing recursive sequence of recursive reals need not converge to a recursive limit. In contrast, our theory converts this counterexample into a *positive result*: ACA_0 is equivalent to the statement that bounded increasing sequences of reals are convergent. This emphasis on positive results is typical of our subject. For another discussion of the difference between our work and recursive analysis see §5 of [**13**].

There are many other examples of reverse mathematics. In particular, there are many other ordinary mathematical theorems which are provably equivalent to ACA_0 over RCA_0. We list a few such theorems:

(1) König's lemma: *Any infinite, finitely branching subtree of* $\omega^{<\omega}$ *has an infinite path* [**2**]. (Contrast weak König's lemma which deals with $2^{<\omega}$.)

(2) Ramsey's theorem for subsets of $[\omega]^3$ (cf. Jockusch [**7**]).

(3) Every countable vector space has a basis.

(4) Every countable commutative ring has a maximal ideal.

(5) Every countable abelian group has a unique divisible closure.

Here the equivalence of ACA_0 with statements (3), (4) and (5) is proved in a paper by Friedman, Simpson and Smith [**5**]. The same paper contains a discussion of the development of countable algebra within RCA_0 and other subsystems of second order arithmetic. See also Simpson [**14**].

There are a number of examples of reverse mathematics in connection with the system WKL_0. In the realm of real analysis we have

THEOREM 7 (RCA_0). *The following are pairwise equivalent*:

(1) WKL_0;

(2) *the Heine-Borel theorem*: *Every covering of* $[0,1]$ *by a countable sequence of open intervals has a finite subcovering*;

(3) *every continuous function on* $[0,1]$ *is uniformly continuous*;

(4) *every continuous function on* $[0,1]$ *is bounded*;

(5) *every continuous function on* $[0,1]$ *has a supremum*;

(6) *every uniformly continuous function on* $[0,1]$, *which has a supremum, attains it.*

In particular, WKL_0 is equivalent over RCA_0 to the maximum principle: Every continuous function on $[0,1]$ attains a maximum value. This result, as well as the equivalence of (1), (3), (4), (5) and (6) above, is due to Simpson [**14**]. the equivalence of (1) and (2) is due to Friedman [**2**] (see also Simpson [**14**]). We also have

THEOREM 8 (RCA_0). WKL_0 *is equivalent to the local existence theorem for solutions of ordinary differential equations* (*Simpson* [**13**]).

From the realm of countable algebra and logic we have

THEOREM 9 (RCA_0). *The following are equivalent*:
(1) WKL_0;
(2) *every countable commutative ring has a prime ideal* [**5**];
(3) *every countable formally real field can be ordered* [**5**],
(4) *every countable formally real field has a real closure* [**5**];
(5) *Gödel's completeness theorem for predicate calculus* [**2**, **14**].

These results suggest that WKL_0 is quite powerful from the viewpoint of what mathematics can be done in the system. Indeed, the elements of Riemann integration and complex variable theory can also be developed handily in WKL_0. On the other hand, it turns out that WKL_0 is not very powerful from the standpoint of logical strength: Kirby and Paris [**8**] have given a simple, model-theoretic argument which shows that WKL_0 is a conservative extension of primitive recursive arithmetic with respect to Π^0_2 sentences. (The observation that their argument gives this result is apparently due to Friedman [**3**]. For a related result see Minc [**10**].) In addition, Harrington has shown that for every model of RCA_0 there is a model of WKL_0 with the same integers. Thus WKL_0 is conservative over RCA_0 with respect to Π^1_1 sentences. (For proofs see Simpson [**14**].)

The above-mentioned result of Kirby, Paris, Friedman and Minc is of great interest in that it is a successful realization of *Hilbert's Program*! Hilbert wanted to show that "ideal" concepts and methods can be eliminated in favor of finitistic ones. The Kirby-Paris-Friedman-Minc result shows that the powerful methods which are available in WKL_0 (including Riemann integration, Gödel's completeness theorem, and ideal theory in countable commutative rings) are indeed eliminable in this sense.

We now turn to the system ATR_0. In recursion-theoretic terms, the principal axiom of ATR_0 asserts that for all $X \subseteq \omega$ and for all relativized ordinal notations $e \in \mathcal{O}^X$, the relativized H-set H^X_e exists. This axiom permits the development of a good theory of countable ordinals and Borel and analytic sets. Within ATR_0 we can prove that every analytic set is Lebesgue measurable, has the property of Baire, and either contains a perfect set or is countable. Moreover, several key results of classical descriptive set theory are, in fact, provably equivalent to ATR_0 over a weak base theory. Namely, we have

THEOREM 10 (RCA_0). *The following are equivalent*:
(1) ATR_0;
(2) *every closed set either contains a perfect set or is countable*;
(3) *determinacy of open games in* ω^ω;
(4) *Ramsey's theorem for open subsets of* $[\omega]^\omega$.

Here equivalents (2), (3) and (4) are due to Friedman [**2**], Steel [**15**] and Simpson [**12**, **4**], respectively. In addition, Friedman has shown that ATR_0 is equivalent (over RCA_0) to the assertion that, of any two countable well-orderings, one is isomorphic to an initial segment of the other. In the realm of countable algebra, it is shown in the Friedman-Simpson-Smith paper **[5]** that ATR_0 is equivalent to Ulm's theorem for reduced abelian p-groups.

It is perhaps interesting to note that ATR_0 is not very strong from the viewpoint of logical strength. Namely, Friedman **[4]** has shown that ATR_0 proves the same Π^1_1 sentences as Feferman's system IR of "predicative analysis". However, as shown above, ATR_0 is much stronger than IR from the standpoint of what mathematics can be done in the system.

Finally, we turn to Π^1_1-CA_0. In recursion-theoretic terms, Π^1_1 comprehension means that, for all $X \subseteq \omega$, the hyperjump of X exists. This principle is much stronger than arithmetical transfinite recursion from the viewpoint of mathematical practice. Each of the following mathematical statements is equivalent to Π^1_1-CA_0 over a weak base theory:

(1) Silver's theorem on coanalytic equivalence relations;

(2) Silver's theorem specialized to closed equivalence relations on ω^ω;

(3) Kondo's theorem : uniformization for coanalytic relations;

(4) the Cantor-Bendixson theory: every closed set is the union of a perfect set and a countable set;

(5) Ulm's theorem for countable abelian groups;

(6) every countable abelian group has a maximal divisible subgroup;

(7) Ramsey's theorem for finitely Borel subsets of $[\omega]^\omega$;

(8) Ramsey's theorem for $F_\sigma \cap G_\delta$ subsets of $[\omega]^\omega$.

Here (1) and (2) are due to Harrington and Sami; (3) is due to Simpson; (4) is due to Kreisel **[9]** and Friedman **[2]**; (5) and (6) appear in the Friedman-Simpson-Smith paper **[5]**; and (7) is closely related to an unpublished result of Solovay. (See also Simpson **[14]**.)

We hope that this paper has conveyed an idea of which parts of ordinary mathematical practice can be carried out in which subsystems of second order arithmetic. For further information on this and other topics, see the author's forthcoming monograph **[14]**, several chapters of which have already been written.

Note added in proof. Recently Douglas Brown and the author have shown that the Hahn-Banach theorem for separable Banach spaces is equivalent to WKL_0 over RCA_0.

REFERENCES

1. H. Friedman, *Some systems of second order arithmetic and their use*, Proc. Internat. Cong. Math. (Vancouver, 1974), Vol. 1, Canadian Mathematical Congress, 1975, pp. 235–242.

2. ______, *Systems of second order arithmetic with restricted induction* (abstracts), J. Symbolic Logic **41** (1976), 557–559.

3. ______, *On fragments of Peano arithmetic*, preprint, May 1979, 7 pages.

4. H. Friedman, K. McAloon and S. G. Simpson, *A finite combinatorial principle which is equivalent to the* 1-*consistency of predicative analyis*, Patras Logic Symposion (G. Metakides, ed.), North-Holland, Amsterdam, 1982, pp. 197–230.

5. H. Friedman, S. G. Simpson and R. L. Smith, *Countable algebra and set existence axioms*, Ann. Pure Appl. Logic **25** (1983), 141–181.

6. D. Hilbert and P. Bernays, *Grundlagen der Mathematik*, Vols. 1, 2, Springer-Verlag, Berlin and New York, 1934, 1939.

7. C. G. Jockusch, Jr., *Ramsey's theorem and recursion theory*, J. Symbolic Logic **37** (1972), 81–89.

8. L. Kirby and J. Paris, *Initial segments of models of Peano's axioms*, Set Theory and Hierarchy Theory.V (Bierutowice, Poland, 1976), Notes in Math., Vol. 619, Springer-Verlag, Berlin and New York, 1977, pp. 211–226.

9. G. Kreisel, *Analysis of the Cantor-Bendixson theorem by means of the analytic hierarchy*, Bull. Acad. Polon. Sci. **7** (1959), 621–626.

10. G. Minc, *What can be done in PRA*? Zap. Nauchn. Sem. Leningrad. Otdel. Mat. Inst. Steklov. Vol. 60, Leningrad, 1976, pp. 93–102.

11. S. G. Simpson, *Notes on subsystems of analysis*, lecture notes, Berkeley, 1973, 38 pages.

12. ______, *Sets which do not have subsets of every higher degree*, J. Symbolic Logic **43** (1978), 135–138.

13. ______, *Which set existence axioms are needed to prove the Cauchy/Peano theorem for ordinary differential equations*?, J. Symbolic Logic **49** (1984), 783–802.

14. ______, *Subsystems of second order arithmetic*, in preparation.

15. J. Steel, *Determinateness and subsystems of analysis*, Ph.D. thesis, Berkeley, 1977.

16. H. Weyl, *Das Kontinuum: Kritische Untersuchungen über die Grundlagen der Analysis*, Berlin, 1917, iv + 84 pages; reprinted by Chelsea, New York, 1960, 1973.

DEPARTMENT OF MATHEMATICS, PENNSYLVANIA STATE UNIVERSITY, UNIVERSITY PARK, PENNSYLVANIA 16802

Proceedings of Symposia in Pure Mathematics
Volume **42**, 1985

Infinite Fixed-Point Algebras

ROBERT M. SOLOVAY

Abstract. Smoryński has recently introduced the notion of the fixed-point algebra of a theory. We prove in this paper that the fixed-point algebras of any two "reasonable" theories are isomorphic. In particular, Peano arithmetic, Zermelo-Fraenkel set theory, and Gödel-Bernays set theory all have isomorphic fixed-point algebras. This confirms one of Smoryński's conjectures and refutes another.

1. Introduction. Let T be a consistent first order theory in a countable language L_T. The *Lindenbaum algebra* of T, B_T, is defined as follows. Say that two sentences, ϕ and ψ, are *equivalent* iff $T \vdash \phi \leftrightarrow \psi$. Let $[\phi]$ be the equivalence class of ϕ. Then the underlying set of B_T, $|B_T|$, is the collection of all $[\phi]$, for ϕ a sentence of T. B_T has a natural Boolean algebra structure whose operations arise from the propositional connectives of L_T:

$$\sim [\phi] = [\sim \phi]; \qquad [\phi] \vee [\psi] = [\phi \vee \psi].$$

If T is essentially undecidable, then B_T is a countable totally nonatomic Boolean algebra. (This will happen (for example) if Peano arithmetic P is relatively interpretable in T; cf. [**4** or **2**].) Since any two countable totally nonatomic Boolean algebras are isomorphic, any two consistent, essentially undecidable theories have isomorphic Lindenbaum algebras.

We now specialize to the following class of first order theories T:

(1) L_T is a recursive language equipped with a fixed Gödel numbering. (It is necessary to assume various syntactic operations are primitive recursive. This will hold for all the usual Gödel numberings.)

(2) P is relatively interpretable in T, and a particular interpretation is fixed, once and for all.

(3) T is recursively axiomatizable and consistent.

For brevity, we refer to a first order theory T satisfying (1)–(3) as *reasonable*. Certainly P is reasonable. If ZF (Zermelo-Fraenkel set theory) is consistent, then

1980 *Mathematics Subject Classification*. Primary 03G05; Secondary 03B45, 03F30.

both ZF and GB (Gödel-Bernays set theory) are reasonable. (In each case, one takes a natural Gödel numbering and relative interpretation of P.)

Let $\phi(x)$ be a formula of T with one free variable. Then $\phi(x)$ determines, for each $n \in \omega$, an element $[\phi(\mathbf{n})]$ of B_T as follows: Using the relative interpretation of P in T, we construct a recursive sequence $\theta_n(x)$ of formulas of T such that $\theta_n(x)$ says "x denotes the integer n" (with respect to the interpretation of P in T). Then $[\phi(\mathbf{n})]$ is the element

$$[(\forall x)(\theta_n(x) \to \phi(x))]$$

of B_T.

Let $\phi(x)$ be a formula of T having one free variable. We say that $\phi(\ )$ is *functional* if it satisfies the following condition:

Let m_1, m_2 be Gödel numbers of sentences ψ_1, ψ_2 of L_T. Then if $[\psi_1] = [\psi_2]$, $[\phi(\mathbf{m}_1)] = [\phi(\mathbf{m}_2)]$.

If $\phi(\ )$ is functional, then $\phi(\ )$ determines a map f_ϕ: $|B_T| \to |B_T|$ via the equation

$$f_\phi([\psi]) = [\phi(\#\psi)].$$

(Here $\#\psi$ is the numeral corresponding to the Gödel number of ψ.)

The basic example of a functional ϕ is $Bew_T(x)$ which expresses (relative to some recursive axiomatization of T)"x is the Gödel number of a theorem of T".

We let A_T be the collection of all maps f: $|B_T| \to |B_T|$ arising from functional ϕ's in the manner just indicated. This construction is due to Smoryński [3] who refers to the pair $\langle A_T, B_T\rangle$ as the fixed-point algebra of T. Smoryński goes on to give a definition of an abstract fixed-point algebra (which we need not recall) and to question whether or not various familiar theories have fixed-point algebras isomorphic to that of P.

Let S, T be reasonable theories. An isomorphism of fixed-point algebras is a bijection f: $B_S \to B_T$ which is an isomorphism of Boolean algebras and which gives rise to a bijection f_*: $A_S \to A_T$ via the formula

$$f_*(h) = f \circ h \circ f^{-1}.$$

THEOREM 1. *If S and T are reasonable theories, then $\langle A_S, B_S\rangle$ and $\langle A_T, B_T\rangle$ are isomorphic fixed-point algebras.*

Our next result will give an alternative description of A_T. But first we need some preliminary definitions. Let $\langle \phi_i, i \in \omega\rangle$ be the enumeration of the sentences of L_T in order of increasing Gödel numbers. Let h_T: $\omega \to |B_T|$ be defined by

$$h_T(i) = [\phi_i].$$

We say that a map f: $|B_S| \to |B_T|$ (which need not be a Boolean algebra homomorphism) is *recursive* if there is a recursive map g: $\omega \to \omega$ such that the following diagram is commutative:

$$\begin{array}{ccc} \omega & \overset{g}{\rightarrow} & \omega \\ \downarrow h_S & & \downarrow h_T \\ |B_S| & \overset{f}{\rightarrow} & |B_T| \end{array}$$

(In other words, from the Gödel number of a representative of b, we can effectively produce the Gödel number of some representative of $f(b)$).

We say that a subset X of $|B_T|$ is *small* if for some formula $\phi(x)$ of L_T,

$$X \subseteq \{[\phi(\mathbf{n})] : n \in \omega\}.$$

THEOREM 2. *Let T be a reasonable theory. Let B_T be the Lindenbaum algebra of T, and let f: $|B_T| \to |B_T|$. Then the following are equivalent*:

(1) $f \in A_T$;

(2) *f is recursive and* range(f) *is small.*

THEOREM 3. *Let S, T be reasonable theories, and let f: $B_S \to B_T$ be an isomorphism of Boolean algebras that induces an isomorphism of fixed-point algebras. Then f is recursive.*

COROLLARY. *Let T be reasonable. Then the set of automorphisms of the fixed-point algebra $\langle A_T, B_T\rangle$ is countable.*

THEOREM 4. *Let T be a reasonable theory. Let ϕ be an automorphism of B_T. Then the following are equivalent*:

(1) *ϕ induces an automorphism of $\langle A_T; B_T\rangle$*;

(2) *ϕ is recursive and for all $X \subseteq |B_T|$, X is small iff $\phi[X]$ is small.*

2. Proof of Theorem 1. Let S, T be reasonable theories. We seek an isomorphism ϕ of B_S with B_T that will prolong to an isomorphism ϕ^* of $\langle A_S, B_S\rangle$ with $\langle A_T, B_T\rangle$. Our first step is to describe a sufficient condition (smoothness) for ϕ to so prolong. I do not know whether or not this condition is necessary; I suspect it is not.

DEFINITION. An isomorphism of Boolean algebras, ϕ: $B_S \to B_T$, is *smooth* if there are recursive functions g_1, g_2: $\omega \to \omega$ such that:

(a) If r is the Gödel number of a formula $\theta(x)$ of L_S, then $g_1(r)$ is the Gödel number of a formula $\psi(x)$ of L_T, and for all $n \in \omega$,

$$\phi([\theta(\mathbf{n})]) = [\psi(\mathbf{n})].$$

(b) (Similar to (a). It is obtained from (a) by replacing ϕ, g_1, S and T by ϕ^{-1}, g_2, T and S.)

The definition is arranged so that if ϕ is smooth, so is ϕ^{-1}.

LEMMA 1. *Let ϕ: $B_S \to B_T$ be smooth. Then ϕ prolongs to an isomorphism of $\langle A_S, B_S\rangle$ with $\langle A_T, B_T\rangle$.*

PROOF. We have to show that if h: $|B_S| \to |B_S|$, then $h \in A_S$ iff $\phi \circ h \circ \phi^{-1}$ lies in A_T.

First assume that $h \in A_S$. We show that $\phi \circ h \circ \phi^{-1}$ lies in A_T. Since $h \in A_S$, there is a formula $\theta_1(x)$ of L_S such that if χ is a sentence of L_S with Gödel number r, then

$$h([\chi]) = [\theta_1(\mathbf{r})].$$

Since ϕ is smooth, there is a recursive function g_3 such that if ψ is a sentence of L_T with Gödel number r, then $g_3(r)$ is the Gödel number of some representative of $\phi^{-1}([\psi])$.

Since P is relatively interpretable in S, we can find a formula $\theta_2(x)$ of L_S which expresses:

$$x \text{ is a number} \quad \text{and} \quad (\exists y)(y = g_3(x) \wedge \theta_1(y)).$$

We may easily arrange things so that if $r \in \omega$ and $s = g_3(r)$, then $[\theta_2(\mathbf{r})] = [\theta_1(\mathbf{s})]$.

Finally, since ϕ is smooth, we can find a formula $\theta_3(x)$ of L_T such that for all $r \in \omega$, $[\theta_3(\mathbf{r})] = \phi([\theta_2(\mathbf{r})])$.

Let $h_1 = \phi \circ h \circ \phi^{-1}$. We show that if ψ is a sentence of L_T with Gödel number r_0, then $h_1([\psi]) = [\theta_3(\mathbf{r}_0)]$. Indeed, let $r_1 = g_3(r_0)$. Then r_1 is the Gödel number of ψ^* with $[\psi^*] = \phi^{-1}([\psi])$. $[\theta_2(\mathbf{r}_0)] = [\theta_1(\mathbf{r}_1)] = h([\psi^*]) = h \circ \phi^{-1}([\psi])$. Finally,

$$[\theta_3(\mathbf{r}_0)] = \phi([\theta_2(\mathbf{r}_0)]) = \phi \circ h \circ \phi^{-1}([\psi]) = h_1([\psi]).$$

Thus θ_3 is functional and h_1 is the element of A_T determined by θ_3.

Conversely, suppose $h_1 \in A_T$. We show $h \in A_S$. Since ϕ is smooth, so is ϕ^{-1}. Applying the preceding argument to ϕ^{-1}, h_1, we see that $h = \phi^{-1} \circ h_1 \circ \phi$ lies in A_S. This completes the proof of Lemma 1.

To prove Theorem 1, it suffices to construct a smooth isomorphism ϕ of B_S with B_T. Such a ϕ can be readily constructed by the usual "back-and-forth" method once the following lemma is proved.

LEMMA 2. *Let $\theta_0(x), \ldots, \theta_r(x)$ be formulas of L_S, and $\psi_0(x), \ldots, \psi_{r-1}(x)$ formulas of L_T. Let B_S^* be the Boolean subalgebra of B_S generated by elements of the form $[\theta_i(\mathbf{k})]$ (where $i < r$ and $k \in \omega$). Similarly, let B_T^* be the Boolean subalgebra of B_T generated by the $[\psi_i(\mathbf{k})]$ ($i < r$, $k \in \omega$). We suppose, given an isomorphism of Boolean algebras, ϕ^*: $B_S^* \simeq B_T^*$ such that $\phi^*([\theta_i(\mathbf{k})]) = [\psi_i(\mathbf{k})]$.*

Then there is a formula $\psi_r(x)$ (which we can effectively find from the data $\theta_0, \ldots, \theta_r, \psi_0, \ldots, \psi_{r-1}$) such that ϕ^* prolongs to an isomorphism ϕ^{**}: $B_S^{**} \cong B_T^{**}$ such that $\phi^{**}([\theta_i(\mathbf{k})]) = [\psi_i(\mathbf{k})]$ (for $i \leqslant r$, $k \in \omega$). (Here B_S^{**} (resp. B_T^{**}) is the Boolean subalgebra of B_S (resp. B_T) generated by elements of the form $[\theta_i(\mathbf{k})]$ (resp. $[\psi_i(\mathbf{k})]$) for $i \leqslant r$, $k \in \omega$.)

In applying Lemma 2 we take for the S and T of Lemma 2 alternately the S and T of Theorem 1 and the T and S of Theorem 1. This is allright since the hypotheses of Theorem 1 apply equally to the pair $\langle S, T \rangle$ and the pair $\langle T, S \rangle$.

As our next reduction, we show that if Lemma 2 holds for $r = 1$, it holds for all r. The case $r = 0$ is easy. To prove Lemma 2 for the input θ_0, apply Lemma 2 for the data $\langle \theta_0', \theta_1', \psi_0' \rangle$, where $\theta_0'(x)$ and $\psi_0'(x)$ are the formula "$x = x$", and $\theta_1'(x)$ is $\theta_0(x)$. If $r \geqslant 2$, we prove that Lemma 2 is valid for r by induction on r. We prove Lemma 2 for the input $\langle \theta_0, \ldots, \theta_r; \psi_0, \ldots, \psi_{r-1} \rangle$ by applying Lemma 2 for the input $\langle \theta_0', \ldots, \theta_{r-1}'; \psi_0', \ldots, \psi_{r-2}' \rangle$. Here for $1 \leqslant j \leqslant r-1$, θ_j' is θ_{j+1}. $\theta_0'(2k)$ holds iff $\theta_0(k)$ holds, while $\theta_0'(2k+1)$ holds iff $\theta_1(k)$ does. The formulas ψ_j', $0 \leqslant j \leqslant r-2$, are defined analogously.

We come now to the proof of Lemma 2 for the case of $r = 1$. (This will complete the proof of Theorem 1.) We shall first sketch the proof under the following additional assumption on T: Every instance of the induction axiom expressible in L_T is a theorem of T. (This assumption is valid for P and ZF; it is not valid for GB.) We then indicate how to eliminate this assumption by slightly complicating the argument. The assumption will not be used in our detailed proof.

Since P is relatively interpretable in T, we can carry out various recursive constructions in T. Let $D = \{n \in \omega: \psi_0(n)\}$. Working in T, we will define a function $h: \omega \times \omega \to 2$ primitive recursive in D. The predicate ψ_1 will then be determined from h via the stipulation that $\psi_1(n)$ holds iff $n \in \omega$ and $\text{limit}_{s \to \infty} h(s, n) = 1$.

The use of ω in the preceding definition should be taken as referring not to the standard integers but to the integers of some model of T. Thus if M is a model of T, the relative interpretation of P in T determines a model $\langle \omega^M; \oplus^M, \otimes^M \rangle$ of P. In general, this model will not be isomorphic to the standard integers. The standard integers will be canonically identified, of course, with an initial segment of ω^M. The term D of the preceding paragraph will determine a subset D^M of ω^M. Similarly, h will determine a function h^M mapping $\omega^M \times \omega^M$ into 2.

Since we are assuming, for the moment, that all instances of induction expressible in L_T are provable in T, the existence and uniqueness of functions defined by primitive recursion relative to an oracle is unproblematical.

We think of $h(s, n)$ as our guess as to the truth value of $\psi_1(n)$ at stage s of our construction. As s increases, more and more identities about the $\theta_i(\mathbf{k})$'s ($i \leqslant 1$, $k \in \omega$) will have been proved by S. We will, in so far as possible, choose $h(s, n)$ so that if $\psi_1(n)$ were set equal to "$h(s, n) = 1$", these identities would hold true of the $\psi_0(\mathbf{k})$'s and $\psi_1(\mathbf{k})$'s.

We must also insure that each identity proved by T about the $\psi_0(\mathbf{k})$'s and $\psi_1(\mathbf{k})$'s is also an identity proved by S about the corresponding $\theta_0(\mathbf{k})$'s and $\theta_1(\mathbf{k})$'s. Now by use of the recursion theorem, we may assume during the course of the construction that we know the Gödel number of ψ_1. If some new identity is proved by T about ψ_0 and ψ_1, we see if it is possible (given our previous commitments and the known truth values of $\langle \psi_0(\mathbf{k}), k \in \omega \rangle$) to falsify this identity. If so, we stop the construction, and set $h(j, n)$ for $j \geqslant s$ to this falsifying assignment. If the construction stopped at a standard integer stage, s, T would be inconsistent. The upshot will be that T proves precisely the identities about the ψ's that S proves about the θ's.

In general, we will not have full induction in T. We get around this as follows. Let $J = \{n \in \omega: (\exists s \in \omega)\ (s \text{ codes } D \cap n)\}$. It is clear that J is an initial segment of ω, containing 0 and closed under the successor operation. The function h will now map $J \times J$ into 2. Finally, $\psi_1(n)$ will express "$n \in J$ and $(\exists x \in J)(\forall y \in J)(x \leqslant y \to h(y, n) = 1)$". In this way, we will be able to use the fact that induction holds in T for arithmetic formulas to carry through the argument.

This completes our sketch of the ideas behind our proof of Lemma 2. We now begin our formal proof.

We let B_0 be the set of formal Boolean polynomials in the variables $\nu_{i,j}$ (for $i \leqslant 1, j \in \omega$). We Gödel number the elements of B_0 in some reasonable manner. In particular, if n is the Gödel number of b, and $\nu_{i,j}$ appears in b, then $i, j \leqslant n$. We write $b(\theta)$ for the sentence of L_S obtained from b by replacing $\nu_{i,j}$ by $\theta_i(\mathbf{j})$. $b(\psi)$ is defined analogously, and is a sentence of L_T.

We shall define a certain primitive recursive function $k: \omega^4 \to 2$. Our desired function h will be obtained from k as follows: Let $s, n \in J$. Let $t \in \omega$ code $D \cap (\max(s, n) + 1)$. Let e be a Gödel number for k. Then $h(s, n)$ will equal $k(s, n, t, e)$. Note that from e, one can primitive recursively compute a Gödel number for ψ_1 (where ψ_1 is derived from h, J as described above).

We shall talk in what follows as if we are defining h directly, and during the construction we know the Gödel number of ψ_1. The crucial thing to observe about the construction is that in defining $h(s, n)$ one only need know about $D \cap (\max(s, n) + 1)$. (So h can be viewed as arising from a k as in the preceding paragraph.) In fact, it will be true that $h(s, n) = 0$ if $s < n$, $s, n \in J$, so $h(s, \cdot)$ is essentially a finite object.

We define $h(s, n)$ by induction on s for $s \in J$.

Case 1. $s = 0$. Set $h(s, n) = 0$, all $n \in J$.

Case 2. $s > 0$, $s \in J$, and for some $s_0 < s$, the construction was frozen at s_0. (The construction will be frozen at most once.) Then set $h(s, n) = h(s_0, n)$.

Case 3. Otherwise.

We first construct a list $\langle c_i, i < s\rangle$ of elements of B_0. The list will record, roughly speaking, those identities about the θ's proved in S before stage s.

If $i = 2^m 3^n$, m is the Gödel number of an element b of B_0 and n is the Gödel number of a proof in S of $b(\theta)$, then we set $c_i = b$. Otherwise, set $c_i = 1$.

Let $t \leqslant s$ be maximal such that there is a truth assignment to the variables appearing in

$$c^t = \bigwedge_{i<t} c_i$$

which makes it true, and which, for each $\nu_{0,j}$ appearing in c^t, gives it the truth value of $\psi_0(\mathbf{j})$. (Here we use the oracle D, but only for $j \leqslant s$.)

We now study whether or not we can spoil the theorem proved by T at stage s.

Case 3A. $s = 2^n 3^m$; n is the Gödel number of a $b \in B_0$; m is the Gödel number of a proof, in T, of $b(\psi)$, and there is a truth assignment τ which makes c^t true, makes b false, and gives each $\nu_{0,j}$ the same truth value as $\psi_{0,j}$.

In this case, we will take the least such assignment (with respect to an ordering to be defined in a moment), say τ_0 and set $h(s, j) = \tau_0(\nu_{1,j})$. If $\nu_{1,j}$ does not appear in b or c^t, set $h(s, j) = 0$. (So, in particular, if $j > s$, $h(s, j) = 0$.) We also freeze the construction at stage s.

The ordering on assignments we use is simply the lexicographic ordering. If τ_0 and τ_1 are distinct assignments, $\tau_0 < \tau_1$, if, for the least j such that $\tau_0(\nu_{1,j}) \neq \tau_1(\nu_{1,j})$, we have $\tau_0(\nu_{1,j}) < \tau_0(\nu_{1,j})$.

Case 3B. Case 3A does not apply.

Let τ_0 be the least truth assignment (to the variables appearing in c^t) such that:

(1) $\tau_0(\nu_{0,j})$ has the same truth value as $\psi_0(\mathbf{j})$;

(2) $\tau_0(c^t) = 1$.

For j such that $\tau_0(\nu_{1,j})$ is defined, set $h(s, j) = \tau_0(\nu_{1,j})$. If $\nu_{1,j}$ does not appear in c^t, set $h(s, j) = 0$.

This completes the definition of h.

We now prove a series of claims which will eventually culminate in the proof of Lemma 2.

Claim 1. The following are provable in T:

(1) For $s < n$, $n \in J$, we have $h(s, n) = 0$.

(2) The set $\{\langle i, h(s, i)\rangle : i \leqslant s\}$ has an integer code for $s \in J$.

(3) Let $s_1 < s_2$, with $s_2 \in J$. Assume there exists an n for which $h(s_1, n) \neq h(s_2, n)$. Then there is a least such n and, for this least n, $h(s_1, n) < h(s_2, n)$ (i.e., $h(s_1, \cdot)$ lexicographically precedes $h(s_2, \cdot)$.)

PROOF. (1) is evident from the construction. (2) is clear since $h(s, \cdot)$ is definable in arithmetic from a code for $D \cap (s + 1)$ (which exists since $s \in J$.) In proving (3), we may assume that $0 < s_1$ and that the construction has not been frozen at a stage prior to s_2. That there is a least n such that $h(s_2, n) \neq h(s_1, n)$ is evident from (1) and (2) of this claim. Suppose $h(s_2, n) < h(s_1, n)$. Let t_1, t_2 be the analogues at stages s_1, s_2 of the t in the description of our action in Case 3. (Clearly $t_1 \leqslant t_2$.) Let τ_1, τ_2 be the corresponding least assignments. Since τ_2 makes c^{t_2} true, it makes c^{t_1} true. But then the restriction of τ_2 to the domain of τ_1 is less than τ_1. This contradicts our choice of τ_1 as the least assignment agreeing with D on the $\nu_{0,j}$'s making c^{t_1} true.

Claim 2. For each $n \in \omega$, T proves

$$(\exists s_n)(s_n \in J \wedge (\forall s \geqslant s_n))(s \in J \rightarrow h(s, n) = h(s_n, n)).$$

REMARK. This asserts that

$$\lim_{s \in J, s \rightarrow \infty} h(s, n)$$

exists for each *standard* n. I do not know how to prove, in T, that this limit exists for all n in J.

PROOF. We prove the claim, in the metatheory, by induction on n. By our induction hypothesis, T proves

$$(\exists s^* \in J)(\forall s \in J)(s \geqslant s^* \rightarrow h(s, i) = h(s^*, i) \text{ for } i < n).$$

By (3) of Claim 1, if $s^* \leqslant s_1 < s_2$, with $s_2 \in J$, $h(s_1, n) \leqslant h(s_2, n) \leqslant 1$. It is now easy to see that the desired limit exists.

Claim 3. For each *standard* s, T proves: The construction does not freeze at stage s.

PROOF. Suppose not. Fix $s \in \omega$ least such that the claim is false. Let M be a model of T in which the construction freezes at stage s. Then Case 3A applies at s. So the following sentence will hold in M:

"$s = 2^m 3^n$, n is the Gödel number of a $b \in B_0$, and n is the Gödel number of a proof in T, of $b(\psi)$".

The phrase in quotes will be primitive recursive (provided our various implicit Gödel numberings are done in the usual (primitive recursive) way). Hence it is really true of s. So $b(\psi)$ holds in M, since M is a model of T. On the other hand, our choice of $h(s, \cdot)$ and the fact that the construction was frozen at stage s insures the following:

(1) $\psi_1(k)$ has the truth value of "$h(s, k) = 1$";

(2) $\sim b(\psi)$ holds in M.

But (2) contradicts a previous remark. Thus M cannot exist, and Claim 3 is proved.

Claim 4. Let $b \in B_0$. Suppose that $b(\theta)$ is a theorem of S. Then $b(\psi)$ is a theorem of T.

PROOF. Let m be the Gödel number of b, and let n be the Gödel number of a proof in S of $b(\theta)$. Let $s = 2^m 3^n$. By Claim 3, T proves: The construction does not freeze at any stage $s_0 \leqslant s$.

We follow the notation used in the description of Case 3 of our construction. So $b = c_s$. We claim T proves that $t(s) = s$. Suppose not. Let M be a model of T such that, in M, $t(s) < s$. We may find a model, N, of S such that for $0 \leqslant i \leqslant s$,

$$M \models \psi_0(\mathbf{i}) \Leftrightarrow N \models \theta_0(\mathbf{i}).$$

(Here is where we use the key hypothesis that ϕ^* exists.) Now $c^s(\theta)$ is a theorem of S. Hence $N \models c^s(\theta)$. It follows that the assignment τ which gives each variable $\nu_{i,j}$ appearing in c^s the truth value of $\theta_i(\mathbf{j})$ makes c^s true and is compatible with D. This contradicts our choice of M and establishes that T proves $t(s) = s$.

Now argue in T. It follows easily by induction on $s' \in J$ that $s' \geqslant s$ implies that the assignment $\tau_{s'}$, which sets $\nu_{0,i}$ equal to the truth value of $\psi_0(\mathbf{i})$ and sets $\nu_{1,i}$ equal to $h(s', i)$, makes b true. (Since for $s' \geqslant s$, $t(s') \geqslant t(s) = s$ and $c_s = b$.) It is important that only arithmetic induction, which is available in T, is used here. By Claim 2, it follows that $b(\psi)$ is true.

Claim 5. Let $b \in B_0$ and let T prove $b(\psi)$. Then S proves $b(\theta)$.

PROOF. Let m be the Gödel number of b, and let n be the Gödel number of a proof in T of $b(\psi)$. Let $s = 2^m 3^n$.

The free Boolean algebra generated by $\{\nu_{0,j}, j \leqslant s\}$ is finite. It follows that there is an element b^* of B_0 with the following properties: (1) If $\nu_{i,j}$ appears in b^*, then $i = 0$, and $j \leqslant s$. (2) T proves $b^*(\psi)$. (3) If $b_1 \in B_0$ satisfies the analogue of

(1) and T proves $b_1(\psi)$, then $b^* \to b_1$ is a tautology. (b^* is equivalent to the conjunction of all b_1 such that T proves $b_1(\psi)$, and b_1 contains $v_{i,j}$ only if $i = 0$, $j \leqslant s$.) It follows from the existence of the partial isomorphism ϕ^* that S proves $b^*(\theta)$.

By the argument used in Claim 4, T proves $t(s) = s$. I claim $(c^s \wedge b^*) \to b$ is a tautology. If not, there is a truth assignment τ making $c^s \wedge b^*$ true and making b false. Since $\tau(b^*) = 1$, we can find a model M of T such that

$$M \vDash \psi_0(\mathbf{i}) \Leftrightarrow \tau(v_{0,i}) = 1.$$

But then $M \vDash$ "The construction is frozen at stage s". This contradicts Claim 3.

Now clearly S proves $c^s(\theta)$ and we have already remarked that S proves $b^*(\theta)$. Since $(c^s \wedge b^*) \to b$ is a tautology, S proves $b(\theta)$.

The existence of the partial isomorphism ϕ^{**} asserted in Lemma 2 is immediate from Claims 4 and 5. Thus Lemma 2 (and hence Theorem 1) is completely proved.

3. Sets with recursive structure. It will be convenient to develop some foundational material on "sets with recursive structure" before turning to the proofs of Theorems 2 through 4. The definitions we present work well for the examples that we have in mind, namely, ω and the underlying set of the Lindenbaum algebra, B_T, of a reasonable theory T. (They would not be appropriate for ω_1^{CK}, the least nonrecursive ordinal.)

A *prerecursive structure on a set* X is a collection of functions, $\mathbf{F} \subseteq \{f: f: \omega \to X\}$. We require that whenever $f \in \mathbf{F}$ and $h: \omega \to \omega$ is recursive, then $f \circ h \in \mathbf{F}$. (We think of $\mathbf{F}$ as the "recursive functions" from ω to X.)

Let $\langle X, \mathbf{F}\rangle$, $\langle Y, \mathbf{G}\rangle$ be sets with prerecursive structures. A map $\Phi: X \to Y$ is *recursive* iff whenever $f \in \mathbf{F}$, the composition $\Phi \circ f \in \mathbf{G}$.

A prerecursive structure on X, $\mathbf{F}$ say, is *recursive* if there is an $h \in \mathbf{F}$ such that: (1) h is surjective; (2) $\mathbf{F} = \{h \circ g: g \text{ is a recursive map from } \omega \text{ to } \omega\}$; (3) $\{(i, j): h(i) = h(j)\}$ is recursively enumerable (r.e.).

PROPOSITION 1. *Let $\langle X, \mathbf{F}\rangle$ be a set with recursive structure. Let $h_0 \in \mathbf{F}$ satisfy* (1)–(3) *of the preceding definition. Let $h_1 \in \mathbf{F}$ map ω onto X. Then h_1 also satisfies* (1)–(3) *of the preceding definition.* (*We say in this case that h_1 is a structure map for X.*)

PROOF. Clearly h_1 satisfies (1). By (2) applied to h_0, there is $g_0: \omega \to \omega$ such that $h_1 = h_0 \circ g_0$. Let $E_i = \{(m, n): h_i(m) = h_i(n)\}$. Then

$$E_1 = \{(m, n): \exists\, m_1, n_1 \in \omega: m_1 = g_0(m), n_1 = g_0(n) \text{ and } (m_1, n_1) \in E_0\}.$$

By assumption, E_0 is r.e. It follows that E_1 is r.e. Hence h_1 satisfies (3). By the definition of prerecursive structure, it is clear that if $g: \omega \to \omega$ is recursive, $h_1 \circ g \in \mathbf{F}$. Thus it suffices to prove that if $f \in \mathbf{F}$, then $\exists\, g: \omega \to \omega$, recursive, such that $f = h_1 \circ g$. A moments reflection shows that it suffices to prove this when f is h_0. The general case will follow since (2) is true of h_0. Now $E^* = \{(i, j): (g_0(i), j) \in E_0\}$ is r.e. and, since h_1 is surjective, $(\forall j)(\exists\, i)((i, j) \in E^*)$. But we need only take $g(j)$ to be the i such that (i, j) is the first pair to appear in E^* with second component j (relative to some fixed recursive enumeration of E^*). (For then $h_1 \circ g(j) = h_0 \circ g_0 \circ g(j) = h_0(j)$.) This completes the proof.

It follows that if h: $\omega \to X$ is surjective, and $\{(i, j): h(i) = h(j)\}$ is r.e., then h determines a recursive structure on X (for which $\mathbf{F} = \{h \circ g$: g: $\omega \to \omega$ *is recursive*$\}$). Any $h' \in \mathbf{F}$ which is surjective will be, by the preceding proposition, a structure map for this structure.

PROPOSITION 2. *Let* $\langle X, \mathbf{F}\rangle$, $\langle Y, \mathbf{G}\rangle$ *be sets with recursive structure and let* f: $\omega \to X$, g: $\omega \to Y$ *be structure maps for* X *and* Y. *Let* h: $X \to Y$. *Then the following are equivalent*:

(1) h *is recursive*.

(2) *There is a recursive* H: $\omega \to \omega$ *such that the following diagram commutes*:

$$\begin{array}{ccc} \omega & \xrightarrow{H} & \omega \\ \downarrow f & & \downarrow g \\ X & \xrightarrow{h} & Y \end{array}$$

PROOF. Left as an exercise for the reader.

PROPOSITION 3. *Let* $\langle X, \mathbf{F}\rangle$ *be a recursive structure on* X. *Let* g: $\omega \to X$, $g \in \mathbf{F}$, *map onto* $X_0 \subseteq X$. *Then* X_0 *inherits a recursive structure from* X *for which* g *is a structure map. The corresponding collection of maps* $\mathbf{F}_0$ *is precisely* $\{f \in \mathbf{F}$: range$(f) \subseteq X_0\}$.

PROOF. Left as an exercise.

PROPOSITION 4. *Let* $\langle X, \mathbf{F}\rangle$ *be a set with recursive structure. Let* f: $X \to X$ *be recursive and a bijection. Then* f^{-1} *is recursive*.

PROOF. Let h: $\omega \to X$ be a structure map for $\langle X, \mathbf{F}\rangle$. Then $f \circ h$ is recursive and surjective. Hence $f \circ h$ is also a structure map for $\langle X, \mathbf{F}\rangle$, by Proposition 1. It follows that $\exists\, g$: $\omega \to \omega$ recursive such that $f \circ h \circ g = h$.

Let $y = f(x)$, $h(n) = y$. Then $y = h(n) = f(h(g(n)))$. So, since f is 1-1, $h(g(n)) = x = f^{-1}(y)$; i.e., g: $\omega \to \omega$ is such that $h(g(n)) = f^{-1}(h(n))$. Since g is recursive, so is f^{-1} by Proposition 2.

PROPOSITION 5. *Let* X, Y *and* Z *be sets with recursive structure, and let* f: $X \to Y$ *and* g: $Y \to Z$ *be recursive. Then the composition* $g \circ f$: $X \to Z$ *is recursive*.

PROOF. Left to the reader.

4. Proofs of Theorems 2 through 4. The usual Gödel numbering gives rise to a map h_S of ω onto the Lindenbaum algebra of S. $h_S(i)$ is obtained by finding the ith number, n_i, which is the Gödel number of a sentence of L_S, say ψ_i, and setting $h_S(i) = [\psi_i]$. Clearly, h_S determines a recursive structure on $|B_S|$ for which h_S is a structure map. The map g: $\omega \to |B_S|$ is recursive iff $\exists\, G$: $\omega \to \omega$ such that, for all i, $G(i)$ is the Gödel number of a representative of $g(i)$.

Moreover, Proposition 2 of §3 just says that the notions of a map $f\colon |B_S| \to |B_T|$ being recursive in the sense of §3, and the definition offered in the Introduction (just prior to the statement of Theorem 2) are equivalent.

Note next that the direction (1) → (2) of Theorem 2 is immediate from the definitions.

For the converse, let $f\colon |B_T| \to |B_T|$ be recursive, and let $\phi(x)$ be a formula of L_T with one free variable such that $\mathrm{Range}(f) \subseteq X_\phi = \{[\phi(\mathbf{n})]\colon n \in \omega\}$. Then X_ϕ inherits a recursive structure from B_T as discussed in Proposition 3 of §3, and the map $f\colon |B_T| \to X_\phi$ is recursive. Moreover, a structure map for X_ϕ is the map $\{n \to [\phi(\mathbf{n})]\}$. It follows from Proposition 2 of §3 that there is a recursive function $g\colon \omega \to \omega$ such that if ψ is a sentence of L_T with Gödel number n, then $f([\psi]) = [\phi(\mathbf{g(n)})]$.

We may easily construct a formula $G(x, y)$ that numeralwise represents g in the following sense:

(1) T proves "$(\forall x, y)G(x, y) \to x$ and y are integers". (The phrase "integer" is expressed in L_T using the given relative interpretation of P in T);

(2) T proves "$\forall$ integer $x\ \exists !y\, G(x, y)$";

(3) let $k \in \omega$, and $m = g(k)$. Then T proves $G(\mathbf{k}, \mathbf{m})$.

Let $\psi(x)$ be the following formula:

$$(\exists y)[G(x, y) \wedge \phi(y)].$$

Then, clearly, if $k \in \omega$, $T \vdash \psi(\mathbf{k}) \leftrightarrow \phi(\mathbf{g(k)})$. So $[\psi(\mathbf{k})] = [\phi(\mathbf{g(k)})]$. If k is the Gödel number of a sentence θ of L_T, then by choice of g, $[\psi(\mathbf{k})] = f([\theta])$. So ψ is functional, and the associated map of B_T into itself is the f we started with. The proof of Theorem 2 is complete.

Before beginning the detailed proof of Theorem 3, we sketch the main ideas behind it. Let $\langle k_1, k_2, k_3\rangle$ be a triple of maps from A_T. Then under favorable circumstances (which we shall detail in a moment) this triple will determine a surjection of ω onto $|B_T|$. This surjection will turn out to be recursive; it will also turn out that the favorable circumstances do occur. Since the "favorable circumstances" depend only on the algebraic structure of the fixed-point algebra, it will follow that the algebraic structure determines the recursive structure. The conditions that the triple must satisfy are as follows:

(1) k_1 is an injective map;

(2) define $g\colon \omega \to |B_T|$ from k_2 as follows: $g(0)$ is the zero of the Boolean algebra B_T and $g(n + 1)$ is $k_2(g(n))$; then g is injective;

(3) define a map $h\colon \omega \to |B_T|$ as follows: $h(n) = y$ iff $k_1(y) = k_3(g(n))$; then h is defined on all of ω and maps onto $|B_T|$.

(The map h is the surjection determined by the triple.)

This completes our sketch of the proof of Theorem 3. The detailed proof will be preceded by a series of lemmas.

LEMMA 1. *Let T be a reasonable theory. Then there is a subalgebra B_T^* of B_T and an $h \in A_T$ such that h is an isomorphism of Boolean algebras mapping B_T onto B_T^*.*

PROOF. In view of Theorem 1, it suffices to prove Lemma 1 for any one reasonable theory T_0. It will then follow for all reasonable theories. We take T_0 to be Zermelo set theory. We let $B^*_{T_0}$ be the set of equivalence classes of arithmetical formulas. Since arithmetical truth is definable in Zermelo set theory, $B^*_{T_0}$ is small. By a back-and-forth argument similar to that used to prove Theorem 1, we get an isomorphism $h\colon B_{T_0} \to B^*_{T_0}$ which is "recursive in the codes". By Theorem 2, h lies in A_{T_0}. This proves Lemma 1.

LEMMA 2. *Let* $f \in A_T$. *Define* $g\colon \omega \to B_T$ *by* $g(0) = 0$, $g(n+1) = f(g(n))$. *Then* g *is recursive. We can choose* f *so that* g *is an injective map.*

PROOF. By picking a formula ϕ which is functional and such that $f = f_\phi$, we see easily that there is a $G\colon \omega \to \omega$ such that $G(n)$ is the Gödel number of a representative of $g(n)$. Hence g is recursive. In proving the last claim, by Theorem 1, it suffices to prove it for a particular theory T_0. We take T_0 to be P, and take for f the map f_ϕ for $\phi(x)$ the formula "x is a theorem of P". In this case, the map g is clearly injective. (See, for example, [**1**].)

LEMMA 3. *Let* $f\colon B_T \to B_T$ *be as in Lemma* 2 (*so that* $g\colon \omega \to B_T$ *is injective*). *Let* $S \subseteq B_T$ *be small. Let* $H\colon \omega \to S$ *be recursive. Then there is an* $h \in A_T$ *such that* $h(g(n)) = H(n)$.

PROOF. Since S is small, there is a formula $\phi(x)$ (with one free variable) of L_T such that $S \subseteq \{[\phi(\mathbf{n})]\colon n \in \omega\}$. It follows from Propositions 3 and 2 of §3 that there is a recursive function $h_1\colon \omega \to \omega$ such that

$$H(n) = [\phi(\mathbf{h}_1(\mathbf{n}))].$$

Let $G\colon \omega \to \omega$ be a recursive function such that $G(n)$ is the Gödel number of a representative of $g(n)$. Since g is 1-1, so is G. We may assume G is monotone increasing. It then follows that range G is recursive.

We may assume that the Gödel numbering of proofs of T has the following property: If n is the Gödel number of a proof of ψ, and ψ_1 is a subformula of ψ, then ψ_1 has Gödel number less than n.

Let S_n be the set of $m < n$ which are Gödel numbers of sentences of L_T. Let R_n be the smallest equivalence relation on S_n such that if m_1, m_2 in S_n are Gödel numbers of sentences ψ_1, ψ_2, and $\psi_1 \leftrightarrow \psi_2$ has a proof in T with Gödel number less than n, then $m_1 R_n m_2$. Clearly S_n, R_n can be effectively computed from n.

Define a partial recursive function h_2 as follows: $h_2(m)$ will be defined only if m is the Gödel number of a sentence of L_T. For such an m, h_2 searches for the least r such that for some $n \in S_r$, $m R_r n$, and $n \in \text{range}(G)$. It then takes the least such n and determines the unique k such that $n = G(k)$. Then the value of $h_2(m)$ will be $h_1(k)$.

We let $\psi_2(x, y)$ be a formula of L_T expressing: "x and y are integers and $y = h_2(x)$" (relative to the given relative interpretation of P in T).

Then we may assume $\psi_2(x, y)$ has the following properties:

(1) T proves "$(\forall x)(\forall y)[\psi_2(x, y) \to x$ and y are integers]".

(2) T proves "$(\forall x)(\forall y)(\forall z)[\psi_2(x, y) \wedge \psi_2(x, z) \to y = z]$".

(3) For each $m \in$ domain h_2, if $k = h_2(m)$, then T proves "$\psi_2(\mathbf{m}, \mathbf{k})$".

(4) Let m_1, m_2 be Gödel numbers of sentences χ_1 and χ_2 of L_T. Suppose $[\chi_1] = [\chi_2]$. Then T proves

$$(\forall y)[\psi_2(\mathbf{m}_1, y) \leftrightarrow \psi_2(\mathbf{m}_2, y)].$$

We sketch the proof of (4). Let r be the least integer which is the Gödel number of a proof, in T, of $\chi_1 \leftrightarrow \chi_2$. Then $m_1 R_r m_2$ and hence, for all $s \geqslant r$, $m_1 R_s m_2$. Now since $g\colon \omega \to B_T$ is injective, $G(k) R_r m_1$ for at most one $k \in \omega$. If there is such a k, then clearly $h_2(m_1) = h_2(m_2) = k$. If not, T can reason as follows: if at any later stage s, $m_1 R_s G(k)$ for some k, then also $m_2 R_s G(k)$ (and conversely) since $m_1 R_s m_2$. Hence $h_2(m_1)$ is defined iff $h_2(m_2)$ is defined, and if defined they have the same value.

Now consider the formula $\psi(x)$ displayed below:

$$(\exists y)[\psi_2(x, y) \wedge \phi(y)].$$

It follows from (4) that $\psi(\ )$ is functional, and so determines an element h of A_T. We show that $h(g(n)) = H(n)$. This will complete the proof of Lemma 3. Now $h(g(n)) = [\psi(\mathbf{G(n)})]$. If we examine the definition of h_2, we see that $h_2(G(n)) = h_1(n)$. Using the definition of ψ and properties (2) and (3) of ψ_2, we see that T proves "$\psi(\mathbf{G(n)}) \leftrightarrow \phi(\mathbf{h}_1(\mathbf{n}))$". But h_1 was chosen so that $[\phi(\mathbf{h}_1(\mathbf{n}))] = H(n)$. Thus $h(g(n)) = H(n)$, and the proof of Lemma 3 is complete.

We turn now to the proof of Theorem 3. Let $f\colon B_S \to B_T$ be an isomorphism that prolongs to an isomorphism of A_S with A_T. Let $h_S\colon \omega \to B_S$, and $h_T\colon \omega \to B_T$ be the surjections that give the canonical recursive structures on B_S, B_T. We must show that there is a recursive map $F\colon \omega \to \omega$ such that $h_T(F(n)) = f(h_S(n))$ for all $n \in \omega$.

Let $k_1\colon B_S \cong B_S^*$, $k_1 \in A_S$ be provided by Lemma 1. Let $k_2\colon B_S \to B_S$, $k_2 \in A_S$ be provided by Lemma 2 such that if $g(0) = 0$, and $g(n+1) = k_2(g(n))$, then g is injective. Let $k_3 \in A_S$ be provided by Lemma 3 such that $k_3(g(n)) = k_1(h_S(n))$.

Let $f_*\colon A_S \cong A_T$ be the isomorphism induced by f. Let k_1', k_2' and k_3' be $f_*(k_1)$, $f_*(k_2)$ and $f_*(k_3)$, respectively. Let g' be $f \circ g$. (Note that g' is recursive since it is defined from k_2' the way that g is defined from k_2.)

Now $k_1'(f(h_S(n))) = f(k_1(h_S(n))) = f(k_3(g(n))) = k_3'(g'(n))$.

Thus given n, we can effectively find a Gödel number of a representative of $k_1'(f(h_S(n)))$ (since k_3' and g' are recursive). Let $F_1\colon \omega \to \omega$, be such that $F_1(n)$ is the Gödel number of a representative of $k_1'(f(h_S(n)))$. Let $F_2\colon \omega \to \omega$, recursive, be such that $F_2(n)$ is the Gödel number of a representative of $k_1'(h_T(n))$. Define $F_3\colon \omega \to \omega$ thus: $F_3(n)$ is the least number of the form $2^r 3^s$, where s is the Gödel number of a proof in T of the equivalence of the sentences with Gödel numbers $F_1(n)$ and $F_2(r)$.

F_3 is clearly recursive, if total. But if $f(h_S(n)) = h_T(r)$, then $F_1(n)$ and $F_2(r)$ are both Gödel numbers of representatives of the same element of B_T (to wit $k_1'(f(h_S(n)))$). It follows that $F_3(n)$ is defined for all n. Let $F(n)$ be the unique r

such that for some s, $F_3(n) = 2^r3^s$. Then, by the definition of F_3, $k_1'(h_T(F(n))) = k_1'(f(h_S(n)))$. Since k_1' is injective, it follows that $h_T \circ F = f \circ h_S$. Since F is clearly recursive, Theorem 3 now proved.

The Corollary to Theorem 3 is immediate. Any automorphism of $\langle B_S, A_S \rangle$ is a recursive map from $|B_S|$ to $|B_S|$ and there are only countably many such maps (as is clear from Proposition 2 of §3). An alternative way to see the Corollary is to note that from k_1, k_2 and k_3 we can canonically define a surjection of ω onto $|B_S|$. Thus any automorphism that fixes k_1, k_2 and k_3 must be the identity. Since A_S is countable, it follows that the automorphism group of $\langle B_S, A_S \rangle$ is countable.

It is now quite easy to prove Theorem 4. We first prove that (1) $\to$ (2). By Theorem 3, if ϕ is an automorphism of $\langle B_T, A_T \rangle$ then ϕ is recursive. To see that $X \subseteq B_T$ is small iff $\phi[X]$ is small, it suffices to define the family of small sets in terms of the structure $\langle B_T, A_T \rangle$. But from Theorem 2 and Lemma 3, it follows that $X \subseteq B_T$ is small iff there is an $f \in A_T$ such that $X \subseteq \text{range}(f)$. So (1) $\to$ (2).

Conversely, let ϕ be an automorphism of B_T which satisfies the conditions of (2). Then (by Proposition 4 of §3), ϕ^{-1} also satisfies (2). It suffices to show that if $h \in A_T$, then so is $\phi \circ h \circ \phi^{-1}$. For then, an analogous argument will show: If $h \in A_T$, so is $\phi^{-1} \circ h \circ \phi$. It will follow that $\phi \circ A_T \circ \phi^{-1} = A_T$, and hence that ϕ satisfies (1).

So assume that $h \in A_T$. As the composition of three recursive maps, $\phi \circ h \circ \phi^{-1}$ is recursive. Since $\text{range}[\phi \circ h \circ \phi^{-1}] \subseteq \phi[\text{range}(h)]$, $\text{range}[\phi \circ h \circ \phi^{-1}]$ is small. By Theorem 2, $\phi \circ h \circ \phi^{-1} \in A_T$. The proof of Theorem 4 is complete.

REFERENCES

1. G. Boolos, *On deciding the truth of certain statements involving the notion of consistency*, J. Symbolic Logic **41** (1976), 779–781.
2. J. R. Shoenfield, *Mathematical logic*, Addison-Wesley, Reading Mass., 1967.
3. C. Smoryński, *Fixed point algebras*, Bull. Amer. Math. Soc. (N.S.) **6** (1981), 317–356.
4. A. Tarski, A. Mostowski and R. M. Robinson, *Undecidable theories*, North-Holland, Amsterdam, 1953.

DEPARTMENT OF MATHEMATICS, UNIVERSITY OF CALIFORNIA, BERKELEY, CALIFORNIA 94720

Proceedings of Symposia in Pure Mathematics
Volume 42, 1985

The "Slow-Growing" Π_2^1 Approach to Hierarchies

S. S. WAINER

The aim of recursion-theoretic hierarchies is to assign ordinal notations to functions in such a way as to reflect, as closely as possible, their computational complexity—whether this be measured in terms of absolute computability or computability relative to some higher-type oracle such as ${}^2E$. We develop here a new "refined" approach to this problem, by means of which each Kleene-computation in the maximal type-structure over the integers N occurs as a natural collapse (under the "slow-growing" function G) of an identical computation defined over abstract ordinal notations. A consequence of this is that Kleene's **O** appears as the collapse of a Π_2^1-complete set of "notations over notations". Thus the partial recursive type-3 operator which generates hierarchies (we follow Girard [5] in calling it Λ although our version arises somewhat differently from his) will, when applied over abstract ordinal notations, generate Π_2^1 hierarchies which under G collapse down to the corresponding Π_1^1 hierarchies over N. As an immediate application of this we can read of Girard's result that the transfinite Grzegorczyk hierarchy is the collapse of the Bachmann hierarchy so that, for example, the ordinal naturally assigned in this way to the ε_0th Grzegorczyk function is now the Howard ordinal [6]—here, $\phi_{\varepsilon_{\omega_1+1}+1}(0)$. Following Girard's first proof of this result, various other elementary treatments of it were found and it is the idea of our [4] which underlies the more general framework set out here. We hope that this framework will provide a useful and more down to earth alternative approach to at least part of Girard's theory of Π_2^1-logic.

1. The collapsing map G. The main obstacle to any direct assignment of functions to ordinals is simply that the ordinal successor function is not continuous. The first lesson we learn from Girard [5] is that this obstacle can be overcome by working instead with *direct* limits inside the category of ordinals

1980 *Mathematics Subject Classification*. Primary 03D55; Secondary 03D65, 03F15.

with increasing functions as morphisms. Then each countable ordinal α can be expressed as the direct limit of various ω-sequences of integers (i.e. functions from N to N), depending on the way in which α is presented. Thus if α is the direct limit of a sequence $(n_x^\alpha)_{x \in N}$ then $\alpha + 1$ is the direct limit of the sequence $(n_x^\alpha + 1)_{x \in N}$ and if $\alpha_0 < \alpha_1 < \cdots$ is a given fundamental sequence with limit α such that the direct systems representing $\alpha_0, \alpha_1, \ldots$ mesh together in an appropriate way, then α can be expressed as the direct limit of the sequence $(n_x^{\alpha_x})_{x \in N}$. There is a clear dependence, here, on the choice of fundamental sequences used in representing a given ordinal so it is more natural for us to regard α above as ranging over ordinal *notations* rather than merely the set-theoretic ordinals $|\alpha|$ which they represent. If we now let $G(\alpha)$ denote the function $x \mapsto n_x^\alpha$ constructed above, then, loosely speaking, α *is* the direct limit of $G(\alpha)$, and $G(\alpha)$ satisfies the recursive definition

$$\begin{aligned} G(0)(x) &= 0, \\ G(\alpha+1)(x) &= G(\alpha)(x) + 1, \\ G(\alpha)(x) &= G(\alpha_x)(x) \quad \text{if } |\alpha| = \lim_x |\alpha_x|. \end{aligned}$$

G is the simplest, and the "slowest growing", hierarchy of functions that we could naturally envisage writing down—note that, in contrast with faster growing hierarchies based on iterations, etc., its definition is given *pointwise* for each x. To illustrate the slow growth of G consider ordinals written in Cantor normal form to the base ω together with their standard fundamental sequences induced by $\omega_x = x$. Then for each x, it is easy to check that

$$G(\omega^{\alpha_1} + \omega^{\alpha_2} + \cdots \omega^{\alpha_k})(x) = x^{G(\alpha_1)(x)} + x^{G(\alpha_2)(x)} + \cdots + x^{G(\alpha_k)(x)},$$

so in particular, we have

$$G(\omega^n)(x) = x^n, \qquad G(\omega^\omega)(x) = x^x,$$

$$G(\omega^{\omega^\omega})(x) = x^{x^x}, \ldots, \quad \text{and} \quad G(\varepsilon_0)(x) = x^{x^{\cdot^{\cdot^{\cdot^x}}}} \Big\} x-1 \text{ times}.$$

Thus the first nonelementary function in this hierarchy occurs at the first "elementary-closed" ordinal ε_0 and similarly we shall later see that the first nonprimitive recursive function occurs at the first prim-closed ordinal—in our notation, $\phi_{\omega+1}(0)$.

The foregoing is all by way of introduction. We are going to develop a more general version of G by a different approach, so that all functions from N to N will appear in its range. Thus we will need to adopt a very general notion of "ordinal notation"—in fact, we shall simply take all countable well-founded trees having a natural ordinal-like structure. (It must be admitted that to call the resulting G "slow-growing" is something of a misnomer, since our general development will at first have nothing to do with rate-of-growth. However, our later applications and the initial motivation for G came out of such considerations and the name seems to have stuck.)

DEFINITION 1. The set Ω of *abstract ordinal notations* or *tree ordinals* is the collection of infinitary terms $\alpha, \beta, \gamma, \ldots$, inductively defined by

$$0 \in \Omega, \quad \alpha \in \Omega \Rightarrow \alpha +_0 1 \in \Omega, \quad \forall x \in N(\alpha_x \in \Omega) \Rightarrow (\alpha_x)_{x \in N} \in \Omega.$$

Each $\alpha \in \Omega$ represents an ordinal $|\alpha|$ in the obvious way:

$$|0| = 0, \quad |\alpha +_0 1| = |\alpha| + 1, \quad |(\alpha_x)_{x \in N}| = \sup_x |\alpha_x|.$$

We often denote sequences $(\alpha_x)_{x \in N}$ by "$\sup \alpha_x$".

Now suppose C is any map from Ω to N^N. It is useful to analyse the action of C in a trivial sheaf-theoretic context as follows.[1] Let $\underline{\underline{\Omega}} \times N$, $\underline{\underline{N}} \times N$ be the sheaves over N whose stalks are copies of the structures $\underline{\underline{\Omega}} = (\Omega, +_0 1, 0)$, $\underline{\underline{N}} = (N, +1, 0)$ respectively and let $\Gamma(\underline{\underline{\Omega}} \times N)$, $\Gamma(\underline{\underline{N}} \times N)$ denote their respective sets of cross-sections. Identifying functions with their sequences of values we then have $\Gamma(\underline{\underline{\Omega}} \times N) = \{(\alpha_x)_{x \in N} | \forall x \in N(\alpha_x \in \Omega)\}$ and $\Gamma(\underline{\underline{N}} \times N) = N^N$.

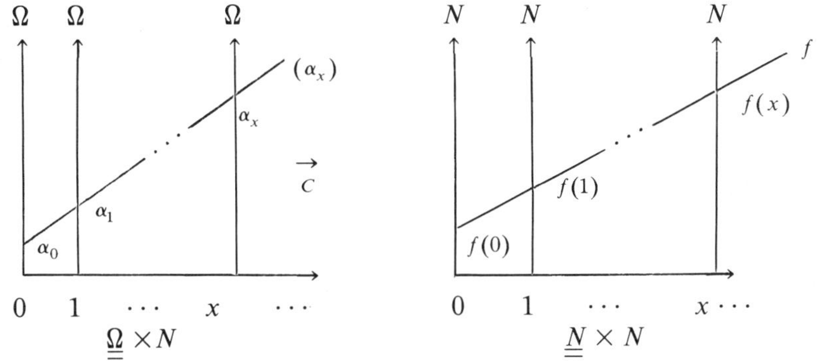

There are two ways in which C defines a map from $\Gamma(\underline{\underline{\Omega}} \times N)$ to $\Gamma(\underline{\underline{N}} \times N)$:

(i) *Locally*. Let $C_x(\alpha) = C(\alpha)(x)$ for each $x \in N$. Then $C_x \colon \Omega \to \overline{N}$ for each x, so C induces the standard sheaf-theoretic map

$$\Gamma(C)\colon \Gamma(\underline{\underline{\Omega}} \times N) \to \Gamma(\underline{\underline{N}} \times N)$$

given by $\Gamma(C)((\alpha_x)_{x \in N}) = f$ where $f(x) = C_x(\alpha_x)$.

(ii) *Globally*. Since each cross-section $(\alpha_x)_{x \in N}$ of $\underline{\underline{\Omega}} \times N$ is itself a member of Ω, $C((\alpha_x)_{x \in N}) \in N^N$ and so $C\colon \Gamma(\underline{\underline{\Omega}} \times N) \to \Gamma(\underline{\underline{N}} \times N)$.

DEFINITION 2. G is that sheaf-morphism C from $\underline{\underline{\Omega}} \times N$ to $\underline{\underline{N}} \times N$ such that, on cross-sections, $C = \Gamma(C)$, i.e. G is the map from Ω to N^N defined, pointwise at x, by

$$G_x(0) = 0, \quad G_x(\alpha +_0 1) = G_x(\alpha) + 1, \quad G_x(\sup \alpha_x) = G_x(\alpha_x).$$

Clearly G collapses $\underline{\underline{\Omega}} \times N$ onto $\underline{\underline{N}} \times N$ and so for each $f \in N^N$ there will be many $\alpha \in \Omega$ such that $G(\alpha) = f$. Each such α is therefore a simple kind of computation-tree for f: to evaluate $f(x) = G_x(\alpha)$ one takes the path through α obtained by choosing the xth branch issuing from each sequence $(\beta_x)_{x \in N}$ encountered. Then $f(x) =$ the number of successors occurring along that path.

[1] We thank S. Feferman for first suggesting this approach.

EXAMPLE 1. Given knowledge of the values of f, choose $\alpha = \widehat{(f(x))}_{x\in N} \in \Omega$ where $\hat{n} = 0 +_0 1 +_0 1 +_0 \cdots +_0 1$. Then $|\alpha| \leqslant \omega$ and $G(\alpha) = f$ since $G_x(\hat{n}) = n$.

EXAMPLE 2. Define addition on Ω by the obvious "continuous" extension of the usual primitive recursive definition of addition on N

$$\alpha +_0 0 = \alpha, \quad \alpha +_0 (\beta +_0 1) = (\alpha +_0 \beta) +_0 1, \quad \alpha +_0 \sup \beta_x = \sup(\alpha +_0 \beta_x).$$

Then $G(\alpha +_0 \beta) = G(\alpha) + G(\beta)$ and similarly for multiplication and exponentiation: $G(\alpha \odot \beta) = G(\alpha) \cdot G(\beta)$ and $G(\alpha^{0^\beta}) = G(\alpha)^{G(\beta)}$. Thus starting with $\omega = (\hat{x})_{x\in N}$ we can build up standard notations for ordinals below ε_0 so that $G_x(\omega) = x$, $G_x(\omega +_0 \omega) = x + x$, $G_x(\omega^2) = x^2$, $G_x(\omega^\omega) = x^x$, $G_x(\omega^{\omega^\omega}) = x^{x^x}$, etc.

For an interesting application of this see Cichon's [**3**] alternative proof of Kirby and Paris's most recent independence result for arithmetic.

2. Type-structures induced by G. We now extend G (and our sheaf-theoretic picture of it) to higher types. Let $\{T_n(N)\}_{n\in N}$ denote the maximal type-structure over N. Again there will be both local and global approaches to this extension and G is the map under which they coincide. First the local approach, developed pointwise at x for $x \in N$:

DEFINITION 3. For each $x \in N$, $\{T_n(\Omega, x)\}_{n\in N}$ denotes the type-structure induced by G_x over Ω as follows:

(0) $T_0(\Omega, x) = \Omega$ and $\alpha \sim_x \beta \Leftrightarrow G_x(\alpha) = G_x(\beta)$.

$(n + 1)$ $T_{n+1}(\Omega, x) = \{\Phi\colon\ T_n(\Omega, x) \to \Omega \mid\ \forall\phi, \psi \in T_n(\Omega, x)(\phi \sim_x \psi \Rightarrow \Phi(\phi) \sim_x \Phi(\psi))\}$ and $\Phi \sim_x \Psi \Leftrightarrow \forall\phi \in T_n(\Omega, x)(\Phi(\phi) \sim_x \Psi(\phi))$.

DEFINITION 4. For each x and each type-level n define $G_x\colon T_n(\Omega, x) \to T_n(N)$ and its inverse $\hat{}\colon T_n(N) \to T_N(\Omega, x)$ as follows:

(0) $G_x(\alpha)$ is already defined for $\alpha \in \Omega$, and for each $m \in N$ set $\hat{m} = 0 +_0 1 +_0 1 +_0 \cdots +_0 1$ (m times).

$(n + 1)$ Assume G_x and $\hat{}$ already defined at level n. Then for $\phi \in T_{n+1}(\Omega, x)$ define, for all $f \in T_n(N)$,

$$G_x(\Phi)(f) = G_x(\Phi(\hat{f})).$$

For $F \in T_{n+1}(N)$ define, for all $\phi \in T_n(\Omega, n)$,

$$\hat{F}(\phi) = \hat{m} \quad \text{where } m = F(G_x(\phi)).$$

It follows from part (i) of the lemma below that $\hat{F} \in T_{n+1}(\Omega, x)$.

LEMMA 1. *For each x and each type-level n,*

(i) $\phi \sim_x \psi \Leftrightarrow G_x(\phi) = G_x(\psi)$,

(ii) $G_x(\hat{f}) = f$ *and hence* $\widehat{G_x(\phi)} \sim_x \phi$.

PROOF. By a simple induction on n.

THEOREM 1. *For each x and each type-level n, G_x induces an isomorphism from $T_n(\Omega, x)/\sim_x$ onto $T_n(N)$ so that if $\phi \in T_n(\Omega, x)$ and $\Phi \in T_{n+1}(\Omega, x)$ then*

$$G_x(\Phi(\phi)) = G_x(\Phi)(G_x(\phi)).$$

PROOF. Immediate from Lemma 1.

DEFINITION 5. For each n let $\Gamma_n(\Omega)$ be the set of cross-sections of the sheaf over N whose stalks are the $T_n(\Omega, x)$, $x \in N$, i.e.

$$\Gamma_n(\Omega) = \{(\phi_x)_{x\in N} | \forall x \in N(\phi_x \in T_n(\Omega, x))\}.$$

Then $\{\Gamma_n(\Omega)\}_{n\in N}$ forms a type-structure with equivalence and application defined pointwise:

$$(\phi_x)_{x\in N} \sim (\psi_x)_{x\in N} \Leftrightarrow \forall x \in N(\phi_x \sim_x \psi_x)$$

and

$$(\Phi_x)_{x\in N}((\phi_x)_{x\in N}) = (\Phi_x(\phi_x))_{x\in N}.$$

DEFINITION 6. For each n let $\Gamma_n(N)$ be the set of cross-sections of the sheaf over N, each of whose stalks is a copy of $T_n(N)$, i.e.

$$\Gamma_n(N) = \{(f_x)_{x\in N} | \forall x \in N(f_x \in T_n(N))\}.$$

Then $\{\Gamma_n(N)\}_{n\in N}$ forms a type-structure of "pointwise-defined" functionals with application:

$$(F_x)_{x\in N}((f_x)_{x\in N}) = (F_x(f_x))_{x\in N}.$$

Note that at level 0 we now have $\Gamma_0(N) = N^N$.

THEOREM 2. *G induces an isomorphism from $\{\Gamma_n(\Omega)\}_{n\in N}$ onto $\{\Gamma_n(N)\}_{n\in N}$, i.e.*

$$(\phi_x)_{x\in N} \sim (\psi_x)_{x\in N} \Leftrightarrow (G_x(\phi_x))_{x\in N} = (G_x(\psi_x))_{x\in N}$$

and

$$G((\Phi_x)_{x\in N}((\phi_x)_{x\in N})) = (G_x(\Phi_x))_{x\in N}((G_x(\phi_x))_{x\in N}).$$

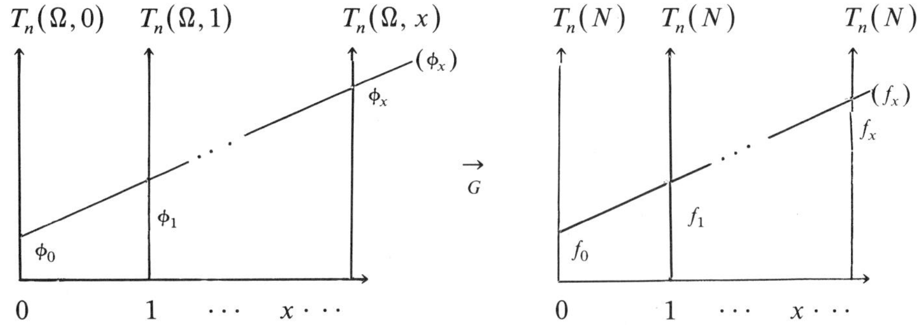

The global approach to $\{\Gamma_n(\Omega)\}_{n\in N}$ runs as follows:

DEFINITION 7. $\{\mathbf{C}_n(\Omega)\}_{n\in N}$ is the "G-continuous" type-structure over Ω given by:

(0) $\mathbf{C}_0(\Omega) = \Omega$ and $\alpha \approx \beta \Leftrightarrow G(\alpha) = G(\beta)$.

$(n+1)$ Call Φ: $\mathbf{C}_n(\Omega) \to \Omega$ *G-continuous* if $\phi \approx \psi \Rightarrow \Phi(\phi) \approx \Phi(\psi)$ and $\Phi(\sup_x \phi^x) \approx \sup \Phi(\phi^x)$. Then $\mathbf{C}_{n+1}(\Omega) = \{\Phi: \mathbf{C}_n(\Omega) \to \Omega | \Phi \text{ is } G\text{-continuous}\}$ and $\Phi \approx \Psi \Leftrightarrow \forall\phi \in \mathbf{C}_n(\Omega)(\Phi(\phi) \approx \Psi(\phi))$ and, given $\Phi^0, \Phi^1, \Phi^2, \ldots \in \mathbf{C}_{n+1}(\Omega)$, $\sup_x \Phi^x$ is defined by $(\sup_x \Phi^x)(\phi) = \sup \Phi^x(\phi)$. It follows easily from the pointwise-at-x definition of G that $\sup_x \Phi^x$ is G-continuous.

DEFINITION 8. The maps p_n: $\mathbf{C}_n(\Omega) \to \Gamma_n(\Omega)$ and q_n: $\Gamma_n(\Omega) \to \mathbf{C}_n(\Omega)$ are defined as follows:

(0) $p_0(\alpha) = (\alpha, \alpha, \alpha, \ldots) \in \Gamma_0(\Omega)$, $q_0((\alpha_x)_{x\in N}) = \sup \alpha_x \in \mathbf{C}_0(\Omega)$.

$(n+1)$ $p_{n+1}(\Phi) = (\Phi_x)_{x\in N} \in \Gamma_{n+1}(\Omega)$ where, for each $\psi \in T_n(\Omega, x)$,

$$\Phi_x(\psi) = \Phi(q_n((0, 0, \ldots, 0, \psi, 0, 0, \ldots))).$$

Here the 0's denote the constant 0 functionals belonging to each $T_n(\Omega, y)$ for $y \neq x$.

$$q_{n+1}((\Phi_x)_{x\in N}) = \Phi \in \mathbf{C}_{n+1}(\Omega)$$

where, for each $\phi \in \mathbf{C}_n(\Omega)$, $\Phi(\phi) = (\Phi_x)_{x\in N}(p_n(\phi)) = \sup \Phi_x(\phi_x)$. To see that $\Phi = q_{n+1}((\Phi_x)_{x\in N})$ is G-continuous, note first that if $\phi^0, \phi^1, \phi^2, \ldots \in \mathbf{C}_n(\Omega)$ then, by the definition of p_n, if $\psi \in T_{n-1}(\Omega, x)$, $(\sup_m \phi^m)_x(\psi) = \sup_m \phi^m_x(\psi) \sim_x \phi^x_x(\psi)$, so $(\sup_m \phi^m)_x \sim_x \phi^x_x$. Thus

$$\Phi\Big(\sup_m \phi^m\Big) = \sup_x \Phi_x\Big(\Big(\sup_m \phi^m\Big)_x\Big) \approx \sup_x \Phi_x(\phi^x_x)$$
$$\approx \sup_m \sup_x \Phi_x(\phi^m_x) = \sup_m \Phi(\phi^m).$$

LEMMA 2. *For each type-level* n,

(i) $\phi \approx \psi \Rightarrow p_n(\phi) \sim p_n(\psi)$,

(ii) $(\phi_x)_{x\in N} \sim (\psi_x)_{x\in N} \Rightarrow q_n((\phi_x)_{x\in N}) \approx q_n((\psi_x)_{x\in N})$,

(iii) $p_n(q_n((\phi_x)_{x\in N})) \sim (\phi_x)_{x\in N}$,

(iv) $q_n(p_n(\phi)) \approx \phi$,

(v) $\phi \approx \sup_x q_n((0, 0, \phi_x, 0, \ldots))$.

PROOF. By induction on n. The case $n = 0$ is easily checked. For $n+1$, (i) and (ii) follow straightforwardly from the definitions and

(iii) $p_{n+1}(q_{n+1}((\Phi_x)_{x\in N})) = (\Psi_x)_{x\in N}$ where, for $\psi \in T_n(\Omega, x)$,

$$\Psi_x(\psi) = (\Phi_x)_{x\in N}(p_n(q_n((0, \ldots, 0, \psi, 0, \ldots))))$$
$$\sim (\Phi_x)_{x\in N}((0, \ldots, 0, \psi, 0, \ldots))$$
$$= (\Phi_0(0), \ldots, \Phi_{x-1}(0), \Phi_x(\psi), \Phi_{x+1}(0), \ldots) \sim_x \Phi_x(\psi),$$

so $\Psi_x \sim_x \Phi_x$ for each x and therefore

$$p_{n+1}(q_{n+1}((\Phi_x)_{x\in N})) = (\Psi_x)_{x\in N} \sim (\Phi_x)_{x\in N}.$$

(iv)

$$q_{n+1}(p_{n+1}(\Phi))(\phi) = \sup_x \Phi(q_n((0, \ldots, 0, \phi_x, 0, \ldots)))$$
$$\approx \Phi\Big(\sup_x q_n((0, \ldots, 0, \phi_x, 0, \ldots))\Big) \approx \Phi(\phi)$$

for each $\phi \in \mathbf{C}_n(\Omega)$ where $(\phi_x)_{x\in N} = p_n(\phi)$. So $q_{n+1}(p_{n+1}(\Phi)) \approx \Phi$.

(v) For each $\Phi \in \mathbf{C}_n(\Omega)$, if $p_{n+1}(\Phi) = (\Phi_x)_{x\in N}$ then

$$\begin{aligned}\sup_x q_{n+1}((0,\ldots,0,\Phi_x,0,\ldots))(\Phi) &= \sup_x\big(q_{n+1}((0,\ldots,0,\Phi_x,0,\ldots))(\phi)\big)\\ &= \sup_x(0,\ldots,0,\Phi_x(\phi_x),0,\ldots)\\ &\approx \sup \Phi_x(\phi_x)\\ &= q_{n+1}(p_{n+1}(\Phi))(\phi) \approx \Phi(\phi) \quad \text{by (iv).}\end{aligned}$$

So $\Phi \approx \sup_x q_{n+1}((0,\ldots,0,\Phi_x,0,\ldots))$.

In the following we shall ambiguously allow ϕ_x to denote either an element of $T_n(\Omega, x)$ or its image $q_n((0,\ldots,0,\phi_x,0,\ldots))$ in $\mathbf{C}_n(\Omega)$. Thus to each $\phi \in \mathbf{C}_n(\Omega)$, p_n assigns a sequence $(\phi_x)_{x\in N}$ such that $\phi \approx \sup\phi_x$ and, to each sequence $(\phi_x)_{x\in N} \in \Gamma_n(\Omega)$, q_n assigns a $\phi \in \mathbf{C}_n(\Omega)$ such that $\phi \approx \sup\phi_x$. By Lemma 2, p_n and q_n are mutually inverse and the definitions of p_n, q_n are such that

$$p_{n+1}(\Phi)(p_n(\phi)) \sim p_0(\Phi(\phi))$$

and $q_{n+1}((\Phi_x)_{x\in N})(q_n((\phi_x)_{x\in N})) \approx q_0((\Phi_x)_{x\in N}((\phi_x)_{x\in N}))$.

Thus we immediately have

THEOREM 3. *For each type-level n,*

(i) $\phi \approx \psi$ *in* $\mathbf{C}_n(\Omega) \Leftrightarrow (\phi_n)_{x\in N} \sim (\psi_x)_{x\in N}$ *in* $\Gamma_n(\Omega)$,

(ii) $\Phi(\phi) \approx \alpha$ *in* $\mathbf{C}_n(\Omega) \Leftrightarrow (\Phi_x)_{x\in N}((\phi_x)_{x\in N}) \sim \alpha$ *in* $\Gamma_n(\Omega)$,

i.e. $\{\mathbf{C}_n(\Omega)\}_{n\in N}$ *and* $\{\Gamma_n(\Omega)\}_{n\in N}$ *are isomorphic.*

We shall continue to work in $\{\mathbf{C}_n(\Omega)\}_{n\in N}$, defining for each $\phi \in \mathbf{C}_n(\Omega)$,

$$G_x(\phi) = G_x(\phi_x) \in T_n(N) \quad \text{and} \quad G(\phi) = (G_x(\phi_x))_{x\in N} \in \Gamma_n(N),$$

so that for $\phi \in \mathbf{C}_n(\Omega)$ and $\Phi \in \mathbf{C}_{n+1}(\Omega)$,

$$G(\Phi(\phi)) = G(\Phi)(G(\phi)) \in N^N.$$

REMARK. If $\phi \in \mathbf{C}_1(\Omega)$ then since $\alpha \approx \sup\widehat{G_x(\alpha)}$,

$$\phi(\alpha) \approx \phi\big(\sup\widehat{G_x(\alpha)}\big) \approx \sup\phi\big(\widehat{G_x(\alpha)}\big),$$

so up to $\approx$-equivalence, ϕ is determined by its values on the integers. To use Girard's terminology each $\phi \in \mathbf{C}_1(\Omega)$ is a "dilation" of a system $(G_x(\phi_x))_{x\in N}$ of number-theoretic functions just as each α is a "dilation" of a single number-theoretic function $G(\alpha)$.

3. Recursion in $\{\mathbf{C}_n(\Omega)\}_{n\in N}$. We now apply Kleene's schemes S1,...,S9 to the G-continuous functionals over Ω, but with primitive recursion S5 extended "continuously", and with the addition of one further scheme S10 introducing sups.

In order to make the necessary index-constructions, we first need a pairing function on Ω, so define

$$[\alpha, \beta] = 2^{(\alpha+_0 1)+_0(\beta+_0 1)} +_0 2^{\alpha+_0 1} +_0 1,$$

with addition $+_0$ as given in §1 and 2^γ defined by $2^0 = 1$, $2^{\gamma +_0 1} = 2^\gamma +_0 2^\gamma$, $2^{\sup \gamma_x} = \sup 2^{\gamma_x}$. It can be checked that [,] gives a 1-1 map from Ω^2 into Ω and that if we use the same pairing on N, i.e., $[a, b] = 2^{(a+1)+(b+1)} + 2^{a+1} + 1$, then for each x, $G_x([\alpha, \beta]) = [G_x(\alpha), G_x(\beta)]$. We make $[\alpha, \beta]$ a successor for convenience only. Finally, set $[\alpha_1, \alpha_2, \ldots, \alpha_k] = [\ldots[[\alpha_1, \alpha_2], \alpha_3], \ldots \alpha_k]$.

Now define the Kleene-style computation theory Rec(Ω) over $\{\mathbf{C}_n(\Omega)\}_{n \in N}$ inductively according to the following schemes where $\underline{\phi} = \phi_1, \ldots, \phi_n$ is a list of G-continuous functionals of various types.

S1.	$\{\varepsilon\}(\alpha, \underline{\phi}) = \alpha +_0 1$	$\varepsilon = [\hat{1}, \hat{n}]$
S2.	$\{\varepsilon\}(\underline{\phi}) = \beta$	$\varepsilon = [\hat{2}, \hat{n}, \beta]$
S3.	$\{\varepsilon\}(\alpha, \underline{\phi}) = \alpha$	$\varepsilon = [\hat{3}, \hat{n}]$
S4.	$\{\varepsilon\}(\underline{\phi}) = \{\varepsilon_1\}(\{\varepsilon_2\}(\underline{\phi}), \underline{\phi})$	$\varepsilon = [\hat{4}, \hat{n}, \varepsilon_1, \varepsilon_2]$
S5.	$\{\varepsilon\}(0, \underline{\phi}) = \{\varepsilon_1\}(\underline{\phi})$	$\varepsilon = [\hat{5}, \hat{n}, \varepsilon_1, \varepsilon_2]$
	$\{\varepsilon\}(\alpha +_0 1, \underline{\phi}) = \{\varepsilon_2\}(\alpha, \{\varepsilon\}(\alpha, \underline{\phi}), \underline{\phi})$	
	$\{\varepsilon\}(\sup \alpha_x, \underline{\phi}) = \sup_x \{\varepsilon\}(\alpha_x, \underline{\phi})$	
S6.	$\{\varepsilon\}(\underline{\phi}) = \{\varepsilon_1\}(\underline{\phi}')$	$\varepsilon = [\hat{6}, \hat{n}, \hat{j}, \hat{k}, \varepsilon_1]$
S7.	$\{\varepsilon\}(\alpha, \psi, \underline{\phi}) = \psi(\alpha)$	$\varepsilon = [\hat{7}, \hat{n}]$
S8.	$\{\varepsilon\}(\Phi, \underline{\phi}) = \Phi(\lambda\zeta \cdot \{\varepsilon_1\}(\Phi, \zeta, \underline{\phi}))$	$\varepsilon = [\hat{8}, \hat{n}, \hat{j}, \varepsilon_1]$
S9.	$\{\varepsilon\}(\gamma, \underline{\phi}) = \{\gamma\}(\underline{\phi})$	$\varepsilon = [\hat{9}, \hat{n}]$
S10.	$\{\sup_x \varepsilon_x\}(\underline{\phi}) = \sup_x \{\varepsilon_x\}(\underline{\phi})$	

REMARKS. (1) In S6, $\underline{\phi}'$ is the rearrangement of $\underline{\phi}$ obtained by moving the kth type-j variable to the front. In S8, Φ is of type-j and the computation is defined just in case $\{\varepsilon_1\}(\Phi, \zeta, \underline{\phi})$ is defined for every $\zeta \in \mathbf{C}_{j-2}(\Omega)$ and (hence) $\lambda\zeta \cdot \{\varepsilon_1\}(\Phi, \zeta, \underline{\phi}) \in \mathbf{C}_{j-1}(\Omega)$. In S10, the computation is defined just in case $\{\varepsilon_x\}(\underline{\phi})$ is defined for every x. The other schemes are interpreted in the usual fashion.

(2) The S_n^m and fixed-point properties hold in Rec(Ω) by standard arguments.

(3) Note crucially that if the sup clause in S5 and S10 are deleted and if $\alpha, \beta, \gamma, \varepsilon, \ldots$ and $\underline{\phi} = \phi_1, \ldots, \phi_n$, etc. are replaced by number variables $a, b, c, e, \ldots$ and higher-type variables $\underline{F} = F_1, \ldots, F_n$ over $\{T_n(N)\}_{n \in N}$ respectively then the resulting schemes S1, ..., S9 define Kleene-recursion $\{e\}(a, \underline{F})$ in $\{T_n(N)\}_{n \in N}$. We denote this recursion theory by Rec(N).

Thus if $\{e\}(a) = m$ in Rec(N) then $\varepsilon = \hat{e}$ will be an index for an identical computation in Rec(Ω) and $\{\hat{e}\}(\hat{a}) = \hat{m}$.

(4) If the higher-type schemes S7 and S8 are omitted then we obtain a notion of *partial recursive function* $\lambda\alpha \cdot \{\varepsilon\}(\alpha)$ over Ω. If $\alpha, \beta, \gamma, \varepsilon, \ldots$ are then interpreted as *reals* encoding tree-ordinals, $\lambda\varepsilon\alpha \cdot \{\varepsilon\}(\alpha)$ becomes a partial-recursive-in E_1 functional in the usual sense.

(5) *Sequence coding*. Though the pairing function [,] is adequate for constructing indices, it is not at all clear that we can decode α and β from $[\alpha, \beta]$ in Rec(Ω).

Henceforth we shall adopt the following sequence coding and decoding functions in Rec(Ω) with their counterparts in Rec(N) defined similarly:

(i) $\langle \alpha, \beta \rangle = [\hat{5}, 0, [\hat{2}, 0, \alpha], [\hat{2}, \hat{2}, \beta]]$,

(ii) $(\gamma)_i = \{\gamma\}(\hat{i})$ for $i = 0, 1$ so that $(\langle \alpha, \beta s \rangle)_0 = \alpha$ and $(\langle \alpha, \beta \rangle)_1 = \beta$,

(iii) $\langle \alpha_1, \dots, \alpha_k \rangle = \langle \dots \langle \langle \alpha_1, \alpha_2 \rangle, \alpha_3 \rangle, \dots, \alpha_k \rangle$.

Then since the coding and decoding functions in Rec(N) are to be defined by identical recursions, we have, by Theorem 4 below,

$$G_x(\langle \alpha_1, \dots, \alpha_k \rangle) = \langle G_x(\alpha_1), \dots, G_x(\alpha_k) \rangle, \quad G_x((\gamma)_i) = (G_x(\gamma))_i \quad \text{for } i = 0, 1.$$

THEOREM 4. *If* $\{\varepsilon\}(\alpha, \phi_1, \dots, \phi_n)$ *is a defined computation in* Rec(Ω) *then for each* x, $\{G_x(\varepsilon)\}(G_x(\alpha), G_x(\phi_1), \dots, G_x(\phi_n))$ *is a defined computation in* Rec(N) *and*

$$G_x(\{\varepsilon\}(\alpha, \phi_1, \dots, \phi_n)) = \{G_x(\varepsilon)\}(G_x(\alpha), G_x(\phi_1), \dots, G_x(\phi_n)).$$

PROOF. By induction on the definition of $\{\varepsilon\}(\alpha, \phi_1, \dots, \phi_n)$. Let $\underline{\phi} = \phi_1, \dots, \phi_n$ and then $G_x(\underline{\phi}) = G_x(\phi_1), \dots, G_x(\phi_n)$. We shall only consider three of the schemes, the others being mostly trivial.

The easiest of all is S10, for suppose

$$\varepsilon = \sup \varepsilon_x \quad \text{and} \quad \{\varepsilon\}(\underline{\phi}) = \sup_x \{\varepsilon_x\}(\underline{\phi}).$$

Then we can assume, inductively, that for each x

$$G_x(\{\varepsilon_x\}(\underline{\phi})) = \{G_x(\varepsilon_x)\}(G_x(\underline{\phi}))$$

and so,

$$\begin{aligned} G_x(\{\varepsilon\}(\underline{\phi})) &= G_x\Big(\sup_x \{\varepsilon_x\}(\underline{\phi})\Big) = G_x(\{\varepsilon_x\}(\underline{\phi})) \\ &= \{G_x(\varepsilon_x)\}(G_x(\underline{\phi})) = \{G_x(\varepsilon)\}(G_x(\underline{\phi})). \end{aligned}$$

Now suppose $\varepsilon = [\hat{5}, \hat{n}, \varepsilon_1, \varepsilon_2]$ and by S5,

$$\{\varepsilon\}(\alpha +_0 1, \underline{\phi}) = \{\varepsilon_2\}(\alpha, \{\varepsilon\}(\alpha, \underline{\phi}), \underline{\phi}).$$

Then again, by the induction hypothesis

$$\begin{aligned} G_x(\{\varepsilon\}(\alpha +_0 1, \underline{\phi})) &= \{G_x(\varepsilon_2)\}(G_x(\alpha), G_x(\{\varepsilon\}(\alpha, \underline{\phi})), G_x(\underline{\phi})) \\ &= \{G_x(\varepsilon_2)\}(G_x(\alpha), \{G_x(\varepsilon)\}(G_x(\alpha), G_x(\underline{\phi})), G_x(\underline{\phi})) \\ &= \{[5, n, G_x(\varepsilon_1), G_x(\varepsilon_2)]\}(G_x(\alpha) + 1, G_x(\underline{\phi})) \\ &= \{G_x(\varepsilon)\}(G_x(\alpha +_0 1), G_x(\underline{\phi})) \end{aligned}$$

since $G_x(\varepsilon) = [5, n, G_x(\varepsilon_1), G_x(\varepsilon_2)]$.

Finally, suppose $\{\varepsilon\}(\Phi, \underline{\phi})$ is defined by S8,

$$\{\varepsilon\}(\Phi, \underline{\phi}) = \Phi(\lambda\zeta \cdot \{\varepsilon_1\}(\Phi, \zeta, \underline{\phi}))$$

where $G_x(\varepsilon) = [8, n, j, G_x(\varepsilon_1)]$.

By the induction hypothesis, for every $\zeta \in \mathbf{C}_{j-2}(\Omega)$,

$$G_x(\{\varepsilon_1\}(\Phi, \zeta, \underline{\phi})) = \{G_x(\varepsilon_1)\}(G_x(\Phi), G_x(\zeta), G_x(\underline{\phi})).$$

Since this holds for each x it follows from Theorems 2 and 3 that

(i) $\zeta \approx \zeta' \Rightarrow \{\varepsilon_1\}(\Phi, \zeta, \underline{\phi}) \approx \{\varepsilon_1\}(\Phi, \zeta', \underline{\phi})$,

(ii) $\{\varepsilon_1\}(\Phi, \sup \zeta_x, \underline{\phi}) \approx \sup\{\varepsilon_1\}(\Phi, \zeta_x, \underline{\phi})$.

Thus $\lambda\zeta \cdot \{\varepsilon_1\}(\Phi, \zeta, \underline{\phi}) \in \mathbf{C}_{j-1}(\Omega)$ and for each x, $G_x(\lambda\zeta \cdot \{\varepsilon_1\}(\Phi, \zeta, \underline{\phi})) = \lambda g \cdot \{G_x(\varepsilon_1)\}(G_x(\Phi), g, G_x(\underline{\phi}))$ which belongs to $T_{j-1}(N)$ by Theorem 1. Therefore, again using Theorem 1, we have

$$\begin{aligned} G_x(\{\varepsilon\}(\Phi, \underline{\phi})) &= G_x(\Phi(\lambda\zeta \cdot \{\varepsilon_1\}(\Phi, \zeta, \underline{\phi}))) \\ &= G_x(\Phi)(G_x(\lambda\zeta \cdot \{\varepsilon_1\}(\Phi, \zeta, \underline{\phi}))) \\ &= G_x(\Phi)(\lambda g \cdot \{G_x(\varepsilon_1)\}(G_x(\Phi), g, G_x(\underline{\phi}))) \\ &= \{[8, n, j, G_x(\varepsilon_1)]\}(G_x(\Phi), G_x(\underline{\phi})) \\ &= \{G_x(\varepsilon)\}(G_x(\Phi), G_x(\underline{\phi})). \end{aligned}$$

COROLLARY 1. *The type-structure* $\{\mathbf{C}_n(\Omega)\}_{n \in N}$ *is closed under* S1, ..., S10 *recursion, i.e. if* $\underline{\phi} = \phi_1, \ldots, \phi_n$ *are G-continuous and*

$$\psi(\zeta) = \{\varepsilon\}(\zeta, \underline{\phi}) \quad \textit{for every } \zeta \in \mathbf{C}_{j-1}(\Omega),$$

then $\psi \in \mathbf{C}_j(\Omega)$.

PROOF. As in the S8 case of Theorem 4.

COROLLARY 2. *Suppose f is a function computable from functionals $\underline{F}$ in* Rec(N) *via index e. Let $\varepsilon = \hat{e}$ and $\underline{\phi}$ be G-continuous functionals such that, for each x, $G_x(\underline{\phi}) = \underline{F}$. Then if $\{\varepsilon\}(\omega, \underline{\phi}) = \{\varepsilon\}(\sup_x \hat{x}, \underline{\phi})$ is a defined computation in* Rec(Ω),

$$G_x(\{\varepsilon\}(\omega, \underline{\phi})) = \{e\}(x, \underline{F}) = f(x),$$

for each x and so $f = G(\{\varepsilon\}(\omega, \underline{\phi}))$.

Thus the computation of the tree-ordinal $\{\varepsilon\}(\omega, \underline{\phi})$ mirrors the computation of the function $f = \lambda x \cdot \{e\}(x, \underline{F})$ in Rec(N), and so the ordinal $|\{\varepsilon\}(\omega, \underline{\phi})|$ should provide a very refined measure of the complexity of the recursion $\lambda x \cdot \{e\}(x, \underline{F})$ for f. Concrete examples of this ordinal-assignment are given later in §5.

4. Hierarchies in Rec(Ω). The hierarchies will be generated analogously to the usual kind of construction in Rec(N), by iterated application of a given operator over suitable "ordinal notations" but now with Ω as the base rather than N.

DEFINITION 9. The set $\mathbf{O}(\Omega)$ of *notations over* Ω is defined inductively by the clauses:

(i) $0 \in \mathbf{O}(\Omega)$, $\|0\|_\Omega = 0$.

(ii) $\alpha \in \mathbf{O}(\Omega) \Rightarrow \langle \hat{1}, \alpha \rangle \in \mathbf{O}(\Omega)$, $\|\langle \hat{1}, \alpha \rangle\|_\Omega = \|\alpha\|_\Omega + 1$.

(iii) $\forall x(\alpha_x \in \mathbf{O}(\Omega)) = \sup \alpha_x \in \mathbf{O}(\Omega)$, $\|\sup \alpha_x\|_\Omega = \sup_x \|\alpha_x\|_\Omega$.

(iv) $\forall\alpha(\{\varepsilon\}(\alpha) \in \mathbf{O}(\Omega)) \Rightarrow \langle\hat{2}, \varepsilon\rangle \in \mathbf{O}(\Omega)$, $\|\langle\hat{2}, \varepsilon\rangle\|_\Omega = \sup_\alpha\|\{\varepsilon\}(\alpha)\|_\Omega$.

THEOREM 5. *$\mathbf{O}(\Omega)$ is a complete Π_2^1 set.*

PROOF. That $\mathbf{O}(\Omega)$ is Π_2^1 follows from known results about inductive definability over reals, again considering Ω as a set of reals encoding tree-ordinals so that Ω is complete Π_1^1. For example, by Moldestad and Normann [**7**], $\mathbf{O}(\Omega)$ is generated by a positive $\Pi_1^1(\Omega)$ induction and is therefore itself $\Pi_1^1(\Omega)$ and hence Π_2^1.

For the rest, we shall only sketch a proof that $\mathbf{O}(\Omega)$ is complete for Π_2^1 sets of integers. A thorough treatment would obscure the main idea which is really just a generalisation of the usual proof that Kleene's $\mathbf{O}$ is Π_1^1-complete. We use Shoenfield's representation of Σ_2^1 sets, as conveniently set out in Moschovakis [**8**, p. 523]. Suppose $a \in A \Leftrightarrow \exists f(T(a, f)$ is well founded) where $T(a, f)$ is the tree of sequences $\langle c_0,\ldots,c_s\rangle$ such that $\forall m \leqslant s\, R(a, \bar{f}(m), \langle c_0,\ldots,c_m\rangle)$ and R is a recursive relation.

Using the S_n^m and fixed-point properties in Rec(Ω) we can find an index ε such that

$$\{\varepsilon\}(\hat{a}, \langle\hat{b}_0,\ldots,\hat{b}_{k-1}\rangle, \langle\beta_0,\ldots,\beta_{k-1}\rangle, \alpha)$$
$$= \begin{cases} 0 & \text{if } \alpha = 0, \\ \sup_x \{\varepsilon\}(\hat{a}, \langle\hat{b}_0,\ldots,\hat{b}_{k-1}\rangle, \langle\beta_0,\ldots,\beta_{k-1}\rangle, \alpha_x) & \text{if } \alpha = \sup \alpha_x, \\ \sup_x \{\varepsilon\}(\hat{a}, \langle\hat{b}_0,\ldots,\hat{b}_{k-1}, \hat{x}\rangle, \langle\beta_0,\ldots,\beta_{k-1}, \alpha - 1\rangle, \gamma) & \\ & \text{if } \alpha = (\alpha - 1) +_0 1, \end{cases}$$

where

$$\gamma = \begin{cases} \beta_l & \text{if } k+1 \text{ is a sequence } \langle c_0,\ldots,c_s\rangle \text{ such that} \\ & \forall m \leqslant_s R(a, \langle b_0,\ldots,b_m\rangle, \langle c_0,\ldots,c_m\rangle) \text{ and } l = \langle c_0,\ldots,c_{s-1}\rangle, \\ 0 & \text{if } k+1 \text{ is a sequence } \langle c_0,\ldots,c_s\rangle \text{ but} \\ & \exists m \leqslant s \neg R(a, \langle b_0,\ldots,b_m\rangle, \langle c_0,\ldots,c_m\rangle), \\ 1 & \text{if } k+1 \text{ is not a sequence number.} \end{cases}$$

Now one has to check that $\{\varepsilon\}(\hat{a}\,\langle\ \rangle, \langle\ \rangle, \alpha)$ does not belong to $\mathbf{O}(\Omega)$ if and only if there is an infinite sequence $f = \langle b_0,\ldots,b_k,\ldots\rangle$ and an infinite sequence of tree-ordinals $\langle\beta_0,\ldots,\beta_k\ldots\rangle$ such that whenever $k = \langle c_0,\ldots,c_s\rangle \in T(a, f)$ and $l = \langle c_0,\ldots,c_{s-1}\rangle$ then $\beta_k \prec \beta_l \prec \alpha$ where $\prec$ is the subtree relation on tree-ordinals. Therefore

$$a \in A \Leftrightarrow \exists f(T(a, f) \text{ is well founded}) \Leftrightarrow \exists\alpha(\{\varepsilon\}(\hat{a}, \langle\ \rangle, \langle\ \rangle, \alpha) \neq \mathbf{O}(\Omega)).$$

Now let ε_a be an index computable from a such that $\{\varepsilon_a\}(\alpha) = \{\varepsilon\}(\hat{a}, \langle\ \rangle, \langle\ \rangle, \alpha)$. Then,

$$a \notin A \Leftrightarrow \forall\alpha(\{\varepsilon_a\}(\alpha) \in \mathbf{O}(\Omega)) \Leftrightarrow \langle\hat{2}, \varepsilon_a\rangle \in \mathbf{O}(\Omega),$$

and so every Π_2^1 set is reducible to $\mathbf{O}(\Omega)$.

DEFINITION 10. The set $\mathbf{O}(N)$ of *notations over* N is defined inductively by the same clauses as for $\mathbf{O}(\Omega)$ but with the sup clause (iii) omitted:

$(\mathrm{i})_N$ $0 \in \mathbf{O}(N), \|0\|_N = 0$.

$(\mathrm{ii})_N$ $a \in \mathbf{O}(N) \Rightarrow \langle 1, a\rangle \in \mathbf{O}(N), \|\langle 1,a\rangle\|_N = \|a\|_N + 1$.

$(\mathrm{iv})_N$ $\forall a(\{e\}(a) \in \mathbf{O}(N)) \Rightarrow \langle 2, e\rangle \in \mathbf{O}(N), \|\langle 2, e\rangle\|_N = \sup_a\|\{e\}(a)\|_N$.

LEMMA 3. $\alpha \in \mathbf{O}(\Omega) \Rightarrow \forall x(G_x(\alpha) \in \mathbf{O}(N))$.

PROOF. By induction over $\mathbf{O}(\Omega)$ using Theorem 4.

EXAMPLE 3. Let ε be an index for the following primitive recursion in $\mathrm{Rec}(\Omega)$:

$$\{\varepsilon\}(0) = 0, \quad \{\varepsilon\}(\alpha +_0 1) = \langle \hat{1}, \{\varepsilon\}(\alpha)\rangle, \quad \{\varepsilon\}(\sup \alpha_x) = \sup\{\varepsilon\}(\alpha_x).$$

Then for each α, $\{\varepsilon\}(\alpha) \in \mathbf{O}(\Omega)$ and $\|\{\varepsilon\}(\alpha)\|_\Omega = |\alpha|$, so $\langle \hat{2}, \varepsilon\rangle$ is the *standard* notation for $\aleph_1$ in $\mathbf{O}(\Omega)$ and we shall denote it by ω_1.

By Lemma 3, $G_x(\omega_1) = \langle 2, G_x(\varepsilon)\rangle \in \mathbf{O}(N)$ and $\{G_x(\varepsilon)\}(0) = 0$,

$$\{G_x(\varepsilon)\}(a + 1) = \langle 1, \{G_x(\varepsilon)\}(a)\rangle,$$

so $G_x(\omega_1)$ is the standard notation for ω in $\mathbf{O}(N)$—which we shall denote by ω_0 in order to avoid confusion with the tree-ordinal $\omega = \sup_x \hat{x} \in \Omega$.

EXAMPLE 4. By the fixed-point property we can define a partial recursive function $+_\Omega$ in $\mathrm{Rec}(\Omega)$ such that if $\alpha, \beta \in \mathbf{O}(\Omega)$ then $\alpha +_\Omega \beta \in \mathbf{O}(\Omega)$ and $\|\alpha +_\partial \beta\| = \|\alpha\|_\Omega + \|\beta\|_\Omega$. By Theorem 4, each G_x will then collapse the definition of $+_\Omega$ onto a partial recursive definition in $\mathrm{Rec}(N)$ of a function $+_N$ such that if $a, b \in \mathbf{O}(N)$ then $a +_N b \in \mathbf{O}(N)$ and $\|a +_N b\|_N = \|a\|_N + \|b\|_N$. Thus for each x, $G_x(\alpha +_\Omega \beta) = G_x(\alpha) +_N G_x(\beta)$.

Similarly we then obtain partial recursive multiplication functions $\cdot_\Omega$, $\cdot_N$ and exponentiation functions $\exp_\Omega$, $\exp_N$ on $\mathbf{O}(\Omega)$ and $\mathbf{O}(N)$ respectively so that for each x

$$G_x(\alpha \cdot_\Omega \beta) = G_x(\alpha) \cdot_N G_x(\beta), \quad G_x(\exp_\Omega(\alpha, \beta)) = \exp_N(G_x(\alpha), G_x(\beta))$$

In this way one builds up notations $\omega_1 2$, ω_1^2, ω_1^3, $\omega_1^{\omega_1}$, $\omega_1^{\omega_1^{\omega_1}}$, etc. in $\mathbf{O}(\Omega)$ which collapse under each G_x onto notations $\omega_0 2$, ω_0^2, ω_0^3, $\omega_0^{\omega_0}$, $\omega_0^{\omega_0^{\omega_0}}$, etc. in $\mathbf{O}(N)$.

Then we can define $\varepsilon_{\omega_1+1} \in \mathbf{O}(\Omega)$ by

$$\varepsilon_{\omega_1+1} = \sup_x \omega_1^{\omega_1^{\cdot^{\cdot^{\cdot \omega_1}}}} \Big\} x - 1 \text{ times}$$

and a notation for ε_0 in $\mathbf{O}(N)$ by

$$\varepsilon_0 = \langle 2, e_0\rangle$$

where $\{e_0\}(0) = 0$ and $\{e_0\}(x + 1) = \omega_0^{\{e_0\}(x)}$ in $\mathrm{Rec}(N)$, so that for each x,

$$G_x(\varepsilon_{\omega_1+1}) = \omega_0^{\omega_0^{\cdot^{\cdot^{\cdot \omega_0}}}} \Big\} x - 1 \text{ times} = \{e_0\}(x),$$

i.e. $G(\varepsilon_{\omega_1+1})$ is the standard fundamental sequence to ε_0 in $\mathbf{O}(N)$.

We are now in a position to generate recursion-theoretic hierarchies over $\mathbf{O}(\Omega)$ in the usual way. We follow Girard in calling the construction Λ.

LEMMA 4. *There is a partial recursive functional*

$$\Lambda^{\Omega}(\phi, \Phi, \delta_1, \delta_2, \beta, \alpha)$$

such that given $\phi, \delta_1, \delta_2 \in \mathbf{C}_1(\Omega)$ *and* $\Phi \in \mathbf{C}_2(\Omega)$ (*where* Φ *is considered here as a mapping from* $\mathbf{C}_1 \times \mathbf{C}_0$ *into* $\mathbf{C}_0$) *then for* $\beta \in \mathbf{O}(\Omega)$ *the functions* ϕ_β *given by*

$$\phi_\beta = \lambda\alpha \cdot \Lambda^{\Omega}(\phi, \Phi, \delta_1, \delta_2, \beta, \alpha)$$

satisfy

$$\phi_0(\alpha) = \phi(\alpha), \qquad \phi_{\langle \hat{1}, \beta \rangle}(\alpha) = \Phi(\phi_\beta, \alpha),$$
$$\phi_{\sup \beta_x}(\alpha) = \sup \phi_{\beta_x}(\alpha), \qquad \phi_{\langle \hat{2}, \varepsilon \rangle}(\alpha) = \phi_{\{\varepsilon\}(\delta_1(\alpha))}(\delta_2(\alpha)).$$

PROOF. Construct an index for Λ^{Ω} by effective transfinite induction over $\mathbf{O}(\Omega)$ using the fixed-point property.

The counterpart to Lemma 4 in Rec(N) is

LEMMA 5. *There is a partial recursive functional*

$$\Lambda^N(f, F, d_1, d_2, b, a)$$

such that, given $f, d_1, d_2 \in T_1(N)$ *and* $F \in T_2(N)$ (*considered here as a mapping from* $T_1 \times T_0$ *into* T_0), *then for* $b \in \mathbf{O}(N)$ *the functions* f_b *given by*

$$f_b = \lambda a \cdot \Lambda^N(f, F, d_1, d_2, b, a)$$

satisfy

$$f_0(a) = f(a), \quad f_{\langle 1, b \rangle}(a) = F(f_b, a), \quad f_{\langle 2, e \rangle}(a) = f_{\{e\}(d_1(a))}(d_2(a)).$$

REMARK. The functions δ_1, δ_2 in Lemma 4 and d_1, d_2 in Lemma 5 simply determine the particular way in which we wish to diagonalize at limits ("large" limits in the case of Lemma 4). In any given application the method of diagonalization will be fixed and simple, so we henceforth omit mention of the diagonalization functions and write

$$\phi_\beta = \Lambda^{\Omega}(\phi, \Phi, \beta), \qquad f_b = \Lambda^N(f, F, b).$$

THEOREM 6. *For each x and all* $\beta \in \mathbf{O}(\Omega)$

$$G_x(\Lambda^{\Omega}(\phi, \Phi, \beta)) = \Lambda^N(G_x(\phi), G_x(\Phi), G_x(\beta)).$$

PROOF. By Theorem 4 or directly by induction on $\beta \in \mathbf{O}(\Omega)$ to show that for all $\alpha \in \Omega$

$$G_x(\Lambda^{\Omega}(\phi, \Phi, \beta, \alpha)) = \Lambda^N(G_x(\phi), G_x(\Phi), G_x(\beta), G_x(\alpha)).$$

5. Application: The Bachmann and Grzegorczyk hierarchies. We illustrate the use of the foregoing ideas with an application to subrecursive hierarchies first worked out by Girard in the rather different framework of [**5**]. (Many other applications can be envisaged but it remains to be seen just how useful they might be. For example, HYP is obtained by considering the usual jump hierarchy as the collapse under G_x of $\Lambda^{\Omega}(\hat{0}, {}^2\hat{E}, \beta)$ for $\beta \in \mathbf{O}(\Omega)$.)

Define the iteration functional It^{Ω} in $\mathrm{Rec}(\Omega)$ by the recursion

$$\mathrm{It}^{\Omega}(\phi, 0) = \sup_x \phi^{(x)}(1),$$

$$\mathrm{It}^{\Omega}(\phi, \alpha +_0 1) = \sup_x \phi^{(x)}\big(\mathrm{It}^{\Omega}(\phi, \alpha) +_0 1\big),$$

$$\mathrm{It}^{\Omega}(\phi, \sup \alpha_x) = \sup_x \mathrm{It}^{\Omega}(\phi, \alpha_x),$$

where $\phi^{(x)}$ denotes the xth iterate of ϕ.

Then for each x, $G_x(\mathrm{It}^{\Omega}) = \mathrm{It}_x^N$ where It_x^N is the iteration functional in $\mathrm{Rec}(N)$ given by

$$\mathrm{It}_x^N(f, 0) = f^{(x)}(1), \qquad \mathrm{It}_x^N(f, a + 1) = f^{(x)}\big(\mathrm{It}_x^N(f, a) + 1\big).$$

Now if we think of α as ranging over countable (set-theoretic) ordinals and ϕ as an increasing, continuous ordinal function, then $\mathrm{It}(\phi, \alpha)$ is the αth fixed-point of ϕ. Thus

$$\Lambda^{\Omega}(\exp_{\omega}, \mathrm{It}^{\Omega}): \quad \begin{aligned} \phi_0(\alpha) &= \omega^{0^{\alpha}}, \\ \phi_{\langle \hat{1}, \beta \rangle}(\alpha) &= \mathrm{It}^{\Omega}(\phi_{\beta}, \alpha), \\ \phi_{\sup \beta_x}(\alpha) &= \sup \phi_{\beta_x}(\alpha), \\ \phi_{\langle \hat{2}, \varepsilon \rangle}(\alpha) &= \phi_{\langle \hat{1}, \{\varepsilon\}(\alpha) \rangle}(0), \end{aligned}$$

is the $\mathrm{Rec}(\Omega)$-version of the *Bachmann hierarchy* **[1]** of critical functions but now defined over abstract ordinal notations in Ω. In fact,

$$|\phi_{\langle \hat{1}, \beta \rangle}(\alpha)| = \text{Bachmann's } \phi_{\|\beta\|}(|\alpha|).$$

(We henceforth abuse notation by writing $\beta + 1$ instead of $\langle \hat{1}, \beta \rangle$.)

By Theorem 6, G_x collapses the Bachmann hierarchy onto

$$\Lambda^{N}(\exp_x, \mathrm{It}_x^N): \quad \begin{aligned} f_0(a) &= x^{a}, \\ f_{\langle 1, b \rangle}(a) &= \mathrm{It}_x^N(f_b, a), \\ f_{\langle 2, e \rangle}(a) &= f_{\langle 1, \{\varepsilon\}(a) \rangle}(0), \end{aligned}$$

which is just a version of the transfinite *Grzegorczyk hierarchy*—but defined pointwise at each $x \in N$ (see **[4]** for comparisons with the more usual definitions). Strictly, the above functions f_b should carry a further subscript x but we shall collect the hierarchies $\Lambda^N(\exp_x, \mathrm{It}_x^N)$, $x = 0, 1, 2, \ldots,$ together by "diagonalizing" over x to form, for each $b \in \mathbf{O}(N)$,

$$f_b(a, x) = f_{b,x}(a) = \Lambda^N(\exp_x, \mathrm{It}_x^N, b, a),$$

so that $\{f_b | \, b \in \mathbf{O}(N)\}$ is now a 2-variable version of the Grzegorczyk hierarchy.

The following comparisons of the slow-growing and the Grzegorczyk hierarchies are now immediate.

THEOREM 7. (i) *Let $\hat{\omega}_0$ be the standard notation for ω in $\mathbf{O}(\Omega)$, so that $G(\hat{\omega}_0)$ is the fundamental sequence for $\omega_0 \in \mathbf{O}(N)$. Then for all x*

$$G_x\big(\phi_{\hat{\omega}_0 + 1}(0)\big) = f_{\omega_0}(x, x),$$

i.e. the first nonprimitive recursive function in the Grzegorczyk hierarchy is the collapse, under G, of the first prim-closed ordinal.

(ii) *Similarly, for all x,*

$$G_x\left(\phi_{\omega_1^{\hat{\omega}0}+1}(0)\right) = f_{\omega_0^{\omega 0}}(x, x)$$

and so on for $\omega_1^{\omega_1^{\hat{\omega}0}}, \ldots,$ *etc.*

(iii) *For all x,*

$$G_x\left(\phi_{\varepsilon_{\omega 1+1}+1}(0)\right) = f_{\varepsilon_0}(x, x),$$

i.e. the first function not provably recursive in arithmetic is the collapse of the Howard ordinal.

We conclude with a rough indication of how these subrecursive comparisons can be further extended (a more precise and detailed presentation is planned).

Firstly, the inductive definitions of $\mathbf{O}(\Omega)$ and $\mathbf{O}(N)$ can be iterated to yield higher "constructive" number classes over Ω, N respectively. Letting $\mathbf{O}_1(\Omega) = \mathbf{O}(\Omega)$, $\mathbf{O}_1(N) = \mathbf{O}(N)$ the next stage is

DEFINITION 11. (a)(i) $0 \in \mathbf{O}_2(\Omega)$,

(ii) $\alpha \in \mathbf{O}_2(\Omega) \Rightarrow \langle \hat{2}, \hat{1}, \alpha \rangle \in \mathbf{O}_2(\Omega)$,

(iii) $\forall x(\alpha_x \in \mathbf{O}_2(\Omega)) \Rightarrow \sup \alpha_x \in \mathbf{O}_2(\Omega)$,

(iv) $\forall \alpha \in \Omega(\{\varepsilon\}(\alpha) \in \mathbf{O}_2(\Omega)) \Rightarrow \langle \hat{2}, \hat{2}, \varepsilon \rangle \in \mathbf{O}_2(\Omega)$,

(v) $\forall \alpha \in \mathbf{O}_1(\Omega)(\{\varepsilon\}(\alpha) \in \mathbf{O}_2(\Omega)) \Rightarrow \langle \hat{2}, \hat{3}, \varepsilon \rangle \in \mathbf{O}_2(\Omega)$.

(b) $\mathbf{O}_2(N)$ is defined analogously, replacing α, ε, Ω by a, e, N and deleting clause (iii).

LEMMA 6. $\alpha \in \mathbf{O}_2(\Omega) \Rightarrow \forall x(G_x(\alpha) \in \mathbf{O}_2(N))$.

Now in $\mathbf{O}_2(\Omega)$ there is a standard notation (which for want of a better name we shall call) ω_2 for the least ordinal not represented in $\mathbf{O}_1(\Omega)$, and $G_x(\omega_2)$ will be the standard notation in $\mathbf{O}_2(N)$ for the least ordinal not represented in $\mathbf{O}_1(N)$, i.e. ω_1^{CK}. Thus we can build up notations

$$\omega_2, \omega_2^{\omega_2}, \omega_2^{\omega_2^{\omega_2}}, \ldots, \varepsilon_{\omega_2+1} \quad \text{in } \mathbf{O}_2(\Omega)$$

which collapse under G down to notations

$$\omega_1^{\mathrm{CK}}, \omega_1^{\omega_1}, \omega_1^{\omega_1^{\omega_1}}, \ldots, \varepsilon_{\omega_1+1} \quad \text{in } \mathbf{O}_2(N).$$

Using $\mathbf{O}_2(\Omega)$ we can then index the construction in $\mathrm{Rec}(\Omega)$ of a "level-2" Bachmann hierarchy of functions ϕ_α^2 from $\mathbf{O}_1(\Omega)$ into $\mathbf{O}_1(\Omega)$, whose collapse under G will be a version of the level-1 Bachmann hierarchy but defined over $\mathbf{O}_1(N)$ and indexed by notations in $\mathbf{O}_2(N)$—we denote this level-1 version $\{f_a^1 | a \in \mathbf{O}_2(N)\}$.

Therefore, if we set

$$\phi_\alpha^{1,2}(\beta, -) = \phi_{\phi_\alpha^2(\beta)}(-) \quad \text{for } \alpha \in \mathbf{O}_2(\Omega), \beta \in \mathbf{O}_1(\Omega),$$

$$f_a^{0,1}(b, -) = f_{f_a^1(b)}(-) \quad \text{for } a \in \mathbf{O}_2(N), b \in \mathbf{O}_1(N),$$

then, for each x, we will have

$$G_x(\phi_\alpha^{1,2}(\beta,\gamma)) = f^{0,1}_{G_x(\alpha)}(G_x(\beta), G_x(\gamma)),$$

and, in particular,

$$G_x\left(\phi^{1,2}_{\varepsilon_{\omega_2+1}+1}(0,0)\right) = f^{0,1}_{\varepsilon_{\omega_1+1}+1}(0,0),$$

which is the natural generalization of Theorem 7(iii) to the next higher number-class.

Now from Howard **[6]** and Bucholz and Pohlers **[2]** we have

$$|\phi_{\varepsilon_{\omega_1+1}+1}(0)| = \|f^1_{\varepsilon_{\omega_1+1}+1}(0)\|_N = |\mathrm{ID}_1|, \qquad |\phi^{1,2}_{\varepsilon_{\omega_2+1}+1}(0)| = |\mathrm{ID}_2|,$$

where $|\mathrm{ID}_n|$ denotes the ordinal of the theory of n-fold iterated inductive definitions.

Thus the results so far obtained could be stated loosely as

$$G(|\mathrm{ID}_1|) = f_{|\mathrm{ID}_0|}, \qquad G(|\mathrm{ID}_2|) = f_{|\mathrm{ID}_1|},$$

provided we now think of the slow-growing and Grzegorczyk hierarchies as being generated over appropriate primitive recursive systems of unique notations for ordinals below $|\mathrm{ID}_\omega| = \sup_n |\mathrm{ID}_n|$.

By further iterations of Definition 11 and the methods of generating ordinals just described, we would then obtain

(i) for each $n < \omega$, $G(|\mathrm{ID}_{n+1}|) = f_{|\mathrm{ID}_n|}$.

(ii) For each x, $G_x(|\mathrm{ID}_\omega|) = f_{|\mathrm{ID}_\omega|}(x-1, x)$.

Thus $|\mathrm{ID}_\omega|$ is the first ordinal α for which there is no $\beta < \alpha$ such that f_β dominates $G(\alpha)$. But $G(\alpha)$ is "honest" in the sense that its rate of growth reflects its computational complexity, so if we let C_β be the class of functions computable in f_β-bounded time (or space) then f_β dominates $G(\alpha)$ if and only if $G(\alpha) \in C_\beta$. Hence if we are prepared to accept $G(\alpha)$ as being a natural functional representaion of α then $|\mathrm{ID}_\omega|$ is the limit to which we can generate the C_β-hierarchy autonomously with respect to the condition that each α should be previously represented in some C_β with $\beta < \alpha$.

References

1. H. Bachmann, *Die normalfunktionen und das problem der ausgezeichneten folgen von ordnungszahlen*, Vierteljahresschrift der Naturforschung Gesellschaft in Zürich **95** (1950), 115–147.

2. W. Buchholz and W. Pohlers, *Provable wellorderings of formal theories for transfinitely iterated inductive definitions*, J. Symbolic Logic **43** (1978), 118–125.

3. E. A. Cichon, *A short proof of two recently discovered independence results using recursion theoretic methods*, Proc. Amer. Math. Soc. **87** (1983), 704–706.

4. E. A. Cichon and S. S. Wainer, *The slow-growing and the Grzegorczyk hierarchies*, J. Symbolic Logic **48** (1983), 399–408.

5. J.-Y. Girard, Π^1_2*-logic*. part I, Ann. Math. Logic **21** (1981), 75–219.

6. W. A. Howard, *A system of abstract constructive ordinals*, J. Symbolic Logic **37** (1972), 355–374.

7. J. Moldestad and D. Normann, *Models for recursion theory*, J. Symbolic Logic **41** (1976), 719–729.

8. Y. N. Moschovakis, *Descriptive set theory*, North-Holland, Amsterdam, 1980.

School of Mathematics, University of Leeds, Leeds LS2 9JT, England

Proceedings of Symposia in Pure Mathematics
Volume **42**, 1985

Gödel Theorems, Exponential Difficulty and Undecidability of Arithmetic Theories: An Exposition[1]

PAUL YOUNG[2]

Abstract. We present another proof of the exponential difficulty of Presburger arithmetic. The methods used here yield slightly stronger results than earlier proofs: They not only show exponential lower bounds but also show a form of exponential *inseparability*. We give a simple finite set of axioms for addition on the integers and prove that this theory is exponentially inseparable from the logically false sentences of the language. Our aim, however, is primarily expository: We wish to make clear that proving the exponential difficulty of Presburger arithmetic is essentially the same as proving the undecidability of Peano arithmetic. We thus arrange the proof of the exponential difficulty of Presburger arithmetic so that it is obtained essentially as a straightforward substitution into a standard proof of the corresponding Gödel Theorem. By so doing, we hope to convince logicians who teach the undecidability of Peano arithmetic that they should simultaneously teach the exponential difficulty of Presburger arithmetic. Since the proofs can be made essentially the same, very little additional work is required.

An overview. We begin by observing that if $\varphi_0, \varphi_1, \varphi_2, \ldots$ is *any* countable listing of total or partial functions mapping the set, N, of nonnegative integers into N, and if we define the sets

$$S_0 =_{\text{def}} \{i \mid \varphi_i(i) = 0\}, \quad \text{and} \quad S_1 =_{\text{def}} \{i \mid \varphi_i(i) = 1\},$$

1980 *Mathematics Subject Classification*. Primary 03-XX; Secondary 68-XX.

[1] This paper grew out of a talk which I gave at the 1982 Summer Workshop on Recursive Function Theory at Cornell University. In that talk, I presented a method which Michael Machtey and I had worked out earlier [**MY78**] for proving the Fischer-Rabin Theorem on the exponential difficulty of Presburger arithmetic. The proof we present here, however, considerably simplifies these earlier methods, and also yields stronger results. Nevertheless, the intent of this paper is still expository. I hope in a later paper to extend the methods used here to obtain additional new information about lower bounds on the difficulty of Presburger arithmetic.

[2] Supported by NSF Research Grant MCS 76091212A, Purdue University. The research reported here was performed while the author was a visiting professor at the University of California, Berkeley. The support of the Computer Science Division of UCB and the use of their facilities is gratefully acknowledged. I also wish to thank Egon Börger for useful discussions.

0082-0717/85 $1.00 + $.25 per page

then the two sets S_0 and S_1 cannot be separated by any function in the list $\varphi_0, \varphi_1, \varphi_2, \ldots$ That is to say, there is no index e_0 such that φ_{e_0} is total and $[i \in S_0 \Rightarrow \varphi_{e_0}(i) = 1]$ while $[i \in S_1 \Rightarrow \varphi_{e_0}(i) = 0]$.

The above fact is simply an instance of Russell's "paradox", which, in the form we use it, states that no reasonble collection of sets can contain a set which separates those sets which contain their own "name" from those which do not contain their own "name". Since in our context there is nothing paradoxical about this, we shall refer to it as *Russell's paradigm*. The proof is trivial: If we assume that φ_{e_0} is a total function on the list which separates S_0 and S_1, then calculating $\varphi_{e_0}(e_0)$ yields an immediate contradiction.

In spite of the simplicity of this result, it has well known and powerful consequences: In the context of recursive function theory, one takes the list $\varphi_0, \varphi_1, \varphi_2, \ldots$ to be the set of all partial recursive functions (i.e., the set of all computable functions). In this setting we say that the sets S_0 and S_1 are *recursively inseparable* because they cannot be separated by any (total) recursive function. The recursive inseparability of the sets S_0 and S_1 leads directly to a proof of the Gödel Undecidability Theorem. In this formulation, the Gödel Theorem states that the theorems of Peano arithmetic, or some finite axiomatization of arithmetic, are recursively inseparable from the logically false sentences in the language of arithmetic.

To see this, we shall observe that this form of the Gödel Theorem follows very easily from Russells's paradigm together with the fact that any programming system for the partial recursive functions contains a universal program. Pedagogically, this itself is a nice observation because universal programs are, from a contemporary standpoint, merely *interpretors* which, given a program and a potential input to that program, "interpret" the behavior of the program on that input by somehow simulating the action of the program on the input. That such interpretors always exist is a familiar fact for students of programming languages. The construction of interpretors is now a fairly routine, although often tedious and boring task. Thus a very powerful form of the Gödel Theorem follows quite easily from the very simple Russell paradigm together with a fact familiar to students of computer science.

But there is an additional advantage to this point of view. People who are familiar with interpretors are very much aware of the fact that they are useless they are fairly efficient. Once this point is understood, the one additional bit of information from number theory—the not too difficult Fischer-Rabin trick for encoding large portions of multiplication into the (first order) language of addition—not only makes it clear why Presburger arithmetic *should* be exponentially difficult, it makes the proof of exponential difficulty essentially *just a simple substitution* into the proof of the Gödel Theorem.

In outline, if one takes $\varphi'_0, \varphi'_1, \varphi'_2, \ldots$ to be a list, not of all the partial recursive functions, but in some suitable sense of listing of all of the functions computable with at most *exponential difficulty*, then the Russell paradigm still guarantees that the corresponding sets S_0' and S_1' cannot be separated by any characteristic function which occurs in the list. In this setting we say that the sets S_0' and S_1' are *exponentially inseparable*.

Once one knows that a universal program (an interpretor) exists, then the heart of the proof of the Gödel Theorem is to represent the sets S_0 and S_1 in the first order language of addition and multiplication. This is done by showing that programs for all of the partial recursive functions can be correctly translated into the first order language of addition and multiplication, and that the resulting arithmetic formulas represent the functions computed by the translated programs. To carry out the proofs, one needs only a simple finite set of axioms for addition and multiplication. If one now takes the very same translation used in the Gödel proof and simply substitutes the Fischer-Rabin formula for limited multiplications for the formal symbol for multiplication, then one always gets a formula in the language of addition alone, and, not too surprisingly, the resulting formula is correct just exactly to the extent that the Fischer-Rabin formula correctly describes multiplications. Since the latter formula is "correct up to an exponential", the resulting formulas of Presburger arithmetic correctly encode programs whose computational complexity is exponentially bounded. Thus the new formulas no longer encode the recursively inseparable sets S_0 and S_1, but instead encode the *exponentially inseparable* sets S_0' and S_1'. Furthermore, if one takes the same axioms as used in the Gödel proof and simply removes all of those which refer to multiplication, then, in a suitable sense, the new formulas and the more limited axioms successfully represent those programs which do not do more than "exponentially large" multiplications. Thus, by virtually a direct substitution, the same proof which shows that the theorems which are provable from the axioms for addition and multiplication are *recursively inseparable* from the logically false sentences of that language *also* shows that the theorems which are provable without using multiplication are *exponentially inseparable* from the logically false sentences of the language of addition.

There is, of course, a trick to this: one must choose a notion of computational complexity which makes this all work. We begin by looking at a set of axioms for addition and multiplication.

The partial recursive functions and some axioms for arithmetic. We begin by listing a variation of Julia Robinson's axioms for addition and multiplication on the natural numbers:

S1. $\mathbf{0} \oplus \mathbf{1} = \mathbf{1}$.	A1. $x \oplus \mathbf{0} = x$.
S2. $x \oplus \mathbf{1} \neq \mathbf{0}$.	A2. $x \oplus (y \oplus \mathbf{1}) = (x \oplus y) + \mathbf{1}$.
S3. $x \neq \mathbf{0} \Rightarrow \exists z[z \oplus \mathbf{1} = x]$.	M1. $x \otimes \mathbf{0} = \mathbf{0}$.
S4. $x \oplus \mathbf{1} = y \oplus \mathbf{1} \Rightarrow x = y$.	M2. $x \otimes (y \oplus \mathbf{1}) = (x \otimes y) \oplus x$.

T1. $x < y$ or $x = y$ or $y < x$.

The above nine axioms are logically equivalent to the single axiom consisting of their conjunction, which for brevity we will denote by MULTAX. Since we will also be interested in the theory of addition alone, we shall be interested in those axioms obtained by simply omitting the axioms M1 and M2. Since the expression $x < y$ is really just an abbreviation for the more complicated formula

$$(\exists z)[z \neq \mathbf{0} \,\&\, x \oplus z = y],$$

the remaining seven axioms refer to the theory of addition alone. For brevity, we denote the conjunction of these seven axioms by ADDAX. (We will also abuse our notation, occasionally referring to the collection of nine axioms as MULTAX and to the collection of seven axioms as ADDAX. For each n, we let $\mathbf{n}$ be the standard (unary) representation of n in this language; i.e., $\mathbf{n}$ is $(((\mathbf{1} \oplus \mathbf{1}) \oplus \cdots \oplus \mathbf{1}) \oplus \mathbf{1})$, n times.

In order to prove the Gödel Theorem, one needs to know that a few very simple facts are provable from these axioms:

PROPOSITION 1. *For each natural number n and corresponding numeral* $\mathbf{n}$:
(i) ADDAX $\vdash x < \mathbf{n} \oplus \mathbf{1} \Rightarrow x \leqslant \mathbf{n}$,
(ii) ADDAX $\vdash x \leqslant \mathbf{n} \Rightarrow [x = \mathbf{0}$ *or* $x = \mathbf{1}$ *or* $\cdots$ *or* $x = \mathbf{n}]$.
Furthermore, for each fixed set of natural numbers m, n and p:
(iii) *if* $m + n = p$, *then* ADDAX $\vdash \mathbf{m} \oplus \mathbf{n} = \mathbf{p}$,
(iv) *if* $m \neq n$, *then* ADDAX $\vdash \mathbf{m} \neq \mathbf{n}$,
(v) *if* $m \times n = p$, *then* MULTAX $\vdash \mathbf{m} \otimes \mathbf{n} = \mathbf{p}$.

DISCUSSION OF PROOFS. The proofs here are all quite simple. Detailed proofs of all except (iv) may be found on p. 125 of [**MY78**]. (In [**MY78**] all proofs are from MULTAX, so one must observe that in cases (i)–(iv) neither of the multiplicative axioms is actually used.)

It is now necessary to consider the relation of these axioms to the partial recursive functions. For our purposes, it will be convenient to choose a definition of the partial recursive functions which is intimately related to the above axioms. To this end, we define the partial recursive functions to be the smallest class of functions containing the base functions consisting of addition, multiplication, the characteristic function for equality, and a finite collection of projection functions. We close this collection of functions under the operations of functional composition and minimalization. Since we will ultimately be concerned with the complexity of translating programs for the partial recursive functions into the language for arithmetic, we will now carefully give an inductive definition of these functions and of the *programs* which compute them:

The following are programs of two arguments: $+$, $\times$, $\mathbf{c}_=$. In addition, for each n with $1 \leqslant n \leqslant 3$ and each j with $1 \leqslant j \leqslant n$, $\mathbf{P}^n_j$ is a program of n arguments. The *semantics* of these programs is given by specifying the obvious functions which they are intended to compute:[3]

$$\mathbf{c}_=(x, y) \equiv_{\text{def}} 1 \text{ if } x = y, \qquad +(x, y) \equiv_{\text{def}} x + y;$$
$$\mathbf{c}_=(x, y) \equiv_{\text{def}} 0 \text{ if } x \neq y; \qquad \times(x, y) \equiv_{\text{def}} x \times y;$$
$$\mathbf{P}_\mathbf{j}^\mathbf{n}(x_1, \ldots, x_n) \equiv_{\text{def}} x_j.$$

If $\mathbf{G_1}, \ldots, \mathbf{G_m}$ ($m \leqslant 3$) are each programs of n ($n \leqslant 3$) arguments, if $\bar{x}$ is an n-tuple of arguments $x_1, \ldots, x_n$, and if $\mathbf{H}$ is a program of m arguments, then $\mathbf{H}(\mathbf{G_1}, \ldots, \mathbf{G_m})$ is a program of n arguments which computes the function

$$\mathbf{H}(\mathbf{G_1}, \ldots, \mathbf{G_m})(\bar{x}) \equiv_{\text{def}} \mathbf{H}(\mathbf{G_1}(\bar{x}), \ldots, \mathbf{G_m}(\bar{x})).$$

Finally, if $\mathbf{H}$ is a program of $n + 1$ arguments ($1 \leqslant n \leqslant 2$), then $\mathbf{min\,H}$ is a program of n arguments with

$$\mathbf{min\,H}(\bar{x}) \equiv_{\text{def}} \text{the smallest } z \text{ such that } \mathbf{H}(\bar{x}, z) = 0$$

and

$$(\forall z' < z)[\mathbf{H}(\bar{x}, z') \text{ returns some value}].^4$$

Now the key to proving the Gödel Undecidability Theorem is to prove that there is an effective process which, given any of the above programs, produces a formula in the language of arithmetic such that the formula "provably represents" the function which the program computes. To this end, we first give the standard definition of what it means for a formula to represent a function (using a set of axioms AX):

DEFINITION. Let φ be an arbitrary function of k arguments, mapping some subset of N^k to N. Then a formula F_φ with $k + 1$ free variables is said to *represent φ using a set of axioms*, AX, if

$$\mathrm{AX} \vdash \left[F_\varphi(\bar{x}, z) \,\&\, F_\varphi(\bar{x}, z') \Rightarrow z = z'\right],$$

---

[3] By allowing ourselves to use only finitely many projection functions, we have restricted ourselves to functions of only finitely many arguments. This will later make our translation of the partial recursive functions into the language of arithmetic somewhat simpler, and, of course, as long as we have functions of two arguments and pairing functions, then functions of many arguments can always be coded into functions of two arguments. Thus it is sufficient to have only functions of two arguments, but three turns out to be more convenient.

[4] If we wished to prove Tarski's Theorem instead of the Gödel Theorem, we would simply leave off the second clause of this definition, making **min H** essentially just existential quantification, thus making the class of functions the class of all arithmetically definable functions. In this case we would not get representability of these functions from MULTAX, but instead we would have representability of these functions from the *true* statements of arithmetic. With only trivial and obvious changes, the proof we are about to outline would then prove that no arithmetically definable function can separate the true sentences of arithmetic from the false sentences of arithmetic. It would also prove that even with "free infinite quantification" (counting *only* the largest multiplication in the infinite searches), deciding Presburger arithmetic still requires exponentially large multiplications.

and for all k-tuples of natural numbers $\bar{n}$ and natural number p,

$$\text{if } \varphi(\bar{n}) = p, \text{ then AX} \vdash F_\varphi(\bar{\mathbf{n}}, \mathbf{p}).$$

The first condition says that whenever $F_\varphi(\bar{x}, z)$ holds, then the value z is uniquely determined. Thus F_φ determines *some* partial function. The second condition guarantees that whenever the function φ is defined, then it agrees with the function determined by F_φ.

THEOREM. *There is an effective procedure which, given any program,* **P** *for a partial recursive function, produces a formula of arithmetic,* $S(\mathbf{P}, \bar{x}, z)$, *such that the formula* $S(\mathbf{P}, \bar{x}, z)$ *represents the function computed by the program* **P**. *If the program* **P** *does not contain the subprogram* $\times$ *(multiplication), then the representation is accomplished using the axioms* ADDAX *alone. If the program* **P** *does contain the subprogram* $\times$, *then the representation is accomplished using the axioms* MULTAX.

PROOF. The necessary translation S from the programs for the partial recursive functions to formulas in the language of arithmetic is easily given, inductively on the lengths of the programs, as follows:

$$S(+, x, y, z) \text{ is } [x \oplus y = z],$$
$$S(\times, x, y, z) \text{ is } [x \otimes y = z],$$
$$S(\mathbf{P}_\mathbf{j}^\mathbf{n}, \bar{x}, z) \text{ is } [x_j = z],$$
$$S(\mathbf{c}_=, x, y, z) \text{ is } [x = y \,\&\, z = \mathbf{1}] \textit{ or } [x \neq y \,\&\, z = \mathbf{0}].$$

Similarly, $S(\mathbf{H}(\mathbf{G}_1, \ldots, \mathbf{G}_\mathbf{m}), \bar{x}, z)$ is

$$\exists y_1 \cdots \exists y_m [S(\mathbf{G}_1, \bar{x}, y_1) \,\&\, \cdots \,\&\, S(\mathbf{G}_\mathbf{m}, \bar{x}, y_m) \,\&\, S(\mathbf{H}(y_1, \ldots, y_m, z))],$$

and $S(\mathbf{min}\,\mathbf{H}, \bar{x}, z)$ is

$$\forall w [w \leqslant z \Rightarrow \exists y [S(\mathbf{H}, \bar{x}, w, y) \,\&\, (y = \mathbf{0} \text{ iff } w = z)]].$$[5]

That this translation works, i.e. that the formulas $S(\mathbf{P})$ do indeed represent the programs **P** using the axiom sets ADDAX and MULTAX, uses Proposition 1 and is equally easy to prove. (The proof is by induction on the length of the programs, and a proof with essentially all of the details filled in can be found on pp. 127 and 128 of [**MY78**]. Once again, the proof given there is using the axioms

[5] Thus the translation of a program **P** depends not just on the program itself, but in a trivial way also on the free variables chosen in the language of arithmetic. Nevertheless, we sometimes suppress mention of these free variables when they are obvious, e.g., writing $S(\mathbf{P})$ instead of $S(\mathbf{P}, \bar{x}, z)$. Obviously also, in the last two recursive definitions, one must sensibly choose the bound variables y_i and w so that they are distinct from the designated free variables $\bar{x}$ and z in the same formula. Normally, they might be chosen as the *first* variables distinct from the free variables $\bar{x}$ and z, and from each other.

MULTAX, so the reader must observe that the only time the two multiplicative axioms are used is in explicitly dealing with the program $\times$.)

If we now knew that the translation S of the preceding theorem could itself be carried out in the specific programming system which we have given for the partial recursive functions, then we would be very close to being able to complete our proof of the Gödel Theorem. Of course, it is obvious in an intuitive sense that our inductive definition of S provides us with an effective means of producing the translation S, and indeed a reader with only limited programming experience should have little trouble writing a recursive program which computes S. The problem is that we have not yet proven that very much can, in fact, be computed using the programs in this system, and we specifially have not allowed recursion as an explicit primitive of the system.

A second problem is that our programs for the partial recursive functions compute functions from integers to integers, while S is a syntactic transformation carrying strings to strings. But this presents no real problem. Both the domain and range of S are strings in some fixed alphabet. Without loss of generality, we can assume that these alphabets have fewer than k symbols, where k is some sufficiently large (prime) number. We will now consider numbers written base k, identifying each *character* in our alphabet with some nonzero digit less than k. Thus for every program **P** and n-tuple of variables $\bar{x}$, both the syntactic string **P**, $\bar{x}$ and its image $S(\mathbf{P}, \bar{x})$ in the language of arithmetic can be regarded simply as integers written in base k notation, and it is now reasonable to ask whether this mapping S of integers to integers can be carried out in our programming system for the partial recursive functions.

Proving *this* is the hardest part of both Gödel's results on the undecidability of full arithmetic and of the Fischer-Rabin result on the exponential difficulty of Presburger arithmetic. For the moment, we shall ignore this difficulty, explaining instead how the Gödel Theorem is an easy consequence of the fact that the translation S can be carried out in our programming system.

To this end, define a sequence of partial functions $\varphi_0, \varphi_1, \varphi_2, \ldots$ by taking $\varphi_{\mathbf{i}}(x) \equiv_{\text{def}} \mathbf{i}(x)$ if **i** is a program of one argument, with $\varphi_{\mathbf{i}}(x)$ undefined otherwise. We know by the Russell paradigm that the corresponding sets S_0 and S_1 cannot be separated by any characteristic function in this list, i.e., they cannot be separated by any characteristic function computed by any of our partial recursive programs of one argument. If the transformation S was computable by a program in our system, then one would certainly expect the syntactic transformation S' defined by

$$S'(i) \equiv_{\text{def}} (\forall x)[x = \mathbf{i} \Rightarrow S(\mathbf{i}, x, \mathbf{0})]^5$$

to also be computable by some program in our system.

But our representability theorem then guarantees that

$$i \in S_0 \text{ implies MULTAX} \vdash S'(i),$$

while

$$i \in S_1 \text{ implies MULTAX} \vdash (\forall x)[x = \mathbf{i} \Rightarrow S(\mathbf{i}, x, \mathbf{1})],^{6}$$

thus

$$i \in S_1 \text{ implies MULTAX} \vdash \text{not } S'(i).$$

If we now take $T(i) \equiv_{\text{def}} [\text{MULTAX} \,\&\, S'(i)]$ we clearly get that

$$i \in S_0 \text{ implies that MULTAX} \vdash T(i),$$

while

$$i \in S_1 \text{ implies that } T(i) \textit{ is logically false}.$$

Thus we see immediately that if the transformation T can be carried out by a program in our system for the partial recursive functions, then the theorems of MULTAX and the *logically false sentences* of the language of plus and times cannot be separated by any characteristic function computably in this programming system for the partial recursive functions: If these sets *were* recursively separable, then the sets S_0 and S_1 would also be separable, contradicting the Russell paradigm.

As immediate corollaries, one, of course, gets that *any* axiomatization for arithmetic at least as powerful as MULTAX must be undecidable, that there can be no decision procedure for the true sentences of arithmetic (and hence by standard arguments, no recursively presented set of axioms for the true sentences of arithmetic), and that there can be no decision procedure for the logically valid sentences in this language (Church's Theorem).

To actually complete this proof of the Gödel Theorem, one must "merely" prove that the translation S, and hence also its variant T can indeed be carried out by a program in our programming system. This is not so easy, so we postpone it until the end. But we now make some related, motivating, pedagogical comments. Once they have accepted our programming system as a reasonable programming system, students with experience with real programming systems will certainly believe that they can write an "interpretor" for the system by a program in the system. Such an "interpretor" will be a program which accepts as input any program in the system and a potential input to the program, and "interprets" the program on that input by somehow "simulating" its behavior. The building of interpretors is standard in computer science. But an interpretor is simply what a logician calls a "universal" program, i.e., it is a program **univ** of two arguments which, given any program **i** and input x, computes the same thing as $\mathbf{i}(x)$. I.e., in our φ notation,

$$\varphi_{\mathbf{univ}}(\mathbf{i}, x) = \varphi_{\mathbf{i}}(x).$$

[6] We have here an unfortunate double use of notation. In this context, the first occurrence of the symbol **i** stands for the *numeral* used in the language of arithmetic to represent the integer i. This is the string $((\cdots(\mathbf{1} \oplus \mathbf{1}) \cdots) \oplus \mathbf{1})$ (taken i times). However, in the formula $S(\mathbf{i}, x, z)$, we use the symbol **i** to denote a program, which we then interpret as a number written in base k notation. Thus if we think of a program **i** as an integer and then name that same program **i** by the corresponding numeral we have to do a shift from base k notation to what amounts to unary notation.

But we know that *every* program, hence also the universal program, is representable from MULTAX. If we wished, we could thus redefine T to

$$T'(i) \equiv_{\text{def}} \text{MULTAX} \,\&\, (\forall x)[x = \mathbf{i} \Rightarrow S(\mathbf{univ}, x, x, \mathbf{0})].$$

Clearly, exactly as before, we again get

$$x \in S_0 \text{ implies that MULTAX} \vdash T'(\mathbf{i}),$$

while

$$x \in S_1 \text{ implies that } T'(\mathbf{i}) \textit{ is logically false}.$$

Now for the purpose of proving the Gödel Theorems, $T'(i)$ is especially pretty. If we ask whether we can compute $T'(i)$ using our programming system, then we see that, since **univ** is *constant*, viewed as a string of characters, $T'(i)$ is simply one constant string w_1 followed by the string which represents the number i, followed by another constant string w_2. Since, as we shall later see, the numerical operation corresponding to concatenation is not too hard to compute in our system, T' really is not very hard to compute in our programming system.

All of this is fine for motivation, but it does no good unless we know that a universal program actually exists for our system. As we have already mentioned, the way one writes a universal program is to write an interpretor which faithfully mirrors the semantics of the programming system. But our original translation S of the programming system into the language of arithmetic was chosen as it was precisely because it, too, mirrored the semantics of these programs. Furthermore, since the programming system itself is chosen deliberately to be "close" to the language of arithmetic, writing a program which effects the translation S is almost exactly the same as writing a universal program for the system. Technically, the two tasks are almost identical. Intellectually, it is perhaps cleaner to demonstrate the existence of a universal program. This is true because it locates all of the real difficulty in proving the Gödel Theorem in the difficulty of writing an interpretor for a (perhaps somewhat bizarre) programming system. That such interpretors can be written is by now a thoroughly familiar, although perhaps technically tedious and boring, task for students of programming languages. Furthermore, this approach makes clear that, contrary to some current popularizations (e.g. [**H79**]), no very deep self-referencing is required to prove the Gödel Theorem. The *only* self-referencings in this proof are in the very trivial application of the Russell paradigm showing that the sets S_0 and S_1 are inseparable, and the by now thoroughly understood self-referencing in the ability of programming languages to possess programs which interpret their own language.[7]

[7] Unlike the proof we give here and most other proofs of the Fischer-Rabin Theorem, the proof given in [**MY78**] totally avoids diagonalization and the use of a universal function, depending instead on a simple application of the recursion theorem, a point discussed extensively in [**MY81**]. By way of contrast, the proof of the Fischer-Rabin Theorem which we give here avoids using either the recursion theorem or a universal function, although the latter is used as a motivating concept.

We now wish to go on to explain how to transform the outline which we have just given of a proof of the Gödel Theorem into a proof of the exponential difficulty of Presburger arithmetic. For this purpose, it seems better to use the translation S to get T rather than the existence of a universal program to get T'. But universal functions and interpretors provide one more motivating observation. The heart of T is the transformation S, which we are assuming can be carried out by a program in our programming system. We have already observed that writing a program for S is essentially like writing an interpretor for our programming system. Just as we believe that we can always write interpretors, we believe that we can always write interpretors which are not too complicated to execute. Since S is essentially like an interpretor, we should believe that we can write a program which computes S, and hence also one which computes T, which is "not too difficult to run." Let us now accept on faith for a little while *not only that we can write a program which will compute* T (Church's Thesis), but also that when this program is run, its difficulty will not be too outrageously difficult; when we have decided on a suitable measure of the difficulty of computing, the difficulty of running our program which computes T on input *i should not be worse than doubly exponential in i*.

Exponential difficulty of theories of addition. We come now to the real point of this paper: exponential difficulty proofs for Presburger arithmetic.

When Gödel proved his undecidability theorems, it had already been shown by Presburger that the first-order theory of addition was decidable. In 1974 Fischer and Rabin proved that, in suitable measures, any such decision procedure must, in fact, be doubly exponentially difficult. We will show how to obtain the Fischer-Rabin result for single exponential difficulty by essentially just a substitution into the proof outlined above for the Gödel Theorem. As a side benefit, we will prove a slightly more general result. We will show not just that Presburger arithmetic is exponentially difficult, but, in fact, that in a suitable sense, the theorems of the very weak theory ADDAX are exponentially inseparable from the logically false sentences of the theory of addition.[8]

To prove this and to establish a suitable meaning of "exponentially difficult," we begin by asking why the preceding proof of Gödel Theorem cannot be used to prove that the theory of addition alone is undecidable. The essential idea of Gödel proof was to use our programming system for the partial recursive functions to map programs in this system to representing formulas in the language of arithmetic, i.e. into the language of addition and multiplication. Obviously, this translation, the translation S given earlier, cannot work directly if

[8] Our methods, in fact, will also yield the Fischer-Rabin results for double exponential difficulty. To do this, one uses the Fischer-Rabin techniques to encode triply exponential portions of multiplication into the theory of addition. These techniques use Chebychev's Theorem and go beyond the simple expository account which we give here, although the same methods will work to yield double exponential difficulty in a simple variation of the multiplicative measure we introduce in this section.

there is no symbol for multiplication in the target language. Nevertheless, it is worth observing that the *only* place the translation S used the fact that the target language had a symbol for multiplication was in the direct translation of the program $\times$, the base program for multiplication. In this case we defined $S(\times)$ to be the formula $[x \otimes y = z]$. Without having the symbol $\otimes$ in the target language, this portion of the translation S is simply meaningless, but all other portions clearly work.

Naively, one might hope to overcome this difficulty by using addition to define multiplication, and then substituting this definition for the symbol $\otimes$. But this would achieve a translation of the programs in our programming language for the partial recursive functions into the language of Presburger arithmetic. If we *could* do this, the preceding proof of the Gödel Theorem would then show that Presburger arithmetic is undecidable, which it is not.

The key idea in the Fischer-Rabin proof is that although one cannot (using only "first-order" notions) define multiplication from addition, one can nevertheless define "exponentially large portions" of multiplication using the language of addition alone. What we shall see is that in a certain sense, once this is done, the above proof of the Gödel Theorem works exactly to the extent that one can successfully define multiplication from addition.

Before further exploring the Fischer-Rabin ideas, it is now time to consider a question which we have so far left deliberately vague. What shall we mean by saying that a problem is exponentially difficult? More precisely, how shall we measure the complexity of a calculation?

In the context in which we are now working, we are trying to map programs for the partial recursive functions into the language of addition alone, and we have seen that what is difficult about *this* translation is that we have no obvious place to map the base program for multiplication. The translation is, however, easy for programs which do not contain multiplication. This suggests that for this application, the difficulty of the translation is solely centered around the multiplications which programs perform.

Now we have defined the programs for the partial recursive functions by starting with certain simple programs for the base functions and combining these programs under composition and minimalization. This means that, intuitively, as these programs compute, they unwind the compositions and the minimalizations in their definitions until ultimately they arrive at the base programs, which do, in a certain sense, all of the real computation. Since the only function which is causing us difficulty is multiplication, this suggests that we make all components of our calculations free *except* for multiplications: We define the *complexity of computation* of a program **P** to be the function

$$\Phi_{\mathbf{P}}(\bar{x}) =_{\text{def}} \max\{\min\{a, b \mid \text{program } \mathbf{P} \text{ multiplies } a \text{ by } b \text{ while it performs its computation on input } \bar{x}\}\}.$$

That is, every time program **P** performs a multiplication, we measure the difficulty of this multiplication by the *smaller* of the two numbers multiplied, and we then measure the complexity of the entire computation by taking the largest of all of these individual minimums.[9]

With this as background, we are now ready to describe the Fischer-Rabin trick for encoding super-exponentially difficult multiplications into addition:

LEMMA. (A) (*Fischer and Rabin*) *For each natural number* k, *there is a formula* $M_k(x, y, z)$ *in the language of addition such that the formula* M_k *is true iff* $x \times y = z$ *and* x *is less than the constant* $2^{2^{2k}}$. *Moreover, there is a positive integer constant* c *such that* $|M_k(x, y, z)| \leqslant c(k + 1)$.[10]

(B) *In fact, there are fixed strings of characters* X *and* P (*which are* ***independent of*** k) *such that* ***formula*** $M_k(x, y, z)$ *is just the string* $X^k M_0(x, y, z) P^k$. *Also, for any natural numbers* m, n *and* p, ADDAX $\vdash M_k(\mathbf{m}, \mathbf{n}, \mathbf{p})$ *iff* $m \leqslant 2^{2^{2k}}$ *and* $m \times n = p$.

OUTLINE OF PROOF. The proof depends on a variety of elementary number-theoretic and logical tricks, all of which are explained in detail in [**FR74**] and on 194 and 195 of [**MY78**]. We merely outline them here.

We begin by letting $M_0(x, y, z)$ be the formula

$$[x = \mathbf{0} \,\&\, z = \mathbf{0}] \quad \text{or} \quad [x = \mathbf{1} \,\&\, z = y].$$

Since in this case, we need only consider $x < 2^{2^{2k}} = 2$, it is trivial to verify the correctness of M_0, establishing the basis of an induction.

We next observe that, since $(n + 1)^2 - 1 = n^2 + 2n$, any number less than $(n + 1)^2$ can be written as a sum $u_1^2 + u_2 + u_3$, where u_1, u_2 and u_3 are all less than $n + 1$. Using this, we can obtain a formula $M(x, y, z)$ which describes multiplications of x and y for values of x less than $2^{2^{2k+1}}$ (instead of less than $2^{2^{2(k+1)}}$): Using $\bar{u}$ for a triple of variables and $\bar{p}$ for a quadruple of variables, the formula $M(x, y, z)$ describes such multiplications:

$$\begin{aligned} \exists \bar{u} \exists s \exists \bar{p} \big[M_k(u_1, u_1, s) \,\&\, x = (s \oplus (u_2 + u_3)) \\ \& M_k(u_1, y, p_1) \,\&\, M_k(u_1, p_1, p_2) \,\&\, M_k(u_2, y, p_3) \\ \& M_k(u_3, y, p_4) \,\&\, z = (p_2 \oplus (p_3 \oplus p_4)) \big]. \end{aligned}$$

Using the number-theoretic facts just mentioned, one easily sees that if M_k correctly describes multiplications for $x < 2^{2^{2k}}$, that M correctly describes multiplications for $x < 2^{2^{2k+1}}$.

The first difficulty with the formula M is that, because it has five occurrences of the subformula M_k in it, its size (as a function of k) has growth rate of order 5^k.

---

[9] One might more naturally have taken the larger of the two numbers multiplied as the individual complexity, but this would make the proof no easier and would yield a weaker, less useful, result, since we are establishing a *lower* bound on the complexity of the decision procedure. (But this is necessary to establish double exponential difficulty.)

[10] For any string or integer, $|w|$ is, in its obvious meaning, the *length* of w.

Following Fischer and Rabin, we overcome this difficulty by considering the related formula $M'(x, y, z)$ defined by:

$$\begin{aligned}&\exists \bar{u} \exists s \exists \bar{p} \forall x' \forall y' \forall z' \\ &\quad [x = (s \oplus (u_2 \oplus u_3)) \,\&\, z = (p_2 \oplus (p_3 \oplus p_4)) \\ &\quad\quad \&([(u_1 = x' \,\&\, u_1 = y' \,\&\, s = z') \text{ or } (u_1 = x' \,\&\, y = y' \,\&\, p_1 = z') \text{ or}) \\ &\quad\quad\quad (u_1 = x' \,\&\, p_1 = y' \,\&\, p_2 = z') \text{ or } (u_2 = x' \,\&\, y = y' \,\&\, p_3 = z') \text{ or} \\ &\quad\quad\quad\quad (u_3 = x' \,\&\, y = y' \,\&\, p_4 = z')] \Rightarrow M_k(x', y', z'))].\end{aligned}$$

If we simply examine the formula M' directly and assume inductively that $M_k(\mathbf{m}, \mathbf{n}, \mathbf{p})$ is true iff ADDAX $\vdash M_k(\mathbf{m}, \mathbf{n}, \mathbf{p})$, iff $m \times n = p$, and $m < 2^{2^{2k}}$, then we easily get that if $m \times n = p$ with $m < 2^{2^{2k+1}}$, then ADDAX $\vdash M'(\mathbf{m}, \mathbf{n}, \mathbf{p})$. On the other hand, soundness together with the induction hypothesis yields that if ADDAX $\vdash M'(\mathbf{m}, \mathbf{n}, \mathbf{p})$, then we must have $m < 2^{2^{2k+1}}$ and $m \times n = p$.

Since M' contains only one occurrence of the subformula M_k, its length is only linear in k. We may next use the formula M' in place of M_k in the above process to obtain the desired formula M_{k+1} which describes multiplications when the first variable is less than $2^{2^{2(k+1)}}$. As we do this, we must take care to reuse bound variables so that even as k grows, we have a fixed bound on the number of variables actually used, thus keeping the lengths of the variables (i.e. their *subscripts*) strictly bounded. Once this is done, we may unwind the recursive definition of M_{k+1} from M_k to obtain an explicit formula describing each M_k directly from M_0. It turns out that we are very lucky: The formula $M_k(x, y, z)$ is of the desired form $X^k M_0(x, y, z) P^k$. Complete details may be found on p. 195 of **[MY78]**, and we omit them.

This Lemma now easily yields the representation of the limited multiplications which we want:

PROPOSITION 2.(A) (*Fischer and Rabin*) *For each natural number k there is a formula* $\otimes_k$, *in the language of addition, which represents the function* $x \times y$ *for values in which either x or y are less than* $2^{2^{2k}}$. *Then length of* $\otimes_k$ *is linear in k.*

(B) *In fact,* $\otimes_k$ *represents this function using only the axioms* ADDAX, *and* $\otimes_k$ *has the very simple form*

$$\begin{aligned}&w_0 X^k M_0(x, y, z) P^k w_1 X^k M_0(x, y, z) \\ &\quad \cdot P^k w_2 X^k M_0(x, y, z) P^k w_3 X^k M_0(x, y, z) P^k w_4,\end{aligned}$$

where all of the w_i*'s, X and P are fixed strings of characters which are independent of k.*

PROOF. The formula M_k of the Lemma comes very close to being the formula we need. There are only two problems. First, one needs a formula which is adequate for multiplications with *either* variable bounded. Second, one needs to be able to prove the *uniqueness* of the functional values from the axioms

ADDAX. Using Proposition 1(ii), both problems are obviously overcome by the following definition of $\otimes_k(x, y, z)$:

$$[M_k(x, y, z) \text{ or } M_k(y, x, z)] \,\&\, \forall z' < z[\text{not } M_k(x, y, z') \,\&\, \text{not } M_k(y, x, z')].$$

(The strings X and P are exactly as from the Lemma, and the strings w_i are extremely short and trivial.)

We now ask: What happens if we take the translation S used in the Gödel proof and define a new translation S_k which is obtained from S simply by replacing our translation $S(\times, x, y, z) \equiv_{\text{def}} [x \otimes y = z]$ by $S_k(\times, x, y, z) \equiv_{\text{def}} \otimes_k(x, y, z)$, otherwise letting S_k be *identical* with S?

Several things are now immediate. We already know that the formula $\oplus_k$ represents the partial function of multiplication of x and y restricted to the case where at least one of x or y is less than $2^{2^{2k}}$. Furthermore, we have also already observed that in verifying that for *any* program **P** the formula $S(\mathbf{P})$ represents (using the axioms MULTAX) the function computed by program **P**; we used the multiplicative axioms *only* in verifying that $S(\times)$ represents multiplication. It thus follows that, for any program **P**, using the axioms ADDAX alone, the formula $S_k(\mathbf{P})$ represents the partial function computed by **P** for those computations in which all of the multiplications multiply only pairs m, n for which at least one of m and n is less than $2^{2^{2k}}$.

What happens if we next take the transformation T used in the proof of the Gödel Theorem and replace it by

$$T_{\text{ADD}}(i) \equiv_{\text{def}} \text{ADDAX} \,\&\, (\forall x)[x = \mathbf{i} \Rightarrow S_i(\mathbf{i}, x, \mathbf{0})]?^{11}$$

In this case we see that for any program **i** of a single argument, if program **i** on input i produces 0 without multiplying numbers bigger than $2^{2^{2i}}$, then ADDAX $\vdash T_{\text{ADD}}(i)$. On the other hand, if program **i** on input i produces 1 without multiplying numbers bigger than $2^{2^{2i}}$, then exactly as with the Gödel proof we see that the sentence $T_{\text{ADD}}(i)$ is logically false.

We now take $\varphi_i'(x)$ to be the result of running program **i** on input x provided **i** is a program of one argument and provided $\Phi_{\mathbf{i}}(x) < 2^{2^{2x}}$. Thus $\varphi_0', \varphi_1', \varphi_2', \ldots$ is a list of all the functions of the argument computed by programs in our language whose difficulty *in our multiplicative measure* is never greater than double exponential in their (actual) inputs. We know by the Russell paradigm that the corresponding sets S_0' and S_1' cannot be separated by any program which has less than double exponential difficulty in our multiplicative measure.

At this point, in the proof of the Gödel Theorem, we needed to know that the transformation T could be carried out by a program in our programming system. At this same point in our proof of the exponential difficulty of Presburger arithmetic, we need to know not only that the corresponding transformation T_{ADD}

[11] Note here that there is again an unfortunate multiple occurrence of '**i**'. In its first occurrence **i** represents the numeral for i in the language of arithmetic; i.e., as explained earlier, it is essentially i written in unary notation. In its second occurrence, inside S, **i** is the *program* **i**.

can be carried out by a program in our programming system, but also that the difficulty of excecuting this program is at worst double exponential in its inputs. But the heart of the transformation T was the transformation S which carried programs to their representing formulas. To prove the Gödel Theorem, one *must* give an explicit program for the transformation S, and we have already observed that once one does this, since S is essentially just an interpretor, the execution of this program for S *should not* be horrendously difficult. But the transformation S_k which is at the heart of the transformation T_{ADD}, is *identical* to the transformation S *except* that in computing $T_{\mathrm{ADD}}(i)$ for a program **i**, one may be required to compute $S_i(\times, x, y, z)$, which is $\otimes_i(x, y, z)$ instead of the simpler formula $[x \otimes y = z]$ used in defining $S(\times, x, y, z)$. But we have already observed that, syntactically, the formula $\otimes_i(x, y, z)$ is always just a concatenation of words of the form $w_j X^i M_0(x, y, z) P^i w_{j+1}$; i.e. its hardest component is just a fixed string X, *independent of* i, concatenated with itself i times, followed by the fixed formula $M_0(x, y, z)$, followed by a fixed string of right parenthesis P, concatenated with itself i times. We have already commented that computing the numerical functions which correspond to concatenation of strings should not be too difficult. Since in our measure of complexity we simply take maximums of individual computations, the difficulty of computing T_{ADD} in this measure will, therefore, be bounded by the difficulty of computing S, which is the hardest part of the Gödel proof. So for the moment, let us assume that in our multiplicative measure the difficulty of computing $T_{\mathrm{ADD}}(\mathbf{i})$ is bounded by $2^{2^{2i}}$. We also ask, how *big* is $T_{\mathrm{ADD}}(\mathbf{i})$? It turns out that $T_{\mathrm{ADD}}(\mathbf{i})$ is of order 2^i for two reasons: First, a quick induction on the definition of $S_i(\mathbf{P})$ shows that the only nonlinear increase in the growth of this formula occurs in the final calculations of $\otimes_i(x, y, z)$, where we must concatenate strings together i times, causing single exponential growth. Second, in calculating the formula $T_{\mathrm{ADD}}(\mathbf{i})$ from $S_i(\mathbf{i})$, we had to explicitly write down the numeral **i** for *i in the language of arithmetic. But in this language i is expressed by* $((((\cdots((\mathbf{1} \oplus \mathbf{1}) \oplus \mathbf{1}) \cdots))))$ (i times).[12] Since the numeral for i and the formula for $S_i(\mathbf{i})$ are essentially just concatenated, we get single exponential growth in passing from i to $T_{\mathrm{ADD}}(i)$. The difficulty of computing $T_{\mathrm{ADD}}(i)$ is less than doubly exponential in i. If $i \in S_0'$, then $T_{\mathrm{ADD}}(i)$ is provable from ADDAX. If $i \in S_1'$, then $T_{\mathrm{ADD}}(i)$ is logically false. If we now had a characteristic function in the list φ_0', φ_1', φ_2' which with *single* exponential complexity separated the theorems of ADDAX from the logically false sentences in the language of addition, then the computation of the function $\mathrm{T}_{\mathrm{ADD}}$ with *this* function would give us a double exponential separation of S_0' and S_1', violating the Russell paradigm. Thus, modulo proving that the transformation S required for the Gödel Theorem

[12] We could avoid this exponential increase in growth by adding an operation of concatenation and axioms relating this operation to our arithmetic operations in the language of arithmetic. But obviously the resulting language would be no easier to solve, and the formula $T_{\mathrm{ADD}}(\mathbf{i})$ would not be essentially smaller because the formula $\otimes_i$ alone introduces single exponential growth.

and its easy derivative T_{ADD} required for the Fischer-Rabin Theorem are not only computable by programs in our programming system, but also at most doubly exponentially difficult in the associated multiplicative measure, we have proven

THEOREM. *Any program in our given system for the partial recursive functions which computes a characteristic function separating the theorems provable from* ADDAX *from the logically false sentences in the language of addition must, on some inputs i, multiply together two numbers each of which are bigger than* 2^i.

Final details: computability and efficiency of the transformations S and T_{ADD}. Anyone who proves the Gödel Theorem must at some point face the problem of showing that one can somehow express recursions by encoding sequences of recursive calls as single finite objects in a language based on addition and multiplication, but with no explicitly given means of doing recursions. That is the problem we are faced with here. The transformation S, and hence also the transformations S_k and T_{ADD}, are explicitly built up recursively from their definitions on the base programs. Our programming system itself has no explicit means of doing recursions, yet we must use it to carry out these transformations.

We have no particularly nice way of doing this. It is a familiar problem for anyone who has taught the Gödel Theorem, and we certainly invite the readers to supply their own proofs. The only point we need mention is that when one does so, one must verify, after writing the programs for S and these other derivative transformations, that the computational complexity of "running" these programs is at worst doubly exponential in our multiplicative measure. Of course, while doing this one must also verify that the cost of doing the concatenations which we have described earlier is also at most exponential. We outline one way of doing this.

We begin by observing that any program $\mathbf{P}$ consists simply of finitely many base programs, strung together by the operations of "composition" and "min". The unwinding of the recursive definition of S basically tells us how to attach specific variables to *each* occurrence of each base program which appears *explicitly* in $\mathbf{P}$. Indeed, the transformation S takes one explicitly given program $\mathbf{P}$ and transforms it by explaining how in one step, to write $S(\mathbf{P})$ in terms of $S(\mathbf{P}_1),\ldots,S(\mathbf{P}_r)$, where each $\mathbf{P}_i$ is already explicitly given as a *subprogram* of $\mathbf{P}$. Since each of the programs $\mathbf{P}_i$ is explicitly given as a subprogram of $\mathbf{P}$, the length of the ultimately unwound program $S(\mathbf{P})$ must be linear in the length of $\mathbf{P}$, and if $\mathbf{P}$ has length n (so as a *number* $\mathbf{P}$ has value of order 2^n) then there can be at most n steps to this unwinding.

Assume for the moment that we hve some computationally nice way of getting from $\mathbf{P} = w'_1$ to each succeeding reduction $\mathbf{P} = w'_1 \to w'_2 \to w'_3 \to \cdots \to w'_m = S(\mathbf{P})$ ($m \leqslant n$) with each w'_i having length of order n and value of order 2^n. One common method for getting a way of going *directly* from $w'_1(= \mathbf{P})$ to w'_m $(= S(\mathbf{P}))$ is to use Gödel's beta function β to encode the sequence $w'_1,\ldots,w'_m$.

Recall that $\beta(i, x, y) = x \bmod (i + 1) \cdot y + 1$. We need to arrange that w_1' be **P** (or at least "nearly" **P**) and that w_{i+1}' be a "next" reduction of w_i' so that w_m' must be S(**P**). Assuming that we can make computational sense of what a "next reduction" should be, we first ask: How big must x and y be, and how difficult is it to compute each $\beta(i, x, y)$?

Recalling that β utilizes the Chinese Remainder Theorem, we need to choose y such that

$$w_i' \leqslant (i + 1) \cdot y + 1$$

with $m!$ dividing y, so that the terms $(i + 1) \cdot y + 1$ and $(i' + 1) \cdot y + 1)$ are pairwise relatively prime. The requirement $w_i' < (i + 1)y + 1$ forces y to be of order 2^n, as does the requirement that $m!$ divides y. In our multiplicative measure, one can show without too much work that the difficulty of computing $w \bmod z$ is $\min\{z, w \div z\}$, where "$\div$" represents integer divison. Thus the difficulty of computing $\beta(i, x, y)$ will be about $(i + 1) \cdot y$ (assuming the worst case, i.e., that x is much bigger than y). Thus, if **P** (as a number) has size about n and value about 2^n, the complexity of computing any value $\beta(i, x, y)$ used in describing any of the intermediate terms in calculating S(**P**) will be about $2^n \cdot n$, i.e., it will be about **P** log **P**. Thus the standard use of the Gödel β function used in the standard method of coding finite sequences has multiplicative complexity about **P** log **P**, far less than the double exponential complexity required for the functions S_i and T_{ADD}.

We now ask how one uses the recursive definition of S to go in individual steps from w_i' to w_{i+1}'. To this end, we introduce a new symbol, S_o, and we treat such symbols as "S_o", "(", ")", ",", "$\times$", "$\mathbf{P}_j^n$" $(1 \leqslant j \leqslant n \leqslant 3)$, etc. as formal symbols. We then reinterpret the recursive definition of S as formal *production rules*. For example: For arbitrary "comma free" *words* w_1, w_2, w_3, w_4, we have the rule

$$S_o(\mathbf{P}_2^3, w_1, w_2, w_3, w_4) \to [w_2 = w_4],$$

and also, for words z_1, z_2, z_3, all of whose commas occur inside parentheses, and, for all comma free words w_1, w_2, w_3, w_4, we have the rule

$$\begin{aligned} S_0(z_1(z_2, z_3), w_1, w_2, w_3, w_4) \to (\exists y_1)(\exists y_2)[S_0(z_2, w_1, w_2, w_3, y_1) \\ \& S_0(z_3, w_1, w_2, w_3, y_2) \& S_o(z_1, y_1, y_2, w_4)]. \end{aligned}$$

(Here, y_1 and y_2 must be chosen as the first variables which are distinct from w_1, w_2, w_3, w_4 and from each other. Thus one must have a syntactically defined notion of a variable.)

Clearly, proceeding in this manner, one can replace the recursive definition of S by a *finite* set of production rules which tell how to implement S "one step at a time", hence how to obtain the particular sequence

$$S_o(\mathbf{P}) = w_0' \to w_1' \to \cdots \to w_m' = S(\mathbf{P}).$$

Clearly, since the *syntactic* choice of what constitutes a variable is somewhat arbitrary, we can so choose our variables that deciding what is a variable has low multiplicative complexity. Since all other aspects of defining the finite set of production rules which implement S involve concatenation of strings, thought of numerically as integers base k, all that remains is to show that operations of string concatenation have relatively low multiplicative complexity.

But *numerically*, the concatenation of words w_1 and w_2 is given by $w_1 \cdot k^{|w_2|} + w_2$, and $k^{|w_2|} = \min z[z$ is a power of k and $z > w_2]$. It is easy to verify that if k is a fixed prime, then, in our multiplicative measure, testing that z is a power of k has complexity at most $\sqrt{z}$. Since one can easily verify that testing $z > w_2$ is free in this measure we see that in this measure the difficulty of concatenating w_1 and w_2 is at most $k^{|w_2|}$, i.e. it is at most about w_2 and is less than $w_1 w_2$.

Once one knows how to concatenate words, one can test that one word is an initial substring of another, that one is a terminal substring of another, and that one is simply a substring of another, all with complexity bounded by the words themselves.

In summary, by paying close attention to the details of the proof *of the Gödel Theorem itself*, one sees that one can describe the smallest pair x, y such that $\beta(0, x, y) = S_o(\mathbf{P})$ and, for all i, if $\beta(i, x, y) = w'$ and the symbol S_o occurs in w', then there exists w'' such that $\beta(i+1, x, y) = w''$, and w'' follows from w' by applying one of the finite production rules. One then obtains $S(\mathbf{P})$ by looking for the smallest i such that S_o does not appear in $\beta(i, x, y)$. For this choice of i, we must have $S(\mathbf{P}) = \beta(i, x, y)$. All of this is standard in any proof of the Gödel Theorem. All that is now is the observation that, in describing the x and y which encode the finite sequence $S_o(\mathbf{P}) = w_i' \to w_2' \to \cdots \to w_m' = S(\mathbf{P})$ and in finding the i such that $\beta(i, m, n) = S(\mathbf{P})$, we never need multiply numbers as big as $2^{2^{\mathbf{P}}}$. This is hardly a surprise, and it is not a difficult observation.

Now to pass from S to the transformation T used in the proof of the Gödel Theorem, one basically needs to, given i, produce the *numeral* for i in the language of multiplication. Clearly, one can again use the β function to produce the string $((\cdots(\mathbf{1} \oplus \mathbf{1}) \cdots \oplus \mathbf{1}))$ (i times) and the multiplicative complexity of doing this will not be significantly larger than the output (which has *value* of order 2^i). From the numeral for i and the formula for $S(i)$, one easily produces the formula $T(i)$ used in the Gödel proof by concatenation. This completes our outline of the proof of the Gödel Theorem. We hope that the reader is now convinced that once all the details are filled in, it is a simple matter to verify that the multiplicative complexity of computing $T(i)$ is well less than 2^{2^i}.

What is different in proving the exponential complexity of Presburger arithmetic? The transformation $T_{\text{ADD}}(i)$ is essentially identical to $T(i)$ except that in unwinding the recursion for S, the production rule

$$S_o(\times, w_1, w_2, w_3) \to [w_1 \otimes w_2 = w_3]$$

is replaced by

$$S_o(\times, w_1, w_2, w_3) \to \bigotimes_i (w_1, w_2, w_3).$$

But $\otimes_i(w_1, w_2, w_3)$ is essentially just $X^i M_o(w_1, w_2, w_3) P^i$ concatenated with itself four times. Thus to produce $\otimes_i(w_2, w_2, w_3)$ we must merely concatenate the strings X and P with themselves i times and then concatenate the results with $M_o(w_1, w_2, w_3)$. But the concatenation of a string with itself i times was exactly the problem we faced earlier in finding the numeral for i. Thus no new idea, and no new complexities are introduced in computing $T_{\mathrm{ADD}}(i)$ instead of $T(i)$. Indeed computing $T_{\mathrm{ADD}}(i)$ is, as advertised, essentially a simple substitution into the computation of $T(i)$. Thus we have proved a version of the Fischer-Rabin Theorem on the exponential difficulty of Presburger arithmetic by a simple substitution into a proof of the Gödel Theorem.

Some concluding remarks. As remarked in an earlier footnote, Fischer and Rabin proved not just that Presburger arithmetic is exponentially difficult, but that it is doubly exponentially difficult. To do this they used more sophisticated methods, using Chebychev's Theorem, to encode triple exponential portions of multiplication in place of the double exponential multiplications encoded by their formulas M_k described here. Clearly, the formulas they construct when substituted for the formulas M_k will yield double exponential difficulty of Presburger arithmetic in the multiplicative measure which we have used here. That is, we obtain double exponential difficulty for the true sentences of the language of addition. However, because of the technical details in doing this, the complexity measure used must count the complexity of multiplying to be the larger, rather than the smaller, of the two numbers multiplied.

One may also wish to observe that in most treatments of exponential difficulty of Presburger arithmetic, one proves that deciding the true sentences of Presburger arithmetic is exponentially or doubly exponentially difficult for the standard Turing machine time measure, for either deterministic or nondeterministic time. To derive this result one must prove that Turing machines can be simulated by our partial recursive functions with no significant complexity loss as we pass from Turing machine time to our multiplicative measure. Since in our measure, existential quantification is essentially an application of minimalization, which in essence is free in our measure, and since nondeterministic computations essentially just involve an existential quantification over deterministic computations, we see immediately that if we can simulate deterministic computations without significant loss as we pass from Turing machine time to our multiplicative measure, then we can also efficiently "simulate" nondeterministic computations. In fact, since in our multiplicative measure both existential *and* universal searches are essentially free, we get not only that Presgurger arithmetic is (doubly) exponentially difficult for nondeterministic (i.e., for "existential") Turing machine time, but also that Presburger arithmetic is (doubly) exponentially difficult for arbitrary alternating Turing machine time computations, a result recently proved in [**B80**].

One way to "simulate" these Turing machine computations with our multiplicative measure is to use the Gödel beta function, much as we did when encoding the translation S, of partial recursive programs. However, other, more efficient, simulations seem to be possbile. We hope to return to the details of this concluding section in a later paper.

References

[**B80**] L. Berman, *The complexity of logical theories*, Theoret. Comput. Sci. **11** (1980), 71–77.

[**BM80**] A. Bruss and A. Meyer, *On time-space classes and their relation to the theory of real addition*, Theoret. Comput. Sci. **11** (1980), 59–69.

[**FR74**] M. Fischer and M. Rabin, *Super-exponential complexity of Presburger arithmetic*, SIAM-AMS Proc. , No. 7, Amer. Math. Soc., Providence, R.I., 1974, pp. 27–41.

[**H79**] D. Hofstadter, *Gödel, Escher, Bach: An Eternal Golden Braid*, Basic Books, New York, 1979.

[**MY78**] M. Machtey and P. Young, *An introduction to the general theory of algorithms*, Elsevier North-Holland, New York, 1978.

[**MY81**] ______, *Remarks on recursion versus diagonalization and exponentially difficult problems*, J. Comput. and System Sci. **22** (1981), 442–453.

Department of Computer Sciences FR - 35 University of Washington, Seattle, Washington, 98195.

Appendix I

LIST OF PARTICIPANTS

NAME	AFFILIATION
Klaus Ambos-Spies	University of Dortmund, West Germany
Arthur Apter	Rutgers University
Ferdinando Arzarello	University of Torino, Italy
Jos Baeten	University of Minnesota, Minneapolis
Joshua D. Bernoff	Massachusetts Institute of Techonology
Mark Bickford	University of Wisconsin
Robert Birmingham	University of Connecticut
Jaime Bohorquoz	Cornell University
Ronald Book	University of California, Santa Barbara
Jerald Bope	Bard College
Egon Borger	University of Dortmund, West Germany
Sheryl Brady	Cornell University
Tim Carlson	University of California, Berkeley
Douglas Cenzer	University of Florida
John C. Cherniavsky	National Science Foundation, Washington, D.C.
Chi Chong	National University, Singapore
Eugeniusz Cichon	Pennsylvania State University, University Park
Peter Clote	University of Paris VII, France
Daniel E. Cohen	London University, Queen Mary College, England
Barry Cooper	Leeds University, England
John Cowles	University of Wyoming
James Cremer	University of Wisconsin
John Crossley	Monash University, Australia
René David	Toulouse University, France
Martin Davis	NYU, Courant Institute of Mathematical Sciences
Jacob Dekker	Rutgers University
Robert A. DiPaola	CUNY, Graduate Center
Carlos A. DiPrisco	Inst. Venezolano de Investigaciones Cientificas, Venezuela

Hans-Dieter Donder	University of Boon, West Germany
Solomon Feferman	Stanford University
Peter Fejer	Cornell University
Matthew Foreman	University of California, Los Angeles
Harvey Friedman	Ohio State University
Sy D. Friedman	Massachusetts Institute of Technology
Jean-Yves Girard	University of Paris VII, France
Warren Goldfarb	Harvard University
Edward Griffor	Uppsala University, Sweden
Marcia J. Groszek	Massachusetts Institute of Technology
Leon Harkleroad	University of Notre Dame
Leo Harrington	University of California, Berkeley
Juris Hartmanis	Cornell University
Louise Hay	University of Illinois at Chicago
James Henle	Smith College
Christer Hennix	SUNY, College at New Paltz
Harold Hodes	Cornell University
Bernard R. Hodgson	University Laval, Canada
Steven Homer	DePaul University
Wen Qi Huang	Huashong Institute of Technology, Peoples' Republic of China
Michael Ingrassia	Western Illinois University
Ljubomir, Ivanov	Sofia University, Bulgaria
John Jayne	University College, England
Thomas Jech	Pennsylvania State University, University Park
Carl Jockusch	University of Illinois
Iraj Kalantari	Western Illinois University
Akihiro Kanamori	CUNY, Baruch College
Miodrag Kapetanovic	Mathematics Institute, Yugoslavia
Alexander Kechris	California Institute of Technology, Pasadena
Clement Kent	Lakehead University, Thunder Bay, Canada
Dennis Kfoury	Boston University
Stephen C. Kleene	University of Wisconsin
Phokion Kolaitis	Occidental College, California
Spiros Kourouklis	University of California, Los Angeles
Evangelos Kranakis	University of Heidelberg, West Germany
Ralph M. Krause	National Science Foundation, Washington, D. C.
Antonin Kucera	Charles University, Czechoslovakia
Stuart Kurtz	University of Chicago
Robert LaGrange	University of Wyoming
Anne Leggett	Western Illinois University
Stephen Legrand	Pennsylvania State University, University Park
Manuel Lerman	University of Connecticut, Storrs

Alain Louveau	CNRS, France
Robert Lubarsky	Massachusetts Institute of Technology
Wolfgang Maass	University of California, Berkeley
Menachem Magidor	Hebrew University, Israel
Mark Manasse	Bell Telephone Laboratories, New Jersey
Alfred Manaster	University of California, San Diego, LaJolla
Richard Mansfield	Pennsylvania State University, University Park
David Marker	Yale University
Donald Martin	University of California, Los Angeles
Pierre Matet	Pensylvania State University, University Park
Adrian R. D. Mathias	Peterhouse, England
Kenneth McAloon	CUNY, Brooklyn College
Charles McCarthy	Wolfson College, Oxford University, England
George Metakides	University of Patras, Greece
Robert Mignone	College of Charleston
Terrence Millar	University of Wisconsin
Arnold Miller	University of Texas
Douglas Miller	Schlumberger-Doll Research, Connecticut
William Mitchell	Pennsylvania State University, University Park
Thomas Moran	Columbia University
Michael Morley	Cornell University
Yiannis N. Moschovakis	University of California, Los Angeles
Kanji Namba	University of Tokyo, Japan
Anil Nerode	Cornell University
Yosaku Nishiwaki	Keio University, Japan
Dag Normann	University of Oslo, Norway
Iain Phillips	Oxford University, England
Marian Pour-El	University of Minnesota, Minneapolis
William Powell	Swampscott, Massachusetts
Hilary Putnam	Harvard University
Nicholas Reingold	Massachusetts Institute of Technology
Jeffrey Remmel	University of California, San Diego, LaJolla
Ian Richards	University of Minnesota, Minneapolis
David Rosenthal	Ithaca College
John Rosenthal	Ithaca College
Gerald Sacks	Harvard University
Ramez Sami	Cairo University, Egypt
Andrej Scedrov	University of Pennsylvania, Philadelphia
Steven Schwarz	Massachusetts Institute of Technology
Philip Scowcroft	Cornell University
Abraham Seidenberg	University of California, Berkeley

Juichi Shinoda	Nagoya University, Japan
Joseph Shoenfield	Duke University
Richard A. Shore	Cornell University
Stephen Simpson	Pennsylvania State University, University Park
Theodore Slaman	University of Chicago
Kay Smith	St. Olaf College
Rick L. Smith	University of Florida
Robert I. Soare	University of Chicago
Robert Solovay	University of California, Berkeley
Dieter Spreen	Technical High School, West Germany
Lee Stanley	Dartmouth College
John Steel	University of California, Los Angeles
Michael Stob	Calvin College
Viggo Stoltenberg-Hanson	Uppsala University, Sweden
It Tan	Massachusetts Institute of Technology
Gregory Taylor	Columbia University
Simon Thompson	Oxford University, England
Tosiyuki Tugue	Nagoya University, Japan
Daniel Velleman	University of Texas, Austin
Catherine Wagner	Cornell University
Stanley S. Wainer	Leeds University, England
Shiqiang, Wang	Beijing Normal University, People's Republic of China
Richard Watnick	University of Connecticut, Stamford
Klaus Weihrauch	University of Hagen, West Germany
Scott Weinstein	University of Pennsylvania, Philadelphia
Galen Weitkamp	Western Illinois University
Lawrence Welch	Western Illinois University
Philip Welch	Oxford University, England
W. Hugh Woodin	California Institute of Technology, Pasadena
Li Xiang	Kweichow University, People's Republic of China
Dong-Ping Yang	Institute of Computing Technology, People's Republic of China
Alexander C. Yessenin-Volpin	Boston, Massachusetts
Paul Young	Purdue University

Appendix II

Short Courses

The fine structure of L, SY. D. FRIEDMAN, Massachusetts Institute of Technology (3 lectures)
Complexity theory, JURIS HARTMANIS, Cornell University (4 lectures)
Descriptive set theory, A. KECHRIS, California Institute of Technology and YIANNIS MOSCHOVAKIS, University of California, Los Angeles (2 lectures each)
Effective mathematics, ANIL NERODE, Cornell University (4 lectures)
E-recursion theory, GERALD E. SACKS, Massachusetts Institute of Technology and Harvard University (4 lectures)
The degrees of unsolvability, RICHARD A. SHORE, Cornell University (4 lectures)
R. E. sets and degrees, ROBERT I. SOARE, University of Chicago (4 lectures)

Invited Hour Addresses

Absolute Π^1_2 *real singletons*, RENÉ DAVID, Toulouse University, France
Strong $\square$*-principles*, HANS-DIETER DONDER, University of Bonn, West Germany
Constructive presentations, SOLOMON FEFERMAN, Stanford University
Π^1_2 *logic and generalized recursion theory*, JEAN-YVES GIRARD, University of Paris VII, France
A gentle approach ot priority arguments, LEO HARRINGTON, University of California, Berkeley
Pseudo-jump operations, CARL G. JOCKUSCH, JR., University of Illinois, Urbana
Recursive functionals and quantifiers of finite types, STEPHEN C. KLEENE, University of Wisconsin, Madison
Inductive definitions and recursion in **E**, PHOKION G. KOLAITIS, Occidental College
The embedding problem for the recursively enumerable degrees, MANUEL LERMAN, University of Connecticut
Some effective results in measure theory, ALAIN LOUVEAU, CNRS, France
Isomorphisms in the lattice of R. E. sets, WOLFGANG MAASS, University of California, Berkeley
Countable decomposable admissible sets: Kleene structures, MENACHIM MAGIDOR, Hebrew University, Israel

Homogeneous trees, DONALD A. MARTIN, University of California, Los Angeles

Semantic methods in complexity theory, KENNETH MCALOON, Brooklyn College, CUNY

Decidable Ehrenfeucht theories, TERRENCE MILLAR, University of Wisconsin, Madison

Aspects of the continuous functionals, DAG NORMANN, University of Oslo, Norway

Constructions in Noetherian rings, ABRAHAM SEIDENBERG, University of California, Berkeley

The fundamental problem of infinite injury, JOSEPH R. SHOENFIELD, Duke University

Mathematical practice and subsystems of second order arithmetic, STEPHEN G. SIMPSON, Pennsylvania State University

The E-recursively enumerable degrees are dense, THEODORE SLAMAN, University of Chicago

Lindebaum algebras with additional structures, ROBERT M. SOLOVAY, University of California, Berkeley

An introduction to gap-one morasses, LEE STANLEY, Dartmouth College

Scales in $L(R)$, JOHN R. STEEL, University of California, Los Angeles

Major subsets and the lattice of recursively enumerable sets, MICHAEL STOB, Calvin College

Reducibility of substitutions, LESLIE G. VALIANT, Harvard University

The slow growing approach ot hierarchies, STANELY S. WAINER, Leeds University, England

Determinacy of real games, scales, and inner models, W. HUGH WOODIN, California Institute of Technology

Subrecursive computational complexity, PAUL YOUNG, Purdue University

ABCDEFGHIJ - AMS/SP - 898765